MARINE NAVIGATION

MARINE NAVIGATION
Piloting and Celestial and Electronic Navigation

Fourth Edition

Richard R. Hobbs
Commander, U.S. Naval Reserve

Naval Institute Press
Annapolis, Maryland

Library of Congress Cataloging-in-Publication Data

Hobbs, Richard R.
    Marine navigation : piloting and celestial and electronic
navigation / Richard R. Hobbs — 4th ed.
        p.   cm.
    ISBN 1-55750-381-8 (alk. paper)
    1. Navigation.   2. Pilots and pilotage.   3. Nautical astronomy.
I. Title.
VK555.H67   1997
623.89—dc21   97-34676

Printed in the United States of America on acid-free paper ∞

05 04 03 02 01 00 99 98   9 8 7 6 5 4 3 2
First Printing

CONTENTS

PART 2. CELESTIAL NAVIGATION

FOREWORD

Throughout the history of warfare at sea, navigation has been an important basic determinant of victory. Occasionally, new members of the fraternity of the sea will look upon navigation as a chore to be tolerated only as long as it takes to find someone else to assume the responsibility. In my experience, such individuals never make good naval officers. Commander Hobbs has succeeded in bringing together the information and practical skills required for that individual who would take the first step down the road toward becoming a competent marine navigator.

At the outset of this book, the author stresses the necessity for safe navigation, but there is another basic tenet of sea warfare that this book serves. The best weapons system ever devised cannot function effectively unless it knows where it is in relation to the real world, where it is in relation to the enemy, and where the enemy is in relation to the real world. Not all defeats can be attributed to this lack of information, but no victories have been won by those who did not know where they were.

W. P. MACK
VICE ADMIRAL, U.S. NAVY

Although these words were written by Admiral Mack for the first edition of this book twenty-five years ago, they are certainly no less valid today than they were then. The modern student of navigation could well take the last sentence to heart, not only as it applies to the practice of navigation, but also to life in general.

THE AUTHOR
ANNAPOLIS, 1997

ACKNOWLEDGMENTS

Marine Navigation was originally published in 1973 as a two-volume text to introduce U.S. Naval Academy and NROTC midshipmen to the shipboard navigation department organization and to principles of piloting and celestial and electronic navigation as practiced in the U.S. Navy. It was consolidated into a single volume with the publication of the third edition in 1990. Over the years since first publication, it has found many additional users both in the military and in the civilian communities in the United States and abroad.

Because the primary purpose of the book remains the introduction of basic marine navigation techniques to naval students, the reader will continue to find a definite U.S. Navy orientation to much of the material presented within. Nevertheless, most of the principles discussed should be just as applicable to navigators of private and commercial craft as to navigators of Navy surface vessels, though the practical application will vary with the type of vessel and her crew.

In order to promote familiarity with sources, excerpts from many of the navigation publications produced by the Defense Mapping Agency and the National Ocean Service that support the practice of seagoing navigation appear throughout this book. Wherever possible, specific year dates have been deleted from these excerpts and from the example problems using them, so that the effect of the passage of time on the text can be minimized. All excerpts, however, reflect the latest formats and contents of their sources as of the publication date of this book. When dealing with actual or theoretical problems and situations other than those presented in this text, the student is cautioned always to obtain and use the correct edition covering the specific dates in question.

I remain greatly indebted to Commander F. E. Bassett, USN (Retired), under whose guidance and assistance the first edition of this text was prepared while he served as chairman of the Navigation Department of the U.S. Naval Academy in the early 1970s, during which time I was a navigation instructor there. Many individuals contributed numerous valuable suggestions incorporated in the various editions over the years, including Lieutenant Commanders

J. D. L. Backus, RN, and J. L. Roberts, USN; Commanders F. Ricci, Italian Navy, and A. Tuttle, USN (Retired); Captains F. Olds, USN, and R. Smith, RN; Colonel E. S. Maloney, USMC (Retired), and many others. Special thanks are also due to Forest Gibson of the U.S. Power Squadron in San Pedro, California, and Lieutenant Commander Robert Irving, USN (Retired), of Marine Research Corporation in Sun Valley, California, both of whom have provided many helpful advisories on material in need of updating in the various editions over the years.

Special thanks for advice and for reviewing critical portions of the manuscript for this fourth edition are due to Captain William Craft, USN, current chairman of the Navigation Department at the Naval Academy. Thanks also to the navigation instructors and chartroom staff who served during academic year 1996–97, most of whom had a hand in reviewing the manuscript or offering valuable suggestions concerning it.

Several publications, technical reports, papers, and Internet sites were consulted for technical facts and details incorporated in this edition. The 1996 edition of *American Practical Navigator (Bowditch)*, published by the Defense Mapping Agency, Washington, D.C., contains a wealth of information on the technical details of all major navigation systems currently operational. The DMA publication *Pub. 117, Radio Navigational Aids,* was also a particularly helpful reference on this subject. Magnavox Corporation provided much technical and illustrative material consulted for the GPS portion of the text. The Magnavox technical reports "GPS Accuracy for Civil Marine Navigation" by G. W. Zackmann, and "GPS Perspectives" by T. A. Stansell, Jr., were particularly helpful in this regard. The 1996 Internet report "Glonass Performance in 1995: A Review," by P. Misra, M. Pratt, R. Muchnik, and B. Burke of MIT Lincoln Laboratory, Lexington, Massachusetts, contains a wealth of up-to-date technical information about the Russian Glonass system; the Russian Internet Glonass site also provided much useful data on system parameters. In regard to electronic charts, the article "Electronic Charts: An Overview of Evolving Chart Standards and Technologies" by Nigel Calder, which was published in the March/April 1996 issue of *Ocean Navigator* magazine, was a very comprehensive source of information concerning the latest developments in electronic chart technology.

Finally, I wish to extend special appreciation to the staff members of the Naval Institute Press and freelance editor Carol Kromminga, whose superb support in the editorial, design, and production aspects of this book is evident throughout. Thanks to you all.

MARINE NAVIGATION

PART 1.
PILOTING

THE ART OF NAVIGATION

1

The word "navigate" is derived from the Latin words *navis,* meaning ship, and *agere,* meaning to move or direct. Navigation is generally defined as the process of directing the movements of a vessel from one place to another. For the contemporary navigator, however, this definition is incomplete, as it lacks two essential modifying terms—the words *safely* and *efficiently*. In today's world of rampant inflation and increasingly serious energy shortages, the cost of replacing a vessel lost through negligent navigation can often be completely prohibitive, quite apart from the consideration of any attendant injuries or deaths among the vessel's crew, loss of cargo, or costs of oil cleanup in the case of tankers. Revenue losses caused by inefficient navigation with consequent increases in fuel bills and other operating costs can be almost if not equally as severe over time. Thus, modern navigation may be more properly defined as the process of directing the movements of a vessel safely and efficiently from one place to another.

It is often said that modern navigation is both an art and a science, with ancient navigators initiating the practice of navigation as an art, and modern navigators developing it into a science. Anyone who has seen the chart of a professional navigator after a day's work in a difficult operating area would certainly be impressed with the artistry displayed in the navigational plot. When one considers the wide range of electronic aids and other sophisticated devices routinely used by the modern navigator, the scientific aspects become equally as apparent. In the last analysis, it is rather difficult to differentiate between navigation as an art and navigation as a science. In fact, the art of navigation probably represents one of the first instances of the practical use of science by humans. The annual migration of early people from one hunting ground to another with the changing seasons is an established archaeological fact. Today, this type of movement from one place to another over land areas by reference to landmarks is called *land navigation*. When people extended their wanderings to coastal waters and rivers by means of primitive boats and river craft, the division of navigation by land or seamarks now known as *piloting* was born.

As humankind continued in its quest for knowledge of unknown territories, mariners began to venture to sea beyond the range of piloting aids, necessitating the development of a new form of navigation: *dead reckoning*. In its initial form, this process was simply concerned with keeping a record of the estimated distances and directions traveled so as to enable the mariner to return to familiar surroundings. As the length and duration of ocean voyages became ever more extended, certain instruments were developed to assist in the determination of course, speed, depth, distance traveled, and ultimately, the position of the vessel at sea. Among these primitive instruments were the lodestone and other early forms of the magnetic compass, the chip log, the hand leadline, the hourglass, and eventually the backstaff, quadrant, early chronometers, the astrolabe, and the early sextant (Figure 1-1).

As improvements continued to be made in navigation instruments, improved techniques of using them were also developed, which ultimately culminated in the mid-1800s in determining position at sea by observing celestial bodies, or *celestial navigation*. With the development of radio and the electromagnetic wave in the early 1900s came *radionavigation* and the subsequent development of more sophisticated means of position-finding now referred to as *electronic navigation systems*. With the advent of the airplane came the practice of *air navigation*; the navigation of a submarine fostered new techniques of *submarine navigation*; and the beginning of the space age necessitated a new branch of navigation called *space* or *astral navigation*.

Marine surface navigation is subdivided into three basic areas—piloting, and celestial and electronic navigation. The first part of this book is concerned with the practice of piloting in waters contiguous to U.S. and foreign shores, and in inland rivers, bays, and lakes in which piloting techniques apply. Although the main orientation is toward surface navigation in these waters, much of the basic information set forth is applicable to all other types of navigation as well.

Throughout seafaring history, the mariner who knew how to navigate was always held in high esteem by his fellows, as without his expertise disaster would surely befall the ship and crew. Until comparatively recent times, the practice of navigation was an art based largely on mathematics and interpretation of written sailing directions, often drafted in Latin or some other foreign language. Usually, only an educated ship's officer or captain could master the subject, and this knowledge was often closely guarded, both as a means of enhancing personal prestige under normal circumstances, and of ensuring control over an unruly crew in times of peril. Thus, an aura of mystique became associated with the art of navigation, which has persisted even to the present day to some extent.

Modern navigation, however, is no longer an art that only an educated few can master. Today's mass-produced charts, easily understood reference publications and sailing directions, and automated electronic receivers have made it

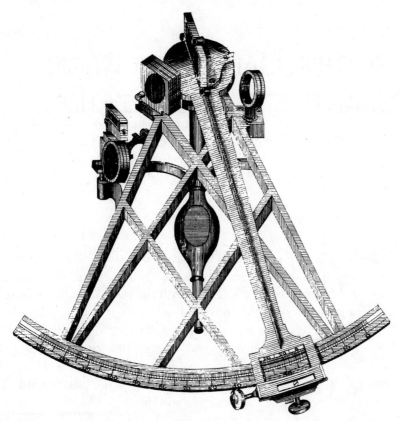

Figure 1-1. *An early sextant.*

possible for just about anyone with basic reading and math skills and some measure of manual dexterity to navigate effectively, once he or she has learned the basic principles involved. Professional standards of accuracy and effectiveness, however, as in any other field of human endeavor, can usually be achieved only by those who are willing to spend the extra time, energy, and constant attention to detail required for above-average success. And the evident skill with which professional navigators use the sextant and other navigational equipment, and the accuracy and neatness of their position plots, still command the respect of their fellow mariners at sea.

THE SHIPBOARD NAVIGATION DEPARTMENT ORGANIZATION

2

While the basic techniques of safe navigation of a seagoing surface vessel are essentially independent of the size of the vessel, the organization that carries out navigational responsibilities will vary with the size and type of vessel. The entire crew of a small boat may consist of only one or two people, who carry out navigational responsibilities in addition to all other responsibilities of operating the vessel. Aboard most merchant vessels, the navigation department usually consists of a single deck officer, typically a second mate, who performs the assignment of ship's navigator as a collateral duty, and who is assisted when the ship is under way by the ship's master and the deck officer on watch.

On board smaller U.S. Navy ships commanded by officers of the rank of lieutenant commander or junior, the duties of the ship's navigator are normally performed by the ship's executive officer as a collateral duty assignment. Larger ships and afloat staffs may have a navigator assigned as a part of their established allowance of officers provided by the Department of the Navy. Since the duties of navigator on a warship are generally considerably more complex than on a merchant ship, Navy navigators are usually assisted by enlisted personnel of the quartermaster rating, with the exact number dependent on the size and mission of the ship. On larger Navy ships such as a cruiser or carrier, these personnel usually constitute a separate division or ship's department, with a junior officer assigned as assistant navigator and N-division officer.

Because this text is oriented primarily toward the practice of marine surface navigation in the U.S. Navy, the remainder of this chapter will discuss the duties of the navigator of a Navy ship in some detail. Students of navigation not affiliated with the Navy should nevertheless find the following discussions of interest, because even though the functional relationships may be unique to Navy warships, the basic responsibilities of navigator on board almost every seagoing vessel are similar.

a. GENERAL DUTIES. The head of the navigation department of a ship will be designated the navigator. The navigator normally will be senior to all watch and division officers. The Chief of Naval Personnel will order an officer as navigator aboard large combatant ships. Aboard other ships, the commanding officer will assign such duties to any qualified officer serving under his command. In addition to those duties prescribed by regulation for the head of a department, he will be responsible, under the commanding officer, for the safe navigation and piloting of the ship. He will receive all orders relating to his navigational duties directly from the commanding officer and will make all reports in connection therewith directly to the commanding officer.

b. SPECIFIC DUTIES. The duties of the navigator will include:

1. Advising the commanding officer and officer of the deck as to the ship's movements; and if the ship is running into danger, as to a safe course to be steered. To this end he will:

(a) Maintain an accurate plot of the ship's position by astronomical, visual, electronic, or other appropriate means.

(b) Prior to entering pilot waters, study all available sources of information concerning the navigation of the ship therein.

(c) Give careful attention to the course of the ship and depth of water when approaching land or shoals.

(d) Maintain record books of all observations and computations made for the purpose of navigating the ship, with results and dates involved. Such books shall form a part of the ship's official records.

(e) Report in writing to the commanding officer, when underway, the ship's position at 0800, 1200, and 2000 each day and at such other times as the commanding officer may require.

(f) Procure and maintain all hydrographic and navigational charts, sailing directions, light lists, and other publications and devices for navigation as may be required. Maintain records of corrections affecting such charts and publications. Correct navigational charts and publications as directed by the commanding officer and, in any event, prior to any use for navigational purposes. Corrections will be made in accordance with such reliable information as may be supplied to the ship or as the navigator is able to obtain.

2. The operation, care, and maintenance of the ship's navigational equipment. To this end he will:

(a) When the ship is underway and weather permits, determine daily the error of the master gyro and standard magnetic compasses and report the result to the commanding officer in writing. He will cause frequent comparisons of the gyro and magnetic compasses to be made and recorded. He will adjust and compensate the magnetic compasses when necessary, subject to the approval of the commanding officer. He will prepare tables of deviations and keep correct copies posted at the appropriate compass stations.

(b) Ensure that the chronometers are wound daily, that comparisons are made to determine their rates and error, and that the ship's clocks are properly set in accordance with the standard zone time of the locality or in accordance with the orders of the senior officer present.

(c) Ensure that the electronic navigational equipment assigned to him is kept in proper adjustment and, if appropriate, that calibration curves or tables are maintained and checked at prescribed intervals.

3. Advise the engineer officer and the commanding officer of any deficiencies observed in the steering system, and monitor the progress of corrective actions.

4. The preparation and care of the deck log. He will daily, and more often when necessary, inspect the deck log and will take such corrective action as may be necessary and within his authority to ensure that it is properly kept.

5. The preparation of such reports and records as are required in connection with his navigational duties, including those pertaining to the compasses, hydrography, oceanography, and meteorology.

6. The required navigational training of all personnel such as junior officers, boat coxswains, and boat officers; the training of all quarterdeck personnel in the procedures for honors and ceremonies and of all junior officers in Navy etiquette.

7. Normally, assignment as the officer of the deck for honor and ceremonies and other special occasions.

8. The relieving of the officer of the deck as authorized or directed by the commanding officer (in writing).

c. DUTIES WHEN PILOT IS ON BOARD. The duties prescribed for a navigator in these regulations will be performed by him whether or not a pilot is on board.

d. ORGANIZATIONAL RELATIONSHIPS. The navigator reports to the commanding officer concerning navigation and to the executive officer for the routine administration of the navigation department. The following officers report to the navigator:

1. The engineer officer concerning the steering engine and steering motors.

2. The assistant navigator, when assigned.

Figure 2-1. Duties of the navigator, OPNAVINST 3120.32.

DUTIES OF THE NAVIGATOR

The responsibilities assigned the U.S. Navy navigator are the same, regardless of the navigator's rank or the size of the ship. They are set forth in *OPNAV INSTRUCTION 3120.32, Standard Organization and Regulations of the U.S. Navy* (see Figure 2-1), and are amplified by Navy instructions such as the joint *COMNAVSURFLANT/-SURFPAC/-AIRPAC/-AIRLANT INSTRUCTION 3530.4, Surface Ship Navigation Department Organization and Regulations Manual.* The remainder of the discussion in this chapter is based on the regulations set forth in these documents, and they are alluded to in various places throughout this book, as they set the standards for the performance of Navy navigators.

RELATIONSHIP OF THE NAVIGATOR TO THE COMMAND STRUCTURE

At this point, it would be well to consider the relationship of the Navy navigator to the commanding officer (CO), the executive officer (XO), and the officer of the deck (OOD). Note that *OPNAVINST 3120.32* specifies that the navigator reports directly to the CO on all matters pertaining to the navigation of the ship. Although this seems like a break in the chain of command, it is not, in reality. *Navy Regulations, 1973*, assigns ultimate responsibility for the safe navigation of the ship to the commanding officer:

> The commanding officer is responsible for the safe navigation of his ship or aircraft, except as prescribed otherwise for ships at a naval shipyard or station, in drydock, or in the Panama Canal. . . .

The CO delegates responsibility for safe navigation to the navigator, and the navigator reports directly to the CO only on those matters pertaining to the navigation of the ship. The navigator reports to the XO in all other matters having to do with administrative functions as a department head, such as the navigation department administration and the training of junior officers in navigation. When separately assigned, the navigator assists the senior watch officer in training watch officers, especially the OODs. The navigator may be empowered to relieve the OOD, if necessary, in a dangerous situation and if so authorized in writing by the CO; this authorization must be in the form of a letter inserted in the service record of the navigator. When so authorized, the navigator can relieve the OOD when such action is necessary for the safety of the ship, and the CO or XO is not on deck. Normally, however, the navigator advises the OOD of a safe course to steer, and the OOD may regard this advice as authority to change the course; the OOD should then report this change to the CO.

THE NAVIGATOR'S STAFF

The navigator does not personally have to perform all of the tasks indicated in *OPNAVINST 3120.32*, but he or she is still responsible for seeing that they are carried out. The leading petty officer (LPO) in the navigation division (N-division) assists the navigator and the assistant navigator, if assigned, in carrying out many of the navigator's duties. On smaller ships the LPO may be a

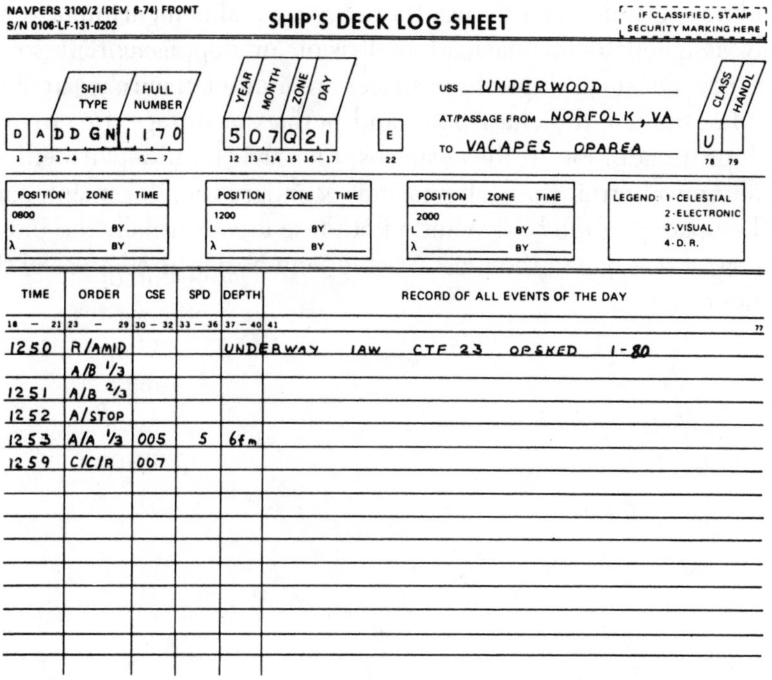

SHIP'S DECK LOG SHEET

IF CLASSIFIED, STAMP
SECURITY MARKING HERE

SHIP TYPE	HULL NUMBER	YEAR	MONTH	ZONE	DAY				CLASS	HANDL
D A D D G N	1 1 7 0	5 0	7	Q	2 1	E			U	

USS __UNDERWOOD__

AT/PASSAGE FROM __NORFOLK, VA__

TO __VACAPES OPAREA__

POSITION	ZONE	TIME	POSITION	ZONE	TIME	POSITION	ZONE	TIME	LEGEND: 1 - CELESTIAL
0800			1200			2000			2 - ELECTRONIC
L _____ BY ____			L _____ BY ____			L _____ BY ____			3 - VISUAL
λ _____ BY ____			λ _____ BY ____			λ _____ BY ____			4 - D. R.

TIME	ORDER	CSE	SPD	DEPTH	RECORD OF ALL EVENTS OF THE DAY
1250	R/AMID				UNDERWAY IAW CTF 23 OPSKED 1-80
	A/B ⅓				
1251	A/B ⅔				
1252	A/STOP				
1253	A/A ⅓	005	5	6fm	
1259	C/C/R	007			

Figure 2-2. A portion of the Ship's Deck Log.

first class quartermaster (QM1). On larger ships a chief quartermaster (QMC) may be assigned.

The navigator and the LPO will detail certain specific areas of responsibility to the more junior of the quartermasters, such as chart petty officer, clock petty officer, and training petty officer. The responsibilities associated with these functions will be discussed later. One of the most important collateral duties of all of the junior petty officers and strikers is standing the quartermaster of the watch (QMOW) duty while the ship is under way. They draft the *Ship's Deck Log* on the bridge. This running chronology of all events of significance occurring during their watch is specified by Navy directives to be one of the legal records of the events occurring on the ship (Figure 2-2).

On smaller ships, the QMOW has several other important functions, in addition to keeping this log. When the navigator is not on deck, the QMOW assists the OOD in all matters pertaining to safe navigation, including the maintenance of the ship's position plot. On ships that do not have a separate meteorological organization, the QMOW is also responsible for filling out an hourly weather observation sheet. A report of these observations is transmitted by radio to the Naval Oceanographic Command Center in whose area of responsibility the ship is operating.

In addition to functional duties, there are several compartments or spaces normally assigned to the navigation division or department personnel for maintenance. On small ships, these spaces are limited in number to the bridge and pilothouse area, the chartroom, and perhaps a storage area or void. On larger ships, in addition to these spaces, the navigation department may be assigned its own berthing area, head, and secondary conn for maintenance. Of these, the upkeep of the bridge and pilothouse area is usually the most time-consuming, as it is one of the most "visible" and heavily used areas of the ship when under way.

THE PILOTING TEAM

3

Unless the navigator is directing the movements of a very small vessel, he or she cannot effectively navigate in confined waters without some assistance from other personnel. The direction of movements of a vessel in these circumstances by reference to land and seamarks is called *piloting*. Piloting may be done by visual methods or by the use of electronics, or by a combination of the two.

Merchant vessels almost always will bring a local pilot on board to take responsibility for conning the ship in piloting waters and in making a berth. The second or third mate deck watch officer in these circumstances will assist the pilot by taking occasional fixes, and by ensuring that the helmsman interprets and carries out all orders to the helm correctly. Naval vessels, on the other hand, tend to rely much less on pilots than do merchant ships, since by their nature warships are intended to proceed in harm's way on their own resources, especially in wartime. Accordingly, the Navy piloting organization is much more extensive than that found on board most merchant ships.

PILOTING TEAM MEMBERS

Aboard Navy ships, the persons who assist the navigator in the piloting environment are customarily referred to as the *piloting*, or *navigation, team*. Its members are normally N-division personnel, if the division is large enough; on small ships having only a few quartermasters on board, members are drawn from related rates, such as operations specialists or signalmen.

The duties of each member of the piloting team on a typical Navy ship are described on the following pages. Again, even though the organization described is unique to the Navy, the functions carried out by the various positions are common to all vessels, so the following descriptions should be of interest to all students of marine surface navigation. The number of personnel assigned to each one of these positions and their ratings varies with the size and type of ship; on smaller ships, one person may carry out the functions of two or more

Figure 3-1A. *The navigator (right) is in charge of the piloting or navigation team. The navigator and/or the navigation evaluator coordinates the team effort through the bearing recorder (left) who communicates with the other team members over the 1JW sound-powered phone circuit, and records each round of bearings in the* Bearing Book *as it is received.*

positions, with the ultimate case being the single-handed small craft whose operator must carry out all functions.

The Navigator, Evaluator, and Plotter

The *navigator* is in overall charge of the piloting/navigation team on board a Navy ship. On large ships he or she may be assisted by a leading quartermaster referred to as the *navigation evaluator*, who in turn supervises and coordinates the actions of all team members, under the navigator's supervision. In the piloting environment, the navigator normally supervises the team from the bridge in proximity to the chart table, the captain, and the officer of the deck. The navigator does not normally do much if any plotting on the navigational chart. Most navigators find it easier to move about and keep the total picture if they assign one of their more senior petty officers, called the *navigation plotter*, to do the actual plotting on the chart. This allows the navigator to monitor the navigation plot and at the same time to be cognizant of oncoming shipping, weather, and any other factors that might affect the ship and its movements.

As the ship proceeds along her intended track during piloting, after each position determination, or *fix* (see chapter 8), the navigator or evaluator makes course and speed recommendations, position reports, and other required navigation advisories verbally directly to the officer of the deck in a standardized format. The standardized advisories are given in italics in the list below.

1. Fix time and accuracy: *Based on a(n) excellent* (or *fair/poor*) *fix at time 1000 . . .*
2. Fix position in relation to intended track: *Navigation holds us on track* (or ___ *yards to right/left of track*).
3. Nearest hazard to navigation: *Nearest hazard to navigation is shoal water 200 yards off the starboard bow.*
4. Nearest aid to navigation: *Nearest aid to navigation is buoy 8, off the port beam.*
5. Depth of water in comparison to chart: *Fathometer reads 45 feet* (or *meters*) *beneath the keel; concurs with charted depth.*
6. Time to next turn and turn bearing and/or slide line (once per leg plus updates): *Next turn will be at time 1020; turn bearing is 100°T* (spoken one-zero-zero degrees true) *off the water tower to starboard.*
7. Distance to turn: *Distance to turn is 2,500 yards.*
8. Course after turn (once per leg): *Next course is 075°T.*
9. Recommendations to regain/maintain track: *Navigation recommends coming left to course 040°T to regain track* (or *recommends maintaining course and speed*).
10. Computed current set and drift (once on each track leg if less than 1,000 yards (meters) and every third fix on legs greater than 1,000 yards (meters): *Set and drift is 090°T at 1 knot.*
11. Comparison with CIC radar navigation plot information passed to the navigator via the 1JS circuit: *CIC concurs* (or *CIC does not concur*).

The complete report for this example would be as follows:

Based on an excellent fix at time 1000, Navigation holds us 50 yards to right of track. Nearest hazard to navigation is shoal water 200 yards off the starboard bow. Nearest aid to navigation is buoy 8, off the port beam. Fathometer reads 45 feet beneath the keel; concurs with charted depth. Next turn will be at time 1020; turn bearing is 100°T off the water tower to starboard. Distance to turn is 2,500 yards. Next course is 075°T. Navigation recommends coming left to course 040°T to regain track. Set and drift is 090°T at 1 knot. CIC concurs.

The piloting information on which these reports are made is gathered by the remaining members of the piloting/navigation team as described in the following sections.

The Bearing Recorder

Although the navigator is the overall director of the piloting/navigation team, the real coordinator of the team is the *bearing recorder*. He or she is stationed on the bridge, alongside the navigator, evaluator, and plotter, and is in constant communication with the other positions on the team via the navigation sound-powered circuit, the 1JW. The navigator or evaluator keeps the bearing recorder informed as to which objects are to be used for obtaining lines of position (LOPs) as the ship proceeds along its track, as well as how often to ask for these LOPs. The bearing recorder keeps track of the passage of time and apportions the objects among the bearing takers and radar operator as applicable. When LOPs are received back from the various positions, the bearing recorder then simultaneously reports them to the navigator or evaluator and plotter and records them in the *Bearing Book* (see Figure 3-4 on page 18). A depth sounding taken at this time is also recorded in the book, for comparison with the charted depth after the ship's position, or fix, is produced from a plot of the LOPs.

Figure 3-1B. *A bearing taker on the bridge wing of a Navy training ship passes visual bearings to the bearing recorder via the 1JW sound-powered phone circuit.*

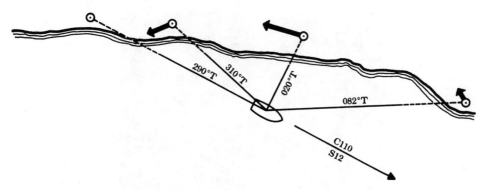

Figure 3-2. *Bearing drifts of objects relative to a moving ship. The bearing drift rate is equal to the ship's speed times the sine of the relative bearing angle to the object observed.*

The Bearing Taker

On small ships, there may be only one *bearing taker* stationed at a central position from which objects on all sides of the ship can be observed. On larger ships, the configuration of the superstructure and placement of the gyro repeaters requires the stationing of several bearing takers. Usually, one person is stationed on the port side or "wing" of the bridge, and one on the starboard side; some ships station a third bearing taker astern as well. The bearing recorder assigns objects to these individuals according to the side of the ship on which the objects are located—port, starboard, or astern—and they then "shoot" the bearings to their assigned objects and report them to the bearing recorder whenever a round (set) of bearings is called for.

It often becomes necessary to assign a single bearing taker more than one object to shoot for a given round of visual bearings. In such cases the bearing to the object that is closest to the ship's beam should be obtained first, as it has the greatest angular velocity relative to the ship, and consequently its bearing is drifting most rapidly. Those objects appearing toward the ship's bow or stern should be shot last; because their bearings change quite slowly, if at all, in the time it takes to shoot a round. In Figure 3-2, the bearing drifts of several objects off the port side of a ship steering course 110°T are depicted, with the lengths of the arrows representing the drift rates. As can be seen from the figure, all bearings except the one directly astern will drift to the left, with the bearing to the object directly abeam at 020°T having the greatest rate of change. Because failure to obtain the bearings in the proper order could result in an erroneous fix position, the bearing takers should be reminded of the proper procedure each time the piloting team is stationed.

Figure 3-3. *A radar operator on the bridge or in CIC provides radar ranges and bearings for the navigation team.*

The Radar Operator

These visual bearings may be supplemented as desired by radar ranges and, to a lesser extent, radar bearings supplied by the *radar operator*. Because of space limitations on the bridge, this individual is often stationed at a radar repeater located in the ship's CIC, or in the case of some larger ships so equipped, in the chart house; he or she is connected to the bearing recorder via the 1JW circuit. When the ship gets beyond visual range to land, the radar ranges and bearings then become a primary source of lines of position. Prior to sailing, various prominent landmarks are identified and labeled with letters of the alphabet on the navigator's chart; the radar operator is then briefed and given a copy of the lettered chart. The bearing recorder requests radar information from the radar operator by referring to point A, B, C, and so on, as the ship progresses along its track. Since radar ranges to objects directly ahead or astern are changing most rapidly, the radar operator should "shoot" these first, then give ranges to objects near the beam last—a procedure exactly opposite that used for visual bearings.

Electronic Navigation System Operator

If the ship is equipped with an electronic navigation system such as Omega, Loran-C, or GPS, and readouts from these systems are not proximate to the bridge chart table, an operator is stationed near the navigation set to relay, as often as desired, position information obtained from this source to the plotter via the 1JW circuit.

The Echo Sounder Operator

Navy instructions require that the ship's echo sounder be energized and recording whenever the ship is in restricted waters or operating in depths less than 100 fathoms. If the ship is not equipped with a remote readout of echo sounder readings near the chart table on the bridge, an *echo sounder operator* is stationed near the console for this instrument and is also on the 1JW circuit. When the operator hears a request being given to the other team members for LOPs, he or she notes the depth reading at that time, and passes this information to the bearing recorder. As was mentioned earlier, the recorder then logs this information into the *Bearing Book* along with the other information obtained for that time, so that the charted depth may be compared with the actual depth of water at the time of each fix.

Quartermaster of the Watch

Although not really a member of the piloting/navigation team per se, the *quartermaster of the watch* (QMOW) serves a vital function during piloting by maintaining the *Ship's Deck Log*, previously described in chapter 2 (see Figure 2-2 page 9). The QMOW also performs any other navigational tasks as required by the OOD or navigator.

THE PILOTING TEAM ROUTINE

The function of the piloting or navigational team is to maintain an accurate and timely plot of the ship's position by all available means, including both visual and electronic information. A cardinal rule is that *no single source of information should be relied upon to the exclusion of others*, to preclude the possibility of unknown error being present.

During piloting, the navigator's plot on the bridge is designated the primary plot, and the radar navigation team in CIC acts as a secondary plot, providing continual positional information to supplement and be compared with the navigator's primary plot. Because the nominal positional accuracy of even the best of standard electronic navigation systems currently available is only ±100

AREA	DISTANCE FROM NEAREST LAND	FIX ACCURACY	RECOMMENDED INTERVAL
Restricted Waters	Less than 2 n. miles	50 yards	2 minutes
Piloting Waters	2-10 n. miles	100 yards	3-15 minutes as conditions warrant
Coastal Waters	10-30 n. miles	500 yards	15 minutes
Enroute Navigation	Over 30 n. miles	.75 n. miles	As conditions warrant, but not greater than 30 minutes

Figure 3-4. *U.S. Navy fix interval guidelines.*

meters (the exception being differential electronic navigation or precise positioning GPS), visual piloting methods are mandated in most piloting situations because of the greater accuracy they can provide—±10 meters or better in many cases.

Although some of the newest state-of-the-art equipment such as GPS-based position-plotters with electronic chart backgrounds is capable of automating the piloting routine to a great extent, current Navy and civil legal requirements specify that vessels under way should maintain a current position plot at all times on the largest scale up-to-date traditional paper chart of the area available, primarily because of the accuracy considerations mentioned above. Nevertheless, such equipment if available can serve as an extremely valuable backup to positional information obtained by traditional methods.

The interval at which the ship's position is determined or fixed depends on the judgment of the navigator, taking into account how constrained the ship is by water depth, current, bottom contour, weather, and navigational hazards in proximity. Navy directives recommend the intervals shown by the table in Figure 3-4.

In practice, most Navy navigators, in all but extreme situations, fix the ship's position every three minutes in most normal piloting environments. This not only allows the use of the "three-" and "six-minute" rules (described in chapter 7) while maintaining the DR plot, but also gives the bearing recorder time to issue new instructions to the team as the ship moves. To let the team know that the time for a round is approaching, the bearing recorder will usually pass the word, "Stand by for a round of bearings," about ten seconds prior to the time of the desired observation. As the time arrives, the command "Stand by— Mark!" is given. The instant the word "Mark" is spoken, the bearing takers should note the bearings to their objects, the radar operator should mark the radar ranges or bearings, and the echo sounder operator should read the depth recorded. All these readings are then recorded in the *Bearing Book*, in the format that appears in Figure 3-5. Note that the gyro error is figured and recorded on the top of the page. Normally, the plotter takes this into account when adjusting the plotting instrument. Bearings recorded are those read by the bearing taker or radar operator from gyro repeaters and radarscopes. The

RECORD GYRO BEARINGS						
Date: *23 June 1969* Place: *Entering Norfolk, Va.*				Gyro Error *1°W*		
Time	Cape Henry Light	Cape Charles Light	Thimble Shoals Light	Lynn-Haven Bridge Tower	Checkered Tank	Echo Sounder Reading
0946	205.5	010.0	287.0			4
0949	201.0	012.5	287.0			
0952	197.0	016.0	287.0			3
			10870			
0955	192.5		287.0			3
0958	186.0		287.0			
1001	178.0		287.0	222.5		
1004	171.0		286.5	217.0		3
1007	164.0		287.0	211.0		
1010	155.0		287.0	206.5		3

Figure 3-5. The Bearing Book.

Bearing Book, in conjunction with the *Ship's Deck Log*, forms a legal record of a vessel's track. If corrections to an entry are required in either record, an erasure must not be made; a single line is drawn through the entry and initialed by the author, and the correct information is inserted above it.

Although not usually considered part of the piloting/navigation team, two additional billets on a ship's special sea detail bill (a listing of personnel assignments for stations manned when operating in shallow or restricted waters) can provide useful information to the navigator in a piloting situation. On most smaller Navy ships, there is a provision for a leadsman on the bow as part of the deck division's area of responsibility. This individual can provide valuable backup to determine whether or not the echo sounder is operating correctly. The leadsman is equipped with a marked line fitted with a weight on the end called a *leadline* (pronounced *lĕd*), which can be lowered into the water until it strikes bottom, thereby making a depth reading possible. Communication with this station can be established via the 1JV circuit on the bridge.

On Navy ships having a gunfire control director, the navigator should request that this equipment be manned whenever the special sea detail is set. Its operator can be patched into the 1JW circuit, and its optical and radar systems can be used to provide very precise ranges and bearings to prominent landmarks. On occasion its radar can even be made to lock onto a suitable navigation aid in order to provide continuous range and bearing LOPs from this object to the navigator—a most valuable backup to supplement conventional radar information.

In conditions of low visibility such as fog, snow, or rain squalls, the normal piloting team routine may be disrupted somewhat because of inability to record visual bearings. In such situations, especially when in constricted waters or ship channels, the primary responsibility for maintaining the navigation plot may be shifted to the radar piloting team in CIC. Even in these circumstances, however, the navigator is still responsible for the ship's safe navigation. The navigator, therefore, acts as a backup to the CIC plot; he or she attempts to verify all CIC recommendations by using all positional information of opportunity. If doubt arises as to the ship's position, the navigator should immediately recommend taking all way off the ship, and perhaps even dropping anchor, until the doubt can be resolved.

To ensure that the CIC radar navigation team will be ready should it be needed, it has become standard procedure, whenever the special sea detail is set, for the CIC team to act as a backup for the piloting team in the piloting environment. The navigation information is usually passed between the bridge and CIC via phone talkers on the 1JA circuit. The navigator should always make sure, therefore, that the cognizant CIC personnel and all other individuals with a need to know are briefed each time prior to getting under way or entering port as to the ship's track and all other matters concerning safe navigation through piloting waters.

More about the prepiloting navigation brief will be discussed in chapter 15 in connection with voyage planning.

CONCLUSION

In conclusion, it must be remembered that the piloting/navigation team, like any other team, needs practice to operate smoothly and efficiently together. The navigator should make it a routine procedure to assemble the team prior to the time the ship enters or leaves port to brief them on the ship's route, the appearance of landmarks and lights, the expected visibility conditions, and any other unusual or pertinent circumstances associated with that particular piloting environment. It must be borne in mind too that it is one thing to try to visualize a navigation aid or landmark from its appearance on a chart or description in a book, and quite another task to pick out the actual object hidden among its natural surroundings. Even an experienced bearing taker occasionally confuses one desired landmark with another when shooting a bearing. The problem is compounded many times for a relatively inexperienced observer operating in poor visibility conditions. For this reason, the bearing taker is considered the "weak link" in the team. With practice and a good idea of what to expect, however, erroneous input from this source can be minimized if not completely eliminated.

THE NAUTICAL CHART

4

The nautical chart is historically the most important and certainly the most frequently used tool employed by the navigator in the execution of his functional responsibilities. Maps, charts, and written sailing directions were probably in use by Egyptian and Greek mariners in the Mediterranean Sea well before the birth of Christ. Ptolemy, a Greek astronomer and mathematician, constructed many maps in the second century A.D., among which was a world map based upon an earlier calculation of the earth's circumference as 18,000 miles. His works remained a standard until the Middle Ages; Columbus believed he had reached the East Indies in 1492 in part because he used the Ptolemaic chart as a basis for his calculations of position. In the Pacific, the natives of the South Sea islands constructed and used crude yet effective charts from palm leaves and sea shells, representing islands, ocean currents, and angles of intersection of ocean swells. Gerhardus Mercator, a Flemish cartographer who produced a world chart in 1556 by a type of projection bearing his name, is considered to be the father of modern cartography. As more and more mariners recorded extended voyages throughout the world, the accuracy of charts continued to improve. Until the invention of the printing press, however, they were done entirely by hand, and the mariner considered them much too scarce and valuable to be used for plotting. This led to wide use of mathematical techniques for calculating position, which was known as deduced reckoning or sometimes simply as "the sailings." These methods of determining approximate position continued in use until the late nineteenth century, when charts came to be mass produced and the system of geometric "dead reckoning," as it is practiced today, came into widespread use.

There is a difference between a map and a nautical chart. A map is a representation of a land area on the earth's surface, showing political subdivisions, physical topography, cities and towns, and other geographic information. A nautical chart, on the other hand, is primarily concerned with depicting navigable water areas; it includes information on the location of coastlines and harbors, channels and obstructions, currents, depths of water, and aids to navigation. Aeronautical charts, designed for use by the aviator, show elevations, obstructions, prominent landmarks, airports, and aids to air navigation. Like maps, they usually depict land areas, but they differ in that they emphasize landmarks, restricted areas, and other features of special importance to the air navigator.

This chapter will discuss the nautical chart in some detail, including the terrestrial coordinate system; the major types of chart projections used by the surface navigator; chart datums; chart interpretation; determination of position, distance, and direction; and chart production, numbering, and correction systems.

THE TERRESTRIAL COORDINATE SYSTEM

Prior to any discussion of charts or their methods of projection, it is first necessary to understand the nature of the terrestrial sphere and its coordinate system. Our earth is basically round, but it is not quite a perfect sphere, being somewhat flattened at the poles and bulged at the equator. The polar diameter has been calculated to be 6,864.57 miles, while the equatorial diameter is about 6,887.91 miles; the earth is therefore often referred to as a *spheroid*, a close approximation to a sphere. For most navigational purposes, it is considered to be a perfect sphere, with a circumference of exactly 21,600 nautical miles.

On a sphere at rest, any point on its surface is similar to every other point, and all points on the surface are defined as being equidistant from the center. To make measurements on the surface, there must be some point or set of points designated as a reference or references to which all other points can be related. As soon as rotation is introduced, two such reference points are immediately defined—the points at which the spin axis pierces the surface of the sphere. On the earth these points are called the *north* and *south poles*; the axis of the earth, together with its poles, constitutes the basic references on which the terrestrial coordinate system is based.

If a straight line is drawn connecting two points on the surface of a sphere, the line drawn actually represents a locus of points formed by the intersection of a plane with the surface of the sphere. Moreover, if the plane passes through the center of the sphere, as well as through the two points of interest on its surface, it can be shown by spherical trigonometry that the resulting line drawn between the two points represents the shortest possible distance between them, as measured across the surface of the sphere. Any line of this type on the surface of a sphere, formed by the intersection of a plane passing through its center, is termed a *great circle*, so named because it is the largest circle that can be formed on the earth's surface. The shortest distance between any two points on the earth lies along the shorter arc of the great circle passing through them. Figure 4-1 illustrates three great circles, of which two, the equator and a meridian, have special significance in the terrestrial coordinate system.

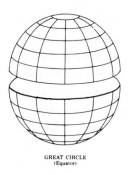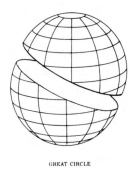

GREAT CIRCLE
(Equator)

GREAT CIRCLE
(Meridian)

GREAT CIRCLE

Figure 4-1. *Examples of great circles.*

   Any other circle formed on the surface of a sphere by the intersection of a plane *not* passing through the center of the sphere is termed a *small circle*. Figure 4-2 depicts three small circles, of which one, the parallel of latitude, has major importance in the earth's coordinate system.

   The great circle formed by passing a plane perpendicular to the earth's axis halfway between its poles is known as the *equator*. The equator, which divides the earth into the *northern* and *southern hemispheres*, is of major importance because it is one of the two great circles from which all locations on the earth's surface are referenced. Any great circle formed by passing a plane through the center of the earth at right angles to the equator is called a *meridian*. There are an infinite number of meridians that could be so formed, but the one that constitutes the second reference line for the terrestrial coordinate system is termed the *prime meridian*. The prime meridian, to which all other meridians are referenced, has been defined by all nations of the world to be the meridian that passes through the original position of the Royal Greenwich Observatory near London, England; although the original structure no longer exists, the spot is marked by a monument visited by thousands of tourists each year. The prime meridian divides the earth in an east-west direction into the *eastern* and the *western hemispheres*.

SMALL CIRCLE

SMALL CIRCLE
(Parallel)

SMALL CIRCLE

Figure 4-2. *Examples of small circles.*

All meridians, because of their construction, lie in a true north and south direction, and are bisected by the earth's axis. That half of a meridian extending from the north to the south pole on the same side of the earth as an observer is called the *upper branch* of the meridian, and the other half, on the other side of the earth from the observer, is referred to as the *lower branch*. The upper branch of the prime meridian is frequently called the *Greenwich meridian*, while its lower branch is the 180th meridian. In common usage, the word "meridian" always denotes the upper branch, unless otherwise specified.

Since there are an infinite number of meridians, all points on the earth's surface have a meridian passing through them. The angular distance between the Greenwich meridian and the meridian of a particular point on the earth's surface is the *longitude* of the point; longitude is measured in degrees of arc from 0° to 180°, either in an easterly or westerly direction from the Greenwich meridian. If a point lies from 0° to 180° east of Greenwich, it is described as being in the eastern hemisphere and having *east longitude*; if it is from 0° to 180° west of Greenwich, it is in the western hemisphere and it has *west longitude*. By convention, the degrees of longitude are always written using three digits, with zeros used as necessary to fill them out, as for example 006° West.

Two meridians are separated by only an infinitely small distance; for the sake of clarity, therefore, meridians of longitude are generally drawn on globes and world maps at intervals of fifteen degrees or so. Figure 4-3 shows the prime meridian of the earth and some other meridians equally spaced around the earth from it.

Any small circle perpendicular to the earth's axis formed by passing a plane parallel to the plane of the equator is termed a *parallel of latitude*. As was the case with the meridian, there are an infinite number of parallels of latitude that can be formed in this manner; therefore, every point on the earth's surface has a parallel of latitude passing through it. The angular distance between the equator and the parallel of latitude passing through a particular point is referred to as the *latitude* of that point. Latitude is measured in degrees of arc from 0° to 90°, either in a northerly or southerly direction from the equator; 90° north latitude is the location of the north geographic pole, and 90° south latitude is the location of the south pole. By convention, the degrees of latitude are always written using two digits, with zeros used to fill them out, as for example 08° North. Parallels of latitude are always at right angles with all meridians that they cross, and thus they lie in an east-west direction. Several parallels of latitude, superimposed on the globe of Figure 4-3, are shown in Figure 4-4.

It should be apparent by now that any point on the earth's surface can be exactly located by specifying its latitude and longitude. When the coordinates of a given location are specified, it has become standard procedure to list the latitude, usually abbreviated by the letter L first, and the longitude, abbreviated either by the Greek letter lambda (λ) or the abbreviation Lo, second. In the

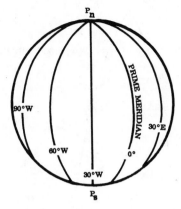

Figure 4-3. The meridians of the earth.

measurement of arc, one degree is made up of 60 minutes of arc, and one minute of arc is made up of 60 seconds. Both latitude and longitude are normally measured to the nearest tenth of a minute of arc, but on some larger-scale charts, the two quantities may be read accurately to the nearest second. After the amount of the latitude or longitude has been measured, each must be suffixed by the proper letter indicating the hemisphere in which the given point is located, either N (north) or S (south) for latitude, and E (east) or W (west) for longitude. If these letters were omitted, a given set of latitude and longitude numbers could refer to any one of four different locations on the earth's surface.

Since latitude is measured in a north-south direction, a convenient location to place a latitude scale is alongside a meridian. In practice, the latitude scale is normally printed on a chart alongside the meridians making up its left and right side boundaries. Since latitude is measured along a meridian, which is a great circle, the length of one degree of latitude is the same everywhere on the

Figure 4-4. Parallels of latitude.

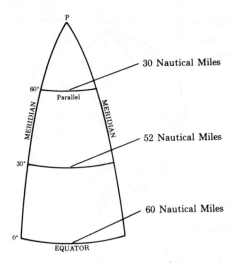

Figure 4-5. *Contraction of the longitude scale.*

earth. It is equal to the earth's circumference, 21,600 nautical miles, divided by 360°, or 60 nautical miles. One minute of latitude, therefore, is equal to one nautical mile of distance, which by strict definition is equal to 1,852 meters or 6,076.1 feet. In the U.S. Navy, however, for distances under 20 miles or so, the nautical mile is usually considered as equivalent to 2,000 yards for most practical purposes.

Longitude is measured in an east-west direction; a logical place for its scale, then, is along a parallel of latitude. On a Mercator chart, the longitude scale is always found alongside the two parallels of latitude constituting the upper and lower boundaries of the chart. There is one very important difference between the latitude and longitude scales. Remember that, with the exception of the equator, parallels of latitude are small circles. One degree of longitude, therefore, will not equal 60 nautical miles except when measured along the equator. At all other locations, as the distance from the equator increases, the length in miles of one degree of longitude decreases. This fact is illustrated in Figure 4-5, where one degree of longitude is shown to be 60 miles wide at the equator, 52 miles at 30° north or south, 30 miles at 60° north or south, and zero at the two poles. For this reason, distance can never be directly measured along the longitude scale of the Mercator chart.

CHART PROJECTIONS

In order for a globe of the world to be of practical use to the navigator when operating in restricted waters, it would have to be thousands of feet in diameter to be large enough to show all the necessary details of importance. Since no vessel could carry a globe of this size, even if one could be made, the navigator must rely instead on flat representations of areas of interest on the globe. Unfortunately, this leads to some problems. Experiment would prove that no considerable portion of a rubber ball can be spread out flat without some stretching or tearing. Likewise, because the earth is spherical in shape, its surface cannot be represented on a flat piece of paper without some distortion. The smaller the portion of the globe to be mapped, however, the less the distortion will be.

The surface of a sphere or spheroid is termed *nondevelopable* because of this fact—no part of it can be spread flat without some distortion. Through centuries of experimentation, however, the cartographer has learned to get around this problem by projecting the surface features of the terrestrial sphere onto other surfaces that are developable, in that they can be readily unrolled to form a plane. Two such surfaces are those of a cone and a cylinder. It is also true that a limited portion of the earth's surface can be projected directly onto a plane surface while keeping distortion within acceptable limits, if the area is small in relation to the overall size of the globe. Projections are termed *geometric* or *perspective* if points on the sphere are projected from a single point that may be located at the center of the earth, at infinity, or at some other location. Most modern chart projections are not projected from a single point, but rather they are derived mathematically.

The desirable properties for any projection include the following:

- True shape of physical features
- Correct angular relationships
- Representation of areas in their correct proportions relative to one another
- True scale, permitting accurate measurement of distance
- Rhumb lines (lines on the surface of the earth that cross all meridians at the same angle) represented as straight lines
- Great circles represented as straight lines

It is possible to preserve any one or even several of these desirable properties in a given projection, but it is impossible to preserve them all in any one type. Although there are several hundred kinds of projections possible, only about half a dozen have ever been used for nautical charts. Of these six, only two have come into general use by the seagoing surface navigator—the Mercator and the gnomonic projections.

The Mercator Projection

As was mentioned in the introductory material of this chapter, the Mercator projection gets its name from a Flemish cartographer, Gerhardus Mercator, who developed it some four hundred years ago. It is the most widely used pro-

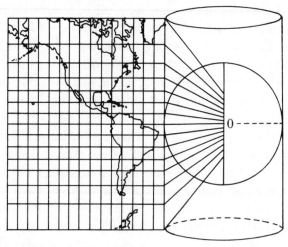

Figure 4-6. *The Mercator conformal projection.*

jection in marine navigation. Position, distance, and direction can all be easily determined, and rhumb lines plot as straight lines; it is also *conformal*, meaning that all angles are presented correctly, and, for small areas, true shape of features is maintained.

The Mercator is a cylindrical projection. To envision the principles involved, imagine a cylinder rolled around the earth, tangent at the equator, and therefore parallel to the earth's axis, as shown in Figure 4-6. Meridians, because they are formed by planes containing the earth's axis, appear as straight vertical lines when projected outward onto the cylinder from within the earth. As has been discussed earlier, however, the distance between successive meridians on the terrestrial sphere lessens as the distance from the equator increases, and finally becomes zero at the poles. On the cylinder, then, as the distance from the equator increases, the amount of lateral distortion steadily increases, and would approach infinity near the poles, projected to the ends of the cylinder. To maintain conformality—true shape—as distance increases from the equator on the cylinder, the latitude scale must be expanded as well. In the Mercator conformal projection, this is done mathematically; the expansion of the latitude scale approximates the secant of the latitude. It should be evident that the greater the distance from the equator, the greater is the distortion of this projection. An example often cited is Greenland, which when completely shown on a Mercator projection, appears to be larger than South America, although in actuality it is about one-ninth as large. For this reason, most Mercator projections of the world are cut off at about 80° north and south latitudes.

The distortion of the true size of surface features, increasing continually as distance increases from the equator, constitutes the major disadvantage of this type of projection. Although there are Mercator charts that depict the polar regions by using a meridian as the circle of tangency of the cylinder, the gnomonic projection, described below, is generally preferred for this purpose. Another disadvantage of a Mercator chart depicting a large area is that great circles, other than a meridian or the equator, appear as curved lines. For conventional methods of navigation in the midlatitudes of the world, however, the advantages of easy measurement of position, distance, and direction on the Mercator projection far outweigh its disadvantages, especially when relatively small areas are depicted.

The Gnomonic Projection

The gnomonic projection, in contrast to the mathematically derived Mercator conformal projection, is a geometrical projection in which the surface features and the reference lines of the sphere are projected outward from the center of the earth onto a tangent plane, as illustrated in Figure 4-7.

There are three general types of gnomonic charts, based on the location of the point of tangency. It may be on the equator (equatorial gnomonic), at either pole (polar gnomonic), or at any other latitude (oblique gnomonic). An oblique gnomonic using a point of tangency located in the central North Atlantic is illustrated in Figure 4-8.

The gnomonic projection has been adapted to a number of different applications, but in surface navigation it is chiefly used because it shows every great circle as a straight line. Rhumb lines on the gnomonic projection appear as curved lines. Figure 4-9 contrasts the appearance of a rhumb line and a great circle line on a Mercator chart with their appearance on a gnomonic chart.

In all three types of gnomonic projection, distortion of shape and scale increases as the distance from the point of tangency increases. Within about

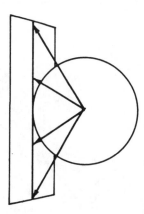

Figure 4-7. *A gnomonic projection.*

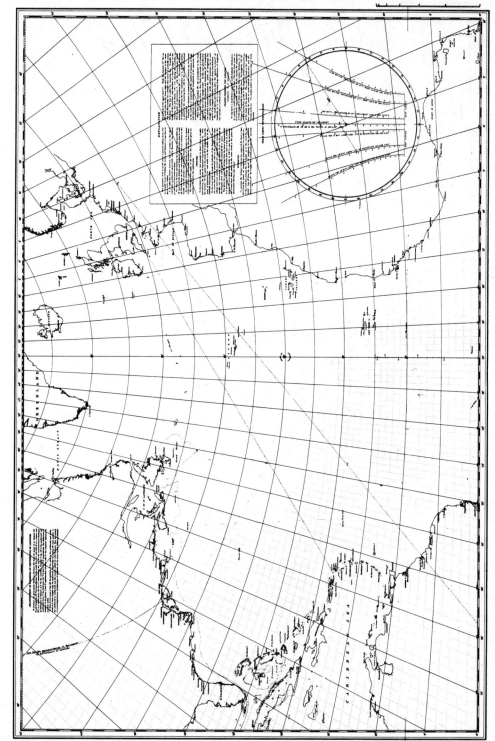

Figure 4-8. An oblique gnomonic chart of the North Atlantic.

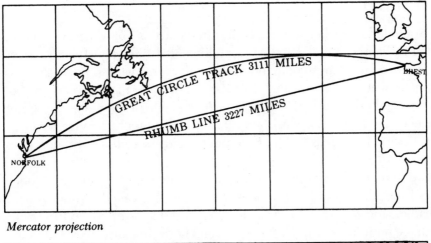

Mercator projection

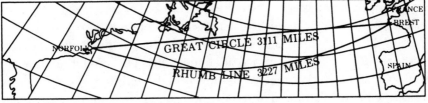

Figure 4-9. *Appearance of a rhumb line and a great circle on a Mercator versus a gnomonic projection.*

1,000 miles of the point of tangency the amount of distortion is not greatly objectionable, but beyond that, it increases rapidly. It is not possible to include as much as a hemisphere in a single gnomonic chart, because at 90° away from the point of tangency, a point would be projected parallel to the plane of the projection.

Distance and direction cannot be measured directly from a gnomonic projection, although it is possible to determine great circle distances by means of a nomogram printed on the chart. This type of chart projection is used mainly to plot the optimum great circle route as a straight line from one place to another. Coordinates of points along the route are then picked off and transferred to a Mercator projection for further use, as described in chapter 15 of this text, which deals with voyage planning. The gnomonic chart is useless as a working chart for normal plotting of navigational data.

There is a third type of projection, called a conic projection, which is based on the projection of a portion of the earth's surface onto a cone. With the exception of a series of charts depicting the Great Lakes, this type of projection finds its most extensive use in aeronautical charts; because the marine surface navigator rarely if ever uses this type of chart, it will not be discussed in detail here.

CHART DATUMS

As alluded to earlier in this chapter, the earth is not a perfect sphere, but rather is slightly flattened at the poles and bulged at the equator. Moreover, the surface of earth is not smooth, but instead is covered with irregular land masses having mountains, valleys, and plateaus, and by water masses of varying bottom contour and tidal heights. This poses a formidable problem for cartographers, namely, what to use as a consistent reference plane for reckoning distances and elevations in land areas and bottom profiles and associated water depths in water areas.

The solution for cartographers has been to devise mathematical model surfaces from which to reference their measurements, called *datums*. A datum is a mathematical model of the earth's shape used as a reference from which to calculate position coordinates, heights, and horizontal distances. One such reference model is based on a shape called an *ellipsoid of revolution*, consisting of an ellipse rotated about its vertical axis, as indicated in Figure 4-10A.

A major difficulty with this model, however, is that no single ellipsoid can adequately model the earth's surface at all locations. This limitation gave rise to cartographers in different regions of the world using different ellipsoids as the basis for their chart datums. These provided good results for their own parts of the world, but poor results if applied to other areas. Until the advent of long-range navigation systems in the latter part of the twentieth century, and more recently precision guided weaponry and worldwide navigation satellite systems, this was not of much consequence for navigators. They simply used locally produced charts that covered their local areas of operation, referenced to the local datum. Figure 4-10B shows several of the more important ellipsoid-based chart datums developed for use in various regions of the world.

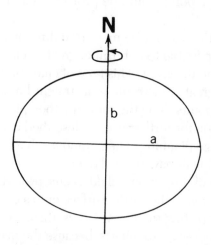

Figure 4-10A. An ellipsoid of revolution, with minor axis b and major axis a.

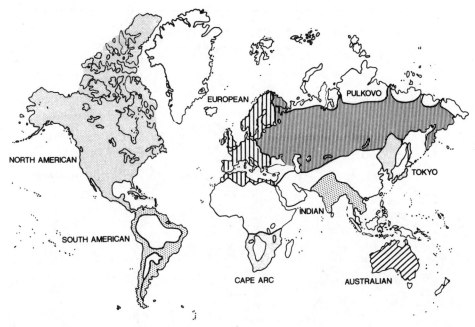

Figure 4-10B. *Major regional datums presently in use.*

The developing need in the 1950s for a better, more comprehensive datum capable of being used worldwide led to the development of an improved model of the earth's shape called a *geoid*. *Geodesy* is the branch of mathematical science that deals with the formulation of geodetic models of the earth.

Basically, a geoid is a theoretical surface of the entire earth to which the mean sea level of the oceans would conform if they were free to adjust to the combined effects of varying gravitational attraction and the centrifugal force of the earth's rotation. The value of gravity is the same everywhere on a geoid surface; the geoid surface is somewhat higher than an ellipsoid surface within land areas, and lower in ocean basin areas. A gravitational plumbline is everywhere perpendicular to a geoid, but not always vertical, as with an ellipsoid. These differences are illustrated in Figure 4-10C.

Because of their far-flung interests, the United States and Great Britain became world leaders in the development of geoid datum systems, eventually resulting in a succession of universally accepted world geodetic systems designated by the year in which they were promulgated: 1960, 1966, 1972, and most recently, 1984. All U.S.-produced nautical charts are now being referenced or converted to the datum of the *World Geodetic System of 1984*, abbreviated WGS-84, as is the U.S.-operated satellite-based Navstar Global Positioning System (GPS), described in detail in chapter 30.

Reference datums are of great importance to the modern navigator as well as to any military personnel involved with precision long-range weaponry and

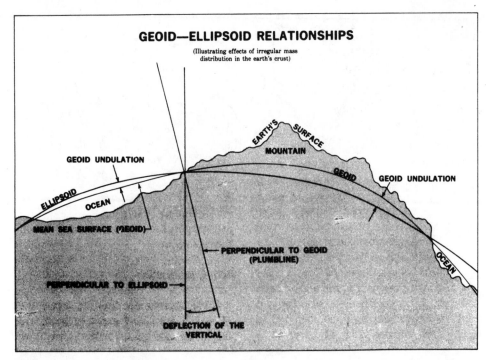

Figure 4-10C. *The geoid, ellipsoid, and topographic surfaces of earth, and deflection of the vertical due to differences in mass distribution within the earth.*

anyone else using long-range electronic or satellite navigation systems. This is because the same position on earth may have different locations when referenced to different geodetic reference planes. An extreme example is in the region of Korea, where locally produced charts have long used a reference plane called the Tokyo Datum. When located using U.S.-produced charts or the GPS system using WGS-84 as the reference plane, a shift in position of some 730 meters occurs—a potential disaster for any mariner, aviator, or weapons director operating in that area! Other less extreme but still very significant deviations can happen in other areas of the world whenever an electronic navigation system based on one datum is used in conjunction with charts based on another. Figure 4-10D shows the corrections that would have to be applied to three commonly used regional chart datums to convert their coordinates to coordinates referenced to the WGS-84 datum.

	NAD 27	ED 50	Tokyo
X	−8 m	−87 m	−146 m
Y	160 m	−98 m	507 m
Z	176 m	−121 m	687 m

Figure 4-10D. *Coordinate shift corrections from North American Datum 27, European Datum 50, and Tokyo Datum to WGS-84.*

In recognition of this problem, many electronic navigation receivers give the user the capability of specifying any one of several different reference datums if desired, with WGS-84 as the default datum. Within the receiver a mathematical conversion subroutine converts the WGS-84-based lat-long output position to new coordinates referenced to the specified datum. The navigator, therefore, should always check the chart datum (provided in the title block) to be certain it is consistent with the datum used by any electronic navigation system or weapons direction system in use.

CHART INTERPRETATION

In order to be able to use any chart effectively, the navigator must first be able to interpret the chart, in regard to both its scale and also its symbols. The scale of a chart refers simply to the ratio between the actual dimensions of an area that the chart depicts and the size of the area as it appears on the chart. A scale of 1:80,000, for example, would mean that one unit of distance measured on the chart would represent 80,000 of the same units in the real world. One inch on the chart would represent 80,000 inches, or one foot would represent 80,000 feet.

The terms *large scale* and *small scale* often cause much confusion to those who are not accustomed to working with charts. A ratio can be written as a fraction: the larger the value of a chart scale ratio, the larger is the scale of the chart. A chart having a large scale such as $\frac{1}{80,000}$ can represent only a relatively small area of the earth's surface without becoming prohibitively large. But a chart having a small scale like $\frac{1}{800,000}$ could represent an area 10 times as large on the same size chart. There is no firm definition of the terms large scale and small scale; the two terms are relative. Thus a chart of scale 1:800,000 would be a small-scale chart compared with one of scale 1:80,000; but the same chart of scale 1:800,000 might be called a large-scale chart in comparison with one having a scale of 1:8,000,000. The phrases

Large scale—small area
Small scale—large area

are often used as memory aids when discussing the subject of chart scale.

When a navigator gets out a nautical chart, it should be examined in detail. All explanatory and cautionary notes appearing on the chart should be read and understood. The scale of the chart, its reference datum, and the date of issue of the chart as well as the date of the survey on which it is based should all be checked. It should be mentioned here that there are some charts in common use that are based on survey data collected some time ago, especially in the Indian, South Atlantic, and South Pacific oceans. Old survey data should be regarded with caution. The units in which water depths are recorded should be

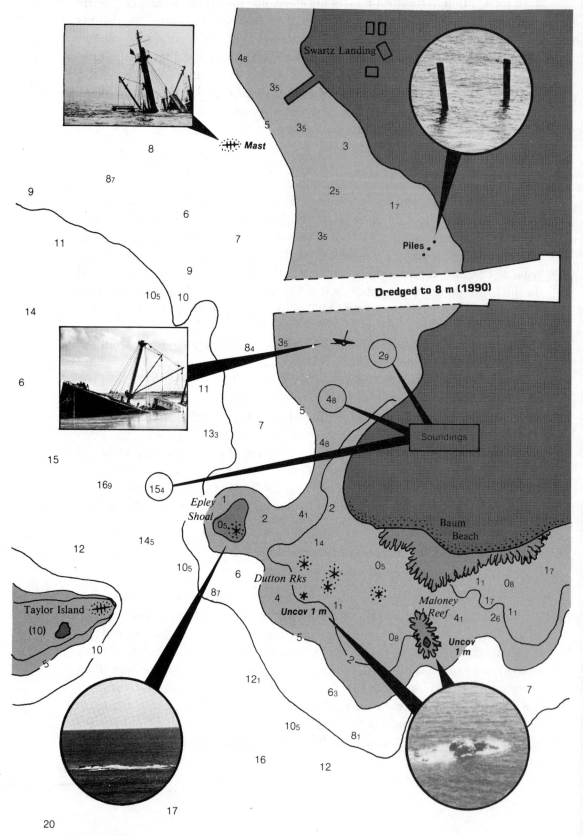

Figure 4-11. A fictitious nautical chart illustrating symbology for various common hazards to navigation.

checked, and areas of the chart in which depth information is either completely missing or widely spaced should also be regarded with extreme caution, particularly in coastal regions.

Many chart symbols and abbreviations are used on a chart to describe features of interest and possible use to the navigator. These constitute a kind of shorthand, and make it possible to insert a great deal of information in a small space on a chart. The symbols used are standardized, but some variations do exist, especially in the shading of shallow water, depending on the scale of the particular chart or chart series. It should be noted that it is not possible to provide the same amount of detailed information on all aids to navigation, whether natural or artificial, on a small-scale chart depicting a given area as would be possible on a large-scale chart. The navigator, therefore, should always keep the master navigation plot on the largest-scale chart practical in a given area. Chart symbols and abbreviations employed to present data on modern charts are contained in a publication entitled *Chart No. 1* published by the Defense Mapping Agency and National Ocean Service, excerpts of which are included as appendix G in the back of this volume. The symbols therein, especially the ones in Section K pertaining to dangers to navigation, should be studied until complete familiarity is attained. Figure 4-11 is a portion of a mock chart depicting chart symbols for some of the more commonly encountered dangers to navigation.

As a final note on chart interpretation, it should be mentioned that a changeover in charted depths and heights from the customary (English) to the international (SI) metric system has been contemplated for some time, in order to conform to bilateral chart reproduction agreements with other nations. On many new charts produced, water depths and heights of lights are shown in meters; land contours are also shown in meters, except where the source data are expressed in feet. As a consequence of this changeover, many shipboard echo sounders are now equipped with dual scales for use with either the customary English or international metric system units.

DETERMINATION OF POSITION, DISTANCE, AND DIRECTION ON A MERCATOR CHART

As was stated earlier, the easy determination of position, distance, and direction is one of the chief advantages of the Mercator projection. In most cases, the tools needed to plot on this type of chart are few, consisting of a plotting compass, a pair of dividers, a parallel rule or some other type of plotter, and a sharp pencil.

A position of known latitude and longitude can be quickly plotted on a Mercator projection using only a compass. Since latitude is usually the first coordinate given in a position, it is natural to plot this coordinate first. To accomplish

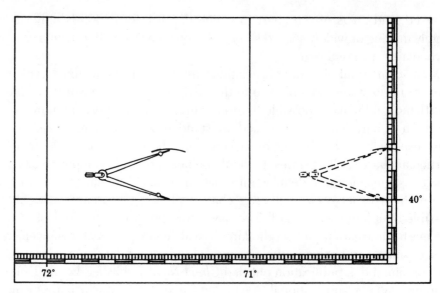

Figure 4-12A. *Swinging an arc to represent a given latitude.*

this, the latitude scale on the most convenient side of the chart is referred to, and the two parallels of latitude that bracket the given latitude are located. The pivot point of the compass is placed on the closest line and the compass is spread across the scale until the leaded point rests at the given latitude. Without the spread thus established being changed, the pivot point is then moved horizontally across the printed parallel of latitude until approximately at the correct longitude; the inexperienced navigator will often lay a straightedge or parallel rule vertically on the chart to assist in this endeavor. An arc is then swung, with the crest representing the proper latitude. Figure 4-12A illustrates this process.

To complete the plot, the same procedure is followed to plot the longitude, except that the compass spread is taken off the top or bottom scale of the chart.

The desired position is located where the crests of the two arcs intersect, as illustrated in Figure 4-12B.

The reverse problem, that of determining the latitude and longitude of a position on a chart, is also easily accomplished. The compass is also a good instrument to use for this purpose, although the dividers can be used as well. Once again, inasmuch as the latitude is written first in a position, it is picked off first from the chart. The pivot point of the compass or dividers is placed on the nearest printed parallel of latitude directly above or below the given position, and the instrument is spread until the other point rests on the position. Without changing the spread thus established, the instrument is then shifted to the most convenient side of the chart, with the pivot point still on the chosen printed parallel of latitude. If the compass is used, a small arc is swung across the scale, as pictured in Figure 4-12C, and the latitude is read off.

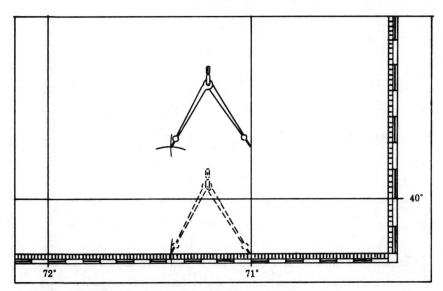

Figure 4-12B. *Completing the plot of a given position on a chart.*

In similar fashion, the longitude is picked off using the compass or dividers set for the distance between the given position and the nearest meridian printed on the chart, as shown in Figure 4-12D.

If dividers are used for this application, chance of error in reading the correct value of latitude and longitude from the chart scale is increased, because there is no definite line drawn across the scale to use as a reference mark.

It has already been explained that one minute of arc on the latitude scale is considered to be equivalent to one nautical mile. Because of this fact, distance

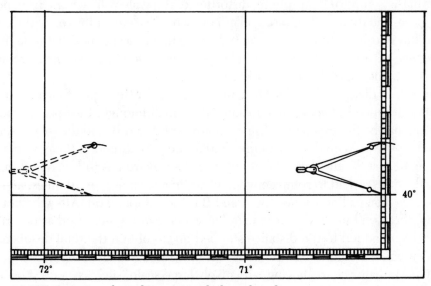

Figure 4-12C. *Swinging a latitude arc across the latitude scale.*

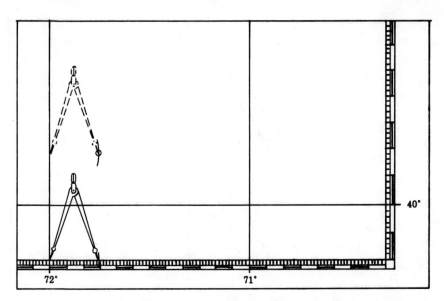

Figure 4-12D. *Swinging a longitude mark across the longitude scale.*

measurement on a Mercator chart is very simple, because the latitude scale on either the left or right border can be used as a distance scale. There is only one caution that must be observed. Because the latitude scale of a Mercator projection expands in length with increased distance from the equator, the length of a mile on the latitude scale of a Mercator chart is not constant. If the chart depicts an area in the northern hemisphere, there are more miles per inch in the southerly portion (the bottom) of the chart. Conversely, if the charted area lies in the southern hemisphere, distortion of the scale will increase from north to south, and there will be more miles per inch of scale near the top of the chart. For this reason, that part of the latitude scale that is at the mean latitude of the distance to be measured should always be used for distance measurement. Consider the example in Figure 4-13.

A pair of dividers is the best instrument to use for this type of measurement. If the distance between points A and B is small enough, the span of the dividers may be increased until its points are over A and B, and the total distance can be measured on the side of the chart at the midlatitude. If the points are so far apart that the dividers cannot reach all the way from A to B in one step, the dividers are spread to some convenient setting, such as 5 or 10 miles, and the distance along a line connecting A and B is then stepped off. After the last full step, there will usually be a remainder, which can be measured as described above and added to the distance stepped off, to obtain the total distance between points A and B. The procedure is illustrated in Figure 4-13.

On Mercator charts having a relatively large scale, such as those showing a harbor or river mouth, the distortion over the small amount of latitude covered

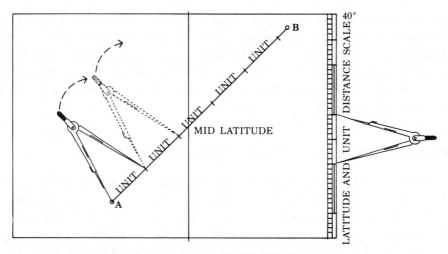

Figure 4-13. *Measurement of distance on a Mercator chart.*

by the chart is negligible. If this is the case, the chart will usually have separate miles and yards and kilometers and meters bar scales printed on it, which are very useful for precise navigation and piloting. The scales, which are based on the midlatitude of the chart, will usually appear over a land area near the chart title block (Figure 4-14).

Measurement of direction on a Mercator projection is, like distance, quite simple because of the conformality of the Mercator. All straight line directions measured on a Mercator chart are *rhumb line* directions; a rhumb line is a line making a constant angle with all meridians it crosses. Since true directions are given with respect to a meridian, it follows that all straight rhumb lines drawn on a Mercator chart are true directions. As soon as a ship is steadied on a given true direction or heading, her course can be represented by a rhumb line drawn across the Mercator chart. Measurement of a rhumb line direction can be made with reference to any convenient meridian or parallel of latitude using any one of several instruments used for measuring direction that incorporates a protractor. A description of the more common instruments of this type is given in chapter 7. Measurement of direction on a Mercator projection can also be accomplished by using a parallel rule or universal drafting machine to transfer the direction of a rhumb line to a nearby *compass rose*. A Mercator

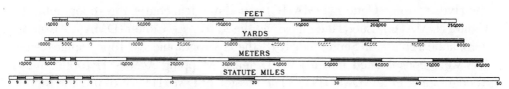

Figure 4-14. *Typical chart bar scales.*

chart compass rose is nothing more than a 360° directional scale referenced to true north by means of a meridian; most roses also contain an inner magnetic direction scale for use with a magnetic compass. Whenever direction is measured on a Mercator chart, care must be taken to read the proper direction off the protractor or true scale of the compass rose; it is very easy to read a direction 180° away from the one desired.

PRODUCTION OF NAUTICAL CHARTS

In the United States, almost all nautical charts, except those covering some inland rivers and lakes, are produced by one of two government agencies—the Defense Mapping Agency (DMA) and the National Ocean Service (NOS). The former agency is concerned mainly with the production and upkeep of charts and related navigational publications covering all ocean areas of the world outside U.S. territorial waters, while NOS produces charts covering inland and coastal waters of the United States and its possessions. This latter organization is also charged with survey responsibilities in support of DMA. In cases where U.S. chart coverage of various ocean and foreign coastal areas is only superficial or lacking entirely, the DMA will often obtain and reproduce applicable charts of various foreign chart-production agencies, especially the British Admiralty.

Charts of some inland waterways, most notably the Mississippi, Ohio, and Missouri rivers and their tributaries, are prepared by the U.S. Army Corps of Engineers. The charts are referred to as "navigational maps," and are available for purchase from district offices of the Corps of Engineers. Because these charts are of relatively minor importance to seagoing navigators, they will not be discussed further in this text.

The Defense Mapping Agency

Until 1972, there were a number of independent U.S. government organizations that produced and supplied charts and maps to the armed services and other government and private users. In 1972, several of these organizations were combined into the *Defense Mapping Agency*, which prints and distributes maps, charts, and supporting publications for all users within the Department of Defense and also for private sale. Between 1972 and late 1978, there were three production centers within DMA, each of which issued its own products: the Hydrographic Center (DMAHC) issued nautical charts, the Topographical Center (DMATC) issued land maps, and the Aerospace Center (DMAAC) issued aeronautical charts. In September 1978, the Hydrographic and Topographic Centers were consolidated into a joint Defense Mapping Agency Hydrographic/Topographic Center (DMAHTC). The mission of DMAHTC is to provide topographic, hydrographic, navigational, and geodetic data, maps, charts, and related products to all armed services, other federal agencies, the Merchant Marine, and private and commercial mariners.

In 1996, DMA was incorporated into a new agency called the National Imagery and Mapping Agency (NIMA), a move designed to consolidate the capabilities of several formerly independent government agencies concerned with chart and map production in both the defense and intelligence communities.

The National Ocean Service

The National Ocean Service, formerly the U.S. Coast and Geodetic Survey (C&GS), is an independent activity within the National Oceanic and Atmospheric Administration (NOAA) in the Department of Commerce. The Coast and Geodetic Survey was established by Congress in 1807 and charged with survey responsibilities for all U.S. coastal waters, harbors, and off-lying island possessions. Since its formation in 1973, NOS has retained the old C&GS responsibility for maintaining accurate surveys of all U.S. coastal waters, and it performs surveys of other areas upon request in support of DMA. As previously mentioned, the NOS also produces a series of large-scale charts covering coastal and certain intracoastal waters of the United States, including the Great Lakes. NOS charts are cataloged in both the DMA chart catalog and a separate series of NOS catalogs, and are distributed by NOAA sales offices and representatives.

THE CHART NUMBERING SYSTEM

Because of the tremendous quantity and variety of modern nautical charts in existence, the need for a logical chart numbering system became apparent to both producers and users of charts. At one time all charts produced by the various U.S. agencies engaged in chart production were simply numbered by series in the order in which they were printed, but in 1974 all chart numbers were converted to the system now in effect, whereby midocean charts, coastal approach charts, and harbor charts can be differentiated from one another by the number of digits in their designations, as described in further detail in the following section.

The U.S. Chart Numbering System

In the U.S. chart numbering system, all commonly used nautical charts produced by both DMAHTC and NOS are assigned a number consisting of from one to five digits, according to their scale and the area they depict. The relationship between the number of digits appearing in the chart number and the chart scale is shown below:

Number of Digits	Scale
1	No scale involved
2	1:9,000,001 and smaller
3	1:2,000,001 to 1:9,000,000
4	Miscellaneous and special nonnavigational charts
5	1:2,000,000 and larger

Because of the chart scale required to represent a given area, it happens that only charts with two- or three-digit identification numbers are of suitable scale to depict large ocean basins or their subdivisions. Likewise, only charts having a five-digit number (a scale larger than 1:2,000,000) are suitable for charting coastal regions with the great detail necessary for piloting applications. A chart number, therefore, not only classifies a chart as to its scale, but it also indicates the size of the geographic region it represents.

Charts bearing a single digit are in reality not nautical charts at all, but rather they are various supporting publications that have no scale. The booklet of nautical chart symbols (*Chart No. 1*) previously mentioned in this chapter is one such publication. Others include chart symbol sheets for other nations, two sheets illustrating national flags and symbols, and a sheet showing the international signal flags and their meanings.

Charts labeled with two- or three-digit numbers are relatively small-scale charts, which for the most part depict either ocean basins or their subdivisions. For the purposes of this type of chart, all ocean areas of the world have been included in one of nine designated basins, numbered as shown in Figure 4-15. The first digit of a two- or three-digit chart number, with three exceptions, denotes the ocean basin in which the area represented by the chart is located. Because of the small size of the Mediterranean (basin 3), the Caribbean (basin 4), and the Indian Ocean (basin 7), there can be no useful two-digit charts of scale smaller than 1:9,000,000 covering these areas. The two-digit numbers 30 through 49 and 70 through 79 are therefore available for other purposes; they are used for large charts that, because of their nature, would not refer to a single ocean basin. Charts of this type include the magnetic dip chart (No. 30), the magnetic variation chart of the world (No. 42), and the standard time zone chart of the world (No. 76).

Charts identified by four-digit numbers are so-called nonnavigational plotting charts and sheets. Examples of this type of chart include large wall and planning charts with scales ranging from 1:1,096,000 to 1:12,000,000, Omega and Loran-C plotting charts, and special gnomonic or azimuthal equidistant charts produced for communications planning purposes. There are over 4,000 different special-purpose charts in existence.

Five-digit charts represent coastal areas; these charts range in scale from

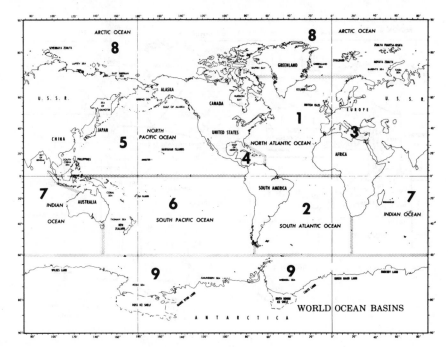

Figure 4-15. *Ocean basins of the world.*

1:2,000,000, which might be used for a chart depicting an entire coastline, to scales as large as 1:8,000, which could depict a river entrance or small harbor. All coastal areas in the world are divided into nine coastal regions, as illustrated in Figure 4-16. Note that these coastal regions are independently numbered in comparison to the ocean basin numbering concept discussed earlier, which forms the basis of the numbering system for two- and three-digit charts. The coastal regions do not have the same numbers as the ocean basins of which they are a part. Where possible, all coastal waters surrounding a major continent are located in the same-numbered coastal region.

Each of the nine coastal regions is further subdivided into a number of subregions; altogether there are 52 different coastal subregions throughout the world. The first two digits of a five-digit coastal chart identify the coastal region and subregion in which the charted area is located. The last three digits place the chart in geographic sequence through the subregion. Charts are not numbered consecutively within a subregion, so that future charts not yet in existence may be placed in their proper geographic order as they are produced, without having to change any numbers on existing charts.

Most of the charts produced by DMAHTC and NOS and used by the navigator for routine marine surface navigation will be marked as described above. This numbering system is also applied to nautical charts produced by foreign

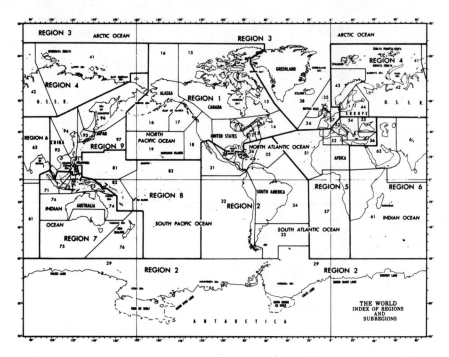

Figure 4-16. *Coastal regions and subregions of the world.*

countries that the DMAHTC maintains within its distribution system. Standard five-digit numbers are assigned to these charts so that they may be filed in a logical sequence with charts produced by the United States. The numbering system provides many benefits to the navigator; not only does it indicate a chart's scale and the area it portrays, but also it facilitates the arrangement of charts into *portfolios*, which are aggregations of charts grouped for the most part by coastal subregion for ease of indexing and storage. Altogether there are 55 chart portfolios, each containing anywhere from about 30 to over 250 charts. Fifty-two of these correspond to the 52 coastal subregions of the world (see Figure 4-16), and three contain general charts of the Atlantic, the Pacific, and the world. Complete listings of the charts comprising each portfolio are contained in the *DMA Catalog of Maps, Charts, and Related Products* described in chapter 5. Navy destroyer-type and larger ships might carry as many as twenty to thirty portfolios of charts on board, depending on the fleet assignment, in order that they might be prepared to operate anywhere within their assigned fleet area on short notice. Most merchant ships of the larger steamship companies, on the other hand, carry far fewer portfolios, as they are usually assigned to specific runs for fairly long periods of time. When necessary, new portfolios of charts can be ordered from the DMA Office of Distribution Services.

DMAHTC Charts outside the Five-Digit Numbering System

There are some types of DMAHTC charts with which the surface navigator will occasionally come into contact that are not covered by the standard five-digit numbering system.

Combat charts are printed in special grid patterns for use in offshore coastal bombardment. Although they are almost all drawn to a scale of 1:50,000 and describe portions of a coastal region, it is nevertheless desirable to distinguish them from normal coastal region navigational charts because of their special military use. They are identified by the letters *COMBT* before a 6-digit chart number. Most combat charts carry a security classification, and so must be kept locked up in a secure storage area.

Other charts outside the five-digit system include a series of world planning charts, pilot charts, and ice surveillance charts.

ELECTRONIC CHARTS

One of the more interesting and certainly most significant developments in marine cartography in recent years has been the introduction of the electronic chart. Electronic charts are not printed on paper, but rather are stored on electronic media such as CD-ROMs for use in electronic plotters and computers. Hybrid electronic receiver/plotters are also coming available that combine the output of an electronic navigation system such as Loran-C, Decca, or GPS (all described in part 3 of this text) with a digital electronic chart on-screen display. Some computer-based displays such as the Navy's NAVSSI (Navigation Sensor System Interface) system can now incorporate all of the foregoing with depth sounder, gyrocompass, and radar input, to present an integrated navigation display not only of the navigator's ship's position but also other ship traffic, weather, and a variety of other navigational information. Electronic chart technology is currently improving at such a pace that it may soon revolutionize the whole manner in which routine marine surface navigation is done.

There are two methods by which electronic charts are being made; each has its advantages and disadvantages. One, called *raster scanning*, is very similar to scanning a document in a copier or computer scanner. An "electronic photo" is made of a master paper chart, composed of hundreds of thousands of tiny color squares called pixels, whose electronic signatures are then stored in an electronic storage device. The chart image is then presented on a computer screen. The main advantage of this method is that all the features of the original paper chart are faithfully and exactly reproduced on the electronic version. Once it is scanned, however, the charted information displayed on this type of electronic chart can be corrected or updated only by rescanning a revised master paper chart, or by overlaying an electronic correction on the existing electronic chart in a way very similar to making a manual correction on a paper chart. Moreover, a separate interface program is required to electronically overlay a lat-long grid on the image, so that inputs from electronic navigation systems, radar, and the ship's gyrocompass can be superimposed on the chart presentation.

The other method of electronic chart production is called *vectorizing*. This too usually starts out with a paper chart. An operator or compiler traces, point by point, the outlines of all the key features on the chart. Sophisticated software stores this information in layers, each one of which contains one type of data, such as all 10-meter depth contours or all aids to navigation, with every point given lat-long coordinates. A complete chart may contain as many as several hundred layers. With a vectorized electronic chart, it is easy to apply corrections simply by specifying the coordinates of a feature and then changing it.

Vectorized charts consume much less memory than do raster charts, but they are far more labor-intensive and therefore costly to produce. Because of the large memory requirements, raster charts can currently be run only on a personal computer (PC) or larger, whereas vector charts can be easily subdivided and zoomed in upon to whatever magnification is required in smaller less-costly plotter-type devices. Distortion and the placement of legend data limit this ability with raster charts. However, the accuracy of vector charts can be questionable, since they are essentially only as reliable as the person and organization producing them. Datums can be a potential problem with raster charts as opposed to vectorized ones, because the former are locked in to the datums of the source charts, which may often be different from the datum used by a WGS-84–based electronic navigation system such as GPS. This could lead to potentially disastrous positioning errors, unless the operator switches the GPS receiver to the datum used by each electronic raster chart sequentially displayed, a situation laden with the possibility of error. In contrast, all vectorized charts currently being produced use WGS-84 as the reference datum, regardless of the source chart datum, since it is relatively easy to do a datum conversion, if required, during the vector production process.

Because of the factors outlined above, initially most of the effort of the large governmental agencies that make charts, such as the U.S. National Ocean Service (NOS) and the British Admiralty, was put into producing rasterized versions of their existing paper charts. (NOS/NOAA has contracted with a private company, BSB Electronic Charts, to produce raster-scanned versions of all 1016 NOS nautical charts.) As demand for the more versatile but costly vectorized charts grew in the mid-1990s, an increasing number of private companies began concentrating on their production for use in a large variety of plotter-type devices now becoming available. In addition, DMA has embarked upon the production of vector-based electronic charts referred to as digital nautical charts (DNC), which will fully conform to the accuracy specifications of the Electronic Chart Display and Information System (ECDIS) of the International Maritime Organization. The first phase of this project, which will eventually yield "seamless" worldwide coverage, is scheduled to be completed about the year 2000.

A complicating factor in regard to electronic charts is the fact that current Navy directives as well as civil legal requirements mandate that a corrected up-to-date paper chart of appropriate scale be used to navigate all major vessels operating at sea, at the risk of otherwise being found negligently liable if an accident or grounding occurs. So, at this time, electronic charts may legally be used only to complement, not replace, traditional paper charts. Nevertheless, electronic charts of one type or the other are experiencing rapidly increasing use, especially on small craft and on larger vessels capable of supporting computer-based integrated navigation displays. Undoubtedly, as soon as a full set of DMA-produced vectorized digital nautical charts becomes available, it will be only a matter of time until they will be accepted by naval authorities and the legal community. Once this happens, the use of paper nautical charts as we know them today will diminish greatly. Indeed, the day may come when the entire world will be covered by an accurate easily correctable electronic chart data base, such that a navigator will need only a few memory chips to replace the drawers full of paper charts now required aboard most naval and merchant vessels.

U.S. Navy and other governmental users may order available electronic charts from NIMA as is now done with paper charts and publications. Civil users may order them directly from the various private companies that produce them. Information on new editions of DMA and NOS electronic charts is promulgated via an *Electronic Notice to Mariners* accessible on the NAVINFONET.

THE CHART CORRECTION SYSTEM

All charts published by both the Defense Mapping Agency and the National Ocean Service are edited and corrected to reflect the latest information available at the time the chart was printed. But a considerable amount of time often elapses between successive printings of a given chart or other navigational publication. As a result, provisions must be made for keeping mariners apprised of changes in hydrographic conditions that will affect the accuracy of their charts and publications as soon as possible after the changes occur.

The principal means by which necessary periodic corrections to DMAHTC and NOS charts and publications are disseminated are the *Notice to Mariners* and *Local Notice to Mariners*. These are bulletins in pamphlet form distributed by mail each week that contain all corrections, additions, or deletions to all DMAHTC and NOS charts and most DMAHTC publications reported during the week preceding issue of the two notices. The *Notice to Mariners* published by NIMA contains all changes relating to oceanic and coastal areas worldwide, while the *Local Notice to Mariners* issued by each of twelve U.S. Coast Guard

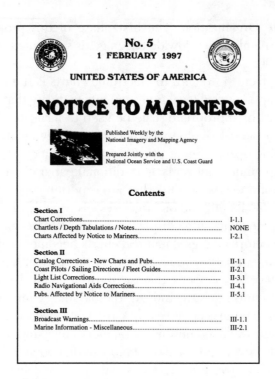

No. 5
1 FEBRUARY 1997

UNITED STATES OF AMERICA

NOTICE TO MARINERS

Published Weekly by the
National Imagery and Mapping Agency

Prepared Jointly with the
National Ocean Service and U.S. Coast Guard

Contents

Figure 4-17A. *Cover of a* Notice to Mariners.

districts contains changes pertaining to U.S. inland waters within each district. Navigators of seagoing vessels that also operate in the inland waterways of one or more Coast Guard districts will need to obtain and use the *Local Notice* issued for each district, as well as the weekly *Notice to Mariners*. Corrections reported in the notices are generated both within and outside of NIMA and NOS. All mariners everywhere are urged to report to the applicable agency any recommended corrections whenever changes are observed. A form for this purpose is included inside the back cover of each *Notice*.

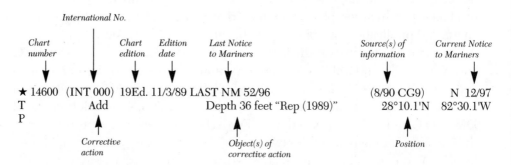

Figure 4-17B. *Format of a* Notice to Mariners *correction.*

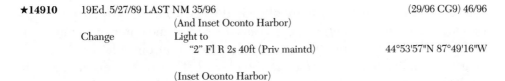

★14910	19Ed. 5/27/89 LAST NM 35/96		(29/96 CG9) 46/96
		(And Inset Oconto Harbor)	
	Change	Light to	
		"2" Fl R 2s 40ft (Priv maintd)	44°53'57"N 87°49'16"W
		(Inset Oconto Harbor)	
	Delete	Light	44°53'49"N 87°49'32"W

Figure 4-17C. *Excerpt from a correction page of the* Notice to Mariners.

Format and Use of *Notice to Mariners*

Each *Notice to Mariners* is made up of three sections, as shown in Figure 4-17A. Within each section, corrections are listed sequentially by chart/publication/item number in the format shown in Figure 4-17B. As noted in the figure, those corrections preceded by a star (★) are based on original U.S. source information; a T indicates a temporary correction; and a P indicates a preliminary correction. Figure 4-17C illustrates a typical correction from a *Notice to Mariners*.

The navigator or quartermasters are not obliged to enter every correction into every chart and publication to which it applies week by week. To conserve the use of nautical charts and publications, and to reduce the amount of chart correction work aboard ship, the chart/publication correction card system was established.

Under this system, every chart and publication kept on board a ship should have a correction card on file (Figure 4-18). As each weekly *Notice to Mariners*

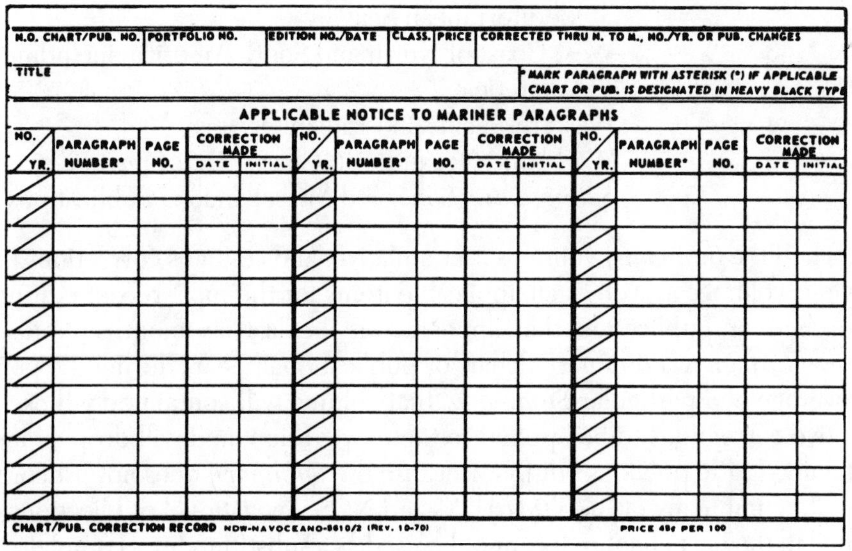

Figure 4-18. *A chart correction card.*

is received, the navigator, or more often in the Navy one of the quartermasters designated as the chart petty officer, scans the *Notice* and enters its number together with the page number bearing the correction onto the cards of any charts or publications that are carried on board. The corrections are actually applied only to those frequently used charts and publications that cover the area in which the ship is operating, plus any additional charts and publications that the ship's commanding officer might specify. After the required changes are made, the *Notice to Mariners* is then inserted into a file for possible future reference.

When a ship is scheduled to begin operating in a new area, the cards of all the charts and publications that will be used therein are pulled from the card file. Corrections are made as necessary using the file of old *Notice to Mariners* to bring the charts and publications up to date.

Summary of Corrections

For convenience in cases where older editions of charts and publications have been newly obtained, such as when a new ship is initially outfitted or when a ship is to proceed to a new area of operations, or where charts and publications stowed on board have not been updated for some time, the DMAHTC issues semiannually a set of five volumes called *Summary of Corrections*, with areas of coverage as follows:

Volume Number	Area Covered
1	East Coast of North and South America
2	Eastern Atlantic and Arctic Oceans, including the Mediterranean Sea
3	West Coast of North and South America, including Antarctica
4	Western Pacific and Indian Oceans
5	World and Ocean Basin Charts, U.S. *Coast Pilots*, *Sailing Directions*, and Miscellaneous Publications

Each of the *Summary* volumes is cumulative, and contains corrections from previous volumes as well as all applicable items for the most recent edition of each chart or publication that appeared in the last six months' *Notice to Mariners*. If a new edition of a chart or publication appears during the period of coverage of a particular *Summary*, that volume will contain only those cumulative corrections to be applied to the new edition, and will drop all items pertaining to the previous edition. Because the *Summary* does not list corrections affecting many of the DMAHTC and NOS navigational publications described in the next chapter, it cannot be used as a substitute for a complete *Notice to Mariners* file.

RADIO BROADCAST WARNINGS

Occasionally it becomes necessary to promulgate changes affecting the safe navigation of a body of water more rapidly than can be done by means of the weekly DMAHTC *Notice to Mariners* or Coast Guard *Local Notice to Mariners*. Radio broadcasts are used for this purpose. Within each of the twelve Coast Guard districts covering U.S. waters, *Broadcast Notices to Mariners* are transmitted as required by various Coast Guard, Navy, and commercial radio stations. The information disseminated in these broadcasts is included in the next *Local Notice to Mariners* issued, if still valid. Information pertaining to littoral and midocean areas plied by oceangoing vessels is broadcast by the Worldwide Navigational Warning System operated by member nations of the International Hydrographic Organization. For purposes of the system, the oceanic areas of the world have been divided into 16 so-called *NAVAREAs*, as shown in Figure 4-19. A different nation or nations are assigned responsibility for coordinating the long-range broadcast of all urgent navigational warnings pertaining to their assigned NAVAREAs. The United States is area coordinator for NAVAREAs IV and XII.

In addition to the NAVAREA region IV and XII broadcasts, DMAHTC also transmits as required special messages called *HYDROLANTS* and *HYDRO-PACS* that provide somewhat redundant coverage to that offered by the

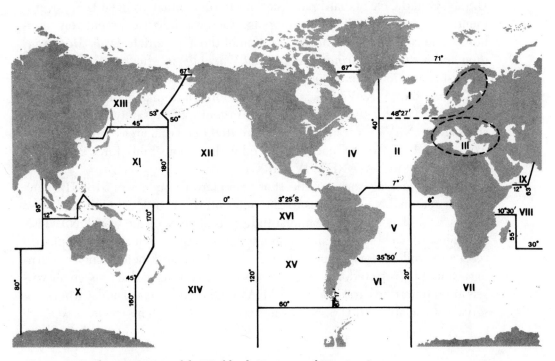

Figure 4-19. *The NAVAREAs of the* Worldwide Navigational Warning System.

NAVAREA system. HYDROLANTS cover the Atlantic Ocean, Gulf of Mexico, and Caribbean Sea, and HYDROPACS cover the Pacific and Indian Oceans. As in the case of the *Broadcast Notices* covering U.S. waters, if it is of continuing interest the information disseminated in the NAVAREA and HYDROLANT and HYDROPAC broadcasts is included in the subsequent issue of the DMAHTC *Notice to Mariners*, as well as in the DMAHTC *Daily Memoranda* described in the following section.

Full details of all navigational warning broadcasts by both U.S. and foreign stations are contained in the DMAHTC publication *Radio Navigational Aids, No. 117*.

Other Printed and Broadcast Warnings

The DMAHTC publishes a set of *Daily Memoranda* each working day in two editions, one for the Atlantic and one for the Pacific. Each *Daily Memorandum* contains the text of appropriate NAVAREA IV and XII warnings, as well as any HYDROLANT or HYDROPAC messages broadcast in the past 24 hours or since the previous working day. The *Daily Memoranda* are sent to naval operating bases, naval stations, customs houses, and major shipping company offices, where they may be picked up for use by navigation personnel of both naval and merchant vessels in port.

Another source of information is *Special Warnings*, occasional broadcasts by U.S. Navy and Coast Guard radio stations to disseminate official U.S. governmental proclamations affecting shipping. Each warning is consecutively numbered, and is included in the next edition of the appropriate *Daily Memorandum* and *Notice to Mariners*. All *Special Warnings* in effect are published in the first *Notice to Mariners* of each year, which is issued in January.

A third source, *NAVTEX*, is an international automated broadcast direct printing service for the promulgation of urgent marine safety information to ships at sea. NAVTEX transmitters are located in coastal areas throughout the world. The U.S. Coast Guard is charged with responsibility for U.S. stations that broadcast in NAVAREAs IV and XII.

NAVTEX messages are broadcast at a standard frequency of 518 kHz. Coverage is reasonably continuous to 200 miles off the U.S. East, Gulf, and West Coasts, as well as most other midlatitude coastal areas of the world. NAVTEX messages carry codes that allow some to be selectively rejected by shipboard receivers, while others containing vital navigational and meteorological warnings cannot be rejected. All cargo vessels over 300 tons and all passenger vessels are required to carry operational NAVTEX receivers in all areas where the service is available. All U.S. Navy ships receive NAVTEX messages as well.

NAVINFONET FLOW DIAGRAM

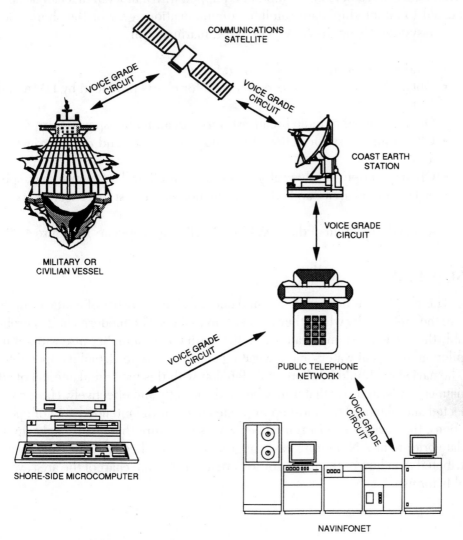

COMMUNICATIONS
SATELLITE

VOICE GRADE
CIRCUIT

VOICE GRADE
CIRCUIT

COAST EARTH
STATION

VOICE GRADE
CIRCUIT

MILITARY OR
CIVILIAN VESSEL

VOICE GRADE
CIRCUIT

PUBLIC TELEPHONE
NETWORK

VOICE GRADE
CIRCUIT

SHORE-SIDE MICROCOMPUTER

NAVINFONET

Figure 4-20. *NAVINFONET flow diagram.*

THE NAVIGATION INFORMATION NETWORK

In the late 1980s, the DMA initiated an automated service designed to give mariners real-time access to its navigational database. Called the *Navigation Information Network* (NAVINFONET), it enables ships in port or at sea anywhere in the world to send queries to and receive navigational data from DMA via voice-grade telephone circuits and communication satellite links such as the commercial INMARSAT.

To use the NAVINFONET, a mariner needs a user ID number obtainable from DMA, and a personal computer equipped with a modem that can be connected to either shipboard satellite communications gear or the shore telephone system. Using the NAVINFONET, mariners can

- Send electronic mail (e-mail) to DMA
- Obtain *Notice to Mariners* corrections for charts produced by DMA and NOS
- Obtain all current radio broadcast warnings and other special warnings
- Obtain the latest corrections to catalogs, light lists, and radio navigation aids
- Obtain a variety of other information such as oil drill rig movements, special operation areas, and other information of interest to navigators.

A schematic diagram of the NAVINFONET system appears in Figure 4-20.

SUMMARY

This chapter has described the fundamental characteristics of nautical charts and the methods by which they are kept up to date. The modern chart is probably the single most important aid to the marine navigator, because without it, piloting in coastal waters and navigating at sea would be virtually impossible. The navigator should be thoroughly familiar with this most fundamental of all nautical tools, in order that it may be used correctly and effectively. Having selected and obtained the most appropriate chart for the intended area of operations, the navigator must remember always to ensure that it is brought up to date prior to use. Numerous cases of grounding have been at least partially attributed to failure of the navigator to have an up-to-date chart of the area available for use.

NAVIGATIONAL PUBLICATIONS

5

In addition to nautical charts, the navigator has several other sources of navigational information about any area of the world in which his or her ship may be operating. This information is contained in a variety of publications, which may be broadly classified by type as chart supplemental publications, manuals, navigation tables, and almanacs. Most of these publications are produced by the Defense Mapping Agency Hydrographic/Topographic Center (DMAHTC) and distributed by the Office of Distribution Services in the same manner as charts, with the exception of certain manuals and almanacs obtained from the U.S. Coast Guard, the National Ocean Service (NOS), and the Naval Observatory. On board ship, the publications are usually stowed in the chartroom, for easy access by the navigator, the quartermasters, and the ship's deck watch officers.

This chapter will describe the contents and use of the more important navigational publications usually carried on board almost all seagoing vessels, and particularly those of the U.S. Navy. Some of the information given here is reiterated in the chapters of this text dealing with applications of the various publications described.

THE *DMA CATALOG OF MAPS, CHARTS, AND RELATED PRODUCTS*

The *Defense Mapping Agency Catalog of Maps, Charts, and Related Products* is an illustrated catalog of various maps, nautical charts, and navigational publications and other related products produced by the DMA. It contains serialized graphic drawings of the world that allow the navigator to locate visually the numbers of all charts and applicable *Sailing Directions* that cover areas of interest. The *Catalog* consists of a set of eleven chapters, the first nine of which catalog the various charts by geographic region. A summary of their contents follows. Each region equals a chapter.

REGION 1	United States and Canada	REGION 5	West Africa and the Mediterranean
REGION 2	Central and South America and Antarctica	REGION 6	Indian Ocean
REGION 3	Western Europe, Iceland, Greenland, and the Arctic	REGION 7	Australia, Indonesia, and New Zealand
REGION 4	Scandinavia, Baltic, and Russia	REGION 8	Oceania
		REGION 9	East Asia

Chapter 10, *Miscellaneous Charts and Publications*, contains information on miscellaneous and special-purpose navigational charts, plotting sheets, and tables. Chapter 11, *Classified Nautical Charts and Publications*, carries a SECRET security classification and is available only to authorized U.S. government users, as are the products it catalogs.

A new edition of the *Catalog* is published every year. Interim corrections are disseminated via change pages and the *Notice to Mariners*.

Use of the *DMA Catalog*

Each of the nine regional chapters in the *DMA Catalog* contains a series of interrelated graphics that indicate the numbers of all available DMAHTC charts and *Sailing Directions* volumes covering that region (the latter publications are described later in this chapter). The first page of each chapter shows the subdivided region marked with the page number on which the large-scale chart coverage for that area is depicted. *Sailing Directions Enroute* and *Planning Guide* volume numbers (see pages 59–60) applicable within the subdivision are also specified; the two types of volumes are differentiated by a rectangle printed around *Planning Guide* numbers. The next page has a graphic depicting the intermediate-scale chart coverage throughout the subregion. The remaining pages depict all other large-scale charts available within each of the subdivisions of page 1.

As an example of the use of the *Catalog*, suppose that a navigator wished to find the numbers of all charts he or she would need in entering Rio de Janeiro, Brazil. First, the navigator ascertains from a diagram of the coastal regions of the world that the eastern coast of South America lies in Region 2, so he or she refers to the Region 2 booklet. Opening the booklet to page 1, reproduced in Figure 5-1A, the navigator notes that Rio de Janeiro lies within a subdivision marked with page number 25, and a box containing *Sailing Directions* volume number 121, which is the *Planning Guide* volume describing the coastal area of South America within which Rio is located. Turning next to page 3, shown in Figure 5-1B, the navigator obtains and records the numbers of the intermediate-scale charts covering the approaches to Rio, nos. 24004 and 24008. Finally, turning to page 25, reproduced in part in Figure 5-1C, he or she obtains and

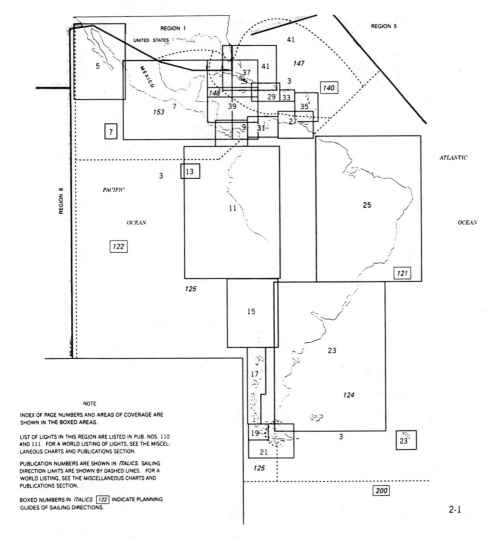

REGION 1
UNITED STATES

MEXICO

REGION 5

REGION 8

PACIFIC

OCEAN

ATLANTIC

OCEAN

NOTE
INDEX OF PAGE NUMBERS AND AREAS OF COVERAGE ARE
SHOWN IN THE BOXED AREAS.

LIST OF LIGHTS IN THIS REGION ARE LISTED IN PUB. NOS. 110
AND 111. FOR A WORLD LISTING OF LIGHTS, SEE THE MISCEL-
LANEOUS CHARTS AND PUBLICATIONS SECTION.

PUBLICATION NUMBERS ARE SHOWN IN *ITALICS*. SAILING
DIRECTION LIMITS ARE SHOWN BY DASHED LINES. FOR A
WORLD LISTING, SEE THE MISCELLANEOUS CHARTS AND
PUBLICATIONS SECTION.

BOXED NUMBERS IN *ITALICS* [122] INDICATE PLANNING
GUIDES OF SAILING DIRECTIONS.

2-1

Figure 5-1A. *Page 1 of* Region 2 *of the* DMA Catalog of Charts.

records the numbers of the large-scale charts covering the approaches and harbor of Rio de Janeiro. They are nos. 24160, 24161, and 24162, in order of increasing scale.

After having thus determined the numbers of the charts and the *Sailing Directions* volume needed, the navigator then goes to the appropriate portfolio storage location, pulls the needed charts, and corrects them up to date by use of the chart correction card system explained in the preceding chapter. In this case, all the charts would be located in Portfolio 24. He or she would obtain the volume of *Sailing Directions* from the location where they are stored, normally the charthouse.

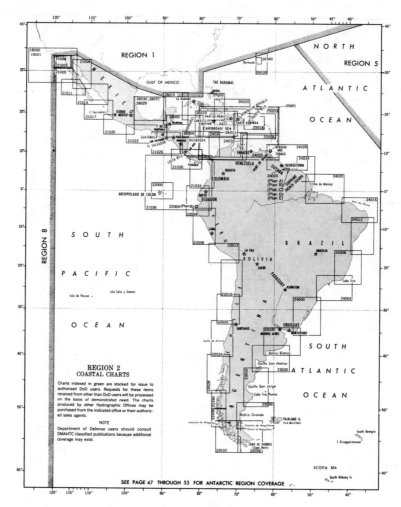

Figure 5-1B. *Page 3 of* Region 2 *of the* DMA Catalog of Charts.

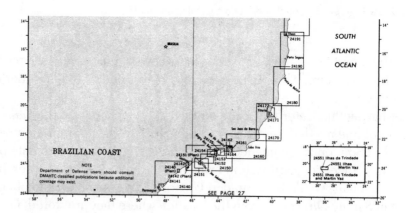

Figure 5-1C. *A portion of page 25 of* Region 2 *of the* DMA Catalog of Charts.

The chart and *Sailing Directions* numbers applying to any location throughout the world can be found by using the *DMA Catalog* in the manner described above.

U.S. Navy Ship Chart Portfolio and Publication Requirements

As mentioned in the previous chapter, U.S. Navy ships normally carry only those chart portfolios associated with areas in which they operate or to which they may deploy. Allowance lists for Navy ships and staffs are promulgated in the form of fleet commanders' instructions, usually in the *3140* series of instruction numbers. These instructions specify for each activity the chart portfolio numbers and publications that they should carry on board. As needs dictate, the Navy navigator can order new or additional charts or publications from NIMA, using procedures specified in the *DMA Catalog*.

As an example, if a Navy navigator's ship were attached to the Second Fleet home-ported in Norfolk, Virginia, he or she might determine from the fleet commander's instruction that all ships of that type in the Atlantic Fleet are required to carry on board, among other portfolios, all charts of Portfolio 14. He or she would then consult Chapter 1 of the *Catalog* to ascertain the numbers and latest edition dates of all charts in Portfolio 14. Any charts in Portfolio 14 presently on board could be checked against this list to determine whether they were the latest editions available. Any charts superseded by new editions and any missing charts could then be ordered from NIMA to bring the ship's supply of charts into conformance with the allowance requirements for that ship.

NOS *NAUTICAL CHART CATALOGS*

The National Ocean Service/NOAA *Nautical Chart Catalogs* consist of five separate folded sheets printed with graphic drawings similar to those in the *DMA Catalog*. Each sheet contains the numbers of all NOS charts and related publications that pertain to the area covered, as well as descriptions of other NOS publications of general interest, and ordering information. The areas of coverage of each of the five NOS *Catalog* sheets are indicated below:

Nautical Chart Catalog Number	Area Covered
1	Atlantic and Gulf Coasts
2	Pacific Coast and Pacific Islands
3	Alaska and the Aleutian Islands
4	Great Lakes
5	Topographic/Bathymetric Maps and Fishing Maps

A portion of *Nautical Chart Catalog 1* covering the Chesapeake Bay appears in Figure 5-2.

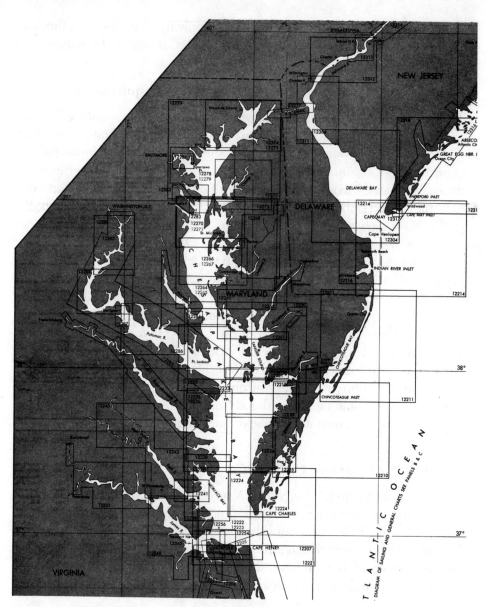

Figure 5-2. *Chesapeake Bay area, NOS Nautical Chart Catalog 1.*

As mentioned in the preceding chapter, all NOS coastal charts are assigned a standard five-digit chart number and most are also cataloged in Region 1 of the *DMA Catalog*.

COAST PILOTS

The U.S. *Coast Pilots* is a series of nine bound volumes containing a wide variety of supplemental information concerning navigation and piloting in the coastal and intracoastal waters of the United States and its possessions, including the Great Lakes. It is published by the National Ocean Service and distributed to Navy users by NIMA. Civil users may purchase them at any NOS distribution agency.

The nine volumes of the *Coast Pilots* are arranged as follows:

Volume Number	Area Covered
	Atlantic Coast
1	Eastport, Maine to Cape Cod
2	Cape Cod to Sandy Hook
3	Sandy Hook to Cape Henry
4	Cape Henry to Key West
5	Gulf of Mexico, Puerto Rico, and Virgin Islands
	Great Lakes
6	Great Lakes and their Connecting Waterways
	Pacific Coast
7	California, Oregon, Washington, and Hawaii
	Alaska
8	Dixon Entrance to Cape Spencer
9	Cape Spencer to Beaufort Sea

The *Coast Pilots* are of great value to the navigator during the planning stages of any voyage through the coastal waters of the United States. Typical information in this series includes the appearance of coastlines, topographical features, navigation aids, normal local weather conditions, tides and currents, local rules of the road, descriptions of ports and harbors, pilot information, general harbor regulations, and many other items of interest to the navigator. Black-and-white photos of major river entrances, harbors, and anchorages are also included. (A sample page from Volume 3 of the *Coast Pilots* appears in Figure 5-3.) Some of the more important areas of concern in which the *Coast Pilots* are especially valuable references are the following:

- Recommended tracks
- Physical descriptions of navigation aids and prominent landmarks
- Description and boundary of navigation hazards
- Procedures for obtaining pilots
- Local rules of the road and local speed limits
- Normal weather, tide, and current information
- Berthing information

New editions of *Coast Pilots 1* through 7 are published annually, with each edition containing all changes reported during the preceding year. *Coast Pilots*

13. CHESAPEAKE BAY, PATUXENT AND SEVERN RIVERS

This chapter describes the western shore of Chesapeake Bay from Point Lookout, on the north side of the entrance to Potomac River, to Mountain Point, the northern entrance point to Magothy River. Also described are Patuxent River, Herring Bay, West River, South River, Severn River, and Magothy River, the bay's principal tributaries; the ports of Solomons Island, Benedict, Chesapeake Beach, Shady Side, Galesville, and Annapolis; and several of the smaller ports and landings on these waterways.

COLREGS Demarcation Lines.–The lines established for Chesapeake Bay are described in 82.510, chapter 2.

Charts 12230, 12263, 12273.– From Potomac River to Patuxent River, the western shore of Chesapeake Bay is mostly low, although the 100-foot elevation does come within 1 mile of the water midway between the two rivers. Above Patuxent River, the ground rises and 100-foot elevations are found close back of the shore along the unbroken stretch northward to Herring Bay. Above Herring Bay, the 100-foot contour is pushed back by the tributaries. Except for the developed areas, the shore is mostly wooded.

The bay channel has depths of 42 feet or more, and is well marked by lights and buoys.

The **fishtrap** areas that extend along this entire section of the western shore are marked at their outer limits and are shown on the charts.

Ice is encountered in the tributaries, particularly during severe winters. When threatened by icing conditions, certain lighted buoys may be replaced by lighted ice buoys having reduced candlepower or by unlighted buoys, and certain unlighted buoys may be discontinued. (See Light List.)

Tidal Current Charts, Upper Chesapeake Bay, present a comprehensive view of the hourly speed and direction of the current northward of Cedar Point, at the south entrance to Patuxent River. The series of 12 charts may be obtained from NOS sales agents and from the National Ocean Survey, Distribution Division (C44), 6501 Lafayette Avenue, Riverdale, Md. 20840.

Weather.–Storm warning display locations are listed on the NOS charts and shown on the Marine Weather Services Charts, published by the National Weather Service.

Chart 12230.–The **danger zone** of an aerial gunnery range and target area begins off Point Lookout and extends northward to Cedar Point. (See 204.42, chapter 2, for limits and regulations.)

A middle ground with depths of 10 to 18 feet is about 8 miles eastward of Point Lookout; the area is about 7 miles long in a north-south direction and 2 miles wide. The stranded wreck near the middle of the shoal is marked by lighted buoys.

Chart 12233.–St. Jerome Creek, 5 miles north of Point Lookout, is entered by a marked channel. In 1966-1971, the controlling depth was 7 feet. There are general depths of 8 to 4 feet above the marked channel. The creek is used principally as an anchorage for oyster and fishing boats.

There are several small wharves along St. Jerome Creek. The landing at Airedele, on the south side just above the entrance, has depths of about 5 feet at the channel face; gasoline is available.

Point No Point, on the west side of Chesapeake Bay 6 miles north of Point Lookout, has no prominent natural marks. **Point No Point Light** (38°07.7'N., 76°17.4'W.), 52 feet above the water, is shown from a white octagonal brick dwelling on a brown cylinder, in depths of 22 feet, 1.6 miles southeastward of the point; a seasonal fog signal is sounded at the light. The light is 1.7 miles due west of a point on the bay ship channel 76.4 miles above the Capes.

An aerial bombardment **prohibited area** is 5.5 miles north-by-west of Point No Point Light. (See 204.42, chapter 2, for limits and regulations of the prohibited area.) The 200-yard-square target area has rock and concrete piers at the corners and in the center, all in depths of 37 feet. Each pier is 50 feet in diameter and 12 feet high; lighted buoys are moored east and west of the target. The steel piling of a Navy radar target (38°14'15"N., 76°20'25"W.) is about 1.6 miles northwestward of the center of the aerial target. The piling is marked by a light; mariners are advised to exercise caution when transiting the area.

Hooper Island Light (38°15.4'N., 76°15.0' W.), 63 feet above the water, is shown from a white conical tower on a brown cylindrical base, in depths of 18 feet near the outer edge of the shoals, 3 miles westward from Hooper Islands; a seasonal fog signal is sounded at the light. The light is 2.8 miles due east of a point on the bay ship channel 84.4 miles above the Capes.

Charts 12264, 12265.–The enclosed Navy seaplane basin 8.5 miles north-northwestward of Point No Point and 2 miles southwestward of Cedar Point has depths of about 10 feet. The entrance to the basin is between two breakwaters, each marked at their outer ends by a light.

Cedar Point (38°17.9'N., 76°22.5'W.) is 10 miles north-by-west of Point No Point. The ruins of an abandoned lighthouse are on the tiny islet 0.3 mile off the point. The shoal extending 0.5 mile eastward from the islet is marked at its outer end by a lighted buoy.

Figure 5-3. A sample page from Coast Pilot 3.

8 and 9 are published every two years. Interim corrections to all volumes are published in the *Notice to Mariners* and *Local Notice to Mariners*.

SAILING DIRECTIONS

The *Sailing Directions* consist of ten geographic groups of loose-leaf volumes that provide information about foreign coasts and coastal waters similar to that found in the *Coast Pilots* for U.S. coastal areas, as well as information on adjacent midocean areas. The *Sailing Directions* are published and kept up to date by DMAHTC.

The *Sailing Directions* currently consist of a set of 47 volumes; 10 of these are *Planning Guides* for ocean transits, and 37 are *Enroute* directions for piloting in coastal waters throughout the world. In addition to these volumes, one other publication, the *World Port Index, Publication No. 150*, provides information on the facilities available in some 7,000 seaports throughout the world.

Each of the *Planning Guides* contains five chapters, their contents being as follows:

1. COUNTRIES	Governments
	Regulations
	Search & Rescue
	Communications
2. OCEAN BASIN ENVIRONMENT	Oceanography
	Climatology
	Magnetic Disturbances
3. DANGER AREAS	Operating Areas, Firing Areas
	Reference Guide to Warnings
	—NM No. 1
	—NEMEDRI, DAPAC
	—Charts
4. OCEAN ROUTES	Route Chart & Text
	Traffic Separation Schemes
5. NAVAID SYSTEMS	Electronic Navigation Systems
	Systems of Lights & Buoyage

By using the appropriate *Planning Guide* in conjunction with the applicable small-scale ocean charts, the navigator should find most of the midocean navigational information needed to plan intelligently for a prospective ocean voyage in the area covered.

The 37 *Enroute* publications are designed to be used in conjunction with applicable large-scale DMAHTC coastal charts to provide all information required for piloting in foreign coastal and intracoastal waters. The *Enroute* volumes are divided into ten geographic groups, corresponding with the ocean areas covered by the *Planning Guide* volumes. The number of volumes in each group varies with the size of the geographic area covered. The Mediterranean area, for example, is covered by two *Enroute* volumes, while the Northern Pacific area requires six.

Each *Enroute* publication is divided into a number of geographic subdivisions called sectors; each sector contains the following information:

CHART INFORMATION—Index: Gazetteer
COASTAL WINDS & CURRENTS
OUTER DANGERS
COASTAL FEATURES
ANCHORAGES (COASTAL)
MAJOR PORTS
    —Directions; Landmarks; Navaids; Depths;
    —Limitations; Restrictions; Pilotage; Regulations;
    —Winds; Tides; Currents; Anchorages

Three of the more valuable features of the *Enroute* volumes are panoramic photographs of coastal features of the more heavily traveled coasts and harbors; graphic keys to charts within sectors, as illustrated in Figure 5-4A; and "graphic direction" plates that combine an annotated chartlet of an area with an orientation photograph and line drawings of prominent navigational features, as in the example shown in Figure 5-4B on page 67. After transiting a midocean area using the appropriate *Planning Guide*, the navigator should then be able to brief all personnel concerned as to the characteristics of the piloting waters to be entered at the end of the voyage, using the appropriate volume of the *Enroute* publications, together with the complementary large-scale coastal charts of the area.

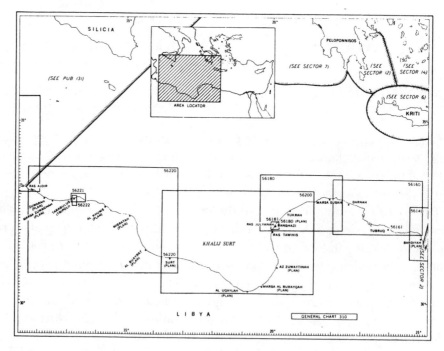

Figure 5-4A. *Graphic key to charts within Sector 1, "Coast of Libya," from the* Enroute *volume* Eastern Mediterranean Sea *of the* Sailing Directions.

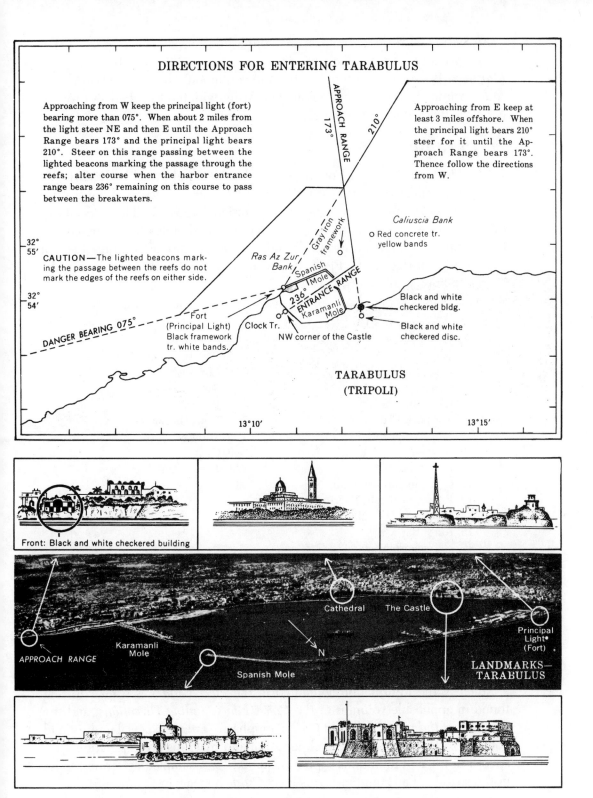

Figure 5-4B. *Graphic directions for entering Tarabulus,* Enroute Sailing Directions.

All *Sailing Directions* volumes, once published, are kept current by the periodic issuance of sets of changed pages, supplemented by the *Notice to Mariners*. Because its content is subject to continuous change, the *World Port Index* volume is updated and reissued annually, with interim changes promulgated via the *Notice to Mariners*.

FLEET GUIDES

The *Fleet Guides* consist of two sets of booklets published by DMAHTC, containing information primarily designed to acquaint incoming naval ships with pertinent command, navigational, operational, repair, and logistical information on frequently visited ports in both the United States and foreign countries. One set is intended for use by ships of the Atlantic Fleet, the other for use by ships of the Pacific Fleet. Following is a list of ports described in the current edition of the *Fleet Guides*:

Atlantic Area	**Pacific Area**
1. Portsmouth, N.H	1. Puget Sound
2. New London	2. San Francisco
3. New York	3. Port Hueneme, Cal.
5. Hampton Roads	4. Los Angeles–Long Beach
6. Canal Zone	5. San Diego
7. Charleston, S.C.	6. Panama Canal
8. Mayport, Fla.	7. Pearl Harbor
9. Port Everglades, Fla.	8. Midway
10. New Orleans, La.	9. Guam
11. Bermuda	11. Tokyo Wan
12. Guantanamo Bay, Cuba	12. Adak
13. Roosevelt Roads, Puerto Rico	
14. Morehead City, N.C.	
15. Kings Bay, Ga.	
16. Narragansett Bay	
17. Port Canaveral	
18. Key West	
19. Pensacola	
21. Ingleside	

Missing volume numbers are for ports no longer having any U.S. Navy facilities. All *Fleet Guides* are restricted for use only by agencies of the U.S. government, and are not issued to private users or to the Merchant Marine.

Much of the information contained in the *Fleet Guides* is similar to that found in applicable volumes of the *Coast Pilots, Sailing Directions*, and the *World Port Index*, but the *Fleet Guides* emphasize areas of special interest and concern to U.S. Navy ships, such as command relationships, operational responsibilities, and munitions support capabilities. A sample page from the logistics section of a Hampton Roads *Fleet Guide* appears in Figure 5-5.

5-8 LOCAL SUPPLY ACTIVITIES

NAVY REGIONAL FINANCE CENTER (NAV REGFINCEN), NORFOLK

Mission

Pay dealers' bills; examine and consolidate disbursing returns; review and consolidate property returns; perform accounting functions; maintain accounts of, and make payment to, personnel of the regular Navy and Naval Reserve on active duty and training duty and personnel of assigned Organized Naval Reserve Units; prepare and pay military and civilian travel claims; receive miscellaneous cash collections; issue transportation requests and meal tickets; settle accounts of disestablished activities as directed; act as field representative of the Comptroller of the Navy; and perform such other functions as may be assigned by the Controller of the Navy.

Command and Support

NAVREGFINCEN Norfolk, administered by a Commanding Officer, is under the command of, and receives primary support from, the Comptroller of the Navy.

Commander, Naval Base, Norfolk, exercises area coordination authority.

Location

The NAVREGFINCEN and the Military Pay Department are in Building E (Figure 3), NAVSTA Norfolk.

Finance Services

Ships without disbursing officer, and desiring to be paid, should submit requests to the Navy Regional Finance Center, Norfolk, immediately upon arrival. Ships in the Little Creek area should contact the Disbursing Office, Building 3015, which is a Branch Office of the NAVREGFINCEN Norfolk. Services include payments for paydays, travel and reenlistments. (See NAV PHIBASE "Logistical Services Bulletin.")

Finance services for ships in the Norfolk Naval Shipyard are provided by the Finance Office, Building 11 (Figure 15).

Official funds required by ship's disbursing officers can be obtained at the office of the Virginia National Bank in Building Z-133 (Figure 3). In an emergency, funds may be arranged for by telephoning the Navy Regional Finance Center, Norfolk.

NAVY PUBLICATIONS AND PRINTING SERVICE OFFICE (NAVPUBPRINTSERVO), FIFTH NAVAL DISTRICT, NORFOLK

Mission

Serve as the publications and printing service for the Fifth Naval District in accordance with the Navy Industrial Fund Charter and assure the economical and efficient provision of the publications and printing requirements for the area.

Tasks

See NAVPUB INST 5450.5 series.

Command and Support

NAVPUBPRINTSERVO Norfolk (Building K-BB, Figure 3), administered by a civilian Director, is under the command of, and receiving primary support from, the Chief of Naval Material (NAVSUPSYSCOM).

5-9 FUEL AND WATER

FUEL

Fueling Arrangements

Requests for fuel to be delivered at piers or in the stream should be placed with the Bulk Fuel Scheduler, Fuel Department, Naval Supply Center, Norfolk, by message, requisition, or telephone, giving as much advance notice as practicable and informing him of complete accounting information, quantity, desired time and place of delivery. Have requisitions prepared and ready to be presented to the Fuel Supervisor at time of fueling.

Navy Special Fuel Oil (NSFO) is not available at the Naval Amphibious Base; however, NSFO can be delivered to ships by barge (YOG). Arrangements may be made with SOPA(ADMIN) Little Creek Sub-Area, at least 72 hours in advance, specifying requirements. In an emergency, NSFO can be obtained at the Craney Island Fuel Terminal at any time with arrangements made through SOPA(ADMIN) and NSC, Norfolk Duty Officer.

Ships in the Norfolk Naval Shipyard should request fuel oil from the Fuel Division, NSC Norfolk, by requisition, message, or telephone, giving as much advance notice as practicable. Deliveries must be reduced to a minimum and must be limited to actual emergencies. Ships should notify the Shipyard Operations Officer of any fueling arrangements proposed to take place in the Shipyard.

Figure 5-5. A sample page from the Hampton Roads *volume of the* Atlantic Fleet Guide.

New editions of the various *Fleet Guide* chapters are published at intervals as conditions warrant, while interim changes are promulgated via the *Notice to Mariners*.

LIGHT LIST

The *Light List* is a series of seven bound volumes describing lighted aids to navigation, and unlighted buoys, daybeacons, fog signals, radiobeacons, and Loran-C coverage in the coastal and intracoastal waters of the continental United States and the islands of Hawaii. (A radiobeacon is a navigation aid incorporating a radio transmitter, to be used by vessels equipped with radio direction finders; see part 3 of this text for further information.) The *Light List* is published annually and distributed by the U.S. Coast Guard; its seven volumes are arranged as indicated:

Volume Number	Contents
I	Atlantic Coast, St. Croix River, Maine, to Ocean City Inlet, Maryland
II	Atlantic Coast, Ocean City Inlet, Maryland, to Little River, South Carolina
III	Atlantic and Gulf Coasts, Little River, South Carolina, to Econfina River, Florida
IV	Gulf Coast, Econfina River, Florida, to Rio Grande, Texas
V	Mississippi River System
VI	Pacific Coast and Pacific Islands
VII	Great Lakes

Information for each aid to navigation described is arranged in eight columns as illustrated in Figure 5-6. Data in the *Light List* are arranged sequentially in the order in which the mariner would encounter the lights and buoys when approaching from seaward. For the purposes of the *Light List*, "from seaward" is defined as proceeding in a clockwise direction around the continental United States and in a northerly direction up the Chesapeake Bay and the Mississippi River.

The range given for lighted aids in column 6 is the nominal range of the light, defined as the distance the light should be seen in "clear" meteorological visibility conditions, disregarding the height of eye of the observer (the nominal range of a navigation light and other related topics are covered in detail in chapter 6). Differentiation of the various kinds of aids to navigation covered is done by the use of different type sizes, as indicated at the top of the following page:

(1) No.	(2) Name and location	(3) Position	(4) Characteristic	(5) Height	(6) Range	(7) Structure	(8) Remarks
			CHESAPEAKE BAY (Maryland) – Fifth District				
	SEVERN AND MAGOTHY RIVERS (Chart 12282)	N/W					
	Severn River						
	Annapolis Harbor **(Main Channel)**						
17970	TRITON LIGHT	38 58.9 76 28.6	Fl (4+5) G 30s 0.3s fl. 1.3s ec. 0.3s fl. 1.3s ec. 0.3s fl. 1.3s ec. 0.3s fl. 3.4s ec. 0.3s fl. 1.3s ec. 0.3s fl. 1.3s ec. 0.3s fl. 1.3s ec. 0.3s fl. 1.3s ec 0.3s fl. 14.8s ec.	25	8	Bronze strucure.	Maintained by U.S. Navy Academy.
17975	NAVAL ACADEMY LIGHT		Q W (R sector)	10	W 7		Red from 220° to 259.2° white from 259.2° to 273.2°. Maintained by U.S Navy Academy.
17980	Naval Station Ramp Obstruction Buoy	38 58.8 76 27.9				White with orange bands and diamond worded DANGER ROCKS.	
17985	Severn River Restricted Area Buoy B	38 59.0 76 28.4				Yellow can.	Marks southwest corner of restriced area.
17990	Severn River Restricted Area Buoy C	38 59.0 76 28.7				Yellow can.	
17995	Severn River Restricted Area Buoy D	38 59.0 76 28.8				Yellow can.	
18000	– Channel Buoy 18					Red nun.	
	Lake Ogleton Entrance						
18005	– LIGHT 1	38 57.1 76 27.5	Fl G 4s	15	4	SG on pile.	
18010	– Daybeacon 2					TR on pile.	
18015	– Daybeacon 3					SG on pile.–	
18020	– Daybeacon 4					TR on pile.	
18025	– LIGHT 5		Fl G 4s	15	4	SG on pile.	
	Chesapeake Harbor						
18030	– ENTRANCE LIGHT 2	38 57.6 75 28.0	Fl R 4s	14		TR on pile.	Maintained from Mar. 1 to Dec. 1. Private aid.
18035	– Buoy 3	38 57.6 76 28.1				Green can.	Maintained from Mar. 1 to Dec. 1. Private aid.
18040	– Buoy 4	38 57.6 76 28.1				Red nun.	Maintained from Mar. 1 to Dec. 1. Private aid.
18045	– Buoy 5	38 57.6 76 28.2				Green can.	Maintained from Mar. 1 to Dec. 1. Private aid.
18050	– Buoy 6	38 57.6 76 28.3				Red nun.	Maintained from Mar. 1 to Dec. 1. Private aid.

Figure 5-6. A sample page of Volume 1 of the Light List.

Place-names are indexed alphabetically in the back of each volume for ease of reference if the name of the location of a particular navigation aid is known. Each volume also contains a luminous range diagram, described in the next chapter, for use in light visibility computations. Much of the data in the *Light List* also appears in abbreviated form near the symbol for the nav aid on a chart, but the *Light List* is much more complete. Information on the *Light List* is updated as changes occur by means of the *Notice to Mariners*.

LIST OF LIGHTS

The *List of Lights* is a series of seven bound publications containing detailed information on the location and characteristics of lighted navigational aids, fog signals, and radiobeacons located in foreign and selected U.S. coastal areas. The *List of Lights* volumes are produced and supported by DMA, and distributed by NIMA. A major point of difference between the *Light List* and *List of Lights* is that the latter does not contain descriptions of unlighted buoys. Only those nav aids incorporating either a light, fog signal, or radiobeacon are described in the *List of Lights*.

Arrangement of the data for each light is basically similar to the *Light List*, except that the visibility given for a light is always expressed as the distance that a light can be seen in clear weather with the eye of the observer 15 feet above water. Column 1 contains numbers assigned to each light by DMAHTC, with any international numbers beneath. Names of lights having a geographic range of 15 miles or more, intended for making a landfall, are printed in bold-faced type; italics are used for all floating aids; and ordinary roman type is used for all other lights. Heights of lights are given in feet, followed by meters in bold-faced type. A sample page from the *List of Lights* appears in Figure 5-7.

An updated set of volumes is published annually, with interim changes being promulgated as they occur via weekly *Notice to Mariners*.

TIDE AND TIDAL CURRENT TABLES

Tide Tables are published annually in four volumes covering the following areas:

Europe and the West Coast of Africa, including the Mediterranean Sea;
East Coast, North and South America, including Greenland;
West Coast, North and South America, including the Hawaiian Islands;
Central and Western Pacific Ocean and the Indian Ocean.

At one time the *Tide Tables* were published by the National Ocean Service, but now they are also published by a private publisher, using data provided by

(1) No.	(2) Name and location	(3) Position	(4) Characteristic	(5) Height	(6) Range	(7) Structure	(8) Remarks
						CANADA– ST. LAWRENCE RIVER	
5156	lle St. Ours Buoy.	N/W 45 56.8 73 13	Fl.(2+1)R. period 6ˢ			Red and green bands marked "Ile St. Ours".	
5160	– Buoy M7.	45 56.5 73 12.7	Fl.G.			Green buoy marked "M7".	
5164	– Buoy M10.	45 55.8 73 13	Fl.R.			Red buoy marked "M10".	
5168	– Buoy M9.	45 55.8 73 12.7	Fl.G.			Green buoy marked "M9".	
5172 H 2452	lle St. Ours Course Range, front.	45 53.1 73 13.0	F.G.	59 18		White skeleton tower, fluorescent orange daymark, black stripe; 45.	Visible in line of range. Emergency light.
5176 H 2452.1	–Rear, 671 meters 182°20′ from front.		F.G.	108 33		Square skeleton tower, flourescent orange slatwork daymark, black stripe; 60.	Visible in line of range, also visible 066°20′-112°23′. Emergency light.
5180	– lle St. Ours Buoy M16.	45 55 73 13	Fl.R.			Red buoy marked "M16".	
5184	Bellmouth Curve Buoy M20, middle of curve.	45 54.5 73 13.1	Q.R.			Red buoy marked "M20".	Radar reflector.
5188	Bellmouth Curve Buoy M19.		Fl.G.			Green buoy marked "M19".	
5192 H 2454	Petite Traverse Range, front.	45 54.7 73 12.5	F.G.	53 16		White circular tower, fluorescent orange rectangular daymark; 16.	Visible in line of range. Also visible 045°30′ to 144°59′.
5196 H 2454.1	– Rear, 518 meters 045° 40′ from front.		F.G.	108 33		Red skeleton tower, fluorescent orange slatwork daymark, black stripe; 53.	Visible in line of range. Also visible 045°30′ to 164°59′.
5200	– Buoy M23.	45 54.1 73 13.1	Fl.G.			Green buoy marked "M23".	
5204	– Buoy M26, N. side of range line.	45 54 73 13.6	Fl.R.			Red buoy marked "M26".	
5206 H 2453	lle St. Ours, S. end.	45 54.3 73 13.5	Q.W.	46 14		6 White square skeleton tower; 21.	Radar reflector.
	RACON		M(—–)		10		
5208 H 2456	Bellmouth Curve.	45 55.2 73 12.5	F.G.	69 21		5 White skeleton tower, fluorescent red and white rectangular daymark; 34.	Maintained from Nov. to May. Visible on a bearing of 024°47′.
5212	Contrecoeur Bend Buoy M29.	45 53.6 73 14	Fl.G.			Green buoy marked "M29".	
5216	– Buoy M32.	45 53.5 73 14.5	Fl.R.			Red buoy marked "M32".	
5220	Contrecoeur Buoy M33.	45 53.1 73 14.5	Fl.G.			Green buoy marked "M33".	
5224 H 2458	Contrecoeur Course Range, front.	45 55.3 73 12.6	F.G.	56 17		White tower, fluorescent orange front face and daymark, black stripe; 22.	Visible in line of range.

Figure 5-7. A sample page from the List of Lights.

NOS. Together they contain daily tide time predictions for some 190 reference ports and listings of time and height difference data for an additional 5,000 locations, referred to as substations. The navigator can construct a daily tide table for any one of the 5,000 substations by applying the time difference data for that substation to the tabulated daily data for its reference port. The *Tide Tables* and their use are described in greater detail in chapter 11.

Tidal Current Tables are published annually in two volumes, one for the Atlantic Coast of North America, and the other for the Pacific Coast of North America and Asia. Like the *Tide Tables*, they were originally published by the National Ocean Service, but are now also published privately, using NOS data. They are arranged somewhat like the *Tide Tables*, with one part of each volume containing daily tidal current predictions for a number of reference stations, and a second part containing time and maximum strength difference data for a large number of subordinate locations. The *Tidal Current Tables* and their use are thoroughly described in chapter 12.

Both sets of tables are distributed to U.S. governmental users by NIMA.

PILOT CHARTS

Pilot charts are special charts of portions of the major ocean basins, designed to assist the navigator during voyage planning. Their name is misleading, because they are small-scale charts of ocean areas, of little use in the actual practice of piloting as defined in this text. Nevertheless, they are invaluable to the navigator. They present, in color-coded graphic form, a complete forecast of the hydrographic, navigational, and meteorological conditions to be expected in a given ocean area during a given time of year. Included is information concerning average tides, currents, and barometer readings; frequency of storms, calms, or fogs; possibility of the presence of ice, including the normal limits of iceberg migration; and a great variety of other meteorological and oceanographic data. Lines of equal magnetic variation—*isogonic* lines—are given for each full degree of variation. The shortest and safest routes between principal ports are also indicated.

Pilot charts are published by DMAHTC in an annual atlas format for certain years for the following regions: Central American waters and the South Atlantic; South Pacific and Indian oceans; the northern North Atlantic; and the Indian Ocean separately.

A portion of a pilot chart of the North Atlantic is shown in Figure 5-8.

DISTANCES BETWEEN PORTS, PUBLICATION NO. 151

The publication *Distances Between Ports, Publication No. 151*, is a useful reference publication made available to the navigator by DMAHTC for use in the preliminary planning stages of a voyage. It is simply a tabulated compendium of great-circle distances calculated along the most frequently traveled sea routes between U.S. and foreign ports, and between foreign ports. A sample page, showing distances from Norfolk, Virginia, to various locations throughout the western hemisphere is shown in Figure 5-9.

This publication finds its greatest use during the preplanning phase of voyage planning to determine estimated total distances between the points of de-

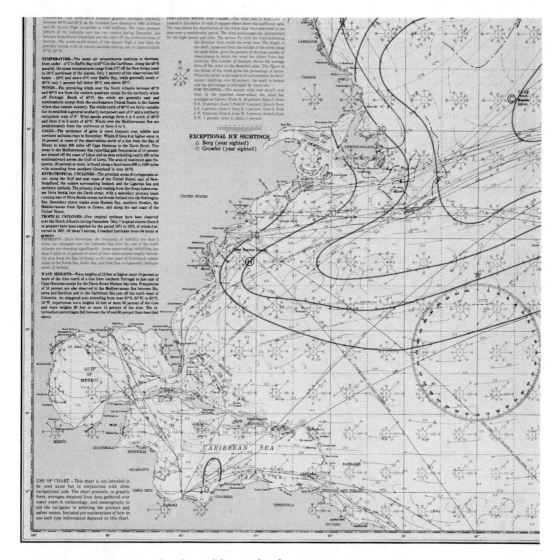

Figure 5-8. *A portion of a pilot chart of the North Atlantic.*

parture and arrival, measured along the most common routes between them. An example of its use to find the initial estimate of the distance for a typical voyage from Norfolk, Virginia, to Naples, Italy, is given in chapter 15.

ALMANACS

Books giving the positions of the various celestial bodies used for celestial navigation, times of sunrise and sunset, moonrise and moonset, and other astronomical information of interest to the navigator are called *almanacs*. There

Iskenderun, Turkey, 1,553
Istanbul, Turkey (south of Greece), 1,321
Istanbul, Turkey (via Corinth Canal), 1,279
Izmir, Turkey, 1,160
Kerch, Ukraine (south of Greece), 1,756
Kerch, Ukraine (via Corinth Canal), 1,714
Larnaca, Cyprus, 1,457
Livorno, Italy, 134
Marsaxlokk, Malta, 596
Mersin, Turkey, 1,495
Messina, Sicily, 515
Milos, Greece, 970
Napoli, Italy, 369
Patrai, Greece, 825
Sant Antioco Ponti, Sardinia, 310
Thessaloniki, Greece, 1,210
Valletta, Malta, 592

NIIGATA, JAPAN
(37°58'00"N., 139°04'00"E.) to:

Junction Points*

Panama, Panama, (via Tsugaru-Kaikyo), 7,648
Punta Arenas, Chile (via Tsugaru Kaikyo), 9,687
Singapore, 2,976

Ports

Busan, Republic of South Korea, 516
Hakodate, Japan, 247
Incheon, Republic of South Korea, 894
Kawasaki, Japan (via Balintang Channel), 762
Maizuru, Japan, 256
Muroran, Japan (via Balintang Channel), 293
Nagoya, Japan (via Balintang Channel), 906
Nevel'sk, Russia, 549
Osaka, Japan (via Shimonoseki Strait), 743
Shimonoseki, Japan, 492
Vladivostok, Russia, 453

NIKISKI, ALASKA, U.S.A.
(60°40'30"N., 151°24'30"W.) to:

Junction Points*

Panama, Panama, 5,055
Punta Arenas, Chile, 7,947

Ports

Antofagasta, Chile, 6,572
Callao, Peru, 5,794
Manzanillo, Mexico, 3,351
Mollendo, Peru, 6,213
Quepos, Costa Rica, 4,665
Victoria, Canada, 1,311

NIKOLAYEVSK-NAAMURE, RUSSIA
(53°07'30"N., 140°43'00"E.) to:

Junction Points*

Panama, Panama (via Unimak Pass), 7,296

Ports

Aleksandrovsk-Sakhalinskiy, Russia (via South Channel), 162
Bukhta Shelikhova, Kuril Islands, Russia (north of Sakhalin), 677
Dagu, China, 1,961
Hong Kong, B.C.C. (via South Channel), 2,394
Matsugahama, Japan, 741
Pago Pago, American Samoa (via Nemuro Strait and La Perouse Strait), 4,913
Qingdao, China, 1,750

San Francisco, California, U.S.A., (via Unimak Pass), 4,099
Seattle, Washington, U.S.A., (via Unimak Pass), 3,748
Shanghai, China, 1,742
Sovetskaya Gavan, Russia (via South Channel), 276
Tomari, Japan, 694
Xiamen, China (via South Channel), 2,138
Zaliv Urkt, Russia (north of Sakhalin), 209

NOME, ALASKA, U.S.A.
(64°29'00"N., 165°25'00"W.) to:

Junction Points*

Panama, Panama, 5,834

Ports

Apra, Guam, 3,686
Auckland, New Zealand, 6,159
Callao, Peru, 6,578
Cordova, Alaska, U.S.A., 1,409
Hakodate, Japan, 2,347
Hong Kong, B.C.C., 4,145
Honolulu, Hawaii, U.S.A., (via Unimak Pass), 2,674
Los Angeles, California, U.S.A., 2,968
Manila, Philippines, 4,379
Pago Pago, American Samoa, 4,748
Petropavlovsk-Kamchatskiy, Russia, 1,317
Pevek, Russia, 700
Prince Rupert, Canada, 1,846
San Francisco, California, U.S.A., 2,636
Sasebo, Japan (via Osumi Kaikyo), 3,373
Seattle, Washington, U.S.A., (via Juan de Fuca Strait), 2,288
Sydney, Australia, 6,280
Valparaiso, Chile, 7,728
Vancouver, Canada, 2,304
Yokohama, Japan, 2,696

NORD-OSTSEE KANAL (EAST ENTRANCE), GERMANY
(54°22'00"N., 10°09'00"E.) to:

Junction Points*

Bishop Rock, England, 727
Ile d'Ouessant, France, 716
Pentland Firth, Scotland, 565
Skagens Odde, Denmark, 224

Ports

Aalborg, Denmark, 207
Aarhus, Denmark, 131
Abenra, Denmark, 66
Amsterdam, Netherlands, 291
Antwerp, Belgium, 398
Assens, Denmark, 65
Baltiysk, Russia, 353
Bremerhaven, Germany, 129
Calais, France, 408
Cherbourg, France, 554
Den Helder, Netherlands, 263
Dover, England, 410
Dundee, Scotland, 501
Dunkerque, France, 386
Emden, Germany, 197
Esbjerg, Denmark, 372
Fecamp, France, 482
Flensburg, Germany, 52
Fredericia, Denmark, 85
Frederikshavn, Denmark, 200
Gavle, Sweden, 615
Gdansk, Poland, 344
Gdynia, Poland, 339
Goteborg, Sweden, 230
Grena, Denmark, 140
Haderslev, Denmark, 72
Halmsted, Sweden, 188
Hamburg, Germany, 91
Hamina, Finland, 747

Hango, Finland, 564
Helsingborg, Sweden, 176
Helsingor, Denmark, 175
Helsinki, Finland, 625
Horsens, Denmark, 131
Kalmar, Sweden, 289
Kalundborg, Denmark, 111
Karlskrona, Sweden, 235
Klaipeda, Lithuania, 400
Kobenhavn, Denmark (via Store Baelt), 202
Kobenhavn, Denmark (via The Sound), 162
Koge, Denmark, 153
Kolding, Denmark, 86
Korsor, Denmark, 73
Kotka, Finland, 694
Kyndbyvaerkets, Denmark, 164
Landskrona, Sweden, 167
Le Havre, France, 522
Limhamn, Sweden, 159
London, England, 439
Lubeck, Germany, 99
Lulea, Sweden, 875
Malmo, Sweden, 205
Masnedsund, Denmark, 149
Middlefart, Denmark, 84
Musko Island, Sweden, 473
Naestved, Denmark, 94
Nakskov, Denmark, 46
Newcastle, England, 439
Norrkoping, Sweden, 459
Odense, Denmark, 117
Oostende, Belgium, 364
Ornskoldsvik, Sweden, 733
Oulu, Finland, 918
Oxelosund, Sweden, 420
Peenemunde, Germany, 155
Pitea, Sweden, 864
Plymouth, England, 639
Porkkala, Finland, 647
Portland, England, 565
Portsmouth, England, 515
Randers, Denmark, 188
Rauma, Finland, 678
Riga, Latvia, 555
Ronne, Denmark, 175
Rostock, Germany, 92
Rotterdam, Netherlands, 323
Ruch'i, Russia, 766
Sankt-Peterburg, Russia, 779
Sonderborg, Denmark, 46
Southampton, England, 533
Stockholm, Sweden, 503
Stralsund, Germany, 104
Sundsvall, Sweden, 684
Swinoujscie, Poland, 186
Szczecin, Poland, 222
Tallinn, Estonia, 603
Travemunde, Germany, 88
Trelleborg, Sweden, 141
Turku, Finland, 576
Umea, Sweden, 751
Uusikaupunki, Finland, 624
Vaa, Finland, 744
Vejle, Denmark, 106
Ventspils, Latvia, 434
Wilhelmshaven, Germany, 126
Wismar, Germany, 84
Zeebrugge, Belgium, 354

NORDENHAM, GERMANY
(53°29'00"N., 8°29'40"E.) Add 5 1/2 miles to Bremerhaven distances.

NORFOLK, VIRGINIA, U.S.A.
(36°51'00"N., 76°18'00"W.) to:

Junction Points*

Bishop Rock, England, 3,101
Cape of Good Hope, Republic of South Africa, 6,802
Fastnet Rock, Republic of Ireland, 2,979
Inishtrahull, Republic of Ireland, 3,083
Montreal, Canada (via Cabot Strait), 1,700
Panama, Panama (via Windward Passage), 1,825
Pentland Firth, Scotland, 3,280

Punta Arenas, Chile, 6,900
Strait of Gibraltar, 3,335
Straits of Florida (via outer route), 980
Yucatan Channel, 1,165

Ports

Argentia, Canada, 1,189
Arkhangel'sk, Russia, 4,399
Banes, Cuba (via Crooked Island Passage), 1,018
Belem, Brazil, 2,832
Belize, Belize, 1,503
Bocas del Toro, Panama (via Crooked Island Passage), 1,853
Bordeaux, France, 3,405
Buenos Aires, Argentina, 5,786
Cadiz, Spain, 3,303
Cape Town, Republic of South Africa, 6,790
Cartagena, Colombia (via Crooked Island Passage), 1,658
Charlotte Amalie, Virgin Islands, 1,296
Cienfuegos, Cuba (via Windward Passage), 1,482
Coatzacoalcos, Mexico, 1,736
Colon, Panama (via Crooked Island Passage), 1,779
Dakar, Senegal, 3,408
Elizabeth Harbor, Bahamas, 848
Fort de France, Martinique, 1,597
Freetown, Sierra Leone, 3,821
Funchal, Madeira Island, 2,907
Georgetown, Guyana, 2,090
Guantanamo, Cuba, 1,117
Halifax, Canada, 790
Hamilton, Bermuda, 683
Ivittuut, Greenland, 2,116
Kingston, Jamaica (via Crooked Island Passage), 1,279
La Guaira, Venezuela, 1,687
La Habana, Cuba (southbound, outside), 985
Lagos, Nigeria, 4,941
Las Palmas, Canary Islands, 3,130
Limon, Costa Rica (via Crooked Island Passage), 1,852
Lisboa, Portugal, 3,129
Liverpool, Canada, 724
Livingston, Guatemala, 1,595
Maracaibo, Venezuela, 1,682
Montevideo, Uruguay, 5,710
Nassau, Bahamas, 758
Nuevitas, Cuba, 1,076
Parrsboro, Canada, 784
Pointe a Pitre, Guadeloupe, 1,527
Ponta Delgada, Azores, 2,401
Port au Prince, Haiti, 1,178
Port Antonio, Jamaica (via Crooked Island Passage), 1,228
Port Castries, St. Lucia, 1,720
Port of Spain, Trinidad, 1,799
Porto Grande, Cape Verde Islands, 2,971
Puerto Barrios, Guatemala, 1,603
Puerto Cortes, Honduras (via Straits of Florida), 1,568
Recife, Brazil, 3,651
Reykjavik, Iceland, 2,677
Rio de Janeiro, Brazil, 4,723
Salvador, Brazil, 4,042
San Juan del Norte, Nicaragua (via Straits of Florida), 1,846
San Juan del Norte, Nicaragua (via Windward Passage), 1,837
San Juan, Puerto Rico, 1,252
Santa Cruz de Tenerife, Canary Islands, 3,057
Santa Marta, Colombia (via Crooked Island Passage), 1,588
Santiago de Cuba, Cuba, 1,167
Santo Domingo, Dominican Republic, 1,329
Santos, Brazil, 4,910
Sint Nicolaashaven, Aruba (via Mona Passage), 1,600
Sint Nicolaashaven, Aruba (via Windward Passage), 1,611
St. John, Canada, 731 St. John's, Canada, 1,277
Sydney, Canada, 996

Figure 5-9. A sample page from DMAHTC Publication No. 151, Distances Between Ports.

are several of these published in the United States both by commercial presses and by the government. The two most commonly used by American marine navigators are the *Nautical Almanac* and the *Air Almanac*, both of which are prepared by the U.S. Naval Observatory and published annually in the United States by the Government Printing Office. They contain data pertaining to celestial navigation from either surface vessels or aircraft, with the *Air Almanac* specially designed to facilitate air navigation. These publications, used in conjunction with several different types of standard navigation celestial sight reduction tables, are discussed in detail in part 2 of this text on celestial navigation.

REFERENCE TEXTS AND MANUALS

There are two reference texts found on board most Navy and other seagoing ships that deserve mention. The first of these is considered the bible for marine navigation—*The American Practical Navigator*, usually called *Bowditch* after its original author. This book was originally published in 1802, and has been popular with mariners ever since. It is not only an encyclopedic compendium of navigational information and technique, but it also contains many tables that are useful in solving navigational problems; one of these, the table of horizon distances for various heights above sea level, is used in chapter 6, which deals with light visibility computation. *Bowditch* is published by the DMAHTC in two volumes as *Publication No. 9.*

The second text is *Dutton's Navigation and Piloting*, published by the Naval Institute Press, Annapolis, Maryland. It has long been regarded as the standard textbook for the professional mariner on all phases of marine navigation.

In addition to the texts mentioned above, DMAHTC and NOS publish a number of other manuals and reference publications of interest to the navigator. Some of the more important of these are described below.

The DMAHTC *Handbook of Magnetic Compass Adjustment and Compensation*, more familiarly known as *Publication No. 226*, is a simply written and well-organized technical manual that provides step-by-step instruction for magnetic compass adjustment and compensation and the procedures for swinging ship. It has been considered the standard reference on its subject by generations of Navy, Merchant Marine, and private navigators.

Radio Navigational Aids, Publication No. 117, provides information on a wide variety of radio navigational aids, warnings, time signals, and other broadcasts throughout the world. It is produced in annual editions by DMAHTC.

The DMAHTC *Radar Navigation Manual, Publication No. 1310*, is an excellent treatment of the use of radar at sea for both navigation and collision avoidance. Although oriented primarily toward the merchant mariner, much of the material is also applicable to Navy and civilian navigators whose vessels are equipped with radar.

PUBLICATION CORRECTION SYSTEM

The *Notice to Mariners* system of correcting charts and publications has already been described in some detail in chapter 4. The system as it applies to publications is briefly reviewed below.

Each navigational publication supported by NIMA and NOS, like every chart, should have a correction card made up for it and entered into the central chart/publication correction card file when the publication is initially received on board. Each time a *Notice to Mariners* is received and found to contain a correction affecting an on-board publication, the navigator or an assistant enters the number of the *Notice* containing the correction onto the correction card of the affected publication. If the publication is currently being used, the correction is then immediately entered into the publication. If, as happens in a great many cases, the publication is not in current use, nor anticipated to be used in the near future, the correction is not entered into the publication. In the latter case, after the *Notice* number has been written on the correction card, the card is returned to the central file until such time as the publication is needed. At that time, the card is pulled out of the file, along with the file copies of old *Notices to Mariners*, and each correction annotated on the card is entered into the publication. In this way, only a few of the many publications on board need be corrected week by week; the rest are not corrected until an occasion for their use arises.

Corrections for many NIMA and NOS publications can also be accessed through the NAVINFONET described in chapter 4.

SUMMARY

This chapter has discussed and described the major navigational publications usually found on board almost all seagoing naval, commercial, and private vessels. The prudent navigator will make extensive use of these and all other applicable publications both in the piloting environment and while at sea.

Comments concerning their publications are always welcomed by NIMA and NOS. Many observers contribute valuable data from time to time concerning currents, aids and dangers to navigation, port facilities, and related suggestions that help keep all publications described in this chapter in agreement with actual conditions. Much of this type of data could be otherwise obtained only by large expenditures of time and money, if at all. NIMA and NOS solicit such cooperation and greatly appreciate the receipt of any data that may increase the accuracy and completeness of their publications.

VISUAL NAVIGATION AIDS

6

In the earliest days of sail the need for lights and buoys to aid the mariner in the navigation of coastal waters and rivers was recognized. It has been established that a lighthouse was built at Sigeum, near Troy, in the seventh century B.C., and in Alexandria in Egypt in the third century B.C. Wood fires furnished their illumination; wood and sometimes coal continued to be used for this purpose until the invention of the electric light in the nineteenth century. The first lighthouse in the United States was built at Boston in 1716, and logs and empty kegs were used to mark channels in the Delaware River as early as 1767.

An aid to navigation, sometimes abbreviated and called a "nav aid," is defined as any device external to a vessel or aircraft intended to assist a navigator in determining position and safe course, or to warn of dangers or obstructions to navigation. The modern visual aids to be discussed in this chapter are those local to the area in which the navigator is operating and which appear on the chart. These include permanent structures attached to the shore or bottom, such as lighthouses, light towers, automated lights and beacons, as well as floating aids, such as lightships and buoys.

Before any aid to navigation can be used, it must first be positively identified. Once having identified the aid, the navigator can then correlate its position on the navigational chart, and proceed to use it to assist in the determination or verification of the vessel's position. By day, the general location, shape, color scheme, auxiliary features, and markings of a navigation aid can all assist in its identification. By night, most of these identifying characteristics become indistinguishable, and the navigator must then obtain all identifying information primarily from the light shown by lighted navigation aids. Fortunately, over the years a rather extensive codelike system of light colors, patterns of flashes, and time cycles has been developed and standardized, which allows a great deal of identifying information to be imparted by the light itself. By carefully analyzing the various visual characteristics of aids to navigation by day, and their lights at night, in conjunction with the nautical chart covering the area and the applicable volumes of the *Light List* and *Coast Pilots* in U.S. waters, and the *List of Lights* and *Sailing Directions* in foreign waters, even an inexperienced navigator should have no undue difficulty in identifying most aids.

In this chapter, the characteristics, identification, and use of lighted and un-lighted aids to navigation will be discussed. Since almost all modern aids of importance are lighted so that they may be used both by day and by night, the chapter will begin with a discussion of the characteristics and functions of navigation lights. The last half of the chapter will describe in some detail the major U.S. and foreign systems of buoyage. As will be seen, the majority of buoys incorporate a light, so a general understanding of navigation light characteristics is also of fundamental value in the study of buoyage systems.

CHARACTERISTICS OF LIGHTED NAVIGATION AIDS

Lights intended to function as navigational aids are broadly classified into two groups—*major* and *minor* lights. Major lights have high intensity and reliability, and are normally placed in lightships, lighthouses, and highly automated light towers and other permanently installed structures. They are intended to indicate key navigational points along seacoasts and channels, and in harbor and river entrances. Major lights are further subdivided into *primary* and *secondary* lights, with the former being very strong lights of long range established for the purpose of making landfalls or coastal passages, and the latter being those lights of somewhat lesser range established at harbor entrances and other locations where high intensity and reliability are required.

Minor lights are automated lights of low to moderate intensity, placed on fixed structures, and intended to serve as navigational aids for harbors, channels, and rivers, and to mark isolated dangers. They usually have the same numbering, coloring, light, and sound characteristics as the buoyage system in the surrounding waters.

Lighthouses and lightships were, until the early 1900s, the principal types of major navigation lights. Today, however, most lightships and manned lighthouses have been replaced by highly reliable automatic light towers and large navigational buoys (LNBs). Most of these and other primary and secondary lights incorporate radio monitoring devices that give warning at centrally located manned stations if their light should fail.

Whether manned or not, the essential requirements for any major light structure are as follows: it should be placed at the best possible location dependent on the physical conditions at the site, should be built up to a sufficient height for the location, and should incorporate a rugged support for the lantern and a housing for the power source, usually electricity. Most structures are painted in order to make them readily distinguishable from their background by day.

Figure 6-1. *(Top) The Ambrose Lightship off New York Harbor was retired in 1967 after having been replaced by a manned light tower. (Bottom) The Portland Lightship off Cape Elizabeth, Maine, was replaced by a large navigation buoy in 1975. (Courtesy U.S. Coast Guard)*

As was mentioned in the introductory section of this chapter, the first objective of the navigator upon sighting a navigation aid is its identification. In the case of a lighted navigation aid at night, the navigator uses three attributes of the light to do this—the light phase characteristic, the duration of its period, and its color. It is important that each of these attributes be understood. A light *phase characteristic* is defined as the light sequence or pattern of light shown within one complete cycle of the light; the light *period* is defined as the length of time required for the light to progress through one complete cycle of changes; and the light *color* refers to the color of the light during the time it is shining.

Light Phase Characteristics

The following is a summary of the more common light phase characteristics seen for lights both in permanent structures and on buoys. Their standard abbreviations appear in parentheses alongside the characteristic.

Fixed (F.) This is a light that shines with steady, unblinking intensity.

Flashing (Fl.) This light appears as a single flash at regular intervals; the duration of the light is always less than the duration of darkness. Normally, a flashing light should not flash more than 30 times per minute.

Quick Flashing (Qk.Fl.) Basically similar to a flashing light, the quick flashing light shows more frequently to indicate a greater degree of cautionary significance. The duration of the flash is less than the duration of darkness, and the light will flash at least 60 times per minute.

Interrupted Quick Flashing (I.Qk.Fl.) By convention, this is a light that quick flashes six times, followed by a time of darkness, with a standard period of ten seconds.

Group Flashing (Gp.Fl.) This light shows groups of two or more flashes at regular intervals. Its period usually appears on the chart, and may be of any duration.

Morse (Mo.[A]) A light showing a pattern of flashes comprising a Morse Code character, by convention normally the letter "A."

Equal Interval (E.Int.) Sometimes called an *Isophase* (Iso.) light, this is a light having equal durations of light and darkness. Its period may be of any length, and will usually be written on the chart.

Occulting (Occ.) Any light that is on longer than it is off during its period is termed an occulting light.

Group Occulting (Gp.Occ.) This is an occulting light, broken by groups of eclipses into two or more flashes. The pattern of the eclipses is indicated on a chart enclosed by parentheses following the basic abbreviation, as for example, Gp.Occ. (2 + 3), which indicates a light interrupted by a group of 2, then 3, eclipses.

Composite This refers to a light showing two or more distinct light sequences within its period. There is no standard abbreviation for a composite light; the fact that it is composite is indicated on a chart by a set of parentheses placed after the basic light characteristic, which contains the number of flashes or occultations (eclipses) within each period. For example, Gp.Fl. (2 + 3) indicates a light that flashes a total of 5 times during its period, in two groups composed of 2, then 3, flashes.

Occasionally a light combines two or more of the preceding characteristics, so as to be distinctive when several lights are located in the same general vicinity. A light abbreviated F.Gp.Fl (2 + 3) on a chart would be a composite fixed

Illustration	Symbols and meaning		Phase description
	Lights which do not change color	Lights which show color variations	
	F.= Fixed...	Alt.= Alternating.	A continuous steady light.
	F.Fl.=Fixed and flashing	Alt. F.Fl.= Alternating fixed and flashing.	A fixed light varied at regular intervals by a flash of greater brilliance.
	F.Gp.Fl. = Fixed and group flashing.	Alt. F.Gp.Fl = Alternating fixed and group flashing.	A fixed light varied at regular intervals by groups of 2 or more flashes of greater brilliance.
	Fl.=Flashing	Alt.Fl.= Alternating flashing.	Showing a single flash at regular intervals, the duration of light always being less than the duration of darkness.
	Gp. Fl. = Group flashing.	Alt.Gp.Fl.= Alternating group flashing.	Showing at regular intervals groups of 2 or more flashes
	Gp.Fl.(1+2) = Composite group flashing.	....	Light flashes are combined in alternate groups of different numbers.
	Mo.(A) = Morse Code.	....	Light in which flashes of different duration are grouped in such a manner as to produce a Morse character or characters.
	Qk. Fl. = Quick Flashing.	....	Shows not less than 60 flashes per minute.
	I.Qk. Fl. = Interrupted quick flashing.	....	Shows quick flashes for about 5 seconds, followed by a dark period of about 5 seconds.
	E.Int.= Equal interval. (Isophase)	....	Light with all durations of light and darkness equal.
	Occ.=Occulting.	Alt.Occ. = Alternating occulting.	A light totally eclipsed at regular intervals, the duration of light always greater than the duration of darkness
	Gp. Occ. = Group Occulting.	....	A light with a group of 2 or more eclipses at regular intervals
	Gp.Occ.(2+3) = Composite group occulting.	....	A light in which the occultations are combined in alternate groups of different numbers.

Light colors used and abbreviations: W = white, R = red, G = green.

Figure 6-2. *Light phase characteristics.*

flashing light in which the flashes appeared in two groups of 2, then 3, flashes. Figure 6-2 illustrates the different light phase characteristics described above, as well as some of the more common light phase combinations. Those phase characteristics that can be further enhanced by changing or alternating the

(1) No	(2) Name and location	(3) Position	(4) Characteristic	(5) Height	(6) Range	(7) Structure	(8) Remarks
			MAINE – First District				
5395	**The Cuckolds Light**	43 46.8 69 39.0	Fl (2) W 6ˢ 1ˢ fl 1ˢ ec. 1ˢ fl 3.0ˢ ec.	59	12	White octagonal tower on dwelling. 48	Emergency light of reduced intensity if main light is extinguished. HORN: 1 blast ev 15ˢ (2ˢ bl).

Figure 6-3A. *The Cuckolds Light data,* Light List.

light colors are also indicated; light color alternations will be more fully explained later.

When working with charted composite light abbreviations, care must be taken to distinguish between the meanings of the numbers in the parentheses following a composite flashing light, and similar numbers following a composite occulting light. In the case of the composite flashing light, the numbers refer to the pattern of the *flashes* of the light. Contrariwise, when the light is a composite occulting light, the numbers within the parentheses denote the pattern of the *eclipses* in the light. This possible confusion is avoided in the *Light List* and *List of Lights*, because the parentheses are not used to describe a composite light. In these publications, the lengths of each eclipse and duration of light of an occulting light are given in tabular form, similar to the entry shown in Figure 6-3A for the Cuckolds Light, a group flashing light. As will be explained in the following section, this information allows the light to be diagrammed in bar-chart form, thus eliminating all doubt as to its appearance.

It should also be mentioned here that on nautical charts, abbreviated phase characteristics and other identifying data are printed in roman type for fixed light structures, and in italics for lighted floating aids, to assist in differentiation between the two types of navigation aids.

Period of a Light

As previously mentioned, the period of a navigational light is defined as the time required for the light to progress through one complete cycle of changes. The periods of all lights except those having either quick flashing or interrupted quick flashing phase characteristics are indicated both on charts depicting them and in the *Light List* and *List of Lights*. For all types of lights, the period is measured from the start of the first flash of one cycle to the start of the first flash of the succeeding cycle.

It is often advantageous to diagram the period of a light in order to aid in its identification. To diagram a light, it is necessary to obtain the length and pattern of its flashes and eclipses from the applicable volume of the *Light List* or *List of Lights*. Once these data are on hand, the light may then be diagrammed in the form of a simple bar-chart, as in the example shown in Figure 6-3B for

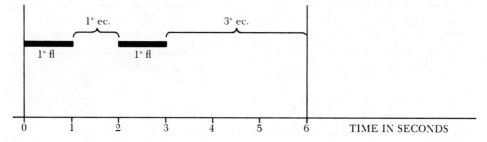

Figure 6-3B. *Diagram of the period of the Cuckolds Light.*

the Cuckolds Light. As can be seen from the diagram, the period of this light is 6 seconds in length. It is a group flashing light, broken by two eclipses of first one then 3 seconds duration.

Color of a Light

There are only three light colors in common use on lighted navigation aids in U.S. and most foreign waters—white, green, and red (some European buoys have yellow lights). All lighted navigation aids, regardless of the color of their light, are symbolized on a chart either by a purple-colored ray in the form of an exclamation point or by a one-eighth-inch purple circle, superimposed over a black dot or small open circle indicating the location of the light. Sections P and Q of *Chart No. 1, Nautical Chart Symbols and Abbreviations*, included as appendix D of this volume, may be referred to for examples. On charts, the color of the light, if other than white, is indicated by the abbreviations "R" for red, and "G" for green, printed near the light symbol. A white light has no abbreviation on a chart. Thus, if a purple light symbol appears on a chart with no color abbreviation nearby, the navigator should assume its color to be white. In the *Light List* and *List of Lights*, however, the color of a white light is indicated by the abbreviation "W."

Alternating Lights

If a light is made to change color in a regular pattern, either by alternately energizing different-colored lights or by passing colored filters around the same light, the light is termed an *alternating* light, abbreviated "Alt." Alternating lights used in conjunction with different phase characteristics, as described in Figure 6-2, show a very distinctive appearance that cannot be easily mistaken. Their use is generally reserved for special applications requiring the exercise of great caution, such as airport beacons, harbor entrance lights, and lighthouses.

Sector Lights

A type of light often confused with alternating lights is the *sector* light. Sector lights are used to warn the navigator of hazards to navigation when

(1) No.	(2) Name and location	(3) Position	(4) Characteristic	(5) Height	(6) Range	(7) Structure	(8) Remarks
			MAINE – First District				
4450	**Two Bush Island Light**	43 57.8 69 04.5	Fl W 5ˢ (R sector)	65	W 25 R 21 42	White square tower.	Red from 061° to 247°. Emergency light of reduced intensity when main light is extinguished. Light displayed continuously. HORN: 1 blast ev 15ˢ (2ˢ bl).

Figure 6-4A. *Two Bush Island Light,* Light List.

approaching the light from certain dangerous arcs or sectors. A sector light may be separated into two or more colored arcs, or rendered invisible in all but one or two narrow arcs, by permanently positioned shields built around the light. If it is multicolored, a color seen in one sector should not be visible in an adjacent sector, except when the observer is located exactly on the border between them. A sector may change the color of a light, as the observer moves around it, but not its phase characteristic or its period. For example, a flashing white light having a red sector will appear flashing red if viewed from within the red sector, but outside of this sector the light will appear flashing white. Sectors may be only a few degrees in width, marking an isolated rock or shoal, or so wide as to mark an entire deep water approach toward shore. Any bearings given to describe the limits of the various sectors are always expressed in degrees true as observed from a hypothetical vessel moving in a clockwise direction around the light. As an example, consider Two Bush Island Light in Penobscot Bay, Maine; its description from the *Light List* appears in Figure 6-4A.

In the Remarks column, this light is described as having a red sector from 061° to 247°, as viewed from seaward. An illustration of this light appears in Figure 6-4B.

Range Lights

Two or more lights in the same horizontal direction, so located that one appears over the other when they are sighted in line, are known as *range* lights.

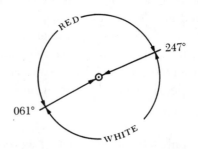

Figure 6-4B. *Sectors for Two Bush Island Light.*

They are usually visible in only one direction, and are mainly used to mark straight reaches of a navigable channel between hazards on either side. The light nearest the observer is called the front light, and the one farthest away, the rear light. By steering the ship in such a way as to keep the front and rear lights always in line, one over the other, the navigator will continually remain on the centerline of the channel.

The lights of ranges may be any of the three standard colors. In U.S. waters, their phase characteristics are standardized, with the front light always quick flashing, and the rear light an equal interval (isophase) light with a period of 6 seconds. Their structures are always fitted with conspicuously colored daymarks for daytime use. Most range lights have very narrow sectors of visibility, and lose brilliance rapidly as the ship diverges from the range. A set of range lights marking the Cape May Harbor entrance in New Jersey, for instance, is visible only a few degrees on each side of the range centerline. When using range lights for navigation, care must be taken to examine the charts beforehand to determine how far a particular set of lights can be safely followed, especially in the vicinity of a bend in the channel.

IDENTIFYING A NAVIGATIONAL LIGHT

As previously mentioned, the navigator should take all three attributes of a navigational light into account when seeking to identify it—the phase characteristic, period, and color. Of the three, the light period is the feature usually weighted most heavily, as it is normally the most unambiguous attribute of a light, especially when other lights with which it might be confused are in close proximity. To eliminate all possible error, no light should ever be considered positively identified until its period is timed by a stopwatch and found to be identical with published information.

One note of caution that should be borne in mind when identifying a sector and especially an alternating light is that the colored segments of the light will almost always be of lesser range than the white segment (see Figure 6-4A). This is so because the same light source is normally used for both the white and colored segments, with one or more colored filters used to produce the latter. In addition to changing the color of the basic white light passing through it, such a filter will also absorb some of the energy of the light, thus decreasing its luminous range. Hence, it is possible at extreme range that only the white portion of an alternating light might be visible, which could lead to false interpretation of the light phase characteristic and an invalid measurement of its period. Likewise, when approaching a sector light on the borderline between a white and a colored sector, only the white sector may be visible initially, until the vessel has drawn within the more limited luminous range of the colored segment of the light.

COMPUTING THE VISIBILITY OF A LIGHT

It frequently happens that the navigator desires to know at what time and position on the ship's track a given light should be expected to be sighted. This is especially important when the ship is making a landfall, as failure to sight certain lights when expected could mean that a serious error had been made in the determination of the ship's position. Moreover, in some circumstances, the navigator will need to determine when the light will be lost from sight, after it has been acquired and used for some time for position-finding, as when making a coastal transit. The basic quantity required for both these determinations is the *computed visibility*, the maximum distance at which the light can be seen in the meteorological visibility conditions in the immediate vicinity. Once this distance is computed, the navigator can use it to swing a corresponding distance arc centered on the light symbol across the projected DR track on the chart, to ascertain the approximate bearing and time at which the light should be seen initially, and if desired, the bearing and time at which the light should no longer be visible.

The following sections describe how to calculate the computed visibility, and how to use the distance thus determined to derive the time and true bearing at which a light may be expected to be sighted, or once acquired, lost from view.

Meteorological Visibility

The concepts of meteorological and computed visibility are often confused by the inexperienced navigator. Meteorological visibility results primarily from the amount of particulate matter and water vapor present in the atmosphere at the location of an observer. It denotes the range at which the unaided human eye can see an unlighted object by day in a given set of meteorological conditions. The more common terms used to describe different meteorological visibility conditions are listed below:

Term	Meteorological visibility range	International visibility codes
Dense to moderate fog	0–500 yards	0–2
Light to thin fog	500 yards–1 mile	3, 4
Haze	1–2 miles	5
Light haze	2–5½ miles	6
Clear	5½–11 miles	7
Very clear	11–27 miles	8
Exceptionally clear	Over 27 miles	9

There are several means whereby the navigator may obtain an estimate of the meteorological visibility range. If in an area covered by marine weather broadcasts, one of these may be tuned in to obtain this as well as other useful weather information (stations broadcasting marine weather are listed in *Publication 117, Radio Navigational Aids*). Probably the most common procedure, however, is simply to determine the meteorological visibility empirically through observations of nearby land and seamarks, other ships in the vicinity, and at night, land lights or minor navigation lights.

In calculating computed visibility distances for lights, it is important to realize that the visibility so computed is not limited to the existing meteorological visibility. A very strong light such as that in a lighthouse might "burn through" even a dense fog and be seen several miles away, yet the prevailing meteorological visibility could be less than 500 yards.

Terms Associated with Light Visibility Computations

In addition to meteorological visibility, there are several other terms that come into use when determining the visibility of a light. These are all illustrated in Figure 6-5.

Horizon distance is the distance measured along the line of sight from a position above the surface of the earth to its visible horizon, the line along which earth and sky appear to meet. The higher the position, the further its horizon distance will be.

Luminous range is the maximum distance at which a light may be seen under the existing meteorological visibility conditions. It depends only on the intensity of the light itself, and is independent of the elevation of the light, observer's height of eye, or the curvature of the earth.

Nominal range is a special case of the luminous range; it is defined as the maximum distance at which a light may be seen in clear weather (considered for this computation to be meteorological visibility of 10 nautical miles). Like luminous range, it takes no account of the elevation, height of eye, or the earth's curvature, and depends only on the light intensity.

Charted range is the range printed on the chart near the light symbol as part of the data describing the light; in some texts, it is referred to as the charted visibility. On U.S. charts edited after June 1973, the charted range is the nominal range rounded to the nearest whole nautical mile. On previously issued charts, and on some current foreign charts, the charted range is the lesser of the nominal range or a geographic range that was then defined as the maximum distance at which a light could be seen in perfect visibility by an observer with a height of eye of 15 feet.

Geographic range is the distance at which a light could be seen in perfect visibility, taking its elevation, the observer's actual height of eye, and the curvature of the earth into account.

Computed visibility (sometimes called the *predicted range*) is the maximum distance at which a light could be seen in the existing visibility conditions, taking the intensity and elevation of the light, the observer's actual height of eye, and the curvature of the earth into account. It is always the lesser of the luminous range or the geographic range.

Figure 6-5 illustrates the relationships among these terms for a situation in which an observer with a height of eye of 100 feet is located at the geographic range to a light having a horizon distance of 12.8 miles and a nominal range of 19.6 miles.

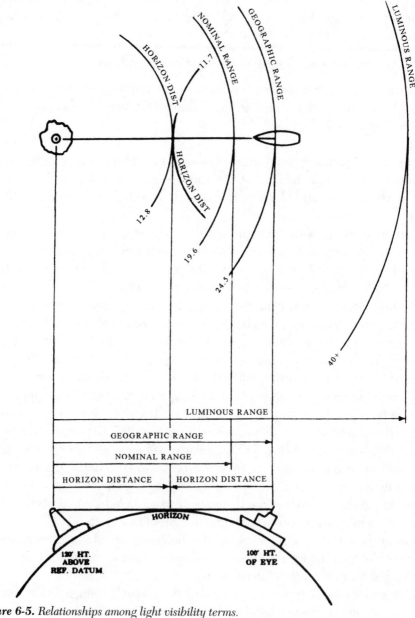

Figure 6-5. *Relationships among light visibility terms.*

Use of the *Light List* and *List of Lights*

In order to compute the visibility of a light, the navigator must first have some information about it, especially its intensity and elevation above the water. The *Light Lists* provide this and other information for lights and buoys in the coastal areas and rivers of the United States and its possessions, and the *Lists of Lights* provide similar information on lights found in foreign waters. The former series is published by the U.S. Coast Guard, and the latter is published by the Defense Mapping Agency Hydrographic/Topographic Center. Descriptions of these publications are given in chapter 5, dealing with navigation publications.

In addition to the information contained in the body of the various volumes, the *Light List* and the *List of Lights* contain another valuable aid for the navigator in the form of the *luminous range diagram*. Figure 6-6 depicts the diagram found in the *Light List* volumes, based on the international visibility code.

By using the diagram, the luminous range of any light of given nominal range can be found for any meteorological visibility condition prevailing. The diagram is entered from either the top or bottom, using the nominal range of the light as an entering argument. The vertical line adjacent to the entry point is followed to the appropriate curve corresponding to the estimated visibility at the time of the observation. The luminous range for the given light under these visibility conditions is then read directly across to the left or right. As an example, consider Two Bush Island Light, from Figure 6-4A on page 86. The nominal range of its white sector is given as 25 miles. Entering the diagram in Figure 6-6 with this number, and going to the 11-mile visibility curve, a luminous range of about 28 miles is obtained from the scale on the sides. For 5½-mile (clear) visibility, the luminous range is about 16 miles, and for 2-mile visibility, the luminous range is just over 7 miles.

If a 10-mile visibility curve were plotted by interpolation on the diagram, it would be seen that the luminous range of any light in 10-mile meteorological visibility conditions would closely agree with the listed nominal range of the light. This would be as expected, since the nominal range and the luminous range in 10-mile visibility conditions are identical by definition.

In situations in which meteorological visibility conditions are different from the curves on the luminous range diagram, an interpolated curve may be drawn in between the two curves bracketing the given value. In practice, however, the navigator usually estimates the visibility conditions by eye, based on experience, and then uses the curve on the diagram closest to the estimated value for luminous range computations.

Procedure for Computing the Visibility of a Given Light

Once the navigator has determined the luminous range of the light through the use of the luminous range diagram, the computed visibility of the light can

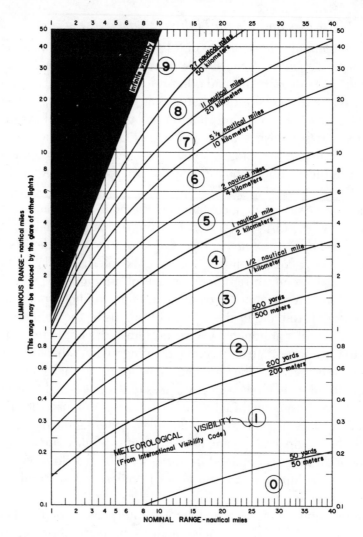

Figure 6-6. *The luminous range diagram,* Light List.

then be found. In addition to the luminous range, the other data that the navigator needs to do this are the height of eye, and the elevation of the light, obtained either from the chart or the applicable *Light List* or *List of Lights* volume. After this information is obtained, the visibility of the light is computed in stepwise fashion as described below.

The first step is to find the horizon distance that a light ray emanating from the light in question would travel before it is cut off by the curvature of the earth, as illustrated in Figure 6-7.

In Figure 6-7 a light ray originating from a light elevated 100 feet above sea level is cut off by the earth's curvature at point P, which lies on the visible hori-

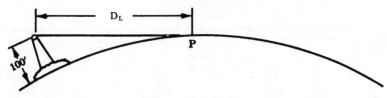

Figure 6-7. Horizon distance for a light of 100-foot elevation.

zon. Although geographic range or horizon distances can be calculated by means of the formula

$$D \text{ (horizon distance in naut. miles)} = 1.169 \sqrt{h \text{ (elevation in feet)}}$$

the usual practice is to look up the desired range or horizon distance from a table of precomputed values, such as shown in Figure 6-8. For an elevation of 100 feet, the table yields a geographic range of 11.7 miles.

The second step is to find the horizon distance for the observer's height of eye, as illustrated by Figure 6-9.

For an observer's height of eye of 50 feet, this distance from the table in Figure 6-8 is 8.3 miles.

As the third step, the horizon distance of the light and the horizon distance of the observer are added together to form the *geographic range* of the light—the distance a light could be seen in perfect visibility, taking the elevation of the light, the observer's height of eye, and the curvature of the earth into account. The geographic range to a light having a 100-foot elevation from an observer with a height of eye of 50 feet is 11.7 + 8.3 = 20.0 miles; it is illustrated in Figure 6-10A.

As an alternative to the computation of the geographic range by the addition of the horizon distances of the observer and the light, there is also included in the *American Practical Navigator (Bowditch)* a table of geographic ranges (Table 13), reproduced in part in Figure 6-10B. Entering arguments are the height of the light in feet or meters on the sides of the table, and the observer's height of eye in feet or meters across the top. The complete table is reproduced in appendix E of this book.

For the fourth and final step in determining the distance at which a given light should be sighted, the navigator figures the *computed* or *predicted visibility* of the light by comparing its luminous range in the existing visibility to the geographic range. If the luminous range is greater than the geographic range, the light should become visible as soon as the vessel comes within the geographic range; under these conditions, the computed visibility is the same as the geographic range. Figure 6-11 is an illustration of this situation for a light having a luminous (nominal) range of 23 miles in 10-mile visibility.

TABLE 12
Distance of the Horizon

Height Feet	Nautical miles	Statute miles	Height meters	Height Feet	Nautical miles	Statute miles	Height meters
1	1.2	1.3	.30	120	12.8	14.7	36.58
2	1.7	1.9	.61	125	13.1	15.1	38.10
3	2.0	2.3	.91	130	13.3	15.4	39.62
4	2.3	2.7	1.22	135	13.6	15.6	41.15
5	2.6	3.0	1.52	140	13.8	15.9	42.67
6	2.9	3.3	1.83	145	14.1	16.2	44.20
7	3.1	3.6	2.13	150	14.3	16.5	45.72
8	3.3	3.8	2.44	160	14.8	17.0	48.77
9	3.5	4.0	2.74	170	15.3	17.6	51.82
10	3.7	4.3	3.05	180	15.7	18.1	54.86
11	3.9	4.5	3.35	190	16.1	18.6	57.91
12	4.1	4.7	3.66	200	16.5	19.0	60.96
13	4.2	4.9	3.96	210	17.0	19.5	64.01
14	4.4	5.0	4.27	220	17.4	20.0	67.06
15	4.5	5.2	4.57	230	17.7	20.4	70.10
16	4.7	5.4	4.88	240	18.1	20.9	73.15
17	4.8	5.6	5.18	250	18.5	21.3	76.20
18	5.0	5.7	5.49	260	18.9	21.7	79.25
19	5.1	5.9	5.79	270	19.2	22.1	82.30
20	5.2	6.0	6.10	280	19.6	22.5	85.34
21	5.4	6.2	6.40	290	19.9	22.9	88.39
22	5.5	6.3	6.71	300	20.3	23.3	91.44
23	5.6	6.5	7.01	310	20.6	23.7	94.49
24	5.7	6.6	7.32	320	20.9	24.1	97.54
25	5.9	6.7	7.62	330	21.3	24.5	100.58
26	6.0	6.9	7.92	340	21.6	24.8	103.63
27	6.1	7.0	8.23	350	21.9	25.2	106.68
28	6.2	7.1	8.53	360	22.2	25.5	109.73
29	6.3	7.3	8.84	370	22.5	25.9	112.78
30	6.4	7.4	9.14	380	22.8	26.2	115.82
31	6.5	7.5	9.45	390	23.1	26.6	118.87
32	6.6	7.6	9.75	400	23.4	26.9	121.92
33	6.7	7.7	10.06	410	23.7	27.3	124.97
34	6.8	7.9	10.36	420	24.0	27.6	128.02
35	6.9	8.0	10.67	430	24.3	27.9	131.06
36	7.0	8.1	10.97	440	24.5	28.2	134.11
37	7.1	8.2	11.28	450	24.8	28.6	137.16
38	7.2	8.3	11.58	460	25.1	28.9	140.21
39	7.3	8.4	11.89	470	25.4	29.2	143.26
40	7.4	8.5	12.19	480	25.6	29.5	146.30
41	7.5	8.6	12.50	490	25.9	29.8	149.35
42	7.6	8.7	12.80	500	26.2	30.1	152.40
43	7.7	8.8	13.11	510	26.4	30.4	155.45
44	7.8	8.9	13.41	520	26.7	30.7	158.50
45	7.8	9.0	13.72	530	26.9	31.0	161.54
46	7.9	9.1	14.02	540	27.2	31.3	164.59
47	8.0	9.2	14.33	550	27.4	31.6	167.64
48	8.1	9.3	14.63	560	27.7	31.9	170.69
49	8.2	9.4	14.94	570	27.9	32.1	173.74
50	8.3	9.5	15.24	580	28.2	32.4	176.78
55	8.7	10.0	16.76	590	28.4	32.7	179.83
60	9.1	10.4	18.29	600	28.7	33.0	182.88
65	9.4	10.9	19.81	620	29.1	33.5	188.98
70	9.8	11.3	21.34	640	29.5	34.1	195.07
75	10.1	11.7	22.86	660	30.1	34.6	201.17
80	10.5	12.0	24.38	680	30.5	35.1	207.26
85	10.8	12.4	25.91	700	31.0	35.6	213.36
90	11.1	12.8	27.43	720	31.4	36.1	219.46
95	11.4	13.1	28.96	740	31.8	36.6	225.55
100	11.7	13.5	30.48	760	32.3	37.1	231.65
105	12.0	13.8	32.00	780	32.7	37.6	237.74
110	12.3	14.1	33.53	800	33.1	38.1	243.84
115	12.5	14.4	35.05	820	33.5	38.6	249.94

Figure 6-8. *Horizon Distance Table 12, from* Bowditch.

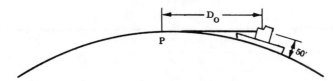

Figure 6-9. Horizon distance for an observer's height of eye of 50 feet.

If the luminous range is less than the geographic range, the light will not become visible until the ship approaches within the luminous range, regardless of the height of eye of the observer. In this situation, the luminous range becomes the computed visibility of the light, as depicted in Figure 6-12, where meteorological visibility has decreased to 5½ miles.

Thus, the distance at which a light should be seen, the computed visibility, is always the lesser of the geographic range or the luminous range. In the absence of a predicted or expected visibility condition in the vicinity of the light, nominal range is normally used as the luminous range for purposes of the light visibility computation. The small boatman not equipped with a *Light List* or *List of Lights* will usually use the charted nominal range as the computed visibility.

Plotting the Computed Visibility of a Light on the Chart

Having determined the computed visibility distance at which a given light should be sighted, the navigator can then find the approximate position and time at which the light should be sighted by referring to the dead reckoning plot. To do this, the navigator swings an arc centered on the charted position of the light, with radius equal to the computed visibility, across the projected DR course line, as shown in Figure 6-13.

The intersection of the arc and the projected DR course line is the approximate position at which the ship should be when the light becomes visible on the horizon. The bearing of the light from this position should be the approximate bearing at which the light will appear; it can be expressed in both degrees

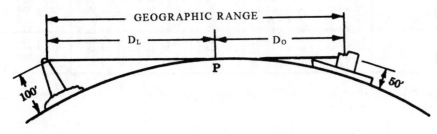

Figure 6-10A. Geographic range for a 100-foot light and an observer's height of eye of 50 feet.

TABLE 13
Geographic Range

Object height Feet	Meters	39 / 12 Miles	43 / 13 Miles	46 / 14 Miles	49 / 15 Miles	52 / 16 Miles	56 / 17 Miles	59 / 18 Miles	62 / 19 Miles	66 / 20 Miles	69 / 21 Miles	Meters	Feet
0	0	7.3	7.7	7.9	8.2	8.4	8.8	9.0	9.2	9.5	9.7	0	0
3	1	9.3	9.7	10.0	10.2	10.5	10.8	11.0	11.2	11.5	11.7	1	3
7	2	10.4	10.8	11.0	11.3	11.5	11.9	12.1	12.3	12.6	12.8	2	7
10	3	11.0	11.4	11.6	11.9	12.1	12.5	12.7	12.9	13.2	13.4	3	10
13	4	11.5	11.9	12.2	12.4	12.7	13.0	13.2	13.4	13.7	13.9	4	13
16	5	12.0	12.4	12.6	12.9	13.1	13.4	13.7	13.9	14.2	14.4	5	16
20	6	12.5	12.9	13.2	13.4	13.7	14.0	14.2	14.4	14.7	15.0	6	20
23	7	12.9	13.3	13.5	13.8	14.0	14.4	14.6	14.8	15.1	15.3	7	23
26	8	13.3	13.6	13.9	14.2	14.4	14.7	15.0	15.2	15.5	15.7	8	26
30	9	13.7	14.1	14.3	14.6	14.8	15.2	15.4	15.6	15.9	16.1	9	30
33	10	14.0	14.4	14.7	14.9	15.2	15.5	15.7	15.9	16.2	16.4	10	33
36	11	14.3	14.7	15.0	15.2	15.5	15.8	16.0	16.2	16.5	16.7	11	36
39	12	14.6	15.0	15.2	15.5	15.7	16.1	16.3	16.5	16.8	17.0	12	39
43	13	15.0	15.3	15.6	15.9	16.1	16.4	16.7	16.9	17.2	17.4	13	43
46	14	15.2	15.6	15.9	16.1	16.4	16.7	16.9	17.1	17.4	17.7	14	46
49	15	15.5	15.9	16.1	16.4	16.6	16.9	17.2	17.4	17.7	17.9	15	49
52	16	15.7	16.1	16.4	16.6	16.9	17.2	17.4	17.6	17.9	18.2	16	52
56	17	16.1	16.4	16.7	16.9	17.2	17.5	17.7	18.0	18.3	18.5	17	56
59	18	16.3	16.7	16.9	17.2	17.4	17.7	18.0	18.2	18.5	18.7	18	59
62	19	16.5	16.9	17.1	17.4	17.6	18.0	18.2	18.4	18.7	18.9	19	62
66	20	16.8	17.2	17.4	17.7	17.9	18.3	18.5	18.7	19.0	19.2	20	66
72	22	17.2	17.6	17.9	18.1	18.4	18.7	18.9	19.1	19.4	19.6	22	72
79	24	17.7	18.1	18.3	18.6	18.8	19.2	19.4	19.6	19.9	20.1	24	79
85	26	18.1	18.5	18.7	19.0	19.2	19.5	19.8	20.0	20.3	20.5	26	85
92	28	18.5	18.9	19.2	19.4	19.7	20.0	20.2	20.4	20.7	20.9	28	92
98	30	18.9	19.3	19.5	19.8	20.0	20.3	20.6	20.8	21.1	21.3	30	98
115	35	19.9	20.2	20.5	20.7	21.0	21.3	21.5	21.8	22.1	22.3	35	115
131	40	20.7	21.1	21.3	21.6	21.8	22.1	22.4	22.6	22.9	23.1	40	131
148	45	21.5	21.9	22.2	22.4	22.7	23.0	23.2	23.4	23.7	24.0	45	148
164	50	22.3	22.7	22.9	23.2	23.4	23.7	24.0	24.2	24.5	24.7	50	164
180	55	23.0	23.4	23.6	23.9	24.1	24.5	24.7	24.9	25.2	25.4	55	180
197	60	23.7	24.1	24.4	24.6	24.9	25.2	25.4	25.6	25.9	26.1	60	197
213	65	24.4	24.7	25.0	25.3	25.5	25.8	26.1	26.3	26.6	26.8	65	213
230	70	25.1	25.4	25.7	25.9	26.2	26.5	26.7	27.0	27.2	27.5	70	230
246	75	25.7	26.0	26.3	26.5	26.8	27.1	27.3	27.6	27.9	28.1	75	246
262	80	26.2	26.6	26.9	27.1	27.4	27.7	27.9	28.2	28.4	28.7	80	262
279	85	26.8	27.2	27.5	27.7	28.0	28.3	28.5	28.8	29.0	29.3	85	279
295	90	27.4	27.8	28.0	28.3	28.5	28.9	29.1	29.3	29.6	29.8	90	295
312	95	28.0	28.3	28.6	28.9	29.1	29.4	29.7	29.9	30.2	30.4	95	312
328	100	28.5	28.9	29.1	29.4	29.6	29.9	30.2	30.4	30.7	30.9	100	328
361	110	29.5	29.9	30.2	30.4	30.7	31.0	31.2	31.4	31.7	31.9	110	361
394	120	30.5	30.9	31.2	31.4	31.7	32.0	32.2	32.4	32.7	32.9	120	394
427	130	31.5	31.8	32.1	32.4	32.6	32.9	33.2	33.4	33.7	33.9	130	427
459	140	32.4	32.7	33.0	33.3	33.5	33.8	34.1	34.3	34.6	34.8	140	459
492	150	33.3	33.6	33.9	34.1	34.4	34.7	34.9	35.2	35.5	35.7	150	492
525	160	34.1	34.5	34.7	35.0	35.2	35.6	35.8	36.0	36.3	36.5	160	525
558	170	34.9	35.3	35.6	35.8	36.1	36.4	36.6	36.9	37.1	37.4	170	558
591	180	35.7	36.1	36.4	36.6	36.9	37.2	37.4	37.7	37.9	38.2	180	591
623	190	36.5	36.9	37.1	37.4	37.6	38.0	38.2	38.4	38.7	38.9	190	623
656	200	37.3	37.6	37.9	38.2	38.4	38.7	39.0	39.2	39.5	39.7	200	656
722	220	38.7	39.1	39.4	39.6	39.9	40.2	40.4	40.7	40.9	41.2	220	722
787	240	40.1	40.5	40.8	41.0	41.3	41.6	41.8	42.0	42.3	42.5	240	787
853	260	41.5	41.8	42.1	42.4	42.6	42.9	43.2	43.4	43.7	43.9	260	853
919	280	42.8	43.1	43.4	43.7	43.9	44.2	44.5	44.7	45.0	45.2	270	919
984	300	44.0	44.4	44.6	44.9	45.1	45.5	45.7	45.9	46.2	46.4	300	984

Figure 6-10B. Excerpt from Table 13, Geographic Range, Bowditch.

G.R. = GEOGRAPHIC RANGE   C.V. = COMPUTED VISIBILITY

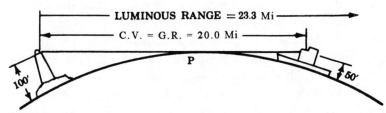

Figure 6-11. *Computed visibility in clear weather for a 100-foot, 23-mile light and an observer's height of 50 feet.*

L.R. = LUMINOUS RANGE   C.V.=COMPUTED VISIBILITY

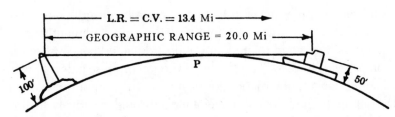

Figure 6-12. *Computed visibility for a 100-foot, 23-mile light in 5½-mile visibility, for an observer at 50 feet.*

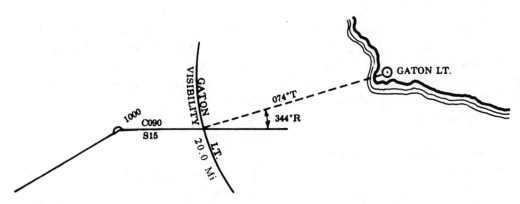

Figure 6-13. *Predicting the ship's position, time, and bearing for a light sighting.*

true and degrees relative, for the benefit of the lookouts. Finally, the approximate time of the sighting can be obtained by computing how long it should take the ship to reach the intercept position.

In practice, the navigator computes the expected distance and time and bearing for each light the ship should sight during a night's steaming by swinging computed visibility arcs for all lights that should be sighted across the ship's intended track for the night. On Navy ships and on most merchant vessels, this information is entered in the night orders for the nighttime deck watch officers, and the watch officer on duty is required to notify the commanding officer or master and navigator if the actual sightings differ too greatly from the precomputed times and bearings. Because of unanticipated changes in the local meteorological visibility conditions, a small amount of variance between actual and expected sighting data is normally to be expected. As mentioned earlier, on some occasions the navigator will also swing the visibility arc for each light across the intended exit track to determine when the ship can be expected to pass out of the range of the light.

BUOYS AND BEACONS

The second category of visual aids to navigation to be described in this chapter, of equal importance to the fixed light structures predominantly described up to this point, is buoys and related immobile aids called beacons (more fully described on pages 103–104). The primary function of these aids is to warn the mariner of some danger, obstruction, or change in the contour of the sea bottom, and to delineate the limits of safe channels through relatively shallow water. A secondary function is to assist the navigator to a limited degree in the determination of the ship's position.

Systems of Buoyage

There are two general systems of buoyage in use throughout the world—the *lateral* system and the *cardinal* system. In the lateral system, the location of each buoy or beacon indicates the direction of the danger it marks relative to the course that should normally be followed; it is best suited for well-defined channels. In the cardinal system, the location of each buoy indicates the approximate true bearing of safe water from the danger it marks. The cardinal system is best suited for marking offshore rocks, shoals, islets, and other dangers in and near the open sea.

Over the last two centuries, most of the major maritime nations of the world began to recognize the need to standardize both of these systems of buoyage. Although the cardinal system became standardized through a number of international accords into what is now the Uniform Cardinal System, the same cannot be said of the lateral system. In 1889, an international marine conference held in Washington, D.C., recommended that under the leadership of the United States, in the lateral system right-hand channel buoys should be painted red, and left-hand buoys black. With the subsequent introduction of lighted aids to navigation about the turn of the century, the logical extension of this pattern led to the use of red or white lights on the right side and green or white lights on the left side of channels. This system became known as the U.S. Lateral System. In 1936, however, a League of Nations subcommittee recommended a system virtually opposite to that of the U.S. Lateral System. The 1936 system became known as the Uniform Lateral System. In this system, black buoys with green or white lights are used to mark the right side of a channel, and red buoys with red or white lights are used for left-side markings.

For the next forty years until the mid-1970s, almost all foreign maritime countries used the Uniform Lateral System to mark their coastal waters and navigable rivers, and the Uniform Cardinal System to mark dangers in offshore areas, while the United States and its possessions used the U.S. Lateral System. In 1977, however, most western European nations began adopting a new system called the IALA (International Association of Lighthouse Authorities) Combined Cardinal and Lateral System, which combined features of both the old uniform systems. By 1980, this system had been implemented throughout western Europe. Then in the spring of 1982, some eighty of the major maritime nations, including the United States, signed an international agreement to implement the IALA system, henceforth called the IALA Maritime Buoyage System, in one of two forms worldwide. The original European system, designated IALA System "A," was prescribed for Europe, Africa, and Asia, and a new IALA System "B," for North, Central, and South America, Korea, and the Philippines. The most significant new feature of both systems is that green instead of black buoys are used as channel markers; in system "A," green buoys mark the starboard side, while in system "B," green buoys mark the port side.

Implementation of IALA System "B" in U.S. waters began in 1983, and was completed in 1989. The main changes from the old U.S. Lateral System were the repainting of black channel buoy colors to green, and the use of new vertically striped red-and-white midchannel buoys and new yellow special buoys.

The remainder of this chapter will discuss the IALA buoyage system in detail.

The IALA System of Buoyage

The buoyage system used in the territorial waters of the United States and its possessions is IALA System "B." It employs an arrangement of colors, shapes, numbers, letters, and light phase characteristics to indicate the side on which the buoy or beacon should be left when proceeding in a given direction with respect to the sea. As was mentioned in chapter 5 in connection with the *Light List*, "from seaward" is defined as proceeding in a clockwise direction around the continental United States, in a northerly direction up the Chesapeake Bay, and in a northerly and westerly direction on the Great Lakes (except southerly on Lake Michigan).

Types of Buoys

Buoys can be classified both by their construction and by their function into several types. Five types of buoy construction are used in U.S. waters:

1. *Can Buoys.* Always unlighted, these buoys are so named because they are built in the cylindrical shape of a can.

2. *Nun Buoys.* Like can buoys, this kind of buoy is always unlighted. They have the shape of a truncated cone above water, resembling the old-time habit of a nun.

3. *Lighted Buoys.* This type of buoy consists of a metal float on which a short skeleton tower of any shape is mounted. The tower supports a lantern powered by electric batteries in the body of the buoy.

4. *Sound Buoys.* These are of framelike construction and are fitted with some type of sound-producing device. There are four main kinds of sound buoys:

Bell Buoys. These are steel floats of any shape, surmounted by a structure able to accommodate a single-toned bell. Most bell buoys are sounded by the motion of the sea, with a very few powered by bottled gas or electricity.

Gong Buoys. Similar in construction to a bell buoy, these sound buoys are fittted with a series of gongs, each having a distinctive sound, rather than with a bell.

Whistle Buoys. Like bell and gong buoys, this type of sound buoy can be of any shape. They are fitted with a whistle or horn powered by a system of bellows actuated by the motion of the sea.

Horn Buoys. Similar to a whistle buoy, a horn buoy is powered by an electric battery or land line rather than the random motion of the sea. Its horn therefore sounds in a regular cadence, rather than in the irregular cadence of most motion-actuated bell, gong, and whistle buoys.

5. *Combination Buoy.* This is a descriptive term that applies to any buoy in which a light and a sound signal are combined, such as a lighted bell, lighted gong, or lighted whistle or horn buoy.

Figure 6-15, page 107, pictures several of the various types of buoys described above.

In addition to the foregoing, *spar buoys*, shaped like long upright poles, are frequently used in European waters, but they are not frequently seen in U.S. waters.

Buoys can also be grouped by function into three main categories:

1. *Channel buoys* are those used to mark the sides, centers, and junctions of channels.

2. *Special buoys* are those used to indicate special areas, dangers, and other features such as anchorages, cable runs, or shoal contours.

3. *Cardinal buoys* are used to indicate special dangers or safe passages, especially offshore. They have not been used much in U.S. waters.

Each of these will be described in the following sections.

Radar reflectors that enhance the buoy as a radar target are now being incorporated in all buoys in U.S. waters to assist in their location and use in radar navigation. All buoys serve as daytime navigation aids; those fitted with lights are available at night, while those equipped with sound signals are more readily located in times of poor visibility, whether in darkness or in fog, rain, or snow.

Chart No. 1. in appendix G of this volume contains various colored plates depicting lighted and unlighted aids to navigation found on the navigable coastal waters, intracoastal waterways, western rivers, and state waterways of the United States. These, especially the ones depicting the IALA systems, should be referred to throughout the remainder of this discussion of buoys. The chart symbol, possible light characteristics, and sample chart descriptive data appear by each type of aid (see also Section Q of *Chart No. 1* for a complete listing of all buoy and beacon symbols). Note that for buoys, the abbreviated chart phase characteristic data are printed in *italics*, as opposed to the roman type used for fixed light structures.

All U.S. coastal waters accessible from the sea are now marked by buoys and daymarks conforming to IALA System "B." Aids on the intracoastal waterway that runs the length of the lower East Coast of the United States follow the same basic color scheme as other U.S. coastal waterways, with the addition of a yellow stripe or border on the various markers. Western rivers and state waterways, however, still use systems unique to them, which are different from either IALA system; since oceangoing mariners will seldom encounter them, only the IALA systems will be described here.

Channel Buoys

As previously mentioned, buoys that are used in and around channels to mark their limits are collectively called *channel buoys*. In U.S. waters, buoys that mark the right side of the channel when returning from sea are always painted red, and they carry even numbers. Buoys that mark the left side are always painted green, and they carry odd numbers. If neither lighted nor equipped with a sound device, right-hand channel buoys must be in the truncated cone shape of the nun buoy, and left-hand channel buoys must be can buoys. If sound buoys are used to mark a channel, bell buoys are placed on the right, and gong buoys on the left. The color-coding scheme of right- and left-hand channel buoys can be remembered by means of an old mnemonic aid— *red right returning*.

Table 6-1.

Characteristics	Type of Channel Buoy			
	Port Hand	*Starboard Hand*	*Safe Water*	*Preferred Channel*
Color	Green	Red	Red/white vertical stripes	Green/red horizontal bands
Shape	Can or combination	Nun or combination	Sphere or combination	Can, nun, or combination
Markings	Odd numbers	Even numbers	May be lettered	May be lettered
Light characteristics	Any [except Gp. Fl. (2 + 1)]	Any [except Gp. Fl. (2 + 1)]	Mo.(A), Iso, Occ, L. Fl. 10s	Gp. Fl. (2 + 1)
Light color	Green	Red	White	Red or green

Note: for preferred channel buoy, topmost band by day and light color by night indicates to which side of vessel the buoy should be left.

Buoys that mark the center of navigable channels, formerly called midchannel buoys in the U.S. Lateral System, are called *safe water marks* in the IALA systems. They are easily recognized by their vertical red-and-white-striped color scheme by day, and by their white Morse "A" (short-long) light at night. Some, particularly in European waters, can also be distinguished by a spherical red "topmark" placed atop the buoy structure. If unlighted, these buoys are spherical in shape, and are therefore often referred to as "beach-ball" buoys by mariners.

Buoys that are used to mark junctions or splits in a channel, formerly called junction buoys in the U.S. Lateral System, are called *preferred channel marks* in the IALA systems. If the preferred channel, when entering from seaward, requires that the buoy be left to port, in order to proceed down the preferred starboard channel, the top band is green; conversely, if the preferred channel calls for leaving the buoy to starboard, the top band is red. Table 6-1 summarizes channel buoys and their various characteristics.

Special Buoys

Special buoys are those used for marking prohibited areas, limits of fishtrap areas, cable crossings, and the like. In the IALA systems, they are painted yellow, and if lighted they carry a yellow light that may have any characteristic other than those used for white lights. In most U.S. waters, they are usually unlighted, and are painted white with orange trim, with lettering on them that identifies their purpose. There is also an *isolated danger mark* in the IALA systems, painted with horizontal black and red bands, and fitted with a group flashing 2 white light if lighted. It is not used much in U.S. waters.

Cardinal Buoys

The preceding types of channel and special-purpose buoys are usually found in and near inshore waters. One other type of buoy is used to indicate dangers and safe passages in offshore waters, especially in Europe and Asia—the *cardinal buoy*. The cardinal buoy is painted in yellow and black horizontal bands. There is a different cardinal mark for each of the four cardinal points of the compass—north, south, east, and west—hence their name. The buoy indicates the direction of safe water from the point of interest marked by the buoy. A cardinal mark may be used to indicate the direction of deepest water or the safe side on which to pass a danger, or to draw attention to a feature in a channel such as a bend, junction, bifurcation, or end of a shoal.

An important feature of a cardinal mark is the two black double-cone topmarks always associated with it, which by their arrangement, coupled with the color scheme of the buoy, identify by day the quadrant indicated by the mark. The shape of the buoy itself does not have significance.

When lighted, the cardinal mark is fitted with a very distinctive white light having a continuous or periodic very quick flashing (i.e., 100 or 120 flashes per minute), quick-flashing, or quick followed by long-flashing (i.e., not less than two seconds in duration) phase characteristic, depending on the orientation of the buoy.

Beacons

In addition to buoys, there is another type of aid to navigation in U.S. waters that is extensively used to delineate channels and mark hazards to navigation. Called *beacons*, they are not floating aids, but rather are rigidly attached to the bottom or to the shore. Beacons may be as large as a lighthouse or as small as a single piling marking a channel or obstruction (see Figure 6-14). If unlighted, they are referred to as *daybeacons*. All beacons exhibit a *daymark* of some sort. In the case of a lighthouse or light tower, the color and shape of the structure constitute the daymarks. For other smaller beacons, particularly those marking channels, these daymarks consist of colored geometric shapes called *dayboards*, affixed to the beacon to convey identifying information and to denote the kind of marker it is (see Figure 6-14).

Dayboards are colored and numbered in basically the same manner as buoys, with red triangular dayboards indicating a right-side boundary, and green square dayboards, a left-side channel boundary, progressing from seaward. The shapes and color schemes for channel, preferred channel, and safe water dayboards, along with their chart symbols, are shown in *Chart No. 1*. Dayboards are often fitted with reflective tape to aid in their identification at close range by searchlight at night.

The basic chart symbols for daybeacons in water areas are squares and triangles. Left-side channel daybeacons are represented by either a green square or an open triangle and the letter "G," and right-side daybeacons by a purple triangle. Safe water and preferred channel daybeacons are symbolized by an open triangle with the color abbreviations "RW" for a red-and-white safe water

Figure 6-14. *Two kinds of typical beacons found in U.S. waters are pictured off the Coast Guard base at Miami, Florida. The traditional type of wooden structure on the left is gradually being replaced by concrete piles fitted with plastic or metal dayboards. Note the minor light atop each beacon. (Courtesy U.S. Coast Guard)*

mark, or "RG" (red uppermost) or "GR" (green uppermost) for preferred channel marks.

In addition to dayboards, *minor lights* are commonly mounted on smaller beacons to facilitate their use at night. These minor lights have the same color and phase characteristics as similarly placed lighted buoys. On charts, they are differentiated from lighted buoys by the use of the standard symbols for structural navigational lights, and by the use of roman instead of italic type to describe their phase characteristics and color.

PART I. PILOTING

Identification of Buoys and Beacons

As can be gathered from the preceding discussion, the size, shape, coloring, signaling equipment, and markings on a buoy or beacon all help to identify it. Note especially the various light phase characteristics and colors associated with the different types of markers, as illustrated in *Chart No. 1*. As mentioned, the *Light List* contains full data on each buoy and beacon in U.S. coastal waters and rivers, regardless of whether it is lighted or unlighted. The *List of Lights* does the same for lighted aids in foreign and U.S. coastal waters. The location of each marker is also shown on charts of the area, with an appropriate symbol indicating its position. In the case of buoys, a small open circle or dot marks the location of its anchor. If the marker is lighted, its location symbol is overprinted by a purple circle or ray. Abbreviated descriptive data are given by the symbol, indicating its color scheme, sound signal, markings, and light color and phase characteristic, if any. Buoy data are italicized, while beacon data are in roman type. Buoys fitted with a radiobeacon are indicated by the abbreviation "Bn" printed nearby.

Light Colors and Phase Characteristics. The light color and phase characteristic schemes used in the IALA buoyage systems are well standardized. Red lights are used only on red channel markers and on preferred channel markers with the topmost band red. Green lights are used only on green channel markers and preferred channel markers with the top band green. White lights are used only on cardinal markers and on safe water channel markers, isolated danger markers, and some U.S. special-purpose buoys, with standardized yellow lights slowly replacing the latter. As was the case for land-based navigation light structures, if the light is colored, the abbreviations "R," "G," or "Y" appear as appropriate on the chart alongside the purple light symbol, printed in italics in the case of buoys. A white light has no abbreviation, so that any lighted buoy or beacon symbol without a color abbreviation nearby should be considered to be fitted with a white light.

The light phase characteristics associated with the various types of buoys and beacons are illustrated in *Chart No. 1*. Port- and starboard-hand channel marker lights may have any characteristic other than that used for preferred channel markers, including fixed, flashing, occulting, quick-flashing, or equal interval. Preferred channel markers are readily identified by their composite group flashing (2 + 1) characteristic. Safe water marks in U.S. waters show a distinctive Morse code pattern, by convention the letter "A." The light phase characteristic always appears in abbreviated form alongside the charted buoy or beacon light symbol, along with the color designation, if any.

Numbering System. The standard numbering scheme used on buoys and daymarks, like light colors and characteristics, greatly facilitates the identification and location of the aid on the chart. In the U.S. IALA System "B," odd numbers are used only on green port channel buoys or green daymarks, while even numbers are found only on red starboard channel buoys or daymarks. The numbers on both sides increase sequentially from seaward. Thus, buoys marked with the numbers 1, 3, 5, *etc.* might mark the port side of a channel entrance, while the numbers 2, 4, 6, *etc.* would appear on starboard-hand buoys. If there are more buoys or daymarks on one side of a channel than the other, some numbers are omitted, so that a buoy marked "4" would never appear opposite one marked "9." If a buoy or daymark is added after a particular system is established, it is marked with the same number as the channel marker preceding it, plus a letter suffix, as for example "1A." Letters without numbers are applied in some cases to red and white vertically striped safe water buoys or daymarks, red and green horizontally banded preferred channel buoys or daymarks, and other special-purpose buoys or daymarks. The markings on a buoy or daymark are indicated on the chart by the placement of quotation marks around the letters or numbers, with italics used for the buoy data. Care must he taken not to confuse an identification mark with the symbol for a color or with a nearby sounding. For example, a charted buoy symbol with the abbreviation RN "2" printed nearby would indicate a red nun buoy marked with the number 2.

Figure 6-15 is a portion of a mock chart illustrating typical chart symbols for various types of visual aids to navigation that might be encountered in U.S. waters.

Differences between U.S. and Foreign Buoyage Systems

Because many Navy and merchant marine navigators frequently find themselves operating in foreign waters, it follows that they should be aware of the major differences between IALA System "B," used to mark most U.S., Central, and South American waters, and System "A," used in Europe, Africa, and Asia. From the point of view of the American navigator, probably the greatest difference is that in IALA System "A," red channel markers and lights are used to indicate the *left* side of channels rather than the right. Even numbers are used to mark these left-side aids, opposite to the practice in the U.S. system. In addition to numbers and letters, many buoys in European waters are stenciled with a name related to their locale or the danger they mark, as for example, "*CASTLE*" buoy or "*SPIT REFUGE*" buoy.

Topmarks. Another major difference between buoys of the U.S. IALA system and those of the old Uniform Lateral and Cardinal Systems, and IALA System "A," is that in many cases, buoys of the latter systems are fitted with a type of visual indicator called a *topmark*. Somewhat similar in concept to the U.S. daymark, topmarks are intended to convey information as to the type of buoy on which they are placed, primarily by their shape, and in the case of cardinal marks, by their orientation. These topmarks are a necessity in the Uni-

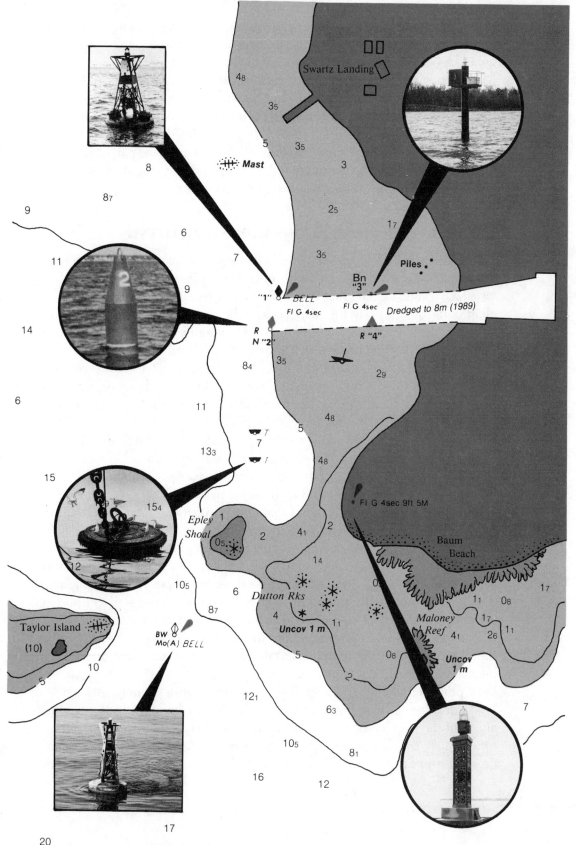

Figure 6-15. *A fictitious nautical chart illustrating symbology for various aids to navigation found in U.S. waters.*

20

form Cardinal, Uniform Lateral, and IALA "A" Systems because, with a few exceptions in the case of the old Uniform Lateral System, the shape of unlighted as well as lighted buoys does not have navigational significance. The different topmarks associated with the various kinds of buoys in the IALA systems are shown in the IALA plates of *Chart No. 1.*

If it is desired, further information on foreign buoyage systems can be obtained from *The American Practical Navigator (Bowditch)* and from the applicable volumes of the *Enroute Sailing Directions* covering particular foreign areas of interest. In the case of lighted buoys, full descriptive data on them are also contained in the appropriate *List of Lights* volume.

USE OF BUOYS AND BEACONS DURING PILOTING

Regardless of the buoyage system in use where a navigator is operating, the first objective upon sighting a buoy or beacon in piloting waters is, as with any visual navigation aid, to identify it and correlate its position on the navigational chart of the area. By day, the shape, color scheme, top- or daymarks, and if distinguishable, any markings can all be used to identify the buoy or beacon, and determine its position relative to the channel or hazard it is marking. At night, the light on a lighted buoy, or minor light on a beacon, will usually serve as the only means of identifying it from any distance. Once the buoy or beacon has been identified, the navigator can then make full use of it as a valuable aid to navigation, subject to the restraints discussed below.

Because a buoy is moored in position by a cable, with enough slack left to allow for rising and falling tide levels, it does not always maintain an exact position directly over its anchor. For this reason, the navigator should never make exclusive use of buoys to determine the ship's position. Buoys should always be regarded as warnings or aids rather than fixed navigation marks, especially during winter mouths or when they are moored in exposed waters. Rough weather, ice, or even collision with a ship may carry a buoy completely away or drag it to a new location. Moreover, the light on a lighted buoy may fail, or motion-actuated sound devices may not sound in calm seas or in heavy icing conditions. A smaller buoy called a station or marker buoy is sometimes placed in close proximity to important navigation aids, such as a sea buoy, to mark its location in case the regular aid is accidentally shifted from station. Beacons, because they are permanently mounted at a stationary location, are much preferred for lines of position (LOPs) if the need arises.

In practice, whenever a ship is proceeding down a channel marked with buoys or beacons, true bearings should continually be taken to the next aids ahead of the ship. If the ship were steaming in a channel marked with buoys, for example, true bearings to buoys on the right side of the channel should be growing ever larger, or in other words, the buoy should show a right-bearing drift. Left-hand buoys should show a left-bearing drift. Initially, the bearings will not drift very rapidly, if at all, while the ship is relatively distant from the buoy, but as she approaches, the drift rate should increase. If the bearings remain unchanged as the ship draws near, the ship is proceeding down a line of bearing to the buoy. Unless the course is adjusted so as to establish a bearing drift, the ship will collide with the buoy—a very distressing event for all concerned, especially the navigator.

When navigating in a narrow channel, buoys and especially beacons can be extremely helpful in the safe piloting of the ship. Because of the time lag inherent in the dead reckoning plot, which will be explained in chapter 8, piloting in a narrow channel with many twists and turns would be very difficult without buoys or beacons to mark the channel boundaries. Although these aids cannot be exclusively relied upon to keep the vessel within the channel, use of buoys and beacons in conjunction with independently obtained fixes and an accurate dead reckoning plot can make piloting in a narrow channel no more difficult than driving an automobile down a city street.

SUMMARY

This chapter has described the characteristics of both lighted and unlighted navigation aids in both U.S. and foreign waters that assist the navigator in the safe direction of the movements of the ship in the piloting environment. The first objective of the navigator upon sighting any visual aid to navigation is its positive identification, in order that its position relative to the channel or a hazard it may be marking can be correlated on the chart. By day, the overall appearance of an aid will usually serve to identify it, and by night, the various attributes of its light, particularly its period, will serve this function. When piloting by means of navigation lights at night, the navigator should be able to predict the time and direction in which a light should be sighted, or once sighted, lost from sight, by means of the computed visibility of the light. Piloting in a winding or narrow channel would be extremely awkward without the availability of buoys and beacons to assist. Piloting would be very difficult if not impossible without the visual aids to navigation discussed herein.

7 NAVIGATIONAL INSTRUMENTS

Almost all professions have a set of tools uniquely associated with them, and the practice of navigation is no different in this respect. This chapter will describe many of the more common instruments used by U.S. Navy, merchant marine, and civilian navigators. Although navigational instruments may be classified in a number of ways, they will be considered here in groups according to the following purposes: instruments to measure direction, distance, and speed; instruments to measure depth; instruments for plotting; and instruments for miscellaneous use.

THE MEASUREMENT OF DIRECTION

The horizontal direction of one terrestrial point from another, expressed as an angle from 000° clockwise to 360°, is termed a *bearing*. There are three different types of bearings with which the surface navigator is concerned, depending on which direction is used as the basic reference, or 000° direction. If a bearing is measured with reference to the ship's longitudinal axis, it is termed a *relative bearing*; if measured with respect to a magnetic compass needle aligned with magnetic north, it is a *magnetic bearing*; and if measured with respect to a gyrocompass repeater having zero gyro error, or a magnetic compass corrected to true north, the angle is termed a *true bearing*. Terminology associated with direction and bearing is fully discussed in chapter 9 of this text, which describes shipboard compasses. In this section, some of the more common instruments used in conjunction with the shipboard compass to determine one of the three types of bearings—relative, magnetic, and true—will be described.

The Azimuth Circle

The term "azimuth" is often used interchangeably with the word "bearing," although technically the former term refers to the bearing of a celestial body,

Figure 7-1. *Azimuth circle for the standard Navy 7½-inch gyrocompass repeater.*

while the latter pertains to the bearing of a terrestrial object. Perhaps the most common device used for obtaining either is the *azimuth* or *bearing circle*. Figure 7-1 depicts the azimuth circle found most often on Navy ships, which is designed to fit on the standard 7½-inch gyro repeater.

The azimuth circle consists of a nonmagnetic brass ring, formed to fit snugly over the repeater face. It can be turned to any desired direction by means of two finger lugs provided on the ring. A pair of sighting vanes, consisting of a peep vane at one end and a vertical wire at the other, are mounted on one diameter of the ring. A reflector of dark glass is also attached to the vane containing the vertical wire, for use in observing the azimuth of celestial bodies.

To observe a bearing, an observer looks through the peep vane, sometimes called the "near" vane, toward the object to be observed or "shot." The observer then rotates the ring until the object appears beyond the vertical wire of the opposite or "far" vane. A reflecting mirror is built into the circle to bring the portion of the compass card directly beneath the far vane into the field of vision; it will also bring the relative bearing scale inscribed around the perimeter of the compass card binnacle into view. The bearing of the observed object is then read by the position of the vertical wire on the compass card, if a true bearing is required, or by its position on the relative bearing scale, if a relative bearing is desired. Care must be taken here to be sure to read the correct bearings, because it is easy for the inexperienced bearing taker to read the wrong scale. The *true* bearings on the compass card always appear in the far vane mirror *above* the relative bearings. As an alternative method of measuring relative bearing, the ring of the azimuth circle is inscribed with bearings in counterclockwise order, from 000° at the far vane; the relative bearing to an object may by obtained by sighting it through the sight vanes and then reading the bearing that appears on the ring directly over the gyro repeater's lubber's line mark representing the ship's head.

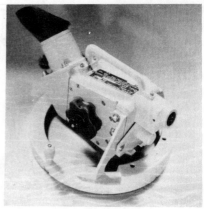

Figure 7-2. Two types of telescopic alidades found on board Navy ships.

A second set of sighting devices designed for observing the azimuth of the sun is attached to the ring at right angles to the sight vanes. One of these devices is a mirror, and the other is a housing containing a triangular reflecting prism. To observe the azimuth of the sun, the azimuth circle is rotated until the sun is above the prism. Its rays strike the mirror and are then reflected onto the prism, which in turn reflects a narrow ray across the graduations on the compass card, where the bearing is read.

Both the far sight vane and the prism housing incorporate leveling bubbles for the purpose of horizontal alignment of the azimuth circle at the moment the bearing or azimuth is shot; if the ring is not level at this time, error may be introduced into the bearing obtained.

The Telescopic Alidade

Another piece of equipment used for observing bearings is the *telescopic alidade*. It is quite similar in construction to the azimuth circle, except that it is fitted with a telescopic observation device rather than a set of sight vanes. The optical system simultaneously projects the image of approximately 25° of the compass card and a view of a built-in level into the field of view of the telescope. The object to which a bearing is to be obtained is sighted in the telescope and its bearing is read off from the compass card. Older models of the telescopic alidade have a straight telescope attached, and newer models have the eyepiece inclined at an angle for ease in viewing. Both types are pictured in Figure 7-2.

In general, most observers seem to prefer the azimuth circle for most applications when observing a bearing, as the field of vision in the telescopic alidade is comparatively narrow. Accurate observations of a lighted navigation aid at night, however, are often facilitated by the use of the telescopic alidade.

Figure 7-3A. *The Fisk stadimeter.*

MEASUREMENT OF DISTANCE

There are two instruments commonly used by the shipboard navigator for the measurement of distance. One of these, radar, is discussed in a separate chapter of this text; the other instrument is the hand-held stadimeter.

The Stadimeter

The stadimeter is normally used most frequently on board Navy ships by the OOD to obtain precise ranges from the OOD's ship to others in a formation. For shorter ranges up to 2,000 yards, it is generally conceded to be more accurate than most surface search radars for this purpose. In piloting, however, the stadimeter is also used as a navigational instrument by the navigator to ascertain accurate distances to navigation aids of known height above the water.

There are two kinds of stadimeters currently in use on board most larger vessels. The Fisk type, pictured in Figure 7-3A, is probably the more common of the two; the other, the Brandon sextant type, is shown in Figure 7-3B.

Figure 7-3B. *The Brandon sextant stadimeter.*

Both stadimeters incorporate two scales. One, located on the index arm of the frame, is the object height scale, graduated in logarithmic form for object heights between 50 and 200 feet. The other, inscribed around the index drum beneath the frame, is the distance scale; it is graduated in a spiral logarithmic scale for distances between 50 and 10,000 yards. Both types of instruments are equipped with a removable telescope fitting in the rear view finder, a reflecting mirror in the right side of the forward view finder, and an index mirror under the rear view finder. To use either instrument, the height of the object to be observed is first set on the index arm scale. Then the object is sighted in the telescope; turning the index drum causes a reflected image in the mirror in the right side of the forward view finder to move up or down relative to the direct image observed through the left side. When the top of the reflected image is superimposed alongside the bottom of the direct image, the distance to the object is read directly from the index drum scale. If the stadimeter is used to obtain the distance to an object having a height less than 50 feet, such as a YP training craft, the index arm height scale is set to some integral multiple of the object's actual height, between 50 and 200 feet. The distance obtained from the index drum scale is then divided by the amount of the height multiple to yield the actual distance. If the stadimeter were used to obtain the distance between two YPs of masthead height 35 feet, for example, the instrument height scale would be set to 70 feet ($2 \times 35'$). If a distance of 620 yards were indicated on the index drum, the actual distance between them would be 310 yards ($620 \div 2$).

Because of the logarithmic scale, distances read on the index drum scale can be read with great precision up to 2,000 yards, but beyond that the accuracy progressively decreases. Prior to use, alignment of the instrument should always be checked by observing the sea horizon. When the reflected and direct images of the horizon appear side by side, the index drum distance scale should read "infinity." If it does not, the mirrors must be adjusted before further use. The stadimeter is a delicate instrument; if handled and stowed properly, however, it is of great value to the navigator in the piloting situation.

MEASUREMENT OF SPEED

There are two kinds of speed with which the navigator of a surface ship is concerned—"true" speed, or speed relative to the earth (often called "speed over the ground" [SOG]), and ship's speed through the water. True speed is normally calculated empirically by measuring the time required for the ship to traverse a known distance. Speed through the water in which the ship is floating is measured both mechanically and empirically by methods to be discussed in this section.

One of the earliest methods developed to measure a ship's speed relative to the water in which she floats is to time the passage of a wood chip or retrievable float along the ship's length. This procedure is still in limited use today by some navigators as a backup to their instruments. A team is set up to periodically throw the chip or float over the bow and time its transit along the side until it reaches the stern, or until a marker knot is reached in the uncoiling retrieving line of the float. Ship's speed is then read from a precalculated table.

Fortunately, the modern navigator has several more sophisticated means of determining speed available. Primary among these are the impeller and pit logs, the doppler speed log, and some recent developments in the determination of speed by the use of satellite and terrestrial electronic navigation systems.

The Impeller and Pit Logs

All marine instruments designed for direct measurement of speed through the water are known as *logs*. Many smaller sail and power boats are equipped with a type of log called an *impeller log* that consists of a sensing device incorporating a small propeller or paddle wheel located beneath the waterline just outside the hull. The speed of rotation of the impeller caused by the water flow past it is mechanically or electrically translated into vessel speed through the water, similar to the speedometer on an automobile.

Probably the most common of all instruments installed in larger vessels to measure speed through the water is the *pitometer* ("pit") log, so called because it incorporates a *pitot tube*. This is a three-foot-long tube generally located near the keel, which can be extended through the ship's hull. It contains two orifices, one of which measures dynamic pressure, and the other, static pressure. Through either a system of bellows or mercury tubes, depending on the type of equipment installed, the difference between the dynamic and static pressure is continually monitored. This pressure difference is proportional to speed. A control unit converts the pressure difference to speed units, and transmits this information to remote locations wherever required on the ship. When using the pit log, the navigator must always remember that when extended the pitot tube increases the ship's draft by about three feet. For this reason, most ships make it a standard procedure to raise the pit log tube whenever the ship is about to transit relatively shallow water.

The Doppler Speed Log

A comparatively recent outgrowth of advances in solid state electronics and sonar technology in recent years is the *doppler speed log*. This instrument depends for its operation on one or more sonar beams projected into the water by a transducer mounted on the bottom of the hull of a vessel. By electronically analyzing the return of the sonar beam pattern reflected back either from the sea bottom or from the water itself in deeper areas, a very accurate determination of speed can be made. Depending on the model of equipment and number of beams, not only fore-and-aft speeds, but also athwartship speed can be measured to the nearest tenth of a knot. Moreover, when bottom echos are being used, the speed determined is the true speed over the ground (SOG), a very desirable quantity to have for navigational purposes. Many models of doppler speed logs provide a readout of depth to the nearest foot, meter, or fathom whenever bottom echos are returned, and most feature a distance-run-since-last-reset odometer.

Doppler sonar systems and associated equipment are discussed in more detail in part 3 of this text.

Electronic Navigation Systems

Perhaps the most esoteric of all means of determining vessel speed is that afforded by the NAVSTAR satellite *Global Positioning System* (GPS). Although the primary purpose of the system is to provide continuous three-dimensional position-fixing capability everywhere on earth, many receivers also have the capability of continuous derivation of speed over the ground. The GPS system and receivers are described more fully in part 3.

Two other long-range electronic navigation systems also provide continuous speed-determination capability, though not as accurately as that possible with GPS. The *Omega Navigation System* is a hyperbolic radionavigation system that makes continuous position-fixing possible everywhere on earth through the generation of an extensive lattice of hyperbolic lines of position that crisscross the entire globe. *Loran-C* is an older radionavigation system, which, like Omega, generates a hyperbolic grid pattern covering most of the littoral zones of North America, Europe, and the Far East. Although both these systems are intended primarily for position-fixing, many Omega and Loran-C receivers now available incorporate a continuous readout of vessel speed over the ground as an important auxiliary feature. Both Omega and Loran-C are covered in detail in part 3.

Use of Shaft RPM to Estimate Speed

In addition to the foregoing methods of direct measurement of vessel speed, another fairly simple yet effective means of estimating approximate speed long familiar to navigators of screw-driven ships is the use of *shaft revolutions per minute* (RPM). For all larger constant-draft vessels having nonvariable pitch propellers, there is a fairly consistent relationship between shaft RPM and speed through the water. When a ship is commissioned, one of her sea trials consists of the preparation of a graph showing speed versus RPM. From this graph, tables are constructed indicating the number of RPMs necessary on the ship's shafts for each knot of speed desired. In fact, on many Navy ships not having any direct speed-measuring devices except a pit log, the navigator will often consider the estimation of speed by shaft RPMs of superior accuracy to this instrument, especially when the pit log has not been calibrated for some

Figure 7-4. *The AN/UQN-4 echo sounder.*

time. Since the hull form and its resistance may change over time because of planned alterations and marine fouling, most navigators will try to schedule their ships to run a measured mile at least once a year, to verify and modify the shaft RPM data as required. Measured miles are marked off by range markers on shore; most major ports have a measured mile laid off in close proximity.

MEASUREMENT OF DEPTH

Measurement of water depth is accomplished on most modern ships not fitted with a doppler speed log primarily by means of an electronic depth finder called the *echo sounder*, or *fathometer*, a now common-use name applied to an early Raytheon model. An echo sounder installed in many Navy ships is the AN/UQN-4, pictured in Figure 7-4. Most echo sounder devices consist of a fixed transducer mounted on the underside of the vessel's hull, and an operating and display console remotely located where needed, usually in the chartroom. The output display may be either a small CRT, an LED display, a strip chart recorder, or a combination of these. In operation, the echo sounder transmits a sound pulse vertically into the water, and computes the depth by measuring the time interval from transmission of the sound signal until the return of its echo from the bottom. Most newer models of echo sounders allow the operator to display the depths thus measured in feet, fathoms, or meters, using any one of several scales.

Echo sounders and their operation, including the AN/UQN-4, and the procedures for navigation by use of the water depth measurements they provide, are elaborated upon in part 3.

When using echo sounder depths, the navigator must always remember that for most models the depths recorded are those from the position of the sonar transducer to the bottom. For actual water depths, the navigator must add the transducer depth of the vessel to all readings.

The auxiliary function of the doppler speed log as a depth finder in shallower water was mentioned in the preceding section. It should be mentioned here that on most models this feature only operates to maximum depths of about 1,000 feet. Beyond this limit, the navigator must revert to the ship's echo sounder for depth determinations.

An alternative mechanical method of measuring depth in piloting waters is the *hand leadline* (pronounced lĕd). The leadline was a reliable means of depth measurement even before the birth of Christ. The standard Navy leadline consists of a lead weight attached to a 25-fathom line marked as follows:

2 fm	2 strips of leather	13 fm	same as 3 fm
3 fm	3 strips of leather	15 fm	same as 5 fm
5 fm	white rag	17 fm	same as 7 fm
7 fm	red rag	20 fm	line with two knots
10 fm	leather with hole	25 fm	line with one knot

An indentation in the bottom of the lead weight allows the application or "arming" of the weight with tallow for the purpose of obtaining a sample of the composition of the bottom.

On vessels making use of the hand leadline, a "leadsman" is stationed on a platform called "the chains," usually located about two-thirds of the distance aft from the bow to the bridge. In reporting depths obtained by use of the leadline, it is customary to refer to the markings simply as marks; reports such as "By the mark five," or "Mark twain," are given.

PLOTTING INSTRUMENTS

Plotting instruments are the least sophisticated yet most fundamental of all of the navigator's tools. Of these, the ordinary *pencil* is the most basic. Most navigators prefer either a no. 2 or 3 pencil; it should be sharpened regularly so as to write all lines clearly and sharply, but lightly enough to facilitate easy erasure. A gum eraser is generally used on nautical charts, since it is less destructive to chart surfaces than the pencil-tip eraser.

The *dividers* and *drawing compass* are standard plotting instruments second only to the pencil in simplicity. The use of these instruments to measure distance on the nautical chart has already been described. Details of their use will not be repeated here, except to mention that the navigator should become practiced enough with the instrument that it can be manipulated with one hand.

The drawing compass is quite similar to the dividers, except that one of the two points is leaded to allow the drawing of an arc. The use of a drawing compass to measure the latitude and longitude of a position and to plot a given position on a chart has already been described in chapter 4. In addition to this, the drawing compass is employed whenever it is necessary to swing a distance arc. The navigator should become proficient in the manipulation of this instrument, like the dividers, with one hand.

There are several types of instruments used for plotting direction, of which the simplest is the *parallel rulers*. The rulers consist of two parallel bars with cross pivot braces of equal length so attached that the bars are always kept parallel as they are opened and closed. In operation, the rulers are laid on the chart's compass rose with a leading edge lying across the center of the rose in the desired direction as indicated on the periphery of the rose. By holding first one bar and then the other, the ruler can be "walked" over the surface of the chart to the location at which a line is to be drawn (Figure 7-5).

Because parallel rules are somewhat slow and are difficult to use if a compass rose is not printed nearby on the chart, several plotting devices have been

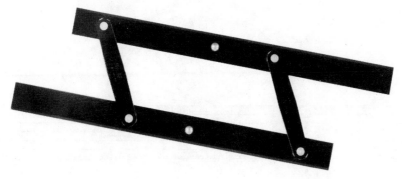

Figure 7-5. *The parallel ruler.*

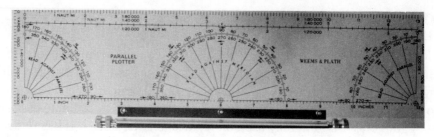

Figure 7-6. *The Weems Parallel Plotter. (Courtesy Weems & Plath, Inc.)*

developed that tend to eliminate the disadvantages of the parallel ruler by incorporating a protractor on the instrument. One of these is the *Weems Parallel Plotter*, shown in Figure 7-6. Printed on its surface are a semicircular protractor for measuring courses and bearings against a charted meridian, and two quarter-circle protractors for use in aligning the plotter against printed parallels of latitude. Once aligned with a meridian or latitude line, the instrument is transferred to the site of the line to be plotted by rolling it over the chart surface using two rollers mounted along one edge. Another somewhat simpler and very popular device is the *Weems Navigation Plotter No. 641*, shown in Figure 7-7. Originally designed for air navigation, this plotter incorporates a straightedge on one side and a protractor on the other that can be quickly aligned with any convenient meridian or parallel of latitude printed on a Mercator chart. Once aligned, the plotter is shifted along the longitude or latitude line until the straightedge is in the desired position for plotting the desired course or bearing line.

Another simple instrument for plotting direction is the *Hoey Position Plotter*, pictured in Figure 7-8. This plotter consists of a clear plastic protractor, with a drafting arm attached and pivoted at the center. The protractor is imprinted with a grid system permitting alignment with any convenient meridian or parallel of latitude on a Mercator chart. A direction line is plotted with this

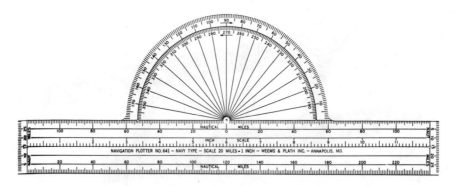

Figure 7-7. *The Weems Navigation Plotter No. 641. (Courtesy Weems & Plath, Inc.)*

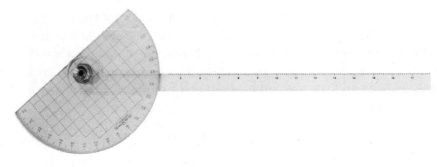

Figure 7-8. *The Hoey Position Plotter.*

instrument by steadying the drafting arm in the desired position with the thumb and forefinger of one hand, while aligning the protractor grid with the other hand. Because it can be aligned with any convenient meridian or parallel of latitude, plotting direction lines with this instrument is quite fast and easily done with a few minutes of practice.

A more complex yet very fast instrument for plotting direction is a type of drafting machine called a *parallel motion protractor*, or PMP; it is shown in Figure 7-9. It is made up of a rotatable protractor, graduated in degrees from 0 to 360, with a drafting arm affixed thereon, which is moved across the chart by a parallel motion linkage fastened to the chart board. The protractor can be aligned and set in position with reference to a chart meridian or latitude line, and the linkage permits movement of the protractor and its drafting arm to any part of the board without any change in the orientation of the protractor disc.

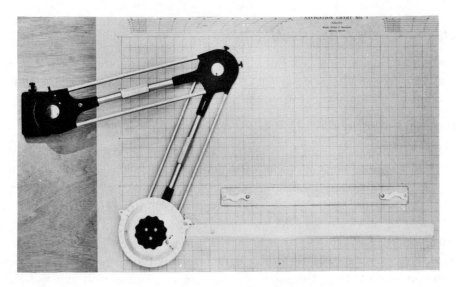

Figure 7-9. *The parallel motion protractor.*

The drafting arm can either be held in position on the protractor by thumb pressure or locked in place by a set screw. Drafting arms marked with various distance scales can be inserted into the protractor, for rapid distance measurement on certain scale charts and plotting sheets. The instrument is a great convenience in laying off courses and bearings and for transferring lines from one location to another on the chart.

It is possible to obtain a very accurate position by measuring the two angles between three adjacent objects on shore. The device used to represent the two angles on a chart is called a *three-arm protractor.* It consists of a central circular disc graduated in 360 degrees, with a central fixed arm and two movable arms attached at the center. After an observation of the two horizontal angles is made, usually by a sextant, the three-arm protractor is set to represent the two angles, one on the right and the other on the left of the fixed arm. The protractor is then set onto the chart, and the central fixed arm is aligned with the central of the three shore objects, as they are represented on the chart. The protractor is moved around, always with the central arm in place on the central object, until the other two arms fall on their respective objects. The center of the protractor then represents the position of the observer. The resultant fix is of very high accuracy, and for this reason, the instrument is extensively employed in chart survey work. Under normal conditions of piloting, however, the three-arm protractor is used primarily as a backup instrument for determining a fix if the ship's gyro fails.

With the exception of the parallel motion protractor, all of these instruments are inexpensive, with their prices for the most part being under $10. They are available commercially from almost every nautical supply store.

MISCELLANEOUS INSTRUMENTS

Several miscellaneous types of instruments usually found on board ship are of great value to the surface navigator. These can be broadly classified as weather instruments, speed-time-distance calculators, and timing devices.

Weather Instruments

On Navy ships not having a separate meteorological division on board, the duties of observing and recording the weather fall to the navigator and the quartermasters. A complete round of weather observations is made hourly on board ship when it is under way, and a synoptic report of these conditions is made to a Naval Oceanographic Command Center four times daily. The basic instruments installed on Navy ships for the purpose of weather observation are the *barometer*, for measuring atmospheric pressure; the *thermometer*, for measuring air temperature; the *psychrometer*, for measuring dry- and wet-bulb temperatures from which the relative humidity and, if desired, the dew point can be calculated; and the *anemometer*, for measuring relative wind speed and direction from which the true wind can be derived.

Complete descriptions and directions for use of all these instruments can be found in the U.S. Navy *NAVOCEANCOMINST* 3144.1 series, *Manual for Ship's Surface Weather Observations*, and in the naval training manual *Quartermaster 3 & 2*. Another excellent reference book for both military and civilian navigators on the subject of weather instruments and their use is *Weather for the Mariner* by W. J. Kotsch (Naval Institute Press, 1983), available at most nautical supply stores.

Speed-Time-Distance Calculators

The navigator has occasion for numerous calculations of speed, time, and distance, based on the formula $D = S \times T$, where D is distance in nautical miles, S is speed in knots, and T is time expressed in hours. If each calculation had to be worked out mathematically, the navigator would have time for little else. Fortunately, several devices have been developed that greatly facilitate the calculation of one of these quantities if the other two are known. These are the 3- and 6-minute rules, the logarithmic speed-time-distance scale, the nautical slide rule, and the electronic calculator.

The first of the "devices" mentioned, the 3- and 6-minute rules, are not really devices at all, but rather are simply special case mathematical formulas for calculating the amount of distance a vessel will traverse at a given speed in knots in 3 or 6 minutes.

The *3-minute rule* can be simply stated as follows:

Distance traveled in yards in 3 minutes = Ship's speed in knots × 100.

The rule can be used to compute quickly the distance a ship proceeding at a given speed will travel in 3 minutes, or, if distance traveled in 3 minutes is given, it will easily yield the ship's speed in knots. For example, if a ship is proceeding at 15 knots, it will travel $15 \times 100 = 1,500$ yards in 3 minutes. If a ship traverses 600 yards in 3 minutes, its speed is $600 \div 100 = 6$ knots.

The *6-minute rule* can be stated as follows:

Distance traveled in miles in 6 minutes = Ship's speed in knots × $\frac{1}{10}$.

Like the previous rule, the 6-minute rule is of great value because it instantaneously yields the number of miles a vessel traveling at a given speed will traverse in 6 minutes. For instance, if a ship were traveling at an ordered speed of 17 knots, it would traverse a distance of $17 \times \frac{1}{10} = 1.7$ miles in 6 minutes. The navigator makes extensive use of both rules, especially during the practice of dead reckoning in the piloting environment, which is discussed in the next chapter.

The *logarithmic speed-time-distance scale* is an extremely useful tool for finding any one of these quantities if the other two are given; it can be thought

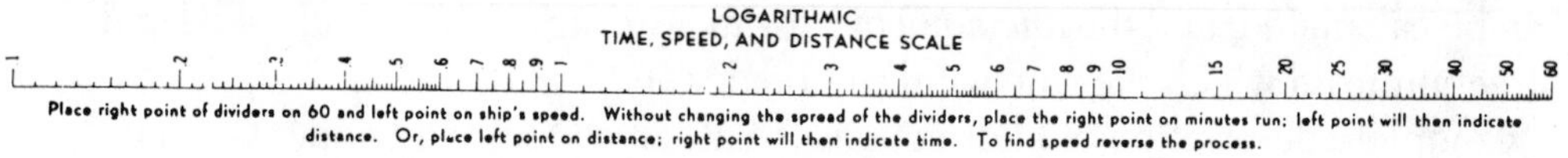

Place right point of dividers on 60 and left point on ship's speed. Without changing the spread of the dividers, place the right point on minutes run; left point will then indicate distance. Or, place left point on distance; right point will then indicate time. To find speed reverse the process.

Figure 7-10. *Logarithmic speed-time-distance scale.*

of as a kind of "paper computer." The scale is usually incorporated on all large-scale charts and plotting sheets, and is also found in nomogram form on the maneuvering board. Figure 7-10 illustrates a typical logarithmic speed-time-distance scale.

This type of scale is always used in conjunction with a pair of dividers. In using it, speed is always represented as the distance a ship would travel in 60 minutes; the distance can be expressed either in yards or nautical miles. If the time in minutes to traverse a certain distance at a given speed is required, for instance, one point of the dividers—the "time" point—is placed at 60 on the scale, and the other—the "distance" point—is set at the distance the ship would proceed in 60 minutes. Without changing the divider spread, the distance point is then moved to the desired distance, and the time required in minutes is read off under the other divider point. Distance is expressed either in miles or yards; once the units of time and distance are chosen for any given problem, they must be consistently used throughout.

As another example, suppose that the distance in miles that a ship will travel in 10 minutes at a speed of 21 knots is required. First, the divider spread is established by placing the "time" point at 60, and the "distance" point on 21 miles. Then the time point is moved to 10 on the scale, and the corresponding distance, 3.5 miles, appears under the distance point. If the distance in yards were required, the distance point could have been initially placed at 42, representing 42,000 yards traveled in 60 minutes at 21 knots. After moving the time point to 10, the number "7" representing 7,000 yards would have appeared under the distance point. Try these manipulations on the scale shown in Figure 7-10.

For a final example, imagine that the speed of a ship that has traveled 4,000 yards in 10 minutes is required. The divider spread may be established in either of two ways. One point may be set at 2, representing 2 miles traveled in 10 minutes, and the other at 10. Or one point may be set at 4, representing 4,000 yards, and the other on 10. After the spread has been thus established, the dividers are then moved so that the time point is placed on 60. If the distance point were originally placed on 2 miles, the speed, expressed as 12 miles per 60 minutes, or 12 knots, can be read directly under the distance point. If the distance point had been originally set on 4,000 yards, however, the speed will be expressed as 24,000 yards traveled in 60 minutes. Speed in knots may be obtained simply by dividing this figure by 2,000 yards. As an alternative method in this last case, the properties of the log scale may be taken advantage of to

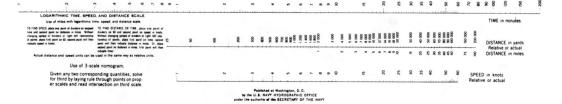

Figure 7-11. *The logarithmic speed-time-distance nomogram on a maneuvering board.*

eliminate the division by 2,000 by setting the time point on 30 instead of 60. The digits appearing under the distance point then represent the ship's speed in knots.

By using the logarithmic speed-time-distance scale in conjunction with a pair of dividers in the fashion described above, any one of the three quantities of the speed-time-distance equation can be found if the other two quantities are given.

The maneuvering board contains a set of three logarithmic scales used in conjunction with one another to solve the equation; the arrangement is referred to as a *nomogram*. The topmost of the three scales is for time, the second is for distance, and the bottom scale is for speed, as shown in Figure 7-11.

To use the nomogram, pencil marks are placed over the two given quantities on the appropriate scales, and the desired third quantity is read by placing a straightedge over the two marks and observing the point of intersection on the third scale. As an alternative method, the topmost time scale can be used as a single logarithmic speed-time-distance scale, as previously explained. Use of the nomogram in this alternative way is considered to produce more accurate results than those obtained by using all three scales in conjunction with one another.

The *nautical slide rule* is an inexpensive plastic device very widely used by surface navigators for the purpose of obtaining a rapid solution of the speed-time-distance equation. In principle it is much like the maneuvering board nomogram described above, except that the three scales have been bent into circular form on a plastic base and covered by a faceplate. Figure 7-12 depicts a nautical slide rule. Distance is expressed both in yards and miles, time is expressed in minutes and hours, and speed is expressed in knots. To use the instrument, the two known factors are set by rotating the slide rule to the appropriate positions, and the third factor appears by the appropriate arrow.

The speed-time-distance equation can be solved using the nautical slide rule in the time it takes to set the two known quantities on it. Its speed, and the fact that it requires no other instrument manipulations, makes it a favorite with all navigators.

Newer devices that are beginning to be used more and more by mariners for performing speed-time-distance conversions, as well as for solving many other types of quantitative navigational problems, are the hand-held *electronic cal-*

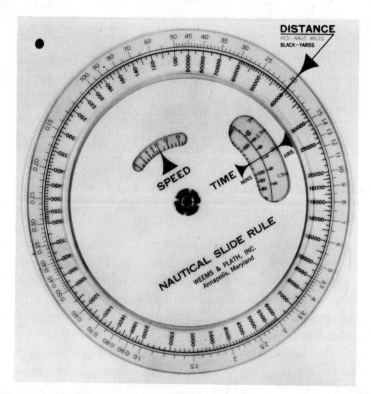

Figure 7-12. *A nautical slide rule. (Courtesy Weems & Plath, Inc.)*

culator and the personal computer. With a few minutes' practice, the navigator can quickly and accurately solve for any one quantity of interest in the basic speed-time-distance formula, given the other two.

More information on the electronic calculator and computer and their steadily increasing marine navigation applications appears in part 2 of this text.

Shipboard Timepieces

The accurate determination of time is of great importance to the navigator both for piloting and celestial navigation. The ship's speed, distance made good, and position are all functions of time, as are most aspects of the daily shipboard routine at sea. It follows, therefore, that the navigator must have reliable timepieces readily available on board ship.

The *chronometer*, considered one of the most accurate mechanical timepieces ever devised, is the principal navigational timepiece normally found on board most oceangoing vessels. Older chronometers such as the Size 85 Hamilton chronometer still found on board some ships (Figure 7-13) contain spring-driven mechanical movements of extremely high precision. Newer chronometers contain movements built around a quartz crystal oscillator powered by a flashlight battery. Regardless of their mechanisms, marine chro-

Figure 7-13. *The Hamilton chronometer.*

nometers are built to withstand shock, vibration, and variations of temperature. Commercial models range in price from about $150 to $600 or more, depending on their size, casing materials, and optional features.

Not even a chronometer can keep absolutely correct time, but the feature that distinguishes a chronometer from ordinary watches and clocks is its rate of gain or loss of time, which is constant over long periods of time. This characteristic allows the chronometer error to be determined with precision, by methods presented in part 2, so that the navigator can arrive at the precisely correct time when desired. The correct time precise to the nearest second is an important requirement for optimal accuracy in celestial navigation.

Most navigators who perform a great deal of celestial navigation at sea will keep at least two chronometers on board for comparison purposes. Most Navy ships have an allowance for three; they are usually kept in the chartroom.

Other timepieces normally found on board most vessels at sea are the *stopwatch* and the *wall clock*. A stopwatch is a necessity for several purposes, among which are the timing of the periods of navigational lights, as discussed in the preceding chapter of this text, and the recording of the time of observation of a celestial body. A special type of stopwatch called a *comparing watch* is sometimes used for this latter purpose; its use is explained in part 2.

Since it would be inconvenient to have to go to the storage location of the ship's chronometers every time a crew member desired the correct time, wall clocks are a common feature on board almost every type of oceangoing vessel. Although not as accurate as a chronometer, many newer battery-powered quartz-crystal models are nearly so. There is a wide variety of marine wall clocks available commercially. Two fairly standard models found on board Navy ships are a spring-driven 24-hour wall clock that must be periodically hand wound with a key, and a newer electric model that will operate for several months on a single flashlight battery.

SUMMARY

This chapter has described many of the more common navigational instruments found on board most seagoing Navy, commercial, and private vessels. The navigator should become very familiar with all of them, as well as with the other more complex navigational aids described elsewhere in this text, such as radar, the gyro, the magnetic compass, the sextant, and the various pieces of electronic gear associated with electronic navigation. The computer is also finding more and more use in the everyday practice of marine navigation, especially aboard U.S. Navy vessels. Just as a surgeon would be lost without a scalpel, the navigator will be quite literally lost without skill in the use of all of the foregoing navigational instruments.

DEAD RECKONING

8

The major concern of the navigator operating at sea is the accurate determination of position. Not only must he or she be continually concerned with the present position of his or her ship, but equally as important, the navigator must also be able to calculate the ship's probable position at future times. To put it in simple terms, the navigator continually seeks answers to the following questions: "Where is the ship now?" and, "Where is the ship going to be in the next minute; the next few minutes; the next hour; or the next day?"

In the eighteenth and nineteenth centuries, and for the first twenty-five years of this century, the navigator calculated the ship's current and future positions by a mathematical process known as deduced—often abbreviated to "ded"—reckoning. This process involved the use of laborious trigonometric computations based on a known point of departure; the technique was necessitated by the inaccuracy and the scarcity of contemporary ocean charts. Although graphic methods made possible by inexpensive yet highly accurate mass-produced modern charts have for the most part replaced the earlier mathematical solutions for the ship's position, the slightly altered term *dead reckoning* is still applied to the process.

Dead reckoning as it is practiced today may be defined as the process of determining a ship's approximate position geometrically by applying to her last established charted position a vector or series of vectors representing all true courses and speeds subsequently ordered. The dead reckoning technique is essential because it is often impossible to obtain an accurate determination or "fix" of the ship's position when required, despite the growing availability of various radionavigaton and satellite navigation systems. Moreover, even after a fix has been obtained and plotted on a chart, it does not represent the location of a moving ship either at the present time or a future time; rather, it only reflects where the ship was at a certain time in the recent past. To obtain the ship's approximate present or future position, the navigator must rely on the dead reckoning plot.

In this chapter, the procedures used by the navigator in maintaining the dead reckoning, or "DR," plot will be examined in detail. Inasmuch as the initial element in every plot is the fix, the first few sections will describe the fix and the techniques for obtaining it. Next, the principles of keeping the DR plot will be fully discussed and illustrated, and finally a variation of the DR plot called "the track" will be explained.

DETERMINING THE FIX

The initial element of the ship's DR plot in any environment is the *fix*, which may be simply defined as the ship's position on the earth's surface at some given point in time. A fix is determined by the intersection of at least two simultaneous *lines of position* (LOPs), each of which may be thought of as a locus of points along which a ship's position must lie. Although the intersection of two such LOPs would be sufficient to obtain a discrete position, it is usually the practice to obtain at least three simultaneous LOPs in determining a fix to guard against the possibility of one or more of them being in error. These LOPs may be obtained in any one of several ways, including but not limited to visual observation of land- or seamarks, observation of celestial bodies, and use of electronic equipment. The remaining sections of this chapter will focus on visual fix techniques; parts 2 and 3 will cover electronic and celestial methods.

The Line of Position

The most accurate LOP possible is obtained by visually observing two or more objects in line, as in Figure 8-1. LOPs of this type are called *visual ranges*; two or more objects are described as being "in range" if they are sighted in line with one another. The visual range LOP is plotted on the chart by placing a straightedge along an imaginary line drawn through the objects sighted in line, and drawing a short segment of the line near the approximate position of the ship on the chart at the time of the observation. The dashed construction lines appearing in Figure 8-1 and in other figures in this chapter are included only for clarity, and should not appear on the chart in practice. LOPs are never extended to the object or navigation aid observed in order to avoid the possibility of the land- or seamark being erased with repeated use of the chart. These, as well as all other single lines of position, are always labeled with the time of observation above the line segment, as shown in Figures 8-1 and 8-2.

Unfortunately, it is seldom possible to observe a range at the precise time at which a fix is desired or required. Consequently, the visual LOP is usually plotted by observing a bearing to a single object by means of the gyro repeater, as in the examples in Figure 8-2.

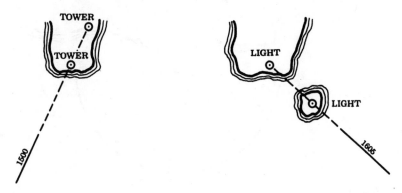

Figure 8-1. *Two illustrations of a visual range LOP.*

The procedure for plotting the visual bearing LOP is basically the same as that used to plot the visual range, except that the straightedge is laid down on the chart along the true line of bearing to the object sighted or "shot"; this bearing is plotted with reference to true north on the chart by means of the chart compass rose. If it is not possible to obtain the true bearing to an object from a gyro repeater, the bearing taker may shoot a relative or magnetic compass bearing to the object. In such cases, the navigator must first convert the relative or magnetic bearing to a true bearing before it can be plotted. As before, only the segment of the line near the ship's approximate charted position is drawn; it is labeled above the line with its time of observation.

If the distance to an object can be found either visually by use of a stadimeter, or electronically by the use of radar, the ship must lie somewhere on a circle centered on the object, with radius equal to the distance measured. An LOP obtained in this manner is technically termed a *distance line of position*, but in normal usage it is generally referred to simply as a "range." A sample range LOP is plotted in Figure 8-3. The range LOP is plotted using a drawing compass set to the appropriate distance on the chart scale. The pivot point of the instrument is placed over the object or landmark shot, and an arc is drawn with the other leaded point. Again, only a small portion of the arc in the vicin-

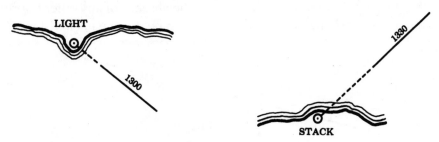

Figure 8-2. *Two illustrations of a visual gyro bearing LOP.*

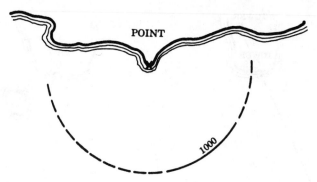

Figure 8-3. *A distance or range LOP.*

ity of the approximate position of the ship on the chart is drawn; it is labeled with the time of the observation above the line.

If a single line of position is obtained by observation of a celestial body, by methods explained in part 2, it should be labeled with the time of the observation, and the name of the body observed, as in Figure 8-4.

The Fix

As was mentioned earlier, if two or more simultaneous lines of position are plotted, the point at which they intersect is the ship's position for that time. Furthermore, it makes no difference how the LOPs were obtained–radar ranges may be crossed with visual bearings, several radar ranges may be crossed, or a radar range and radar bearing to the same object may be used, though this latter technique is not recommended because of the possible radar bearing inaccuracies discussed in chapter 10.

Even though only two simultaneous LOPs are necessary and sufficient to obtain a fix, it is the usual practice to obtain at least three LOPs whenever a fix is to be determined. The third LOP can not only resolve possible ambiguities in selecting the proper fix location when a range LOP is plotted, but also immediately point out any possible errors in the selection, observation, and plotting of the objects shot. Consider the example in Figure 8-5, in which a radar range is simultaneously taken and plotted with a visual bearing.

Is the ship at position A or position B? In the absence of any other information, the navigator should assume the position to be the one in closest proxim-

Figure 8-4. *A single celestial LOP.*

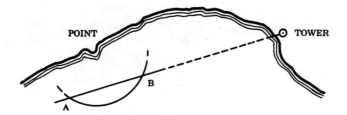

Figure 8-5. *Ambiguity resulting from crossing only two LOPs.*

ity to any navigational hazards present, and make a recommendation for a safe course accordingly. If a third LOP had been obtained at the same time, however, there would be no doubt as to the ship's position (Figure 8-6).

Occasionally, due to uncompensated bearing errors in the gyro repeater, ambiguity will arise even when three simultaneous lines of position are plotted. In such situations, the navigator should again assume the ship to be in the worst possible position and make recommendations accordingly. The ambiguity should then be resolved by immediately shooting another round of bearings. In the situation illustrated in Figure 8-7, the navigator should assume that the ship is located at position C, and recommend a bold alteration of course to starboard to avoid the shoal. If the ship were in fact at position A or B, the margin of safety would thereby be increased.

In order to minimize the effect of possible errors in observed bearings when plotting a fix, the navigator should attempt to optimize the angular spread of the objects shot. If two objects are used, they should be as close as possible to 90° apart; if three objects are shot, they should optimally be 120° apart. The illustrations in Figure 8-8 depict the reasoning behind this rule of thumb by showing the effects of a possible ±5° error in the bearings of two objects 30°, 90°, and 120° apart. The solid lines represent the true bearings to the objects, and the dashed lines, the bearings with a ±5° error applied. As the angle between the objects increases or decreases from the 90° optimum, the shaded areas of uncertainty become larger; at an angle of 90°, the effect of the 5° error is minimized. In similar fashion, it could be shown that for three bearings the optimum angular spread is 120°; if the three bearings were shot to objects all

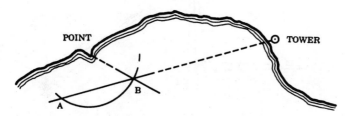

Figure 8-6. *Resolving ambiguity with a third LOP.*

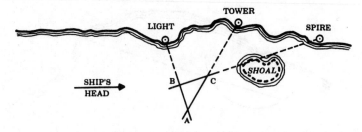

Figure 8-7. *Ambiguity caused by an unknown gyro error.*

located within a 180° arc, as might be the case when transiting a coast, the optimum angle of intersection between adjacent bearings would be 60°.

The symbol used to represent a fix on the navigator's dead reckoning plot is a circle, about one-eighth inch in diameter, placed over the intersection of the LOPs used to determine the fix. Since it can be assumed that these lines of position were shot simultaneously at the time of the fix, it is customary to label only the fix circle with the time at which the LOPs were shot, written horizontally as shown in Figure 8-9. The individual LOPs on which the fix is based are not labeled with times. If the fix was shot on the half minute, a prime (′) is used to denote this fact. If the LOPs were obtained by means of observations of celestial bodies, the names of the bodies should be printed above or below the individual LOP segments.

PRINCIPLES OF THE DEAD RECKONING PLOT

After obtaining and plotting a fix, the navigator is ready to proceed with the DR plot. Through the years, the principles of keeping the dead reckoning plot have been formalized into a set of rules known as the *Six Rules of DR*. If faithfully followed, these rules, used in conjunction with the standard labeling pro-

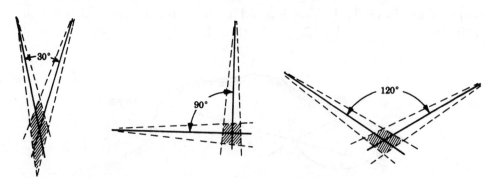

Figure 8-8. *Effect of a ±5° gyro error on two LOPs.*

 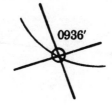 

Figure 8-9. *Examples of fix labels.*

cedures described in this chapter, will result in the navigator's plot being understood by anyone familiar with navigation:

1. A DR position will be plotted every hour on the hour.
2. A DR position will be plotted at the time of every course change.
3. A DR position will be plotted at the time of every speed change.
4. A DR position will be plotted for the time at which a fix or running fix (described later in this chapter) is obtained.
5. A DR position will be plotted for the time at which a single line of position is obtained.
6. A new course line will be plotted from each fix or running fix as soon as it is plotted on the chart.

Navy regulations further specify that the DR plot should be maintained on the largest scale chart available depicting the area in which the ship is operating. For the purposes of the DR plot, each time a fix or running fix is plotted, a vector representing the ordered course and speed is originated from it; in practice, this vector is usually referred to simply as a "course line." The direction in which this vector is drawn represents the ordered true course, referenced to the chart compass rose. Its length represents the distance that the ordered speed would have carried the ship in the time interval under consideration, measured against the chart distance scale. A DR position for any time is always labeled with a semicircle, one-eighth inch in diameter, and its time is printed nearby at an oblique angle. As a simple example, consider the plot in Figure 8-10.

Here, the navigator has obtained and plotted a fix at 0830, and desires to plot a 0900 DR position based on it. A DR course line or vector was drawn in the direction 085°T from the fix, representing the ship's true course to be steered between 0830 and 0900. The ship's ordered speed during this period is to be 15 knots, so the length of the vector measured on the chart distance scale is 7½ miles. The resulting DR position is thus located on the course line and labeled with a semicircle. Note that the course line is labeled with the ordered true course above the line and the ordered speed below. Each time a new course line is laid down in accordance with one of the Six Rules of DR, it should be so labeled.

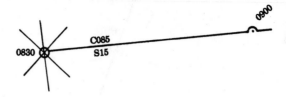

Figure 8-10. Laying down a DR plot.

Both by custom and by Navy directives, once a DR course line has been laid down from a fix, it should be extended at least to two fix intervals. Thus, in the example above, if the navigator were obtaining fixes every half hour, he or she should extend the course line to 0930 and plot another DR position on it for that time.

As a more complicated example of the utilization of the rules of DR in maintaining the DR plot, consider the following excerpt from a ship's *Deck Log*:

0800. With Pollock Rip Light bearing 270°T at 6 miles, departed
      anchorage L-21 on course 090°T, speed 15 knots.
0930. Changed speed to 10 knots to avoid a sailboat.
1000. Changed course to 145°T, increased speed to 20 knots.
1030. Changed course to 075°T.
1106. Obtained radar fix on lighthouse, bearing 010°T, range 7 miles.
1115. Changed course to 090°T, changed speed to 18 knots to arrive
      at operating area S-2 on time at 1215.

The corresponding DR track appears in Figure 8-11.

As previously mentioned, fix positions are shown by one-eighth-inch diameter circles, and DR positions are indicated by semicircles. Fixes and DR positions are further differentiated by printing the times for fixes horizontally, while times for DR positions are printed at an oblique angle.

Beginning at the 0800 position, the navigator plotted the ship's true course line in the direction 090°T, the ordered course. This line was labeled with the ordered course above the line, and the ordered 15-knot speed below. Since the ship traveled 15 miles in one hour at 15 knots, the prescribed 0900 DR position was plotted on the 090° course line 15 miles as measured on the chart distance scale from the 0800 position. Due to the change in speed at 0930, a DR position for this time was plotted 7.5 miles further on. At 1000 the navigator plotted the 1000 DR position, changed the direction of the course line to 145°T, and labeled it with the new ordered course and speed. At 1030 the direction of the course line was changed to 075°T; the 1100 DR position was then plotted to scale on this line. At 1106, a new fix was obtained; hence the new DR course line commenced at that point. Note that the old course line

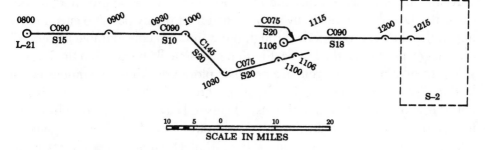

Figure 8-11. *An extended DR plot.*

ends at the 1106 DR position. To enhance the neatness of the plot, any portion of the old course line extending beyond the 1106 DR position should be erased. To complete the plot, the DR position for the 1115 course and speed change was plotted, and the plot was then extended to 1215 to obtain an estimate of the ship's position for that time.

Notice in the example above that it took the navigator 9 minutes from the fix at 1106 to decide on the new course and speed to make the rendezvous position on time. It was not until 1115 that the ship was placed on its new heading and speed. Because of the time lag from the time of obtaining and plotting a fix until the information it depicts can be acted upon, the navigator should lay out a DR course line from any new fix or running fix as soon as it is plotted and, as previously mentioned, extend it at least two fix intervals.

As explained in chapter 3, in confined waters, the navigator usually updates the ship's position by taking a fix at least every 3 minutes (see Figure 3-4); a DR course line is then run out for 6 minutes, and two 3-minute DR positions are plotted on it. This particular interval is chosen in order to allow the navigator to use the 3-minute rule described in the preceding chapter to instantly calculate the distance the ship will proceed along the DR course line from each new fix in 3 minutes. Plotting 3-minute DR positions in this manner not only allows the navigator to see where the ship's approximate position is at frequent intervals, but it also allows a comparison to be made between the calculated and true ship's position every 3 minutes. If the ship is being set to the right or left of its DR position by the effects of a current, this fact becomes immediately apparent and the navigator can recommend a small change of course or speed to compensate for it. In less-restricted waters, the navigator may plot a fix only once every 15 minutes; each time a new fix is plotted, a DR course line is run out from it for 30 minutes, with two 15-minute DR positions plotted on it.

If for some reason the navigator cannot obtain a new fix when expected, the dead reckoning plot is extended until such time as a new fix is obtainable.

Proper labeling and neatness are always of paramount importance in keeping the DR plot. A sloppy plot almost always leads to errors in the plot.

It must be emphasized that the DR position is only an approximation, because the DR plot intentionally ignores the effect of any possible current acting on the ship as it proceeds through the water. If a considerable amount of time has elapsed since the last fix, the ship may actually be far from the DR position. In piloting waters, reliance on a DR plot under these conditions could lead to disaster. When operating far at sea, however, the frequency of fixes is not nearly so critical. Often in the case of vessels lacking any useable electronic navigation equipment, many hours or even days may elapse between successive fixes. Under these circumstances, the DR plot assumes especially great importance, for it may be the only means of estimating the ship's position with any degree of accuracy.

THE RUNNING FIX

It sometimes happens, especially when piloting in low visibility, that the navigator can obtain a line of position from only one object at a time as the vessel proceeds. In such circumstances the navigator may desire to advance a line of position obtained earlier to the time at which a later LOP was obtained. The position thus produced is termed a *running fix*, because the ship has proceeded or "run" a certain distance during the time interval between the two LOPs. The earlier LOP is advanced to the time of the later LOP by using the ship's DR positions plotted for the two times. This is the reason for Rule 5 of the Six Rules of DR, which requires a DR position to be plotted at the time of obtaining a single LOP.

Two examples of the procedure for plotting a running fix are described below. The first involves advancing a bearing LOP, with no intervening course change; and the second concerns the advancement of a range LOP, with a course and speed change occurring in the interval between the times of the first and second LOPs.

For the first example, consider the DR plot in Figure 8-12A. In this figure, a single line of bearing was obtained and plotted on the chart at 1805, together with a DR position for this time, as shown. At 1830, a second line of bearing was obtained. To produce a running fix, the inexperienced navigator will normally draw a construction line connecting the two DR positions involved. In this example the DR course line may be used, since there was no course change between 1805 and 1830. Next, a point on the 1805 LOP is selected and advanced parallel to the line connecting the two DR positions, an amount of distance equal to the distance between them. At this stage, the plot appears as in Figure 8-12B. Finally, the remainder of the advanced 1805 line of bearing is drawn through this advanced point, parallel to the original LOP. To denote the fact that it is an advanced LOP, it is labeled as in Figure 8-12C with the original time and the time to which it was advanced. The running fix is located at

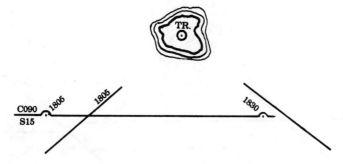

Figure 8-12A. *A 25-minute DR plot, with two LOPs. The 1830 LOP, plotted to form the 1830 running fix, is not labeled.*

the point at which the advanced LOP and the second LOP intersect; it is labeled with a one-eighth-inch diameter circle and the time, as shown. To ensure differentiation between a regular fix and a running fix, the abbreviation "R Fix" is always placed near the running fix symbol in addition to the time, which is written horizontally.

Since the time of the running fix and the time of the second LOP are the same, it is unnecessary to label the second LOP. After plotting the running fix, a new DR course line should be originated from it. The old course line ends at the DR position plotted for the time of the second LOP.

As an example of the technique used to obtain a running fix by advancing a range LOP, consider the DR plot shown in Figure 8-13A. In Figure 8-13A, a radar range was obtained and plotted at 1805, along with the corresponding DR position, as shown. At 1820 the ship changed course and speed; at 1830 a second LOP was obtained, this time a line of bearing. The same basic procedure previously described is used to obtain a running fix. A construction line may first be drawn connecting the two DR positions for 1805 and 1830. Since at least three points on the range arc would have to be advanced in order to reconstruct it, the simpler technique of advancing its center is used instead.

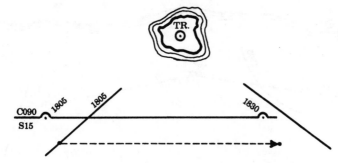

Figure 8-12B. *Advancing a point on the original line of bearing, in preparation for advancing the line itself.*

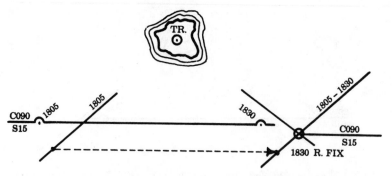

Figure 8-12C. Completed and labeled running fix using an advanced line of bearing.

Since the arc is centered on the tower symbol, this symbol is advanced parallel to the construction line connecting the two DR positions, a distance equal to the distance between them. The plot now appears as in Figure 8-13B. To complete the running fix, the arc is reproduced using the advanced center and the same radius that was used in constructing the original range arc. As was the case in the first example, only the advanced LOP and the running fix symbol are labeled with the time. The completed running fix for this example appears in Figure 8-13C.

Notice in the second example that the basic procedure for obtaining the running fix did not vary, even though there was a course and speed change between the times of the two LOPs. In plotting the running fix, only the DR positions for the times of the two LOPs are considered, regardless of the number of intervening course or speed changes. The earlier LOP is always advanced to the time of the later LOP through a distance parallel and equal to the distance between the two DR positions.

The fact that the DR plot plays a significant part in the determination of each running fix leads to two important considerations. First, the plot must be kept as accurately as possible during the interval between the times at which

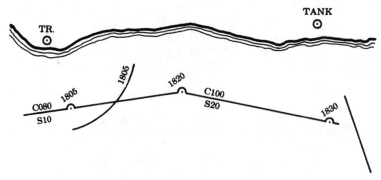

Figure 8-13A. A 25-minute plot, with a course and speed change. Again, the 1830 LOP is not labeled, as it was plotted to form the 1830 running fix.

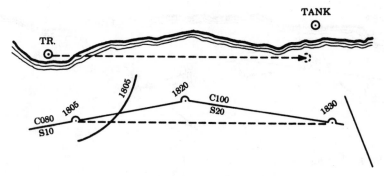

Figure 8-13B. *Advancing the center of a distance arc.*

the two LOPs are obtained, since the DR positions for these times determine the distance and direction through which the earlier LOP will be advanced. Second, in the piloting environment if more that 30 minutes have elapsed after a single LOP is obtained, it should not normally be advanced to form a running fix. Even with a comparatively weak one-knot current flowing, a DR position could be as much as a half-mile from the actual ship's position after a half-hour of travel, since the DR plot does not take current into account. Because of the precision required, in piloting any DR position recorded more than 30 minutes after the time of obtaining a single LOP is considered too inaccurate to use as a basis for advancing a line of position. Even when a line of position is advanced for lesser amounts of time, the resulting running fix is merely a better approximation of position than the corresponding DR position—better, because the ship must lie somewhere along the second LOP. In the confined piloting environment, if more than 30 minutes elapse without the navigator being able to obtain a second LOP or fix, a recommendation should be made to stop the ship until such time as a good fix or estimated position (described in the following section) can be obtained.

If an ambiguity develops when working with a range arc as either the advanced or subsequent LOP, the running fix position chosen should be the one

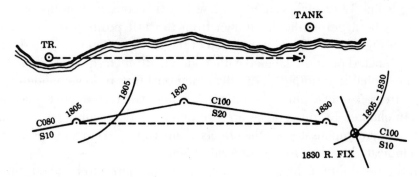

Figure 8-13C. *Completed running fix using an advanced distance arc.*

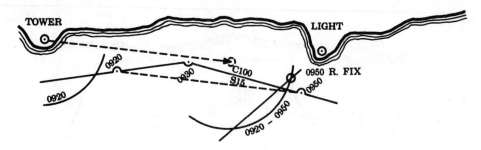

Figure 8-14. *Choosing the most hazardous position.*

that places the ship closest to any existing navigation hazards. The navigator should then recommend the best course of action for this position, just as was the case in similar circumstances when simultaneous LOPs resulted in a large triangle. An illustration of a situation of this type appears in Figure 8-14.

THE ESTIMATED POSITION

As alluded to in the preceding section, sometimes in the routine practice of piloting, it proves impossible to obtain more than one LOP at a time within the recommended half-hour time limit required to form a running fix. Situations of this nature can often arise when piloting at night, in foul weather, or in fog. In such circumstances, it may be possible for the navigator to determine an *estimated position* (EP) based on whatever incomplete positioning information might be available, in connection with the DR plot. For example, a single LOP of good reliability may be obtainable in the form of a bearing to an aid to navigation, or distance to the nearest land. Or a series of echo sounder depth readings might be recorded that would tend to corroborate (or in some cases conclusively invalidate) the position indicated by the DR plot.

One fairly well-established method of obtaining an estimated position when a single LOP is available is to draw a construction line from the DR position corresponding to the time of the LOP to the closest point on it. In the case of a straight-line LOP such as a line of bearing, this construction line would by definition be a perpendicular drawn from the DR position to the LOP, as shown in Figure 8-15. The intersection of the construction line with the LOP is the estimated position, labeled with a one-eighth-inch square and the time.

A questionable running fix, or a fix based on LOPs of low confidence, can also be treated as an estimated position, depending on the circumstances of the case. In chapter 13, procedures are presented for determining an estimated position that compensates for the effects of any expected current upon the position indicated by the DR plot or a running fix.

On the reliability scale, an EP is normally considered to be about midway between a good running fix and an unsubstantiated DR position. The exact de-

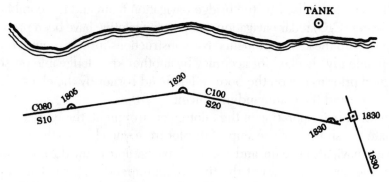

Figure 8-15. *Forming an estimated position based on a single LOP.*

gree of confidence that can be placed in an EP is a matter of judgment based on experience and the circumstances of each case. A new DR plot is normally not originated from an EP during piloting, although in some cases a line representing the estimated course and speed may be extended from it, to determine whether any hazards might be encountered if the vessel were in fact at the position estimated.

THE TRACK

Whenever the ship is to get under way, the navigator always lays down on the chart a kind of preplanned DR plot called a *track*, which is the intended path that the ship should follow over the ground. This track is in fact a form of DR plot, with its course and speed vectors representing the intended course and planned speed, rather than the ordered true course and speed.

Navy instructions require the following information to be included on the plot of the ship's track:

- The course in both degrees true and magnetic; the intended speeds (called the speed of advance, or SOA); distances of each track leg; distances to the next turn
- Danger bearings and ranges to navigation hazards not marked by navigation aids
- Turn bearings in degrees true and relative
- A statement for each turn stating the speed/rudder combination upon which the turn bearing computation was based
- Sound signal characteristics of all navigational aids to be used, if not printed on the chart
- Annotations for shoal water and other hazards or dangers including overhead obstructions, with danger bearings laid out for hazards not marked by a navigation aid

After all charts to be used by the bridge navigation team and the combat information center (CIC) radar navigation team during piloting have been prepared in accordance with these requirements, Navy instructions further require that they be independently checked for accuracy by another knowledgeable person and then signed prior to use on the bottom left-hand corner by the chart preparer, the navigator, and the commanding officer.

When the ship is operating in the piloting environment, the navigator should coordinate the shifting of the ship's DR plot from one chart to the next so that the bridge navigation team and the CIC navigation team do not both shift charts at the same time, nor at the time of an impending turn. The navigator should ensure that a good fix has been plotted on the new chart by one team before the other team follows suit on its new chart.

Figure 8-16 shows a portion of a track that a navigator has laid down to exit Norfolk, Virginia. Note that each of the legs of the track are labeled on the top with the 3-digit true direction of the intended track preceded by the abbreviation TR, followed by the direction in degrees magnetic in parentheses, with the intended speed of advance labeled by the letters SOA on the bottom.

In addition to plotting an exit track from a port or anchorage, the navigator also plots a track to show the planned course across open water, or through the entrance to a port or approach to an anchorage. In these circumstances, the navigator may use as the initial point of the track some preplanned point of departure or arrival positioned outside the confines of the harbor or anchorage area. The technique of planning a track across open water will be discussed in detail in chapter 15 dealing with voyage planning; the procedure for laying down an approach track to an anchorage is described in chapter 14.

In practice, a DR plot is always plotted in conjunction with and relative to some preplanned track; rarely, if ever, does a navigator lay out a DR plot with no preconceived notion of where the ship is supposed to proceed. If a fix places the ship's position 50 yards to the right of her intended track at some time, for instance, the navigator will extend the DR course line from the fix for a 2- or 3-minute distance in the direction of the ordered true course, and plot a DR position. From this future DR position, a new course line may be laid down in a direction suitable to bring the ship back on track, usually by aiming directly for the next junction point. If the ship arrives at a junction point early, the navigator may recommend that speed be decreased so as to arrive at the next point on time. Conversely, if the ship arrives late, the navigator may recommend an increase in speed.

Danger Bearings

In conjunction with plotting the intended track, the navigator should clearly mark the safe limits of navigable water on either side of a channel by means of a precomputed visual bearing to a prominent landmark or navigation aid

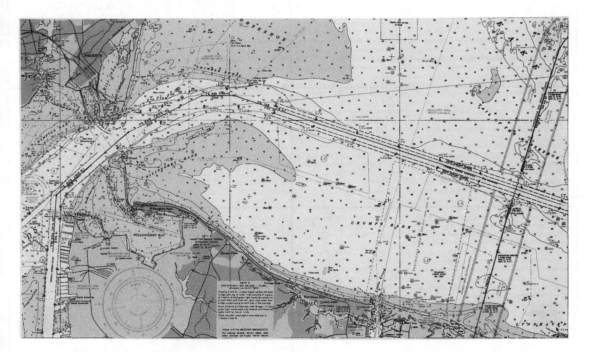

Figure 8-16. *An exit track from Norfolk, Virginia.*

known as a *danger bearing*. Hatching is always applied on the hazardous side of the bearing, and the side on which the hazard exists is indicated by labeling the bearing NLT for "not less than," or NMT for "not more than," the indicated bearing. Consider the track shown in Figure 8-17. In this example, the navigator has laid a track down the center of a narrow channel, along with danger bearings marking the shallow water to the right of the 075° leg and the

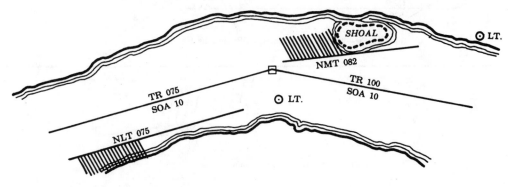

Figure 8-17. *Two examples of danger bearings.*

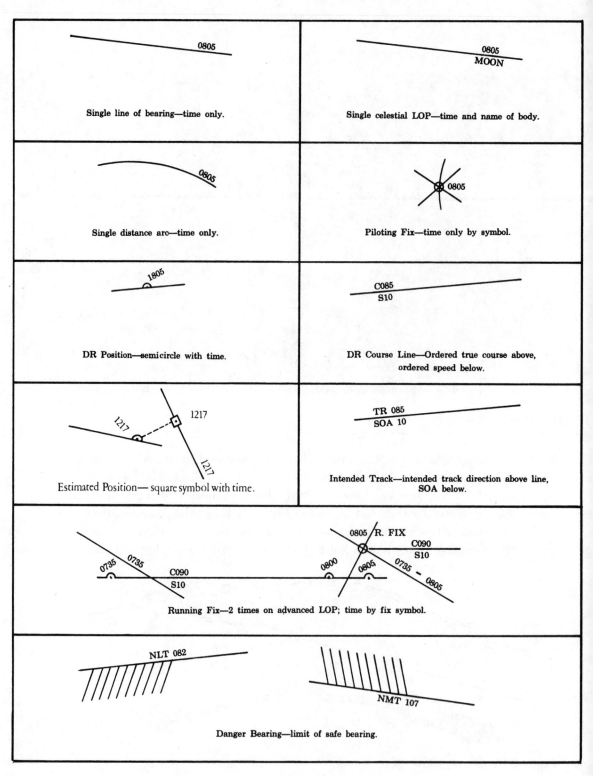

Figure 8-18. *Summary of standard labeling procedures.*

shoal to the left of the 100° leg. If the actual bearing to the first light were any-thing less than 075°T as the ship approached on the 075° track leg—say 070°T—she would be in danger of running through the shallow water. On the other hand, if she were so far to the left of track as she approached the junction point of the two legs that the actual bearing to the second light were greater than 082°T, she would be in danger of running onto the shoal. As long as bearings to the first light remain greater than 075°T and bearings to the second light remain less than 082°T as the ship approached, she would be proceeding in safe water.

When danger bearings are drawn in relation to an intended track, those marking dangers ahead and to the left of track are always labeled NMT, and those to the right, NLT. If danger bearings are required relative to a ship that is under way, they may be either NMT or NLT, depending on whether the ob-jects upon which they are based lie ahead or astern of the ship. If the chart on which the danger bearings are drawn is intended for daytime use only, the dan-ger bearings may be drawn on the chart with a red pencil. If the possibility ex-ists that the chart will be used at night, some dark color other than red should be used, such as blue or black, so that the danger bearings will be visible when viewed under a red light.

In addition to drawing danger bearings, some navigators take the additional precaution of shading all water areas too shallow for safe navigation by their ship with a blue pencil, so that navigable areas of water will stand out.

SUMMARY

This chapter has examined in some detail the procedures used by the navi-gator to maintain the dead reckoning plot and to plot the intended track. As an aid in labeling the various components of both, Figure 8-18 presents in visual form a summary of all of the different labeling procedures described herein.

The secret of success in dead reckoning can be summed up in two words: practice and neatness. Although the basic principles of plotting can be learned and practiced in the classroom environment, it is quite another thing to actu-ally do it on a moving platform in the middle of the night. While working with sample problems in the classroom cannot ensure success, the standards of la-beling and neatness that can be acquired and practiced there can prepare the inexperienced navigator, in large measure, for the practical application at sea.

SHIPBOARD COMPASSES
9

On board a vessel at sea, there are three principal references for direction: the ship's longitudinal axis, the magnetic meridian, and the true or geographic meridian. The horizontal direction of one terrestrial point from another, expressed as an angle from 000° clockwise to 360°, is termed a *bearing*. Bearings measured using the ship's longitudinal axis as the reference direction are called *relative bearings*, indicated by the abbreviation "R" following the bearing. Those based upon the magnetic meridian, determined by use of the magnetic compass, are referred to as *magnetic bearings*, abbreviated "M." And bearings given with reference to the geographic meridian, determined by the use of a ship's gyrocompass of known error, are *true bearings*, abbreviated by the letter "T." The ship's head, or heading, can be thought of as a special bearing denoting the direction in which the ship is pointing; it can be expressed either with reference to magnetic or true north, or with respect to the north axes of the magnetic or gyrocompasses. No matter what reference direction the navigator uses for the ship's head and other bearings, however, they must first be converted to true bearings before they can be used in the navigation plot.

In practice, relative bearings are not normally used for navigation purposes. They find their most extensive use in relating an object's position relative to the ship's bow, for purposes of visualizing the physical relationships involved. Ordinarily they are estimated visually, but if the ship's gyro system becomes inoperative, exact relative bearings to land or seamarks can be shot with a bearing circle or alidade. In order to use relative bearings in a navigation plot, the navigator must have a method of determining the ship's true head when the gyro is inoperative, so as to be able to convert the relative bearings to true.

THE MAGNETIC COMPASS

Virtually all vessels from the smallest of recreational craft to the jumbo tanker and aircraft carrier, are fitted with at least one magnetic compass. On

Figure 9-1A. *The Navy standard No. 1, 7-inch magnetic compass.*

most small boats and vessels that operate mainly in inland and coastal waters, the magnetic compass is the primary reference for direction and course headings. Most oceangoing vessels, including almost all Navy warships, have one or more gyrocompass systems installed to serve this purpose, but even on these ships, the magnetic compass serves as an important backup in case of gyro failure, and as a primary means of checking the accuracy of the gyrocompass at regular intervals while under way. Even though the modern gyrocompass is extremely accurate, highly reliable, and easy to use, it is nevertheless a highly complex instrument requiring periodic expert maintenance, dependent on an electrical power supply, and subject to electronic and mechanical failures of its component parts. The magnetic compass, on the other hand, is a comparatively simple, self-contained mechanism that operates independent of any electrical power supplies, requires little or no maintenance, and is not easily damaged.

Most older commercial ships and Navy warships having a secondary conning station carry two magnetic compasses. The main one, located in close proximity to the helmsman's station in the pilot house or bridge, is called the *steering* (abbreviated "stg") *compass*. The other, usually located in or near the secondary conning station, is often called the *standard* ("std") *compass*. They are usually the same configuration of magnetic compass, differing only in name for reference purposes. Many newly constructed merchant and Navy ships, however, carry only one steering compass, because they are fitted with two redundant gyrocompass systems for which the probability of simultaneous failure is so low that a second magnetic compass is considered to be unnecessary. Figure 9-1A is a photograph of a U.S. Navy standard No. 1, 7-inch magnetic compass. The compass is mounted in a stand called a binnacle, pictured in Figure 9-1B.

Figure 9-1B. *Binnacle for a standard Navy magnetic compass.*

The compass itself consists of a circular card, graduated with 360 degrees around the face; this card floats within a bowl containing a compass fluid. A pair of magnets is attached to the underside of the card, beneath its north-south axis. The card is suspended within the fluid, thereby reducing friction on the pivot about which the card rotates and damping vibrations and oscillations of the card as the ship moves. The compass bowl assembly is supported externally by a set of gimbals, hinged on both the longitudinal and athwart-ships axes, which permits the compass to remain nearly horizontal at all times.

In operation, the compass magnets tend to align themselves with the earth's magnetic lines of force existing at that location. If these lines coincided exactly with the earth's meridians, the north position on the compass card would always point toward true north, and any compass directions read with reference to the card would be true directions. Unfortunately, this ideal situation seldom occurs, because of two effects that come into play—variation and deviation.

Variation

Variation can be defined as the angle between the magnetic line of force or magnetic meridian, and the geographic meridian, at any location on the earth's surface. It is expressed in either degrees east or west to indicate on which side of the geographic meridian the magnetic meridian lies. Variation is caused primarily by the fact that the earth's magnetic and geographic poles do not coin-

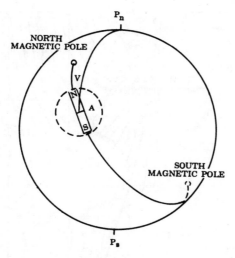

Figure 9-2. Illustration of the variation angle.

cide, and to a lesser extent, by certain magnetic anomalies in the earth's crust. Thus, if we are at position A on the earth's surface in Figure 9-2, our compass would tend to align itself with the magnetic lines of force flowing from the south to the north magnetic pole, as shown. But true north, as represented by the geographic meridian passing through A, would lie off to the right. The angle formed at position A between the magnetic and geographic meridians, angle V, is the variation. In this case the variation is west, since the north axis of the compass card is pointing to the west of geographic or true north.

If we were to change our position on the diagram in a random manner, we would find that the angle V would in most cases change each time we moved— at some locations, the angle would become smaller, while at others it would become larger. If we moved to the back side of the globe, the magnetic lines of force would for the most part tend off to the right of the geographic meridian. The variation would now be east. Finally, if we moved near to the geographic north pole, we would observe that the lines of force flowed in a southerly direction. For this reason, the magnetic compass provides only a general estimate of direction in high latitudes.

As we shifted our position about on the globe, we would find that at some locations the values of the variation would be the same as at other locations. If we moved in such a way that the angle V between the meridians remained constant, we would trace a path known as an *isogonic line*—a line along which the measured variation is the same. Figure 9-3 depicts a series of isogonic lines drawn on a simplified world chart.

There is another characteristic associated with variation that concerns the navigator. The magnetic field of the earth does not remain constant, but rather

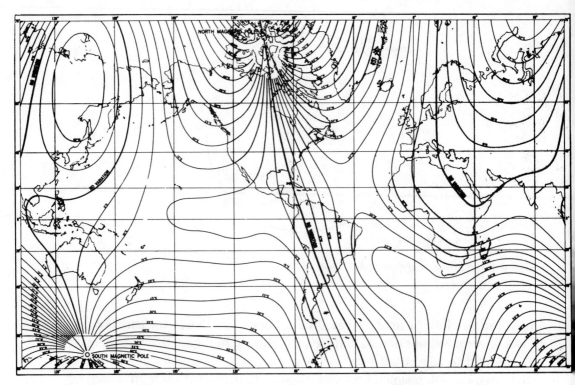

Figure 9-3. *Simplified chart of the world with selected isogonic lines.*

is continually slowly changing both in direction and in intensity. The magnetic poles actually wander slightly over the earth's surface from year to year. Because of this phenomenon, the variation at any given location on the earth changes a small amount from year to year. The navigator can determine the value for the variation at any given position by referring to a chart of the area in which the ship is located. In the centers of the compass roses inscribed on every chart, the value of the variation as measured at that location is given for the date on which the chart was printed. In addition, the annual rate of change of the variation is also supplied; this allows the navigator to compute the variation for the date of concern, if different from the year in which the chart was printed. The variation information contained in the compass rose closest to the ship's DR position on the chart should be used for computations. As an example, consider the compass rose in Figure 9-4.

The variation at that location in 1996, when the chart was printed, was 14° 45′ West, with a 2′ annual increase. If the navigator wished to compute the variation for 1999, the 2′ annual increase would be multiplied by 3 (1999 minus 1996) to get 6′ of increase from 1996 to 1999. The navigator would then add this 6′ to the 1996 value, 14° 45′ West, to get the 1999 variation, 14° 51′ West.

Figure 9-4. *A sample chart compass rose.*

In practice, the navigator always rounds off the value thus computed to the nearest half degree; in the example above, therefore, the 1999 value would be rounded to 15.0° West. The reason for this practice will be discussed later. Note also that the inner bearing circle of the compass rose has been rotated 14° 45′ to the west, corresponding to the variation. This is done for the convenience of the small boatman, who navigates for the most part by the use of a magnetic compass. Magnetic bearings can instantly be converted to true bearings or a true heading to a magnetic heading by laying a straightedge across the compass rose.

Deviation

In addition to variation, the navigator must take one other effect into account when working with the magnetic compass—*deviation*. Deviation is defined as the angle between the magnetic meridian and the north axis of the magnetic compass card. Like variation, it is expressed in either degrees east or west to indicate on which side of the magnetic meridian the compass card north lies. The effect is caused by the interaction of the ship's metal structure and electrical currents with the earth's magnetic lines of force and the compass magnets. Essentially, a ship is like a metal bar moving through a magnetic field; the magnetic effect on the bar varies depending on the angle it makes with respect to the lines of force comprising the field. Likewise, the deviation on a ship varies with the ship's heading. The deviation will also change periodically as metal is moved from one place to another on the ship, as when making equipment alterations. It can even be affected by semipermanent magnetism induced in the ship as it remains alongside a pier for an extended time, such as during overhaul. A Navy ship's degaussing system also has an effect. The degaussing system consists of a number of horizontal wire coils wrapped around a ship on the underside of its hull. The system is designed to lessen the

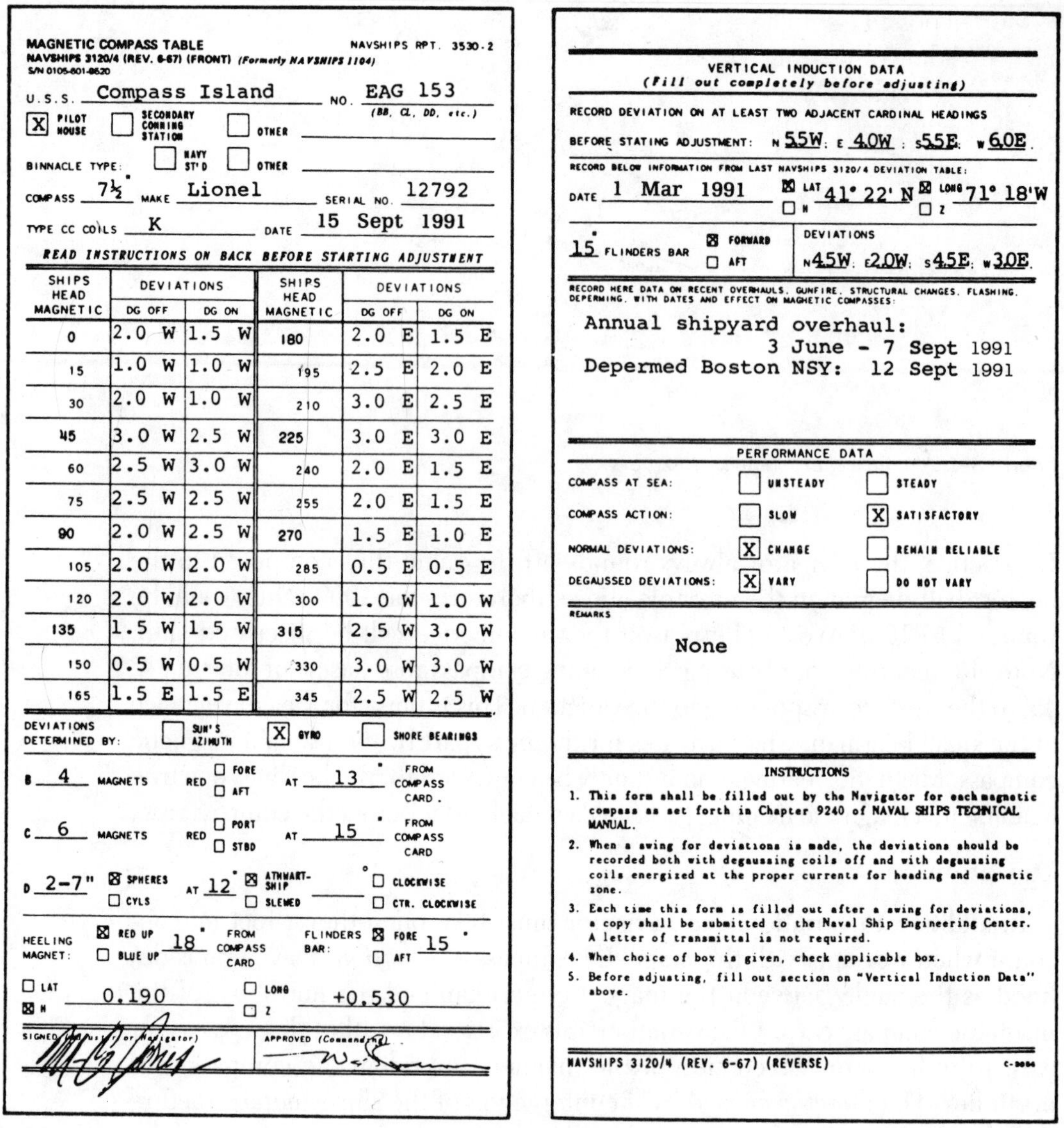

Figure 9-5. *A sample ship's deviation table.*

Form left side:

MAGNETIC COMPASS TABLE
NAVSHIPS 3120/4 (REV. 6-67) (FRONT) *(Formerly NAVSHIPS 1104)*
S/N 0105-801-8620   NAVSHIPS RPT. 3530.2

U.S.S. **Compass Island**   NO. **EAG 153**
                                      (BB, CL, DD, etc.)

[X] PILOT HOUSE   [ ] SECONDARY CONNING STATION   [ ] OTHER

BINNACLE TYPE: [ ] NAVY STD   [ ] OTHER

COMPASS **7½** MAKE **Lionel**   SERIAL NO. **12792**

TYPE CC COILS **K**   DATE **15 Sept 1991**

READ INSTRUCTIONS ON BACK BEFORE STARTING ADJUSTMENT

SHIPS HEAD MAGNETIC	DEVIATIONS DG OFF	DEVIATIONS DG ON	SHIPS HEAD MAGNETIC	DEVIATIONS DG OFF	DEVIATIONS DG ON
0	2.0 W	1.5 W	180	2.0 E	1.5 E
15	1.0 W	1.0 W	195	2.5 E	2.0 E
30	2.0 W	1.0 W	210	3.0 E	2.5 E
45	3.0 W	2.5 W	225	3.0 E	3.0 E
60	2.5 W	3.0 W	240	2.0 E	1.5 E
75	2.5 W	2.5 W	255	2.0 E	1.5 E
90	2.0 W	2.5 W	270	1.5 E	1.0 E
105	2.0 W	2.0 W	285	0.5 E	0.5 E
120	2.0 W	2.0 W	300	1.0 W	1.0 W
135	1.5 W	1.5 W	315	2.5 W	3.0 W
150	0.5 W	0.5 W	330	3.0 W	3.0 W
165	1.5 E	1.5 E	345	2.5 W	2.5 W

DEVIATIONS DETERMINED BY: [ ] SUN'S AZIMUTH   [X] GYRO   [ ] SHORE BEARINGS

B **4** MAGNETS RED [ ] FORE [ ] AFT AT **13** ° FROM COMPASS CARD

C **6** MAGNETS RED [ ] PORT [ ] STBD AT **15** ° FROM COMPASS CARD

D **2-7"** [X] SPHERES [ ] CYLS AT **12** ° [X] ATHWART-SHIP [ ] SLEWED   [ ] CLOCKWISE [ ] CTR. CLOCKWISE

HEELING MAGNET: [X] RED UP [ ] BLUE UP **18** ° FROM COMPASS CARD   FLINDERS BAR: [X] FORE [ ] AFT **15** °

[ ] LAT [X] N **0.190**   [ ] LONG [ ] Z **+0.530**

SIGNED (Aide or Navigator)   APPROVED (Commanding)

Form right side:

VERTICAL INDUCTION DATA
(Fill out completely before adjusting)

RECORD DEVIATION ON AT LEAST TWO ADJACENT CARDINAL HEADINGS

BEFORE STATING ADJUSTMENT: N **5.5W** E **4.0W** S **5.5E** W **6.0E**

RECORD BELOW INFORMATION FROM LAST NAVSHIPS 3120/4 DEVIATION TABLE:

DATE **1 Mar 1991**   [X] LAT **41° 22' N** [ ] N   [X] LONG **71° 18'W** [ ] Z

15 FLINDERS BAR [X] FORWARD [ ] AFT   DEVIATIONS N **4.5W** E **2.0W** S **4.5E** W **3.0E**

RECORD HERE DATA ON RECENT OVERHAULS, GUNFIRE, STRUCTURAL CHANGES, FLASHING, DEPERMING, WITH DATES AND EFFECT ON MAGNETIC COMPASSES:

Annual shipyard overhaul:
                3 June – 7 Sept 1991
Depermed Boston NSY: 12 Sept 1991

PERFORMANCE DATA

COMPASS AT SEA:	[ ] UNSTEADY	[X] STEADY
COMPASS ACTION:	[ ] SLOW	[X] SATISFACTORY
NORMAL DEVIATIONS:	[X] CHANGE	[ ] REMAIN RELIABLE
DEGAUSSED DEVIATIONS:	[X] VARY	[ ] DO NOT VARY

REMARKS   None

INSTRUCTIONS

1. This form shall be filled out by the Navigator for each magnetic compass as set forth in Chapter 9240 of NAVAL SHIPS TECHNICAL MANUAL.
2. When a swing for deviations is made, the deviations should be recorded both with degaussing coils off and with degaussing coils energized at the proper currents for heading and magnetic zone.
3. Each time this form is filled out after a swing for deviations, a copy shall be submitted to the Naval Ship Engineering Center. A letter of transmittal is not required.
4. When choice of box is given, check applicable box.
5. Before adjusting, fill out section on "Vertical Induction Data" above.

NAVSHIPS 3120/4 (REV. 6-67) (REVERSE)

strength of the ship's magnetic "signature" as a countermeasure against magnetic mines. On most headings, the deviation of a ship will normally change if its degaussing system is energized.

The effect of deviation on a shipboard compass can be regulated somewhat by adjusting or "compensating" the compass for the influence of the ship's metal structure, but deviation can never be completely eliminated. It is necessary, therefore, to have some means of determining what value of deviation to use for all possible ship headings, so as to be able to convert from a compass to a true heading or bearing. In practice, the navigator constructs a table in which values of deviation are entered for every fifteen degrees of the ship's head magnetic, starting with 000°M. The table is based on the ship's head magnetic so that it can be used regardless of the variation at any particular location. An example of a Navy ship's magnetic compass deviation table appears in Figure 9-5. Note that in the table there is one column for degaussing off (DG OFF) and a second for degaussing on (DG ON), a necessity because of the effect of the system on the compass. If the ship's head magnetic lies between two tabulated entries, the value of the deviation must be interpolated by methods discussed below. The result, like variation, is normally rounded off to the nearest half degree.

It usually takes about three hours to make up the table, as the ship must be "swung" around 15 degrees at a time while under way in order to record the data. *Swinging ship*, as the procedure is called, is required by Navy instructions whenever the deviation on any heading exceeds 3 degrees. Since the deviation on a ship is always changing because of the causes previously mentioned, it is usually necessary to swing ship several times each year.

Having described variation and deviation, it is now possible to discuss the technique for converting a steering compass heading or bearing to a true heading or bearing, and vice versa. Conversion of a ship's heading will be stressed here, because, as has been explained, the deviation depends on the ship's magnetic head. To convert any other bearing obtained by use of a shipboard magnetic compass to a true bearing, the ship's head magnetic must first be determined as an intermediate step. In practice, it is necessary to convert a steering or standard compass heading to a true heading primarily when it is desired to convert relative bearings to true bearings, for purposes of the navigation plot, with the gyrocompass system inoperative. Occasionally, it is desirable to compute a true ship's head from a steering or standard compass heading in order to obtain the error of an operable gyro. Situations that necessitate going the other way, from true to magnetic compass headings, include the recommendation of a suitable steering compass course to steer in order to achieve a desired true course when the gyro is inoperative, and the calculation of the proper magnetic headings on which to set the ship during the process of swinging ship.

Converting from Compass to True

To convert a steering compass heading to a true heading, the navigator must take both variation and deviation into account. The amount by which a steering compass direction differs from a true direction is often referred to as the *compass error*, sometimes abbreviated "CE," which is nothing more than the algebraic sum of the variation and deviation. The sequence of operations in converting a steering or standard compass heading to true is as follows: first, apply the deviation to the steering compass heading to obtain the magnetic heading;

second, apply the variation to the magnetic heading to produce the desired true heading. In the process, westerly errors must be subtracted and easterly errors added. Through the years, mariners have used the following memory aid to keep the proper sequence in mind when making such conversions:

Can	Dead	Men	Vote	Twice	At	Elections
Compass head	Deviation	Magnetic head	Variation	True head	Add	Easterly error

The most difficult part of the conversion of a magnetic compass heading to a true heading is to obtain the proper value of deviation for the given steering compass heading. Recall that the standard deviation table of the kind shown in Figure 9-5 is based on the ship's head magnetic, or M in the sequence above. Since this is the case, when converting from compass to true headings it is necessary to interpolate *twice* to be certain of obtaining the correct value of the deviation D. For the first interpolation, the steering compass heading can be considered an approximation of the magnetic head. The deviation so obtained allows a better approximation to the magnetic head to be made, by applying this deviation to the original steering compass heading. In so doing, easterly deviation is added, westerly subtracted.

The magnetic head so computed is now a better approximation than the steering compass heading. It remains only to interpolate a second time to extract the correct deviation; this result is then rounded to the nearest half degree. As an example of this process, consider the following question:

A ship's head is 305° per steering compass (p stg c). What is the ship's magnetic heading? Degaussing is OFF.

Referring to the deviation table in Figure 9-5, the steering compass heading is bracketed by 300°M and 315°M, for which the corresponding deviations are 1.0°W and 2.5°W. Writing them in tabular form, we have:

$$305° \rightarrow \begin{matrix} 300° & 1.0°W \\ \\ 315° & 2.5°W \end{matrix}$$

The desired deviation, therefore, must be $\frac{5}{15}$ = ⅓ of the difference between 1.0°W and 2.5°W:

$$\frac{5}{15} \times (2.5 - 1.0) = \frac{1}{3} \times 1.5° = 0.5°; D = 1.5°W.$$

The first estimate of deviation is, therefore, 1.5°W.

Applying 1.5°W to 305° p stg c, the result 303.5°M is obtained. Now the second interpolation must be performed:

$$\frac{3.5}{15} \times (2.5 - 1.0) = \frac{3.5}{15} \times 1.5 = 0.4; \ D = 1.4°W.$$

The required deviation, rounded to the nearest 0.5°, then, is 1.5°W.

This results in a ship's head magnetic of 305° – 1.5° = 303.5°M.

In this case the difference between successive deviations was small enough that no change in computed deviation resulted from the second interpolation. However, when differences are large, this will often result in a change of 0.5° or more after the second interpolation.

After having determined M, the ship's head magnetic, the next step in the conversion procedure is to apply the variation. Its value, rounded to the nearest half degree, is either added to or subtracted from the magnetic heading to obtain the true heading.

The values for variation and deviation are rounded to the nearest half degree because the magnetic compass graduations are so small that it is impossible to read a smaller subdivision. Thus, it would be meaningless to calculate a value for either the variation or deviation to the nearest tenth or beyond.

Because of the construction and placement of the magnetic compass on most ships, it is virtually impossible to shoot directly a steering compass bearing to an external object. Rather, relative bearings are shot, then these are applied to the steering compass heading to obtain magnetic bearings "per steering compass (p stg c)." In converting such a magnetic compass bearing, other than the ship's head, to a true bearing, care must be exercised to use only the ship's magnetic head in determining the proper value for the deviation. Deviation depends only on the ship's head, and not on any other bearing shot relative to that compass. Once the compass error is determined for a given ship's head, it can be applied to any bearing obtained by the use of that compass, as long as the ship remains on the given course. When the ship changes course, a new compass error incorporating the new value of deviation must be figured. An example of this type of problem follows:

While steaming on a heading of 305° p stg c, the following bearings were observed:

| Lighthouse | 102° p stg c |
| Reef Light | 329° p stg c |

What are the true bearings of the lighthouse and reef light? Degaussing is OFF, and variation is 9°E.

From the previous example, the deviation on course 305° p stg c is 1.5°W, and ship's head magnetic is 303.5°M. Knowing the variation, the conversion formula can be written thus:

C	D	M	V	T	A	E
305°	1.5°W	303.5°	9°E			

The true heading, T, is 303.5° + 9° = 312.5°T. The compass error on this heading is the algebraic sum of the deviation and the variation:

$$(-)\ 1.5°W\ (+)\ 9°E = 7.5°E$$

This compass error (CE) can now be applied to the observed bearings to convert them to true:

Object	°p stg c	CE	°T
Lighthouse	102°	(+) 7.5E	109.5°
Reef Light	329°	(+) 7.5E	336.5°

If the ship changed course, the compass error would change, because of the resulting change in the deviation. If further magnetic compass bearings were observed and converted to true, a new value of the compass error would have to be computed.

Converting from True to Compass

To convert a true heading to a steering compass heading, for the purpose of steering some true course through the water with the gyro inoperative, for example, corrections are applied in reverse order according to this sequence:

$$T \quad V \quad M \quad D \quad C \quad A \quad W$$

The last two letters indicate the rule to add westerly errors, and subtract easterly errors, when converting from true to compass headings. Since in this type of problem the ship's head magnetic is an intermediate step found by applying the variation to the true heading, only one interpolation for deviation is required. As a typical example of a problem requiring this methodology, consider the following:

> While steaming on a heading of 149°T, the ship's gyro suddenly tumbled. What steering compass course should be steered to keep the ship on the same true course? Assume a variation of 9°E, with degaussing OFF.

Writing down the known quantities in the format given above, we obtain the following:

T	V	M	D	C
149°	9°E			

Since easterly variation is subtracted when proceeding from true to compass heading, the ship's head magnetic, M, is determined to be:

$$149° (-) 9° = 140°M.$$

Referring to the deviation table of Figure 9-5, this magnetic heading is bracketed by 135°M and 150°M, for which the respective deviations with degaussing OFF are 1.5°W and 0.5°W. Writing them in tabular form, we have

$$
\begin{array}{ll}
135° & 1.5°W \\
140° \rightarrow & \\
150° & 0.5°W
\end{array}
$$

A one-step interpolation, rounded to the nearest half degree, yields a value of 1.0°W for the deviation. The conversion can now be completed as follows:

T	V	M	D	C
149°	(−)9°E	140°	(+)1.0°W	141°

Thus, the ship must steer 141° p stg c in order to remain on a heading of 149°T.

THE GYROCOMPASS

A gyrocompass is essentially a north-seeking gyroscope. It is encased in a housing fitted with various electronic components that keep the spin axis of the gyro aligned with terrestrial meridians, and sense the angle between the ship's head and the gyro spin axis. A detailed explanation of the gyroscope and the theory of its operation are beyond the scope of this text, but the navigator should be aware of the basic principles by which the gyro operates.

Basically, a classical gyro consists of a comparatively massive, wheel-like rotor balanced in gimbals that permit rotation in any direction about three mutually perpendicular axes through the center of gravity of the rotor. The three axes are pictured in Figure 9-6; they are called the spin axis, the torque axis, and the precession axis.

Once a gyroscope rotor is made to rotate, its spin axis would remain forever oriented toward the same point in space unless it were acted upon by an outside force. In the marine gyrocompass, the spin axis is kept aligned with a terrestrial meridian in a plane tangent to the earth's surface by a directive force derived from the tangential velocity component of the earth's rotational motion. Because this tangential velocity component is a maximum at the equator and diminishes to zero at the poles, the directive force is great in lower and mid-latitudes, but diminishes in strength as the earth's poles are approached. In latitudes beyond 70° north or south, the ship's velocity may become so great in relation to the earth's tangential velocity that large errors can be introduced into the directive force of the gyrocompass. For this reason, the gyrocompass must be continually checked for error beyond 70° north and south latitudes. Beyond

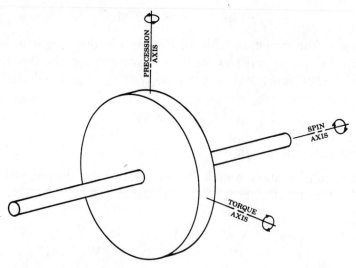

Figure 9-6. *The three axes of a gyroscope.*

75° to 80° latitude, most standard shipboard gyrocompass systems become so slow to respond to correcting forces that extreme errors on the order of 10 to 20 degrees or more are routinely experienced. Most shipboard gyrocompasses become virtually useless beyond about 85° latitude.

Fortunately, however, ice conditions in the polar regions are such that even in summer it is extremely rare that either commercial or Navy surface ships would venture much beyond 70° north or south latitudes in the normal course of operations, so most marine navigators will never be confronted with this problem. Should an occasion to operate in these polar regions arise, navigation is accomplished by employing a number of rather specialized piloting techniques and instruments developed over the years for use there. If required, further information on polar navigation can be found in *Dutton's Navigation and Piloting* and the *American Practical Navigator (Bowditch)*.

As was stated earlier, many newer Navy and merchant ships have two redundant gyrocompass systems installed. Even on older ships, the reliability of the gyrocompass is great enough that the magnetic compass is infrequently used for the most part. The gyrocompass itself on all but the very smallest ships is usually placed well down in the interior of the ship's hull, where it receives minimum exposure to roll, pitch, and yaw, and in the case of Navy ships, maximum protection from battle damage. It is connected by cables to *gyrocompass repeaters*, positioned where required throughout the ship. These repeaters use electronic servo-mechanisms that reproduce the master gyrocompass readings at their remote locations.

A gyrocompass repeater in design is similar to a magnetic compass. It consists basically of a movable card, graduated in 360°, which is mounted within a

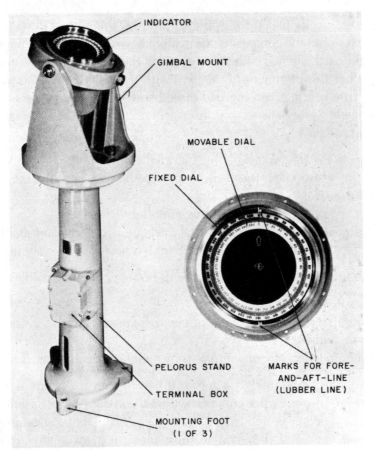

Figure 9-7. *A gyrocompass repeater.*

case called a binnacle. Gyrocompass repeaters located on bridge wings and other locations at which they will be used for obtaining visual bearings for navigational and other purposes are mounted in gimbals that keep the repeater horizontal, atop a stand of sufficient height to allow comfortable observation of bearings. A picture of a gyro repeater and its stand appears in Figure 9-7. The ship's head is indicated by a mark called the lubber's line on the rim of the binnacle; relative bearings are inscribed around the interior of the binnacle alongside the compass card for use if the gyro fails. To aid in shooting bearings with the repeater, it is fitted with an azimuth circle (see Figure 7-1 on page 111), which can be rotated about the face of the binnacle; its use was previously described in chapter 7.

In operation, the gyro is made to spin in such a way that its axis is nearly always aligned with the earth's geographic meridians, with small errors caused by the combined effects of the ship's motion, friction, and malfunctions of the gyro's electronic circuitry. Electronic systems within the gyro translate the angular difference between the spin axis and the ship's head into an electronic signal that drives the compass cards in all the gyro repeaters, keeping the north axis on the cards aligned with the spin axis of the gyro. As long as there is no error in either the alignment of the gyro with true north or in the transmission system, true bearings can be read directly off the gyro repeaters and used in the navigation plot. The gyrocompass, then, has several advantages over the magnetic compass:

- It seeks the true or geographic meridian instead of the magnetic meridian.
- It can be used near the earth's magnetic poles, where the magnetic compass is useless.
- It is not affected by surrounding material.
- Its signal can be fed into inertial navigation systems, automatic steering systems, and, in warships, into integrated navigational systems.

Being an intricate electronic instrument, however, it is also subject to certain disadvantages:

- It requires a constant source of electrical power and is sensitive to power fluctuations.
- It requires periodic maintenance by qualified technicians.

As has been mentioned, the gyro is inherently a very accurate instrument, but there are several sources of error within the system. Consequently, it is rare that a gyro repeater is free from all error, especially that caused by the transmission network, but the error in most modern gyrocompass systems is so small (less than 0.1°–0.2°) as to be considered insignificant for most applications. Even if the error is relatively large, it causes no particular problem for the navigator, as long as it remains within manageable limits and is constant. The navigator must always take any significant error (greater than ~0.3°) into account while keeping the DR plot and when making a course recommendation to the conning officer. Navy instructions require the gyro error to be determined at least once a day. In practice, the navigator will probably make this determination much more frequently, especially in piloting waters. As explained in other chapters of this text, the Navy navigator must record the error on the 0800, 1200, and 2000 position reports that are submitted to the commanding officer each day when under way; the error is entered on each page of the *Bearing Book*, and, as is discussed in this section, the error is applied each time a gyro bearing is plotted or a course recommendation is made.

Methods of Determining Gyrocompass Error

There are several methods of determining gyro error. In the following paragraphs, only those procedures applicable to pilot waters will be described; the determination of gyro error on the open sea will be covered in part 2 as part of

Figure 9-8A. A 5° westerly gyro error illustrated.

celestial navigation. Gyro error, like compass error, is expressed in degrees east or west. If the gyro is rotated to the west of true north, the error is west, and if rotated to the east, the error is east. The diagrams in Figures 9-8A and 9-8B may help to visualize this. In the situation depicted below, a gyro error caused the gyro repeater card to be rotated away from true north toward the west by an amount of 5°. Any bearings shot on this repeater, then, would be too high. Here, the tower at a right angle to true north should bear 090°T; with reference to this repeater, however, it would bear 095° per gyro compass (pgc).

Now let the card be rotated by an easterly error 5° to the east of true north, as in Figure 9-8B. Now the same object would bear 085° pgc, not 090°T. Any other bearings shot using this repeater would, in similar fashion, be 5° too low.

One of the best methods of determining gyro error is to observe an artificial or natural visual range. The range is first shot visually by use of the azimuth circle on the gyro repeater, and its gyro bearing is noted. Then the observed bearing is compared to the true bearing of the visual range, as determined from a chart showing the objects, or in the case of an established lighted range, from the *Light List* or *List of Lights*. The difference in the two bearings is the amount of the gyro error. If the gyro repeater bearing is higher than the actual true bearing, the error is west; if lower, the error is east.

If the ship is at a known location, such as a pier or an anchorage, a gyro error can be obtained by comparing a known bearing to an object ashore, as measured on a chart, with the bearing as observed from the gyro repeater. Once

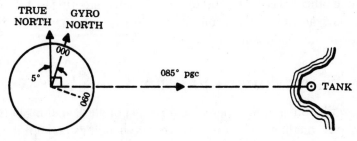

Figure 9-8B. A 5° easterly gyro error illustrated.

again, the difference in bearings is the amount of the error; if the gyro bearing is too high, the error is west; if too low, the error is east.

In the same manner, the ship's heading while tied up alongside and parallel to a pier can be compared with a known pier heading. Any difference is the gyro error.

If the ship is not under way, a favorite method of obtaining gyro error is trial and error adjustment of three or more simultaneous lines of position until a point fix results. If the three LOPs meet at a point when initially plotted, there is no gyro error. If they form a triangle, the lines are adjusted by successive additions or subtractions of 1°, then if necessary, 0.5° to the bearings, until they meet at a point fix. The total correction thus applied to any one LOP is the gyro error. If the correction had to be subtracted, the error is west; if added, the error is east.

Finally, if no other method is available, the gyro can be compared to another compass of known error. If, for example, the navigator had a reliable deviation table available, the ship's magnetic compass could be used for this purpose by converting the compass heading to true and then comparing the true ship's head to the gyro heading. If a second gyrocompass system of known error were installed, the headings of the two gyros could be compared. If the gyro heading in question is too high, the error is west; if too low, the error is east.

There are two memory aids to help the navigator decide whether the gyro error is east or west:

> If the compass is best (higher), the error is west;
> if the compass is least (lower), the error is east,

and,

G.E.T.—Gyro + East = True.

When plotting a gyro bearing or series of bearings, the navigator must always remember to subtract the amount of any westerly gyro errors, and add the amount of any easterly errors prior to plotting them, since only true bearings should be plotted on a chart. In practice, Navy navigators usually take gyro error into account by rotating the compass rose attached to the parallel motion protractor (PMP), described in chapter 7, to compensate for the error. Since the conning officer is conning and the helmsman is steering by reference to the gyro, the navigator must also remember to incorporate any gyro error into any course recommendations made.

THE *MAGNETIC COMPASS RECORD*

Navy regulations require that a legal record of all deviations between magnetic compass readings and comparative true headings be kept in a log called the *Magnetic Compass Record*. Gyro errors are also entered in this log. Every day the ship is under way, the gyro error and any navigational and conning gyro repeater errors are required to be computed and recorded in the log. Comparisons between the magnetic compass and the helm gyro repeater in use for steering are required to be made and recorded in the log every half hour and every time a new ordered course is steered, when practical. Helm repeater error is required to be computed and recorded every four hours, by comparing the master gyro or ship's inertial navigation system (INS) heading to the helm steering repeaters. A compass check is also required any time a gyro compass alarm occurs. A need for maintenance is indicated any time the steering gyrocompass repeater and gyrocompass or INS headings are out of agreement by more than one degree. The ship's officers of the deck (OODs) should also be informed whenever this condition is found to exist.

The ship's navigator reviews and signs the *Magnetic Compass Record* each day the ship is under way, and submits it to the commanding officer for signature on the last day of each calendar quarter.

SUMMARY

There are three references for horizontal direction on board a vessel at sea: the ship's head (relative bearings), the ship's magnetic compass (magnetic bearings), and the ship's gyrocompass (gyrocompass bearings, equivalent to true bearings if the gyro error is negligible). Because most modern oceangoing ships are fitted with one or more gyrocompass systems, the gyroscope is normally the main reference for direction for the surface navigator. When properly used, serviced, and maintained, the modern gyrocompass is an extremely reliable and accurate instrument, but as is the case with all electronic instruments, it is subject to error and damage. In the event of complete gyro failure, the navigator has an excellent backup system available in the form of the magnetic compass. Even when the gyro is fully operational, however, the navigator must be constantly aware of the possible existence of error. It is the unknown, unobserved error that contributes to marine disasters. There are on file numerous reports of vessels having been put aground and lost because of a navigator's adherence to a course laid in safe waters, while the actual track was an unknown path leading to danger. With the ever-increasing volume of supertanker and liquid natural gas carrier traffic at sea, the modern consequences of such a mishap can range, and all too often in the recent past have ranged, far beyond the loss of the ship herself. Now, more than ever, the watchword of the navigator must be constant vigilance at sea.

RADAR

10

Radar, a word derived from the terms *r*adio *d*etection *a*nd *r*anging, is of great practical value to the navigator in the piloting environment. Since its original development in World War II as a detection device for enemy ships and aircraft, improved technology in electronics and electronic circuitry has made possible quantum jumps in the state of the art of this device, so that today there are many different applications. Almost all modern oceangoing vessels, and many smaller craft, are equipped with some variety of navigational radar. It is, in fact, required by Coast Guard regulations for all commercial vessels over 1,600 tons operating in U.S. waters; vessels over 10,000 tons must have two independent radar systems. Naval surface warships will often have as many as four different varieties of radar equipment, depending on the size and type of ship. They are air-search, fire-control, surface-search, and navigational radars. Often, a single surface-search radar will also perform the function of a separate navigational radar, especially on ships the size of a destroyer and smaller. As an important auxiliary function, navigational radars can be used not only to locate navigational aids and perform radar navigation, but also for tracking other vessels in the vicinity so as to avoid risk of collision.

This chapter will briefly examine the more important characteristics of a navigational or surface-search radar and the fundamental techniques of its operation with which the navigator must be familiar in order to use this device effectively. An excellent reference for a more detailed discussion of this subject is *Publication No. 1310*, the *Radar Navigation Manual*, published by the Defense Mapping Agency Hydrographic/Topographic Center. This publication contains a very thorough yet uncomplicated discussion of the theory and application of radar as it applies to all aspects of surface navigation.

CHARACTERISTICS OF A SURFACE-SEARCH/ NAVIGATIONAL RADAR

The basic principle of radar is the determination of range to an object or "target" by the measurement of the time required for an extremely short burst or "pulse" of radio-frequency (RF) energy, transmitted in the form of a wave, to travel from a reference source to a target and return as a reflected echo. Most surface-search and navigational radars use very short high-frequency electromagnetic waves formed by an antenna into a beam very much like that of a searchlight. The antenna is usually parabolic in shape, and rotates in a clockwise direction to scan the entire surrounding area. Bearings to the target are determined by the orientation of the antenna at the moment the reflected echo returns.

In marine surface radars, the beam is fairly narrow in horizontal width, but may be quite broad in vertical height. Figure 10-1 illustrates a typical radar beam, and two side lobes, which are additional beams of low energy unavoidably radiated out in many types of surface radars due to limitations in the antenna size and shape.

The standard radar set is made up of five components—the transmitter, modulator, antenna, receiver, and indicator. The transmitter consists of an oscillator that produces the RF waves. The modulator is essentially a timing device that regulates the transmitter so that it sends out relatively short pulses of energy, separated by relatively long periods of rest. The antenna performs two functions—it forms the outgoing pulse train into a beam during the time the transmitter is on, and it collects returning target echoes during the interval the transmitter is at rest. In the receiver, the reflected radio energy collected by the antenna is converted into a form that may be presented visually on an output display device or indicator, usually a form of cathode ray tube.

A radar system that operates in this way is termed "pulse-modulated"; if the system radiated energy continuously, the strong output signal would completely mask the weaker incoming reflected echo, making range determination impossible. This gives rise to the military electronic countermeasure (ECM) technique known as "jamming," whereby an enemy radar set is neutralized by

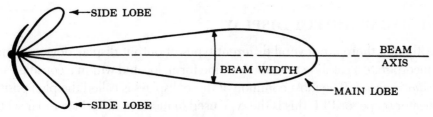

Figure 10-1. *A navigation or surface-search radar beam and its side lobes, from above.*

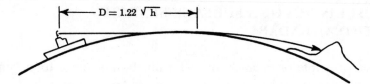

Figure 10-2. *Refraction of a radar beam.*

flooding it with continuous RF energy having the same frequency as the set's radar pulse.

Because it is made up of high-frequency RF energy of very short wave length, usually between 3 and 10 centimeters long for most surface-search and navigational radars, the radar beam acts much like a beam of light. Energy comprising the beam travels out and is reflected back at the speed of light; for this reason, pulse time lengths and rest periods between them are measured in microseconds. Radars are commonly described by listing their frequency, number of pulses per second (pulse repetition rate), length of their pulse in microseconds (pulse length), and antenna rotation rate.

Radar beams are also like light waves in that they travel in straight lines for the most part, although refraction does cause them to bend downward and follow the curvature of the earth to some extent, especially at lower frequencies. For this reason, radars are limited in range to the distance of their *radar horizon*, which for most standard surface radars can be closely approximated by the formula

$$D = 1.22 \sqrt{h}$$

where D is the horizon distance in miles, and h is the height of the radar antenna in feet. Objects beyond the horizon can be detected only if they are of sufficient height above the earth's surface, as is the case with the mountain range in Figure 10-2. In this figure, the low shore line located under the radar horizon would not be detected, but the high mountain range located inland would be.

Radar antennas are usually placed high on the superstructure of the ship to extend the radar horizon as far as possible.

THE RADAR OUTPUT DISPLAY

Although there are several different types of output display devices, almost all incorporate a cathode ray tube in some form, located within a console called a *radar repeater*. The most common of these displays is called the plan position indicator scope, or PPI; this is the type used in most naval surface radar sets. In this presentation, the observer's ship is located at the center of a circular scope,

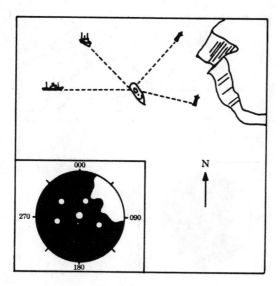

Figure 10-3A. A PPI presentation oriented to true north.

and external objects within the range of the radar presentation are depicted at scaled distances outward from the center. Bearing on the PPI scope is indicated around the periphery of the screen, from 000° at the top clockwise to 360°. On ships having a gyrocompass, the scope has a gyro input, and the presentation is oriented so that the true north direction lies under the 000° mark. If gyro failure occurs, the radar presentation automatically reorients to a relative picture, with the ship's head at the 000° position at the top of the scope. Figure 10-3A shows a typical PPI presentation with the scope oriented by gyro input to true north, and Figure 10-3B depicts the same presentation oriented relative to the ship's head or longitudinal axis.

As the antenna rotates, its beam is represented on the scope by a thin line that "sweeps" around the center, much like a spoke of a turning wheel. This line, called the *sweep*, illuminates or "paints" any objects within range of the radarscope onto its face. This presentation of an object on the radarscope is called a "pip" or "blip." On many radarscopes, the sweep can also be made to illuminate range rings at selected intervals outward from the center of the scope, thus allowing range estimates to be made. Bearing estimates can be made by referring to the bearings surrounding the scope face. A drawing of a PPI scope with its sweep and range rings appears in Figure 10-4.

To assist in the determination of range and bearing, most radar repeaters are fitted with devices called a *bearing cursor* and a *range strobe*. The bearing cursor is a narrow radial beam of light, usually centered on the ship's position on the scope; it can be rotated around by the operator through 360°. The range strobe is a pinpoint of light, which is moved in and out along the bearing cur-

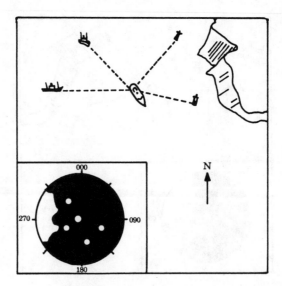

Figure 10-3B. *A PPI presentation oriented relative to ship's head.*

sor. On some older radar repeaters, the range strobe moves in and out along the sweep, and creates a variable range ring as the sweep rotates around the scope. Each indicator is controlled by a separate hand crank incorporating an appropriate distance or bearing readout.

To obtain a "mark"—a range and bearing to a target—the operator first rotates the bearing cursor so that it bisects the pip representing the object on the

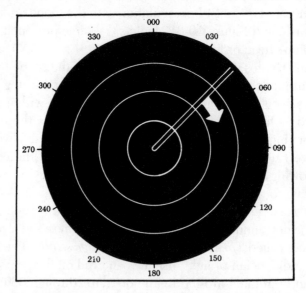

Figure 10-4. *A PPI radarscope, with range rings illuminated.*

scope, and then positions the range strobe so that it touches the inner edge of the pip. The range and bearing are read off from dials near the scope.

The size of the physical area to be depicted on the scope is variable on most radar repeaters, and is determined by the operator by the selection of a suitable range scale. Some repeaters only allow selection of specific scales such as 2, 4, 8, or 10 miles, whereas others allow selection of any scale between an arbitrary upper and lower limit, such as 1 to 50 miles. The scale number refers to the radius of the presentation; if a scale of 8 miles were selected, for example, the most distant object that could be shown on the scope would be 8 miles away, and it would appear near the perimeter of the presentation.

INTERPRETATION OF A RADARSCOPE PRESENTATION

Interpretation of information presented on a radarscope is not always easy. Much experience is often required on the part of the operator to obtain correct readings, especially during unfavorable meteorological conditions or when the radar is operating in a degraded state owing to partial failure of its electronic components. Even in the best of conditions with a finely tuned radar set, many factors tend to produce errors in interpretation of radar information. Among these are bearing resolution, range resolution, radar shadows, multiple echoes, and false echoes.

Bearing resolution is the minimum difference in bearing between two targets at the same range that can be discerned on the radarscope. The radar beam width causes a target to appear wider than it is in actuality. If two or more objects are close together at about the same range, their pips may merge together, giving the impression to the operator that only one target is present. Such erroneous presentations often appear in coastal areas, where numbers of rocks, piles, and boats located offshore may present a false impression of the location of the shoreline, called a "false shoreline" (Figure 10-5A).

Range resolution is the minimum difference in range between two objects at the same bearing that can be discerned by the radar. Pulse length and frequency both affect the range resolution of a particular radar; they can be adjusted on some sets to improve resolution at either long or short ranges. False interpretation of a radar presentation may occur as a result of this cause if two or more targets appear as one, or if objects such as rocks or small boats near the shore are merged with the shoreline (Figure 10-5B).

Radar shadows occur when a relatively large radar target masks another smaller object positioned behind it, or when an object beyond the radar horizon is obscured by the curvature of the earth. The shoreline in Figure 10-2 on page 168 lies in such a radar shadow zone.

The *multiple echo* is caused by pulses within a radar beam bouncing back and forth between the originating ship and a relatively close-in target, espe-

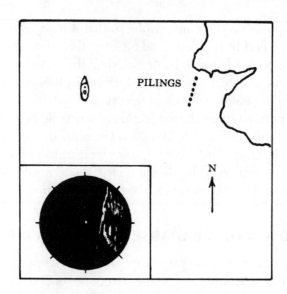

Figure 10-5A. *A false shoreline caused by lack of bearing resolution.*

cially another ship. A multiple echo is a false pip that appears on the scope at the same bearing as the real target but at some multiple of the actual target range (Figure 10-6). If only one false pip appears at twice the actual range, it is often termed a *double echo*. If a second false pip appears at three times the range, it is a *triple echo*. Ordinarily, not more than one or two false pips appear in this fashion.

Another type of false pip sometimes encountered is the *false echo*. Like the multiple echo, this is a pip that appears on the scope where there is no target

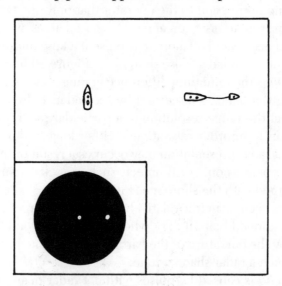

Figure 10-5B. *Tug and tow merged because of lack of range resolution.*

Figure 10-6. *A multiple echo.*

in actuality. One type of false echo results when a portion of the reflected energy from a target returns to the antenna by bouncing off part of the ship's structure, as illustrated in Figure 10-7. The resulting false pip, sometimes called an *indirect echo*, always appears at the same range as the actual target, but at the bearing of the intermediate reflective surface.

A second type of false echo is the *side-lobe effect*. As explained in the first part of this chapter, the parabolic antenna usually radiates several side lobes, in addition to the main lobe of the radar beam. If energy from these side lobes is reflected back by a target, an echo will appear on the radarscope on each side of the main lobe echo. If the target is close enough, a semicircle or even a complete circle may be produced, with a radius equal to the target range. Nor-

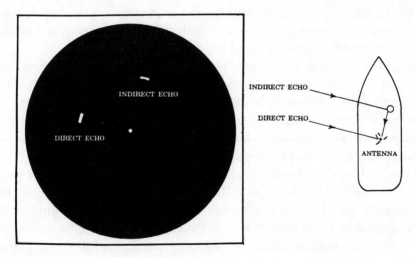

Figure 10-7. *An indirect echo.*

Figure 10-8. *Side-lobe effects.*

mally, a target must be fairly close to produce a side-lobe effect, because of the low energy of the side lobes (Figure 10-8).

USE OF RADAR DURING PILOTING

Because of the way in which a radar set is constructed, most marine surface radar bearings are accurate only to within 3° to 5°. Accuracy in range is usually much better, however. A well-tuned radar should give ranges precise to within ±100 yards out to the radar horizon, with slowly increasing inaccuracy beyond that point to the limits of the extreme range.

It follows that for piloting applications, radar range lines of position (LOPs) are much preferred over radar bearings. In fact, some navigators make it a rule never to use radar bearings unless absolutely no other information is available, and then only with a high degree of suspicion.

Almost any object that is fixed in position, appears in symbolic form on the chart, and is visible on the radarscope presentation can be used to obtain a radar range LOP. Care must be taken, however, especially when using land-marks, to be certain of which features are being painted on the radar set. The operator must be constantly aware of the pitfalls associated with radar range resolution and radar shadow zones, described in the preceding section.

As was explained in chapter 3, which describes the operation of the piloting team on board a Navy ship, the CIC radar navigation team is usually relied upon to guide the ship when poor visibility conditions make piloting by visual LOPs impossible. Even then, however, the navigator is still ultimately respon-sible to the commanding officer for the safe navigation of the ship. The navi-

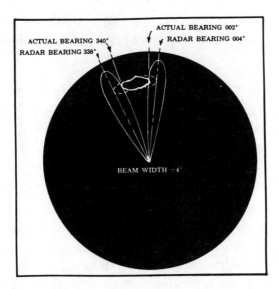

Figure 10-9. *Effect of beam width on radar tangent bearings.*

gator will normally keep a parallel plot on the bridge, using radar information from the pilot house repeater and any visual LOPs of opportunity.

The technique of plotting and labeling a range LOP has already been discussed in detail in chapter 8, and will not be repeated here. A radar range LOP may be crossed with a simultaneous LOP from any other source to obtain a fix, or two or more simultaneous radar ranges may be used for this purpose. If radar information alone is being used for plotting fixes, a minimum of three LOPs should always be obtained at any given time to guard against the ever-present possibility that one of them was misinterpreted by the radar operator.

If, as a last resort, radar bearings must be used in the navigation plot, normally only relatively small, discrete objects should be used for LOPs. When none are available, bearings should then be taken on tangents. Since these bearings will be exaggerated because of the beam widths, it is necessary to decrease right tangent bearings by half the beam width, and to increase left tangents by the same amount. Consider the example in Figure 10-9. In the figure, the actual tangent bearings to the island pictured should be 340°T and 002°T. Because of the 4° radar beam width, however, energy begins to be reflected from the left tangent of the island when the centerline of the beam is directed toward 338°T. Thus, the left edge of the blip representing the island on the radarscope presentation would be located at 338°T—2° too low. Likewise, energy would still be reflected back from the right tangent when the radar beam is 2° beyond it, leading to a bearing for the right tangent on the radarscope of 004°, which is 2° too high.

If the radarscope presentation is oriented to true north by gyro input, all bearings obtained should also be corrected for any known gyro error.

SUMMARY

Radar is an extremely important tool for the navigator operating in any environment, and it is especially valuable during piloting. It is the only instrument widely available that is capable of providing precise LOPs even in conditions of poor visibility, such as fog or during hours of darkness. The primary advantages associated with radar from the navigator's point of view may be summarized as follows:

- Radar can be used at night and during periods of reduced visibility when visual means of navigation are limited or impossible to use.
- It is possible (though not recommended) to obtain a fix from a single object, because both range and bearing are obtainable.
- Radar is available for use at greater distances from land than are most other navigational methods.
- Radar fixes are obtained quickly and accurately.
- Radar can be used to locate and track other shipping and storms.

Like any other highly sophisticated equipment, however, radar also has its limitations. Among the more serious of these are the following:

- It is a complex electronic instrument, dependent upon a power source, and is subject to mechanical and electrical failure.
- There is a minimum range limitation, resulting from returning echoes from nearby wave crests (sea return), and a maximum range limitation.
- Interpretation of the radarscope display is difficult at times, even for a trained and experienced operator.
- LOPs from radar bearings are inaccurate.
- Radar is susceptible to both natural and deliberate interference.
- Radar shadows and sea return may render objects undetectable by radar.

Although radar is not a panacea for the navigator, intelligent use of its capabilities certainly will aid in safely directing the movements of a vessel.

TIDE

11

The rise and fall of the surface of a body of water on earth is called *tide*. This phenomenon is of great interest and importance to the navigator, especially when piloting in relatively shallow coastal waters, because the height of tide determines the total depth of water at any given location and time. In some places, passages that could be safely negotiated by a deep-draft seagoing vessel during periods of high tide levels would be difficult or impossible to transit during periods of low water. This chapter will discuss the causes and effects of tide and will describe the procedures used by the navigator in computing and allowing for them during piloting.

CAUSES OF TIDE

Tide is caused primarily by the interaction of the gravitational forces of the moon, the sun, and the rotating and revolving earth upon the waters comprising the earth's oceans. Because the moon is many times closer to the earth than the sun, the effect of its gravitational pull is some two-and-a-quarter times more pronounced, even though the sun has a mass thousands of times greater. For purposes of illustration, the earth can be visualized as a spherical core covered by water. Although a detailed theoretical explanation of tide is beyond the scope of this text, in essence the strong gravitational pull of the moon on the side of the earth nearest the moon, together with the strong outward centrifugal force generated by the earth-moon system on the opposite side of the earth, cause the surrounding water to bulge out in the form of high tides on both sides, as depicted in an exaggerated fashion in Figure 11-1. The same effect on a much smaller scale also takes place with respect to the earth-sun system, but because the sun is so much farther away, the effect is insignificant in comparison to that of the earth-moon system.

The moon revolves completely around the earth once each month. Since the earth rotates beneath it in the same direction as the moon revolves, it takes 24

Figure 11-1. *Effect of the moon on the earth's oceans.*

hours and 50 minutes for the earth to complete one revolution with respect to the moon. It follows that every location on earth should experience four tides every 24 hours and 50 minutes: two high tides, and two intervening low tides. This is in fact the usual tidal pattern. The period of 24 hours and 50 minutes is called a *tidal day*, and each successive high and low tide is said to constitute one *tide cycle*. At some locations, however, the tide patterns that actually occur are distorted from this norm because of the effects of land masses, constrained waterways, friction, the Coriolis effect (an apparent force acting on a body in motion on the earth's surface, caused by the rotation of the earth), and other factors. There are three major tidal patterns experienced at various locations on earth. Each will be described in the following section.

As the moon rotates about the earth, it happens that at certain times its effects on the earth's oceans are reinforced by those of the sun, but at other times the effects are opposed. The tidal effects of the sun and moon act in concert twice each month, once near the time of the new moon, when this body is on the same side of the earth as the sun, and again near the time of the full moon, when it is at the opposite side of the earth from the sun. Tides produced at these times are abnormally high and unusually low, and are called *spring tides*. Figure 11-2A illustrates the two possible relationships of the earth, sun, and moon during spring tide. The tidal effects of the sun and the moon are in opposition to one another when the moon is at quadrature—the first and last quarter—at which times the moon is located at right angles to the earth-sun line. At these times, high tides are lower and low tides are higher than usual;

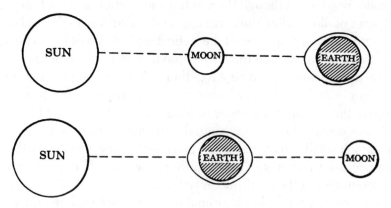

Figure 11-2A. *The earth-sun-moon system at spring tide.*

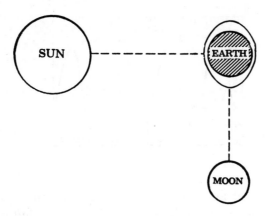

Figure 11-2B. *The earth-sun-moon system at neap tide.*

these are referred to as *neap tides*. Figure 11-2B depicts the relationship of the earth, sun, and moon at neap tide.

TYPES OF TIDES

Before progressing further with the discussion of tide, it is necessary to define some of the terms associated with its description:

Sounding datum: An arbitrary reference plane to which both heights of tides and water depths expressed as soundings are referenced on a chart. The sounding datum planes in most common use will be discussed later in this chapter.

High tide or *high water*: The highest level normally attained by an ascending tide during a given tide cycle. Its height is expressed in feet or meters relative to the sounding datum.

Low tide or *low water*: The lowest level normally attained by a descending tide during a given tide cycle. Like high tide, its height is expressed in feet or meters relative to the sounding datum.

Range of tide: The vertical difference between the high and low tide levels during any given tide cycle. It is normally expressed in feet or meters.

Stand: The brief period during high and low tides when no change in the water level can be detected.

As was mentioned in the preceding section, the physical laws governing movement of the large water masses located in the ocean basins and constricted by the geography of the world's land masses result in three major types of tides experienced at various locations throughout the world. They are classified according to their characteristics as semidiurnal, diurnal, and mixed.

The *semidiurnal* tide (Figure 11-3A) is the basic type of tide pattern observed over most of the world. There are two high and two low tides each tidal

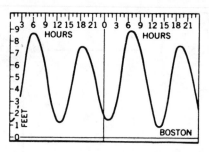

Figure 11-3A. *Semidiurnal tide pattern.*

day, and these occur at fairly regular intervals a little more than 6 hours in length. Usually there are only relatively small variations in the heights of any two successive high or low waters. Tides at most locations on the U.S. Atlantic coast are of this variety.

The *diurnal* tide (Figure 11-3B) is a pattern in which only a single high and a single low water occur each tidal day. High and low tide levels on succeeding days usually do not vary a great deal. Tides of this type appear along the northern shore of the Gulf of Mexico, in the Java Sea, and in the Tonkin Gulf.

The *mixed* tide pattern (Figure 11-3C) is characterized by wide variation in heights of successive high and low waters, and by longer tide cycles than those of the semidiurnal tide. This blend of the semidiurnal and diurnal tides is prevalent on the U.S. Pacific coast, and on many of the Pacific islands.

TIDAL REFERENCE PLANES

In order for water depths, heights, and elevations of topographical features, nav aids, and bridge clearances to be meaningful when printed on a chart, standard reference planes for these measurements have been established. In general, heights and elevations are given on a chart with reference to a standard *high*-water reference plane, and heights of tide and charted depths of water are given with respect to a standard *low*-water reference plane. The reference planes used on charts produced by the United States are listed and described

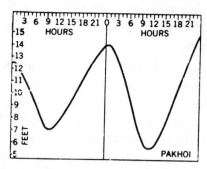

Figure 11-3B. *Diurnal tide pattern.*

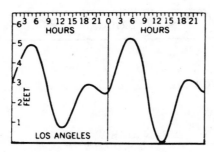

Figure 11-3C. *Mixed tide pattern.*

below, in the order of their relative heights from the sea bottom. All are derived from various computed mean or average water levels, measured with respect to the bottom, as observed and recorded over a number of years.

First, the high-water reference planes:

Mean high-water springs (MHWS): The highest of all high-water reference planes is the average height of all spring tide high-water levels.

Mean higher high water (MHHW): The average of the higher of the high-water levels occurring during each tidal day at a location, measured over a 19-year period.

Mean high water (MHW): The average of all high-tide water levels, measured over a 19-year period. It is the high-water reference plane used on most charts produced by the United States for the basis of measurement of heights, elevations, and bridge clearances.

Mean high-water neaps (MHWN): The lowest of all high-water reference planes in common use; the average recorded height of all neap tide high-water levels.

Next, the low-water reference planes:

Mean low-water neaps (MLWN): The highest of all common low-water reference planes; the average height of all neap tide low-water levels.

Mean low water (MLW): The average height of all low-tide water levels observed over a period of 19 years. Until 1980 it was the low-water reference plane used on charts of the U.S. Atlantic and Gulf coasts.

Mean lower low water (MLLW): The average of the lower of the low-water levels experienced at a location over a 19-year period. It has been the low-water reference plane used for many years for charts of the U.S. Pacific coast, and after 1980, for charts of the U.S. Atlantic and Gulf coasts, as a basis of measurement of charted depth and height of tide.

Mean low water springs (MLWS): The lowest of all low-water reference planes; the average of all spring tide low-water levels. It is the sounding datum on which most water depths of foreign charts are based.

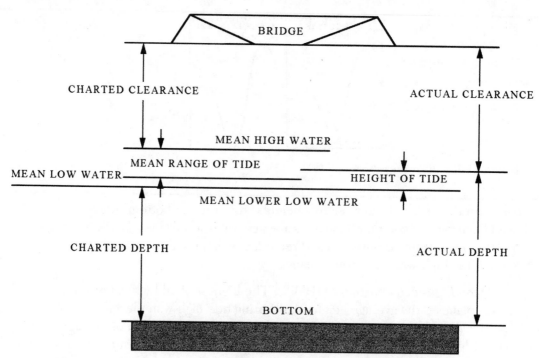

Figure 11-4. *Relationship of terms applying to water depth and vertical clearance.*

As a example to help visualize the relationship between these reference planes, the charted depth, height of tide, and clearance under a bridge, suppose the navigator had a chart that used mean high water for the high-water reference plane from which charted heights and bridge clearances were reckoned, and mean lower low water, for water depths and heights of tide. The physical relationships existing would then be as illustrated in Figure 11-4.

By definition, the mean high-water and mean low-water reference planes represent the average limits within which the water level would normally be located. The vertical distance between these two planes is called the *mean range of tide*; it would represent the average range of tide at that location. Midway in the range of tide, halfway between mean high and mean low water, is a reference plane called the *mean tide level* (MTL). It is not to be confused with another tidal datum, called the *mean sea level*, which is simply the average height of tide at a location over a 19-year observation period.

The navigator must always remember that the actual water level will occasionally fall below the low-water reference plane, particularly around the time of spring tides. Height of tide in this situation is *negative*, denoted by placing a minus sign in front of the height of tide figure. In these extreme cases, the actual water depth will be *less* than that indicated on the chart. Likewise, actual vertical clearance will sometimes be smaller than that indicated on a chart using mean high water as its reference plane for heights and clearances. On the other hand, if the chart is based on water levels at mean high- and mean low-water springs, the navigator will tend to regard charted clearances and depths with more confidence, since the actual water level will seldom, if ever, exceed these limits.

PREDICTING HEIGHT OF TIDE

There are many situations that arise during the practice of piloting that require the navigator to predict the height of tide at a given time in the future. When a ship is entering port, the navigator must know the minimum depth of water through which the ship will pass; some ports must be entered over sandbars or shoals that can be safely transited by a deep draft vessel only at high water. If a ship must pass under a bridge, the navigator will want to know how much vertical clearance exists at the expected time of passage. The ship's deck officers will want to know the depth of water at an intended anchorage so they can calculate the proper scope of chain to be let out, or the height and range of tide expected at a pier mooring so as to be able to allow the proper amount of slack in the mooring lines. For these and other purposes, the navigator refers to a set of four publications known as the *Tide Tables*, previously introduced in chapter 5.

The *Tide Tables* are arranged geographically, with one volume covering each of the following four areas: East Coast, North and South America; West Coast, North and South America; Europe and West Coast of Africa; and Central and Western Pacific and Indian Oceans. A new set of tables is published each year by a private publisher utilizing NOS data, and distributed to Navy users by the National Imagery and Mapping Agency (NIMA). They are available to merchant and civil mariners at most nautical supply stores. Collectively, the *Tide Tables* contain daily predictions of the times of high and low tides at some 190 major reference ports throughout the world, and, in addition, they list differences in times of tides from specific reference ports for an additional 5,000 locations referred to as *subordinate stations*, sometimes abbreviated to *substations*. The navigator can construct a daily tide table, compute the height of tide at a given time, or compute the time frame within which the tide will be above or below a desired level, for each one of the 5,000 subordinate stations, by using the tabulated tide-difference data in conjunction with the daily predictions for the proper reference port.

Layout of the *Tide Tables*

Each volume of the *Tide Tables* is made up of eight tables. The first three are of primary interest for making tide predictions. Tables four through seven consist of astronomical tables for predicting the times of sunrise and sunset and moonrise and moonset at selected latitudes and reference ports, a feet-to-meters conversion table, and a tidal prediction accuracy table.

New York (The Battery), N.Y.

Times and Heights of High and Low Waters

October

Day	Time	ft	cm
1 W ●	0148	0.2	6
	0753	5.2	158
	1406	0.3	9
	2003	5.0	152
2 Th	0226	0.2	6
	0826	5.2	158
	1446	0.3	9
	2036	4.9	149
3 F	0302	0.3	9
	0858	5.2	158
	1524	0.3	9
	2107	4.7	143
4 Sa	0335	0.4	12
	0927	5.1	155
	1601	0.4	12
	2138	4.5	137
5 Su	0404	0.5	15
	0958	5.0	152
	1638	0.6	18
	2215	4.4	134
6 M	0431	0.7	21
	1035	5.0	152
	1717	0.7	21
	2301	4.2	128
7 Tu	0502	0.9	27
	1123	4.9	149
	1808	0.9	27
	2357	4.1	125
8 W	0547	1.0	30
	1218	4.9	149
	1919	1.0	30
9 Th ◐	0057	4.1	125
	0711	1.2	37
	1318	4.9	149
	2030	0.8	24
10 F	0201	4.2	128
	0842	1.0	30
	1424	5.0	152
	2131	0.5	15
11 Sa	0309	4.5	137
	0948	0.7	21
	1534	5.1	155
	2226	0.2	6
12 Su	0416	4.8	146
	1046	0.3	9
	1641	5.3	162
	2317	-0.2	-6
13 M	0517	5.3	162
	1142	-0.1	-3
	1741	5.6	171
14 Tu	0007	-0.5	-15
	0610	5.7	174
	1236	-0.4	-12
	1834	5.8	177
15 W ○	0057	-0.7	-21
	0659	6.1	186
	1329	-0.7	-21
	1924	5.9	180
16 Th	0146	-0.9	-27
	0746	6.3	192
	1421	-0.8	-24
	2014	5.8	177
17 F	0234	-0.8	-24
	0834	6.3	192
	1511	-0.8	-24
	2106	5.6	171
18 Sa	0322	-0.7	-21
	0925	6.1	186
	1600	-0.6	-18
	2201	5.3	162
19 Su	0409	-0.4	-12
	1019	5.8	177
	1651	-0.2	-6
	2300	5.0	152
20 M	0457	0.0	0
	1117	5.5	168
	1744	0.1	3
	2358	4.7	143
21 Tu	0550	0.5	15
	1213	5.2	158
	1843	0.5	15
22 W ◐	0055	4.5	137
	0649	0.9	27
	1309	4.9	149
	1946	0.7	21
23 Th	0151	4.3	131
	0755	1.1	34
	1404	4.6	140
	2047	0.8	24
24 F	0247	4.3	131
	0857	1.1	34
	1501	4.5	137
	2141	0.7	21
25 Sa	0344	4.3	131
	0953	1.0	30
	1559	4.4	134
	2228	0.6	18
26 Su	0438	4.5	137
	1043	0.9	27
	1653	4.5	137
	2312	0.5	15
27 M	0527	4.7	143
	1129	0.7	21
	1741	4.6	140
	2354	0.4	12
28 Tu	0609	4.9	149
	1214	0.5	15
	1823	4.7	143
29 W	0034	0.3	9
	0647	5.1	155
	1257	0.3	9
	1901	4.8	146
30 Th	0114	0.2	6
	0722	5.2	158
	1340	0.2	6
	1936	4.7	143
31 F ●	0153	0.1	3
	0754	5.3	162
	1422	0.1	3
	2009	4.7	143

November

Day	Time	ft	cm
1 Sa	0231	0.2	6
	0825	5.3	162
	1502	0.1	3
	2042	4.6	140
2 Su	0307	0.2	6
	0855	5.3	162
	1542	0.1	3
	2116	4.4	134
3 M	0341	0.3	9
	0928	5.2	158
	1621	0.2	6
	2158	4.3	131
4 Tu	0415	0.5	15
	1010	5.1	155
	1704	0.3	9
	2250	4.2	128
5 W	0453	0.6	18
	1102	5.0	152
	1753	0.4	12
	2349	4.2	128
6 Th	0543	0.8	24
	1202	4.9	149
	1855	0.5	15
7 F ○	0049	4.2	128
	0702	0.9	27
	1303	4.9	149
	2003	0.4	12
8 Sa	0149	4.4	134
	0824	0.8	24
	1407	4.8	146
	2104	0.2	6
9 Su	0251	4.6	140
	0931	0.5	15
	1513	4.9	149
	2200	-0.1	-3
10 M	0355	4.9	149
	1030	0.2	6
	1620	5.0	152
	2252	-0.4	-12
11 Tu	0456	5.3	162
	1125	-0.2	-6
	1722	5.2	158
	2342	-0.6	-18
12 W	0550	5.7	174
	1219	-0.5	-15
	1816	5.3	162
13 Th	0032	-0.8	-24
	0640	6.0	183
	1312	-0.7	-21
	1907	5.4	165
14 F ○	0122	-0.8	-24
	0727	6.1	186
	1403	-0.8	-24
	1957	5.3	162
15 Sa	0211	-0.8	-24
	0814	6.1	186
	1453	-0.8	-24
	2047	5.1	155
16 Su	0259	-0.6	-18
	0902	5.9	180
	1541	-0.7	-21
	2140	4.9	149
17 M	0345	-0.4	-12
	0953	5.6	171
	1629	-0.4	-12
	2236	4.7	143
18 Tu	0432	0.0	0
	1047	5.2	158
	1717	-0.1	-3
	2333	4.4	134
19 W	0520	0.4	12
	1142	4.9	149
	1810	0.3	9
20 Th	0028	4.3	131
	0614	0.8	24
	1235	4.6	140
	1907	0.5	15
21 F ○	0120	4.2	128
	0715	1.0	30
	1327	4.4	134
	2005	0.7	21
22 Sa	0212	4.1	125
	0819	1.1	34
	1419	4.2	128
	2100	0.7	21
23 Su	0305	4.1	125
	0918	1.1	34
	1513	4.1	125
	2149	0.6	18
24 M	0358	4.2	128
	1010	0.9	27
	1610	4.0	122
	2234	0.5	15
25 Tu	0448	4.4	134
	1058	0.7	21
	1703	4.1	125
	2316	0.3	9
26 W	0534	4.7	143
	1144	0.4	12
	1750	4.2	128
	2358	0.2	6
27 Th	0615	4.9	149
	1229	0.2	6
	1831	4.3	131
28 F	0040	0.1	3
	0651	5.1	155
	1314	0.0	0
	1909	4.4	134
29 Sa ●	0121	0.0	0
	0726	5.2	158
	1358	-0.2	-6
	1945	4.4	134
30 Su	0203	0.0	0
	0759	5.3	162
	1441	-0.3	-9
	2022	4.4	134

December

Day	Time	ft	cm
1 M	0243	-0.1	-3
	0833	5.3	162
	1523	-0.4	-12
	2101	4.3	131
2 Tu	0323	0.0	0
	0912	5.3	162
	1605	-0.3	-9
	2147	4.3	131
3 W	0404	0.0	0
	0958	5.2	158
	1649	-0.3	-9
	2241	4.2	128
4 Th	0448	0.2	6
	1053	5.0	152
	1737	-0.2	-6
	2340	4.3	131
5 F	0542	0.3	9
	1152	4.9	149
	1832	-0.1	-3
6 Sa	0037	4.4	134
	0651	0.5	15
	1251	4.8	146
	1935	0.0	0
7 Su	0134	4.5	137
	0805	0.5	15
	1351	4.7	143
	2037	-0.1	-3
8 M	0232	4.7	143
	0912	0.3	9
	1454	4.6	140
	2134	-0.3	-9
9 Tu	0334	4.9	149
	1013	0.0	0
	1601	4.6	140
	2228	-0.5	-15
10 W	0435	5.1	155
	1109	-0.3	-9
	1704	4.6	140
	2320	-0.6	-18
11 Th	0532	5.4	165
	1203	-0.5	-15
	1801	4.7	143
12 F	0011	-0.7	-21
	0624	5.6	171
	1255	-0.7	-21
	1853	4.8	146
13 Sa ○	0101	-0.7	-21
	0711	5.7	174
	1346	-0.8	-24
	1942	4.8	146
14 Su	0150	-0.7	-21
	0757	5.7	174
	1434	-0.8	-24
	2030	4.7	143
15 M	0238	-0.6	-18
	0842	5.5	168
	1521	-0.7	-21
	2119	4.6	140
16 Tu	0323	-0.4	-12
	0929	5.3	162
	1605	-0.6	-18
	2211	4.4	134
17 W	0407	-0.1	-3
	1018	5.0	152
	1650	-0.3	-9
	2304	4.3	131
18 Th	0451	0.2	6
	1108	4.7	143
	1735	0.0	0
	2355	4.1	125
19 F	0538	0.5	15
	1157	4.4	134
	1823	0.3	9
20 Sa	0043	4.0	122
	0631	0.8	24
	1245	4.1	125
	1915	0.5	15
21 Su ○	0131	3.9	119
	0732	1.0	30
	1332	3.9	119
	2011	0.6	18
22 M	0219	3.9	119
	0835	1.0	30
	1421	3.7	113
	2103	0.6	18
23 Tu	0309	4.0	122
	0932	0.9	27
	1516	3.6	110
	2152	0.5	15
24 W	0401	4.1	125
	1024	0.7	21
	1615	3.6	110
	2237	0.4	12
25 Th	0452	4.3	131
	1113	0.4	12
	1711	3.7	113
	2322	0.2	6
26 F	0539	4.6	140
	1200	0.2	6
	1759	3.9	119
27 Sa	0007	0.1	3
	0620	4.8	146
	1247	-0.1	-3
	1842	4.1	125
28 Su ○	0052	-0.1	-3
	0659	5.1	155
	1334	-0.4	-12
	1922	4.2	128
29 M ●	0138	-0.3	-9
	0737	5.3	162
	1419	-0.6	-18
	2003	4.3	131
30 Tu	0223	-0.4	-12
	0816	5.4	165
	1503	-0.8	-24
	2046	4.4	134
31 W	0308	-0.5	-15
	0900	5.4	165
	1546	-0.9	-27
	2133	4.5	137

Time meridian 75° W. 0000 is midnight. 1200 is noon.
Heights are referred to mean lower low water which is the chart datum of soundings.

Figure 11-5. *A sample daily prediction page for New York,* Tide Tables, *Table 1.*

Table 1 lists the times and heights of tide in both feet and meters at each high water and low water, in chronological order, for each day of the year for every reference location, called *reference stations*, used in that volume. The datum or reference plane from which the predicted heights are reckoned is the same as that used for the major large-scale charts of the area. Figure 11-5 is a typical page from Table 1 of the volume of the *Tide Tables* covering the East Coast of North and South America. High and low tides are identified by a comparison of consecutive heights of tide listed. Usually, two high and two low tides will appear on any given day, and there will be a high or low tide about every 6 hours. Occasionally, because the tidal day is 24 hours and 50 minutes long, there will be only three listings for a given day, as on 8 October in Figure 11-5. Here, a high tide appears just before midnight on the preceding day, and just after midnight on the succeeding day, with the result that only one high tide occurs on the 8th. A negative sign appearing before the height of tide figure indicates that this low tide falls below the tidal reference plane used for this volume. Times of tides given in Table 1 are those of the standard time zone in which the port is located. If the port is keeping daylight savings time, each tide time prediction must be adjusted by adding one hour to the time listed.

Table 2 of each volume of the *Tide Tables* contains a listing of tide time and height-difference data, as well as other useful information, for each of the subordinate stations located within the area of coverage of that particular volume. Data in Table 2 are arranged in geographical sequence, and an alphabetical index of subordinate station names is located in the back of each volume. A sample page from Table 2 of an *East Coast Tide Tables* appears in Figure 11-6 on page 186. As can be seen by an inspection of this table, information given for each subordinate station includes its latitude and longitude, differences in times and heights of tide between the subordinate and reference stations, the ranges of mean and spring tides, and the mean tide level.

In Table 2, time corrections are given as algebraic differences to be applied to the reference station time data, and height of tide corrections are given both as algebraic differences and multiplicative ratios, indicated by the use of an asterisk (*) before the ratio. When such a ratio is given, the height of tide at the subordinate station at a particular time is obtained by multiplying the heights of high and low tide given for the reference station by the respective subordinate station ratios, and then applying any correction factors that may be required (indicated within parentheses surrounding the multiplicative ratio).

Table 3 of the *Tide Tables* is used primarily to find the height of tide at a given time, after daily tide predictions for a given location have been computed. It can also be used to find the time frame within which the tide will be either above or below a desired height. A typical page from Table 3 is depicted in Figure 11-7 on page 187.

TABLE 2 – TIDAL DIFFERENCES AND OTHER CONSTANTS

No.	PLACE	POSITION		DIFFERENCES				RANGES		Mean Tide Level
				Time		Height				
		Latitude	Longitude	High Water	Low Water	High Water	Low Water	Mean	Spring	
		North	West	h m	h m	ft	ft	ft	ft	ft
	CONNECTICUT, Long Island Sound–cont. Time meridian, 75° W			on New London, p.48						
	Thames River									
1205	NEW LONDON, State Pier	41° 22'	72° 06'	*Daily predictions*				2.6	3.0	1.5
1207	Smith Cove entrance	41° 24'	72° 06'	0 00	+0 10	*0.97	*0.95	2.5	3.0	1.4
1209	Norwich	41° 31'	72° 05'	+0 13	+0 25	*1.16	*1.15	3.0	3.6	1.7
1211	Millstone Point	41° 18'	72° 10'	+0 09	+0 01	*1.05	*1.05	2.7	3.2	1.5
	Connecticut River									
1213	Saybrook Jetty	41° 16'	72° 21'	+1 11	+0 45	*1.36	*1.35	3 5	4.2	2.0
1215	Saybrook Point	41° 17'	72° 21'	+1 11	+0 53	*1.24	*1.25	3.2	3.8	1.8
1217	Lyme, highway bridge	41° 19'	72° 21'	+1 25	+1 10	*1.20	*1.20	3.1	3.7	1.7
1219	Essex	41° 21'	72° 23'	+1 39	+1 38	*1.16	*1.15	3.0	3.6	1.7
	Connecticut River									
1221	Hadlyme <7>	41° 25'	72° 26'	+2 19	+2 23	*1.05	*1.05	2.7	3.2	1.5
1223	East Haddam	41° 27'	72° 28'	+2 42	+2 53	*1.12	*1.10	2.9	3.5	1.6
1225	Haddam <7>	41° 29'	72° 30'	+2 48	+3 08	*0.97	*0.95	2.5	3.0	1.4
1227	Higganum Creek	41° 30'	72° 33'	+2 55	+3 25	*1.01	*1.00	2.6	3.1	1.5
1229	Portland <7>	41° 34'	72° 38'	+3 51	+4 28	*0.85	*0.85	2.2	2.6	1.3
1231	Rocky Hill <7>	41° 39'	72° 38'	+4 44	+5 44	*0.78	*0.80	2.0	2.4	1.2
1233	Hartford <7>	41° 46'	72° 40'	+5 30	+6 52	*0.74	*0.75	1.9	2.3	1.1
				on Bridgeport, p.52						
1235	Westbrook, Duck Island Roads	41° 16'	72° 28'	−0 24	−0 32	*0.61	*0.60	4.1	4.7	2.2
1237	Duck Island	41° 15'	72° 29'	−0 26	−0 35	*0.67	*0.68	4.5	5.2	2.4
1239	Madison	41° 16'	72° 36'	−0 21	−0 30	*0.73	*0.72	4.9	5.6	2.6
1241	Falkner Island	41° 13'	72° 39'	−0 14	−0 25	*0.80	*0.80	5.4	6.2	2.9
1243	Sachem Head	41° 15'	72° 42'	−0 11	−0 15	*0.80	*0.80	5.4	6.2	2.9
1245	Money Island	41° 15'	72° 45'	−0 12	−0 23	*0.83	*0.84	5.4	6.2	3.0
1247	Branford Harbor	41° 16'	72° 49'	−0 08	−0 18	*0.88	*0.88	5.9	6.8	3.1
1249	New Haven Harbor entrance	41° 14'	72° 55'	−0 09	−0 14	*0.92	*0.92	6.2	7.1	3.3
1251	New Haven (city dock)	41° 18'	72° 55'	+0 01	−0 01	*0.89	*0.88	6.0	6.9	3.2
1253	Milford Harbor	41° 13'	73° 03'	−0 08	−0 10	*0.98	*0.96	6.6	7.6	3.5
1255	Stratford, Housatonic River	41° 11'	73° 07'	+0 26	+1 01	*0.82	*0.80	5.5	6.3	2.9
1257	Shelton, Housatonic River	41° 19'	73° 05'	+1 35	+2 44	*0.74	*0.72	5.0	5.8	2.7
1259	BRIDGEPORT	41° 10'	73° 11'	*Daily predictions*				6.8	7.7	3.6
1261	Black Rock Harbor entrance	41° 09'	73° 13'	−0 04	−0 03	*1.02	*1.04	6.9	7.9	3.7
1263	Saugatuck River entrance	41° 06'	73° 22'	−0 02	+0 01	*1.04	*1.04	7.0	8.0	3.8
1265	South Norwalk	41° 06'	73° 25'	+0 09	+0 15	*1.05	*1.04	7.1	8.2	3.8
1267	Greens Ledge	41° 03'	73° 27'	−0 02	−0 01	*1.07	*1.08	7.2	8.3	3.9
1269	Stamford	41° 02'	73° 33'	+0 03	+0 08	*1.07	*1.08	7.2	8.3	3.9
1271	Cos Cob Harbor	41° 01'	73° 36'	+0 05	+0 11	*1.07	*1.08	7.2	8.3	3.9
1273	Greenwich	41° 01'	73° 37'	+0 01	+0 01	*1.10	*1.08	7.4	8.5	4.0
1275	Great Captain Island	40° 59'	73° 37'	0 00	+0 01	*1.08	*1.08	7.3	8.4	3.9
	NEW YORK Long Island Sound, north side			on Willets Point, p.56						
1277	Port Chester	41° 00'	73° 40'	−0 03	−0 14	*1.01	*1.01	7.2	8.5	3.9
1279	Rye Beach	40° 58'	73° 40'	−0 22	−0 31	*1.01	*1.01	7.2	8.4	3.9
1281	Mamaroneck	40° 56'	73° 44'	−0 02	−0 13	*1.02	*1.04	7.3	8.6	3.9
1283	New Rochelle	40° 54'	73° 47'	−0 18	−0 21	*1.02	*1.04	7.3	8.4	3.9
1285	Davids Island	40° 53'	73° 46'	+0 04	−0 09	*1.01	*1.00	7.2	8.5	3.9
1287	City Island	40° 51'	73° 47'	+0 03	−0 05	*1.01	*1.00	7.2	8.5	3.9
1289	Throgs Neck	40° 48'	73° 48'	+0 08	+0 12	*0.98	*0.98	7.0	8.2	3.8
	East River									
1291	Whitestone	40° 48'	73° 49'	+0 04	+0 06	*1.00	*1.00	7.1	8.3	3.8
1293	Old Ferry Point	40° 48'	73° 50'	+0 10	+0 14	*0.99	*0.99	7.1	8.3	3.8
1295	College Point, Flushing Bay	40° 47'	73° 51'	+0 14	+0 13	*1.00	*0.95	6.8	7.9	3.7
1297	Northern Boulevard Bridge, Flushing	40° 46'	73° 50'	+0 29	+0 35	*0.95	*0.95	6.8	8.0	3.7
1299	Westchester, Westchester Creek	40° 49'	73° 50'	+0 16	+0 14	*0.98	*0.98	7.0	8.3	3.8
1301	Hunts Point	40° 48'	73° 52'	+0 11	+0 10	*0.97	*1.04	6.9	8.1	3.8
1303	Westchester Avenue Bridge, Bronx River . . .	40° 50'	73° 53'	+0 16	+0 15	*0.97	*0.97	6.9	8.1	3.7
1305	North Brother Island	40° 48'	73° 54'	+0 15	+0 15	*0.92	*0.92	6.6	7.8	3.6
1307	Port Morris (Stony Point)	40° 48'	73° 54'	+0 04	+0 07	*0.89	*1.11	6.3	7.4	3.4
1309	Lawrence Point	40° 47'	73° 55'	+0 03	+0 11	*0.90	*0.90	6.4	7.6	3.5
1311	Wolcott Avenue	40° 47'	73° 55'	+0 03	+0 11	*0.85	*0.86	6.1	7.2	3.2
				on New York, p.60						
1313	Hell Gate, Wards Island	40° 47'	73° 55'	+2 58	+3 45	*1.33	*1.59	6.0	7.3	3.4
1315	Hell Gate, Hallets Point	40° 47'	73° 56'	+2 04	+2 07	*1.12	*1.14	5.1	6.1	2.7
1317	Horns Hook, East 90th Street	40° 47'	73° 56'	+1 52	+1 34	*1.02	*0.91	4.7	5.7	2.6
1319	Roosevelt Island, north end	40° 46'	73° 56'	+1 49	+1 28	*1.05	*1.05	4.8	5.8	2.6
1321	37th Avenue, Long Island City	40° 46'	73° 57'	+1 34	+1 13	*0.99	*1.00	4.5	5.5	2.4
1323	East 41st Street, New York City	40° 45'	73° 58'	+1 03	+0 46	*0.95	*1.09	4.3	5.2	2.4
1325	Hunters Point, Newtown Creek	40° 44'	73° 57'	+1 22	+0 56	*0.89	*0.90	4.1	4.9	2.2
1327	English Kills entrance, Newtown Creek	40° 43'	73° 55'	+1 34	+1 07	*0.90	*0.90	4.2	5.0	2.3
1329	East 27th Street, Bellevue Hospital	40° 44'	73° 58'	+1 12	+1 06	*0.92	*0.91	4.2	5.0	2.3
1331	East 19th Street, New York City	40° 44'	73° 58'	+1 06	+1 01	*0.90	*0.91	4.1	4.9	2.2
1333	North 3rd Street, Brooklyn	40° 43'	73° 58'	+0 59	+0 45	*0.90	*0.90	4.1	4.9	2.2
1335	Williamsburg Bridge	40° 43'	73° 58'	+0 56	+0 41	*0.90	*0.91	4.1	4.9	2.2
1337	Wallabout Bay	40° 42'	73° 58'	+0 32	+0 22	*0.94	*1.05	4.3	5.2	2.4
1339	Brooklyn Bridge	40° 42'	74° 00'	+0 17	+0 10	*0.94	*0.95	4.3	5.2	2.3

Endnotes can be found at the end of table 2.

Figure 11-6. *A sample subordinate data page,* Tide Tables, *Table 2.*

Time from the nearest high water or low water

Duration of rise or fall, see footnote (h. m.)	h. m.	h. m.	h. m.	h. m.	h. m.	h. m.	h. m.	h. m.	h. m.	h. m.	h. m.	h. m.	h. m.	h. m.	h. m.
4 00	0 08	0 16	0 24	0 32	0 40	0 48	0 56	1 04	1 12	1 20	1 28	1 36	1 44	1 52	2 00
4 20	0 09	0 17	0 26	0 35	0 43	0 52	1 01	1 09	1 18	1 27	1 35	1 44	1 53	2 01	2 10
4 40	0 09	0 19	0 28	0 37	0 47	0 56	1 05	1 15	1 24	1 33	1 43	1 52	2 01	2 11	2 20
5 00	0 10	0 20	0 30	0 40	0 50	1 00	1 10	1 20	1 30	1 40	1 50	2 00	2 10	2 20	2 30
5 20	0 11	0 21	0 32	0 43	0 53	1 04	1 15	1 25	1 36	1 47	1 57	2 08	2 19	2 29	2 40
5 40	0 11	0 23	0 34	0 45	0 57	1 08	1 19	1 31	1 42	1 53	2 05	2 16	2 27	2 39	2 50
6 00	0 12	0 24	0 36	0 48	1 00	1 12	1 24	1 36	1 48	2 00	2 12	2 24	2 36	2 48	3 00
6 20	0 13	0 25	0 38	0 51	1 03	1 16	1 29	1 41	1 54	2 07	2 19	2 32	2 45	2 57	3 10
6 40	0 13	0 27	0 40	0 53	1 07	1 20	1 33	1 47	2 00	2 13	2 27	2 40	2 53	3 07	3 20
7 00	0 14	0 28	0 42	0 56	1 10	1 24	1 38	1 52	2 06	2 20	2 34	2 48	3 02	3 16	3 30
7 20	0 15	0 29	0 44	0 59	1 13	1 28	1 43	1 57	2 12	2 27	2 41	2 56	3 11	3 25	3 40
7 40	0 15	0 31	0 46	1 01	1 17	1 32	1 47	2 03	2 18	2 33	2 49	3 04	3 19	3 35	3 50
8 00	0 16	0 32	0 48	1 04	1 20	1 36	1 52	2 08	2 24	2 40	2 56	3 12	3 28	3 44	4 00
8 20	0 17	0 33	0 50	1 07	1 23	1 40	1 57	2 13	2 30	2 47	3 03	3 20	3 37	3 53	4 10
8 40	0 17	0 35	0 52	1 09	1 27	1 44	2 01	2 19	2 36	2 53	3 11	3 28	3 45	4 03	4 20
9 00	0 18	0 36	0 54	1 12	1 30	1 48	2 06	2 24	2 42	3 00	3 18	3 36	3 54	4 12	4 30
9 20	0 19	0 37	0 56	1 15	1 33	1 52	2 11	2 29	2 48	3 07	3 25	3 44	4 03	4 21	4 40
9 40	0 19	0 39	0 58	1 17	1 37	1 56	2 15	2 35	2 54	3 13	3 33	3 52	4 11	4 31	4 50
10 00	0 20	0 40	1 00	1 20	1 40	2 00	2 20	2 40	3 00	3 20	3 40	4 00	4 20	4 40	5 00
10 20	0 21	0 41	1 02	1 23	1 43	2 04	2 25	2 45	3 06	3 27	3 47	4 08	4 29	4 49	5 10
10 40	0 21	0 43	1 04	1 25	1 47	2 08	2 29	2 51	3 12	3 33	3 55	4 16	4 37	4 59	5 20

Correction to height

Range of tide, see footnote (Ft.)	Ft.	Ft.	Ft.	Ft.	Ft.	Ft.	Ft.	Ft.	Ft.	Ft.	Ft.	Ft.	Ft.	Ft.	Ft.
0.5	0.0	0.0	0.0	0.0	0.0	0.0	0.1	0.1	0.1	0.1	0.1	0.2	0.2	0.2	0.2
1.0	0.0	0.0	0.0	0.0	0.1	0.1	0.1	0.2	0.2	0.2	0.3	0.3	0.4	0.4	0.5
1.5	0.0	0.0	0.0	0.1	0.1	0.1	0.2	0.2	0.3	0.4	0.4	0.5	0.6	0.7	0.8
2.0	0.0	0.0	0.0	0.1	0.1	0.2	0.3	0.3	0.4	0.5	0.6	0.7	0.8	0.9	1.0
2.5	0.0	0.0	0.1	0.1	0.2	0.2	0.3	0.4	0.5	0.6	0.7	0.9	1.0	1.1	1.2
3.0	0.0	0.0	0.1	0.1	0.2	0.3	0.4	0.5	0.6	0.8	0.9	1.0	1.2	1.3	1.5
3.5	0.0	0.0	0.1	0.2	0.2	0.3	0.4	0.6	0.7	0.9	1.0	1.2	1.4	1.6	1.8
4.0	0.0	0.0	0.1	0.2	0.3	0.4	0.5	0.7	0.8	1.0	1.2	1.4	1.6	1.8	2.0
4.5	0.0	0.0	0.1	0.2	0.3	0.4	0.6	0.7	0.9	1.1	1.3	1.6	1.8	2.0	2.2
5.0	0.0	0.1	0.1	0.2	0.3	0.5	0.6	0.8	1.0	1.2	1.5	1.7	2.0	2.2	2.5
5.5	0.0	0.1	0.1	0.2	0.4	0.5	0.7	0.9	1.1	1.4	1.6	1.9	2.2	2.5	2.8
6.0	0.0	0.1	0.1	0.3	0.4	0.6	0.8	1.0	1.2	1.5	1.8	2.1	2.4	2.7	3.0
6.5	0.0	0.1	0.2	0.3	0.4	0.6	0.8	1.1	1.3	1.6	1.9	2.2	2.6	2.9	3.2
7.0	0.0	0.1	0.2	0.3	0.5	0.7	0.9	1.2	1.4	1.8	2.1	2.4	2.8	3.1	3.5
7.5	0.0	0.1	0.2	0.3	0.5	0.7	1.0	1.2	1.5	1.9	2.2	2.6	3.0	3.4	3.8
8.0	0.0	0.1	0.2	0.3	0.5	0.8	1.0	1.3	1.6	2.0	2.4	2.8	3.2	3.6	4.0
8.5	0.0	0.1	0.2	0.4	0.6	0.8	1.1	1.4	1.8	2.1	2.5	2.9	3.4	3.8	4.2
9.0	0.0	0.1	0.2	0.4	0.6	0.9	1.2	1.5	1.9	2.2	2.7	3.1	3.6	4.0	4.5
9.5	0.0	0.1	0.2	0.4	0.6	0.9	1.2	1.6	2.0	2.4	2.8	3.3	3.8	4.3	4.8
10.0	0.0	0.1	0.2	0.4	0.7	1.0	1.3	1.7	2.1	2.5	3.0	3.5	4.0	4.5	5.0
10.5	0.0	0.1	0.3	0.5	0.7	1.0	1.3	1.7	2.2	2.6	3.1	3.6	4.2	4.7	5.2
11.0	0.0	0.1	0.3	0.5	0.7	1.1	1.4	1.8	2.3	2.8	3.3	3.8	4.4	4.9	5.5
11.5	0.0	0.1	0.3	0.5	0.8	1.1	1.5	1.9	2.4	2.9	3.4	4.0	4.6	5.1	5.8
12.0	0.0	0.1	0.3	0.5	0.8	1.1	1.5	2.0	2.5	3.0	3.6	4.1	4.8	5.4	6.0
12.5	0.0	0.1	0.3	0.5	0.8	1.2	1.6	2.1	2.6	3.1	3.7	4.3	5.0	5.6	6.2
13.0	0.0	0.1	0.3	0.6	0.9	1.2	1.7	2.2	2.7	3.2	3.9	4.5	5.1	5.8	6.5
13.5	0.0	0.1	0.3	0.6	0.9	1.3	1.7	2.2	2.8	3.4	4.0	4.7	5.3	6.0	6.8
14.0	0.0	0.2	0.3	0.6	0.9	1.3	1.8	2.3	2.9	3.5	4.2	4.8	5.5	6.3	7.0
14.5	0.0	0.2	0.4	0.6	1.0	1.4	1.9	2.4	3.0	3.6	4.3	5.0	5.7	6.5	7.2
15.0	0.0	0.2	0.4	0.6	1.0	1.4	1.9	2.5	3.1	3.8	4.4	5.2	5.9	6.7	7.5
15.5	0.0	0.2	0.4	0.7	1.0	1.5	2.0	2.6	3.2	3.9	4.6	5.4	6.1	6.9	7.8
16.0	0.0	0.2	0.4	0.7	1.1	1.5	2.1	2.6	3.3	4.0	4.7	5.5	6.3	7.2	8.0
16.5	0.0	0.2	0.4	0.7	1.1	1.6	2.1	2.7	3.4	4.1	4.9	5.7	6.5	7.4	8.2
17.0	0.0	0.2	0.4	0.7	1.1	1.6	2.2	2.8	3.5	4.2	5.0	5.9	6.7	7.6	8.5
17.5	0.0	0.2	0.4	0.8	1.2	1.7	2.2	2.9	3.6	4.4	5.2	6.0	6.9	7.8	8.8
18.0	0.0	0.2	0.4	0.8	1.2	1.7	2.3	3.0	3.7	4.5	5.3	6.2	7.1	8.1	9.0
18.5	0.1	0.2	0.5	0.8	1.2	1.8	2.4	3.1	3.8	4.6	5.5	6.4	7.3	8.3	9.2
19.0	0.1	0.2	0.5	0.8	1.3	1.8	2.4	3.1	3.9	4.8	5.6	6.6	7.5	8.5	9.5
19.5	0.1	0.2	0.5	0.8	1.3	1.9	2.5	3.2	4.0	4.9	5.8	6.7	7.7	8.7	9.8
20.0	0.1	0.2	0.5	0.9	1.3	1.9	2.6	3.3	4.1	5.0	5.9	6.9	7.9	9.0	10.0

Obtain from the predictions the high water and low water, one of which is before and the other after the time for which the height is required. The difference between the times of occurrence of these tides is the duration of rise or fall, and the difference between their heights is the range of tide for the above table. Find the difference between the nearest high or low water and the time for which the height is required.

Enter the table with the duration of rise or fall, printed in heavy-faced type, which most nearly agrees with the actual value, and on that horizontal line find the time from the nearest high or low water which agrees most nearly with the corresponding actual difference. The correction sought is in the column directly below, on the line with the range of tide.

When the nearest tide is high water, subtract the correction.

When the nearest tide is low water, add the correction.

Figure 11-7. Table 3 of a Tide Table.

COMPLETE TIDE TABLE

Date:_____

Subordinate Station _____

Reference Station _____

HW Time Difference _____

LW Time Difference _____

Difference in height of HW _____

Difference in height of LW _____

Reference Station   Subordinate Station

HW _____ _____   _____ _____

LW _____ _____   _____ _____

HW _____ _____   _____ _____

LW _____ _____   _____ _____

HW _____ _____   _____ _____

LW _____ _____   _____ _____

HEIGHT OF TIDE AT ANY TIME

Locality:_____ Time:_____ Date:_____

Duration of Rise or Fall: _____

Time from Nearest Tide: _____

Range of Tide: _____

Height of Nearest Tide: _____

Corr. from Table 3: _____

Height of Tide at: _____

Figure 11-8. *A standard form for tide calculations.*

USE OF THE *TIDE TABLES*

To solve problems involving the predicted height of tide, the navigator makes use of the first three tables of the applicable volume of the *Tide Tables*, in conjunction with a standard tide form such as that which appears in Figure 11-8. The top part of the tide form is designed for construction of a daily tide table for a given reference or subordinate station, making use of Tables 1 and 2. The bottom part is for computing the height of the tide at a given time at a designated location, using information from Table 3.

The use of the *Tide Tables* is best demonstrated by means of an example. Suppose that a navigator wished to construct a complete tide table for 8 October at the Brooklyn Bridge, New York. Upon completion, the navigator will then use this table both to draw a graph of tide versus time and, in conjunction with Table 3, to compute the height of tide at 0900, to see if the ship could pass safely underneath the bridge at that time.

Constructing a Complete Tide Table

The first step in constructing the complete tide table for Brooklyn Bridge for 8 October is to locate the subordinate station difference data by referring to the index for Table 2 in the back of the proper *Tide Table* volume, in this case the *East Coast of North and South America*. A portion of the index appears in Figure 11-9 below. In the index, the subordinate station number of Brooklyn Bridge is found to be 1339. Turning to this number in Table 2, illustrated in Figure 11-6, the Brooklyn Bridge time and height-difference data are noted and recorded onto the top of the blank tide form. The reference station for Brooklyn Bridge and other nearby subordinate stations appears in boldfaced type in the "Differences" column above the subordinate station data; it is New York (at the Battery), located on pages 60–63 of Table 1.

Opening Table 1 to the proper page, the daily predictions for the month of October for New York at the Battery are found, as reproduced in Figure 11-5,

INDEX TO STATIONS 289

	No.		No.
Blackbeard Island, Ga.............	3457,3473	Bridgewater, Nova Scotia..............	509
Blackbird Creek, Del..................	1997	Bridport Inlet, Melville Island........	7
Blacks Creek, N. J....................	2113	Brigantine Channel, N. J..............	1817
Blackwater River, Fla.............	4535,4537	Bristol Ferry, R. I....................	1163
Blackwater Sound, Fla.............	3951,3953	Bristol Harbor, R. I...................	1171
Blake Cove, Va........................	2157	Bristol Highlands, R. I................	1161
Blessing Plantation, S. C.............	3075	Bristol Point, R. I....................	1159
Block Island, Great Salt Pond, R. I....	1189	Broad Creek, S. C.................	3339,3341
Block Island (Old Harbor), R. I........	1191	Broad River, S. C.................	3317-3335
Blomsterbugten, Greenland.............	53	Broad Sound, Mass....................	929,931
Bloody Point Bar Light, Md............	2313	Bronx River, N. Y......................	1303
Bloody Point, S. C....................	3359	Brooklyn, North Third Street, N. Y.....	1333
Blount Island Bridge, Fla.............	3613	Brooklyn Bridge, N. Y..................	1339
Bluefields Lagoon, Nicaragua..........	4747	Broomes Island, Md.....................	2419
Blue Hill Bay, Maine...............	677-683	Brosewere Bay, N. Y....................	1503
Blue Hill Harbor, Maine...............	679	Broughton Point, S. C..................	3319
Bluff Island, S. C....................	3217	Broward River, Fla.....................	3619
Bluff Plantation, S. C................	3247	Brown Cove, Va.........................	2747
Bluff Point, S. C.....................	3169	Brown Island, S. C.....................	2995
Bluffton, S. C........................	3355	Browns Bay, Va.........................	2595
Boca Chica Key, Fla...............	4261,4265	Brunswick, East River, Ga..............	3505
Boca Chita Key, Fla...................	3909	Brunswick, Maine.......................	805
Boca Ciega Bay, Fla...............	4387-4395	Buchanan Creek entrance, Va............	2743

Figure 11-9. *A portion of the index to Table 2,* Tide Tables.

COMPLETE TIDE TABLE

Date: **8 OCTOBER**

Substation   **BROOKLYN BRIDGE**

Reference Station   **NEW YORK**

HW Time Difference   **+ 0 17 m**

LW Time Difference   **+ 0 10 m**

Difference in height of HW   *** 0.94**

Difference in height of LW   *** 0.95**

Reference Station			Substation	
HW **2357** **4.1'**	**(7 OCT)**		___	___
LW **0547** **1.0'**			___	___
HW **1218** **4.9'**			___	___
LW **1919** **1.0'**			___	___
HW ___ ___			___	___
LW ___ ___			___	___

Figure 11-10A. *Reference station tide table, 8 October.*

and the daily predictions for 8 October are recorded on the tide form. In addition, the last tide event on the preceding day is also recorded, for reasons that will become apparent later. At this point, the partially completed tide form now appears as illustrated in Figure 11-10A above.

As the final step in preparing a complete tide table for Brooklyn Bridge for 8 October, the high and low tide time differences are added algebraically to the daily predictions recorded for New York, and the multiplicative ratios given for Brooklyn Bridge are applied to the New York tidal height data. At this time it is noted that when the high water time difference, +17 minutes, is added to the last tide on 7 October—a high tide at 2357—the result is another high tide appearing at the bridge on 8 October. The reason for recording the last tide event on the preceding day at the reference station should now be obvious. If the subordinate station tide time differences are positive, they may result in a prior day's tide event at the reference station occurring on the day in question at the subordinate station, when added to the reference station time predictions. Conversely, if subordinate station time differences are negative, then the first

COMPLETE TIDE TABLE

Date: __8 OCTOBER__

Substation __BROOKLYN BRIDGE__

Reference Station __NEW YORK__

HW Time Difference __+0 17__

LW Time Difference __+0 10__

Difference in height of HW __*0.94__

Difference in height of LW __*0.95__

Reference Station			Substation	
HW __2357__ __4.1'__	(7 OCT)		__0014__ __3.9'__	
LW __0547__ __1.0'__			__0557__ __1.0'__	
HW __1218__ __4.9'__			__1235__ __4.6'__	
LW __1919__ __1.0'__			__1929__ __1.0'__	
HW _____ _____			_____ _____	
LW _____ _____			_____ _____	

tide event on the succeeding day at the reference station should be recorded, because the algebraic addition of a negative time correction could well result in a tide occurring on the day in question at the subordinate station.

After all tide time and height conversions have been made, the complete tide table for 8 October at Brooklyn Bridge appears as shown in Figure 11-10B.

The preceding problem was worked under the assumption that the location of interest, Brooklyn Bridge, was observing standard zone time, the time on which the predictions in the *Tide Tables* are based. If daylight savings time were to be taken into account, it would first be necessary to convert all reference station daily predictions to daylight savings time by adding one hour. The complete tide table for the subordinate station would then be computed in the same manner as described above.

Graphing Tidal Data

On many occasions it may be more helpful to the navigator to construct a sinusoidal graph of the height of tide versus time for a particular location, rather than present the data in tabular form only. Such a graph is required to be done by Navy navigators whenever a ship enters piloting waters, for each reference station and at least three substations to be passed enroute. Navy directives further require all such graphs and the computations upon which they are based to be entered in the *Navigation Workbook*, one of the ship's required records.

To construct such a graph, a method called the *one-quarter, one-tenth rule* is used. It is described in the notes to Table 3 in each volume of the *Tide Tables*. The procedure is as follows:

1. On graph paper, plot the high- and low-water points in the order of their occurrence for the day, measuring time on the horizontal axis and height on the vertical axis. These are the basic points to be used to draw the curve.
2. Draw light straight lines connecting these points.
3. Divide each of these straight construction lines into four equal parts. The halfway point of each line is the crossing point for the curve.
4. At the quarter point adjacent to high water, draw a vertical construction line above the point, and at the quarter point adjacent to low water, draw a vertical point below the point. Make the length of these lines equal to one-tenth of the range between the high- and low-water tide heights used. The points at the ends of these construction lines then give two intermediate points for the curve.
5. Draw a smooth curve (perhaps with the help of a drawing instrument called a *French curve*) through the high- and low-water points and the intermediate points, making the curve rounded near high and low water.

The completed curve so plotted will approximate the actual tidal curve that should occur at that location. Heights of tide for any time, as well as time frames within which the tide will be at, above, or below a desired height, can be accurately estimated from it.

As a caution, the navigator should remember that this curve, like the data upon which it is based, represents a forecast of the tide for that particular day, assuming nominal good weather prevails. If the weather is stormy and/or windy, these predictions could vary by several feet. Generally, prolonged onshore winds or a low barometric pressure can produce higher tide levels than those predicted, while the opposite can result in lower levels.

An example tide graph drawn for 8 October using the data previously computed for Brooklyn Bridge appears in Figure 11-10C.

If the navigator has access to a personal computer, tidal graphs can also be produced by various navigation software applications programs, which are discussed in chapter 22.

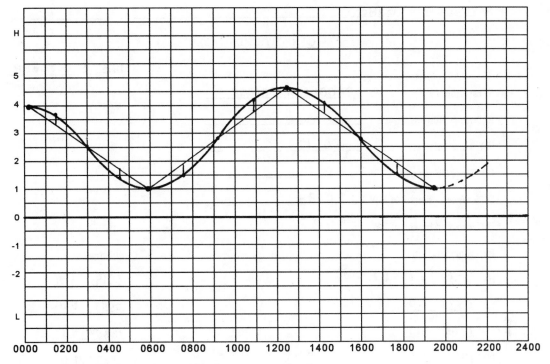

Figure 11-10C. *A tidal graph drawn by the quarter-tenth rule.*

Computing the Height of Tide at Any Time

After either a complete tide table or a graph of tide versus time for a substation has been constructed, the navigator can then determine the tide for any particular time of interest. On the graph of tide versus time, the navigator need only go to the desired time on the time scale, then read up to pick off the corresponding tide on the plotted curve. For the example discussed here, the graph indicates a tidal height of about 2.6 feet at 0900 at Brooklyn Bridge on 8 October.

As an alternative, Table 3 in the *Tide Tables* can be used in conjunction with the complete tide table for Brooklyn Bridge to compute the height of tide at 0900 8 October. To do this, the quantities indicated on the bottom of the tide form are computed for the tide cycle containing the time in question, and recorded on the form for use with Table 3. First, the duration of the rising tide from the 0557 low water to the 1235 high water is computed and found to be 6 hours 38 minutes. Second, the time from the nearest tide comprising the tide cycle—in this case the 0557 low—is figured; it is 3 hours 3 minutes. Third, the range of tide between the 0557 low and the 1235 high water is found—3.6 feet. Fourth, Table 3 in Figure 11-7 is entered to find the correction to height of the nearest tide, as described in the following paragraph.

COMPLETE TIDE TABLE

Date: __8 OCTOBER__

Substation	__BROOKLYN BRIDGE__
Reference Station	__NEW YORK__
HW Time Difference	__+ 0 17__
LW Time Difference	__+ 0 10__
Difference in height of HW	__*0.94__
Difference in height of LW	__*0.95__

Reference Station			Substation	
HW __2357__ __4.1'__	(7 OCT)		__0014__ __3.9'__	
LW __0547__ __1.0'__			__0557__ __1.0'__	
HW __1218__ __4.9'__			__1235__ __4.6'__	
LW __1919__ __1.0'__			__1929__ __1.0'__	
HW ____ ____			____ ____	
LW ____ ____			____ ____	

HEIGHT OF TIDE AT ANY TIME

Locality: __BROOKLYN BRIDGE__   Time: __0900__   Date: __8 OCTOBER__

Duration of Rise or Fall:	__6 38__
Time from Nearest Tide:	__3 03__
Range of Tide:	__3.6__
Height of Nearest Tide:	__1.0__
Corr. from Table 3:	__+ 1.6__
Height of Tide at: __0900__	__2.6__

Figure 11-10D. *Height of tide at a given time.*

Table 3 is entered at the top left with the tabulated duration of rise or fall that is closest to the computed value; in this case, the entering argument is 6 hours 40 minutes. Next, the tabulated time from the nearest high or low water closest to the computed value is found directly across to the right from the entering argument; for this problem, it is 3 hours 7 minutes. The correction sought is in the bottom half of the table, directly beneath this second argument, and opposite the tabulated range of tide that most nearly agrees with the actual range; the correction found is +1.6 feet. Finally, following the instructions on the bottom of Table 3, the correction is added to the 0557 low water to find the height of tide at 0900 of 2.6 feet. The completed calculation appears in Figure 11-10D. In practice, this result would probably be rounded down to 2.0 feet if water depth were critical, or up to 3.0 feet if clearance above high water were critical, to provide a small safety factor to compensate for any difference in the standard conditions on which the tidal predictions are based. Notice that the result, 2.6 feet, agrees well with the value indicated for 0900 on the graph of tide versus time in Figure 11-10C.

THE BRIDGE PROBLEM

During the course of piloting in coastal waters or rivers, it occasionally becomes necessary for a vessel to pass under an overhanging obstruction such as a bridge or cable. In this situation, the navigator can be faced with two problems: first, whether or not the ship can pass safely under the obstruction at the time she is nominally scheduled to reach it; and second, if she cannot, determining the time frame within which the vessel can pass.

As a first step in determining whether a ship can pass under a bridge at a given time, the navigator should refer to a diagram such as Figure 11-4 on page 182. On the diagram it can be seen that the actual clearance can be computed by adding the charted clearance to the distance from the mean high-water plane to the tidal height at the time passage is desired. The charted clearance can be obtained from a small-scale chart of the area in which the bridge is located or from the appropriate *Coast Pilot* or *Sailing Directions* volume. The distance between the mean high-water plane and the tidal height is a quantity equal to the sum of the mean tide level plus half the mean range of tide, minus the tidal height for the time in question; see the diagram in Figure 11-11. If the actual clearance is at least equal to or greater than the masthead height from the waterline to the top of the mast, the ship will be able to pass safely beneath the bridge. In practice an additional safety factor is included as well.

As an example, suppose that the navigator desired to pass beneath the Brooklyn Bridge at the date and time for which the height of tide was computed in the preceding section, that is, 0900 on 8 October. The charted clearance for the bridge is 127 feet, and the navigator's ship has a masthead height of 130 feet. The height of tide for this time was computed to be 2.6 feet. The mean tide level and the mean range of tide can quickly be found by referring to the Brooklyn Bridge data in Table 2 of the *Tide Tables*, which appears in Figure 11-6; they are 2.3 feet and 4.3 feet, respectively. The actual bridge clearance at 0900 is given by the algebraic sum

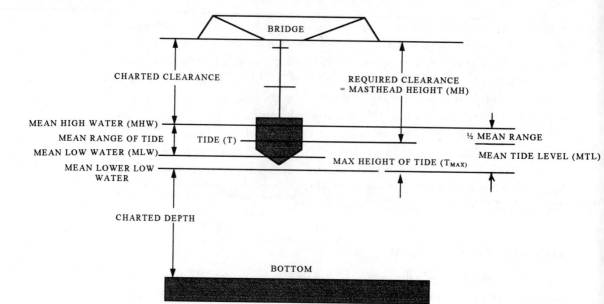

Figure 11-11. *Determining height of tide allowing passage under a bridge. The Maximum Height of Tide (T_{MAX}) ≤ Mean Tide Level + ½ Mean Range of Tide + Charted Clearance − Masthead Height.*

$$127 + [(2.3 + \frac{4.3}{2}) - 2.6] = 129.8 \text{ feet}$$

Since the masthead height of the ship is greater than 129.8 feet, the ship cannot pass under the Brooklyn Bridge at 0900.

Having reached this conclusion, the navigator then needs to determine when the ship can pass beneath the bridge. To do this, the quantitative relationships for the bridge problem must be rearranged to solve for the maximum height of tide that will allow the ship to pass beneath the bridge (see Figure 11-11). The resulting formula looks like this:

> Tide ≤ Mean Tide Level + ½ Mean Range + Charted Clearance
> − Masthead Height

In this example, the formula would be

> Tide ≤ 2.3 + ½ (4.3) + 127 − 130 or Tide ≤ 1.4 feet.

The navigator refers to the tidal graph completed earlier for 8 October, which appears in Figure 11-10C, to determine the times nearest to 0900 when the tide will be equal to or less than 1.4 feet. As this figure is rather low in comparison to the mean range of tide, 4.3 feet, it is obvious that the passage must be made at some time fairly close to the time of a low tide—in this case, the low tide at 0557. Inspection of the diagram indicates a time frame between about 0430 and 0730. If more precise times than these are required, they could

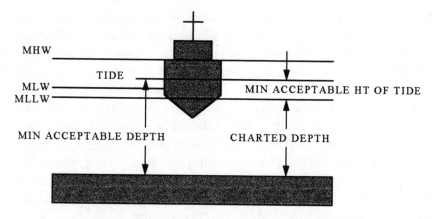

Figure 11-12. *Physical relationships of the shoal problem. The Minimum Acceptable Height of Tide* (T_{MIN}) ≥ *the Minimum Acceptable Depth – the Charted Depth.*

be obtained by working backward through Table 3 in the *Tide Tables*. In most situations, the times as obtained from the tidal graph will suffice.

THE SHOAL PROBLEM

An analogous situation to the bridge clearance problem occurs when a ship is required to pass over a shallow bar or shoal. In this type of problem, it is necessary to compute the times on each side of a high tide between which the tide will be above a certain level. In the shoal problem, the actual depth of water must be equal to or greater than the sum of the ship's draft, measured from the waterline to the keel, plus an appropriate safety factor. The physical relationships existing in the shoal problem are represented in Figure 11-12.

The minimum acceptable height of tide is that which when added to the charted depth yields the minimum acceptable depth for passage, usually the ship's draft (waterline to keel depth) plus an optional safety factor (see Figure 11-12). In formula form, this becomes

Tide ≥ Minimum Acceptable Depth – Charted Depth

Once the minimum acceptable height of tide is determined, the time frame is found by methods identical to those described above for the bridge problem, using either the tide graph or Table 3 of the *Tide Tables*.

SUMMARY

This chapter has examined the causes and effects of tide, as well as the more common types of problems associated with tide predictions. The navigator must be continually aware of the effects of tide, especially when passage under a low overhead obstruction or over shallow ground is scheduled. Water depths may occasionally be less than those indicated on a chart, and overhead clearances may sometimes be lower, especially if the less-extreme reference planes such as mean high and mean low water are used as the basis for charted depths and clearances. To be certain of passing an obstacle safely, the navigator should always include a suitable safety margin in all calculations of computed clearances and depths. If possible, passages should be scheduled when extreme tide conditions exist that will provide the greatest possible margin for safety in the given situation. Finally, the navigator should bear in mind that, although the predictions in the *Tide Tables* will be accurate most of the time, there exists the possibility that actual tide levels may differ from the predictions, particularly under unusual conditions of wind and barometric pressure.

CURRENT

12

In the preceding chapter, tide was defined as a rise or fall in the surface of a body of water. This chapter will examine a related phenomenon known as *current*, which, in contrast to tide, is strictly defined as a horizontal movement of water. As will be explained, this horizontal movement is caused in large measure by tide, but other physical factors such as wind, rain, and the Coriolis effect also come into play. Like tide, current is also of great interest and importance to the navigator, both at sea and especially when piloting in coastal waters, because it continually affects the movements of the ship as it proceeds through the water.

There are two main types of current with which the surface navigator is concerned—ocean and tidal. The first part of this chapter will examine the causes, effects, and methods of prediction of ocean currents, and the second part will discuss similar aspects of tidal current.

OCEAN CURRENT

The oceans and their currents have long been an enigma. Even though the oceans cover over 70 percent of the earth's surface, it has only been in comparatively recent years that we have begun to understand the mysterious forces affecting the oceans and their currents. This section will relate some of the more important facts that have been established concerning major ocean currents.

An ocean is never in a state of equilibrium. It is, instead, always in a state of motion, attempting to attain a forever unattainable balance. The primary reason for this imbalance is the rise and fall of the ocean level caused by the gravitational forces of the sun and moon, but there are several other factors that also have an effect. The waters of the oceans are continually being heated and cooled by the earth's atmosphere, blown by its winds, and salted and diluted by evaporation and rain. All these factors combine to produce flows of both surface and subsurface waters from higher to lower levels, colder to warmer areas,

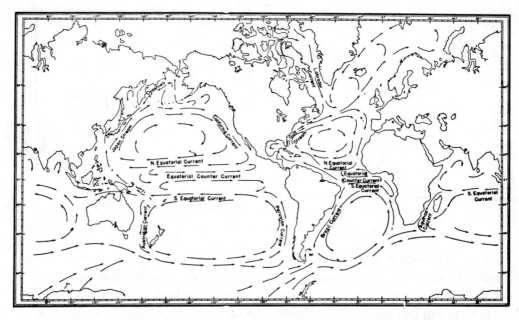

Figure 12-1. *Important ocean currents of the world.*

and higher- to lower-density regions. Once set in motion, the flows of water in the oceans are affected by the Coriolis force associated with the earth's rotation, so that they form giant patterns of rotation called "gyres" in each of the major ocean basins. These gyres rotate in a basically clockwise direction in the northern hemisphere, and in a counterclockwise direction in the southern hemisphere. Moreover, it is these gyres, or portions of them, that form the more well-known and well-defined of the world's great *ocean currents*, some of which are pictured in Figure 12-1.

The major ocean currents are like rivers in the oceans, but they far surpass any continental river in size and strength. The Gulf Stream, for example, is spawned in the western Caribbean and Gulf of Mexico. It surges north through the Straits of Florida, and follows a meandering course off the U.S. Atlantic seaboard to Newfoundland, where it turns eastward, crosses the North Atlantic, and finally dissipates off northern Europe. Where it is strongest off Florida and South Carolina, it is 40 miles wide and 2,000 feet deep, and it carries 100 billion tons of water at a velocity sometimes approaching 5 knots. The fastest known current is the Somali, located in the Indian Ocean; during the summer season of the monsoon wind, it has been measured at speeds up to 7 knots.

Even more interesting than the surface currents are the *deep ocean currents*, caused primarily by water-density differences. Although little was known about these currents until recent years, it is generally accepted today that the deep-ocean current system is equally as extensive as the surface system. It is theorized that these currents have been instrumental in shaping the ocean floor in many areas of the world, particularly near the continental slopes. In the western end of the Mediterranean basin off Gibraltar, there is an incoming surface current that flows at speeds of 3 to 5 knots. At the other end of the basin, the water grows more dense because of evaporation and the influx of salts from the land, and sinks. It forms a deep ocean current that flows westward underneath the incoming surface current, thus maintaining roughly the same volume of water in the Mediterranean. Another amazing deep-sea current is the Cromwell Current, discovered in 1951. It flows at a top speed of 5 knots about 300 feet below and in the opposite direction to a surface current that flows westward south of the equator in the Pacific.

Research into currents is being actively pursued by oceanographers today, using both traditional methods, such as anchored and free-flowing buoys, installed current meters, dyes, and the drift of survey ships, and to an ever-increasing extent, new techniques fostered by space-age technology. Satellites equipped with infrared sensors are being used to chart currents by mapping heat flow patterns, and others have been used to chart oceanographic buoys and even icebergs fitted with a transmitter.

Predicting Ocean Currents

Although the *Planning Guides* of the *Sailing Directions* contain some information on normal locations and strengths of currents encountered on the open sea, the navigator should find the *pilot chart* described in chapter 5 (see Figure 5-8 page 75) of most value in predicting the direction and velocity of ocean currents.

The direction in which a current flows is called its *set,* and its velocity is referred to as its *drift.* The navigator determines set and drift from the pilot chart by referring to the green color-coded arrows on the chart designated for that purpose. The direction of the arrows indicates the average set of the current at that location during the time period covered by the chart, and the figures printed nearby represent the average drift.

When using current data from the appropriate pilot chart, the navigator should remember that these data represent average conditions based on historical records. Actual ocean currents encountered will often vary somewhat from these norms, depending on the meteorological activity taking place at the time.

TIDAL CURRENT

The major ocean currents must be reckoned with by the surface navigator while at sea. Once coastal waters are entered, however, the navigator becomes primarily concerned with another type of current known as *tidal current.* This current is so called because it is caused primarily by the rise and fall of the tide;

coastal waters affected by tidal currents are often referred to as *tidewater* areas. A tidal current that flows toward shore as a result of the approach of a high tide is called a *flood current*, and that which flows away from shore because of a low tide is an *ebb current*. During each tide cycle, there is a moment corresponding to the stand of a tide, when no horizontal movement of the water takes place as the current changes direction; this moment, which may in reality be several minutes long, is called *slack water*.

At first consideration, it might seem that the times of minimum velocity of a flood or an ebb current should coincide with the times of high and low tide stands, but in general this is not the case. The change of direction of a tidal current always lags behind the turning of the tide by a time interval of varying length, which depends on the geographic characteristics of the shoreline. Along a relatively straight coast with few indentations, the interval might be quite small. On the other hand, if a large harbor were connected to the sea by a narrow inlet, the tide and current might be out of phase by as much as 3 hours. In this latter kind of situation, the velocity of current in the connecting channel is usually at or near a maximum when the tide is at its extreme high or low level.

In most inshore waters, the direction of flow, or set, of an incoming flood current is not exactly opposite the direction of the outgoing ebb current, because of irregularities in the shape of the shore. Where the two currents are approximately opposite from one another, they are referred to as a *reversing current*. Offshore, where the direction of flow is not as restricted, it often happens that the effect of semidiurnal tides is to create a current that flows continuously with no clearly defined ebb or flood; its set moves completely around the compass during each tide cycle. A tidal current of this type is called a *rotary current*. As a result of the Coriolis effect, its direction of rotation is usually clockwise in the northern hemisphere, and counterclockwise in the southern hemisphere. In U.S. waters, rotary currents are fairly common off the Atlantic coast, but are rarely observed off the Gulf or Pacific coasts where the diurnal tide pattern prevails.

Prediction of Tidal Current

Prediction of tidal current is of great importance to the navigator during the practice of piloting in coastal tidewater areas. Not only must the navigator take current into account when transiting a channel, in order to remain within its bounds, but he or she must also realize that some approaches into harbors and turning basins are virtually impossible for a deep-draft vessel to negotiate when a current of any appreciable velocity is flowing. A carrier or cruiser, for example, normally enters the U.S. Navy base at Mayport, Florida, at or very near the time of slack water, because of the strong tidal current flowing in the mouth of the St. Johns River at other times.

For making various predictions concerning tidal currents, the navigator has several means available, depending on the area of the world in which the ship is operating. In the coastal waters of the United States, a set of two publications called the *Tidal Current Tables* is extensively used for tidal current predictions; in addition, a series of tidal current diagrams and charts is also available for many of the major U.S. ports. British Admiralty–produced coastal approach charts often contain a graphic known as a *current diamond*, which is used in conjunction with charted tables and the applicable volume of the *Tide Tables* to yield current estimates.

The *Tidal Current Tables*

The *Tidal Current Tables*, often referred to simply as the *Current Tables*, consist of two volumes, one titled the *Atlantic Coast of North America*, and the other the *Pacific Coast of North America and Asia*. Each is published yearly by a private publisher using NOS data. They are distributed to U.S. Navy users by the National Imagery and Mapping Agency (NIMA). They are also available to merchant and civil mariners at most nautical supply stores. In appearance, they are much like the *Tide Tables*. Each volume is divided up into several numbered tables; the West Coast edition has four tables, and the East Coast edition five, the extra table being for rotary current predictions. Like the *Tide Tables*, the *Current Tables* contain daily predictions of the times of slack water and maximum flood and ebb current velocities for each of several reference stations, and they list time and velocity difference ratios for each of several hundred subordinate locations. The navigator can apply these differences to a specified set of reference station data to obtain a complete set of current predictions for a given subordinate location for a day of interest.

Table 1 of the *Current Tables* lists the times and maximum velocities of each flood and ebb current and the time of each slack water in the chronological order of appearance for all reference stations used in that volume. Figure 12-2 is a page from Table 1 of an *Atlantic Coast Current Tables* volume. Inasmuch as the tidal currents are primarily caused by the action of the tides, there are usually four maximum currents for every tidal day period of 24 hours and 50 minutes, with a slack water between each one. As was the case with tide, however, a vacancy does occasionally appear; in these instances, the slack or maximum current that seems to be missing will be found to occur either just before midnight on the preceding day or just after midnight on the following day. For ease of reference, successive slack-water times are placed in one column, and maximum current velocities are arranged in consecutive order in a second column. As in the *Tide Tables*, times given in Table 1 are based on the standard time zone in which the reference station is located.

Table 2 consists of time differences for flood and ebb currents, and for the minimum currents preceding them (usually but not always slack waters), speed

Hell Gate (off Mill Rock), East River, New York

F—Flood, Dir. 050° True   E—Ebb, Dir. 230° True

October

Day	Slack h m	Max h m	knots
1 W ●		0003	4.6E
	0324	0627	3.7F
	0933	1217	4.7E
	1542	1845	3.7F
	2155		
2 Th		0036	4.7E
	0359	0700	3.7F
	1009	1252	4.8E
	1617	1919	3.8F
	2233		
3 F		0111	4.7E
	0434	0734	3.7F
	1046	1329	4.8E
	1653	1955	3.7F
	2310		
4 Sa		0148	4.7E
	0508	0810	3.7F
	1123	1407	4.8E
	1729	2032	3.7F
	2348		
5 Su		0227	4.6E
	0543	0848	3.6F
	1201	1447	4.7E
	1808	2112	3.6F
6 M	0028	0309	4.5E
	0620	0929	3.5F
	1243	1531	4.6E
	1849	2155	3.4F
7 Tu	0112	0353	4.4E
	0703	1015	3.4F
	1330	1618	4.5E
	1937	2244	3.3F
8 W	0201	0443	4.3E
	0753	1106	3.3F
	1423	1711	4.4E
	2032	2339	3.2F
9 Th ☽	0258	0538	4.2E
	0853	1205	3.2F
	1524	1809	4.4E
	2136		
10 F		0040	3.2F
	0359	0638	4.2E
	1001	1310	3.3F
	1629	1911	4.4E
	2243		
11 Sa		0145	3.3F
	0503	0742	4.4E
	1111	1417	3.4F
	1734	2016	4.5E
	2348		
12 Su		0251	3.5F
	0604	0847	4.6E
	1216	1522	3.6F
	1836	2120	4.7E
13 M	0048	0352	3.7F
	0702	0950	4.8E
	1317	1622	3.9F
	1933	2220	4.9E
14 Tu	0144	0448	3.9F
	0757	1048	5.0E
	1413	1718	4.1F
	2028	2314	5.0E
15 W ○	0237	0542	4.1F
	0850	1142	5.2E
	1506	1811	4.2F
	2120		
16 Th		0008	5.1E
	0328	0633	4.2F
	0941	1234	5.3E
	1558	1902	4.2F
	2211		
17 F		0058	5.2E
	0419	0723	4.2F
	1031	1324	5.3E
	1649	1952	4.2F
	2301		
18 Sa		0148	5.1E
	0509	0812	4.1F
	1121	1413	5.2E
	1741	2042	4.0F
	2352		
19 Su		0237	4.9E
	0600	0903	3.9F
	1213	1503	5.0E
	1833	2134	3.8F
20 M	0044	0327	4.7E
	0653	0956	3.7F
	1306	1554	4.7E
	1927	2229	3.5F
21 Tu	0138	0420	4.4E
	0748	1052	3.5F
	1401	1649	4.4E
	2024	2327	3.3F
22 W ◑	0234	0516	4.1E
	0847	1153	3.4F
	1459	1748	4.1E
	2123		
23 Th		0030	3.1F
	0332	0619	3.9E
	0948	1258	3.3F
	1558	1854	4.0E
	2222		
24 F		0133	3.0F
	0429	0727	3.9E
	1048	1401	3.0F
	1655	2001	3.9E
	2318		
25 Sa		0231	3.0F
	0524	0830	3.9E
	1143	1456	3.1F
	1749	2058	4.0E
26 Su	0010	0321	3.1F
	0613	0920	4.1E
	1233	1544	3.2F
	1838	2142	4.1E
27 M	0056	0405	3.2F
	0659	1000	4.2E
	1317	1626	3.3F
	1922	2219	4.3E
28 Tu	0137	0443	3.4F
	0740	1035	4.4E
	1358	1703	3.4F
	2004	2253	4.4E
29 W	0216	0519	3.5F
	0820	1110	4.6E
	1437	1739	3.6F
	2043	2328	4.6E
30 Th	0254	0553	3.6F
	0858	1146	4.7E
	1515	1814	3.6F
	2122		
31 F ●		0004	4.7E
	0330	0628	3.7F
	0936	1222	4.9E
	1552	1850	3.7F
	2200		

November

Day	Slack h m	Max h m	knots
1 Sa		0041	4.8E
	0405	0704	3.7F
	1013	1301	4.9E
	1630	1926	3.7F
	2238		
2 Su		0120	4.8E
	0441	0741	3.7F
	1051	1340	4.9E
	1708	2005	3.6F
	2317		
3 M		0200	4.7E
	0519	0821	3.6F
	1131	1422	4.9E
	1748	2046	3.6F
	2358		
4 Tu		0243	4.7E
	0559	0904	3.5F
	1214	1507	4.8E
	1832	2131	3.4F
5 W	0043	0329	4.6E
	0645	0951	3.4F
	1302	1555	4.7E
	1921	2221	3.3F
6 Th	0133	0420	4.5E
	0738	1044	3.3F
	1357	1648	4.6E
	2017	2316	3.2F
7 F ◑	0230	0516	4.4E
	0840	1143	3.3F
	1458	1746	4.5E
	2120		
8 Sa		0018	3.2F
	0332	0617	4.4E
	0948	1249	3.3F
	1603	1849	4.5E
	2226		
9 Su		0124	3.3F
	0436	0722	4.5E
	1058	1357	3.4F
	1709	1955	4.6E
	2330		
10 M		0230	3.4F
	0539	0828	4.7E
	1203	1503	3.5F
	1811	2100	4.7E
11 Tu	0031	0332	3.6F
	0638	0932	4.9E
	1304	1605	3.7F
	1910	2201	4.9E
12 W	0128	0430	3.8F
	0734	1031	5.1E
	1400	1702	3.9F
	2005	2258	5.0E
13 Th	0221	0524	4.0F
	0828	1126	5.2E
	1454	1755	4.0F
	2058	2351	5.1E
14 F ○	0313	0615	4.1F
	0919	1218	5.3E
	1545	1845	4.0F
	2148		
15 Sa		0041	5.1E
	0403	0705	4.1F
	1009	1307	5.3E
	1635	1935	4.0F
	2238		
16 Su		0129	5.0E
	0452	0754	4.0F
	1058	1355	5.1E
	1725	2023	3.8F
	2327		
17 M		0216	4.9E
	0542	0842	3.8F
	1147	1442	5.0E
	1815	2112	3.6F
18 Tu	0016	0304	4.7E
	0633	0932	3.6F
	1237	1529	4.7E
	1906	2202	3.4F
19 W	0106	0352	4.5E
	0725	1023	3.3F
	1328	1618	4.5E
	1959	2255	3.2F
20 Th	0157	0442	4.3E
	0820	1117	3.1F
	1420	1709	4.3E
	2053	2350	3.0F
21 F ○	0250	0535	4.1E
	0916	1214	2.9F
	1514	1803	4.1E
	2148		
22 Sa		0046	2.9F
	0343	0631	4.0E
	1013	1312	2.8F
	1608	1859	4.0E
	2242		
23 Su		0142	2.8F
	0436	0728	4.0E
	1107	1408	2.8F
	1701	1954	4.0E
	2333		
24 M		0233	2.9F
	0526	0821	4.1E
	1158	1458	2.9F
	1751	2045	4.1E
25 Tu	0020	0320	3.0F
	0614	0908	4.2E
	1245	1544	3.1F
	1839	2130	4.3E
26 W	0104	0402	3.1F
	0659	0952	4.4E
	1328	1625	3.2F
	1923	2212	4.4E
27 Th	0145	0441	3.3F
	0741	1033	4.6E
	1410	1704	3.4F
	2006	2252	4.6E
28 F	0224	0519	3.4F
	0822	1113	4.8E
	1450	1743	3.5F
	2047	2332	4.7E
29 Sa	0303	0557	3.5F
	0902	1154	4.9E
	1530	1822	3.6F
	2127		
30 Su ●		0013	4.8E
	0341	0636	3.6F
	0943	1235	5.0E
	1610	1901	3.6F
	2208		

December

Day	Slack h m	Max h m	knots
1 M		0054	4.9E
	0420	0717	3.7F
	1024	1317	5.1E
	1651	1943	3.8F
	2249		
2 Tu		0137	4.9E
	0501	0759	3.7F
	1106	1401	5.1E
	1734	2026	3.6F
	2333		
3 W		0222	4.9E
	0546	0844	3.6F
	1152	1447	5.0E
	1820	2112	3.5F
4 Th	0019	0310	4.8E
	0634	0933	3.5F
	1241	1537	4.9E
	1910	2203	3.4F
5 F	0110	0401	4.7E
	0729	1027	3.4F
	1336	1630	4.8E
	2005	2258	3.3F
6 Sa	0206	0457	4.7E
	0830	1126	3.3F
	1435	1727	4.7E
	2106	2358	3.3F
7 Su	0307	0557	4.6E
	0937	1230	3.3F
	1539	1829	4.6E
	2210		
8 M		0103	3.3F
	0411	0702	4.6E
	1045	1338	3.3F
	1644	1934	4.6E
	2314		
9 Tu		0210	3.4F
	0514	0809	4.7E
	1150	1446	3.4F
	1748	2040	4.7E
10 W	0015	0314	3.5F
	0615	0915	4.9E
	1252	1549	3.5F
	1848	2144	4.8E
11 Th	0113	0414	3.7F
	0713	1017	5.0E
	1349	1647	3.7F
	1944	2242	4.9E
12 F	0207	0509	3.8F
	0807	1113	5.1E
	1442	1741	3.8F
	2036	2336	5.0E
13 Sa ○	0259	0601	3.9F
	0859	1204	5.2E
	1533	1831	3.8F
	2126		
14 Su		0025	5.0E
	0349	0649	3.9F
	0948	1252	5.2E
	1621	1918	3.8F
	2214		
15 M		0112	5.0E
	0437	0736	3.8F
	1035	1337	5.1E
	1709	2004	3.7F
	2301		
16 Tu		0156	4.9E
	0524	0822	3.7F
	1122	1420	5.0E
	1755	2049	3.5F
	2347		
17 W		0239	4.8E
	0611	0907	3.5F
	1208	1503	4.8E
	1842	2133	3.4F
18 Th	0032	0322	4.6E
	0659	0952	3.3F
	1253	1546	4.6E
	1929	2218	3.2F
19 F	0118	0406	4.5E
	0747	1038	3.1F
	1340	1630	4.4E
	2017	2305	3.0F
20 Sa	0205	0451	4.3E
	0837	1126	2.9F
	1428	1716	4.3E
	2106	2353	2.8F
21 Su ◐	0254	0539	4.2E
	0929	1217	2.8F
	1518	1805	4.2E
	2157		
22 M		0044	2.8F
	0344	0630	4.1E
	1022	1310	2.7F
	1610	1856	4.1E
	2248		
23 Tu		0136	2.8F
	0435	0723	4.2E
	1115	1403	2.8F
	1702	1948	4.1E
	2337		
24 W		0227	2.8F
	0526	0815	4.2E
	1205	1454	2.9F
	1753	2039	4.2E
25 Th	0025	0315	2.9F
	0615	0906	4.4E
	1253	1542	3.0F
	1842	2129	4.4E
26 F	0110	0401	3.1F
	0702	0955	4.6E
	1339	1628	3.2F
	1928	2216	4.5E
27 Sa	0153	0445	3.3F
	0748	1041	4.8E
	1423	1712	3.3F
	2014	2302	4.7E
28 Su	0236	0529	3.5F
	0833	1127	4.9E
	1507	1755	3.5F
	2058	2347	4.9E
29 M ●	0318	0612	3.6F
	0917	1212	5.1E
	1550	1839	3.6F
	2142		
30 Tu		0032	5.0E
	0401	0656	3.7F
	1002	1257	5.2E
	1633	1923	3.6F
	2227		
31 W		0118	5.1E
	0446	0742	3.7F
	1048	1343	5.2E
	1718	2008	3.7F
	2313		

Time meridian 75° W. 0000 is midnight. 1200 is noon.

Figure 12-2. *A sample reference station page,* Atlantic Coast Tidal Current Tables, *Table 1.*

TABLE 2 – CURRENT DIFFERENCES AND OTHER CONSTANTS

No.	PLACE	Meter Depth	POSITION		TIME DIFFERENCES				SPEED RATIOS		AVERAGE SPEEDS AND DIRECTIONS							
			Latitude	Longitude	Min. before Flood	Flood	Min. before Ebb	Ebb	Flood	Ebb	Minimum before Flood		Maximum Flood		Minimum before Ebb		Maximum Ebb	
		ft	North	West	h m	h m	h m	h m			knots	Dir.	knots	Dir.	knots	Dir.	knots	Dir.
	EAST RIVER–cont. Time meridian, 75° W				on Hell Gate, p.32													
	Roosevelt Island																	
3401	west of, off 75th Street		40° 46'	73° 57'	−0 02	−0 04	−0 08	+0 07	1.1	1.0	0.0	−−	3.8	037°	0.0	−−	4.7	215°
3406	east of, off 36th Avenue		40° 46'	73° 57'	−0 08	−0 04	−0 08	−0 11	1.0	0.7	0.0	−−	3.5	030°	0.0	−−	3.4	210°
3411	west of, off 67th Street		40° 45.74'	73° 57.24'	+0 13	−0 08	+0 06	+0 11	1.1	0.9	0.0	−−	3.6	011°	0.0	−−	4.0	230°
3416	west of, off 63rd Street		40° 45.58'	73° 57.27'	−0 10	−0 08	0 00	+0 03	0.8	0.6	0.0	−−	2.8	036°	0.0	−−.	2.9	223°
3421	east of		40° 45.49'	73° 57.08'	0 00	−0 06	+0 02	+0 07	0.8	0.6	0.0	−−	2.8	028°	0.0	−−	2.6	200°
3426	Manhattan, off 31st Street		40° 44.38'	73° 58.17'	+0 09	−0 11	−0 02	+0 36	0.4	0.5	0.0	−−	1.5	000°	0.0	−−	2.1	175°
3431	Newtown Creek entrance		40° 44'	73° 57'	Current weak and variable													
3436	Pier 67, off 19th Street		40° 44'	73° 58'	−0 08	+0 08	−0 08	+0 07	0.5	0.4	0.0	−−	1.8	355°	0.0	−−	1.9	179°
3441	Williamsburg Bridge, 0.3 mile north of		40° 43.08'	73° 58.24'	−0 05	+0 12	−0 01	+0 10	0.8	0.6	0.0	−−	2.7	020°	0.0	−−	2.9	220°
3446	Manhattan Bridge, East of	15	40° 42.5'	73° 59.4'	−0 28	+0 19	−0 13	+0 03	0.7	0.5	0.1	161°	2.5	083°	0.0	−−	2.9	259°
3448	Brooklyn Bridge	15d	40° 42.36'	73° 59.85'	+0 29	+0 41	+0 33	+0 29	0.8	0.6	0.1	324°	2.7	063°	0.0	−−	2.8	253°
3451	Brooklyn Bridge, 0.1 mile southwest of		40° 42.2'	74° 00.0'	−0 18	+0 08	−0 04	−0 07	0.9	0.8	0.0	−−	2.9	046°	0.0	−−	3.5	222°
3456	Buttermilk Channel (SEE CAUTION NOTE)	15	40° 41.3'	74° 00.8'	−0 31	0 00	+0 03	−0 18	0.5	0.6	0.0	−−	1.8	050°	0.0	315°	2.6	221°
3461	Buttermilk Channel		40° 41.15'	74° 00.81'	−0 12	−0 18	−0 06	+0 18	0.5	0.5	0.0	−−	1.8	050°	0.0	−−	2.4	220°
	HARLEM RIVER																	
3466	East 107th Street	15	40° 47.4'	73° 56.1'	−0 08	−0 03	−1 09	−1 39	0.2	0.2	0.0	−−	0.8	206°	0.0	−−	0.8	030°
3471	Willis Ave. Bridge, 0.1 mile NW of		40° 48.3'	73° 55.8'	−0 20	−0 12	−0 12	−0 13	0.4	0.3	0.0	−−	1.2	140°	0.0	−−	1.3	330°
3476	Madison Ave. Bridge		40° 48.8'	73° 56.1'	−0 20	+0 18	−0 21	−0 14	0.5	0.4	0.0	−−	1.8	180°	0.0	−−	1.7	000°
3481	Macombs Dam Bridge		40° 49.7'	73° 56.1'	−0 20	+0 14	−0 22	−0 11	0.5	0.3	0.0	−−	1.7	180°	0.0	−−	1.4	000°
3486	High Bridge		40° 50.5'	73° 55.9'	−0 20	+0 08	−0 23	−0 08	0.6	0.4	0.0	−−	2.0	189°	0.0	−−	2.0	015°
3491	West 207th Street Bridge		40° 51.8'	73° 54.9'	−0 22	+0 05	−0 22	−0 02	0.6	0.4	0.0	−−	2.0	215°	0.0	−−	2.0	035°
3496	Broadway Bridge		40° 52.4'	73° 54.7'	−0 23	+0 08	−0 20	+0 04	0.6	0.5	0.0	−−	2.1	116°	0.0	−−	2.3	299°
3501	Henry Hudson Bridge, 0.7 nmi. SE of	16	40° 52.6'	73° 55.3'	+0 12	+0 31	−0 31	+0 41	0.2	0.3	0.0	−−	1.8	137°	0.0	−−	1.3	326°
	LONG ISLAND, South Coast				on The Narrows, p.36													
3506	Fire Island Lighted Whistle Bouy 2Fl		40° 29'	73° 11'	See table 5.													
3511	Fire Island Inlet, 22 miles S of <17>		40° 16'	73° 16'	See table 5.													
3516	Shinnecock Canal, railroad bridge <18>		40° 53.2'	72° 30.1'				−0 38	−−	0.8	−−	−−	−−	−−	−−	−−	1.5	180°
3521	Ponquogue bridge, Shinnecock Bay		40° 50.7'	72° 30.1'	+0 54	+0 35	+0 27	+0 37	0.5	0.3	0.0	−−	0.8	250°	0.0	−−	0.6	090°
3526	Shinnecock Inlet		40° 50.6'	72° 28.7'	−0 06	−0 21	−0 30	−1 03	1.5	1.2	0.0	−−	2.5	350°	0.0	−−	2.3	170°
3531	Fire I. Inlet, 0.5 mi. S of Oak Beach		40° 37.78'	73° 18.40'	−) 03	−0 01	+0 29	−0 01	1.4	1.2	0.0	−−	2.4	082°	0.0	−−	2.4	244°
3536	Jones Inlet		40° 35.5'	73° 34.0'	−1 15	−0 49	−0 48	−1 05	1.8	1.3	0.0	−−	3.1	035°	0.0	−−	2.6	217°
3541	Long Beach, inside, between bridges		40° 35.7'	73° 39.6'	−0 54	+0 23	+0 32	0 00	0.3	0.3	0.0	−−	0.5	076°	0.0	−−	0.6	277°
3546	East Rockaway Inlet		40° 35.4'	73° 45.3'	−1 46	−1 35	−1 03	−1 38	1.3	1.2	0.0	−−	2.2	042°	0.0	−−	2.3	227°
3551	Ambrose Light		40° 27'	73° 49'	See table 5.													
3556	Sandy Hook App. Lighted Horn Bouy 2A		40° 27'	73° 55'	See table 5.													
	JAMAICA BAY																	
3561	Rockaway Point	15	40° 32.3'	73° 56.8'	−2 36	−2 34	−1 38	−3 02	1.1	0.5	0.2	228°	1.9	301°	0.2	217°	1.1	140°
3566	Rockaway Inlet		40° 33.7'	73° 56.1'	−1 55	−2 20	−1 33	−2 11	1.1	1.3	0.0	−−	1.8	085°	0.0	−−	2.7	244°
3571	Rockaway Inlet	14	40° 34.2'	73° 53.8'	−1 53	−2 00	−1 15	−2 29	0.9	0.8	0.0	−−	1.6	066°	0.1	344°	1.5	261°
3576	Barren Island, east of		40° 35'	73° 53'	−1 59	−2 28	−2 03	−2 19	0.7	0.9	0.0	−−	1.2	004°	0.0	−−	1.7	192°
3581	Canarsie (midchannel, off pier)		40° 37.6'	73° 53.0'	−1 54	−1 38	−1 18	−2 06	0.3	0.4	0.0	−−	0.5	045°	0.0	−−	0.7	222°
3586	Beach Channel (bridge)		40° 35'	73° 49'	−1 18	−1 13	−0 57	−1 25	1.1	1.0	0.0	−−	1.9	062°	0.0	−−	2.0	225°
3591	Grass Hassock Channel		40° 36.6'	73° 47.1'	−1 21	−1 02	−0 57	−0 54	0.6	0.5	0.0	−−	1.0	052°	0.0	−−	1.0	228°
	NEW YORK HARBOR ENTRANCE																	
3596	Ambrose Channel	15	40° 31.0'	73° 58.8'	−0 57	−1 10	−0 25	−0 07	0.9	0.8	0.1	025°	1.6	303°	0.0	−−	1.7	123°
3601	Norton Point, WSW of	16	40° 33.5'	74° 01.5'	−0 13	−1 01	+0 26	+0 27	0.6	0.6	0.3	263°	1.0	341°	0.1	071°	1.2	166°
3606	THE NARROWS, midchannel		40° 36.6'	74° 02.8'	Daily predictions						0.0	−−	1.7	340°	0.0	−−	2.0	160°

Endnotes can be found at the end of table 2.

Figure 12-3. *A typical subordinate location data page,* Atlantic Coast Tidal Current Tables, *Table 2.*

ratios and average speed and true direction of flood and ebb currents, and other useful information for each of the subordinate locations located within the area of coverage of that volume. As in the *Tide Tables*, the subordinate location data are arranged in geographic sequence, with an alphabetic index to Table 2 located in the back of each volume. A sample page from Table 2 of an *Atlantic Coast Current Tables* appears in Figure 12-3. Notice that at some locations the water depth for which the data were determined is given, as in the case of the Hampton Roads Newport News stations in Figure 12-3. If no water depths appear, the data presented represent average data for all depths. At a few locations precise data are supplied for multiple meter depths, as in the case of Hussey Sound, whose data are partially reproduced in Figure 12-4.

Note that the differences in velocity of maximum currents between a subordinate location and its reference station in the *Current Tables* are always expressed in the form of arithmetic speed ratios, and never as algebraic differences sometimes used for tidal height differences in the *Tide Tables*.

		METER DEPTH	POSITION		TIME DIFFERENCES				SPEED RATIOS		AVERAGE SPEEDS AND DIRECTIONS							
NO.	PLACE		Lat.	Long.	Min. before Flood	Flood	Min. before Ebb	Ebb	Flood	Ebb	Minimum before Flood		Maximum Flood		Minimum before Ebb		Maximum Ebb	
		ft	° ' N	° ' W	h. m.	h. m.	h. m.	h. m.			knots deg.		knots deg.		knots deg.		knots deg.	

TABLE 2. - CURRENT DIFFERENCES AND OTHER CONSTANTS

KENNEBEC RIVER
Time meridian, 75°W — on PORTSMOUTH HARBOR ENTRANCE, p.10

NO.	PLACE	METER DEPTH (ft)	Lat.	Long.	Min. before Flood	Flood	Min. before Ebb	Ebb	Flood	Ebb	Min. before Flood (kn)	(deg)	Max. Flood (kn)	(deg)	Min. before Ebb (kn)	(deg)	Max. Ebb (kn)	(deg)
180	Bluff Head, west of		43 51.3	69 47.8	+0 33	+0 53	+0 26	+0 24	1.9	1.9	0.0	- -	2.3	014	0.0	- -	3.4	184
185	Fiddler Ledge, north of		43 52.8	69 47.8	+0 47	+1 12	+0 22	+0 48	1.6	1.4	0.0	- -	1.9	267	0.0	- -	2.6	113
190	Doubling Point, south of		43 52.8	69 48.4	+0 28	+0 49	+0 23	+0 53	2.2	1.7	0.0	- -	2.6	300	0.0	- -	3.0	127
195	Lincoln Ledge, east of		43 53.8	69 48.6	+0 32	+0 45	+0 23	+0 34	1.6	1.6	0.0	- -	1.9	359	0.0	- -	2.8	174
200	Bath, 0.2 mile south of bridge <3>		43 54.5	69 48.5	+0 29	+1 28	+0 43	+0 23	0.8	0.8	0.0	- -	1.0	003	0.0	- -	1.5	177
	CASCO BAY																	
205	Broad Sound, west of Eagle Island		43 42.7	70 03.8	-1 16	-1 05	-1 27	-0 59	0.8	0.7	0.0	- -	0.9	010	0.0	- -	1.3	168
210	Hussey Sound, SW of Overset Island	15	43 40.27	70 10.52	-1 28	-1 18	-0 58	-1 30	0.9	0.6	0.0	- -	1.1	316	0.3	189	1.2	153
	---do.---	25	43 40.27	70 10.52	-1 39	-1 19	-1 06	-1 32	0.9	0.6	0.0	- -	1.1	318	0.3	211	1.1	155
	---do.---	40	43 40.27	70 10.52	-1 58	-1 16	-1 05	-1 32	0.9	0.5	0.1	228	1.1	314	0.3	200	1.0	154
212	Hussey Sound, SE of Pumpkin Nob	40	43 40.45	70 10.78	-2 21	-1 29	-1 32	-1 14	1.0	0.5	0.1	068	1.2	346	0.1	066	0.9	168
215	Hussey Sound, east of Crow Island	40	43 41.33	70 10.79	-2 18	-0 42	-0 55	-1 24	0.7	0.4	0.1	114	0.9	016	0.0	- -	0.8	197
220	Portland Hbr. ent., SW of Cushing I.		43 37.9	70 12.7	-1 43	-1 11	-1 20	-0 58	0.8	0.6	0.0	- -	1.0	322	0.0	- -	1.1	154
225	Diamond I. Ledge, midchannel SW. of		43 39.6	70 13.5	-1 26	-1 12	-1 11	-1 06	0.8	0.5	0.0	- -	0.9	300	0.0	- -	0.9	150
230	Portland Breakwater Light 0.3 mi. NW of <1> <4>		43 39.5	70 14.5	- - -	-0 47	- - -	-1 07	0.3	0.3	0.0	- -	0.4		0.0	- -	0.5	048

Figure 12-4. *Tabulated data for multiple meter depths are given for Hussey Sound in the* Atlantic Coast Current Tables.

Table 3 of the *Current Tables* is used to find the velocity of current for a given time at a reference station or subordinate location, after a complete current table has been constructed by use of the first two tables as applicable. This table is reproduced in Figure 12-5.

Note that Table 3 actually consists of two similarly constructed but separate tables. Table 3A is used for most computations, with Table 3B being reserved for use only with those locations listed in Table 2 that use Cape Cod Canal, Hell Gate, or the Chesapeake and Delaware Canal as their reference stations.

Table 4, pictured in Figure 12-6, is used to find the time frame around a slack-water or minimum current within which the current velocity will be below a desired maximum. Like Table 3, it is made up of two parts, A and B, with B used only for subordinate locations referred to Cape Cod Canal, Hell Gate, and the Chesapeake and Delaware Canal.

Table 5, as mentioned earlier, appears only in the *Atlantic Coast* volume of the *Current Tables*. It is designed for use in predicting the set and drift of an offshore rotary current at 46 locations off the Atlantic seaboard. A page from this table appears in Figure 12-7 on page 209. For each location listed in Table 5, the average direction and velocity of the rotary current are given for each hour after the occurrence of maximum flood current at a specified reference station. The velocities given should be increased by about 15 to 20 percent at the time of a new or full moon, and decreased by the same amount when the moon is at or near quadrature. In practice, this table is not extensively used, inasmuch as the rotary current velocities listed for most locations are so small for the most part as to be considered negligible.

TABLE A

Interval between slack and maximum current

Interval between slack and desired time (h. m.)	1 20	1 40	2 00	2 20	2 40	3 00	3 20	3 40	4 00	4 20	4 40	5 00	5 20	5 40
0 20	0.4	0.3	0.3	0.2	0.2	0.2	0.2	0.1	0.1	0.1	0.1	0.1	0.1	0.1
0 40	0.7	0.6	0.5	0.4	0.4	0.3	0.3	0.3	0.3	0.2	0.2	0.2	0.2	0.2
1 00	0.9	0.8	0.7	0.6	0.6	0.5	0.5	0.4	0.4	0.4	0.3	0.3	0.3	0.3
1 20	1.0	1.0	0.9	0.8	0.7	0.6	0.6	0.5	0.5	0.5	0.4	0.4	0.4	0.4
1 40	-----	1.0	1.0	0.9	0.8	0.8	0.7	0.7	0.6	0.6	0.5	0.5	0.5	0.4
2 00	-----	-----	1.0	1.0	0.9	0.9	0.8	0.8	0.7	0.7	0.6	0.6	0.6	0.5
2 20	-----	-----	-----	1.0	1.0	1.0	0.9	0.9	0.8	0.7	0.7	0.7	0.6	0.6
2 40	-----	-----	-----	-----	1.0	1.0	1.0	0.9	0.9	0.8	0.8	0.7	0.7	0.7
3 00	-----	-----	-----	-----	-----	1.0	1.0	1.0	0.9	0.9	0.8	0.8	0.8	0.7
3 20	-----	-----	-----	-----	-----	-----	1.0	1.0	1.0	0.9	0.9	0.9	0.8	0.8
3 40	-----	-----	-----	-----	-----	-----	-----	1.0	1.0	1.0	0.9	0.9	0.9	0.9
4 00	-----	-----	-----	-----	-----	-----	-----	-----	1.0	1.0	1.0	1.0	0.9	0.9
4 20	-----	-----	-----	-----	-----	-----	-----	-----	-----	1.0	1.0	1.0	1.0	0.9
4 40	-----	-----	-----	-----	-----	-----	-----	-----	-----	-----	1.0	1.0	1.0	1.0
5 00	-----	-----	-----	-----	-----	-----	-----	-----	-----	-----	-----	1.0	1.0	1.0
5 20	-----	-----	-----	-----	-----	-----	-----	-----	-----	-----	-----	-----	1.0	1.0
5 40	-----	-----	-----	-----	-----	-----	-----	-----	-----	-----	-----	-----	-----	1.0

TABLE B

Interval between slack and maximum current

Interval between slack and desired time (h. m.)	1 20	1 40	2 00	2 20	2 40	3 00	3 20	3 40	4 00	4 20	4 40	5 00	5 20	5 40
0 20	0.5	0.4	0.4	0.3	0.3	0.3	0.3	0.3	0.2	0.2	0.2	0.2	0.2	0.2
0 40	0.8	0.7	0.6	0.5	0.5	0.5	0.4	0.4	0.4	0.4	0.3	0.3	0.3	0.3
1 00	0.9	0.8	0.8	0.7	0.7	0.6	0.6	0.5	0.5	0.5	0.4	0.4	0.4	0.4
1 20	1.0	1.0	0.9	0.8	0.8	0.7	0.7	0.6	0.6	0.6	0.5	0.5	0.5	0.5
1 40	-----	1.0	1.0	0.9	0.9	0.8	0.8	0.7	0.7	0.7	0.6	0.6	0.6	0.6
2 00	-----	-----	1.0	1.0	0.9	0.9	0.9	0.8	0.8	0.7	0.7	0.7	0.7	0.6
2 20	-----	-----	-----	1.0	1.0	1.0	0.9	0.9	0.8	0.8	0.8	0.7	0.7	0.7
2 40	-----	-----	-----	-----	1.0	1.0	1.0	0.9	0.9	0.9	0.8	0.8	0.8	0.7
3 00	-----	-----	-----	-----	-----	1.0	1.0	1.0	0.9	0.9	0.9	0.9	0.8	0.8
3 20	-----	-----	-----	-----	-----	-----	1.0	1.0	1.0	0.9	0.9	0.9	0.9	0.8
3 40	-----	-----	-----	-----	-----	-----	-----	1.0	1.0	1.0	0.9	0.9	0.9	0.9
4 00	-----	-----	-----	-----	-----	-----	-----	-----	1.0	1.0	1.0	1.0	0.9	0.9
4 20	-----	-----	-----	-----	-----	-----	-----	-----	-----	1.0	1.0	1.0	1.0	0.9
4 40	-----	-----	-----	-----	-----	-----	-----	-----	-----	-----	1.0	1.0	1.0	1.0
5 00	-----	-----	-----	-----	-----	-----	-----	-----	-----	-----	-----	1.0	1.0	1.0
5 20	-----	-----	-----	-----	-----	-----	-----	-----	-----	-----	-----	-----	1.0	1.0
5 40	-----	-----	-----	-----	-----	-----	-----	-----	-----	-----	-----	-----	-----	1.0

Use table A for all places except those listed below for table B.
Use table B for Cape Cod Canal, Hell Gate, Chesapeake and Delaware Canal and all stations in table 2 which are referred to them.

1. From predictions find the time of slack water and the time and velocity of maximum current (flood or ebb), one of which is immediately before and the other after the time for which the velocity is desired.
2. Find the interval of time between the above slack and maximum current, and enter the top of table A or B with the interval which most nearly agrees with this value.
3. Find the interval of time between the above slack and the time desired, and enter the side of table A or B with the interval which most nearly agrees with this value.
4. Find, in the table, the factor corresponding to the above two intervals, and multiply the maximum velocity by this factor. The result will be the approximate velocity at the time desired.

Figure 12-5. Atlantic Coast Tidal Current Tables, *Table 3.*

TABLE A

Maximum current	Period with a speed not more than -				
	0.1 knot	0.2 knot	0.3 knot	0.4 knot	0.5 knot
Knots	Minutes	Minutes	Minutes	Minutes	Minutes
1.0	23	46	70	94	120
1.5	15	31	46	62	78
2.0	11	23	35	46	58
3.0	8	15	23	31	38
4.0	6	11	17	23	29
5.0	5	9	14	18	23
6.0	4	8	11	15	19
7.0	3	7	10	13	16
8.0	3	6	9	11	14
9.0	3	5	8	10	13
10.0	2	5	7	9	11

TABLE B

Maximum current	Period with a speed not more than -				
	0.1 knot	0.2 knot	0.3 knot	0.4 knot	0.5 knot
Knots	Minutes	Minutes	Minutes	Minutes	Minutes
1.0	13	28	46	66	89
1.5	8	18	28	39	52
2.0	6	13	20	28	36
3.0	4	8	13	18	22
4.0	3	6	9	13	17
5.0	3	5	8	10	13

Figure 12-6. Atlantic Coast Tidal Current Tables, *Table 4.*

USE OF THE *TIDAL CURRENT TABLES*

In solving tidal current prediction problems involving the use of the *Current Tables*, the navigator normally uses a standard current form such as that shown in Figure 12-8 on page 210. The form has three sections: the first is for constructing a complete current table for a given location of interest in conjunction with Tables 1 and 2; the second is for computing the velocity of current for a given time, using Table 3; and the third is for use with Table 4 in finding the earliest and latest times on each side of a slack water within which the velocity of current will be below a specified limit.

As was the case for the *Tide Tables*, the use of the *Current Tables* is best illustrated by an example. Suppose that the navigator desired to compute the set and drift of the current at Brooklyn Bridge, New York, at 0900 on 8 October, the same location and time used for the tide example in chapter 11. To accomplish this, the navigator first constructs a complete current table for Brooklyn

TABLE 5.—ROTARY TIDAL CURRENTS   193

Georges Bank Lat. 41°50′ N., long. 66°37′ W.

Time	Direction (true)	Velocity
	Degrees	Knots
0	285	0.9
1	304	1.1
2	324	1.2
3	341	1.1
4	10	1.0
5	43	0.9
6	89	1.0
7	127	1.3
8	147	1.6
9	172	1.4
10	197	0.9
11	232	0.8

(Time column headed: Hours after maximum flood at Pollock Rip Channel, see page 28)

Georges Bank Lat. 41°54′ N., long. 67°08′ W.

Time	Direction (true)	Velocity
	Degrees	Knots
0	298	1.1
1	325	1.4
2	344	1.5
3	0	1.2
4	33	0.7
5	82	0.8
6	118	1.1
7	138	1.5
8	153	1.2
9	178	1.1
10	208	0.9
11	236	0.8

Georges Bank Lat. 41°48′ N., long. 67°34′ W.

Time	Direction (true)	Velocity
	Degrees	Knots
0	325	1.5
1	332	2.1
2	342	2.0
3	358	1.3
4	35	0.7
5	99	0.8
6	126	1.3
7	150	2.0
8	159	1.9
9	169	1.7
10	197	1.2
11	275	0.9

Georges Bank Lat. 41°42′ N., long. 67°37′ W.

Time	Direction (true)	Velocity
	Degrees	Knots
0	316	1.1
1	341	1.3
2	356	1.0
3	16	0.8
4	43	0.6
5	92	0.8
6	122	1.0
7	146	1.1
8	170	1.1
9	195	1.0
10	215	1.0
11	272	0.9

Georges Bank Lat. 41°41′ N., long. 67°49′ W.

Time	Direction (true)	Velocity
	Degrees	Knots
0	318	1.6
1	320	1.8
2	325	1.4
3	330	0.8
4	67	0.3
5	111	0.9
6	117	1.5
7	126	1.7
8	144	1.7
9	160	1.1
10	242	0.8
11	292	1.2

Georges Bank Lat. 41°30′ N., long. 68°07′ W.

Time	Direction (true)	Velocity
	Degrees	Knots
0	312	1.5
1	338	1.7
2	346	1.5
3	14	1.1
4	59	0.9
5	99	0.9
6	123	1.3
7	144	1.7
8	160	1.6
9	187	1.3
10	244	1.0
11	274	1.1

Georges Bank Lat. 41°29′ N., long. 67°04′ W.

Time	Direction (true)	Velocity
	Degrees	Knots
0	277	1.0
1	302	1.2
2	329	1.4
3	348	1.3
4	15	1.2
5	48	1.1
6	85	1.2
7	122	1.4
8	145	1.5
9	166	1.3
10	194	1.2
11	223	1.1

Georges Bank Lat. 41°14′ N., long. 67°38′ W.

Time	Direction (true)	Velocity
	Degrees	Knots
0	305	1.4
1	332	1.6
2	355	1.6
3	15	1.4
4	38	1.1
5	77	0.9
6	112	1.2
7	141	1.6
8	162	1.6
9	187	1.5
10	214	1.4
11	252	1.2

Georges Bank Lat. 41°13′ N., long. 68°20′ W.

Time	Direction (true)	Velocity
	Degrees	Knots
0	319	1.5
1	332	2.0
2	345	1.4
3	9	0.8
4	42	0.6
5	80	0.7
6	118	1.0
7	138	1.3
8	154	1.4
9	169	1.5
10	188	1.3
11	236	0.9

Georges Bank Lat. 40°48′ N., long. 67°40′ W.

Time	Direction (true)	Velocity
	Degrees	Knots
0	304	0.9
1	340	0.9
2	353	0.8
3	29	0.6
4	56	0.6
5	83	0.6
6	107	0.9
7	140	1.0
8	156	1.0
9	175	0.9
10	202	0.8
11	245	0.8

Georges Bank Lat. 40°49′ N., long. 68°34′ W.

Time	Direction (true)	Velocity
	Degrees	Knots
0	301	1.2
1	326	1.5
2	345	1.4
3	8	1.1
4	36	0.8
5	69	0.8
6	106	1.0
7	139	1.4
8	153	1.5
9	175	1.4
10	201	1.1
11	237	0.9

Great South Channel, Georges Bank Lat. 40°31′ N., long. 68°47′ W.

Time	Direction (true)	Velocity
	Degrees	Knots
0	320	0.7
1	331	0.9
2	342	1.1
3	3	1.0
4	23	0.8
5	63	0.4
6	129	0.7
7	140	0.9
8	164	1.0
9	179	1.0
10	190	0.8
11	221	0.6

Figure 12-7. Atlantic Coast Tidal Current Tables, *Table 5.*

COMPLETE CURRENT TABLE

Locality: _____   Date: _____

Reference Station: _____

Time Difference: Min Bef Flood: _____ Flood: _____
                      Min Bef Ebb: _____ Ebb: _____
     Speed Ratio: Flood: _____
                      Ebb: _____

Maximum Flood Direction: _____
Maximum Ebb Direction: _____

Reference Station: _____ Locality: _____

_____ _____   _____ _____
_____ _____   _____ _____
_____ _____   _____ _____
_____ _____   _____ _____
_____ _____   _____ _____
_____ _____   _____ _____
_____ _____   _____ _____
_____ _____   _____ _____
_____ _____   _____ _____
_____ _____   _____ _____

VELOCITY OF CURRENT AT ANY TIME

Int. between slack and desired time:
Int. between slack and maximum current: _____   (Ebb) (Flood)
Maximum current: _____
Factor, Table 3   _____
Velocity: _____
Direction: _____

DURATION OF SLACK

Times of maximum current: _____ _____
Maximum current: _____ _____
Desired maximum: _____ _____
Period — Table 4:   _____ _____
Sum of periods: _____
Average period: _____
Time of slack: _____
Duration of slack: From: _____ To: _____

Figure 12-8. U.S. Naval Academy current form.

Bridge for 8 October. Then, using this, the navigator can construct a tidal current graph, required by Navy directives whenever a ship is to transit piloting waters. Once completed, the navigator can then use this graph to find the velocity of tidal current at any desired time, or the duration of desired maximum current around the time of slack water. As an alternative, the navigator can also use the complete current table in conjunction with Table 3 in the *Current Tables* to find the same information.

Constructing a Complete Current Table

To construct the complete current table, the Brooklyn Bridge difference data are first located by referring to the index to Table 2 located in the back of the Atlantic Coast volume of the *Tidal Current Tables*. The applicable portion of this index is shown in Figure 12-9. From the index, the numbers of the Brooklyn Bridge subordinate locations are found as 3448 and 3451. Turning to these numbers in Table 2, illustrated in Figure 12-3, the data of interest to the navigator at the bridge itself are noted and recorded on the top of the current form. The reference station for this and the other nearby location appears in boldfaced type above the various difference data; it is Hell Gate, located in Table 1 starting on page 32.

Opening Table 1 to the proper page, the daily predictions for 8 October appearing in Figure 12-2 are located and recorded on the left side of the current form in chronological order. In addition, the first current event on the preceding day is also recorded, for reasons to be explained later. At this point, the partially completed current form now appears as in Figure 12-10A.

INDEX TO STATIONS

	No.		No.
Brenton Point	2211	Cape Fear River	6336-6436
Breton Bay entrance	5736	Cape Fourchu	26,31
Brewer Point	6031	Cape Haze	8106
Brewerton Angle	6071	Cape Henlopen	4021-4036
Brewerton Channel	4946,6066	Cape Henry Light	4436-4461,4771,4491,6176
Brickyard Creek	7081	Cape Lookout Shoals	6331
Bridgeport Harbor entrance	3066	Cape May	3966
Bridgeton	4171	Cape May Canal	3991,3996
Brier Island	51,56	Cape May Channel	4001
Bristol Harbor	2271	Cape May Harbor	3986
Bristol, N.J	4396	Cape May Point	4006-4011
Broad Creek	5951	Cape Neddick	761
Broad River	7091,7096,7111	Cape Poge Light	1826,1836,1876
Broad River Bridge	7111	Cape Porpoise	756
Broad River Entrance	7036	Cape Romain	6511
Broadkill Slough	4051	Cape Sable	6,11
Broad Sound	611	Cape Spencer	76
Broadway Bridge, Harlem River	3496	Capers Inlet	6516
Broken Ground-Horseshoe Shoal, between	1841	Captain Harbor	3201
Bronx River	3366	Captiva Pass	8086
Bronx-Whitestone Bridge	3346	Carrot Island	6321
Brooklyn Bridge	3448,3451	Casco Bay	611-746
Broomes Island	5866	Casco Passage	121
Broughton Island	7631	Castine Harbor	291
Browns Ledge	1981	Castle Hill	2216

Figure 12-9. *A portion of the index to Table 2,* Atlantic Coast Tidal Current Tables.

COMPLETE CURRENT TABLE

Locality: __BROOKLYN BRIDGE__   Date: __8 OCTOBER__

Reference Station: __HELL GATE__

Time Difference:	Min Bef Flood:	+0 29	Flood:	+0 41
	Min Bef Ebb:	+0 33	Ebb:	+0 29
Speed Ratio:	Flood:	0.8		
	Ebb:	0.6		

Maximum Flood Direction: _____ 063°
Maximum Ebb Direction: _____ 253°

Reference Station: __HELL GATE__   Locality: __BROOKLYN BRIDGE__

2244	3.3 F (7 OCT)	
0201	0	
0443	4.3 E	
0753	0	
1106	3.3 F	
1423	0	
1711	4.4 E	
2032	0	
2339	3.2 F	

Figure 12-10A. *A partially completed current table, with reference station data entered.*

As the final step in the production of a complete current table for Brooklyn Bridge on 8 October, the time differences for slack-water minimums and maximum currents are added algebraically to the times at Hell Gate, and the current maximum velocities are multiplied by the appropriate velocity ratios to obtain the corresponding velocities at Brooklyn Bridge. Care must be taken when doing this to be sure to use the flood ratios for flood-current velocity conversions, and the ebb ratios for ebb-current conversions. The converted velocities are always rounded to the nearest tenth.

At this time, the time difference between ebbs at Hell Gate and Brooklyn Bridge is applied to the last current event on the preceding day—the flood at 2244—to determine whether that event will occur on the day in question at the subordinate location. In this case it will not, so it is dropped from further consideration. The reason for recording the last current event on the previous day at the reference station should now be obvious. If the subordinate location time differences are positive, this may result in a previous day's current event being moved forward into the day in question at the subordinate location. On the other hand, if the time differences are negative, the result may be to move a following day's event back into the current day at the subordinate location. Hence the following rule: when the time corrections are *positive*, the last event

COMPLETE CURRENT TABLE

Locality: **BROOKLYN BRIDGE**   Date: **8 OCTOBER**

Reference Station: **HELL GATE**

Time Difference: Min Bef Flood: **+0 29**   Flood: **+0 41**
                    Min Bef Ebb: **+0 33**   Ebb: **+0 29**

Speed Ratio: Flood: **0.8**
                    Ebb: **0.6**

Maximum Flood Direction: **063°**
Maximum Ebb Direction: **253°**

Reference Station: **HELL GATE**   Locality: **BROOKLYN BRIDGE**

2244	**3.3 F (7 OCT)**		
0201	**0**	**0234**	**0**
0443	**4.3 E**	**0512**	**2.6 E**
0753	**0**	**0822**	**0**
1106	**3.3 F**	**1147**	**2.6 F**
1423	**0**	**1456**	**0**
1711	**4.4 E**	**1740**	**2.6 E**
2032	**0**	**2101**	**0**
2339	**3.2 F**	**0020**	**2.6 F**

Figure 12-10B. *Complete current table for Brooklyn Bridge for 8 October.*

on the *preceding day* at the reference station should always be recorded; when the corrections are *negative*, the first event on the *following day* should be recorded.

After all current time and velocity differences have been applied, the complete current table for 8 October at Brooklyn Bridge now appears as shown in Figure 12-10B.

Had daylight savings time been taken into account for this example, one hour would have been added to all times given for the reference station at Hell Gate prior to the application of the Brooklyn Bridge difference data.

Graphing Tidal Current

As an adjunct to the complete current table for a subordinate location, the navigator may construct a tidal current graph for the location. As mentioned earlier, such graphs are required by Navy directives whenever a ship enters piloting waters. Tidal current graphs and the data upon which they are based, like the tide graphs described in chapter 11, are entered into the *Navigation Workbook*, where they become a part of the ship's navigational records.

To construct a tidal current graph the same type graph paper is used as for the tide graph. The procedure is as follows: (1) On properly labeled graph paper, plot

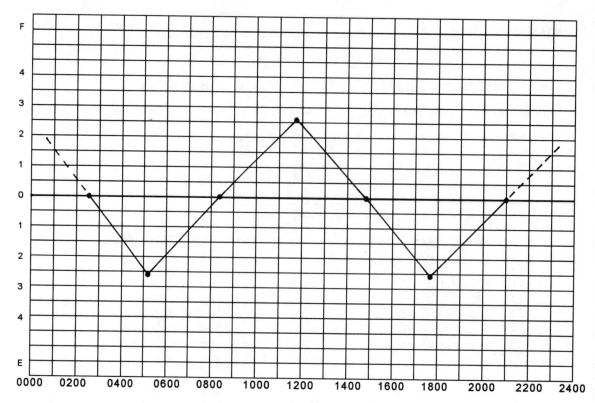

Figure 12-10C. *A tidal current graph, straight-line method, for Brooklyn Bridge for 8 October.*

points for all times of maximum current and slack waters, measuring time on the horizontal axis and current drift on the vertical. Flood maxima are plotted above the reference (zero-velocity) axis, and ebb maxima below. (2) Connect the points with straight line segments.

The completed graph should approximate the current versus time curve to be expected at the subordinate location. As with tidal predictions, the navigator should remember that unusual meteorological conditions could result in the actual current being different from that predicted by the *Tidal Current Tables*.

If the navigator has access to a personal computer, several navigation software application programs now available can generate high-quality tidal current graphs without the necessity to first produce a complete current table for the subordinate location. These programs are discussed in chapter 22.

An example tidal current graph drawn for 8 October using the current data previously computed for Brooklyn Bridge appears in Figure 12-10C.

Computing the Tidal Current at Any Time

After the complete tidal current table and/or tidal current graph has been constructed for a subordinate location, the navigator can then determine the tidal current for any particular time of interest. On the graph of current versus time, the navigator need only go to the desired time on the time scale, then read up or down to pick off the corresponding current velocity on the plotted curve. For the example problem under consideration, it appears from Figure 12-10C that the current at Brooklyn Bridge at 0900 8 October is about 1.0 knot.

If a complete current table is on hand, the navigator can use it in conjunction with Table 3 in the *Current Tables* to determine the same information with more precision. As an example of this procedure to find the velocity of current at 0900 on 8 October at Brooklyn Bridge, first the quantities indicated on the second part of the current form must be figured and recorded for use with Table 3 (see Figure 12-8).

The interval between the nearest slack water—the 0822 slack—and the desired time is 38 minutes; the interval between this slack and the next maximum current, the 1147 flood, is 3 hours 25 minutes; and the maximum current is the 1147 flood, 2.6 knots in velocity. Table B of Table 3 is entered using as entering arguments the tabulated values closest to the computed values for the interval between slack and desired time, and the interval between slack and maximum current. For this example, the table yields a velocity factor of 0.4. The 1147 flood velocity, 2.6, is then multiplied by this factor, resulting in a velocity of current of 1.0 knots at 0900; the product of this multiplication is always rounded to the nearest tenth. Since this is a flood current, its direction from Table 2 is 063°. The current form now appears as in Figure 12-10D.

Notice that the result, 1.0 knot, does not agree very favorably with the value indicated for 0900 on the graph of current versus time in Figure 12-10C. This is due to the straight-line approximation method used for plotting the current graph.

Finding Duration of Slack Water

It often happens that passage through restricted waters or a docking maneuver is greatly facilitated when the current is either slack or at minimum velocity on either side of slack water. Either the tidal current graph or the complete tidal current table can be used to find the time duration around slack water within which the current will be at or below a desired velocity.

To use the tidal current graph for this purpose, the navigator need only enter with the desired maximum current on each side of the slack water (zero-velocity) reference line, and read off the times bracketing each slack water within which the current will be at or below the desired value. For the example situation used in this chapter, suppose that a passage is to be made near the time of the 0822 slack water at Brooklyn Bridge, with a desired maximum current of 0.5 knots. Referring to the graph in Figure 12-10C, the two times obtained would be about 0745 and 0900.

If more precision were required, the complete tidal current table could be used in conjunction with Table 4 in the *Current Tables* and the bottom portion

COMPLETE CURRENT TABLE

Locality: **BROOKLYN BRIDGE**   Date: **8 OCTOBER**

Reference Station: **HELL GATE**

Time Difference: Min Bef Flood: **+0 29**   Flood: **+0 41**
                    Min Bef Ebb: **+0 33**   Ebb: **+0 29**
   Speed Ratio: Flood: **0.8**
                       Ebb: **0.6**

Maximum Flood Direction: **063°**
Maximum Ebb Direction: **253°**

Reference Station: **HELL GATE**   Locality: **BROOKLYN BRIDGE**

2244	3.3 F (7 OCT)			
0201	0		0234	0
0443	4.3 E		0512	2.6 E
0753	0		0822	0
1106	3.3 F		1147	2.6 F
1423	0		1456	0
1711	4.4 E		1740	2.6 E
2032	0		2101	0
2339	3.2 F		0020	2.6 F

VELOCITY OF CURRENT AT ANY TIME

Int. between slack and desired time:	0 38	
Int. between slack and maximum current:	3 25	(Ebb) (Flood)
Maximum current:	2.6	
Factor, Table 3	.4	
Velocity:	1.0	
Direction:	063°	

Figure 12-10D. Computed current at 0900 8 October at Brooklyn Bridge, using the complete current table in conjunction with Table 3 in the Current Tables.

of the tidal current form, to find these same time limits (see Figure 12-8). The maximum currents on either side of the desired slack are the 0512 ebb and the 1147 flood; their times and maximum velocities are recorded on the form, along with the desired maximum velocity of 0.5 knot. Next, Table B of Table 4 is entered twice, using as entering arguments the tabulated maximum currents closest to the predicted values and the desired maximum velocity. A period of 22 minutes is obtained for both currents in this example.

The period in each case represents the total duration of the desired maximum velocity, *assuming the maximum velocities of the current on each side of the slack are identical.* Since consecutive flood and ebb currents may not have the same velocity, it is necessary to find the *average* period; it is this period that the navigator will use to compute the desired time frame. To find the average period, the length of each tabulated period is entered on the form, added together, and added together, and divided by 2. In this example, the average period is:

$$\frac{22 + 22}{2} \text{ or 22 minutes in duration.}$$

To complete the problem, it is further assumed that the slack water will occur in the middle of the average period of low velocity. Thus, the times within which the velocity of the current will be below 0.5 knot are given by the expression

$$0822 \pm (22 \div 2) = 0811, 0833.$$

The completed current form appears in Figure 12-10E. Again, notice that these times vary somewhat from those predicted by means of the tidal current graph (see Figure 12-10C), due to the straight-line approximation method used for plotting the graph.

Current Diagrams and Charts

In addition to the tabulated data in the *Tidal Current Tables*, there are several types of graphic aids available to help the navigator to predict the set and drift of tidal current at certain heavily traveled locations in U.S. waters. Two of the more widely used of these are a set of current diagrams included in the *Atlantic Coast Tidal Current Tables*, and a series of 13 *Tidal Current Charts* issued by NOS covering the more important bays, harbors, and sounds on the east and west coasts of the continental United States.

One example of the current diagrams contained in the *Atlantic Coast Tidal Current Tables*, the one for Chesapeake Bay, appears in Figure 12-11 on page 219. These diagrams are based on the occurrence of specified maximum current or slack-water events at a designated reference station. The diagrams use as their entering arguments the number of hours preceding or following the designated event, and the ship's speed. The diagram in Figure 12-11 is entered by setting one edge of a parallel ruler along the applicable speed line of the nomogram to the right of the diagram, and then expanding the ruler until the opposite edge lies on the vertical line associated with the proper time interval adjacent to the location of interest listed along the left border. The numbers on the horizontal lines indicate current velocities. Those figures that appear alongside the expanded ruler edge on the diagram indicate the current that should be encountered at each successive location as the ship proceeds either northbound or southbound at the given speed. Set of the current is northerly during flood currents and southerly during ebb currents; in the diagram it is determined by whether the edge of the ruler cuts a horizontal line in a shaded flood-current area or an unshaded ebb-current area.

COMPLETE CURRENT TABLE

Locality: __BROOKLYN BRIDGE__   Date: __8 OCTOBER__

Reference Station: __HELL GATE__

Time Difference: Min Bef Flood: __+0 29__ Flood: __+0 41__
                   Min Bef Ebb: __+0 33__ Ebb: __+0 29__
    Speed Ratio: Flood: __0.8__
                        Ebb: __0.6__

Maximum Flood Direction: __063°__
Maximum Ebb Direction: __253°__

Reference Station: __HELL GATE__   Locality: __BROOKLYN BRIDGE__

2244	3.3 F (7 OCT)		
0201	0	0234	0
0443	4.3 E	0512	2.6 E
0753	0	0822	0
1106	3.3 F	1147	2.6 F
1423	0	1456	0
1711	4.4 E	1740	2.6 E
2032	0	2101	0
2339	3.2 F	0020	2.6 F

VELOCITY OF CURRENT AT ANY TIME

Int. between slack and desired time:	0 38	
Int. between slack and maximum current:	3 25	(Ebb) (Flood)
Maximum current:	2.6	
Factor, Table 3	.4	
Velocity:	1.0	
Direction:	063°	

DURATION OF SLACK

Times of maximum current:	0512	1147
Maximum current:	2.6	2.6
Desired maximum:	.5	.5
Period – Table 4:	22	22
Sum of periods:		44
Average period:		22
Time of slack:		0822

Duration of slack: From: __0811__ To: __0833__

Figure 12-10E. *Completed current form for Brooklyn Bridge, showing computed duration of slack water.*

Tidal Current Charts are sets of small-scale chartlets of a particular bay, harbor, or sound, which depict the normal set and drift of the tidal current for each hour after occurrence of a particular current event at a designated reference station. Figure 12-12 shows a portion of one of a set of 11 such chartlets

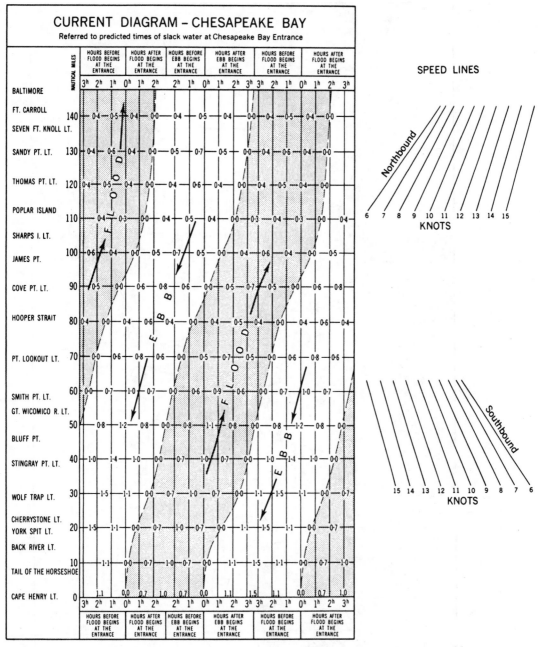

Figure 12-11. *A current diagram for Chesapeake Bay in the* Atlantic Coast Tidal Current Tables.

that show the predicted current at various locations within Narragansett Bay for each hour after high water at Newport, Rhode Island. The direction of the arrows on the chartlet represents the set of the tidal current at that place and time, and the numbers near the arrows represent the drift.

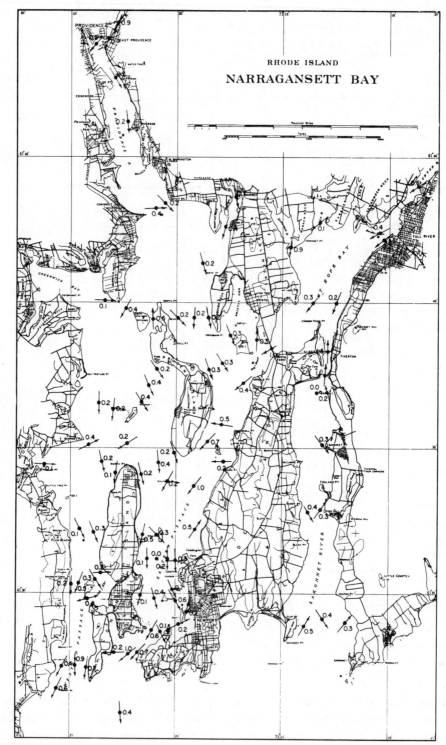

Figure 12-12. *A portion of a tidal current chart for Narragansett Bay.*

Tidal Current Diamonds

For predictions of the set and drift of tidal currents in foreign coastal areas, the navigator can use the applicable *Enroute* volume of the *Sailing Directions*, which contains tidal current diagrams and chartlets for selected ports, foreign-produced equivalents of the *Tidal Current Tables* and *Charts*, or charted current information. On charts produced by the British Admiralty, lettered symbols called *tidal current diamonds* are printed at various locations along major channels of interest to the navigator. The letters within the diamonds refer to small tables printed on the chart, which give hourly predictions of set and drift based on the occurrence of a certain tide event at a designated reference station, usually high water. A portion of a British Admiralty chart showing a current diamond and its associated charted information appears in Figure 12-13. The table shown contains two velocity columns, one for current velocities at spring tides, and the other for velocities at neap tides.

To use this type of charted current information, the navigator must initially determine whether the date in question is at spring tides or neap tides, or somewhere in between. To do this, the applicable volume of the *Tide Tables* is used to determine the heights of high and low water on the day under consideration at the reference station referred to by the charted current diamond table. These heights are then compared to the corresponding heights of spring and neap tides at the reference station; this latter information is usually given on the chart. To predict the tidal current at the site of the current diamond for a given time on the date in question, it is necessary to interpolate first for current strength values for that day of the mouth, and then for current strength and direction for times between the whole hours listed.

As an example, suppose that the high and low water levels at Greenock on a particular day were computed by use of a *Tide Table* to be 9.9 feet and 1.7 feet, respectively. From charted information (not shown in Figure 12-13), it is found that high water at spring tide is 10.8 feet, and at neap tide, 9.0 feet. Low waters at these times are 0.8 and 2.9 feet, respectively. By comparing the computed heights with these charted values, it appears that this day is about midway between the spring tide and neap tide times of the month. Returning to the table for diamond "B" in Figure 12-13, current velocities for each whole hour are thus computed by interpolating halfway between the velocities given for spring and neap tides.

After computing the current velocity for each hour, it is then necessary to complete the set and drift determination by interpolating in both these computed strengths and in the direction column for the values corresponding to the exact time of interest. If, for example, it were 4 hours 30 minutes before high water at Greenock, the interpolated direction would lie halfway between the tabulated directions 026° and 046°, or 036°. The computed current 5 hours

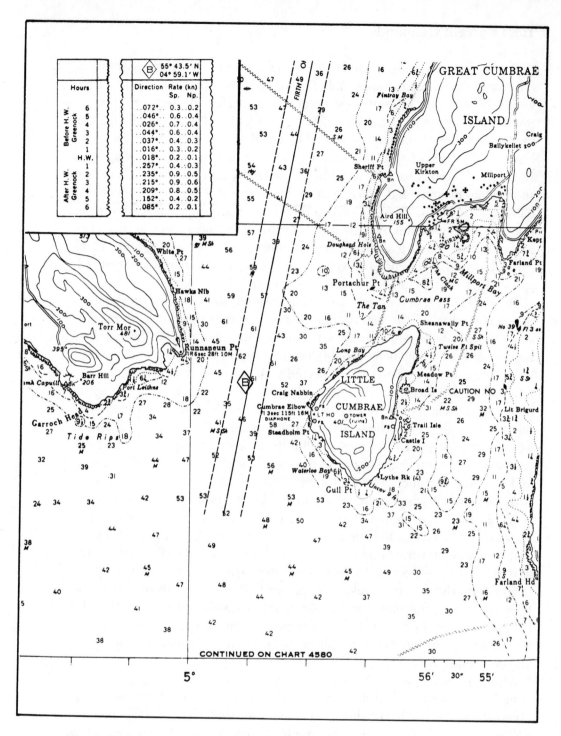

Figure 12-13. *A portion of a British Admiralty approach chart.*

Wind velocity (miles per hour)	10	20	30	40	50
Average current velocity (knots) due to wind at following lightship stations:					
Boston and Barnegat	0.1	0.1	0.2	0.3	0.3
Diamond Shoal and Cape Lookout Shoals	0.5	0.6	0.7	0.8	1.0
All other locations	0.2	0.3	0.4	0.5	0.6

Figure 12-14. *Velocities of wind-driven currents.*

before high water would be 0.5 knots (halfway between 0.6 and 0.4), and, for 4 hours before, 0.55 knots. Thus, for 4 hours 30 minutes, a drift of 0.525 knots is obtained; it would be rounded off to 0.5 knots in practice.

WIND-DRIVEN CURRENTS

When using current predictions from any source, the navigator should always bear in mind that the predictions are subject to error during unusual conditions of wind or river discharge. Figure 12-14 is an excerpt from the *Current Tables* showing the average current velocity that will normally result when the surface of a body of water is subjected to continuing winds of various strength. The set of wind-driven currents is normally slightly to the right of the direction in which the wind is blowing in the northern hemisphere, and slightly to the left in the southern hemisphere. The navigator may wish to combine estimated wind-driven current vectors with predicted ocean or tidal current vectors in order to obtain a more accurate prediction of the actual currents existing at a particular location and time.

SUMMARY

This chapter has described the two major types of surface currents normally encountered by the seagoing surface navigator, and has examined in some detail the methods used to predict their effects at a given place and time. Although the predictions contained in the various reference publications discussed in this chapter are usually accurate most of the time, the navigator must always be alert for those occasions when the actual currents encountered will differ greatly from those predicted as a result of unusual wind or storm conditions.

The following chapter on current sailing will explain how the navigator takes the estimated currents predicted by methods described in this chapter into account in directing the movements of a vessel safely from one point to another.

13 CURRENT SAILING

Chapter 8, "Dead Reckoning," states that a vessel's calculated dead reckoning course and speed line is seldom if ever identical to her actual track over the surface of the earth. It also mentions that the difference between the two is caused by the effect of current. Used in this context, current refers not only to the horizontal movement of the water through which the ship is proceeding, but also to all other factors which, operating singly or combined, might have caused the ship's projected position to differ from her true position at any time. Among these additional factors are the following:

Wind
Steering inaccuracy
Undetermined compass or gyro error
Error in engine or shaft RPM indications
Excessively fouled bottom
Unusual conditions of loading or trim

Current, in the strict oceanographic sense of horizontal water movement, however, usually accounts for the major portion of the discrepancy between a vessel's DR and fix positions.

Current sailing refers to the methods used by the navigator to take the effects of current into account in directing the movements of the vessel. There are two distinct phases of current sailing. During the first phase, called the *presailing* or *planning* phase, the navigator uses methods described in the previous chapter to obtain a predicted or estimated current. This estimate is then applied to the intended track to find the optimum course and speed to order. In the second phase, referred to as *postsailing*, the navigator computes the "actual" current that has acted on the ship during a transit between two points on the track. This computed current is then used as an estimate for the next leg of the track, if the situation is such that the current may be expected to remain unchanged. During presailing procedures, the current is considered to consist entirely of predicted horizontal water movements; however, the current used during the postsailing phase consists not only of water movements, but also all other factors mentioned above that may have affected the ship's movement during the time of the plot.

To work both presailing problems involving an estimated current or postsailing problems with an actual current, the navigator uses a graphic drawing known as a *current triangle*. This is a geometric, scaled drawing that uses vectors to represent the relationships between the ship's ordered course and speed, the current, and the resulting ship's track. There are two types of current triangles. The one employed during the presailing or planning phase of current sailing is called the *estimated current triangle*. The other, used during the postsailing phase, is known as the *actual current triangle*.

THE ESTIMATED CURRENT TRIANGLE

In the estimated current triangle, the three legs represent the ordered course (C) and speed (S), the estimated set (S) and drift (D) of the current, and the resultant track (TR) and speed of advance (SOA). An estimated current triangle appears in Figure 13-1. As can be seen from Figure 13-1, the vector representing the track and speed of advance of the ship is the vectorial sum of the vectors representing the ship's ordered course and speed and the current. Furthermore, if the triangle were drawn to scale using some convenient speed scale and compass rose, it should be apparent that any one of the vectors constituting the triangle could be found if the other two were known. The navigator makes use of the geometric solution of the estimated current triangle to solve the following two basic types of problems. The predetermined set and drift of the estimated current are considered given quantities in each case:

1. To find the expected track and speed of advance when the vessel proceeds at a given course and speed; and
2. To find the course and speed a vessel should order to achieve an intended track and speed of advance.

As a medium on which to work the current triangle, most navigators prefer the maneuvering board. This device incorporates several speed scales, and by locating the foot of the track and ordered course vectors at the center of the

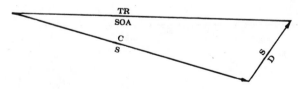

Figure 13-1. *The estimated current triangle.*

maneuvering board, direction measurements are greatly facilitated. The estimated current triangle may also be worked on a portion of the chart isolated from the plot, or within a convenient chart compass rose. It is also possible to construct the triangle directly on the dead reckoning plot, using the DR course line as the ordered course and speed vector for purposes of the triangle. Inasmuch as this practice leads to a proliferation of extra lines in the region of the DR plot, however, it is usually best to work the triangle away from the plot if at all possible.

SOLVING THE ESTIMATED CURRENT TRIANGLE

As an example of the use of the current triangle to solve the first type of estimated current problem mentioned above, suppose that a navigator wished to determine the track and SOA that the ship was making while proceeding on an ordered course of 090°T at an ordered speed of 20 knots. By use of the *Current Tables*, the navigator has estimated that a current having a set of 180°T and a drift of 4 knots is flowing in the area.

To find the solution, the navigator selects a blank portion of the chart away from the dead reckoning plot for the site of the current triangle. Using the distance scale of the chart as a speed scale for vector lengths, a vector 20 units long in the direction 090°T is first drawn to represent the ordered course and speed. From the tip or head of this vector, a second vector 4 units long in the direction 180°T, representing the estimated current, is laid down and labeled. At this stage, the semicompleted current triangle appears as shown in Figure 13-2A. To complete the triangle, a vector representing the expected track and speed of advance is drawn, as in Figure 13-2B. The track is obtained by measuring the direction of vector AB; it is 101°T. Measuring the length of the vector against the chosen speed scale yields the speed of advance, 21 knots.

To solve the second type of estimated current problem, in which the direction of the intended track and the speed of advance are either given or calculated, and the proper ordered course and speed are required, the solution would be found in a similar fashion, except that the vector representing the track and speed of advance would be drawn first, followed by the current vector. The required course and speed would then be the missing vector. Figure 13-3 illustrates the estimated current triangle as it appears in this type of problem. The direction of the intended track vector, AB in the figure, is implicitly

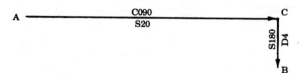

Figure 13-2A. *Determining track and SOA by a current triangle.*

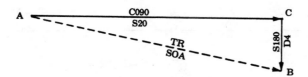

Figure 13-2B. *Completed estimated current triangle.*

determined as soon as the navigator decides to proceed from a point A to a point B across the chart. The speed of advance, or length of vector AB, can be specified in a number of ways. It may simply be given by directive; a certain distance may be required to be traversed in a given amount of time; or it may be required to arrive at a point B from a point A at a given time, as when proceeding to rendezvous. SOAs in the latter two cases must be calculated by means of the speed-time-distance formula.

Chapter 8, "Dead Reckoning," states that the navigator normally lays down an intended track on the chart prior to sailing through a given area, to represent the planned path and speed of the ship. Each leg of the track is labeled with its direction and the intended SOA. It should be apparent that to achieve this planned track and SOA, the navigator must solve a problem similar to the one just discussed. In practice, the navigator will often simply make an intelligent guess as to the proper ordered course and speed, taking into consideration the estimated set and drift, rather than draw an estimated current triangle. The navigator will recommend a course a few degrees to the right or left of the intended track, and a speed close to the SOA; these recommendations are then adjusted based on the trend of the subsequent fix information.

THE ESTIMATED POSITION ALLOWING FOR CURRENT

Returning for a moment to the first estimated current problem discussed in the last section, suppose that the navigator desired to obtain an estimate of the ship's position 30 minutes after proceeding from a 1200 fix position on an ordered course of 090°T and at an ordered speed of 20 knots. To do this, a DR position could simply be plotted for 1230 on the 090°T DR course line. But suppose further that the navigator desired to be more accurate in the determination

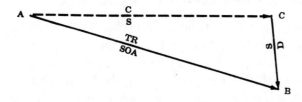

Figure 13-3. *Solving for ordered course and speed using the current triangle.*

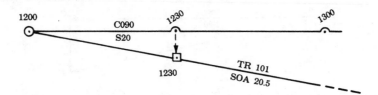

Figure 13-4. *Determining an estimated position (EP) with allowance for current.*

of this position, and it was not possible to obtain a fix or running fix. If the estimated current were applied to the DR position, a more accurate *estimated position* previously described in chapter 8 could well result, assuming that the actual current encountered was close to the predicted estimated current.

To obtain an estimated position (EP) for a given time taking current into account, a DR position is first plotted for the time of interest. Then the DR position is "adjusted" in the direction of the set of the current the amount of distance the drift of the current would have carried the ship in the time since the last fix. If done properly, the adjusted DR should fall on the resultant track previously computed. The estimated position is labeled by the box symbol for an EP, as shown in Figure 13-4. In the figure, the 1230 DR is advanced 2 miles, the distance the current would carry the ship in 30 minutes, in the direction of the set of the current, 180°T. An estimated position can be plotted at whatever time intervals the navigator desires in this way; all EPs should lie on the intended track. This fact leads to an alternative method of locating estimated positions by extending the intended track out from the last good fix, and calculating the distance traveled along the track by using the speed of advance. In this way EPs can be plotted on the intended track similarly to the way in which DR positions are plotted on the DR course line.

As pointed out in chapter 8, the navigator should refrain from placing an unwarranted amount of confidence in any estimated position, including one derived by the method described above. The reliability of a position obtained by any means other than crossing two or more simultaneous LOPs is always open to doubt, and must be decided by the navigator based on the circumstances of each case.

DETERMINING AN EP FROM A RUNNING FIX

During the discussion of the running fix in chapter 8, it was explained that in the piloting environment a line of position, or LOP, was never advanced more than 30 minutes in order to obtain a running fix. Furthermore, in that chapter the effects of any possible current were not considered in advancing the LOP. In certain circumstances, particularly when the navigator has observed that a strong current is flowing, the navigator may improve the potential accuracy of

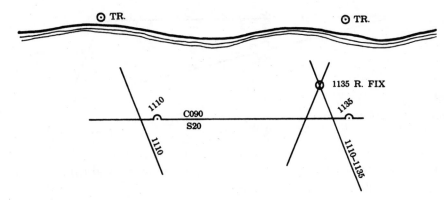

Figure 13-5A. *A 25-minute running fix.*

the running fix by forming an estimated position based on it, that takes into account the estimated current.

To do this, the navigator simply adjusts the final position of the advanced LOP by moving it in the direction of the set a distance the drift of the current would have moved it in the time interval through which the LOP was advanced. Consider the situation in Figure 13-5A, in which an LOP was advanced 25 minutes for a running fix at 1135.

Suppose that the navigator has estimated a current to be flowing with a set of 230°T and a drift of 5 knots. In order to obtain an estimated position incorporating this estimated current, the 1110–1135 LOP must be adjusted for the estimated current by moving it in the direction 230°T a distance of $\frac{25}{60} \times 5$ or 2.1 miles. This is done by advancing a point on the line and then reconstructing the remainder of the LOP through the point, just as was originally done to advance the 1110 LOP. The repositioned LOP and resulting estimated position appear in Figure 13-5B.

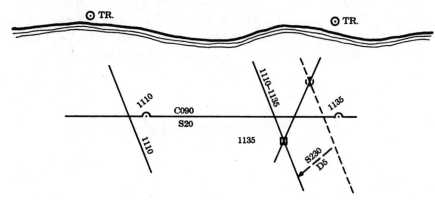

Figure 13-5B. *Obtaining an EP from a running fix.*

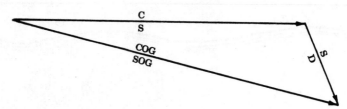

Figure 13-6. The actual current triangle.

As is the case with all estimated positions, the navigator should be wary in placing complete confidence in an estimated position derived from a running fix, which is itself an artificial position.

THE ACTUAL CURRENT TRIANGLE

In the actual current triangle, the three legs represent the ordered course (C) and speed (S), the course over the ground (COG) and speed over the ground (SOG), and the resultant set (S) and drift (D) of the actual current experienced. An illustration of an actual current triangle appears in Figure 13-6.

Note that the intended track and speed of advance of the estimated current triangle have become the course and speed over the ground in the actual current triangle. In practice, this type of current triangle is used mainly to find the actual current vector, with the other two sides of the triangle given or determined from the navigation plot.

As an example, suppose that a ship had been steaming on a course of 270°T at a speed of 15 knots for 60 minutes from a 1200 fix position. At the end of the 60 minutes, a new fix was obtained, which placed the ship 3 miles south of its 1300 DR position, as shown in Figure 13-7A. Why didn't the 1300 fix show the ship to be at the 1300 DR position? The answer to this question, as mentioned at the beginning of this chapter, is that some "current" acted on the ship during its 60 minutes of travel, causing its actual course and speed over the ground to be different from the DR course line, representing the ordered course and speed.

To visualize the situation, the navigator could draw an actual current triangle. A vector representing the ordered course and speed from the 1200 fix is first drawn and labeled, followed by a second vector representing the actual

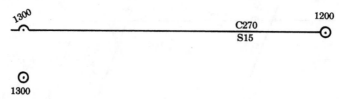

Figure 13-7A. A 60-minute DR track.

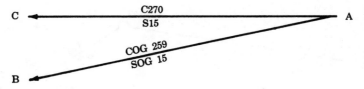

Figure 13-7B. *The known legs of the actual current triangle.*

path the ship must have followed between the 1200 and 1300 fixes. At this point, the triangle appears as in Figure 13-7B. The direction of the COG-SOG vector is the direction of the 1300 fix from the 1200 fix. Its length represents the speed at which the ship must have traveled to traverse the distance between the two fixes in the given 60 minutes of time—the SOG. Since the path of the ship with respect to the ground is the resultant of its ordered course and speed and the actual current, the missing vector in Figure 13-7B must be the actual current vector. Furthermore, it must lie in the direction from C to B in the figure, by the laws of vector mechanics. The completed actual current triangle appears in Figure 13-7C. The set and drift of the actual current experienced between 1200 and 1300 can be obtained by measuring the direction and length of vector CB; they are 180°T and 3 knots. The navigator could then use these values as the estimated current during the planning phase for the next part of the transit.

In practice, it is not usually required to draw a separate current triangle to determine the actual current. A ship can always be thought of as being displaced from its DR course line by the actual current acting on it in the time interval since the last good fix. Moreover, the ship's course and speed do not alter the effect of the current, because the whole body of water in which the ship floats is moving with the set and drift of the current. Thus, the ship's fix position at any time will always be offset from the corresponding DR position in the direction of the set of the current by the distance the drift would have carried the ship during the time period.

It follows, therefore, that the actual set can be determined simply by measuring the direction of a fix from the corresponding DR position at any time; the actual drift can be determined by measuring the distance from the DR to the fix position, and then dividing this distance by the number of hours elapsed

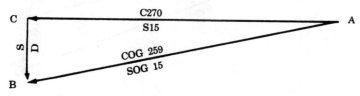

Figure 13-7C. *The completed actual current triangle.*

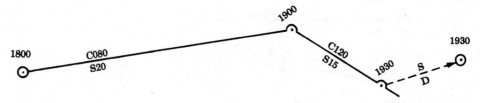

Figure 13-8. *Determining actual set and drift after a course change.*

since the last good fix. In the plot of Figure 13-7A, the set could have been obtained by measuring the direction of the 1300 fix from the 1300 DR position, 180°T. The drift could have been figured by dividing the distance, 3 miles, by the number of hours, 1 in this case, for a drift of 3 knots. As a second example, consider the plot in Figure 13-8. Even though there was a course and speed change at 1900, the effect of the current on the vessel is everywhere the same during the interval from 1800 to 1930. Thus, the set of the actual current can be measured from the 1930 DR to the 1930 fix; it is 070°T. The drift is found to be 4 knots by dividing the distance from the 1930 DR to the 1930 fix, 6 miles, by the number of hours, 1½.

Whenever actual current is to be determined as described above, care must be taken to use only the last two good fixes as a basis for the determination. If one or more running fixes were plotted in the interim between the two fixes, they must be disregarded, since the effect of any current was not taken into account while advancing the earlier LOPs to form the running fix(es). In this situation, the DR course line must be replotted from the time of the first fix to the time of the second, ignoring any intervening running fixes.

For purposes of comparison, the estimated and actual current triangle have been superimposed on each other in Figure 13-9. As can be seen, the ordered course and speed vector is common to both triangles. In the estimated current triangle, the estimated current vector combines with the ordered course and speed vector to form the resultant track and speed of advance. In the actual

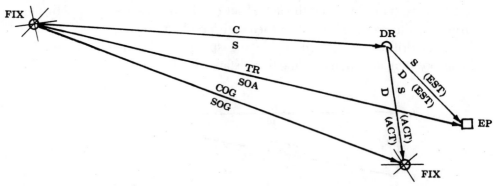

Figure 13-9. *Relationship between the estimated and actual current triangles.*

current triangle, the actual current vector causes the ship to proceed along the vector representing its actual course and speed over the ground.

As a final cautionary note, it should be pointed out that current set and drift often undergo rapid change with both time and location in relatively constricted bodies of water such as straits, bays, rivers, and river mouths. When piloting in such waters, the navigator should not consider the actual current previously determined to be valid as a future estimate for more than a reasonably short time period, usually about 15 minutes.

SUMMARY

This chapter has explained the procedures used by the navigator to take the estimated current into account while directing the movements of a vessel from one point to another. The technique of determining the actual current, which may have been the result of many other factors in addition to water movement, has also been examined in some detail.

The navigator must be cognizant of the fact that current predictions from whatever source are always subject to error, and may not represent the actual conditions existing at a future time. Because of this, the navigator should always regard any estimated position with suspicion, tempered by experience. In the last analysis, the fix is the only certain way to ascertain the ship's position with 100 percent accuracy.

PRECISE PILOTING AND ANCHORING

14

In chapter 8 of this text dealing with dead reckoning, it was assumed that a vessel executed a change of course or speed instantaneously as it proceeded along the intended track laid down on the chart by the navigator. This is not the case, of course, in reality. Generally speaking, the larger the vessel, the more time and distance are required for either a course or speed change to be effected. On a comparatively small-scale chart of a large coastal approach region or ocean area, this time and distance can be disregarded for the most part, and any errors in the dead reckoning track so introduced are insignificant.

During piloting operations in more constricted waters and in narrow channels, however, the increased degree of accuracy required makes it impossible to ignore these factors. The practice of taking the turning diameter, time to turn, and acceleration and deceleration data into account when plotting and directing the movements of a vessel is termed *precise piloting*. The principles of precise piloting come into play whenever the vessel is engaged in any type of maneuvering requiring high precision, especially when following a narrow channel or when anchoring.

The first part of this chapter will explain the techniques involved in precise piloting, and the second part will examine in some detail the procedures used in selecting, plotting, and executing an anchorage.

SHIP'S HANDLING CHARACTERISTICS

The attributes of a particular vessel relating to her performance in making turns at various rudder angles and speeds, and in accelerating and decelerating from one speed to another, are collectively called the vessel's *handling characteristics*. In the case of naval warships, these characteristics are referred to by the more specific term *tactical characteristics*, since these handling qualities bear directly upon the tactics that may be employed by the ship and others of her type and class. Every ship has a set of handling characteristics peculiar to

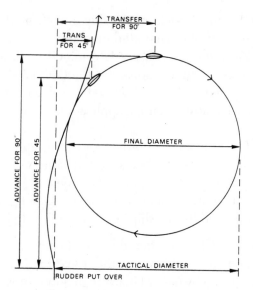

Figure 14-1. *A ship's turning characteristics.*

herself; even warships within the same class often differ to some extent in the manner in which they respond to a given rudder angle or engine speed change, although nominally they are designed to have identical tactical characteristics.

The handling or tactical characteristics pertaining to the ship's performance during turns are called her *turning* characteristics. The following terms are extensively used in describing these characteristics. They are illustrated in Figure 14-1.

Advance is the distance gained in the direction of the original course until the ship steadies on her final course. It is measured from the point at which the rudder is put over, and will be a maximum for a turn of 90°.

Transfer is the distance gained at right angles to the direction of the original course until the ship steadies on her final course.

The *turning circle* is the path followed by the point about which the ship seems to pivot—the pivoting point—as she executes a 360° turn. Every rudder angle and speed combination will normally result in a different turning circle.

Tactical diameter is a naval term referring to the distance gained at a right angle to the left or right of the original course in executing a single turn of 180°. Tactical diameter can be thought of as the transfer for a turn of 180°; it will be different for each rudder angle and speed combination.

Final diameter is the diameter of the turning circle the ship would describe if she were allowed to continue a particular turn indefinitely. For all but a few small ships, the final diameter will always be less than the tactical diameter, due to the initial "kick" of the ship away from the direction of the turn, shown in a somewhat exaggerated manner in Figure 14-1.

Standard tactical diameter is a predetermined tactical diameter established by various tactical publications, most notably *ATP 1-B Volume 1*, for each ship type. It is used to standardize the tactical diameters for all ships by ship type, and finds its most extensive application when maneuvering in formation.

Standard rudder is the amount of rudder necessary to turn a ship in her standard tactical diameter at standard speed. It varies with the ship type, and also with the class of ship within a particular type.

Angle of turn is the horizontal angle through which the ship swings in executing a turn, measured from a ship's original course to her final course.

Each ship will normally have on board a complete list of all turning and acceleration/deceleration data pertaining to the ship. Merchant vessels operating in U.S. waters are required by Coast Guard regulations to have handling characteristics tables readily available to the conning officer. On Navy ships, the ship's tactical data are maintained in the form of a file called the *tactical data folder*. This folder contains a wealth of information in tabular form about the ship and her tactical turning and acceleration/deceleration characteristics for different rudder angle and speed combinations, and engine speed changes. All conning officers, as well as the ship's navigator, should be intimately familiar with their ship's handling and tactical characteristics and data.

One part of these tables of particular interest to the navigator is the turning characteristics for the ship for various rudder and speed combinations. This information usually appears in tabulated form, with the advance and transfer listed for various angles of turn at various rudder angles and speeds. Following is a typical table of this kind, showing the advance and transfer for every 15° of angle of turn using 15° rudder at 15 knots of speed.

Angle of Turn	Advance	Transfer	Angle of Turn	Advance	Transfer
15°	180	18	105°	330	280
30°	230	30	120°	310	335
45°	270	60	135°	270	380
60°	310	110	150°	230	418
75°	330	170	165°	180	470
90°	335	220	180°	100	500

In the case of Navy ships, tables of this type are normally based on the tactical characteristics of the first ship of the class within the type, as the time and expense of making the large number of runs required to generate a similar table for each ship would be prohibitive. When extensive alterations are made on a ship, however, a new set of tactical data should be experimentally determined by procedures set forth in the U.S. Navy publication *NavShips Technical Manual*. Unfortunately, in practice it is not uncommon to find tactical data folders on older ships that have never been updated, even though many extreme modifications may have been made since they were commissioned. A newly reported navigator faced with this situation should recommend that a period of the ship's operating schedule be set aside to update her tactical data folder at the earliest opportunity, and any existing advance, transfer, and acceleration data should be used with caution, until such a time as new data are obtained.

Merchant ships' handling characteristics are usually determined for different conditions of loading at the builder's trials. As is the case for older Navy ships, the data are often outdated or incomplete on many older vessels.

USE OF ADVANCE AND TRANSFER DURING PILOTING

As has been previously stated, the navigator should always take the ship's turning characteristics into consideration during piloting operations in which any degree of precision is necessary. In practice, the professional navigator will use these characteristics whenever the scale of the chart allows, in order to increase the accuracy of his or her dead reckoning plot. Consider the following example.

Suppose that a navigator of a medium-sized vessel has laid down an intended track on the chart to negotiate a 50° bend in a narrow river channel, as illustrated in Figure 14-2A. If the ship's rudder were not put over until she reached the intersection of the old and new track directions, point A in Figure 14-2A, the turning diameter of the ship might cause her to go aground on the left side of the channel as she was making the turn; at the very least, it would be far to the left of the intended track. Obviously, the navigator should recommend a point on the old track, called the "turning point," at which the ship should put her rudder over, taking her turning circle into account, so that the ship will come out of the turn on the new track leg.

To do this, the navigator uses the proper advance and transfer table for the rudder and speed combination utilized in making the turn. For this example, the table just discussed will be used. First, the amount of advance and transfer for an angle of turn of 50° must be calculated by interpolation in the table. The advance for a turn of 45° is 270 yards, and for a turn of 60°, 310 yards. Interpolating between these values, the following expression is set up to yield the advance for a 50° turn:

$$270 + \frac{5}{15}(310 - 270) = 283 \text{ yards.}$$

In similar fashion, the transfer for this turn is calculated to be 77 yards.

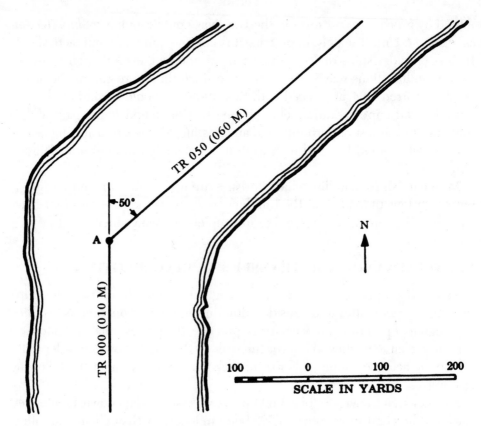

Figure 14-2A. *A track laid down to negotiate a narrow channel.*

After the advance and transfer have been calculated for the given angle of turn, the next step is to apply their values to the track to determine the point at which the rudder should be put over. Initially, there is no point upon which to base the advance, so the transfer is laid off first. The transfer is indicated by drawing a dashed construction line parallel to the original track, at a distance from it equal to the amount of transfer, as shown in Figure 14-2B. The intersection of the dashed transfer line with the new track, labeled point B in Figure 14-2B, defines the point on the new track at which the turn should be completed, taking transfer into account. Next, advance is included by laying its value off from point B back along the dashed transfer line. This locates a second point on the transfer line, labeled point C in Figure 14-2C. To locate the turning point on the original track, a perpendicular is dropped to it from point C. The resulting point, labeled point D, is also shown in Figure 14-2C.

Turn Bearing

To complete the plot, the navigator locates a suitable object along the shoreline to use as the basis for a predicted line of position referred to as a *turn bear-*

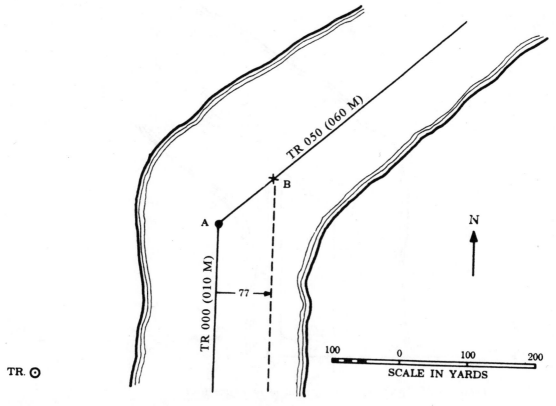

Figure 14-2B. *Laying off the transfer for a turn of 50°.*

ing. It is drawn from the turning point toward but not through the object, and it is labeled with the true and relative bearings of the object from the turning point, as indicated in Figure 14-2C. After determining the turn bearing, the navigator informs the conning officer of its value so that the ship's rudder can be put over when this bearing is sighted on the gyro repeater.

When selecting an object for a turn bearing, the ideal object should be located as nearly as possible at a right angle from the turning point on the original track. The effect of any unknown gyro error on a bearing to an object in this position is minimized, and the rate of change of bearing is maximized, thus enhancing the probability of starting the turn at the proper moment. The object used for a turn bearing in Figure 14-2C is so located. Many navigators prefer to select an object on the same side of the ship as the direction of the turn if available, especially if the conning officer who will be giving greatest attention to that side is to personally sight the turn bearing. In practice, however, there are usually no objects in the ideal position; and in coastal piloting, there may often be only one or two distinguishable landmarks available anywhere along the shore, so the navigator must often select a less than optimum turning bearing.

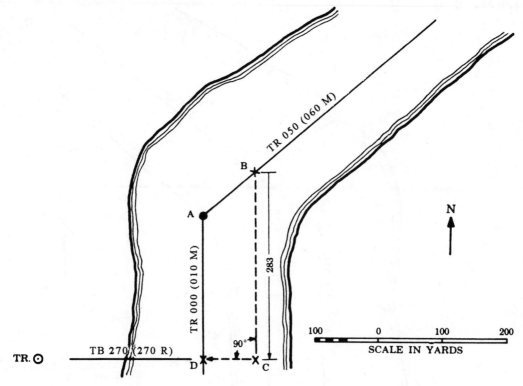

Figure 14-2C. *Locating the turning point on the original track.*

It should be reiterated that in the example discussed above, the advance and transfer table used to determine the turning point was compiled for a rudder angle of 15 degrees and a speed of 15 knots. If either a different rudder angle or different speed had been used, the advance and transfer, and hence the turning point and turn bearing, would vary. An appropriate advance and transfer table consistent with the planned rudder angle and speed must always be used to determine the proper advance and transfer values. When using any advance and transfer tables, the navigator must also bear in mind that the data in the tables were compiled under zero wind and current conditions. When an appreciable wind or current exists, the actual response of a ship in a turn may differ markedly from the tabulated data. Ships with high freeboard such as a container ship or aircraft carrier usually turn into a wind much more rapidly than they would turn under zero wind conditions, and turn out of the wind much more slowly. Ships with a deep draft may be much affected by even a slight current. Thus, the navigator must always be ready to adjust his or her recommendations as to the time to turn and the amount of rudder to use, based on the physical conditions existing at the given location and time.

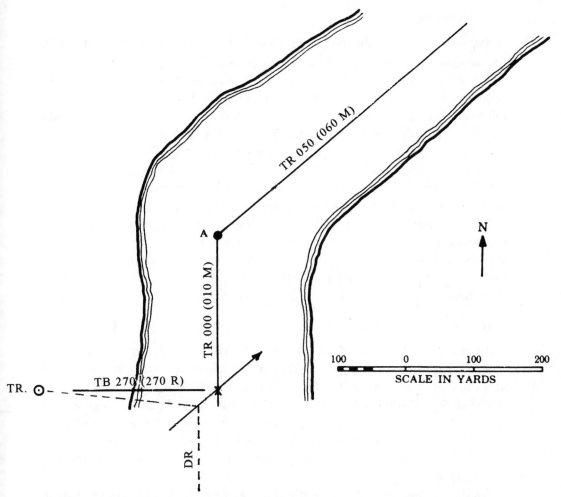

Figure 14-2D. *The slide line technique. The revised turn bearing is the bearing from the intersection of the DR course line and slide line, to the navaid used for the original turn bearing.*

Slide Lines

As an adjunct to the turn bearing, Navy procedures call for the plotting of an additional line called a *slide line* or *slide bar*. As indicated in Figure 14-2D, the slide line is drawn parallel to the new course through the turning point on the original course line. Its function is to assist the navigator in quickly revising a turn bearing if the ship finds itself off track immediately prior to a turn.

To use the slide line for this purpose, a DR course and speed line is projected ahead of the ship's last fix position as usual. The revised turn bearing is then the bearing of the navaid from the intersection point of the DR course line with the slide line, as indicated in Figure 14-2D.

ANCHORING

Among the more critical of all precision piloting evolutions carried out by the navigator is anchoring in the exact center of a predetermined anchorage. It is one of the few instances in which all of the piloting skills of the navigator are brought into play, including plotting, computing tide and current, and precise piloting. Moreover, it is not enough that the navigator ready the piloting team for the evolution. The anchor detail personnel who will handle the ground tackle must also be briefed, as well as the commanding officer or master and sea detail deck watch officers who will have the responsibility for ship control. All other personnel who will be concerned with the anchorage must also be apprised of the overall plan and the part they are to play in it.

From the navigator's point of view, much of the effort in anchoring actually is expended before the evolution ever takes place. There are four stages in any successful anchoring, although they may not be formally recognized as such— selection, plotting, execution, and postanchoring procedures. They will each be examined in the following sections of this chapter.

Selection of an Anchorage

An anchorage position in most cases is specified by higher authority. Anchorages for most ports are assigned by the local port authority in response to individual or joint requests for docking or visit. Naval ships submit a Port Visit (PVST) Request letter or Logistic Requirement (LOGREQ) message well in advance of the ship's scheduled arrival date. Operational anchorages in areas outside of the jurisdiction of an established port authority are normally assigned by the Senior Officer Present Afloat (SOPA) for ships under this officer's tactical command.

If a ship is steaming independently and is required to anchor in other than an established port, the selection of an anchorage is usually made by the navigator and then approved by the commanding officer or master. In all cases, however, regardless of whether the anchorage is selected by higher authority or by the navigator, the following conditions should always apply insofar as possible:

1. The anchorage should be at a position sheltered from the effects of strong winds and currents.
2. The bottom should be good holding ground, such as mud or sand, rather than rocks or reefs.
3. The water depth should be neither too shallow, hazarding the ship, nor too deep, facilitating the dragging of the anchor.
4. The position should be free from such hazards to the anchor cable as fish traps, buoys, and submarine cables.
5. The position should be free from such hazards to navigation as shoals and sand bars.

6. There should be a suitable number of landmarks, daymarks, and lighted navigation aids available for fixing the ship's position both by day and by night.
7. If boat runs to shore are to be made, the anchorage chosen should be in close proximity to the intended landing.

Even when an anchorage has been specified by higher authority, the commanding officer or master, who is in the position of being ultimately responsible for the safety of the ship, has the prerogative of refusing to anchor at the location assigned if it is judged to be unsafe. In such circumstances, an alternate location less exposed to hazard should be requested.

Many of the coastal charts of the United States and its possessions drawn up by the National Ocean Survey contain colored anchorage circles of various sizes for different types of ships, located on the chart in those areas best suited for anchoring, taking into account the factors listed above. These circles are lettered and numbered, allowing a particular berth to be specified. Foreign charts often have anchorage areas specified as well. Amplifying information on possible anchorage sites can be obtained from the applicable volume of the *Coast Pilots* for U.S. waters, from the proper volume of the *Enroute Sailing Directions* for foreign waters, and from the *Fleet Guide* for ports in both foreign and domestic waters frequented by U.S. Navy ships.

When it is desired to anchor at a location other than that shown as an anchorage berth on a chart, the anchorage is normally specified by giving the range and bearing to it from some charted reference point, along with the radius of the berth.

Plotting the Anchorage

After the anchorage position has been determined, the navigator is ready to begin plotting the anchorage. In so doing, reference is often made to the following terms:

The approach track. This is the track along which the ship must proceed in order to arrive at the center of the anchorage. Its length will vary from 2,000 yards or more for a large ship, to 1,000 yards for a ship the size of a Navy destroyer or smaller. Under most circumstances, it should never be shorter than 1,000 yards.

The head bearing. If at all possible, the navigator selects an approach track such that a charted navigational aid will lie directly on the approach track if it were extended up to the aid selected. The bearing to the aid thus described is termed the "head" bearing; it should remain constant if the ship is on track during the approach.

The letting-go circle. This is a circle drawn around the intended position of the anchor at the center of the berth with a radius equal to the horizontal distance from the hawsepipe to the pelorus.

The letting-go bearing. Sometimes referred to as the "drop" bearing, this is a predetermined bearing drawn from the intersection of the letting-go circle with the approach track to a convenient landmark or navigation aid, generally selected near the beam.

Range circles. These are preplotted semicircles of varying radii centered on the center of the anchorage, drawn so that the arcs are centered on the approach track. Each is labeled with the distance from that arc to the letting-go circle.

Swing circle. This is a circle centered at the position of the anchor, with a radius equal to the sum of the ship's length plus the length of chain let out.

Drag circle. This is a circle centered at the final calculated position of the anchor, with a radius equal to the sum of the hawsepipe to pelorus distance and the final length of chain let out. All subsequent fixes should fall within the limits of the drag circle.

The actual radii of both the swing and drag circles will in reality be less than the values used by the navigator in plotting them on the chart, because the catenary of the chain from the hawsepipe to the bottom is disregarded. Thus, a built-in safety factor is always included in the navigator's plot.

Prior to commencing the anchorage plot, it is always wise to draw a swing circle of estimated radius around the designated anchorage site to check whether any charted hazards will be in close proximity to the ship at any time as she swings about her anchor. If any such known hazards are located either within or near the swing circle, an alternate berth should normally be requested.

If the anchorage appears safe, the navigator begins the anchorage plot by selecting the approach track. During this process, due regard must always be given to the direction of the predicted wind and current expected in the vicinity of the anchorage. Insofar as possible, the approach should always be made directly into whichever of these two forces is predicted to be strongest at the approximate time at which the anchorage is to be made. Doing so will not only minimize the side forces that might tend to deflect the ship from the intended approach track, but also make it easier for the ship to back down to deploy the chain and set the anchor after it is let go.

The letting-go circle is drawn with a radius equal to the horizontal distance between the anchor-hawsepipe and the pelorus from which bearings will be observed. If the anchor were not let go until the pelorus was over the center of the assigned berth, the anchor would miss the center by the length of the ship from the hawsepipe to the pelorus. Thus, when the letting-go bearing, measured from the intersection of the letting-go circle and the approach track, is observed on the pelorus, the anchor will be in position directly over the center of the assigned berth.

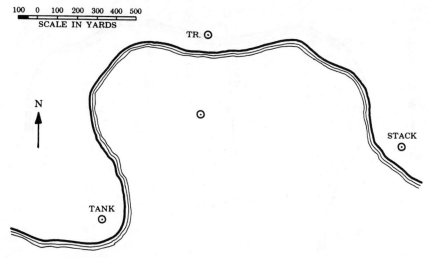

Figure 14-3A. *An anchorage assignment.*

The anchorage plot is completed by laying down the remainder of the intended track leading up to the approach track, and then swinging the range circles across the track. These arcs are normally drawn at 100-yard intervals measured outward from the letting-go circle to 1,000 yards, and at ranges of 1,200, 1,500, and 2,000 yards thereafter. After the anchor has been let go and the chain let out to its final length, a second swing circle is plotted, followed by the drag circle.

The use of these various quantities is best illustrated by an example. Suppose that a ship having a total length of 300 feet (100 yards) and a hawsepipe to pelorus distance of 150 feet (50 yards) has been directed to anchor at the position specified in the bay pictured in Figure 14-3A.

After an estimated swing circle has been plotted and the anchorage has been determined to be safe, the navigator is ready to begin the construction of the anchorage plot. As the first step, the approach track is selected by considering the different objects available for a head bearing, taking into account the expected winds and current in the bay. Assuming negligible current and a northerly wind, the tower in Figure 14-3A is a good choice for a head bearing, especially since it is doubtful that an approach track of sufficient length could be constructed using any other navigation aid shown. The approach track is then laid off from it, as in Figure 14-3B.

As the next step, the intended track leading to the final approach track is laid down, with care being taken to allow for the proper length for the approach track. It is assumed here that the ship will be approaching the bay from the southwest. The advance and transfer for the 60° left turn onto the approach track are obtained, and the turning point is located. A turn bearing is drawn from this point and labeled. At this stage, the plot appears as in Figure 14-3C.

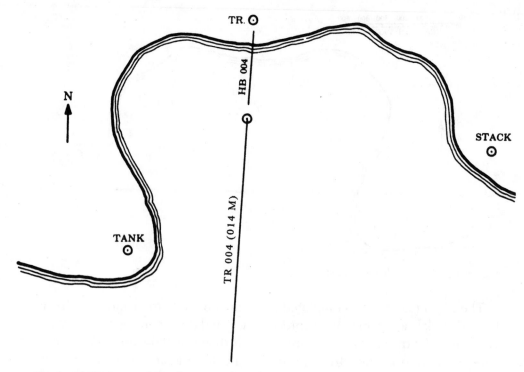

Figure 14-3B. *Laying down the approach track.*

To complete the plot, the letting-go circle is drawn, with a radius equal to the 50-yard hawsepipe to pelorus distance. The letting-go bearing is then constructed using the stack, as it is nearly at a right angle to the approach track. Finally, range circle arcs are drawn and labeled, centered on the middle of the anchorage, with radii measured in 100-yard increments outward from the letting-go circle to 1,000 yards. Arcs are also swung for 1,200-, 1,500-, and 2,000-yard distances to the letting-go circle; the last of these is not shown because of space limitations. The anchorage plot is now complete and appears as shown in Figure 14-3D.

The various arcs, bearings, and tracks are drawn as shown to make the reading of information from the plot fast and easy for the navigator as the ship is approaching its anchorage. He or she will already have briefed the captain, conning officer, and pilot if aboard as to the intended head bearing and letting-go bearing; they could then drop anchor in the proper location without any further word from the navigator, by keeping the head bearing always constant and dropping anchor when the letting-go bearing reached the proper value. The navigator does not rest during this time, however. Quite to the contrary, he or she obtains fixes as often as possible and maintains a running commentary to inform all concerned as to the distance to go to the drop circle and whether the ship is to the right, to the left, or on the approach track as it proceeds to the anchorage.

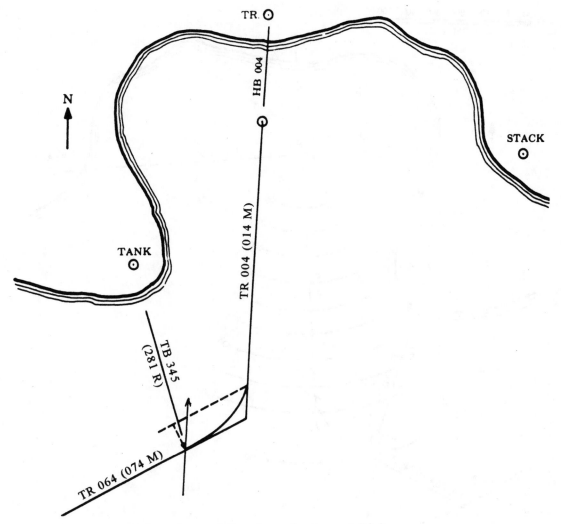

Figure 14-3C. Laying down the intended track, turn bearing, and slide line.

Executing the Anchorage

When executing the actual anchorage, the navigator's dual objective is to keep the ship as nearly as possible on its preplanned approach track, and to have all headway off the ship when the hawsepipe is directly over the center of the anchorage. As mentioned above, frequent fixes are obtained as the ship proceeds along its track, and the bridge is kept continually informed as to the position of the ship in relation to the track and letting-go circle. The navigator recommends courses to get back onto track if necessary. Since every ship has her own handling characteristics, speeds that should be ordered as the ship proceeds along the track are difficult to specify. In general, however, with 1,000

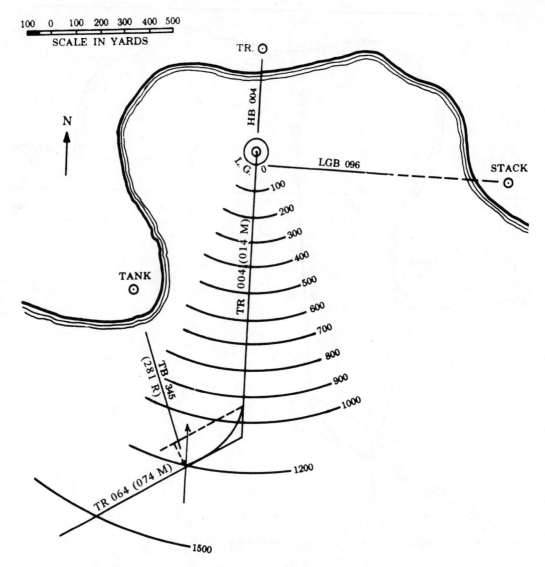

Figure 14-3D. *The completed anchorage plot.*

yards to go, most ships are usually slowed to a speed of 5 to 7 knots. Depending on wind and current, engines should be stopped when about 300 yards from the letting-go circle, and the anchor detail should be instructed to "stand by." As the vessel draws near the drop circle, engines are normally reversed so as to have all remaining headway off the ship as she passes over the letting-go circle. When the pelorus is exactly at the letting-go bearing, the word "Let go the anchor" is passed to the anchor detail, and the anchor is dropped.

As the anchor is let go, the navigator should immediately call for a round of bearings to be taken, and the ship's head should be recorded. After the resulting fix is plotted, a line is extended from it in the direction of the ship's head, and the hawsepipe to pelorus distance is laid off along the line, thus plotting the position of the anchor at the moment that it was let go. If all has gone well, the anchor should have been placed within 50 yards of the center of the anchorage.

Post-anchoring Procedures

After the anchor has been let go, chain is let out or "veered" until a length or "scope" of chain five to seven times the depth of water is reached. During this time the ship may drift backward due to the effects of wind and current, thus preventing any tendency for the anchor chain to pile up on itself on the bottom and thus foul the anchor. When sufficient chain has been veered, it is secured and the engines are backed, causing the flukes of the anchor to dig into the bottom, thereby "setting" the anchor. When the navigator receives the word that the chain has been let out to its full precomputed length and that the anchor appears to be holding with a moderate strain on the chain, a round of bearings is taken and the ship's head is recorded, as well as the direction in which the chain is tending. With this information, another fix is plotted, and the position of the anchor is recomputed by laying off the sum of the hawsepipe to pelorus distance plus the scope of chain in the direction in which the chain is tending. This second calculation of the position of the anchor is necessary because it may have been dragged some distance from its initial position during the process of setting the anchor.

After the final position of the anchor has been thus determined, the navigator draws a second swing circle, this time using the computed position of the anchor as the center, and the sum of the ship's length plus the actual scope of chain let out as the radius. If any previously undetermined obstruction, such as a fish net buoy or the swing circle of another ship anchored nearby, is found to lie within this circle, the ship may have to weigh anchor and move away from the hazard. If the ship is anchored in a designated anchorage area, due care should be taken to avoid fouling the area of any adjacent berths, even though they might presently be unoccupied. If the swing circle intersects another berth, it may be necessary to take in some chain to decrease the swing radius; if this is not possible, a move to a larger berth may be advisable.

If satisfied that no danger lies within the swing circle, the navigator then draws the drag circle concentric with the swing circle, using as a radius the sum of the hawsepipe to pelorus distance plus the scope of chain. All fixes subsequently obtained should fall within the drag circle; if they do not, the anchor should be considered to be dragging. Both the swing circle and drag circle are shown in Figure 14-4, assuming that a scope of chain of 50 fathoms to the hawsepipe has been let out.

After plotting the drag circle, the navigator selects several lighted navigation aids suitable for use in obtaining fixes by day or night, and these are entered in

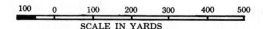

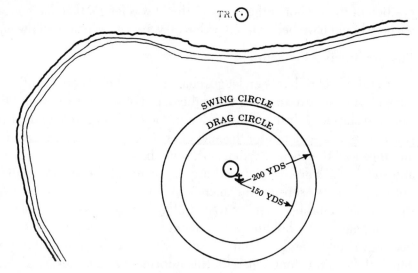

Figure 14-4. *An expanded view of the swing and drag circles.*

the *Bearing Book* for use by the anchor-bearing watch. The anchor-bearing watch is charged with obtaining and recording in the *Bearing Book* a round of bearings to the objects designated by the navigator at least once every 15 minutes, and plotting the resulting fix on the chart each time. Should any fix fall outside the drag circle, another round of bearings is immediately obtained and plotted. If this second fix also lies outside the drag circle, the ship is considered to be dragging anchor, and all essential personnel are alerted.

In practice, if the ship is to be anchored for any length of time, the navigator will usually cover the area of the chart containing the drag circle with a sheet of semiclear plastic, so that the chart will not be damaged by the repeated plotting and erasures of fixes within the drag circle.

The advent of highly automated high-precision GPS receivers in recent years has provided an extremely valuable auxiliary feature for the anchor-bearing watch. Many of today's GPS receivers incorporate an audible alarm that can be set to go off whenever a subsequent fix falls outside of an arbitrary distance from an indicated position, such as the center of an anchorage. Such a receiver can easily be set up to alarm whenever a fix falls outside the drag circle, providing an invaluable backup for the anchor-bearing watch on board Navy ships, and a kind of "automated" anchor-bearing watch for small craft with insufficient crew to post such a watch.

Even with such an aid, however, the importance of maintaining a competent and alert anchor-bearing watch cannot be overemphasized, as many recorded groundings have occurred because the duties of this watch were improperly performed. When a ship is dragging anchor, especially in high wind conditions, there is often no unusual sensation of ship's motion or other readily apparent indication of the fact. The safety of the ship depends on the ability of the anchor-bearing watch to alert all concerned if fixes obtained by whatever means begin to fall outside the drag circle. If conditions warrant, the ship may have to get under way. As interim measures to be taken while the ship is preparing to do this, more chain may be veered to increase the total weight and catenary of chain in the water, and a second anchor may be dropped if the ship is so equipped.

In situations in which high winds are forecast, the ship should assume an increased degree of readiness, with a qualified conning officer stationed on the bridge, and a skeleton engineering watch standing by to engage the engines if necessary. As an example, during a Mediterranean deployment, a U.S. Navy destroyer was anchored off Cannes, France, in calm waters with less than 5 knots of wind blowing. Because high winds had been forecast for later in the day, the officer of the deck was stationed on the bridge, and a skeleton engineering watch was charged with keeping the engines in a 5-minute standby condition. Two hours after anchoring, after two liberty sections had gone ashore, the winds began to increase. In the next 45 minutes, wind force increased to the point where 75-knot gusts were being recorded. The ship got under way and steamed throughout the night until the storm abated the next day.

SUMMARY

This chapter has described the procedures used by the navigator during precise piloting in constricted waters when the accuracy required necessitates the incorporation of the vessel's handling characteristics data in the dead reckoning plot. The navigator should take the effects of wind and current into consideration when using handling data, and should also be aware that subsequent alterations might have changed the ship's actual characteristics somewhat from the listed figures.

Precision anchoring involves almost every skill the navigator possesses in order to drop the anchor successfully within 50 yards of the center of the designated anchorage. A large percentage of the navigator's work in anchoring is accomplished before the actual evolution takes place. The navigator must ensure that all members of the ship's crew who will be involved are prepared to play their roles.

After the anchor has been set and the chain secured at its final length, the navigator must make sure that a properly trained and alert anchor-bearing watch is stationed to plot and record bearings for a fix at least once every 15 minutes to insure that the ship is not dragging anchor. In the event that the anchor is discovered to be dragging, the ship must be prepared to get under way to reset the anchor if necessary.

VOYAGE PLANNING AND INTRODUCTION TO TIME

15

The previous chapters in this text have each dealt in detail with some aspect of either the marine navigator's environment or duties during the practice of piloting. This chapter will conclude the piloting part of this text by examining how these different factors are brought together and employed by the navigator in planning an extended voyage.

Since a knowledge of time and how it is determined is essential during voyage planning for estimating departure and arrival times and dates, for planning time zone adjustments en route, and for many other purposes, an introductory section on time is included at the beginning of this chapter. It will provide a brief overview of the determination of time, time zones, and their use. Subsequent sections of this chapter will examine various aspects of the planning activities of a Navy navigator as he or she prepares for a typical Mediterranean deployment of a Navy ship with the U.S. Sixth Fleet, with the main emphasis being placed on the planning of a voyage from Norfolk, Virginia, to Gibraltar, and thence to Naples, Italy.

Although the particular planning process discussed herein as an example is oriented toward a Navy ship, most aspects of the process will have direct parallels in the case of a navigator planning a similar transoceanic voyage on a commercial ship or private vessel. In fact, the same concerns and general procedures apply to almost any major voyage to be undertaken on almost any type of surface vessel.

TIME

The consideration of time is always of major importance in every voyage-planning process. Almost every planning action of the navigator is concerned in some way with the timely arrival of the ship at her destination and at intermediate points en route. In the case of both commercial and naval ships, the time and date of arrival are normally specified by higher authority in the form of an estimated time of arrival (ETA) that must be met. In order to arrive on time, the navigator must compute the estimated time and date of departure (ETD) as well as the ETAs at several points along the track to check the ship's progress. As the ship moves from one time zone to another as she proceeds along her route, the navigator must recommend when to reset the ship's clocks and in what direction, to keep the ship's time in accord with the local zone time. As will be explained in part 2 of this text, the consideration of time is also very important in the performance of celestial navigation practiced after the ship has passed beyond the limits of piloting waters.

The Development of Zone Time

To understand the concept of time, it is necessary to have a basic knowledge of how time is derived. Until the nineteenth century, time was mainly reckoned according to the apparent motion of the sun across the sky. In the late 1800s, however, the development of comparatively rapid transit systems such as the railroad and the steamship made keeping time according to the motions of the "apparent" sun impractical, inasmuch as timepieces had to be adjusted every time one's position on the earth changed. Moreover, because of the elliptic path of the earth around the sun, the rate at which the apparent sun moves across the sky is not constant, but varies from day to day. To avoid this latter difficulty, the concept of a theoretical "mean" sun passing completely around the earth at the equator once every 24 hours came to be widely used for marking the passage of time. One "mean" solar day is 24 hours in length, with each hour consisting of 60 minutes and each minute, 60 seconds. Since the mean sun completes one circuit of the earth every 24 hours, it follows that it moves at the rate of 15° of arc as measured at the equator—or 15° of longitude—every hour. This fact will assume great significance in part 2, when the procedures to use the sun to obtain celestial lines of position are discussed.

At first, "mean" solar time was reckoned according to the travel of the mean sun relative to the observer's meridian; this was called *local mean time* (LMT). Shortly thereafter, however, to eliminate the necessity for frequent resetting of timepieces, the earth's surface was divided into 24 vertical sectors called *time zones*, each of which is 15° of longitude in width. Time within each zone is reckoned according to the position of the mean sun in relation to the central meridian of the zone. Thus, clocks are changed by one hour increments only when transiting from one 15° longitude zone to another, and every location within the same time zone keeps the same "standard" time. Time reckoned in this manner is called *zone time* (ZT).

Figure 15-1 depicts a standard time zone chart of the world. Because it is based on a Mercator projection, each sector appears as a vertical band 15° of longitude in width. As can be seen by an inspection of the time zone chart, each zone is defined by the number of hours of difference existing between the

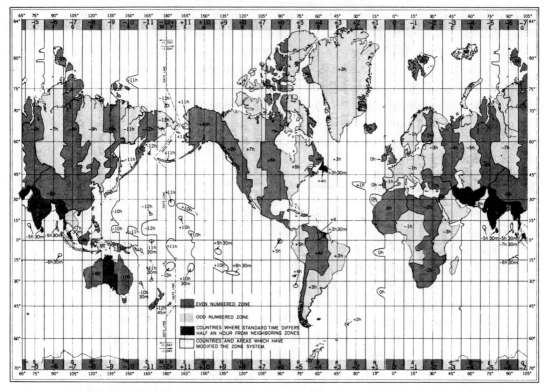

Figure 15-1. *Standard time zone chart of the world.*

time kept within that zone and the time kept within the zone centered on the prime (0°) meridian passing through Greenwich, England. Each zone is also labeled with letters called *time zone indicators* that assist in identification of the zone.

Time based upon the relationship of the mean sun with the prime meridian is called *Greenwich Mean Time*, abbreviated GMT; it is often referred to as ZULU time, because of its time zone indicator letter. The farther to the west of Greenwich that a time zone lies, the earlier will the time kept in that zone be in relation to GMT. This is indicated by placing a plus (+) sign in front of the hourly difference figure, to indicate the number of hours that must be added to the local zone time to convert it to GMT. By the same token, the farther to the east of Greenwich a time zone is located, the later its time will be relative to Greenwich Mean Time. This fact is indicated by minus (–) signs, indicating the number of hours that must be subtracted from local zone time to obtain GMT. The Greenwich zone is centered at the prime meridian, and it extends 7½° to either side. A new time zone boundary lies every 15° thereafter across both the eastern and western hemispheres, resulting in the twenty-fourth zone being split into two halves by the 180th meridian. The half on the west side of this meridian keeps time 12 hours behind GMT, making its difference +12 hours, while the half on the east side keeps time 12 hours ahead, resulting in a difference of –12 hours. Thus, there are in reality 25 different standard time zones, numbered from +1 through +12 to the west of the Greenwich zone, and –1 through –12 to the east. These differences are usually referred to as the *zone description* or *zone difference*, abbreviated ZD.

From the preceding discussion and from an inspection of the time zone chart, it follows that as a general rule the standard time zone in which any particular position on earth is located can be found simply by dividing its longitude by 15. If the remainder from such division is less than 7½°, the quotient represents the number of the zone; if the remainder is greater than 7½°, the location is in the next zone away from the Greenwich meridian. If there is no remainder, the location lies exactly on the central meridian of a time zone. The sign of the zone is determined by the hemisphere in which the position is located. In the western hemisphere the sign is positive, and in the eastern hemisphere the sign is negative. Applying the proper sign to the number found by division yields the zone description (ZD). As an example, if the standard time zone in which Norfolk, Virginia, is located were required, its longitude, 76° 18.0′ West, would be divided by 15 to yield a quotient of 5 with a remainder of 1° 18′. Thus, it is located in the +5 time zone, which has the time zone indicator letter *R* assigned.

For the navigator sailing the world's oceans, the 15°-wide standard time zones are convenient to use, but in practice many places do not always adhere precisely to the time of the zone in which they are physically located, as indicated in Figure 15-1; to do so would in many cases cause a great deal of confusion in conducting business and travel. As a result, time zone boundaries follow state and county boundaries in many cases, while some regions establish their own local "standard" times different from any of the 24 accepted zones. In Antarctica, where all 24 time zones converge at the pole, GMT is used throughout.

Daylight savings time is a device adopted by some countries, including the United States, to extend the hours of daylight in the evening during summer. Locations keeping daylight savings time keep the time of the next zone to the east of the time zone in which they are located. When daylight savings time is being kept in Norfolk, for example, it keeps +4Q time rather than its standard +5R time. For purposes of the voyage described in this chapter from Norfolk to Naples, daylight savings time has been disregarded, even though it probably would be in effect during the time of the year in question. The time kept at any particular location and time of year can normally be found in the applicable volume of the *Sailing Directions* for foreign ports or the *Coast Pilots* for U.S. ports.

Time Conversions

Because of the difficulties inherent in working with the different zone times in calculating times of arrival and departure on long voyages such as the Norfolk to Naples transit of this chapter, the navigator usually first converts all times to GMT prior to the initial planning stages of a voyage. After all ETAs and ETDs have been computed in GMT, certain times of interest can then be converted to local zone times. To do this, extensive use is made of the formula

$$ZT + ZD = GMT,$$

in which ZT is the local zone time for the locations of interest, and ZD is the zone description previously described. When correcting from zone time to GMT, the zone difference is added algebraically to the zone time; but when converting to zone time from GMT, the sign opposite the zone difference must be used.

As an example, suppose that the navigator wished to convert 0800 local zone time on 30 June at Naples, Italy, to Greenwich Mean Time. Since the longitude of Naples is 14° 16′ East, it lies in the −1A time zone, and the local zone time to be converted could be written as 0800A 30 June. Time zone indicators are normally placed after all four-digit time figures in the manner shown here to avoid confusion during time conversion. To convert 0800A 30 June to GMT, the zone difference of −1 hour is added algebraically to yield 0700Z 30 June.

As a second example, suppose that it is desired to convert 1000Z 18 June at Norfolk, Virginia, to local zone time. To accomplish this, the formula above is rearranged to the following form:

$$ZT = GMT − ZD;$$

thus it is necessary to subtract the +5 zone difference from GMT, yielding 0500R 18 June.

ZONE	+12	+11	+10	+9	+8	+7	+6	+5	+4	+3	+2	+1	0	−1	−2	−3	−4	−5	−6	−7	−8	−9	−10	−11	−12
00	01	02	03	04	05	06	07	08	09	10	11	12	13	14	15	16	17	18	19	20	21	22	23	/	01
01	02	03	04	05	06	07	08	09	10	11	12	13	14	15	16	17	18	19	20	21	22	23	/	01	02
02	03	04	05	06	07	08	09	10	11	12	13	14	15	16	17	18	19	20	21	22	23	/	01	02	03
03	04	05	06	07	08	09	10	11	12	13	14	15	16	17	18	19	20	21	22	23	/	01	02	03	04
04	05	06	07	08	09	10	11	12	13	14	15	16	17	18	19	20	21	22	23	/	01	02	03	04	05
05	06	07	08	09	10	11	12	13	14	15	16	17	18	19	20	21	22	23	/	01	02	03	04	05	06
06	07	08	09	10	11	12	13	14	15	16	17	18	19	20	21	22	23	/	01	02	03	04	05	06	07
07	08	09	10	11	12	13	14	15	16	17	18	19	20	21	22	23	/	01	02	03	04	05	06	07	08
08	09	10	11	12	13	14	15	16	17	18	19	20	21	22	23	C	01	02	03	04	05	06	07	08	09
09	10	11	12	13	14	15	16	17	18	19	20	21	22	23	B	01	02	03	04	05	06	07	08	09	10
10	11	12	13	14	15	16	17	18	19	20	21	22	23	A	01	02	03	04	05	06	07	08	09	10	11
11	12	13	14	15	16	17	18	19	20	21	22	23	Z	01	02	03	04	05	06	07	08	09	10	11	12
12	13	14	15	16	17	18	19	20	21	22	23	N	01	02	03	04	05	06	07	08	09	10	11	12	13
13	14	15	16	17	18	19	20	21	22	23	O	01	02	03	04	05	06	07	08	09	10	11	12	13	14
14	15	16	17	18	19	20	21	22	23	P	01	02	03	04	05	06	07	08	09	10	11	12	13	14	15
15	16	17	18	19	20	21	22	23	Q	01	02	03	04	05	06	07	08	09	10	11	12	13	14	15	16
16	17	18	19	20	21	22	23	R	01	02	03	04	05	06	07	08	09	10	11	12	13	14	15	16	17
17	18	19	20	21	22	23	S	01	02	03	04	05	06	07	08	09	10	11	12	13	14	15	16	17	18
18	19	20	21	22	23	/	01	02	03	04	05	06	07	08	09	10	11	12	13	14	15	16	17	18	19
19	20	21	22	23	/	01	02	03	04	05	06	07	08	09	10	11	12	13	14	15	16	17	18	19	20
20	21	22	23	/	01	02	03	04	05	06	07	08	09	10	11	12	13	14	15	16	17	18	19	20	21
21	22	23	/	01	02	03	04	05	06	07	08	09	10	11	12	13	14	15	16	17	18	19	20	21	22
22	23	/	01	02	03	04	05	06	07	08	09	10	11	12	13	14	15	16	17	18	19	20	21	22	23
23	/	01	02	03	04	05	06	07	08	09	10	11	12	13	14	15	16	17	18	19	20	21	22	23	/
ZONE	+12	+11	+10	+9	+8	+7	+6	+5	+4	+3	+2	+1	0	−1	−2	−3	−4	−5	−6	−7	−8	−9	−10	−11	−12

Figure 15-2. Time Comparison Table.

When making time conversions, the navigator must be alert for situations in which the date changes as a result of the conversion process. For instance, suppose in the first example discussed above that the time at Naples to be converted were 0030A 30 June. Subtracting the –1 zone difference yields 2330Z 29 June. Likewise, if the time in the second example at Norfolk had been 0245Z 18 June, conversion to local zone time would result in 2145R 17 June.

Further examples of time conversions are given in the following table:

Location/Position	Local Zone Time (ZT)	Zone Description (ZD)	Greenwich Mean Time (GMT)
Norfolk, Va.	0700 R 30 Jun	+5R	1200Z 30 Jun
Greenwich, Eng.	1200Z 15 Jul	+0Z	1200Z 15 Jul
Tokyo, Japan	1200I 15 Apr	–9I	0300Z 15 Apr
160°W	1200X 15 May	+11X	2300Z 15 May
165°W	1400X 15 Jun	+11X	0100Z 16 Jun
173°W	1200Y 17 Jul	+12Y	2400Z 17 Jun
160°E	1200L 1 Jul	–11L	0100Z 1 Jul
165°E	0800L 1 Jul	–11L	2100Z 30 Jun
173°E	1200M 2 Sep	–12M	0000Z 2 Sep

To assist in conversions of times from one zone to another, particularly those involving date changes, an aid called the Time Comparison Table can be used; it is printed on most standard time zone charts. A sample table of this type appears in Figure 15-2. The table can be used to determine the zone time and date of any zone from a given zone time and date. When crossing the date change line shown in the table from left to right, one day is added to the date; when crossing from right to left, one day is subtracted. As an example of its use, if the time and date at Washington, D.C. (+5R) were 1900R 1 February, the corresponding time and date at Sydney, Australia (–10K) would be 1000K 2 February. As a second example, if the time and date at Tokyo, Japan (–9I) were 0900I 1 February, the time and date in San Francisco (+8U) would be 1600U 31 January. The circled figures in the diagram refer to these two examples.

In labeling the track, as well as in writing messages, times and dates are usually written in an alphanumeric format called a *date-time group*, as illustrated in Figure 15-3. If all dates fall within the same month and year, the month and year abbreviations can be omitted when labeling a track. Several examples of the use of a date-time group in labeling the track are given later in this chapter.

Changing Time Zones and Dates en Route

To avoid undue psychological stress to crew members, ship's clocks are normally set to conform to the time zone in which the ship is operating. The nav-

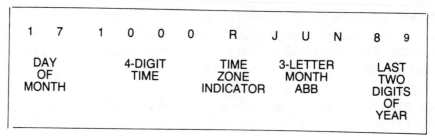

1 7	1 0 0 0	R	J U N	8 9
DAY OF MONTH	4-DIGIT TIME	TIME ZONE INDICATOR	3-LETTER MONTH ABB	LAST TWO DIGITS OF YEAR

Figure 15-3. *Format of a date-time group.*

igator, therefore, should include the schedule of time zone changes in all voyage planning efforts.

When steaming in an easterly direction, as in traveling from Norfolk to Naples, it is necessary to advance ship's clocks by one hour periodically in order to conform to the proper zone time. When proceeding in a westerly direction, the clocks must be continually retarded. In cases requiring clocks to be advanced, the normal procedure in the Navy is to advance them during the morning hours, generally at either 0100 or 0200, so as not to disrupt the normal working day. When clocks are set back, the period during the second dog watch from 1800 to 2000 is usually selected, to avoid the possibility of a 5-hour watch in a 4-hour watch rotation cycle. To assist in determining when the ship's clocks should be reset, the navigator should label the track at intervals with the estimated times of arrival (ETAs) at these points. The times are normally first calculated in Greenwich time and then converted to local zone time. Legs of the track that will require adjustment of the ship's clocks can then be determined by inspection.

The 180th meridian is designated as the International Date Line, because the time kept in the 7½°-wide zones on either side of it differs by 24 hours or one complete day. When crossing the date line on a *westerly* heading, the zone description changes from +12 to –12, so ship's clocks must in effect be *advanced* 24 hours, thereby "losing" (i.e., skipping ahead) one calendar day. Conversely, when crossing the date line on an *easterly* heading, the ship's clocks must in effect be *retarded* 24 hours, since the zone description changes from –12 on the western side of the line to +12 on the eastern side. In this instance, the ship is said to have "gained" a day, since one calendar day is repeated. For convenience, ships crossing the date line will usually effect the date change at night, normally at midnight. In the Navy, if the date change would cause a Sunday or holiday to be repeated or lost, the ship may operate for a time using a zone description of +13 or –13, to allow the date change to be made on either the preceding or the following working day.

THE VOYAGE-PLANNING PROCESS

The key to successful planning of either a single voyage or an extended deployment is early advance preparation by the navigator. In the case of Navy ships, the first notification of a deployment and its associated transits is normally received when the *Yearly Employment Schedule* for the fleet of which the ship is normally a part is published by the cognizant fleet commander. This publication is normally received by the ship's operations officer, who extracts all pertinent information on the ship's schedule, compiles it, and then disseminates it to all officers and personnel concerned. If a deployment is scheduled during the first quarter of the year covered by the *Yearly Employment Schedule*, advance notice is usually received by other means three to six months before. Confirmed deployment dates will be indicated in subsequent published *Quarterly Employment Schedules* or by message.

As soon as the navigator receives word as to the area of the world to which the ship will proceed, the voyage planning process should be commenced. In order not to overlook any of the many facets of an extended voyage and deployment, the navigator should normally lay out all anticipated requirements in the form of a checkoff list. The list below, although compiled for a sample deployment from Norfolk, Virginia, to the Sixth Fleet port of Naples, Italy, is representative of the general items such a list should contain.

1. Publications and Charts
   a. Is the normal operating allowance of publications and charts on board?
   b. Are all charts and publications in normal use corrected and up to date with the latest *Notice to Mariners*?
   c. Which chart portfolios and publications not in normal use, on allowance and on board, will be required for the voyage and which must be brought up to date by the use of chart and publication correction cards listing *Notice to Mariners* corrections? Are any out of date or superseded?
   d. Which charts and publications not normally on allowance have to be ordered?
   e. Do the predeployment checkoff lists and Sixth Fleet operation orders and Fleet instructions require any additional charts and publications?
   f. Do the operation orders require any classified combat charts for gunfire support?
   g. Are the latest pilot charts on board for use in the initial transit?
2. Equipment
   a. Are magnetic compasses within acceptable limits or is a period at sea to swing ship required prior to departure?

b. Is the degaussing system operating properly? When was the last degaussing run completed? Will the ship be required to run the degaussing range prior to departure?

c. Does the ship have a sufficient number of spare PMP parts on board for use by both the navigator and CIC over a six-month deployment?

d. Is there backup plotting equipment on board for emergency use?

e. Are the GPS/Loran/Omega navigation sets working properly? If not, is the electronics material officer aware they need repair? Are spare parts for these receivers on board?

f. Are all chronometers, stopwatches, and comparing watches operating properly and up to allowance? Are the ship's clocks up to allowance and properly calibrated?

g. Are the echo sounder and the pit log operating properly?

3. Personnel

a. Is the department/division up to allowance in key petty officers?

b. Will all key personnel remain on board throughout the cruise?

c. Are sufficient numbers of qualified strikers on board to stand long periods of underway watches?

4. Operations

a. What is the total distance from Norfolk to Naples?

b. What will be the ETD and ETA?

c. Is the voyage at least 1,500 miles so that Optimum Track Ship Routing must be requested from the appropriate Naval Oceanographic Command Center?

d. If the voyage is less than 1,500 miles, is there a possibility of hurricanes or typhoons en route so that OTSR can be requested?

e. Should the track originally recommended be changed based on OTSR recommendations? Will the ETD and ETA change?

f. Have climatological summaries been requested from the appropriate oceanographic center to give frequency of various wind speeds and their directions, wave heights, currents, and the probability of rain and storms en route?

g. Has the operations officer been provided with the final ETD from Norfolk and the ETA at Naples? Has he or she been given the date-time group (DTG) in Greenwich Mean Time (GMT) and the latitude and longitude of all major junction points of the coastal, rhumb line, and great-circle tracks to be employed?

h. Has the operations officer been advised to request weather en route (WEAX) in the movement report?

As can be seen by an inspection of this checklist, the navigator is faced with many areas of concern in planning a voyage of any duration. As the days and weeks pass, and the departure date draws ever closer, the intensity of the planning effort steadily increases until several days before the start of the journey, when virtually all his or her time will probably be filled checking the innumerable and inevitable last-minute details.

The remaining sections of this chapter will focus on the key aspects of planning the transit from Norfolk, Virginia, to Naples, Italy, and will conclude with a discussion of some general points applicable to both this transit and the subsequent deployment.

Obtaining and Updating Charts and Publications

Among the first tasks of the navigator after learning of a scheduled extended voyage or deployment is to ascertain which charts and publications will be required, and of these, which are on board and which are not. To find this information, the applicable allowance lists described in chapter 5 are consulted. Amplifying information on requirements for charts and publications can also be found in applicable standing fleet operation orders and instructions. Any portfolios and charts either not on board or outdated are ordered from the National Imagery and Mapping Agency.

Having ascertained all requirements, the navigator then consults the DMA *Catalog of Maps, Charts, and Related Products* and the NOS *Nautical Chart Catalog 1* to find the numbers of all charts and *Coast Pilot* and *Sailing Directions* volumes that will be of use in planning and executing the initial transit from Norfolk to Naples. The chart petty officer pulls all those charts and publications currently on board from their storage locations, along with their correction cards, and applies to them all the necessary corrections from the file copies of the *Notice to Mariners*.

When the ordered charts and publications are received from NIMA and corrected up to date, the navigator then has on hand all the information needed to successfully and intelligently plan for the forthcoming transit and deployment. Should any material be missing, the navigator does not wait for it to arrive before beginning to plan. Instead, he or she proceeds with the determination of the departure and if necessary the arrival dates and times, as set forth in the following section.

Determination of the Departure and Arrival Dates

When a Navy ship is to deploy, one of two possible situations will exist. She will either deploy in company with other ships in an organized group or she will transit independently to her assigned fleet, joining it on arrival. When a ship is to deploy as part of a group, an operation order is usually made up by the group commander and the commander's staff to cover the transit from the home port to the destination. After completion of the deployment, a second operation order is compiled to govern the return trip. In situations of this type, the navigator of the flagship is normally assigned additional duties as staff navigator, except in those rare instances where a staff is so large as to have a permanently assigned navigator attached to it. In any event, the staff navigator will assume the responsibility of planning the voyage for the entire group, with possible assistance rendered by navigators from the various ships comprising the group. When a ship is to transit independently, on the other hand, all the work of preplanning must be accomplished by the ship's navigator and assigned personnel. For the Norfolk to Naples voyage under discussion, it will be assumed that the ship is to conduct an independent transit.

Regardless of whether the ship is transiting in company or independently, the arrival time at the destination is usually specified by higher authority. For purposes of our mock voyage, it will be assumed that the ship has been directed to arrive at Naples at 0800A 30 June. Should the navigator later conclude that this arrival time is unsafe for some reason, such as a low tide or high current velocity, a request can be initiated to change the ETA; such instances, however, are rare. The navigator, then, is usually concerned mainly with the determination of an estimated time of departure that will ensure the timely arrival of the ship at the destination.

In the case of a ship conducting an independent transit, the date and time of departure normally are specified in the form of a message from the ship's squadron commander, but based on the ETD that the ship's navigator has previously recommended as best fitting the ship's needs for the transit. To determine the estimated time and date at which the ship should depart, the navigator must first determine the approximate distance between Norfolk and Naples along the most suitable route at the given time of year. The total number of hours required for the journey at the approximate speed of advance (SOA) to be used is then computed by the use of the speed-time-distance formula. The distance between Norfolk and Naples can be found easily by use of *Publication No. 151, Distance Between Ports*. The ship can sail a great circle route across the Atlantic to Gibraltar, but sailing a great circle from Gibraltar to Naples is impossible due to the geography of the Mediterranean. Thus, *Publication No. 151* is entered first for the great circle distance from Norfolk to Gibraltar, 3,335 miles. It is entered a second time for the distance from Gibraltar to Messina, on the toe of Italy, 1,049 miles, and a third time for the distance between Messina and Naples, 175 miles. Adding the three together yields a total distance from Norfolk to Naples of 4,559 miles. The maximum SOA is normally specified by higher authority; for most conventionally powered destroyers it is usually set at about 16 knots, the most economical cruising speed for this type of ship. By dividing the 4,559-mile total by 16 knots, a total time in transit of 285 hours, or 11 days and 21 hours, is computed.

At the beginning of this chapter in the section on time, it was shown that Norfolk lies in the +5R time zone, while Naples is in the −1A zone. Thus, the ship will "lose" 6 hours of time en route because of time zone adjustments; this figure must also be added to the initial time requirement figure. The ship must depart Norfolk, therefore, at least 12 days and 3 hours before her scheduled arrival time at Naples. Subtracting this time from 0800A 30 June yields an estimated time of departure from Norfolk of 0500R on 18 June. An alternative and in most cases preferred method of obtaining the same result would be to first convert the designated arrival time at Naples to GMT. The computed transit time, 285 hours, is then subtracted, and finally the resulting time of departure from Norfolk expressed in GMT is converted to Romeo time by subtracting the +5 zone difference. To facilitate departure ceremonies, an ETD for deployments between the hours of 0800 and 1000 local time is usually specified by local authorities. The navigator in this case, therefore, would recommend to the commanding officer that the departure be scheduled for 1000R 17 June. The extra time would be compensated for by specifying a slower SOA on some legs of the transit.

The navigator should also consider the predicted tides and currents at Norfolk on the day of departure and at Naples on the day of arrival. Depending on the size of the ship and the location of the port, this consideration may be the overriding factor in deciding on departure time or entry time. A carrier, for example, will normally only enter or depart when the tide is high and the current is determined to be slack at the Navy piers in Norfolk. The state of the tide and current in Norfolk present no undue difficulties to a destroyer, except possibly during peak ebb or flood and during storm conditions, as all channels are wide, deep, and easily negotiated. In any event, either the navigator or one of the quartermasters will compute complete tide and current tables for the restricted water-piloting portion of the transit on the day of departure. They will also compute a complete tide table for Naples for 30 June using the "Europe" volume of the *Tide Tables*, and obtain tidal current information for this date from charted information such as current diamonds and from the *Enroute* volume of the *Sailing Directions* for the Western Mediterranean. If information from these sources indicates that it is unsafe to enter port at the designated time, a decision may be made to request permission to adjust the arrival time so as to enter port when tide and current conditions are more favorable.

Plotting the Intended Track

Concurrent with the determination of the ETD, the navigator will usually begin to plot the initial estimate of the ship's track for the transit. Before actually beginning to plot, however, the navigator first gathers all available information applicable to a transit of the ocean areas under consideration during the given time of year. A most valuable aid at this stage is a climatological summary for the time of year and area under consideration, which is furnished upon request to Navy navigators by the Naval Oceanography Command Center serving that area. It will provide the normal wind speeds and directions, wave heights, currents, and the probability of rain and storms en route. Other sources of the same type of information are the appropriate editions of the *Sailing Direction Planning Guides*, the *Coast Pilots*, and the pilot chart. For the voyage under consideration, the navigator would need the *Planning Guides* for the Western Mediterranean and for the North Atlantic, the Atlantic Coast volume of the *Coast Pilots*, and the North Atlantic pilot chart for the month of June. The objective in using all these references is to lay down the optimum track from Norfolk to Naples, taking advantage of great circle routes to the fullest extent possible consistent with the meteorological conditions normally prevailing during the month of June. In this regard, it will sometimes be faster when contemplating an extended voyage to choose a longer route, if in so doing known areas of major weather disturbance can be avoided. In addition to these on-board references and the Naval Oceanography Center climatological summary, the Navy navigator should also request another service available to Navy ships known as Optimum Track Ship Routing, or OTSR. In contrast to the aids discussed above, which are based on historical meteorological data, the OTSR is an optimum track for the specific requesting ship based on the actual climatological and hydrographic forecasts covering the time of the voyage. Optimum Track Ship Routing will be described in greater detail later in this chapter.

Since the shortest distance between Norfolk and the Strait of Gibraltar lies along a great circle drawn between them, the navigator begins to plot by constructing the great-circle track as a straight line on a gnomonic projection. In this case, the gnomonic projection with a point of tangency in the North Atlantic is chosen; it is shown in Figure 15-4A. Since this chart cannot be used for navigational purposes, it is then necessary to transfer the great-circle route onto a Mercator projection on which the great-circle track is approximated by a series of rhumb lines. To transfer the track, the navigator selects convenient points along the great circle on the gnomonic chart, about 300 to 500 miles apart, and then transfers these points to the Mercator chart by replotting them at their correct latitude and longitude. Each consecutive point is labeled either with letters or a sequential alphanumeric designation such as R-1, R-2. Rhumb lines connecting these points are then drawn. At this stage, the track appears as shown in Figure 15-4B. Since the small scale of the North Atlantic chart shown in Figure 15-4B makes its use for actual navigation impractical, the points along the track and the rhumb lines connecting them are also laid off on larger-scale "working" charts and plotting sheets of smaller portions of the North Atlantic, which are suitable for plotting celestial and electronic fixes.

It should be mentioned here that it is also possible to determine a series of rhumb lines approximating a great-circle track by use of a modern electronic calculator or a computer to solve a computational algorithm developed for this purpose. More sophisticated calculators designed especially for marine naviga-

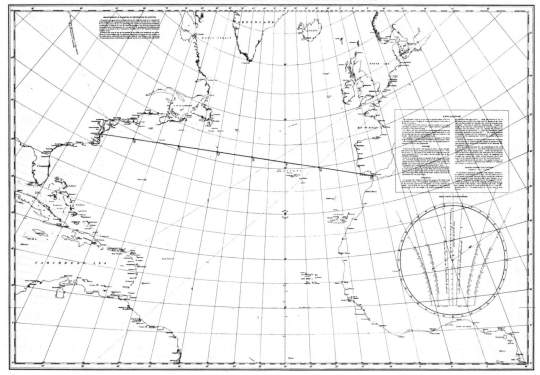

Figure 15-4A. *Great-circle route, Norfolk to Gibraltar.*

tion, and many modern Loran and GPS receivers, come equipped with this algorithm "built in" to their navigation program packages or internal memories. Several commercially available computer navigation programs discussed in part 3 of this text will do this as well.

Completing the Track

After the track from Point A off Hampton Roads to the entrance to Naples has been laid down, the exit track from Norfolk and the entrance track into Naples are plotted in accordance with procedures set forth in chapter 8. Figure 8-16 on page 145 shows a portion of a typical exit track from Norfolk. If the ship is to anchor at a predesignated anchorage at Naples, the anchorage plot may also be completed at this time using procedures described in the preceding chapter.

To complete the plot, distances along the rhumb line segments of the track are measured and totaled to obtain the actual distance measured along the track from Norfolk to Naples. This figure should agree fairly well with the distance found previously from the *Distance Between Ports* publication, but differences of 10 or 20 miles are not uncommon due to slight variations in the

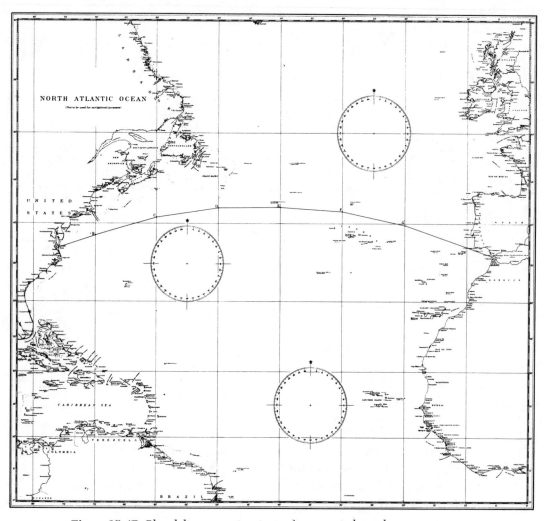

Figure 15-4B. *Rhumb line approximation to the great-circle track.*

tracks upon which they were based. After computing the speeds of advance for the various legs of the track as explained in the following section, most navigators label each junction point on the large-scale chart with the distance remaining to the destination and the computed ETA at each point, expressed in both Greenwich mean and local zone time. Other navigators may record this information in a separate table, while a few do both. In any case, the information should be readily available for use by all concerned. As a minimum, each rhumb line segment on the large-scale plot should be labeled with the track direction and speed of advance, and each junction point should be labeled with the distance remaining and the ETA in local zone time. A portion of the completed track appears in Figure 15-5, and a sample table supporting it follows.

Track Description Norfolk to Gibraltar

Point	Latitude	Longitude	Course to next point	Distance to next point	ETA	Distance to Naples
Pier	36–57.0N	76–20.0W	Various	19	———	4,599
A	36–56.5N	75–56.0W	072	295	171711Z	4,540
B	38–46.0N	70–00.0W	075	465	182041Z	4,245
C	40–40.0N	60–00.0W	082	455	201041Z	3,780
D	41–50.0N	50–00.0W	089	440	211511Z	3,325
E	42–00.0N	40–00.0W	094	455	221841Z	2,885
F	41–30.0N	30–00.0W	102	465	232311Z	2,430
G	39–50.0N	20–00.0W	109	690	250417Z	1,965
H	36–00.0N	06–00.0W	090	51	262323Z	1,275
Gibraltar	35–57.0N	05–45.0W	———	———	270235Z	1,225

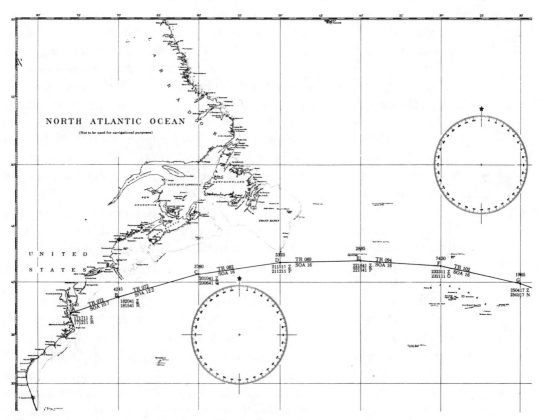

Figure 15-5. Segment of a completed Norfolk-Gibraltar track.

If the ship is to steam within sight of land at night during the transit, as when passing Gibraltar or approaching Naples, the applicable volume of the *List of Lights* should be consulted, and a table should be prepared listing the computed visibility of all lights that should be seen. After the track has been drawn, the visibility arcs for all these lights should be laid down on the large-scale coastal charts of the area in accordance with procedures set forth in chapter 6.

Determining SOAs of the Track Legs

It was shown previously that the ship must depart Norfolk no later than 0500R on 18 June (180500R JUN) to arrive at Naples on time using an average SOA over the entire track of 16 knots. It was further stated, however, that the navigator would probably recommend that the ship actually depart at 1000R on the seventeenth, to facilitate her departure. Thus, if the commanding officer agrees with this recommendation, the ship will have 19 hours of slack time to expend en route. This slack time could be expended in a number of ways. After consulting with the ship's planning board for training, the navigator might plan for the extra time to be used for emergency drills, standard training exercises, and shiphandling drills, or he might simply lower the overall SOA somewhat in order to proceed at a slower speed. In any event, the slack time must be accounted for, because once Navy ships are under way, they are normally required to maintain position within 50 miles or 4 hours of steaming time, whichever is less, from their prefiled planned positions at any time. As can be seen by an inspection of the SOA figures in Figure 15-5 and the accompanying table, in this case the navigator has planned to use up all the slack time during the ship's transit to Point C on the track. Because the ship must proceed at slow speeds while exiting Hampton Roads, the navigator has allowed for an overall SOA from the pier to Point A of only 8.7 knots, which accounts for 1 hour of the extra time. The navigator has planned to use 9 of the remaining 18 hours between Points A and B, and the other 9 hours between Points B and C, thus reducing the required SOAs for these legs from 16 knots to 10.7 knots and 12.2 knots, respectively. In reality, the ship may actually slow down or stop several times during her travel from Point A to Point C while conducting drills and exercises, and then proceed at 20 or 25 knots for a short time to maintain the planned SOAs.

The theoretical position of the ship on the intended track at any time is sometimes referred to as her *position of intended movement*, abbreviated PIM. The PIM moves along the track at the SOA, and the actual ship's position can be described in relation to it. If the ship arrives two hours ahead of the planned ETA at point C, for example, it might be said that she is "two hours ahead of PIM" at that point. At the start of a transit, it is the usual practice to proceed at an ordered speed slightly faster than the SOA, in order to build up some additional slack time in case some unanticipated delay occurs en route, such as a storm or mechanical breakdown. In general, it is always best to proceed slightly ahead of PIM at first, and to decrease speed to account for any time thus gained toward the end of the voyage, rather than being late, after having been delayed en route by some unexpected occurrence.

OPTIMUM TRACK SHIP ROUTING

Optimum Track Ship Routing, or OTSR, is a forecasting service provided by the Naval Oceanographic Command, which seeks to provide an optimum track for naval vessels making transoceanic voyages. This track is selected by applying ocean wave forecasts and climatological information to individual ship loading and performance data. The use of the OTSR recommendations does not guarantee smooth seas and following winds, but it does promise a high probability of sailing the safest and most rapid route consistent with the requirements of the requesting ship. Since the service became fully operational in 1958, ships of the Navy have been able to stay on schedule during transits to a degree seldom seen before the institution of the service, and ship and cargo damage has been significantly reduced.

OTSR is an advisory service only, and use of its recommended track does not relieve the commanding officer and navigator of Navy ships from the responsibility to properly plan the voyage using all other means available. Neither does it seek to limit the prerogative of the commanding officer in deciding upon the route the vessel will follow. If the ship does deviate to any great extent from the OTSR route, the Naval Oceanographic Command Center that performs the service should be informed by message.

Using OTSR Services

Requests for the OTSR service should be submitted for receipt by the Naval Oceanographic Command Center (NOCC) Norfolk, Virginia (Atlantic area), or by the Naval Oceanographic Command Center (NOCC) Pearl Harbor, Hawaii (Pacific area), at least 72 hours prior to the estimated time of departure (ETD). In practice, most navigators submit their requests several weeks before this deadline, and they include a request for route recommendation based on long-range climatological data. The OTSR route recommendation based on the actual short-range weather forecast is normally received by message about 36 hours prior to the ETD.

Information on the OTSR request includes the following:

Name and type of ship
Point of departure and ETD
Destination, including the latest acceptable arrival time
Intended SOA

Draft
Any unusual conditions of loading

Specific instructions for requesting OTSR services can be found in the U.S. Navy *NAVOCEANCOMINST 3140.1* series. After receipt, the navigator should compare the OTSR recommendations with his or her own preplanned track. Usually they will not vary a great deal if there is no major storm activity along the ship's intended track. On those occasions when the two tracks do diverge significantly, the navigator will normally recommend altering the ship's track to conform to the OTSR track, adjusting the SOA if necessary so as to arrive on time at the destination. Should the ship not adhere to the OTSR recommendations and in so doing encounter a storm en route that she would have avoided had she sailed the OTSR route, the commanding officer may be held responsible for any storm damage incurred.

While en route the navigator is required to send a short OTSR weather report at 0800 local zone time each day to the appropriate Naval Oceanographic Command Center. Using this information, the cognizant center continuously surveys and reevaluates the route. The message includes the following information:

Position in latitude and longitude
Course and speed being made good
Wind direction and speed
Wave direction, period, and height
Swell direction, period, and height

MISCELLANEOUS CONSIDERATIONS

The preceding sections of this chapter have each dealt with some aspect of the procedures used by the Navy navigator in deciding upon and laying down the ship's track in preparation for an extended voyage. Although preparation of the track is probably the major concern of the navigator during voyage planning, there are many other areas that require some attention both before and after getting under way on the deployment. Several of the more important of these concerns are highlighted in the following paragraphs.

The *Ship's Position Log*

Navy directives require that a legal document called the *Ship's Position Log* be maintained by both bridge and CIC watchstanders whenever the ship is under way and the *Bearing Book* is not in use. In it, fixes are recorded every half hour from all available sources, with accompanying depth soundings, intermediate DR positions, set and drift of current, and resets of SINS if the ship is so equipped.

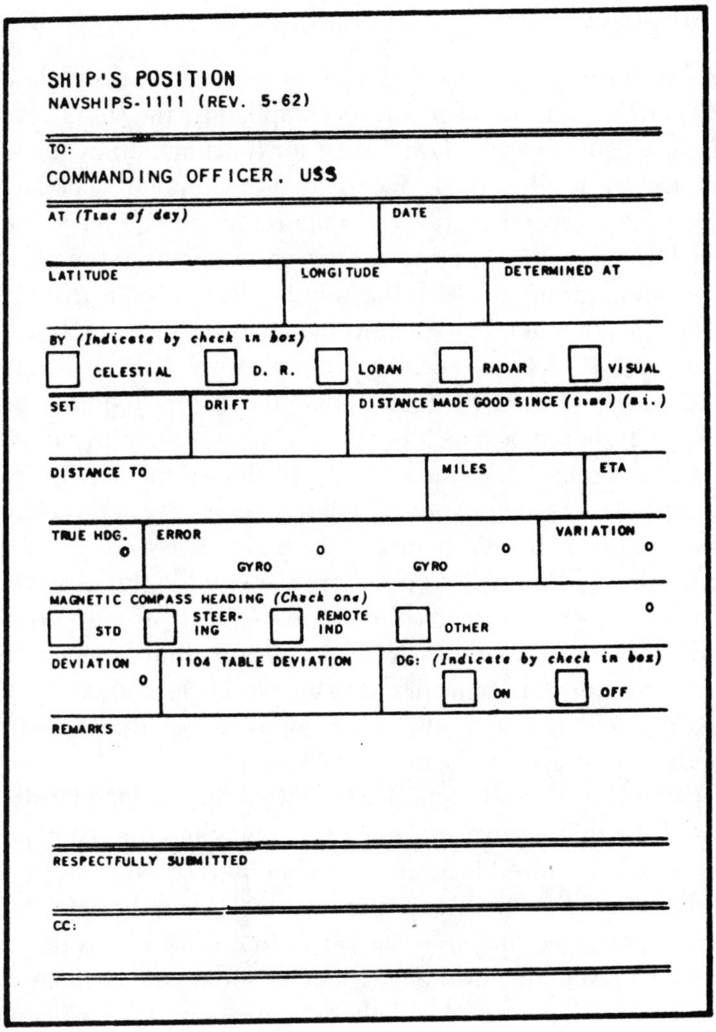

```
SHIP'S POSITION
NAVSHIPS-1111 (REV. 5-62)

TO:
COMMANDING OFFICER, USS
AT (Time of day)                          DATE

LATITUDE              LONGITUDE          DETERMINED AT

BY (Indicate by check in box)
  [ ] CELESTIAL    [ ] D. R.   [ ] LORAN    [ ] RADAR    [ ] VISUAL

SET            DRIFT          DISTANCE MADE GOOD SINCE (time) (mi.)

DISTANCE TO                    MILES              ETA

TRUE HDG.   ERROR                                 VARIATION
     o                    o              o                  o
               GYRO              GYRO

MAGNETIC COMPASS HEADING (Check one)                        o
  [ ] STD   [ ] STEER-   [ ] REMOTE    [ ] OTHER
              ING            IND

DEVIATION   1104 TABLE DEVIATION    DG: (Indicate by check in box)
     o                              [ ] ON    [ ] OFF

REMARKS




RESPECTFULLY SUBMITTED


CC:
```

Figure 15-6. *A ship's position report.*

The Ship's Position Report

By both custom and formal regulations, Navy navigators are required to submit a ship's position report in writing to the commanding officer three times daily when the ship is under way, at 0800, 1200, and 2000 local zone time. This report is submitted on a standard form called a *ship's position report*, shown in Figure 15-6. For the transit from Norfolk to Naples, each report should have entered on it, in addition to the other required data, the distance to and ETA at the next track junction point and, under the remarks section, the amount of time by which the ship is ahead or behind the intended track. All position reports should be signed by the navigator.

Routine Messages

Several days prior to a Navy ship getting under way, a standard message called a *MOVREP* (movement report) is compiled by the ship's operations officer or the navigator in a standard format for transmission by the ship's communication facility. It informs all concerned of the impending departure of the ship, and contains several sections of importance to the navigator, primary among which is a detailed description of every waypoint in the planned track, and the date-time group at which the ship is expected to arrive there. From this, the ship's position of intended movement (PIM), described earlier in connection with OTSR, can be calculated. Ships are required to stay within 4 hours and 50 miles of their PIM; an update message is required to be transmitted if the ship should deviate more than 4 hours from it. From this message the Naval Oceanographic Command Center will generate an Optimum Track Ship Routing (OTSR) recommendation described earlier in this chapter, if this service has not already been requested by separate message.

The navigator of a ship such as the destroyer cited in this chapter that does not have a separate meteorological division on board is responsible for the transmission of synoptic weather observation reports four times each day when under way. The reports are transmitted to the Naval Oceanographic Command Center in whose area of responsibility the ship is located during its transit. The content of these messages is discussed in chapter 2.

Several days prior to entering port, a message apprising the local authorities in that port of the ship's arrival and departure dates and times and any services or logistic support required is normally transmitted by Navy ships. This message, called a LOGREQ (logistics requirements), is, like the MOVREP, a standard format message compiled by the operations officer. Several items of information must be supplied by the navigator, including the estimated dates and times of arrival and departure. Like other departments, the navigator may also request any necessary supplies needed on arrival via the message.

The Captain's Night Orders

It is customary for the commanding officer or master of all ships at sea to set down in writing each night the ship is under way a complete set of instructions concerning the activities to be carried out by the watch during the night. In the Navy, these instructions are called the *Captain's Night Orders*. They are required by Navy directives, and form a permanent part of the ship's records. In practice on most Navy ships, the navigator often writes the rough draft of the night orders for further annotation and signature by the captain. The night orders should contain all the navigation instructions necessary for the OOD to safely conn the ship during the night. They are normally prepared and signed by the captain each night the ship is under way, even if the captain remains on the bridge throughout the night. Night orders may also be written when the ship is at anchorage, if unusual conditions of wind or sea warrant, for use by the in-port OOD and anchor-bearing watch.

A more detailed description of the *Captain's Night Orders* is given in chapter 32 in connection with a day's work in navigation at sea.

Consumable Items

In addition to the foregoing preparations, the navigator should ensure that a large supply of all routinely used navigational materials is on board prior to departure, including maneuvering boards, plotting sheets, spare tide and current forms and smooth deck logs, pencils, and all other miscellaneous items required. It is always preferable to have a bit too much than too little.

THE NAVIGATION BRIEF

As alluded to in chapter 3 on the piloting team, whenever a Navy ship is to enter piloting waters, a navigation brief of the navigation and CIC piloting teams and all other affected crewmembers is held beforehand. Such a brief is required by Navy directives to be held within 24 hours of the time at which the ship expects to station the Special Sea and Anchor Detail.

In addition to the bridge and CIC navigation team members, the other attendees of the navigation brief include the commanding and executive officers; key department heads and division officers; watchstanders on the bridge, in CIC, and in engineering; Sea and Anchor Detail personnel; and any other individuals with a need to know. Typically this brief will be written and chaired by the navigator or assistant navigator, with inputs from the ship's first lieutenant, operations officer, and chief engineer. It covers such things as the expected weather, tides and currents, the planned track, tugs and pilots, and any other anticipated events or conditions of navigational significance. An outline of a typical navigation brief as approved by standing Navy directives is given in appendix C of this book.

SUMMARY

This chapter concludes the piloting section of this text with an overview of the procedures used by the navigator in planning an extended voyage, using as an example a theoretical transit from Norfolk, Virginia, to Naples, Italy. Throughout the process of voyage planning, virtually all of the skills of the navigator are employed in order to be thoroughly prepared for getting under way on the appointed date and time.

During the planning stage of a voyage, it is essential that the navigator make use of all resources available in deciding upon the final intended track. Climatological summaries and Optimum Track Ship Routing provided by the Naval Oceanographic Command Center, *Sailing Direction Planning Guides*, *Coast Pilot* volumes, and pilot charts are especially helpful during the planning process. It is often possible to maintain a faster speed of advance by deviating somewhat from the great-circle route if by doing so areas of major weather disturbance can be avoided. Even after getting under way a cautious navigator can still plan for unexpected delays by traveling at a speed slightly faster than the required SOA during the first part of a voyage.

While proper and sufficient planning cannot ensure success, it nevertheless can do much to make the voyage easier. Insufficient or inattentive planning, on the other hand, can lead to disaster or serious embarrassment. As an example of the latter, consider the predicament of a Navy navigator who one day found that the vital publications and charts needed to go to a small, infrequently visited but highly desirable European port were not on board, even though they were specified as being in the ship's allowance by the fleet commander's instruction.

The subject of time, introduced in this chapter, is covered in more detail in part 2 of this text. A profile of a typical day's work in navigation at sea while en route on the voyage planned herein is also presented.

PART 2.
CELESTIAL NAVIGATION

COORDINATE SYSTEMS OF CELESTIAL NAVIGATION

16

In this first chapter of part 2, the basic facts and theories that constitute the foundations upon which the practice of celestial navigation depends will be set forth. First, the relationship of the earth with the surrounding universe and the other bodies of the solar system will be explored. Next, the terrestrial coordinate system already introduced in part 1 will be reviewed, and two additional coordinate systems necessary for the practice of celestial navigation will be developed: the *celestial system* and the *horizon system*. Finally, the three systems will be combined to form the celestial and navigational triangles; the solution of this latter triangle is the basis of celestial navigation.

THE EARTH AND THE UNIVERSE

In order to simplify the explanation of the relationship of the earth with the surrounding universe, the rotating earth can be likened to a spinning gyroscope suspended in space. As the science of physics tells us, the spin axis of a gyroscope will remain forever oriented toward a particular point in space unless it is subjected to an outside force, in which case it will tend to precess in a direction at a right angle to the spin axis. The rotating earth is, of course, subjected to the gravitational pulls of the sun and the moon, as well as to the minute gravitational pulls of all the remaining bodies of the universe. Hence, the earth is subject to *precession*, which, together with its *rotation* about its axis and *revolution* in its orbit about the sun, constitute the three *major motions of the earth*. The period of the earth's precession is so long—about 25,800 years—that for relatively short periods of time the effects of precession can be disregarded, except when certain types of celestial tables designed for use over extended time periods are used.

By good fortune, it happens that in the present era of time the northern axis of the earth is aligned almost exactly with a star that was given the name Polaris—the pole star—by early Greek astronomers. Since the distance of this star is many thousands of times greater than the diameter of the earth's orbit about the sun, the orientation of the north pole of earth with respect to Polaris does not vary more than one or two degrees throughout the year. The beneficial effects of this situation will be fully developed in chapters 25 and 26.

In addition to the aforementioned major motions, there are three *minor motions of the earth* of which the navigator should be aware— *wandering of the terrestrial poles, variations in the rotational speed of the earth,* and *nutation.* The positions of the north and south terrestrial poles, like the magnetic poles discussed in chapter 9, are not stationary, but rather they move in a circular path approximately 100 feet in diameter. A complete cycle of movement is so slow, however, as to be almost immeasurable, and the effect can be disregarded for most navigational purposes. The rotational speed of the earth is presently slowing gradually at the rate of about 0.001 revolution per century; the effect of this minor motion is also disregarded for most navigational purposes. Nutation is an irregularity in the earth's precession caused by the influence of the moon and to a lesser extent the other bodies of the solar system; this effect, like precession, must be reckoned with when certain types of celestial tables designed for use over extended time periods are used.

Like the star Polaris, the other stars of the universe are so far from the earth that their positions seem to remain constant with respect to one another and to the earth's rotational axis. This fact led to the early misconception among ancient astronomers that the earth was located at the center of a hollow sphere upon which the stars were fixed; this sphere seemed to make one complete rotation around the earth every 24 hours. The concept lingered until 1851, when the rotation of the earth was conclusively proven by Jean Foucault with his now-famous pendulum experiment. The idea of such a *celestial sphere* is still useful even today, however, in visualizing the celestial coordinate system described later in this chapter.

Besides being responsible for their apparently unchanging locations on the celestial sphere, the vast distances of the stars from earth also cause some to appear brighter than others. Usually, but not always, those stars nearer the earth will appear brighter than those more distant. The relative brightness of a celestial body is expressed in terms of *magnitude,* based on the Greek astronomer Ptolemy's division of the visible stars into six groups according to their brilliance and the order in which they appeared at night. The first group of stars to become visible to the naked eye at twilight, called the *first magnitude stars,* are considered to be 100 times brighter than the faintest stars visible to the naked eye in full night, the *sixth magnitude stars.* Hence, a magnitude ratio of the fifth root of 100 or 2.512 exists between each magnitude group. The magnitude of a body with an intensity between two groups is denoted by a decimal fraction, as for example, a 2.6-magnitude star. The two brightest stars in the sky, Sirius and Canopus, are actually more than 100 times brighter than a sixth magnitude body. Consequently, they are assigned *negative* magnitudes of –1.6 and –0.9 respectively. In practice, all stars and planets of magnitude 1.5 or greater are collectively referred to as first magnitude bodies. As a point of interest, the moon varies in magnitude from –12.6 to –3.3, and the magnitude of the sun is about –26.7. The planets Venus, Mars, and Mercury also appear at negative magnitudes at certain times of the year.

The patterns in which many of the brighter stars were arranged reminded ancient people of various common terrestrial forms, and it was natural that these patterns or *constellations* became known by the names of the creatures they seemed to resemble, such as Ursa Major (the Big Bear), Scorpio (the Scorpion), and Aries (the Ram). The prominence of the early Greek and Roman mathematicians and astronomers in establishing astronomy as a science is indicated by the fact that most of the constellations as well as most of the individual celestial bodies visible to the naked eye bear names of Greek or Latin derivation.

In celestial navigation, the earth is usually considered to be a perfect sphere suspended motionless at the center of the universe. All heavenly bodies are assumed to be located on a second celestial sphere of infinite radius centered on the center of the earth. This *celestial sphere* is considered to rotate from east to west, with the rotational axis of the sphere concurrent with the axis of the earth; the north pole of the celestial sphere is designated by the abbreviation P_n, while the south pole is denoted by P_s. The sphere completes one rotation with respect to the earth every 24 hours.

To an observer located on the surface of the earth, it appears that all celestial bodies on the celestial sphere rise in the east, follow a circular path across the heavens, and set in the west. The circular path of each body across the heavens is referred to as its *diurnal circle*; if the heavens were photographed at night by a time-lapse camera, each celestial body would subtend a diurnal circle as it rotated across the heavens. Figure 16-1 depicts the celestial sphere and several diurnal circles.

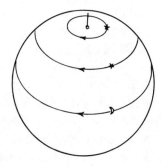

Figure 16-1. *The celestial sphere and several diurnal circles.*

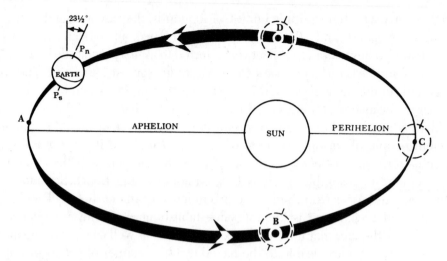

Figure 16-2. *The earth-sun system.*

THE SOLAR SYSTEM

The earth, together with eight other known planets, their moons, and the sun, constitute the *solar system.* Like the other planets of the solar system, the earth follows an elliptical path of revolution about the sun called an *orbit;* it takes about 365 days, or one year, for the earth to complete one circuit around the sun. The earth's axis of rotation is inclined at an angle of 23½° from the perpendicular with respect to the plane of the earth's orbit, called the *plane of the ecliptic.* As will be explained more fully later, this inclination is of special interest as it is primarily responsible for the different seasons experienced on the earth's surface, as well as the ever-changing length of the solar day. Figure 16-2 depicts the relationships between the earth, its orbit, and the sun.

Because of the tilt of the earth's axis with respect to the ecliptic, it seems to an observer on earth as though the sun is continually shifting its position with respect to the earth's equator as the earth moves in its orbit around the sun. When the earth reaches position A in Figure 16-2, the sun appears to be exactly over the 23½° north parallel of latitude. It is at its northernmost point with respect to the earth. This time, called the *summer solstice* in the northern hemisphere, occurs about June 21 of each year, about 12 days before *aphelion,* the orbital point farthest from the sun. Because the angle of incidence of the sun's rays is relatively large at this time in the northern hemisphere and relatively small in the southern hemisphere, the northern hemisphere experiences summer and the southern hemisphere winter at this point in the earth's orbit. As the earth continues in its orbit toward point B, the sun seems to move ever lower, until at position B in the figure it appears to cross over the equator from north to south latitudes. This time is referred to as the *autumnal equinox* in the northern hemisphere, and occurs about September 22. When the earth has proceeded to point C, the sun is at its extreme southern limit, 23½° south latitude; this time, which occurs about December 21, is termed the *winter solstice,* and precedes *perihelion,* the earth's closest approach to the sun, by about 12 days. The northern hemisphere is now experiencing winter, while in the southern hemisphere it is summer. Finally, the sun again crosses the equator at point D, this time from south to north latitudes, about March 21. This event is called the *vernal equinox,* and as noted below, it is of special significance in celestial navigation.

As the early astronomers observed the sun in the course of the earth's revolution about it, it seemed to them as though the sun were an independent ball moving across the heavenly sphere, transiting twelve different constellations in the course of a year. These twelve constellations comprise the *zodiac.* Although with one important exception the zodiac has no navigational significance in modern times, it continues to be of importance even today in the practice of astrology.

When they observed the sun at the time of the vernal equinox, it seemed to the ancients as though the sun were located at a point within the constellation Aries. They chose this point, the *First Point of Aries,* abbreviated by the symbol ♈, as a reference point by which all bodies on the celestial sphere could be located. Although the sun today no longer lies in Aries at the vernal equinox as a result of the effects of the earth's precession, the First Point of Aries continues to be used as the fixed reference point for the location of bodies on the celestial sphere. The sun now appears to be in the constellation Pisces at the time of the vernal equinox.

With the exception of the outermost planet, Pluto, the other known planets of the solar system also revolve about the sun in about the same plane as the plane of the earth's orbit. Although they move at different speeds, all the planets revolve about the sun in the same direction. Mercury, the planet nearest the sun, and Venus, the second planet, are often referred to in celestial navigation as the *inferior planets,* as their orbits lie inside that of the earth, while Mars, Jupiter, and Saturn, the fourth, fifth, and sixth planets, are called the *superior planets* because they are the visible planets whose orbits are outside that of the earth. These three planets, together with Venus, are sometimes referred to as the *navigational planets.* Uranus, Neptune, and Pluto, the seventh, eighth, and ninth planets, are so far distant from the sun as to be of insufficient magnitude for navigational purposes most of the year.

Because of the relative closeness of the other planets of the solar system, when they are viewed from earth they appear to move across the background of the unchanging stars and constellations. As will be discussed later, this motion necessitates the determination of the position of the visible planets and the moon in relation to the stars for each occasion on which they are used for celestial navigation purposes. Because the planets are illuminated by the reflected light of the sun, the inferior planets go through phases similar to the earth's moon, being "full" when on the opposite side of the sun from the earth, and "new" when on the same side. As the superior planets never pass between the earth and the sun, they never appear in the "new" phase.

THE MOON

Just as the earth revolves about the sun, the moon revolves once about the earth in a period of about 30 days. As mentioned in chapter 11 in connection with tide, the moon revolves about the earth in the same direction as the earth rotates, thereby resulting in a time period of about 24 hours and 50 minutes for the earth to complete one rotation with respect to the moon. Since the moon itself completes one rotation every 30 days, the same side of the moon is always facing toward earth. Because the moon, like the planets, is illuminated by the sun, the size and shape of the visible portion of the moon as viewed from the earth is constantly changing, as shown in Figure 16-3. The names given to the various visible portions or *phases* of the moon appear in the figure.

Occasionally the moon passes directly between the sun and the earth, thereby casting a shadow across the surface of the earth, as illustrated in Figure 16-4A. When viewed from within the shadow zone, it seems to the observer as though the sun disappears behind the superimposed moon. This re-

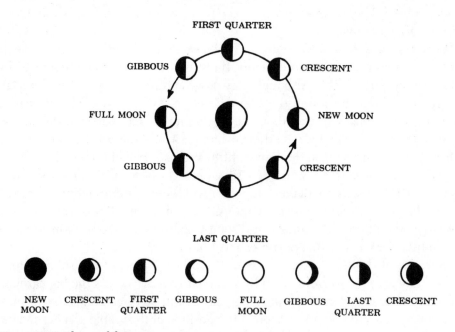

Figure 16-3. *Phases of the moon.*

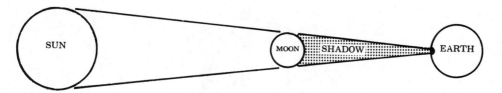

Figure 16-4A. *The earth-sun-moon system during a solar eclipse.*

sult, termed a *solar eclipse,* is of great interest to solar scientists, as it provides an opportunity to study the sun's atmosphere or corona while the moon blocks the sun's surface from view. Locations on the borderline of the shadow zone on the earth's surface experience only a partial blockage of the sun's disk, a *partial solar eclipse.*

At other rather infrequent times, the moon's orbit takes it through the conical shadow zone cast by the earth, thereby creating a *lunar eclipse* visible to observers on the dark side of the earth. This phenomenon has little scientific value, but is interesting to observe. The position of the earth-sun-moon system during a lunar eclipse is illustrated in Figure 16-4B.

REVIEW OF THE TERRESTRIAL COORDINATE SYSTEM

As the celestial coordinate system is based in large measure upon the terrestrial coordinate system, it is necessary to review briefly the terrestrial system introduced in chapter 4 before proceeding with a discussion of the celestial system. The terrestrial system, it may be recalled, is based on the equator and the prime meridian. The *equator* is defined as that great circle on the earth's surface formed by passing a plane perpendicular to the earth's axis midway between the poles. A *meridian* is any great circle on the earth's surface formed by passing a plane containing the earth's axis through the earth perpendicular to the equator: the *prime meridian* is the upper branch of the meridian that passes from pole to pole through the site of the Royal Observatory at Greenwich, England. The location of any position on the earth is specified by stating its location relative to the equator and the prime meridian. The *latitude* of a position is defined as the angular distance measured from the equator north-

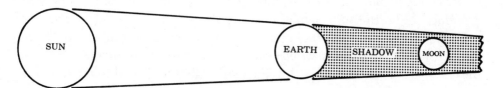

Figure 16-4B. *The earth-sun-moon system during a lunar eclipse.*

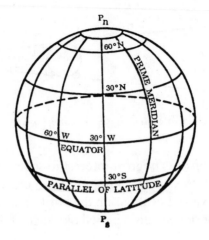

Figure 16-5. *The terrestrial coordinate system.*

ward or southward through 90°; the direction of measurement is indicated by placing a suffix N (north) or S (south) after the angular measure. The *longitude* of a position is the angular distance measured from the prime meridian eastward or westward through 180°; the direction of measurement is indicated by placing a suffix E (east) or W (west) after the angular measure. Each degree of latitude or longitude is subdivided into 60 minutes, and each minute is further subdivided into either 60 seconds or tenths of a minute. Figure 16-5 depicts the equator, prime meridian, and several parallels of latitude and meridians.

As a final note, it was shown in chapter 4 that every degree of arc of a great circle drawn upon the earth's surface is assumed for navigational purposes to subtend 60 nautical miles of distance, and every minute of such a great circle arc therefore subtends 1 nautical mile. This assumption is of great importance in celestial navigation, for reasons to be presented shortly.

THE CELESTIAL COORDINATE SYSTEM

Just as any position on the earth can be located by specifying its terrestrial coordinates, any heavenly body on the celestial sphere can be located by specifying its *celestial coordinates.* To form the *celestial coordinate system,* the terrestrial equator is projected outward onto the celestial sphere to form the *celestial equator,* sometimes called the *equinoctial.* The celestial equator is the reference for north-south angular measurements on the celestial sphere.

In similar fashion, terrestrial meridians can be projected outward to the celestial sphere to form *celestial meridians.* Because of the apparent rotation of the celestial sphere with respect to the earth, these projected celestial meridians appear to sweep continuously across the inner surface of the sphere, making them inconvenient to use as a basis for lateral measurements of position on the celestial sphere. Hence, a separate set of great circles are "inscribed" on the surface of the celestial sphere perpendicular to the celestial equator for use in describing the position of one point on the sphere relative to another. These great circles, called *hour circles,* are defined as follows:

An *hour circle* is a great circle on the celestial sphere perpendicular to the celestial equator and passing through both celestial poles.

Every point on the celestial sphere has an hour circle passing through it. Just as the meridian passing through the observatory at Greenwich, England, was chosen as the reference for the lateral coordinate of a point on the terrestrial sphere, the hour circle passing through the First Point of Aries (♈) forms the reference for the lateral coordinate of a point on the celestial sphere. This hour circle is usually referred to simply as the "hour circle of Aries."

The celestial equivalent of terrestrial latitude is *declination,* abbreviated *dec.*; its definition follows:

Declination is the angular distance of a point on the celestial sphere north or south of the celestial equator, measured through 90°.

Declination is labeled with the prefix N (north) or S (south) to indicate the direction of measurement; prefixes are used to differentiate declination from latitude. Figure 16-6A depicts the declination of a star located 30° north of the celestial equator; the terrestrial sphere is printed in black, while the celestial sphere is shown in blue.

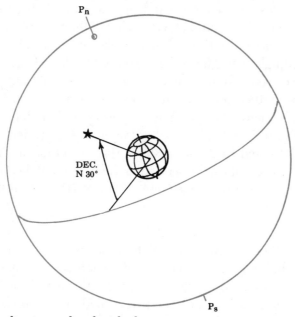

Figure 16-6A. *Declination on the celestial sphere.*

The celestial equivalent of longitude is the *hour angle*; it is defined below:

The *hour angle* is an angular distance measured laterally along the celestial equator in a westerly direction through 360°.

If it is desired to locate a body on the celestial sphere relative to the location of Aries, the hour circle of Aries is used as a reference for the measurement of the hour angle. Such hour angles, measured in a westerly direction from the hour circle of Aries to the hour circle of the body, are called *sidereal hour angles*, abbreviated *SHA*.

For the purposes of celestial navigation, it is not only desirable to locate a body on the celestial sphere relative to Aries, but also it is desirable to locate a body relative to a given position on earth at a given time. To do this, two terrestrial meridians are projected onto the surface of the celestial sphere for use as references for hour angle measurements—the Greenwich meridian and the observer's meridian. The celestial meridians thus projected are termed the *Greenwich celestial meridian* and the *local celestial meridian*, respectively. Hour angles measured relative to the Greenwich celestial meridian are called *Greenwich hour angles*, abbreviated *GHA*, while those measured relative to the local celestial meridian are termed *local hour angles*, abbreviated *LHA*. Both Greenwich hour angles and local hour angles are measured westward from a projected terrestrial meridian to a celestial hour circle moving ever westerly with the rotating celestial sphere. Consequently, both GHA and LHA values grow increasingly larger with time, increasing from 0° to 360° once each 24 hours; they relate the rotating celestial sphere to the meridians of the earth. Sidereal hour angles, on the other hand, are measured between two hour circles on the celestial sphere; although the value of the SHA of the stars changes with time as the stars move through space relative to one another, the rate of change is exceedingly slow. Hence for purposes of celestial navigation, sidereal hour angles are considered to remain constant over a significant period of time.

In Figure 16-6B, the hour circle of Aries and the projected Greenwich and observer's celestial meridians are shown superimposed in blue upon the celestial sphere of Figure 16-6A. The resulting sidereal, Greenwich, and local hour angles (SHA, GHA, and LHA) of the star at this given time are indicated.

As a point of interest, it can be seen from the figure that the GHA of the star (GHA☆) is equal to the sum of the GHA of Aries (GHA♈) plus the SHA of the star (SHA☆):

$$\text{GHA}☆ = \text{GHA}♈ + \text{SHA}☆$$

The importance of this relationship will become apparent later.

For some applications in celestial navigation, it is advantageous to use an alternative angle to LHA to express the angular distance from the observer's meridian to the hour circle of a body. This is called the *meridian angle*, by convention abbreviated with a lowercase *t*. It is defined as follows:

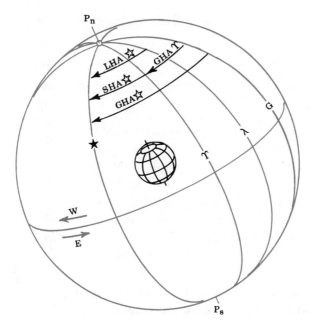

Figure 16-6B. *SHA, GHA, and LHA of a star on the celestial sphere.*

The *meridian angle t* is the angular distance between 0° and 180°, measured at the pole nearest the observer, from the observer's meridian either easterly or westerly to the hour circle of the body.

Because it can be measured either easterly or westerly from the observer's meridian, depending on which direction will result in an angle less than 180°, the meridian angle t is always labeled with the suffix E (east) or W (west) to indicate the direction of measurement; for example, in Figure 16-6B, the meridian angle t might be about 30°W. More will be said about the meridian angle later in connection with the solution of the celestial triangle.

Astronomers use a different coordinate called *right ascension* (RA) to locate bodies on the celestial sphere. Right ascension is similar to SHA, but it is measured easterly from the hour circle of Aries, and is expressed in time units (hours) rather than arc. Right ascension finds only limited application in the practice of celestial navigation.

THE HORIZON COORDINATE SYSTEM

In order to obtain a celestial line of position by observation of a celestial body, a third set of coordinates, called the *horizon system,* is required. It differs from the celestial coordinate system described above in that it is based on the position of the observer, rather than on the projected terrestrial equator and poles. The reference plane of the horizon system corresponding with the plane of the equator in the terrestrial and celestial systems is the observer's celestial horizon, defined as follows:

> The *celestial horizon* is a plane passing through the center of the earth, perpendicular to a line passing through the observer's position and the earth's center.

The line mentioned in the definition, extended outward from the observer to the celestial sphere, defines a point on the sphere directly over the observer called the *zenith*. The observer's zenith is always exactly 90° of arc above the celestial horizon. The extension of the line through the center of the earth to the opposite side of the celestial sphere defines a second point directly beneath the observer called the *nadir*. The observer's zenith and nadir correspond to the terrestrial and celestial poles, while the *zenith-nadir line* connecting the observer's zenith and nadir corresponds to the axis of the terrestrial and celestial spheres.

The equivalent of a meridian in the terrestrial system and an hour circle in the celestial system in the horizon system is the *vertical circle*:

> A *vertical circle* is a great circle on the celestial sphere passing through the observer's zenith and nadir, perpendicular to the plane of the celestial horizon.

The vertical circle passing through the east and west points of the observer's horizon is termed the *prime vertical,* and the vertical circle passing through the north and south points is the *principal vertical.* The observer's principal vertical is always coincident with the projected terrestrial meridian (i.e., the local celestial meridian) passing through his or her position.

The equivalent of latitude in the horizon system is *altitude,* defined as follows:

> *Altitude* is the angular distance of a point on the celestial sphere above a designated reference horizon, measured along the vertical circle passing through the point.

The reference horizon for the horizon coordinate system is the celestial horizon of the observer, defined above as the plane passing through the center of the earth perpendicular to the zenith-nadir line of the observer. Altitude measured relative to the celestial horizon is termed *observed altitude,* abbreviated *Ho.* The observed altitude of a celestial body can be defined as the angle formed at the center of the earth between the line of sight to the body and the plane of the observer's celestial horizon. The other horizon used as a reference for altitude measurements is the observer's *visible* or *sea horizon,* the line along which sea and sky appear to meet. This is the reference horizon from which altitudes are measured by the sextant: such altitudes are called *sextant altitudes,*

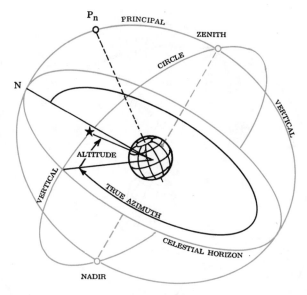

Figure 16-7. *Observed altitude and true azimuth of a star in the horizon coordinate system.*

abbreviated hs. In practice, sextant altitudes must be converted to observed altitudes (*Ho*) to obtain an accurate celestial LOP; this conversion procedure will be explained in chapter 19.

The equivalent of longitude in the horizon system is *true azimuth,* abbreviated *Zn*:

> *True azimuth* is the horizontal angle measured along the celestial horizon in a clockwise direction from 000°T to 360°T from the principal vertical circle to the vertical circle passing through a given point or body on the celestial sphere.

True azimuth can be thought of simply as the true bearing of a celestial body from the observer's position.

Figure 16-7 illustrates the horizon system of coordinates in red, and depicts the observed altitude and true azimuth of a star.

THE CELESTIAL TRIANGLE

For the purposes of celestial navigation, the terrestrial, celestial, and horizon coordinate systems are combined on the celestial sphere to form the *astronomical* or *celestial triangle.* When this triangle is related to the earth it becomes the *navigational triangle,* the solution of which is the basis of celestial navigation.

Each of the three coordinate systems examined in the foregoing sections of this chapter is used to form one of the sides of the celestial triangle. The three

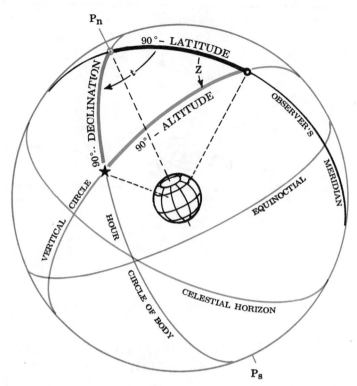

Figure 16-8. *Astronomical or celestial triangle.*

vertices of the triangle are the celestial pole nearest the observer, the observer's zenith, and the position of the celestial body. To illustrate, consider the celestial triangle depicted in Figure 16-8; in the figure, each of the coordinate systems discussed previously has been superimposed on the celestial sphere to form the celestial triangle, with the terrestrial system shown in black, the celestial system in blue, and the horizon system in red. The side of the triangle connecting the celestial pole with the observer's zenith is a segment of a projected terrestrial meridian; the side between the pole and the celestial body is a segment of the celestial hour circle of the body; and the side of the triangle between the observer's zenith and the position of the body is a segment of a vertical circle of the horizon system. In Figure 16-8, both the observer's zenith and the star being observed are located on the northern half of the celestial sphere. Other possible relationships will be discussed later.

In celestial navigation, the lengths of the sides of the celestial (later the navigational) triangle are of paramount importance. In Figure 16-8, the length of the side formed by the projected terrestrial meridian between the north celestial pole P_n and the observer's zenith expressed as an angle is 90° minus the observer's latitude. The length of the side concurrent with the hour circle of the body is in this case 90° minus the declination, but if the body were south of the celestial equator, this length would be 90° plus the declination. The length of the third side, which is measured along the vertical circle from the observer's zenith to the body, is 90° (the altitude of the zenith) minus the altitude of the body. Figure 16-8 should be studied carefully in order to fix these relationships in mind.

Only two of the angles within the celestial triangle are of concern in celestial navigation. The angle marked t in the figure at the north celestial pole is the *meridian angle,* defined earlier as the angle measured east or west from the observer's celestial meridian to the hour circle of the body. The meridian angle, like longitude, is measured from 0° to 180°, and is labeled with the suffix E (east) or W (west) to indicate the direction of measurement. The meridian angle bears a close relationship to the local hour angle (LHA). If the LHA is less than 180°, the meridian angle t is equal to the LHA, and is west. If the LHA is greater than 180°, the meridian angle is equal to 360° minus the LHA, and is east.

The other angle of importance within the celestial triangle is the *azimuth angle,* which is located at the observer's zenith. It is abbreviated Z. The azimuth angle is defined as the angle at the zenith between the projected celestial meridian of the observer and the vertical circle passing through the body; by convention, it is measured from 0° to 180° either east or west of the observer's meridian. It is important to distinguish this azimuth angle of the celestial triangle from the true azimuth of the observed body, which, as mentioned earlier, can be likened to the true bearing of the body from the observer. Azimuth in this latter sense is always abbreviated Zn, and with altitude forms the two horizon coordinates by which a celestial body is located with respect to an observer on the earth's surface.

The third interior angle of the celestial triangle is called the *parallactic angle*; it is not used in the ordinary practice of celestial navigation, and will not be referred to henceforth.

THE NAVIGATIONAL TRIANGLE

In the determination of position by celestial navigation, the celestial triangle just described must be solved to find a *celestial line of position* passing through the observer's position beneath the zenith. While it would be possible to solve the celestial triangle by spherical trigonometric methods, in the usual practice of celestial navigation the solution is simplified by the construction of a second closely related *navigational triangle.* To form this triangle, the observer is imagined to be located at the center of the earth, and the earth's surface is expanded outward (or the celestial sphere compressed inward) until the surface of the earth and the surface of the celestial sphere are coincident.

After the earth's surface has been thus expanded (or the celestial sphere thus compressed), the position of the celestial body being observed becomes the

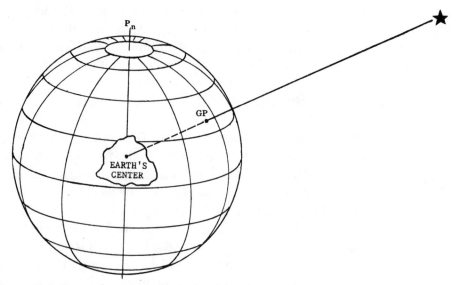

Figure 16-9. *Geographic position (GP) of a celestial body.*

geographic position (GP) of the body on the earth's surface. The concept of this geographic position or GP of a celestial body on the earth's surface is important and should be completely understood. Every celestial body has a GP located on the earth's surface directly beneath it, as shown in Figure 16-9. As the celestial sphere rotates about the earth, all geographic positions of celestial bodies move from east to west across the earth's surface. The GP of the sun is sometimes called the *subsolar point,* the GP of the moon the *sublunar point,* and the GP of a star the *substellar point.* In every case, the diameter of the body is considered compressed to a point on the celestial sphere, located at the center of the body. The GP of the observed celestial body forms one vertex of the navigational triangle.

Since the coordinates of the observer's position on earth are not known, but rather are to be determined, the zenith of the observer in the celestial triangle becomes a hypothesized *assumed position* (AP) of the observer in the navigational triangle. This assumed position or AP forms a second vertex of the navigational triangle; the procedure for selection of the coordinates of the assumed position will be presented later.

The remaining vertex of the navigational triangle, the celestial pole, is termed the *elevated pole,* abbreviated P_n or P_s. The elevated pole is always the pole nearest the observer's assumed position; it is so named because it is the celestial pole above the observer's celestial horizon. Thus, the assumed position of the observer and the elevated pole are always on the same side of the celestial equator, while the geographic position of the body observed may be on either side. A sample navigational triangle appears in Figure 16-10.

292

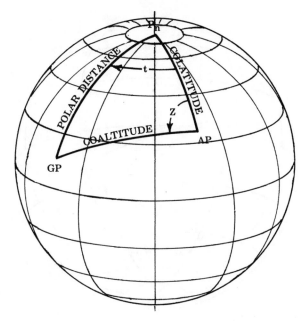

Figure 16-10. *The navigational triangle.*

Inasmuch as the positions of the AP of the observer and GP of the celestial body are unlimited, it follows that an infinite number of shapes are possible for the navigational triangle.

The three sides of the navigational triangle are called the *colatitude,* the *coaltitude,* and the *polar distance.* They are defined as follows:

The *colatitude* is the side of the navigational triangle joining the AP of the observer and the elevated pole. Since the AP is always in the same hemisphere as the elevated pole, the length of the colatitude is always 90° (the latitude of the pole) minus the latitude of the AP.

The *coaltitude,* sometimes called the *zenith distance,* is the side of the navigational triangle joining the AP of the observer and the GP of the body. Inasmuch as the maximum possible altitude of any celestial body relative to the observer's celestial horizon is 90°, the altitude of the zenith, the length of the coaltitude is always 90° minus the altitude of the body.

The *polar distance* is the side of the navigational triangle joining the elevated pole and the GP of the body. For a body in the same hemisphere, the length of the polar distance is 90° minus the declination of the GP; for a body in the opposite hemisphere, its length is 90° plus the declination of the GP.

The interior angles of the navigational triangle bear the same names as the corresponding angles of the celestial triangle, with only the *meridian angle t* at

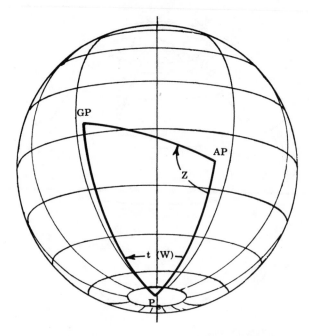

Figure 16-11A. *A navigational triangle with the south pole the elevated pole.*

the elevated pole and the *azimuth angle* Z at the AP of the observer being of any consequence in the solution of the triangle. The measurement and labeling of the meridian angle was discussed in the preceding section. The azimuth angle is always measured from the observer's meridian toward the vertical circle joining the observer's AP and the GP of the body. Since the angle between the observer's meridian and the vertical circle of the body can never exceed 180°, the azimuth angle must always have a value between 0° to 180°. It is labeled with the prefix N (north) or S (south) to agree with the elevated pole, and with the suffix E (east) or W (west) to indicate on which side of the observer's meridian the GP lies. Labeling the azimuth angle in this manner is important, as it may be measured with reference to either the north or south poles, and in either an easterly or westerly direction. It may be recalled here that the meridian angle *t* is also measured either eastward or westward from the observer's meridian to the hour circle of the body, from 0° to 180°. Thus, the suffix of the meridian angle *t* and the suffix of the azimuth angle Z will always be identical.

   In the ordinary practice of celestial navigation it is necessary to convert the azimuth angle of the navigational triangle to the true azimuth or bearing of the GP of the body from the AP of the observer. As an example of the conversion process, consider the navigational triangle in Figure 16-11A in which the south pole is the elevated pole. In this figure, the south pole is the elevated pole be-

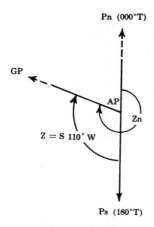

Figure 16-11B. *Directional relationships for an azimuth angle.*

cause the AP of the observer is located in the southern hemisphere. At the time of the observation of the body, its GP has been determined to be north of the equator and west of the observer. Thus, the prefix for the azimuth angle Z in this case is S (south), to agree with the elevated pole, and the suffix is W (west), identical with the suffix of the meridian angle. Hence, if the size of the azimuth angle were 110°, the angle would be written S 110°W. To convert this azimuth angle to a true azimuth, it is helpful to draw a sketch of the directional relationships involved similar to that in Figure 16-11B. From the figure, it should be obvious that to convert the azimuth angle S 110°W to true azimuth, it is necessary only to add 180°. Thus, the true azimuth or bearing of the GP from the AP in this case is 180° + 110° = 290°T.

In practice, an alternative method of conversion is frequently used that makes use of a set of rules based on the relationship of the azimuth angle with the latitude of the observer and the local hour angle of the body. This method will be discussed in a later chapter.

A complete understanding of the celestial and navigational triangles is essential in celestial navigation. By solution of the appropriate triangle, the navigator can determine the ship's position at sea, check compass accuracy, predict the rising and setting of any celestial body, locate and identify any bodies of interest, and much more. It can well be stated that the solution of the celestial and navigational triangle *is* celestial navigation.

THE CIRCLE OF EQUAL ALTITUDE

To illustrate the basic concepts involved in obtaining a celestial line of position, suppose that a steel pole perpendicular to a level surface were raised, and a wire stretched from its top to the surface, such that the angle formed by the

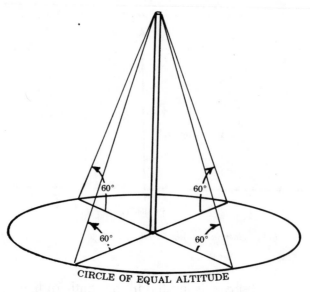

Figure 16-12A. *A circle of equal altitude inscribed about a pole.*

wire and the surface was 60°. If the end of the wire were rotated around the base of the pole, a circle would be described, as shown in Figure 16-12A. At any point on this circle, the angle between the wire and the surface would be the same, 60°. Such a circle is termed a *circle of equal altitude*.

Now, suppose that the end of the pole were extended to an infinite distance. The angle formed by the wire anywhere on the flat surface would approach 90°, since the wire would be nearly parallel with the pole. If the surface were spherical, however, and the measurement made relative to a tangent plane, the angle would vary from 90° at the base of the pole to 0° at all points on the spherical surface 90° away from the location of the base. Figure 16-12B depicts two concentric circles of equal altitude inscribed on such a spherical surface, centered on the location of the base—the "GP"—of the pole. At all points on the circumference of the smaller circle, the angle formed by the wire and the tangent plane is 60°, while the angles measured along the larger circle are all 30°.

In celestial navigation, the situation is analogous to that depicted in Figure 16-12B. Suppose that a celestial body was observed and found to have an altitude of 60° above the observer's celestial horizon, and its GP at the moment of the observation was determined to have been located at 10° south latitude, 30° west longitude. Suppose further that the AP of the observer determined perhaps from the DR plot was 10° north latitude, 10° west longitude. Assuming a globe of sufficient scale were available, a navigational triangle similar to the one in Figure 16-13A could be constructed by plotting the coordinates of the GP and AP.

Now, let us plot a circle of equal altitude about the GP of the body, from which circle an altitude of 60° could be observed. To do this, let us assume for

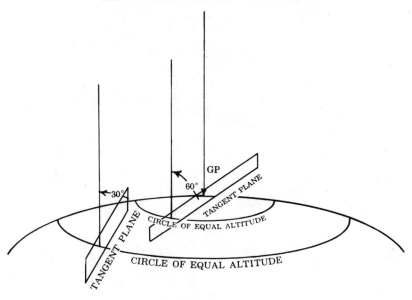

Figure 16-12B. *Two concentric circles of equal altitude.*

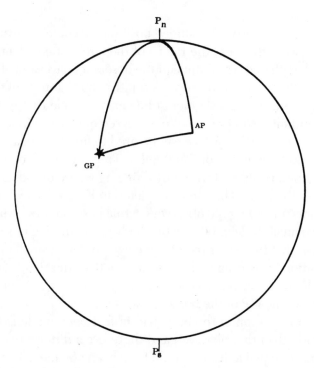

Figure 16-13A. *A navigational triangle.*

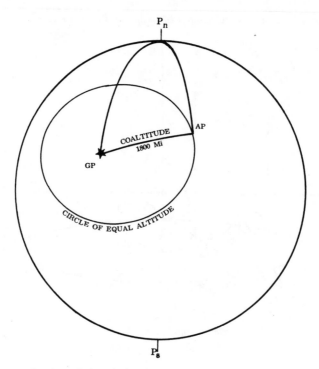

Figure 16-13B. *Circle of equal altitude for Ho 60°.*

the moment that the observer's assumed position AP coincided with the actual position. If this were the case, the radius of the circle of equal altitude would have the same length as the coaltitude of the observer's navigational triangle; expressed as an angle, this is 90° (observer's zenith) –60° (altitude of the body), or 30°. Since the coaltitude is a segment of a vertical circle that is itself a great circle, the important assumption that one degree of a great circle subtends 60 nautical miles on the earth's surface can be used to find the linear length of the coaltitude; it is 30° × 60 miles, or 1,800 miles. Thus, the circle of equal altitude for this observation could be formed by swinging an arc of radius 1,800 miles about the GP of the body. This circle, depicted in Figure 16-13B, represents a locus of all points, including the observer's actual position, from which it is possible to observe an altitude of 60° for this body at the time of observation.

Now suppose that the AP of the observer was in fact a small distance from the actual position, as is usually the case in practice. In this situation, the AP will probably lie off the circle as shown in Figure 16-13C, either closer to or farther away from the GP of the body.

Thus, if the observation of the body had in fact been made from the AP, a different altitude than that observed and therefore a different coaltitude (radius of the circle of equal altitude) would have been obtained. The significance of this difference in the lengths of the two coaltitudes corresponding with the

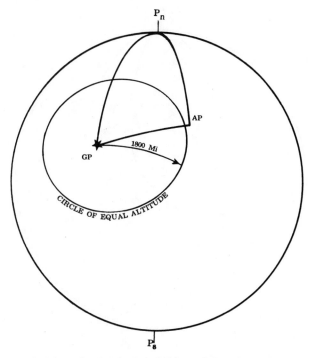

Figure 16-13C. *An AP not on the observed circle of equal altitude.*

observer's actual observation and the assumed position will become apparent in chapter 17.

To find his or her exact position, that is, the *celestial fix*, the observer could observe a second celestial body and plot a second circle of equal altitude around its GP. Normally the two circles would intersect in two places; the observer's position must be located at one of the intersections, probably the one nearest the AP of the observer, as depicted in Figure 16-13D.

If the two intersection points were so close together as to cause confusion, the doubt could be resolved by observing a third body and plotting a third circle of equal altitude.

Because of the large-scale globe or chart that would have to be used to plot a celestial fix of meaningful accuracy by finding the intersection of two or more complete circles of equal altitude, this method is not normally used in the practice of celestial navigation. For bodies of low altitude, the length of the coaltitude is prohibitive. If the altitude of a body were 20°, for example, the coaltitude would be $(90° - 20°) \times 60$, or 4,200 miles long. To obtain any meaningful degree of accuracy, a chart hundreds of feet in length and width would have to be used to plot the resulting circle. Bodies of extremely high altitude are difficult to observe with accurate results using the marine sextant. An alternative method of plotting the celestial LOP, therefore, has been devised wherein only

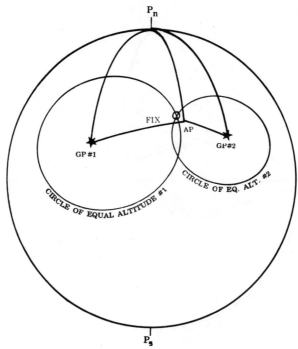

Figure 16-13D. *Determining a celestial fix using two circles of equal altitude.*

a small portion of the coaltitude and circle of equal altitude for each body observed is plotted. This method, called the *altitude-intercept method,* is the subject of the following chapter.

SUMMARY

This chapter has set forth the basic theory upon which the remaining chapters of this book on the practice of celestial navigation are based. The earth's relationship to the universe and the solar system of which it is a part were examined in some detail, and the celestial and horizon systems of coordinates were developed. Finally, the terrestrial, celestial, and horizon systems were combined to form the celestial triangle, from which the navigational triangle is derived. The solution of the navigational triangle is basic to the practice of celestial navigation. Succeeding chapters will deal in detail with various aspects of the use of the navigational triangle in solving problems of particular interest to the navigator.

THE ALTITUDE-INTERCEPT
METHOD

17

As discussed in the last chapter, a celestial line of position can be produced from an observation of a celestial body by plotting a circle on the surface of the earth with radius equal to the coaltitude of the body, centered on the geographical position (GP) of the body at the time of the observation. A celestial LOP of this type, termed a *circle of equal altitude,* is considered impractical for bodies of altitude less than about 87°, because of the unwieldy length of the resulting coaltitude radius.

The *altitude-intercept method* is an alternative method of plotting the celestial LOP that eliminates the disadvantages of the circle of equal altitude. In the manual practice of celestial navigation, this method normally utilizes daily data tabulated in either one of two almanacs, the *Nautical Almanac* or the *Air Almanac,* either in conjunction with one of several different *sight reduction tables* or an electronic calculator to produce a computed altitude Hc to a body being observed from an assumed position (AP) of the observer. The computed altitude Hc is then compared with the observed altitude Ho to determine the position of the celestial LOP.

If a navigation calculator or computer program is being utilized, the almanac data are resident in memory, and the celestial LOP can be calculated without further reference to any almanac or sight reduction tables. This electronic sight reduction is further discussed in chapter 22.

THE ALTITUDE-INTERCEPT METHOD

The altitude-intercept method of plotting a celestial line of position was developed in 1875 by the Frenchman Marc St. Hilaire as an alternative to more cumbersome methods then in vogue. Originally the method employed two trigonometric equations called the "cosine-haversine" formulas, but because they were rather difficult to solve by manual methods, about 1930 Ogura, a Japanese, developed a more convenient solution based on the use of certain

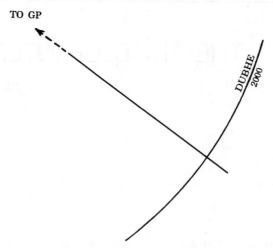

Figure 17-1A. *Segment of a circle of equal altitude, GP nearby.*

sight reduction tables. The use of similar but improved tables predominates to-day, although the advent of the miniature electronic calculator equipped with a capability of handling trigonometric functions has sparked renewed interest in the use of the original cosine-haversine formulas and others derived from them. In its present form, the altitude-intercept method requires little other than the ability to plot on a chart, to look up required data in tables, and to add, subtract, and interpolate between numbers in a column.

In essence, in the altitude-intercept method only small segments of the radius and the circumference of the circle of equal altitude are plotted on a chart or plotting sheet. To do this, an assumed position (AP) for the time of the observation of a celestial body is chosen, and the true azimuth and computed altitude of the body observed are determined for this position by use of one of the sight reduction tables mentioned earlier. A line drawn from the AP of the observer toward the GP of the body in the direction of the true azimuth then represents a segment of the radius of the circle of equal altitude; a segment of the circumference of the circle is positioned along the true azimuth (radius) line by comparing the computed altitude with the observed altitude actually obtained.

Consider for a moment a small segment of the radius and circumference of a circle of equal altitude in a situation in which the AP of an observer is assumed to be fairly close to the GP of the body, as in Figure 17-1A. Since the GP of the body is nearby, the portion of the circumference of the circle of equal altitude in the figure appears curved. If the distance to the GP were gradually increased, the portion of the circumference shown would appear increasingly less curved, finally approaching a straight line as the distance to the GP increased to a few hundred miles or more. This situation is depicted in Fig-

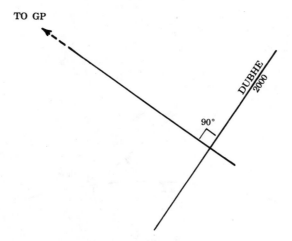

Figure 17-1B. *Segment of a circle of equal altitude, GP distant.*

ure 17-1B. As can be seen from the figure, the circumference of the circle near the radius line has approached a straight line perpendicular to the radius at the point of intersection.

In the altitude-intercept method of plotting the celestial line of position, the LOP is always represented by a short segment of the circle of equal altitude such as that appearing in Figure 17-1B, drawn at right angles to the radius line. Since the radius of the circle of equal altitude always lies in the direction of the GP, the LOP can be thought of as being drawn perpendicular to the bearing or true azimuth (Zn) of the GP of the body. It is the positioning of the LOP along the true azimuth line that constitutes the basic problem solved by the altitude-intercept method.

As was mentioned earlier, the altitude of a given celestial body can be computed for the assumed position (AP) of the observer by use of an almanac in conjunction with a sight reduction table; this computed altitude is abbreviated Hc. If the observed altitude Ho were identical to the calculated altitude Hc, the circles of equal altitude corresponding to the computed and observed altitudes would be coincident and would both pass through the AP of the observer. If the observed altitude Ho were greater than the computed altitude Hc, the radius of the circle of equal altitude corresponding to the observed altitude would be smaller than the radius of the circle for the computed altitude. In this case, the observer must in reality be located closer to the GP of the body than the assumed position. Conversely, if the observed altitude were less than the computed altitude, the observer must be located farther away. Figure 17-2 illustrates the reasoning upon which the foregoing conclusions are based.

In the figure, three segments of circles of equal altitude are depicted, corresponding to an observed altitude Ho greater than, equal to, or less than a com-

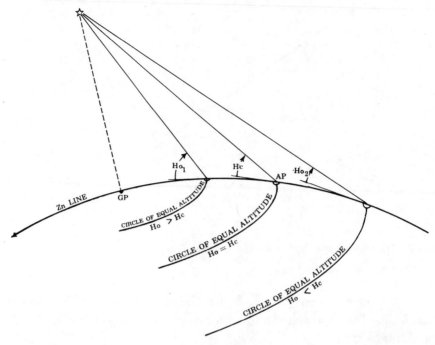

Figure 17-2. *Effect of Ho larger than, equal to, or smaller than Hc. Note: Celestial body shown unrealistically close to earth to emphasize difference in size of angles.*

puted altitude Hc for a given assumed position AP. It should be noted that as the size of the observed altitude increases, the distance from the GP of the body decreases. This is reasonable, since the length of the coaltitude (90° – altitude), which is the radius of the circle of equal altitude, decreases as the altitude increases.

To calculate the distance between the observer and the AP along the true azimuth line, it is only necessary to find the difference between the observed and computed coaltitude distances from the GP of the body. The following algebraic expression represents the distance expressed in degrees of arc between the two coaltitudes:

$$(90° - Ho) - (90° - Hc)$$

where 90° – Ho is the coaltitude of the observed altitude, and 90° – Hc is the coaltitude of the computed altitude. In practice, however, it is not necessary to compute the coaltitudes, since

$$(90° - Ho) - (90° - Hc) = 90° - Ho - 90° + Hc$$
$$= Hc - Ho$$

Thus, to find the distance from the AP along the true azimuth line at which the LOP corresponding with Ho should be drawn, it is only necessary to find the

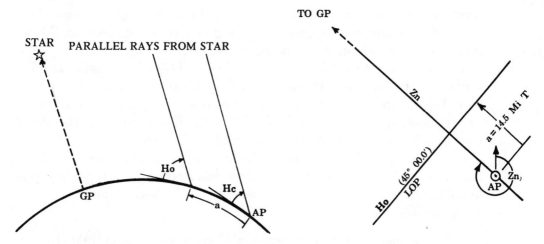

Figure 17-3. *Altitude-intercept method. Ho greater than Hc.*

difference between the computed and observed altitudes. For every minute of arc difference, the intercept point of the LOP with the true azimuth line is moved 1 nautical mile from the AP, either toward the GP of the body if Ho is greater than Hc, or away from the GP if Ho is less than Hc. The distance thus determined between the intercept point and the AP is called the *intercept distance,* symbolized by a lowercase *a.*

Figure 17-3 shows an example of the computation of an intercept distance, in which an observed altitude Ho is greater than the computed altitude Hc.

In this example, a star was observed having an Ho of 45° 00.0′. For an assumed position AP chosen for the time of the observation, an Hc of 44° 45.5′ was computed, as well as the true azimuth of the GP from the AP. To plot the celestial LOP corresponding to the observed altitude, the intercept point must be advanced along the true azimuth line toward the GP of the body (since Ho is greater than Hc) by an intercept distance equal to the arc-difference between Ho and Hc expressed in miles. This intercept distance is given by the following computation:

Ho	45° 00.0′
−Hc	44° 45.5′
a	14.5′T

The computed intercept distance, 14.5 miles in this case, is labeled with the letter *T* to indicate that the intercept point must be moved *toward* the GP of the body. If the intercept distance is to be laid off *away from* the GP, in the case of Ho being less than Hc, the letter *A* is used to indicate this fact. Hence, in this example, the LOP corresponding with the observed Ho is drawn perpendicular to the true azimuth line through a point on the line 14.5 miles toward the GP of the star. Because the resulting LOP represents a portion of the circle of equal altitude along which the angle Ho could be observed, the observer's position must lie somewhere on this LOP.

To find the exact location of the observer on the LOP, the procedure described can be repeated for a second body observed simultaneously with the first. The intersection of the two simultaneous LOPs then defines the observer's position—the celestial fix—at the time of the observation. The plot of a celestial fix and running fix will be the subject of chapter 23.

As a memory aid in determining whether the intercept distance should be laid off toward or away from the GP of the observed body when plotting a celestial LOP by the altitude-intercept method, two phrases have long been used. They are:

"Coast Guard Academy" (Computed Greater Away), and "HoMoTo" (Ho More than Hc, Toward).

In order to achieve accuracy in the plot of the celestial LOP, and to a lesser extent to facilitate the plot by the altitude-intercept method, the assumed position is always chosen in such a way that the intercept distance is kept fairly short. The reason for this is that the true azimuth of the GP of an observed body from the AP of the observer actually lies along a great circle, while the intercept distance laid down on a Mercator chart follows a rhumb line. The longer the intercept distance is, the more the rhumb line representing it on the Mercator projection will diverge from the great circle true azimuth. Thus, an error is introduced into the plot of the celestial LOP, which increases in size in proportion to the length of the intercept distance. For relatively short intercept lengths of 30 miles or less, the error is insignificant, but for intercept distances longer than 60 miles, the effect of the error is considerable.

It may be apparent by now that the altitude-intercept method of plotting the celestial line of position takes its name from the fact that a computed *altitude* is first determined for an assumed position, then compared with the observed altitude to find the *intercept* distance a. The following section will briefly describe how the assumed position is chosen, and how the computed altitude and true azimuth are obtained for it.

THE DETERMINATION OF THE ASSUMED POSITION, Hc, AND Zn

The altitude-intercept method of solving the navigational triangle and plotting the resulting celestial line of position depends on the ability of the observer to determine the computed altitude Hc and the true azimuth Zn for an observed celestial body from a selected assumed position at the time of the observation. As was stated in the introduction to this chapter, to do this determination manually the navigator usually employs one of two almanacs in conjunction with one of several different sight reduction tables. In chapter 20, the use of the *Nautical Almanac* in conjunction with the *Sight Reduction Tables for Marine Navigation, No. 229,* is described, and in chapter 21, the use of the *Air Almanac* with the *Sight Reduction Tables for Air Navigation, No. 249,* will be demonstrated. Chapter 22 describes the use of a concise sight reduction table included in the *Nautical Almanac.*

The coordinates of the assumed position AP are chosen to provide the proper values of the quantities required for entering arguments in the sight reduction tables used, consistent with the restraint that the intercept distance from the AP to the resulting celestial LOP should be as short as possible. The rules for the selection of the assumed latitude and longitude will be presented in chapters 20 and 21.

PLOTTING THE CELESTIAL LINE OF POSITION

After the assumed position, intercept distance, and true azimuth have been determined by the solution of the navigational triangle, the celestial LOP can be plotted. Although the plot could be done on a fairly large-scale "working" chart of the area in which the ship is operating, it is the usual procedure to use a scaled *plotting sheet* for this purpose. The DMA Hydrographic/Topographic Center makes available a Large Area Plotting Sheet, or LAPS, covering several degrees of latitude and longitude, as well as a Small Area Plotting Sheet (SAPS), which covers two degrees of latitude and longitude. Only the latitude lines are labeled, with the labeling of the meridians left to the user. Each sheet can be used for either north or south latitudes by inverting the sheet. A typical SAPS centered on latitude 31° (SAPS-31) appears in Figure 17-4.

To plot a celestial line of position on a plotting sheet, a suitable SAPS for the latitudes involved is first selected, and the longitudes are labeled, increasing from right to left if in west longitudes, and from left to right if in east longitudes. Once the sheet has been properly prepared, the assumed position is plotted on it. From this AP, a dashed line is drawn either in the direction of the true azimuth Zn for "Toward" intercepts, or 180° away from the Zn in the case of "Away" intercepts. The length of the intercept is then laid off along this line, and a point is inscribed on the line to indicate the length of the intercept distance. To complete the plot, the LOP is drawn through this point, perpendicular to the intercept line.

As an example of a plot of a typical celestial LOP, suppose that for an observation of the planet Venus an assumed position at L 34°S, λ 163° 08.4′E had been selected, and the intercept distance and the true azimuth Zn were deter-

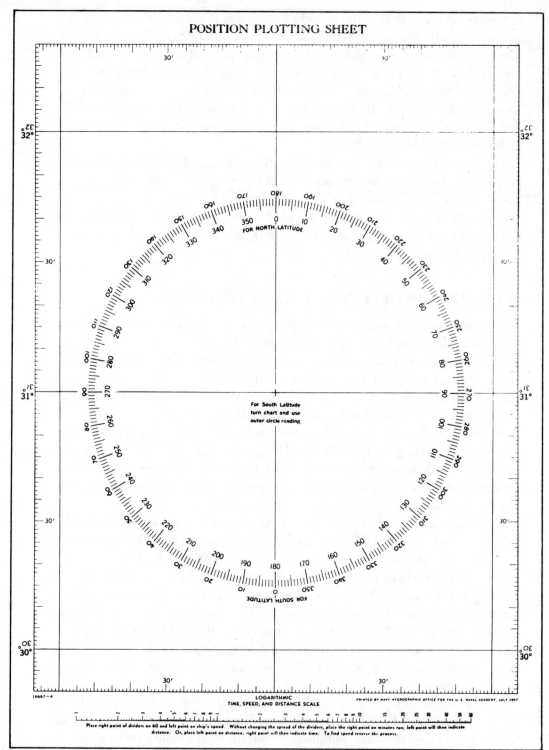

Figure 17-4. SAPS-31.

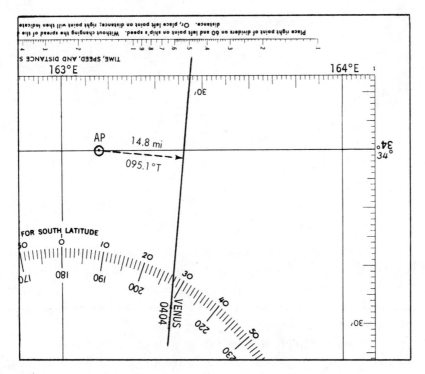

Figure 17-5. *Plot of a Venus LOP.*

mined to be 14.8 miles "toward," and 095.1°T, respectively. The completed plot appears in Figure 17-5. Note that the LOP is labeled with the name of the body and the zone time of the observation to the nearest minute.

The procedure for plotting several celestial LOPs on the same plotting sheet to obtain a celestial fix or running fix will be demonstrated in chapter 23. In practice, after the fix has been plotted on the plotting sheet, its coordinates are then picked off by a pair of dividers or a drawing compass and the fix position is shifted onto the actual working chart.

SUMMARY

The altitude-intercept method of solving the navigational triangle and plotting the resulting celestial LOP is of great use to the navigator engaged in the practice of celestial navigation at sea. After the assumed position has been selected and the computed altitude and true azimuth determined by the use of an almanac in conjunction with one of the several sight reduction tables, the celestial LOP is easily plotted either on an appropriate chart or plotting sheet. Once plotted, two or more celestial LOPs may be combined to form a fix or running fix. If a plotting sheet has been used for the plot, the coordinates of the fix are then transferred to the "working" chart of the area.

TIME

18

The subject of time was introduced in chapter 15 in connection with voyage planning. The fundamental concepts of time were set forth, including the reckoning of time according to two types of mean solar time: Greenwich Mean Time and local zone time. The procedures for time conversions from one time zone to another were also examined in some detail.

This chapter will complete the discussion of time, emphasizing its use and applications during the practice of celestial navigation. The first section will reexamine the various bases by which time is reckoned, including a hitherto unmentioned reference for the passage of time. Next, the relationship between time and longitude will be discussed, and an aid in visualizing these and other time relationships called the time diagram will be introduced. Finally, a discussion of the methods used for the timing of celestial observations will conclude the chapter.

THE BASES OF TIME

From the beginning of recorded history, humans have reckoned time according to the passage of two celestial bodies through the heavens—the sun and the moon. The science of archeology has produced much evidence of the fascination of ancient people with the passage of the sun across the sky by day, and with its migration in declination over the period of a year. It is theorized by many contemporary scientists that the ancient monument at Stonehenge, England, constructed about 1500 B.C., may well be the oldest known solar observatory. Many of the monuments of ancient Egypt were constructed and oriented in such a way that they marked significant points in the annual migration of the sun in declination, and the amazing accuracy of the calendars of the South and Central American Inca and Aztec Indian cultures that flourished in pre-Columbian times is well documented. In fact, almost every known culture since the dawn of history has made extensive use of the sun as the basis for dividing the passage of time into days and years. Use of the moon as the basis for grouping the days of the year into 30-day months was a natural corollary, given the approximately 30-day progression of its phases caused by its period of revolution about the earth.

Apparent Solar Time

Time reckoned according to the passage of the sun observable in the sky is termed *apparent solar time*; used in this context, the sun is referred to as the *apparent sun*. Until the latter half of the nineteenth century, when comparatively rapid means of transportation and communication in the form of the steam locomotive and the telegraph evolved, the passage of the apparent sun across the heavens was almost exclusively used as the basis by which all times were reckoned. When the sun's disk passed over an observer's meridian, all local clocks were adjusted to read noon. Around 1800, it was of little consequence to a man living in Norfolk, Virginia, that his cousin located at the capital in Richmond kept a time that differed from his by about 4 minutes, since it took several days to make the journey between them on horseback.

Mean Solar Time

With the advent of the steam locomotive, however, which by 1890 could complete the Richmond to Norfolk transit in a few hours, maintaining schedules and computing arrival and departure dates and times by reference to the apparent sun became impossible for two reasons. First, if time is reckoned according to apparent solar time, a traveler's watch must be adjusted by 1 minute each time 15 minutes of longitude are traversed, as will be explained later. Second, as the earth proceeds along its elliptical path of revolution about the sun, the speed with which it moves in its orbit continually changes, as does the inclination of the earth's axis with respect to the sun. The net effect of these irregularities is that the length of the solar day varies at different times of the year.

To overcome these disadvantages of nonuniform apparent solar time, the railroads adopted a method of reckoning time, first proposed in the mid-eighteenth century, called *mean solar time*. Mean solar time employs as its reference a hypothesized *mean sun*, the hour circle of which is considered to move at a constant rate along the celestial equator. The rate at which the mean sun moves is the average rate of motion of the apparent sun during each mean solar day over a solar year. Since there are 24 hours in every mean solar day, and 360° in the circumference of the earth, the mean sun moves at the constant rate of 360° ÷ 24 hours, or 15° per hour. This fact is of fundamental importance in celestial navigation, since by this means degrees of arc as measured along a parallel of latitude are directly convertible into time units. As will be seen later, this convertibility of arc to time allows precomputation of the times of rising and setting of any celestial body, particularly the sun, as well as calculation of the times of twilight and the time at which apparent noon will occur. Spurred by railroad interests, mean solar time came into wide use throughout the United States, and from 1900 to the present it has been the predominant means by which the passage of time is reckoned throughout the world.

Equation of Time

At certain times of the year, the apparent sun moves across the heavens at a slower rate than the mean sun, while at other times, the apparent sun travels faster. In the former case, the apparent sun may be as much as 15 minutes behind the mean sun, while in the latter case, the apparent sun may be as much as 15 minutes ahead. The difference between mean and apparent time at any instant is called the *equation of time.* It is tabulated for each day in the *Nautical Almanac,* in the format shown in Figure 18-1. (The daily values of the equation of times vary from year to year, so the values in Figure 18-1 are valid only for the examples in this book.)

As can be seen in the figure, there are three entries in the *Almanac* for each day. The first column gives the number of minutes and seconds by which the apparent sun either lags or leads the mean sun as these bodies pass over the lower branch of the Greenwich meridian—the 180th meridian (00^h)—and the second column contains the time differences applying at the upper branch, or prime meridian (12^h). The figures in the third column are simply the local mean times rounded to the nearest whole minute at which the apparent (actual) sun will pass over or *transit* the Greenwich meridian. During the period 25–27 May, for example, the local mean time at which the apparent sun will transit the Greenwich meridian is approximately 1157. Thus, the apparent sun must be ahead of the mean sun on these dates, by the amount of minutes and seconds listed in the first two columns for each day in the 3-day period covered. Conversely, were the apparent sun lagging behind the mean sun, the times of meridian passage listed in the third column would be 1200 or later, with the figures in the other two columns indicating the exact amount of the lag to the nearest second. Because the equation of time will vary by only a few seconds as the sun continues around the earth in the course of each day, the local

Equation of Time

Day	SUN Eqn. of Time 00^h	12^h	Mer. Pass.
	m s	m s	h m
25	03 13	03 10	11 57
26	03 07	03 03	11 57
27	03 00	02 57	11 57

Figure 18-1. *The equation of time,* Nautical Almanac, 25, 26, 27 May.

mean time of the sun's meridian passage at Greenwich tabulated in the *Almanac* for each day can also be used for all other meridians. On 25 May, for instance, the apparent sun will transit the upper branches of all meridians approximately 3 minutes ahead of the mean sun. Hence, the apparent sun will cross the standard meridians of all time zones at 1157 local mean time (equivalent to 1157 local zone time).

The equation of time is used primarily to determine the local mean times of meridian passage of the apparent sun at Greenwich for tabulation in the *Nautical Almanac*. In chapter 25, the use of these tabulated local mean times of meridian passage to determine the zone time of apparent noon at any position on earth will be demonstrated. The zone time of apparent noon thus determined is referred to as *local apparent noon,* or *LAN.*

Sidereal Time

In addition to apparent and mean solar time, there is one other basis of celestial time that the navigator may occasionally see mentioned in some advanced texts. This time, called *sidereal* time, uses the earth's rotation relative to the stars for its basis, rather than its rotation relative to the sun. Because the earth rotates in the same direction as its direction of revolution about the sun—counterclockwise as viewed from above the plane of the ecliptic—each sidereal day is about 3 minutes and 56 seconds shorter than the mean solar day, as illustrated in Figure 18-2.

In this figure, the earth begins one complete rotation at position A. At position B, the rotation has been completed with respect to the stars, but with respect to the sun, the earth must still rotate an amount equal to the shaded arc.

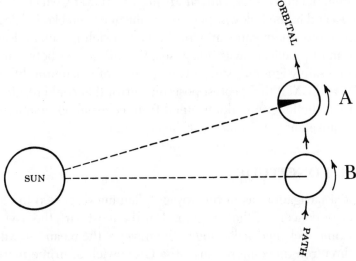

Figure 18–2. Difference between the sidereal and mean solar days.

Thus, the sidereal day is about 23 hours 56 minutes in length, while the mean solar day is exactly 24 hours long.

Sidereal time does not have extensive application in the usual practice of celestial navigation on earth. It is used mainly as the basis by which star charts and star finders are constructed.

Atomic Time Standards

In contrast to the aforementioned bases of time dependent upon the spatial relationship of the earth with the sun and the surrounding universe, there is another basis of time independent of these relationships that is coming into ever wider use today because of the extreme accuracy it can afford, particularly in scientific applications and in electronic navigation systems. This is *Universal Coordinated Time* (UTC), based on the frequency of vibrations of excited atoms of the radioactive element cesium. Because UTC is based on an unchanging atomic time standard, and mean time is affected by variations in the rate of rotation and revolution of the earth, there can be as much as a 0.9-second difference between times reckoned on the two bases. The exact difference at any time is called the *DUT1 correction*: its value is held between ±0.9 seconds by the occasional addition or subtraction of a so-called *leap second* adjustment to UTC applied as required, normally at the end of either June or December.

For some years now, UTC has been exclusively used as the basis of time for the various radio time signals broadcast worldwide by designated radio stations of the various maritime nations. As will be discussed later in this chapter, most navigators rely on these signals to set their chronometers and to compare stopwatches when at sea. The data entries in the various navigational almanacs, however, are tabulated for *Greenwich Mean Time* (GMT), which, as discussed in chapter 15 and in the following section, is mean time reckoned according to the relationship of the mean sun with the Greenwich meridian. Fortunately, though, for most practical navigation at sea, the differences between UTC and GMT may be safely ignored, since even at times of maximum difference between UTC and GMT, the greatest positional error that could result is only 0.2 of a minute of the longitude determined from celestial observations made at that time (equivalent to 0.2 mile near the equator).

TIME AND LONGITUDE

In celestial navigation, as in the voyage planning discussed in chapter 15, time is a major concern of the navigator. For the most part, the navigator uses mean solar time reckoned according to the travel of the mean sun with respect to one of three reference meridians—the Greenwich or prime meridian, the central meridian of the time zone in which the ship is located, and the meridian passing through the ship's position. Time reckoned according to the position of the mean sun relative to the prime meridian is called *Greenwich Mean Time.* Time using the central meridians of the various time zones as reference meridians is called *zone time,* abbreviated *ZT,* and time using the observer's meridian as a reference is referred to as *local mean time,* or *LMT.* The first two types of time, GMT and ZT, are extensively discussed in chapter 15; local mean time, LMT, is explained below.

Local mean time is based on the relationship of the mean sun with the observer's meridian. When the mean sun is at *lower transit* of the observer's meridian, meaning that it is passing over the lower branch of his or her meridian (i.e., that portion of the meridian on the opposite side of the earth from the observer), the LMT is midnight. At *upper transit* of the mean sun, when the upper branch of the observer's meridian (on the same side of the earth as the observer) is crossed, it is noon in LMT. The concept of local mean time by itself does not have much importance in celestial navigation, but the relationship of local mean time with local zone time and with Greenwich mean time is of great importance in determining times of sunrise and sunset, moonrise and moonset, and local apparent noon.

Local mean time differs from zone time by the amount of time required for the mean sun to traverse the difference in longitude between the observer's meridian and the standard meridian of the time zone. As was shown in the preceding section, the mean sun is considered to move around the earth at the constant rate of 15° of arc (longitude) per hour, or 1° every 4 minutes. Thus, differences in longitude, written dλ, can be directly converted into differences in time. To facilitate such conversions of arc to time, the table partially shown in Figure 18-3 is included in the *Nautical Almanac.*

As an example of the use of the table in Figure 18-3, suppose that an observer located at longitude 64° 13.3′W desired to compute the time required for the mean sun to travel from the central meridian of the observer's time zone, 60°W, to the local meridian. The difference in longitude is 64° 13.3′ − 60° = 4° 13.3′. Entering the first column of the table, a value of time of 0 hours 16 minutes is first extracted for 4° of the dλ. To find the additional time increment for the remaining 13.3 minutes of arc, the right-hand portion of the table is used. First, the horizontal row containing entries for 13′ is located, then the value under the column heading closest to the decimal fraction desired is read. Since 0.3 is closest to 0.25, the value for 13.3′ of 0 minutes 53 seconds is extracted. This result is then added to the value of time previously obtained for 4° to yield a final figure of 0 hours, 16 minutes, 53 seconds for the conversion of 4° 13.3′ of arc to time. Rounding off to the nearest minute, it could be stated that it will require about 17 minutes for the mean sun to traverse the indicated arc distance.

It is the constant rate of travel of the mean sun that makes conversions of arc to time, and thence conversions between zone time and local mean time, pos-

CONVERSION OF ARC TO TIME

0°–59°		60°–119°		120°–179°		180°–239°		240°–299°		300°–359°			0ʹ·00	0ʹ·25	0ʹ·50	0ʹ·75
°	h m	°	h m	°	h m	°	h m	°	h m	°	h m	ʹ	m s	m s	m s	m s
0	0 00	60	4 00	120	8 00	180	12 00	240	16 00	300	20 00	0	0 00	0 01	0 02	0 03
1	0 04	61	4 04	121	8 04	181	12 04	241	16 04	301	20 04	1	0 04	0 05	0 06	0 07
2	0 08	62	4 08	122	8 08	182	12 08	242	16 08	302	20 08	2	0 08	0 09	0 10	0 11
3	0 12	63	4 12	123	8 12	183	12 12	243	16 12	303	20 12	3	0 12	0 13	0 14	0 15
4	0 16	64	4 16	124	8 16	184	12 16	244	16 16	304	20 16	4	0 16	0 17	0 18	0 19
5	0 20	65	4 20	125	8 20	185	12 20	245	16 20	305	20 20	5	0 20	0 21	0 22	0 23
6	0 24	66	4 24	126	8 24	186	12 24	246	16 24	306	20 24	6	0 24	0 25	0 26	0 27
7	0 28	67	4 28	127	8 28	187	12 28	247	16 28	307	20 28	7	0 28	0 29	0 30	0 31
8	0 32	68	4 32	128	8 32	188	12 32	248	16 32	308	20 32	8	0 32	0 33	0 34	0 35
9	0 36	69	4 36	129	8 36	189	12 36	249	16 36	309	20 36	9	0 36	0 37	0 38	0 39
10	0 40	70	4 40	130	8 40	190	12 40	250	16 40	310	20 40	10	0 40	0 41	0 42	0 43
11	0 44	71	4 44	131	8 44	191	12 44	251	16 44	311	20 44	11	0 44	0 45	0 46	0 47
12	0 48	72	4 48	132	8 48	192	12 48	252	16 48	312	20 48	12	0 48	0 49	0 50	0 51
13	0 52	73	4 52	133	8 52	193	12 52	253	16 52	313	20 52	13	0 52	0 53	0 54	0 55
14	0 56	74	4 56	134	8 56	194	12 56	254	16 56	314	20 56	14	0 56	0 57	0 58	0 59
15	1 00	75	5 00	135	9 00	195	13 00	255	17 00	315	21 00	15	1 00	1 01	1 02	1 03
16	1 04	76	5 04	136	9 04	196	13 04	256	17 04	316	21 04	16	1 04	1 05	1 06	1 07
17	1 08	77	5 08	137	9 08	197	13 08	257	17 08	317	21 08	17	1 08	1 09	1 10	1 11
18	1 12	78	5 12	138	9 12	198	13 12	258	17 12	318	21 12	18	1 12	1 13	1 14	1 15
19	1 16	79	5 16	139	9 16	199	13 16	259	17 16	319	21 16	19	1 16	1 17	1 18	1 19
20	1 20	80	5 20	140	9 20	200	13 20	260	17 20	320	21 20	20	1 20	1 21	1 22	1 23
21	1 24	81	5 24	141	9 24	201	13 24	261	17 24	321	21 24	21	1 24	1 25	1 26	1 27
22	1 28	82	5 28	142	9 28	202	13 28	262	17 28	322	21 28	22	1 28	1 29	1 30	1 31
23	1 32	83	5 32	143	9 32	203	13 32	263	17 32	323	21 32	23	1 32	1 33	1 34	1 35
24	1 36	84	5 36	144	9 36	204	13 36	264	17 36	324	21 36	24	1 36	1 37	1 38	1 39
25	1 40			145	9 40						40	25	1 40	1 41		
26	1															

Figure 18-3. *Extract from the Conversion of Arc to Time table,* Nautical Almanac.

sible. Differences between *any* two solar mean times based on the travel of the mean sun can be thought of as being equal to the difference of longitude between their reference meridians, converted to units of time. Thus, GMT differs from zone time by the longitude of the standard meridian of the zone converted to time; GMT differs from local mean time by the longitude of the place converted to time; and zone time differs from local mean time by the arc-time difference corresponding to the difference of longitude between the standard meridian of the zone and the meridian of the observer. The only rule that must be remembered when applying arc-time differences is that a location that is east of another has a later time than the more westerly place, and a location that is west of another has an earlier time.

Several examples of conversion of local zone time to Greenwich mean time and vice versa, based on the difference of longitude between the standard meridian of the time zone and the Greenwich meridian, were given in chapter 15. In Figure 18-4A, suppose that observers at three meridians, 57° 45′W, 60°W, and 64°W, wished to convert the local mean time of the sun's meridian passage at their meridians to the local zone times of apparent noon (LAN) at these same

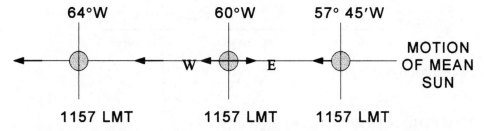

Figure 18-4A. *Local mean time of noon is 1157 at each of three meridians on the +4 time zone on 15 May.*

meridians on 25 May. Earlier it was shown that on this date the local mean time of upper transit of the sun was 1157. Thus, at each of the three meridians, the sun will transit at 1157 local mean time, as depicted in the figure.

Since the 60°W meridian is the standard meridian of this time zone, and local mean time and zone time are equivalent along the standard meridians of all zones, it follows that at 60°W, the LMT and ZT of meridian passage of the sun (LAN) are the same, 1157. All meridians within the zone east of 60°W will experience LAN at an earlier zone time, and all meridians to the west will experience LAN at a later zone time, with the differences being the amount of time it takes the mean sun to travel the arc between the local meridian and the standard meridian of the zone. From the arc-time conversion table in Figure 18-3, the time required for the mean sun to traverse the 2° 15′ of arc between 57° 45′W and 60°W is 9 minutes. Hence, LAN will occur at 57° 45′W at 1157 − 9 = 1148 zone time. In like manner, converting the 4° of arc between 60° and 64°W yields an arc-time difference of 16 minutes, which when applied to 1157 results in the zone time of LAN at 64°W of 1213. These results are indicated in Figure 18-4B.

The procedures for determination of the time of local apparent noon will be discussed in greater detail in chapter 25.

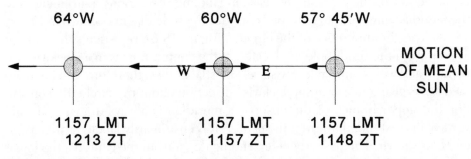

Figure 18-4B. *LMT of apparent noon converted to ZTs at 57° 45′W, 60°W, and 64°W.*

| HOURS | MINUTES | SECONDS |

Figure 18-5. *Format of a 6-digit time.*

THE FORMAT OF A WRITTEN TIME

In celestial navigation, when writing a complete time expressed in hours, minutes, and seconds, the format shown in Figure 18-5 is generally used. Hours, minutes, and seconds are expressed using two digits, and each quantity is separated from the others by a dash.

As an example of a written time, if the apparent sun crossed the central 60°W meridian in the example in the preceding section at exactly 1157 zone time, it would cross an observer's meridian at 64° 13.3′W at 12-13-53 zone time. If any quantities to be written are less than 10 in value, zeros are used to fill in the blank spaces, for example, 01-02-03 for 1 hour, 2 minutes, and 3 seconds. Whenever celestial observations are made, they are always recorded to the nearest second in the format given.

THE TIME DIAGRAM

To assist the navigator in visualizing the time relationships existing among the hour circle of the sun, the hour circle of the celestial body being observed, the Greenwich meridian, and the meridian passing through the assumed position, an aid called the *time diagram* is frequently used. This diagram is especially helpful when converting from either local mean or local zone time to GMT, a conversion of basic importance because GMT is the time used in all almanacs as the basis for tabulation of the coordinates of all celestial bodies.

Essentially, a time diagram is nothing more than a sketch of the earth centered on the north or south pole, with the hour circles and meridians of interest depicted as radial lines. By convention, the south pole is generally selected as the center of the diagram, for ease in labeling the various hour angles and the meridian angle. As an example, consider the time diagram shown in Figure 18-6A. The circumference of the circle in Figure 18-6A represents the equator as seen from the south pole, P_s, located at the center. East is in a clockwise direction, and west counterclockwise; all celestial bodies, therefore, can be imagined as revolving in a counterclockwise direction about the circle. By convention the upper branch of the observer's meridian is always drawn as a solid vertical line extended up from the center; it is customarily labeled with a capital *M*, as shown. The dashed line extending down from the center is the lower branch of the observer's meridian, usually labeled with a lowercase *m*.

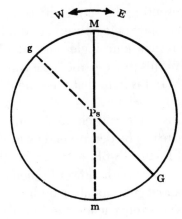

Figure 18-6A. *A time diagram.*

After the observer's meridian has been drawn, the Greenwich meridian is then located, based on the observer's assumed longitude. For this figure the observer's longitude was assumed to be 135° west, so the upper branch of the Greenwich meridian was drawn 135° clockwise from the observer's meridian M. The upper branch of the prime meridian is ordinarily labeled with a capital G, while the lower branch, represented by a second dashed line, is denoted by a lowercase g.

The hour circle of the sun is next located on the diagram by referring to the time of the observation being depicted. In positioning the sun's hour circle, the time diagram may be thought of as the face of a 24-hour clock, with m representing ZT 2400/0000, and M 1200 ZT. Thus, if it were 0600 zone time, the sun would be located 90° clockwise from M; if it were 1800, the sun would be 90° counterclockwise. For this example, suppose that it were 2100 local zone time. The hour circle of the sun would then be located as shown in Figure 18-6B,

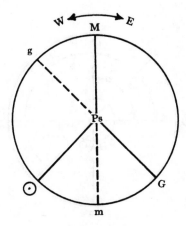

Figure 18-6B. *Locating the sun on the time diagram.*

with the symbol ⊙ representing the sun itself. To complete the diagram, the hour circle of the celestial body being observed, if other than the sun, is plotted. To do this, its Greenwich hour angle must first be obtained from an almanac. As the first step of this process, the local zone time of the observation must be converted to GMT; it is here that the time diagram is of probably the greatest value.

Since the sun is the basis of Greenwich Mean Time as well as zone time, the approximate GMT and date can be determined by inspection from the time diagram. In the diagram of Figure 18-6B, for instance, the hour circle of the sun is located about 90° to the west of the lower branch of the Greenwich meridian. Hence, it must be about 6 hours after midnight at Greenwich, or 0600 GMT. Since the sun has not yet reached the lower branch of the observer's meridian, the date at Greenwich must be one day later than the date at the observer's position.

In general, a difference between the dates at Greenwich and at the location of the observer is always indicated by the time diagram if the hour circle of the sun falls between the angle formed by the lower branches of the Greenwich and the observer's meridians. The meridian whose lower branch is to the west of the sun will always have the earlier date.

After the GMT time and date of the observation have been calculated, the time diagram can then assist in determining the value of the local hour angle of the body observed; the importance of this determination will become apparent in later chapters.

To use the time diagram for this purpose, the hour circle of the body must first be plotted on the diagram. If a body other than a star is observed, the value of its Greenwich hour angle can be obtained from the daily pages of an almanac; its hour circle can then be located on the diagram relative to the Greenwich meridian. Figure 18-7A depicts the hour circles of the moon and

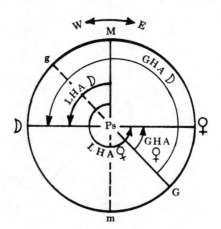

Figure 18-7A. *Time diagram, showing the LHA of the moon and Venus.*

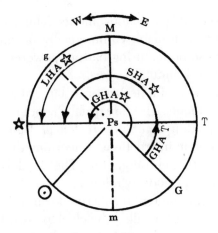

Figure 18-7B. *Time diagram, showing the LHA of a star.*

the planet Venus (note the Venus symbol ♀) positioned on the time diagram of Figure 18-6B by use of their GHAs. The LHAs of the bodies are indicated.

If the body observed is a star, its GHA is not directly obtainable from an almanac, but rather it must be computed from the sum of the GHA of Aries plus the SHA of the star as explained in chapter 16. Once the GHA has been thus determined, the hour circle of the star can be plotted on the diagram. In Figure 18-7B, the hour circles of Aries and a star are shown on the time diagram, and the LHA of the star is identified.

Additional examples of the use of the time diagram as an aid in converting local zone time to GMT and in determining the LHA of an observed body will be presented in chapter 20.

TIMING CELESTIAL OBSERVATIONS

As mentioned earlier, the coordinates of celestial bodies are tabulated in almanacs with respect to the Greenwich Mean Time and date. It is necessary, therefore, that the GMT of each celestial observation be recorded to the nearest second, in order to be able to extract the required data from the almanacs with sufficient precision to make an accurate determination of position possible. There are two methods of determining the correct time to the required accuracy on board ship—the *chronometer* and *radio time signals*. The chronometer and an associated timepiece, the *comparing watch,* will be discussed in the remainder of this section, while radio time signals will be the subject of the balance of this chapter.

The chronometer is described in chapter 7 on navigational instruments; its description will not be repeated here. Figure 7-13 shows the Hamilton chronometer, an older type of spring-driven chronometer; newer models incorporate battery-powered quartz movements. This and a number of other makes of chronometer are available commercially, ranging in price from $150 to $300 or more, with some featuring battery-powered quartz crystal movements accurate to within 1 minute per year.

Most Navy ships the size of a destroyer or larger have an allowance for at least two of these instruments. They are usually kept in the chartroom. Several Navy publications, including *NavShips Technical Manual* and certain other NavShips directives contain instructions concerning the care, winding, transportation, and use of the chronometer. The models like the Hamilton chronometer that require periodic winding are wound once every 24 hours, to ensure that they will never unintentionally run down. This responsibility is usually assigned to a quartermaster called the clock petty officer, who in addition to winding the chronometers, also periodically checks and resets all other wall clocks on board the ship.

Even a chronometer cannot keep absolutely exact time, but the characteristic that distinguishes this instrument from a normal clock or watch is that its rate of gain or loss of time is nearly constant. The difference between the time indicated on the chronometer or *chronometer time*, usually abbreviated *C*, and the correct GMT at any instant is referred to as the *chronometer error*, or *CE*. It is labeled either (F) or (S), depending on whether the chronometer time is ahead (faster than) or behind (slower than) GMT. Because chronometers are normally not reset on board ship, an accumulated error may become large in time. If, for example, the chronometer lost 3 seconds each day, in one 30-day month it would indicate a chronometer time, C, 1 minute and 30 seconds behind or "slower" than GMT. The CE in this case would be written (S) 01-30. Thus, if the chronometer read 01-02-03 at a particular time, the conversion to GMT would be as follows:

$$
\begin{array}{ll}
\text{C} & 01\text{-}02\text{-}03 \\
\text{CE} & +(\text{S})\ 01\text{-}30 \\
\hline
\text{GMT} & 01\text{-}03\text{-}33
\end{array}
$$

Slow chronometer errors are always *added* to chronometer time to obtain GMT, while *fast* errors are always *subtracted*.

Navy directives require that every chronometer be checked for accuracy at least once each day, and the results entered into a log book called the *Navigational Timepiece Rate Book*. This procedure is normally accomplished on board ship by means of the radio time signal described in the next section. The navigator usually arranges for the time signal to be fed into the chartroom by the shipboard radio communications facility at a given time each day. The error of each chronometer relative to the broadcast time signal is observed and recorded into the *Rate Book*. Differences between successive daily chronometer errors are computed and entered for each day, and average daily rates are computed periodically by the following formula:

ADR (Average daily rate) =

$$\frac{\text{Error on last day observed} - \text{error on first day observed}}{\text{Difference in dates of observation}}$$

A sample page of the *Navigational Timepiece Rate Book* from a ship having three chronometers appears in Figure 18-8.

It is humanly impossible to observe a chronometer error much more accurately than to the nearest half second, so the daily errors (successive daily rates) are entered in the *Rate Book* to this precision. This is the reason why the successive daily rates shown in Figure 18-8 are not all the same. For most chronometers, however, the average daily rate should not change much from one computation to another. If it does, it may indicate that the chronometer is in need of overhaul; it should be exchanged for a new chronometer from the chronometer pool when the ship returns to her home port.

Since the chronometer should not be removed and brought topside when celestial observations are to be made, a small handheld watch the size of a stopwatch called a *comparing watch* is often used to time actual observations. A typical comparing watch is shown in Figure 18-9.

DATE	A					B				C					OBSERVATION	
YEAR 19__	MAKE HAMILTON TYPE SIZE 85 SERIAL NO. 192 W					MAKE HAMILTON TYPE SIZE 85 SERIAL NO. 1135				MAKE HAMILTON TYPE SIZE 35 SERIAL NO. 156 B						
MONTH JULY	ERROR RELATIVE TO G.C.T. +=FAST -=SLOW			SUCCESSIVE DAILY RATES		ERROR RELATIVE TO G.C.T. +=FAST -=SLOW		SUCCESSIVE DAILY RATES		ERROR RELATIVE TO G.C.T. +=FAST -=SLOW			SUCCESSIVE DAILY RATES		LOCAL TIME TO NEAREST MINUTE	
DAY	±	MIN.	SECONDS	±	SECONDS	± MIN.	SECONDS	±	SECONDS	±	MIN.	SECONDS	±	SECONDS	TIME	INITIALS
1	+	11	49.0	+	1.5	− 6	02.0	−	1.0	+	16	22.0	+	2.0	1130	RC
2	+	11	51.5	+	2.5	− 6	02.5	−	0.5	+	16	24.0	+	2.0	1130	RC
3	+	11	53.0	+	1.5	− 6	03.0	−	0.5	+	16	25.0	+	1.0	1125	RC
4	+	11	54.5	+	1.5	− 6	04.0	−	1.0	+	16	23.0	−	2.0	1130	GG
5	+	11	55.5	+	1.0	− 6	04.0		0.0	+	16	21.5	−	1.5	1130	GG
24	+	12		+	2.0	−				+	16	20.0	−			
25	+	12	30.5	+	2.5	− 6	13.5		0.0	+					1130	RC
26	+	12	32.0	+	1.5	− 6	13.5	.	0.0	+	16	22.5	+	1.0	1130	RC
27	+	12	34.0	+	2.0	− 6	14.0	−	0.5	+	16	23.0	+	0.5	1135	RC
28	+	12	35.5	+	1.5	− 6	15.0	−	1.0	+	16	24.0	+	1.0	1130	RC
29	+	12	37.0	+	1.5	− 6	16.0	−	1.0	+	16	25.5	+	1.5	1140	GG
30	+	12	39.0	+	2.0	− 6	16.5	−	0.5	+	16	26.5	+	1.0	1130	GG
31	+	12	41.5	+	2.5	− 6	17.0	−	0.5	+	16	27.0	+	0.5	1130	DAH

+1.75 AVERAGE DAILY RATE   −0.5 AVERAGE DAILY RATE   +0.17 AVERAGE DAILY RATE

NOTE: FOR COMPUTATION OF AVERAGE DAILY RATE SEE PARAGRAPH 5 UNDER INSTRUCTIONS.

10—66873-1

Figure 18-8. *Sample page from a* Navigational Timepiece Rate Book.

Figure 18-9. *A comparing watch.*

The comparing watch differs from a normal stopwatch in that it usually has a sweep hand that can be stopped at the moment of observation, while the remainder of the watch movement continues in motion. After the time of the observation has been recorded, the indicator sweep can then be returned to run with the second sweep of the main movement, until such time as it is again stopped to record the next observation. If a comparing watch is not available, an ordinary stopwatch or good-quality wristwatch may be used as the timing device. In either case, it is a good practice to read and record first the seconds, then the minutes, and finally the hours for a celestial observation. This procedure is, of course, particularly desirable if the watch does not have the capability of being stopped to "freeze" the time of the observation.

When a round of celestial observations is to be made, the watch to be used is set as close as possible to either the local zone time (ZT) or to GMT, usually by reference to the chronometer. It is also possible to use a radio time signal to set the watch; many small craft navigators will do this, particularly if they use a quartz timepiece or wristwatch in lieu of a chronometer at sea. In either case, the difference between watch time (W) and ZT or GMT is then determined. This difference is the *watch error,* abbreviated *WE;* like the chronometer error discussed earlier, watch error is labeled either fast (F) or slow (S). After recording the watch time of each celestial observation, fast watch errors are subtracted, and slow watch errors added, in order to obtain the correct ZT or GMT for each observation. If the watch is set to local zone time, the zone time of each observation is converted to GMT by applying the zone difference figure.

In cases in which a comparing watch is set by reference to the chronometer, the usual procedure is to set the watch to the indicated chronometer time. The watch error is then identical to the algebraic sum of any difference between the watch time and chronometer time and the chronometer error.

As an example of the determination and application of watch error, suppose that for a given observation the navigator's watch read 20-11-02. Earlier, the watch had read 19-29-01 when the chronometer read 19-30-50. The chronometer error (CE) on GMT on the date in question is determined to be (F) 00-30-01. To obtain the GMT of the observation, it is first necessary to determine the watch error, which in this case is the difference between the time indicated by the watch and the correct GMT.

When the watch time (W) was compared to the chronometer time (C), the watch time was found to be 1 minute 49 seconds slower than the chronometer time:

$$
\begin{array}{lll}
\text{C} & 19\text{-}30\text{-}50 \\
\text{W} & 19\text{-}29\text{-}01 \\
\hline
\text{C–W} & \text{(S) } 01\text{-}49
\end{array}
$$

To find the watch error (WE), this time difference (C – W) must be added algebraically to the chronometer error (CE):

$$
\begin{array}{lll}
\text{C – W} & (+\text{S}) & 01\text{-}49 \\
\text{CE} & (-\text{F}) & 30\text{-}01 \\
\hline
\text{WE} & (-\text{F}) & 28\text{-}12
\end{array}
$$

Having obtained the watch error, the GMT of the observation can be determined by applying the WE to the watch time of the observation:

$$
\begin{array}{lll}
\text{W} & 20\text{-}11\text{-}02 \\
\text{WE} & (-\text{F})28\text{-}12 \\
\hline
\text{GMT} & 19\text{-}42\text{-}50
\end{array}
$$

Again, a positive "plus" sign is associated with all slow errors, and a negative "minus" sign is used in conjunction with fast errors.

RADIO TIME SIGNALS

Radio time signals are broadcast worldwide by radio stations of many foreign maritime nations and by two U.S. Bureau of Standards stations. The two Bureau of Standards stations are WWV at Fort Collins, Colorado, and WWVH at Kauai, Hawaii. As mentioned earlier in this chapter, the time signals transmitted by the U.S. and most foreign stations are based on Universal Coordinated Time (UTC), which can be considered as equivalent to GMT for most practical navigational purposes.

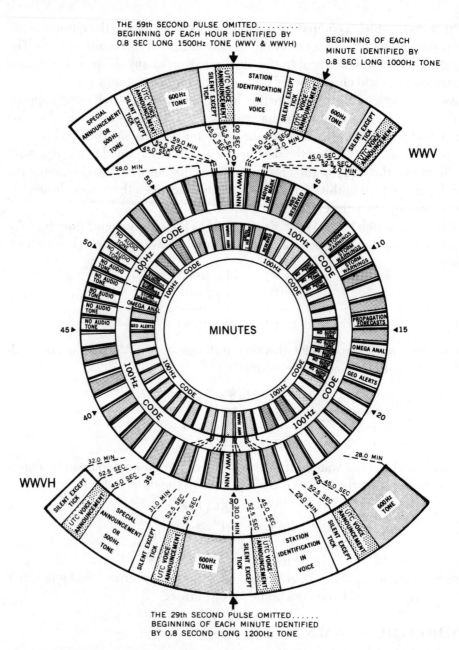

Figure 18-10A. *The WWV and WWVH radio time broadcast format. Detailed information on the various segments of the format is contained in the DMAHTC publication* Radio Navigational Aids.

Minute	Second										
	50	51	52	53	54	55	56	57	58	59	60
55	-		-	-	-						-
56	-	-		-	-						-
57	-	-	-		-	-					-
58	-	-	-	-		-					-
59	-										——

Figure 18-10B. The United States System of time signal tones.

The DMAHTC *Publication No. 117, Radio Navigational Aids,* contains descriptions of the various time signals and the frequencies at which they are transmitted worldwide. There are several standard time signal transmission formats used. The WWV and WWVH format is depicted in Figure 18-10A. As indicated in the diagram, voice announcements of UTC are made each minute, with a female voice on WWVH and a male voice on WWV. These stations, as well as a number of foreign stations, transmit a tone pattern called the *United States System* shown in Figure 18-10B, beginning 5 minutes before each hour of UTC.

The value of DUT1 at the time of each broadcast is indicated by a pattern of either double ticks or emphasized second tones transmitted for each of the 15 seconds immediately following each whole minute tone. The number of second ticks or tones emphasized indicates the numerical value of DUT1. When DUT1 is positive, seconds 1 through 8 are emphasized, corresponding to a DUT1 value of from +0.1 to +0.8 second. When DUT1 is zero, no seconds are emphasized. When DUT1 is negative, seconds 9 through 15 are emphasized, corresponding to a DUT1 value of from −0.1 to −0.7 second. In situations requiring extreme accuracy, the DUT1 correction thus obtained is added algebraically to UTC to yield GMT correct to within 0.1 second.

In practice, when one or more chronometers are carried on board a vessel at sea, a time signal is used to determine the error of the most accurate one once each day. The remaining chronometers, if any, are then compared with this reference chronometer to obtain their individual errors. In the case of small craft that may not have a chronometer on board, the navigator will generally use the time signal much more often to check the accuracy of the boat's timepieces, and to set his or her watch prior to every round of celestial observations

SUMMARY

In this chapter, the fundamental aspects of time as it applies to the practice of celestial navigation have been discussed, including the bases by which time is reckoned, the time diagram, and the use of the chronometer, radio time signal, and watch in recording the time of celestial observations. The material in succeeding chapters will relate the various applications of time and time theory to such diverse problems as the solution of the navigational triangle for a celestial LOP, the determination of times of rising and setting of the sun and moon, the calculation of the duration and commencement of twilight, and the computation of the time of local apparent noon.

THE MARINE SEXTANT

19

The instrument most commonly associated with the practice of celestial navigation is the *marine sextant*. The sextant can be simply described as a hand-held instrument designed to measure the angle between two objects with great precision. In practice, it is usually used to measure the altitudes of celestial bodies above the visible sea horizon, but, as stated in part 1, it can also be employed to measure the horizontal angle between two terrestrial objects in order to obtain a terrestrial line of position.

The name sextant is derived from the Latin *sextans*, meaning the sixth part; its arc is approximately one-sixth of a circle. Because of the optical principles incorporated in the sextant, however, modern instruments can measure angles up to about 145°. The sextant has been a symbol of the practice of navigation at sea for more than 200 years, and the professional navigator takes great pride in being skilled in its use.

CHARACTERISTICS OF THE MARINE SEXTANT

There are many models of sextants of varying quality, precision, and cost; they range in price from about $25 for a plastic model useful for training and as a spare for emergency use, to over $1,000 for one of high-quality construction. In the U.S. Navy, a model similar to that pictured in Figure 19-1 has been in common use for many years, and is still often found on board most Navy ships.

The nomenclature of the principal parts of this sextant, which are identified by the letters in the figure, is representative of nearly all varieties of the micrometer drum sextant:

A The *frame* is usually constructed of either brass or aluminum in the form shown in the figure. It is the basic part of the sextant to which all others are attached.

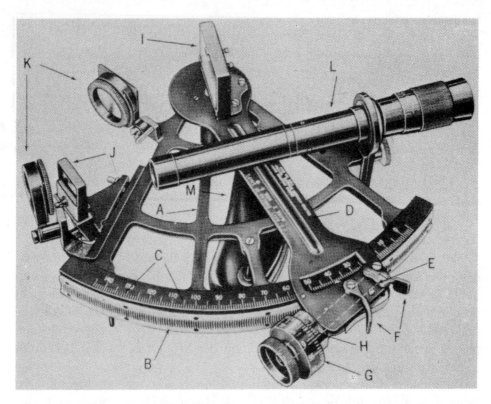

Figure 19-1. *The U.S. Navy standard Mark 2 Mod 0 marine sextant.*

B The *limb* is the bottom part of the frame, cut with teeth on which the micrometer drum rides.

C The *arc* refers to the altitude graduations of arc inscribed on the limb. On many sextants the arc is marked in a strip of brass, silver, or platinum inserted along the side of the limb.

D The *index arm* is a movable bar pivoted about the center of curvature of the limb, on which the index mirror and micrometer drum are fixed.

E The *tangent screw* (not visible in the figure) is a screw gear mounted on the end of the micrometer drum shaft; it engages the teeth of the limb. Turning the *micrometer drum* (G) turns the tangent screw, which in turn has the effect of moving the index arm along the arc of the sextant.

F The *release levers* are spring-actuated clamps that hold the tangent screw in place against the teeth of the limb. Compressing the release levers disengages the tangent screw, allowing rapid movement of the index arm along the arc.

G The *micrometer drum* is graduated in 60 minutes of arc around its circumference; one complete turn of the drum moves the index arm one degree of altitude along the arc, thus allowing readings of minutes of arc between whole degrees to be made.

H The *vernier,* adjacent to the micrometer drum and fixed to the index arm, allows readings to be made to the nearest tenth of a minute of arc.

I The *index mirror* is a piece of silvered glass mounted on the index arm, perpendicular to the plane of the instrument, and centered directly adjacent to the pivot of the index arm.

J The *horizon glass* or *mirror* is constructed of silvered glass and is designed to superimpose the body observed on the visible horizon. There are two types. One, called the traditional type, is divided into two halves, with the half nearest the frame being silvered and the other half either clear or empty. It produces a split image consisting of the horizon on the clear side and the body on the silvered side. This is the type featured in this chapter. The other, a later development, is called the whole horizon mirror. Using specially coated optics, it superimposes both the horizon and the celestial body on the entire mirror with no split image, at the cost of a slight reduction in light transmission. Professional navigators tend to favor the former, while beginners tend to favor the latter, which is considered easier to use.

K *Shade glasses* of variable darkness are mounted on the frame in front of the index mirror and horizon glass. They can be rotated into the line of sight between the mirrors and between the observer and the horizon to reduce the intensity of the light reaching the eye of the observer.

L The *telescope* screws into an adjustable collar in line with the horizon glass, and amplifies both the reflected and direct images observed.

M The *handle,* made of wood or plastic, is designed to be held in the right hand during sextant observations.

During the 1980s there was a Navy-sponsored project to develop an advanced sextant for shipboard use that would incorporate an electronic digital readout instead of a micrometer drum to indicate the angle measured on the sextant. It would have the capability of accepting a night vision telescope for observations using the visible horizon after dark, or a device that could provide an artificial horizon if the natural horizon were obscured by clouds or fog. The electronic output from this sextant could be fed directly into a computer for virtually instantaneous solution for the resulting celestial LOP. Several such LOPs would then be compared by mathematical algorithms to produce a computer solution for the celestial fix. Unfortunately for practitioners of celestial navigation, the increased availability of and tendency for reliance upon modern electronic position-fixing systems and equipment caused this development program to be discontinued, so it may be some time before such a space-age sextant becomes available for general use.

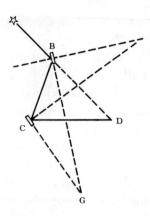

Figure 19-2. *Optical principle of the sextant.*

Optical Principle of the Sextant

The optical principle of the sextant is illustrated in Figure 19-2, with the solid line representing the path of an incoming light ray from a celestial body being observed. The instrument is constructed in such a way that the angle BDC between the body and the horizon is always equal in value to twice the angle between the index mirror and horizon glass, angle BGC, which is measured along the arc of the sextant. Thus, an arc encompassing one-sixth of a circle can be graduated so that angles up to 120° can be read. The arc of most modern sextants is extended slightly beyond one-sixth of a circle, so that angles up to 145° can be measured.

READING THE SEXTANT

When an angle representing either the altitude of a celestial body or the difference in bearing of two terrestrial objects is to be read with the sextant, a 3-step procedure is always used. The sample sextant reading pictured in Figure 19-3 illustrates the steps. First, the number of whole degrees is read by observing the position of the arrow on the index arm in relation to the arc. In this case, the arrow lies between 29° and 30°, so a value of 29° is obtained. Next, the minutes are read by noting the position of the zero mark on the vernier with respect to the graduations of the micrometer drum. In Figure 19-3 the zero falls between 42′ and 43′; hence, a value of 42′ results. Finally, the tenths of a minute are read by noting which of the 10 marks on the vernier is most nearly opposite one of the graduations on the micrometer drum. In this case, a value of 0.5 is indicated. Thus, the angle depicted in the figure is 29° 42.5′.

If the index arm arrow is very close to a whole degree mark on the arc, care must be taken to obtain the correct angle by referring to the micrometer drum.

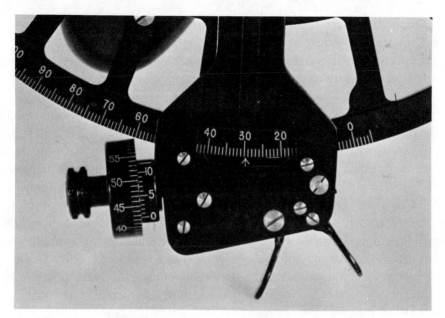

Figure 19-3. *A sextant angle of 29° 42.5'.*

If the arrow were pointing directly at 45° on the arm, for instance, and 57′ on the micrometer drum were opposite the zero of the vernier, the correct reading would be 44° 57′, not 45° 57′. Similarly, any doubt as to the correct minute can be resolved by noting the fraction of a minute indicated on the vernier.

PREPARATION FOR USE OF THE SEXTANT

Prior to going to sea, the navigator should always check the sextant carefully to ensure that all correctable mechanical error that may be present in the instrument is eliminated. There are seven major sources of this so-called *instrument error* in the micrometer drum sextant. Of these seven, four are adjustable, and three are not. The adjustable instrument errors include the following:

Lack of perpendicularity of the frame and index mirror
Lack of perpendicularity of the frame and horizon glass (side error)
Nonparallelism of the index mirror and horizon glass with each other at the zero setting
Nonparallelism of the telescope with the frame (collimation error)

The major nonadjustable instrument errors are:

Prismatic error, occurring because of nonparallelism in the faces of the shade glasses, index mirror, and horizon glass

Graduation error resulting from improper calibration of the scales of the arc, micrometer drum, or vernier

Centering error as a result of the index arm not being pivoted at the exact center of curvature of the arc

In the U.S. Navy it is the usual practice to leave the adjustment of the sextant to the optical repair facility of either a tender or a shipyard, but if necessity requires, an experienced navigator may attempt to adjust the instrument. The *American Practical Navigator (Bowditch)* contains an excellent description of the adjustment procedures, and can be used as a reference if the need arises.

After all adjustable errors have been reduced or eliminated insofar as possible, the sextant will still retain some residual variable adjustable error as well as a small fixed nonadjustable instrument error. Additionally, a small variable error called *personal error* may often be produced as a result of the eye of the observer acting in conjunction with the optical system of the sextant. In practice, the fixed component of instrument error and any personal error are usually considered to be insignificantly small for most practical purposes.

One component of variable instrument error, however, must be taken into account each time sextant observations are to be made. This is *index error,* which results from nonparallelism of the horizon glass and index mirror. Inasmuch as this error continually changes over time as external conditions vary, it can never be completely adjusted out, and it is usually of a significant size. Index error, therefore, must be separately determined on each occasion that the sextant is to be used for a round of observations.

By day, the index error is best determined by an observation of the sea horizon. With the sextant set at 0° 0.0' of arc, the horizon should appear as shown in Figure 19-4A. If the horizon appears as in Figure 19-4B, there is some index error present. To determine the magnitude and direction of the error, the micrometer drum is slowly rotated until the direct and reflected images of the horizon are adjacent. If after so doing the micrometer drum reads *more than* 0.0', the error is *positive.* Conversely, if the reading on the drum is *less than* 0.0', the error is *negative.* In the former case, all angles read on the instrument would be too large, and in the latter case, all angles would be too small. Consequently, when a sextant has a *positive* index error, sometimes referred to as being "on the arc," all subsequent sextant altitudes must be corrected by *subtracting* the amount of the error. When a sextant has a *negative* ("off the arc") index error, all subsequent observations must be corrected by *adding* the amount of the error. This correction is referred to as the *index correction,* abbreviated IC; it is always equal in amount but opposite in sign to the index error. The following mnemonic aid has been coined to help in determining the sign of the index correction:

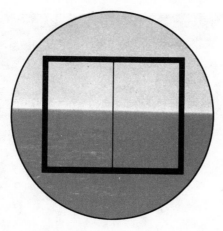

Figure 19-4A. *Sextant set at zero, no index error.*

When it (the index error) is *on* (positive), it (the IC) is *off*; when it's *off*, it's *on*.

As an example, suppose that a navigator observed a visible horizon and found that when it was aligned as in Figure 19-4A, the sextant read as shown in Figure 19-5. Here, the micrometer drum has been rotated on the negative side of the 0° mark to 57.5'. Hence, the index error is –(60.0' – 57.5') or –2.5' "off the arc." All subsequent sextant altitudes, therefore, must be corrected by adding "on" an index correction, IC, of +2.5'.

At night, the index error may be determined by observing a star. The direct and indirect images are either brought into coincidence or aligned directly alongside one another as described above for the horizon, and the amount by

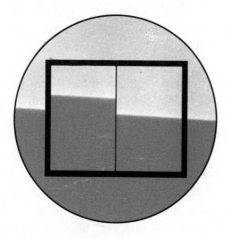

Figure 19-4B. *Sextant set at zero, index error present.*

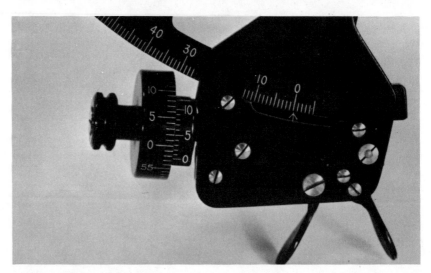

Figure 19-5. *A sextant index error.*

which the micrometer drum was rotated away from the zero setting to achieve this result is the index error.

Other examples of the application of the index correction to the altitude of a celestial body as observed with the sextant will be given later in this chapter.

THE TECHNIQUES OF SEXTANT ALTITUDE OBSERVATIONS

After having determined the index error of the sextant, the navigator is ready to proceed with the altitude observations of the selected celestial bodies. In chapter 16 the observed altitude of a celestial body, Ho, is defined as the angle formed at the center of the earth between the line of sight to the body and the plane of the observer's celestial horizon. This angle is always measured along the vertical circle passing through the observer's zenith and nadir and the body under observation.

Altitude observations made with the handheld marine sextant measure the vertical angle along the vertical circle between the observer's visible sea horizon and the body at the time of observation. The angle thus measured is termed the *sextant altitude* of the body; it is abbreviated *hs*. Because the observer is located on the surface of the earth rather than at its center, the observer's visible sea horizon and the celestial horizon are not coincident. In fact, as will be discussed in more detail later, the line of sight to the observer's sea horizon in most cases is not even parallel to the plane of the celestial horizon, owing to the height of eye of the observer above the earth's surface. Hence, certain corrections to be discussed later in this chapter must always be applied to the sextant altitude hs to obtain the desired observed altitude Ho. In the remainder of this section, the technique of using the sextant to obtain the sextant altitude hs of the sun, moon, planets, and stars will be examined.

When the sun is to be observed, the sextant is initially set at 0°, and then it is held vertically in the right hand with the line of sight directed at the sea horizon below the position of the sun. Suitable shade glasses are moved into position, depending on the brightness of the horizon and of the sun, and the index arm is moved outward by means of the release levers until the reflected image of the sun is brought down and appears in the horizon glass roughly alongside the direct view of the horizon. Next, the micrometer drum is slowly rotated until the sun appears to be resting exactly on the horizon. To check the perpendicularity of the sextant, the instrument should then be tilted slightly to either side around the axis of the telescope, causing the sun to appear to swing in an arc across the horizon glass, as in Figure 19-6A. This process is called *swinging the arc.*

After swinging the arc several times, the image of the sun should be adjusted to its final position at the bottom of the arc tangent to the horizon as shown in Figure 19-6B, and the altitude should be read.

The lower edge of the sun is referred to as its *lower limb,* and the upper edge as its *upper limb.* Although most navigators prefer to observe the lower limb of the sun when measuring the sun's altitude, it is also possible to obtain an altitude by observation of the upper limb. For the upper-limb observation, the sun's image is brought below the horizon until the upper limb is tangent. Upper-limb observations are particularly useful if the lower limb of the sun is obscured by clouds, making only the top portion of the sun's disk sharply defined.

When the moon is observed, the same procedure outlined above is employed, except that the shade glasses are not used. As is the case with the sun, either the upper or lower limb may be brought tangent to the visible horizon, with upper-limb observations being the only technique available if the phase of the moon is such that only the upper limb is illuminated.

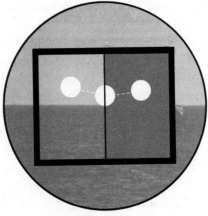

Figure 19-6A. *Swinging the arc of a sextant.*

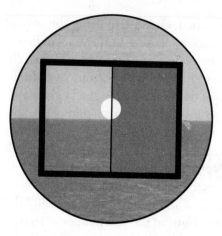

Figure 19-6B. *The sun at the instant of tangency.*

When a star or planet is observed, a different procedure than the one just described for the sun or moon must usually be employed, because of the reduced intensity and small apparent size of these bodies. There are three methods that can be used to observe a star or planet. In the first method, the altitude and true azimuth of the body at the approximate time of observation are first precomputed by means of a set of tables or a device called a *starfinder.* Both of these traditional manual methods of precomputation of the position of a star or planet will be examined in a later chapter. In recent years computer starfinder programs have also been developed for use on personal computers, thus automating the precomputation process to a great extent. However it is done, when the time of observation draws near, the sextant is set to the predetermined altitude of the body and aimed in the direction of the precomputed true azimuth. The body should then appear in the horizon glass approximately on the visible horizon. By swinging the arc and adjusting the micrometer drum, the center of the body is brought into exact coincidence with the horizon as shown in Figure 19-7, and its altitude is read.

The second method of observation of a star or planet used if the altitude and true azimuth have not been precomputed is to set the sextant to zero and elevate it until the body appears in the field of view of the telescope. The release levers are then compressed and the index arm gradually moved forward, while at the same time the sextant is slowly depressed. In this manner, the body is kept always in the telescope field, until the horizon appears and is brought approximately tangent to the star. At this point, the release levers are expanded, and the micrometer drum is used to align the center of the body exactly on the visible horizon.

The third method that can be used to bring a star or planet into the field of view for observation consists of inverting the sextant and sighting the body

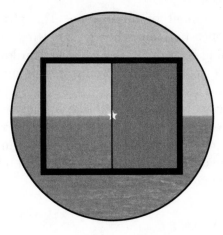

Figure 19-7. *A star at the moment of altitude observation.*

through the telescope, as shown in Figure 19-8. The release levers are then compressed with the fingers of the right hand, and the index arm is rotated forward until the horizon is brought "up" to approximate tangency with the body in the horizon glass. The sextant is then shifted to its normal vertical position in the right hand, with care being taken not to disturb the setting of the index arm. The desired star or planet should appear about on the horizon, and the micrometer drum is then used to adjust the final position of the body.

The choice of the method used for star or planet observations depends on the preference of the individual navigator and the circumstances under which the observations are being conducted.

When a sextant observation is to be made, it is the usual practice for the navigator to make the actual observation and for an assistant to record the exact

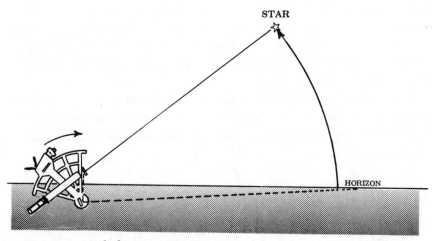

Figure 19-8. *Bringing the horizon up to a star.*

watch time at the moment of observation. The observer should give a "Standby," when the measurement is nearly complete, and a "Mark!" at the instant when the reading is made. The navigator then lowers the sextant and reads off the altitude to the nearest tenth of a minute, while the time is recorded to the nearest second.

Much practice is usually required with the sextant before the observation technique of the inexperienced navigator is perfected. Precision in the observation of the sextant altitude of a celestial body is very important, since every minute of error in the observed altitude of a celestial body causes the plotted celestial line of position to be off by one mile.

CARE OF THE SEXTANT

The modern marine sextant is a well built, very precise optical instrument capable of rendering years of service if it is properly maintained. Its usefulness can be greatly impaired, however, by careless handling or neglect. If the sextant is ever dropped, some error is almost certain to be introduced in all subsequent sightings.

When not in use, the sextant should always be kept secured in its case, and the case itself should be securely stowed in a location free from excessive heat, dampness, and vibration. In particular, the sextant should never be left unattended on the chartroom chart table or other furnishings.

Next to careless handling, moisture is the greatest enemy of the sextant. The sextant mirrors and telescope lens should always be wiped dry with a piece of lens paper after each use; cloth of any type tends to retain dust particles that may scratch the mirror or lens surface. Moisture has an extremely deleterious effect on both the silvering of the mirrors and the graduations of the arc. Should resilvering of the mirrors become necessary, this task, like instrument alignment, is best left to an optical instrument repair facility. Materials can be procured to perform resilvering of the mirrors on board ship, however; the *American Practical Navigator* contains a description of the resilvering procedure. The arc and teeth of the limb should always be kept lightly coated with a thin film of oil or petroleum jelly.

If the sextant is normally stowed in an air-conditioned space and the ship is operating in a humid climate, it is a good practice to bring the sextant in its case out into the open well before use, to prevent condensation from forming on the mirror surfaces.

SEXTANT ALTITUDE CORRECTIONS

For purposes of solving the navigational triangle in order to obtain a celestial line of position, the sextant altitude hs must be corrected to the value that would represent the observed altitude of the center of the body above the celestial horizon for an observer located at the center of the earth. This observed altitude is symbolized Ho. The corrections that must be applied to the sextant altitude hs to obtain the observed altitude Ho may be grouped into five categories, each of which will be described in a following section of this chapter:

1. Corrections for inaccuracy in reading the sextant
2. Corrections for deviation from the horizontal reference plane
3. Corrections for bending of light rays from the body
4. Adjustment to the equivalent reading at the center of the body
5. Adjustment to the equivalent reading at the center of the earth

As will be explained below, not all of these corrections are applied to each sextant observation, but every sextant altitude must be adjusted by certain of them. To aid in the systematic application of the appropriate corrections for each body when this is done manually, standardized *sight reduction forms* are usually employed. Several examples of the application of the various corrections using sight reduction forms developed by the U.S. Navy will be presented in this and the following chapters for each of the types of bodies observed in celestial navigation.

Corrections for the effects listed above that must be applied to the sextant altitude hs to obtain the desired observed altitude Ho are usually obtained from a publication known as an *almanac*. Two such almanacs are used in the normal practice of celestial navigation by U.S. Navy and most civilian navigators. Each will be the subject of a later chapter of this text. In the *Nautical Almanac*, designed primarily for surface navigators, all corrections to the sextant altitude are tabulated precise to the nearest tenth of a minute of arc. In the *Air Almanac*, designed primarily for air navigators, corrections are precise only to the nearest minute. In marine navigation it is usually desirable to take advantage of the increased accuracy afforded by using the more precise tables of the *Nautical Almanac*, whereas in air navigation the higher speeds of travel render this degree of precision impractical. In the following sections, the use of the *Nautical Almanac* to obtain all necessary corrections for each type of celestial body observed will be featured, as this text is oriented primarily to the practice of marine surface navigation.

If a handheld electronic calculator is available, most of the corrections discussed in the following sections can be computed using either preprogrammed algorithms or formulas given in the *Nautical Almanac*. Chapter 22 discusses calculator-based navigation in more detail.

In recent years several computer programs have been developed that automate the entire sight reduction process, including application of the appropriate sextant altitude corrections. These are also discussed in chapter 22.

Corrections for Inaccuracy in Reading the Sextant

Corrections compensating for inaccuracy in reading the sextant must be applied to every sextant altitude observed. The sources of error leading to inaccuracy in reading the sextant have already been discussed earlier in this chapter: fixed instrument error, personal error, and variable index error. Of these three, only the last—index error—is of significance for most sextant observations.

As mentioned previously, the *index correction,* or *IC,* is always equal in amount to the index error, but opposite in sign. Hence, if a sextant micrometer drum setting were 01.5′ on the arc when the horizon was aligned in the horizon glass, the following expression would represent a sextant altitude of 34° 31.6′ corrected for index error:

$$
\begin{array}{ll}
\text{hs} & 34°\ 31.6' \\
\text{IC} & \underline{-01.5'} \\
& 34°\ 30.1'
\end{array}
$$

Once determined, the index correction is considered constant for all angles subsequently read with the sextant on that occasion.

Correction for Deviation from the Horizontal Reference Plane

As is the case with the index correction, corrections for the difference between the line of sight to the observer's sea horizon and a horizontal reference plane parallel to the celestial horizon must be made for every sextant altitude observation. The necessity for this correction is illustrated in Figure 19-9.

In Figure 19-9, the angle measured with the sextant between the incoming light ray from a celestial body and the observer's visible horizon is the sextant altitude hs. As can be seen, it is always larger than the *apparent altitude ha* between the incoming ray and a horizontal reference plane parallel to the observer's celestial horizon, by the amount of dip of the visible horizon beneath the horizontal plane. This *dip angle,* as it is called, results primarily from the height of eye of the observer and the curvature of the earth and is exaggerated somewhat by atmospheric refraction near the surface. As the observer's height of eye above the earth's surface increases, the distance to the visible sea horizon, and therefore the size of the dip angle, also increases. Because the sextant altitude ha is used as the entering argument in the *Air* and *Nautical Almanacs* for all additional corrections required to produce the desired observed altitude Ho, all altitudes observed with the sextant must be decreased by the amount of the dip angle. This correction, known as the *dip correction* or simply *dip,* is therefore always negative; its amplitude slowly increases with the increased height of eye of the observer.

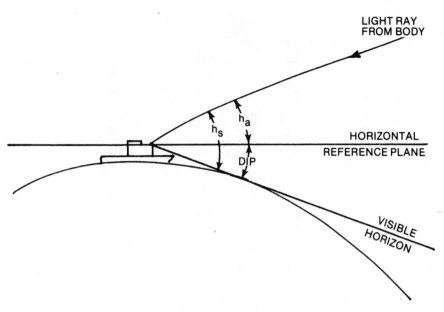

LIGHT RAY FROM BODY

HORIZONTAL REFERENCE PLANE

h_s

h_a

DIP

VISIBLE HORIZON

Figure 19-9. *Necessity for dip correction illustrated. The sextant altitude hs will always be greater than the apparent altitude ha by the amount of the dip angle.*

Although the dip correction could be determined mathematically using a handheld calculator, the usual procedure is simply to look up its value in a table such as that which appears on the inside front cover of the *Nautical Almanac*. An excerpt from this table appears in Figure 19-10.

The values of dip in the table are computed by the formulas

$$D = -0.97\sqrt{h} \quad \text{or} \quad D = -1.76\sqrt{m}$$

where h is the height of the observer in feet, m is the height of the observer in meters, and D is the dip correction in minutes of arc; normal conditions of atmospheric refraction are assumed. It has been found that under unusual atmospheric refraction conditions, such as those existing in polar regions or in midlatitudes after the passage of a squall line, the actual values of dip may be up to 15′ in excess of the tabulated value computed by these formulas. Hence, the navigator should always be aware of the possibility that the actual dip may be greater than the tabulated value when unusual atmospheric conditions prevail.

To use the dip table shown in Figure 19-10, the observer's height of eye expressed either in meters or in feet is used as the entering argument. The basic table is a *critical value table,* meaning that corrections to be extracted from the table are tabulated for ranges of values of the entering argument. The dip correction extracted is the one tabulated for the interval containing the actual

DIP			
Ht. of Eye	Corrⁿ	Ht. of Eye	Ht. of Eye	Corrⁿ

Ht. of Eye	Corrⁿ	Ht. of Eye	Ht. of Eye	Corrⁿ
m		ft.	m	
2·4	−2·8	8·0	1·0 − 1·8	
2·6	−2·9	8·6	1·5 − 2·2	
2·8	−3·0	9·2	2·0 − 2·5	
3·0	−3·1	9·8	2·5 − 2·8	
3·2	−3·2	10·5	3·0 − 3·0	
3·4	−3·3	11·2	See table ←	
3·6	−3·4	11·9		
3·8	−3·5	12·6	m	
4·0	−3·6	13·3	20 − 7·9	
4·3	−3·7	14·1	22 − 8·3	
4·5	−3·8	14·9	24 − 8·6	
4·7	−3·9	15·7	26 − 9·0	
5·0	−4·0	16·5	28 − 9·3	
5·2	−4·1	17·4		
5·5	−4·2	18·3	30 − 9·6	
5·8	−4·3	19·1	32 − 10·0	
6·1	−4·4	20·1	34 − 10·3	
6·3	−4·5	21·0	36 − 10·6	
6·6	−4·6	22·0	38 − 10·8	
6·9	−4·7	22·9		
7·2	−4·8	23·9	40 − 11·1	
7·5	−4·9	24·9	42 − 11·4	
7·9	−5·0	26·0	44 − 11·7	
8·2		27·1		
8·5				1·9

Figure 19-10. *Excerpt from dip table,* Nautical Almanac.

height of eye of the observer. If the observer's height of eye falls exactly on a tabulated height of eye argument, the dip correction corresponding to the preceding interval is extracted.

If the observer's height of eye is less than 2.4 meters or 8.0 feet, or greater than 21.4 meters or 70.5 feet, the inserts to the right of the basic critical value table are used with interpolation to determine the dip correction. Thus, if the observer's height of eye were 2.3 meters, an interpolated dip correction of −2.7′ would be obtained from the table insert; if the height of eye were 4.8 meters, a dip correction of −3.9′ would be extracted from the basic table; and if the height of eye were 26.0 feet, an exact tabulated argument, the dip correction of −4.9′ from the preceding interval would apply.

The inside back cover of the *Nautical Almanac* also contains a small dip table identical to the basic table inside the front cover. It is convenient to use with the adjacent moon Altitude Correction Tables used for observations of the moon.

If a clear view of the sea horizon is not obtainable because of close proximity of a ship or a shoreline, an altitude can be measured by using the waterline of the obstruction as a reference horizon if its distance from the observer is known. In such cases, Table 14 of the *American Practical Navigator (Bowditch)* may be used to obtain the value of the dip correction; this table is partially reproduced in Figure 19-11. As an example of the use of Table 14, if the sea horizon of an observer with a height of eye of 30 feet were obscured by a column of ships 2 miles distant, the dip correction extracted from Table 14 in

Figure 19-11 would be –9.3'. This contrasts with the value of –5.3' determined for a similar height of eye from the dip table of the *Nautical Almanac* for the visible sea horizon.

If a calculator is available, the dip angle for a foreshortened horizon can also be computed using a formula given in chapter 22. After the value of the dip correction has been determined, it is usually added algebraically to the IC. The resulting aggregate correction is then applied to the sextant altitude hs to obtain the apparent altitude ha of the body. As mentioned above, all remaining corrections to be applied in order to obtain the observed altitude Ho are tabulated in the *Air* and *Nautical Almanacs*, using ha as the entering argument.

Corrections for Bending of Light Rays from the Body

Light is assumed to travel through a transparent medium of uniform properties in a straight line at a constant speed. When a light ray passes into a medium of different properties, particularly different density, the speed of light, it is theorized, changes slightly. Moreover, if this ray enters the second medium at an angle, the change of speed does not take place simultaneously across the ray. The effect of this sequential change of speed across the light ray is to cause it to change direction upon entering the second medium; this change in direction is termed *refraction*. A light ray entering a denser medium at an oblique angle is bent toward a line perpendicular to its surface, and a ray entering a medium of less density is bent away from the perpendicular. The greater this *angle of incidence* between the incoming light ray and the perpendicular to the surface of the new medium, the greater will be the angle of change of direction, or *angle of refraction*.

Light from a celestial body travels through the vacuum of space in a relatively straight line until it encounters the earth's atmosphere. Being a denser medium, the atmosphere has the effect of bending incoming light rays toward the earth's surface. Since the atmosphere itself is not uniform, but increases in density as the earth's surface is approached, a light ray emanating from a celestial body striking the atmosphere at an oblique angle is bent in a gradual curving path, as shown in Figure 19-9 and Figure 19-12. This gradual bending of an incoming light ray in the earth's atmosphere is called *atmospheric refraction*. The greater the angle of incidence of the ray with the atmosphere, the greater will be the angle of refraction toward the surface. Hence, if a celestial body has an apparent altitude of near 90°, the effect of atmospheric refraction is negligible, but as its altitude decreases, the refraction effect increases to a maximum of about 34.5' of arc for a body located on the visible horizon.

As is shown in Figure 19-12, the effect of atmospheric refraction is always to cause the celestial body observed to appear to have a greater altitude than is the case in reality. Thus, the refraction correction for all celestial observations would normally always be negative. In the *Nautical Almanac*, however, the re-

TABLE 22

Dip of the Sea Short of the Horizon

Dis-tance	Height of eye above the sea, in feet										Dis-tance
	5	10	15	20	25	30	35	40	45	50	
Miles	′	′	′	′	′	′	′	′	′	′	Miles
0. 1	28. 3	56. 6	84. 9	113. 2	141. 5	169. 8	198. 0	226. 3	254. 6	282. 9	0. 1
0. 2	14. 2	28. 4	42. 5	56. 7	70. 8	84. 9	99. 1	113. 2	127. 4	141. 5	0. 2
0. 3	9. 6	19. 0	28. 4	37. 8	47. 3	56. 7	66. 1	75. 6	85. 0	94. 4	0. 3
0. 4	7. 2	14. 3	21. 4	28. 5	35. 5	42. 6	49. 7	56. 7	63. 8	70. 9	0. 4
0. 5	5. 9	11. 5	17. 2	22. 8	28. 5	34. 2	39. 8	45. 5	51. 1	56. 8	0. 5
0. 6	5. 0	9. 7	14. 4	19. 1	23. 8	28. 5	33. 3	38. 0	42. 7	47. 4	0. 6
0. 7	4. 3	8. 4	12. 4	16. 5	20. 5	24. 5	28. 6	32. 6	36. 7	40. 7	0. 7
0. 8	3. 9	7. 4	10. 9	14. 5	18. 0	21. 5	25. 1	28. 6	32. 2	35. 7	0. 8
0. 9	3. 5	6. 7	9. 8	12. 9	16. 1	19. 2	22. 4	25. 5	28. 7	31. 8	0. 9
1. 0	3. 2	6. 1	8. 9	11. 7	14. 6	17. 4	20. 2	23. 0	25. 9	28. 7	1. 0
1. 1	3. 0	5. 6	8. 2	10. 7	13. 3	15. 9	18. 5	21. 0	23. 6	26. 2	1. 1
1. 2	2. 9	5. 2	7. 6	9. 9	12. 3	14. 6	17. 0	19. 4	21. 7	24. 1	1. 2
1. 3	2. 7	4. 9	7. 1	9. 2	11. 4	13. 6	15. 8	17. 9	20. 1	22. 3	1. 3
1. 4	2. 6	4. 6	6. 6	8. 7	10. 7	12. 7	14. 7	16. 7	18. 8	20. 8	1. 4
1. 5	2. 5	4. 4	6. 3	8. 2	10. 0	11. 9	13. 8	15. 7	17. 6	19. 5	1. 5
1. 6	2. 4	4. 2	6. 0	7. 7	9. 5	11. 3	13. 0	14. 8	16. 6	18. 3	1. 6
1. 7	2. 4	4. 0	5. 7	7. 4	9. 0	10. 7	12. 4	14. 0	15. 7	17. 3	1. 7
1. 8	2. 3	3. 9	5. 5	7. 0	8. 6	10. 2	11. 7	13. 3	14. 9	16. 5	1. 8
1. 9	2. 3	3. 8	5. 3	6. 7	8. 2	9. 7	11. 2	12. 7	14. 2	15. 7	1. 9
2. 0	2. 2	3. 7	5. 1	6. 5	7. 9	9. 3	10. 7	12. 1	13. 6	15. 0	2. 0
2. 1	2. 2	3. 6	4. 9	6. 3	7. 6	9. 0	10. 3	11. 6	13. 0	14. 3	2. 1
2. 2	2. 2	3. 5	4. 8	6. 1	7. 3	8. 6	9. 9	11. 2	12. 5	13. 8	2. 2
2. 3	2. 2	3. 4	4. 6	5. 9	7. 1	8. 3	9. 6	10. 8	12. 0	13. 3	2. 3
2. 4	2. 2	3. 4	4. 5	5. 7	6. 9	8. 1	9. 2	10. 4	11. 6	12. 8	2. 4
2. 5	2. 2	3. 3	4. 4	5. 6	6. 7	7. 8	9. 0	10. 1	11. 2	12. 4	2. 5
2. 6	2. 2	3. 3	4. 3	5. 4	6. 5	7. 6	8. 7	9. 8	10. 9	12. 0	2. 6
2. 7	2. 2	3. 2	4. 3	5. 3	6. 4	7. 4	8. 4	9. 5	10. 6	11. 6	2. 7
2. 8	2. 2	3. 2	4. 2	5. 2	6. 2	7. 2	8. 2	9. 2	10. 3	11. 3	2. 8
2. 9	2. 2	3. 2	4. 1	5. 1	6. 1	7. 1	8. 0	9. 0	10. 0	11. 0	2. 9
3. 0	2. 2	3. 1	4. 1	5. 0	6. 0	6. 9	7. 8	8. 8	9. 7	10. 7	3. 0
3. 1	2. 2	3. 1	4. 0	4. 9	5. 9	6. 8	7. 7	8. 6	9. 5	10. 4	3. 1
3. 2	2. 2	3. 1	4. 0	4. 9	5. 7	6. 6	7. 5	8. 4	9. 3	10. 2	3. 2
3. 3	2. 2	3. 1	3. 9	4. 8	5. 7	6. 5	7. 4	8. 2	9. 1	9. 9	3. 3
3. 4	2. 2	3. 1	3. 9	4. 7	5. 6	6. 4	7. 2	8. 1	8. 9	9. 7	3. 4
3. 5	2. 2	3. 1	3. 9	4. 7	5. 5	6. 3	7. 1	7. 9	8. 7	9. 5	3. 5
3. 6	2. 2	3. 1	3. 8	4. 6	5. 4	6. 2	7. 0	7. 8	8. 6	9. 4	3. 6
3. 7	2. 2	3. 1	3. 8	4. 6	5. 4	6. 1	6. 9	7. 7	8. 4	9. 2	3. 7
3. 8	2. 2	3. 1	3. 8	4. 6	5. 3	6. 0	6. 8	7. 5	8. 3	9. 0	3. 8
3. 9	2. 2	3. 1	3. 8	4. 5	5. 2	6. 0	6. 7	7. 4	8. 1	8. 9	3. 9
4. 0	2. 2	3. 1	3. 8	4. 5	5. 2	5. 9	6. 6	7. 3	8. 0	8. 7	4. 0
4. 1	2. 2	3. 1	3. 8	4. 5	5. 1	5. 8	6. 5	7. 2	7. 9	8. 6	4. 1
4. 2	2. 2	3. 1	3. 8	4. 4	5. 1	5. 8	6. 5	7. 1	7. 8	8. 5	4. 2
4. 3	2. 2	3. 1	3. 8	4. 4	5. 1	5. 7	6. 4	7. 0	7. 7	8. 4	4. 3
4. 4	2. 2	3. 1	3. 8	4. 4	5. 0	5. 7	6. 3	7. 0	7. 6	8. 3	4. 4
4. 5	2. 2	3. 1	3. 8	4. 4	5. 0	5. 6	6. 3	6. 9	7. 5	8. 2	4. 5
4. 6	2. 2	3. 1	3. 8	4. 4	5. 0	5. 6	6. 2	6. 8	7. 4	8. 1	4. 6
4. 7	2. 2	3. 1	3. 8	4. 4	5. 0	5. 6	6. 2	6. 8	7. 4	8. 0	4. 7
4. 8	2. 2	3. 1	3. 8	4. 4	4. 9	5. 5	6. 1	6. 7	7. 3	7. 9	4. 8
4. 9	2. 2	3. 1	3. 8	4. 3	4. 9	5. 5	6. 1	6. 7	7. 2	7. 8	4. 9
5. 0	2. 2	3. 1	3. 8	4. 3	4. 9	5. 5	6. 0	6. 6	7. 2	7. 7	5. 0
5. 5	2. 2	3. 1	3. 8	4. 3	4. 9	5. 4	5. 9	6. 4	6. 9	7. 4	5. 5
6. 0	2. 2	3. 1	3. 8	4. 3	4. 9	5. 3	5. 8	6. 3	6. 7	7. 2	6. 0
6. 5	2. 2	3. 1	3. 8	4. 3	4. 9	5. 3	5. 7	6. 2	6. 6	7. 1	6. 5
7. 0	2. 2	3. 1	3. 8	4. 3	4. 9	5. 3	5. 7	6. 1	6. 5	6. 9	7. 0
7. 5	2. 2	3. 1	3. 8	4. 3	4. 9	5. 3	5. 7	6. 1	6. 5	6. 9	7. 5
8. 0	2. 2	3. 8	3. 8	4. 3	4. 9	5. 3	5. 7	6. 1	6. 5	6. 9	8. 0
8. 5	2. 2	3. 1	3. 8	4. 3	4. 9	5. 3	5. 7	6. 1	6. 5	6. 9	8. 5
9. 0	2. 2	3. 1	3. 8	4. 3	4. 9	5. 3	5. 7	6. 1	6. 5	6. 9	9. 0
9. 5	2. 2	3. 1	3. 8	4. 3	4. 9	5. 3	5. 7	6. 1	6. 5	6. 9	9. 5
10. 0	2. 2	3. 1	3. 8	4. 3	4. 9	5. 3	5. 7	6. 1	6. 5	6. 9	10. 0

Figure 19-11. *Table 14,* American Practical Navigator (Bowditch). *The stepped line gives the unobstructed horizon distance for various heights of eye.*

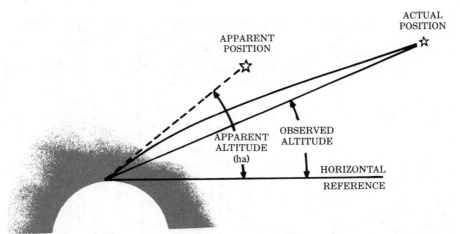

Figure 19-12. *Effect of atmospheric refraction. It will always cause the apparent altitude ha to appear to be greater than the observed altitude Ho to the center of the body.*

fraction correction for the sun and the moon is combined with other predominantly positive corrections described below to form a so-called aggregate *altitude correction*. Except in the case of upper limb and extremely low altitude lower-limb observations of the sun, the absolute value of the various positive corrections is always greater than the negative refraction correction. This results in the net altitude correction being positive for all lunar and most lower-limb solar observations. The term *altitude correction* is also used for the refraction correction for the planets and stars in the *Nautical Almanac*. Being based entirely on refraction in the case of these bodies, it is always negative for them.

The altitude correction tables for the sun, planets, and stars are located inside the front cover of the *Nautical Almanac*. Correction tables for apparent altitudes for these bodies between 10° and 90° are printed on the inside front cover, while corrections for ha less than 10° are located on the facing page. These tables are reproduced in Figures 19-13A and 19-13B on pages 348 and 349. Altitude correction tables incorporating the refraction correction for the moon are located inside the back cover, with corrections for ha between 0° and 35° on the left-hand page, and those for ha between 35° and 90° on the right. Figures 19-14A and 19-14B, pages 350 and 351, depict these tables. The tables on the left side of the inside front cover, shown in Figure 19-13A, are, like the dip table, *critical value tables,* that is, tables in which values of the correction to be found are tabulated for the ranges of values of the entering argument. The tabulated altitude correction is correct for any ha between those printed half a line above and half a line below. Should an apparent altitude fall exactly on a tabulated argument, the correction a half line above is used. Interpolation to the nearest tenth is required in the remaining tables of Figures 19-13B, 19-14A, and 19-14B, for entering arguments of apparent altitude between the tabulated values.

ALTITUDE CORRECTION TABLES 10°–90°—SUN, STARS, PLANETS

OCT.—MAR. SUN APR.—SEPT.						STARS AND PLANETS		DIP			
App. Alt.	Lower Limb	Upper Limb	App. Alt.	Lower Limb	Upper Limb	App. Alt.	Corrⁿ	App. Alt.	Additional Corrⁿ	Ht. of Eye	Corrⁿ	Ht. of Eye	Corrⁿ	Ht. of Eye	Corrⁿ

OCT.—MAR. SUN App. Alt.	Lower/Upper Limb	APR.—SEPT. App. Alt.	Lower/Upper Limb	STARS AND PLANETS App. Alt. Corrⁿ	App. Alt. Additional Corrⁿ	DIP
9 34 +10·8 −21·5		9 39 +10·6 −21·2		9 56 −5·3	**VENUS**	m / ft / 2·4 −2·8 8·0 / m 1·0 − 1·8
9 45 +10·9 −21·4		9 51 +10·7 −21·1		10 08 −5·2		2·6 −2·9 8·6 / 1·5 − 2·2
9 56 +11·0 −21·3		10 03 +10·8 −21·0		10 20 −5·1	Jan.1–Sept.30	2·8 −3·0 9·2 / 2·0 − 2·5
10 08 +11·1 −21·2		10 15 +10·9 −20·9		10 33 −5·0	0° ′	3·0 −3·1 9·8 / 2·5 − 2·8
10 21 +11·2 −21·1		10 27 +11·0 −20·8		10 46 −4·9	42 +0·1	3·2 −3·2 10·5 / 3·0 − 3·0
10 34 +11·3 −21·0		10 40 +11·1 −20·7		11 00 −4·8		3·4 −3·3 11·2 /
10 47 +11·4 −20·9		10 54 +11·2 −20·6		11 14 −4·7	Oct.1–Nov.15	3·6 −3·4 11·9 / See table
11 01 +11·5 −20·8		11 08 +11·3 −20·5		11 29 −4·6	0° ′	3·8 −3·5 12·6 / ←
11 15 +11·6 −20·7		11 23 +11·4 −20·4		11 45 −4·5	47 +0·2	4·0 −3·6 13·3 / m
11 30 +11·7 −20·6		11 38 +11·5 −20·3		12 01 −4·4		4·3 −3·7 14·1 / 20 − 7·9
11 46 +11·8 −20·5		11 54 +11·6 −20·2		12 18 −4·3	Nov.16–Dec.12	4·5 −3·8 14·9 / 22 − 8·3
12 02 +11·9 −20·4		12 10 +11·7 −20·1		12 35 −4·2	0° ′	4·7 −3·9 15·7 / 24 − 8·6
12 19 +12·0 −20·3		12 28 +11·8 −20·0		12 54 −4·1	46 +0·3	5·0 −4·0 16·5 / 26 − 9·0
12 37 +12·1 −20·2		12 46 +11·9 −19·9		13 13 −4·0		5·2 −4·1 17·4 / 28 − 9·3
12 55 +12·2 −20·1		13 05 +12·0 −19·8		13 33 −3·9	Dec.13–Dec.28	5·5 −4·2 18·3 /
13 14 +12·3 −20·0		13 24 +12·1 −19·7		13 54 −3·8	0° ′	5·8 −4·3 19·1 / 30 − 9·6
13 35 +12·4 −19·9		13 45 +12·2 −19·6		14 16 −3·7	11 +0·4	6·1 −4·4 20·1 / 32 − 10·0
13 56 +12·5 −19·8		14 07 +12·3 −19·5		14 40 −3·6	41 +0·5	6·3 −4·5 21·0 / 34 − 10·3
14 18 +12·6 −19·7		14 30 +12·4 −19·4		15 04 −3·5		6·6 −4·6 22·0 / 36 − 10·6
14 42 +12·7 −19·6		14 54 +12·5 −19·3		15 30 −3·4	Dec.29–Dec.31	6·9 −4·7 22·9 / 38 − 10·8
15 06 +12·8 −19·5		15 19 +12·6 −19·2		15 57 −3·3	0° ′	7·2 −4·8 23·9 /
15 32 +12·9 −19·4		15 46 +12·7 −19·1		16 26 −3·2	6 +0·5	7·5 −4·9 24·9 / 40 − 11·1
15 59 +13·0 −19·3		16 14 +12·8 −19·0		16 56 −3·1	20 +0·6	7·9 −5·0 26·0 / 42 − 11·4
16 28 +13·1 −19·2		16 44 +12·9 −18·9		17 28 −3·0	31 +0·7	8·2 −5·1 27·1 / 44 − 11·7
16 59 +13·2 −19·1		17 15 +13·0 −18·8		18 02 −2·9		8·5 −5·2 28·1 / 46 − 11·9
17 32 +13·3 −19·0		17 48 +13·1 −18·7		18 38 −2·8	**MARS**	8·8 −5·3 29·2 / 48 − 12·2
18 06 +13·4 −18·9		18 24 +13·2 −18·6		19 17 −2·7	Jan.1–June14	9·2 −5·4 30·4 /
18 42 +13·5 −18·8		19 01 +13·3 −18·5		19 58 −2·6	0° ′	9·5 −5·5 31·5 / ft. ′
19 21 +13·6 −18·7		19 42 +13·4 −18·4		20 42 −2·5	60 +0·1	9·9 −5·6 32·7 / 2 − 1·4
20 03 +13·7 −18·6		20 25 +13·5 −18·3		21 28 −2·4	June15–Aug.26	10·3 −5·7 33·9 / 4 − 1·9
20 48 +13·8 −18·5		21 11 +13·6 −18·2		22 19 −2·3	0° ′	10·6 −5·8 35·1 / 6 − 2·4
21 35 +13·9 −18·4		22 00 +13·7 −18·1		23 13 −2·2	41 +0·2	11·0 −5·9 36·3 / 8 − 2·7
22 26 +14·0 −18·3		22 54 +13·8 −18·0		24 11 −2·1	75 +0·1	11·4 −6·0 37·6 / 10 − 3·1
23 22 +14·1 −18·2		23 51 +13·9 −17·9		25 14 −2·0	Aug.27–Nov.28	11·8 −6·1 38·9 /
24 21 +14·2 −18·1		24 53 +14·0 −17·8		26 22 −1·9	0° ′	12·2 −6·2 40·1 / See table
25 26 +14·3 −18·0		26 00 +14·1 −17·7		27 36 −1·8	34 +0·3	12·6 −6·3 41·5 / ←
26 36 +14·4 −17·9		27 13 +14·2 −17·6		28 56 −1·7	60 +0·2	13·0 −6·4 42·8 / ft. ′
27 52 +14·5 −17·8		28 33 +14·3 −17·5		30 24 −1·6	80 +0·1	13·4 −6·5 44·2 / 70 − 8·1
29 15 +14·6 −17·7		30 00 +14·4 −17·4		32 00 −1·5	Nov.29–Dec.31	13·8 −6·6 45·5 / 75 − 8·4
30 46 +14·7 −17·6		31 35 +14·5 −17·3		33 45 −1·4	0° ′	14·2 −6·7 46·9 / 80 − 8·7
32 26 +14·8 −17·5		33 20 +14·6 −17·2		35 40 −1·3	41 +0·2	14·7 −6·8 48·4 / 85 − 8·9
34 17 +14·9 −17·4		35 17 +14·7 −17·1		37 48 −1·2	75 +0·1	15·1 −6·9 49·8 / 90 − 9·2
36 20 +15·0 −17·3		37 26 +14·8 −17·0		40 08 −1·1		15·5 −7·0 51·3 / 95 − 9·5
38 36 +15·1 −17·2		39 50 +14·9 −16·9		42 44 −1·0		16·0 −7·1 52·8 /
41 08 +15·2 −17·1		42 31 +15·0 −16·8		45 36 −0·9		16·5 −7·2 54·3 / 100 − 9·7
43 59 +15·3 −17·0		45 31 +15·1 −16·7		48 47 −0·8		16·9 −7·3 55·8 / 105 − 9·9
47 10 +15·4 −16·9		48 55 +15·2 −16·6		52 18 −0·7		17·4 −7·4 57·4 / 110 − 10·2
50 46 +15·5 −16·8		52 44 +15·3 −16·5		56 11 −0·6		17·9 −7·5 58·9 / 115 − 10·4
54 49 +15·6 −16·7		57 02 +15·4 −16·4		60 28 −0·5		18·4 −7·6 60·5 / 120 − 10·6
59 23 +15·7 −16·6		61 51 +15·5 −16·3		65 08 −0·4		18·8 −7·7 62·1 / 125 − 10·8
64 30 +15·8 −16·5		67 17 +15·6 −16·2		70 11 −0·3		19·3 −7·8 63·8 / 130 − 11·1
70 12 +15·9 −16·4		73 16 +15·7 −16·1		75 34 −0·2		19·8 −7·9 65·4 / 135 − 11·3
76 26 +16·0 −16·3		79 43 +15·8 −16·0		81 13 −0·1		20·4 −8·0 67·1 / 140 − 11·5
83 05 +16·1 −16·2		86 32 +15·9 −15·9		87 03 0·0		20·9 −8·1 68·8 / 145 − 11·7
90 00		90 00		90 00		21·4 70·5 / 150 − 11·9
						155 − 12·1

App. Alt. = Apparent altitude = Sextant altitude corrected for index error and dip.

Figure 19-13A. *Altitude Correction Tables, p. A2, Nautical Almanac.*

App. Alt.	OCT.–MAR. SUN Lower Limb	Upper Limb	APR.–SEPT. SUN Lower Limb	Upper Limb	STARS PLANETS
0 00	−18.2	−50.5	−18.4	−50.2	−34.5
03	17.5	49.8	17.8	49.6	33.8
06	16.9	49.2	17.1	48.9	33.2
09	16.3	48.6	16.5	48.3	32.6
12	15.7	48.0	15.9	47.7	32.0
15	15.1	47.4	15.3	47.1	31.4
0 18	−14.5	−46.8	−14.8	−46.6	−30.8
21	14.0	46.3	14.2	46.0	30.3
24	13.5	45.8	13.7	45.5	29.8
27	12.9	45.2	13.2	45.0	29.2
30	12.4	44.7	12.7	44.5	28.7
33	11.9	44.2	12.2	44.0	28.2
0 36	−11.5	−43.8	−11.7	−43.5	−27.8
39	11.0	43.3	11.2	43.0	27.3
42	10.5	42.8	10.8	42.6	26.8
45	10.1	42.4	10.3	42.1	26.4
48	9.6	41.9	9.9	41.7	25.9
51	9.2	41.5	9.5	41.3	25.5
0 54	−8.8	−41.1	−9.1	−40.9	−25.1
0 57	8.4	40.7	8.7	40.5	24.7
1 00	8.0	40.3	8.3	40.1	24.3
03	7.7	40.0	7.9	39.7	24.0
06	7.3	39.6	7.5	39.3	23.6
09	6.9	39.2	7.2	39.0	23.2
1 12	−6.6	−38.9	−6.8	−38.6	−22.9
15	6.2	38.5	6.5	38.3	22.5
18	5.9	38.2	6.2	38.0	22.2
21	5.6	37.9	5.8	37.6	21.9
24	5.3	37.6	5.5	37.3	21.6
27	4.9	37.2	5.2	37.0	21.2
1 30	−4.6	−36.9	−4.9	−36.7	−20.9
35	4.2	36.5	4.4	36.2	20.5
40	3.7	36.0	4.0	35.8	20.0
45	3.2	35.5	3.5	35.3	19.5
50	2.8	35.1	3.1	34.9	19.1
1 55	2.4	34.7	2.6	34.4	18.7
2 00	−2.0	−34.3	−2.2	−34.0	−18.3
05	1.6	33.9	1.8	33.6	17.9
10	1.2	33.5	1.5	33.3	17.5
15	0.9	33.2	1.1	32.9	17.2
20	0.5	32.8	0.8	32.6	16.8
25	−0.2	32.5	0.4	32.2	16.5
2 30	+0.2	−32.1	−0.1	−31.9	−16.1
35	0.5	31.8	+0.2	31.6	15.8
40	0.8	31.5	0.5	31.3	15.5
45	1.1	31.2	0.8	31.0	15.2
50	1.4	30.9	1.1	30.7	14.9
2 55	1.6	30.7	1.4	30.4	14.7
3 00	+1.9	−30.4	+1.7	−30.1	−14.4
05	2.2	30.1	1.9	29.9	14.1
10	2.4	29.9	2.1	29.7	13.9
15	2.6	29.7	2.4	29.4	13.7
20	2.9	29.4	2.6	29.2	13.4
25	3.1	29.2	2.9	28.9	13.2
3 30	+3.3	−29.0	+3.1	−28.7	−13.0
3 30	+3.3	−29.0	+3.1	−28.7	−13.0
35	3.6	28.7	3.3	28.5	12.7
40	3.8	28.5	3.5	28.3	12.5
45	4.0	28.3	3.7	28.1	12.3
50	4.2	28.1	3.9	27.9	12.1
3 55	4.4	27.9	4.1	27.7	11.9
4 00	+4.5	−27.8	+4.3	−27.5	−11.8
05	4.7	27.6	4.5	27.3	11.6
10	4.9	27.4	4.6	27.2	11.4
15	5.1	27.2	4.8	27.0	11.2
20	5.2	27.1	5.0	26.8	11.1
25	5.4	26.9	5.1	26.7	10.9
4 30	+5.6	−26.7	+5.3	−26.5	−10.7
35	5.7	26.6	5.5	26.3	10.6
40	5.9	26.4	5.6	26.2	10.4
45	6.0	26.3	5.8	26.0	10.3
50	6.2	26.1	5.9	25.9	10.1
4 55	6.3	26.0	6.0	25.8	10.0
5 00	+6.4	−25.9	+6.2	−25.6	−9.9
05	6.6	25.7	6.3	25.5	9.7
10	6.7	25.6	6.4	25.4	9.6
15	6.8	25.5	6.6	25.2	9.5
20	6.9	25.4	6.7	25.1	9.4
25	7.1	25.2	6.8	25.0	9.2
5 30	+7.2	−25.1	+6.9	−24.9	−9.1
35	7.3	25.0	7.0	24.8	9.0
40	7.4	24.9	7.2	24.6	8.9
45	7.5	24.8	7.3	24.5	8.8
50	7.6	24.7	7.4	24.4	8.7
5 55	7.7	24.6	7.5	24.3	8.6
6 00	+7.8	−24.5	+7.6	−24.2	−8.5
10	8.0	24.3	7.8	24.0	8.3
20	8.2	24.1	8.0	23.8	8.1
30	8.4	23.9	8.1	23.7	7.9
40	8.6	23.7	8.3	23.5	7.7
6 50	8.7	23.6	8.5	23.3	7.6
7 00	+8.9	−23.4	+8.6	−23.2	−7.4
10	9.1	23.2	8.8	23.0	7.2
20	9.2	23.1	9.0	22.8	7.1
30	9.3	23.0	9.1	22.7	7.0
40	9.5	22.8	9.2	22.6	6.8
7 50	9.6	22.7	9.4	22.4	6.7
8 00	+9.7	−22.6	+9.5	−22.3	−6.6
10	9.9	22.4	9.6	22.2	6.4
20	10.0	22.3	9.7	22.1	6.3
30	10.1	22.2	9.8	22.0	6.2
40	10.2	22.1	10.0	21.8	6.1
8 50	10.3	22.0	10.1	21.7	6.0
9 00	+10.4	−21.9	+10.2	−21.6	−5.9
10	10.5	21.8	10.3	21.5	5.8
20	10.6	21.7	10.4	21.4	5.7
30	10.7	21.6	10.5	21.3	5.6
40	10.8	21.5	10.6	21.2	5.5
9 50	10.9	21.4	10.6	21.2	5.4
10 00	+11.0	−21.3	+10.7	−21.1	−5.3

Additional corrections for temperature and pressure are given on the following page.
For bubble sextant observations ignore dip and use the star corrections for Sun, planets, and stars.

Figure 19-13B. *Altitude Correction Tables*, p. A3, Nautical Almanac.

ALTITUDE CORRECTION TABLES 0°–35°—MOON

App. Alt.	0°–4° Corrⁿ	5°–9° Corrⁿ	10°–14° Corrⁿ	15°–19° Corrⁿ	20°–24° Corrⁿ	25°–29° Corrⁿ	30°–34° Corrⁿ	App. Alt.
00	0° 33.8	5° 58.2	10° 62.1	15° 62.8	20° 62.2	25° 60.8	30° 58.9	00
10	35.9	58.5	62.2	62.8	62.1	60.7	58.8	10
20	37.8	58.7	62.2	62.8	62.1	60.7	58.8	20
30	39.6	58.9	62.3	62.8	62.1	60.7	58.7	30
40	41.2	59.1	62.3	62.8	62.0	60.6	58.6	40
50	42.6	59.3	62.4	62.7	62.0	60.6	58.5	50
00	1° 44.0	6° 59.5	11° 62.4	16° 62.7	21° 62.0	26° 60.5	31° 58.5	00
10	45.2	59.7	62.4	62.7	61.9	60.4	58.4	10
20	46.3	59.9	62.5	62.7	61.9	60.4	58.3	20
30	47.3	60.0	62.5	62.7	61.9	60.3	58.2	30
40	48.3	60.2	62.5	62.7	61.8	60.3	58.2	40
50	49.2	60.3	62.6	62.7	61.8	60.2	58.1	50
00	2° 50.0	7° 60.5	12° 62.6	17° 62.7	22° 61.7	27° 60.1	32° 58.0	00
10	50.8	60.6	62.6	62.6	61.7	60.1	57.9	10
20	51.4	60.7	62.6	62.6	61.6	60.0	57.8	20
30	52.1	60.9	62.7	62.6	61.6	59.9	57.8	30
40	52.7	61.0	62.7	62.6	61.5	59.9	57.7	40
50	53.3	61.1	62.7	62.6	61.5	59.8	57.6	50
00	3° 53.8	8° 61.2	13° 62.7	18° 62.5	23° 61.5	28° 59.7	33° 57.5	00
10	54.3	61.3	62.7	62.5	61.4	59.7	57.4	10
20	54.8	61.4	62.7	62.5	61.4	59.6	57.4	20
30	55.2	61.5	62.8	62.5	61.3	59.6	57.3	30
40	55.6	61.6	62.8	62.4	61.3	59.5	57.2	40
50	56.0	61.6	62.8	62.4	61.2	59.4	57.1	50
00	4° 56.4	9° 61.7	14° 62.8	19° 62.4	24° 61.2	29° 59.3	34° 57.0	00
10	56.7	61.8	62.8	62.3	61.1	59.3	56.9	10
20	57.1	61.9	62.8	62.3	61.1	59.2	56.9	20
30	57.4	61.9	62.8	62.3	61.0	59.1	56.8	30
40	57.7	62.0	62.8	62.2	60.9	59.1	56.7	40
50	57.9	62.1	62.8	62.2	60.9	59.0	56.6	50

H.P.	L	U	L	U	L	U	L	U	L	U	L	U	L	U	H.P.
54.0	0.3	0.9	0.3	0.9	0.4	1.0	0.5	1.1	0.6	1.2	0.7	1.3	0.9	1.5	54.0
54.3	0.7	1.1	0.7	1.2	0.7	1.2	0.8	1.3	0.9	1.4	1.1	1.5	1.2	1.7	54.3
54.6	1.1	1.4	1.1	1.4	1.1	1.4	1.2	1.5	1.3	1.6	1.4	1.7	1.5	1.8	54.6
54.9	1.4	1.6	1.5	1.6	1.5	1.6	1.6	1.7	1.6	1.8	1.8	1.9	1.9	2.0	54.9
55.2	1.8	1.8	1.8	1.8	1.9	1.9	1.9	1.9	2.0	2.0	2.1	2.1	2.2	2.2	55.2
55.5	2.2	2.0	2.2	2.0	2.3	2.1	2.3	2.1	2.4	2.2	2.4	2.3	2.5	2.4	55.5
55.8	2.6	2.2	2.6	2.2	2.6	2.3	2.7	2.3	2.7	2.4	2.8	2.4	2.9	2.5	55.8
56.1	3.0	2.4	3.0	2.5	3.0	2.5	3.1	2.6	3.1	2.6	3.2	2.7	3.2	2.7	56.1
56.4	3.4	2.7	3.4	2.7	3.4	2.7	3.4	2.7	3.4	2.8	3.5	2.8	3.5	2.9	56.4
56.7	3.7	2.9	3.7	2.9	3.8	2.9	3.8	2.9	3.8	3.0	3.8	3.0	3.9	3.0	56.7
57.0	4.1	3.1	4.1	3.1	4.1	3.1	4.1	3.1	4.2	3.1	4.2	3.2	4.2	3.2	57.0
57.3	4.5	3.3	4.5	3.3	4.5	3.3	4.5	3.3	4.5	3.3	4.5	3.4	4.6	3.4	57.3
57.6	4.9	3.5	4.9	3.5	4.9	3.5	4.9	3.5	4.9	3.5	4.9	3.5	4.9	3.6	57.6
57.9	5.3	3.8	5.3	3.8	5.2	3.8	5.2	3.7	5.2	3.7	5.2	3.7	5.2	3.7	57.9
58.2	5.6	4.0	5.6	4.0	5.6	4.0	5.6	4.0	5.6	3.9	5.6	3.9	5.6	3.9	58.2
58.5	6.0	4.2	6.0	4.2	6.0	4.2	6.0	4.2	6.0	4.1	5.9	4.1	5.9	4.1	58.5
58.8	6.4	4.4	6.4	4.4	6.4	4.4	6.3	4.4	6.3	4.3	6.3	4.3	6.2	4.2	58.8
59.1	6.8	4.6	6.8	4.6	6.7	4.6	6.7	4.5	6.7	4.5	6.6	4.4	6.5	4.3	59.1
59.4	7.2	4.8	7.1	4.8	7.1	4.8	7.1	4.8	7.0	4.7	7.0	4.7	6.9	4.6	59.4
59.7	7.5	5.1	7.5	5.0	7.5	5.0	7.5	5.0	7.4	4.9	7.3	4.8	7.2	4.7	59.7
60.0	7.9	5.3	7.9	5.3	7.9	5.2	7.8	5.2	7.8	5.1	7.7	5.0	7.6	4.9	60.0
60.3	8.3	5.5	8.3	5.5	8.2	5.4	8.2	5.4	8.1	5.3	8.0	5.2	7.9	5.1	60.3
60.6	8.7	5.7	8.7	5.7	8.6	5.7	8.6	5.6	8.5	5.5	8.4	5.4	8.2	5.3	60.6
60.9	9.1	5.9	9.0	5.9	9.0	5.9	8.9	5.8	8.8	5.7	8.7	5.6	8.5	5.4	60.9
61.2	9.5	6.2	9.4	6.1	9.4	6.1	9.3	6.0	9.2	5.9	9.1	5.8	8.9	5.6	61.2
61.5	9.8	6.4	9.8	6.3	9.7	6.3	9.7	6.2	9.5	6.1	9.4	5.9	9.2	5.8	61.5

DIP

Ht. of Eye	Corr	Ht. of Eye	Corr	Ht. of Eye	Corr
ft.		ft.		ft.	
4.0		24	−4.9	63	−7.8
4.4	−2.0	26	−5.0	65	−7.9
4.9	−2.1	27	−5.1	67	−8.0
5.3	−2.2	28	−5.2	68	−8.1
5.8	−2.3	29	−5.3	70	−8.2
6.3	−2.4	31	−5.4	72	−8.3
6.9	−2.5	32	−5.5	74	−8.4
7.4	−2.6	33	−5.6	75	−8.5
8.0	−2.7	35	−5.7	77	−8.6
8.6	−2.8	36	−5.8	79	−8.7
9.2	−2.9	37	−5.9	81	−8.8
9.8	−3.0	38	−6.0	83	−8.9
10.5	−3.1	40	−6.1	85	−9.0
11.2	−3.2	41	−6.2	87	−9.1
11.9	−3.3	42	−6.3	88	−9.2
12.6	−3.4	44	−6.4	90	−9.3
13.3	−3.5	45	−6.5	92	−9.4
14.1	−3.6	47	−6.6	94	−9.5
14.9	−3.7	48	−6.7	96	−9.6
15.7	−3.8	51	−6.8	98	−9.7
16.5	−3.9	52	−6.9	101	−9.8
17.4	−4.0	54	−7.0	103	−9.9
18.3	−4.1	55	−7.1	105	−10.0
19.1	−4.2	57	−7.2	107	−10.1
20.1	−4.3	58	−7.3	109	−10.2
21.0	−4.4	60	−7.4	111	−10.3
22.0	−4.5	62	−7.5	113	−10.4
22.9	−4.6	63	−7.6	116	−10.5
23.9	−4.7		−7.7	118	−10.6
24.9	−4.8			120	

MOON CORRECTION TABLE

The correction is in two parts; the first correction is taken from the upper part of the table with argument apparent altitude, and the second from the lower part, with argument H.P., in the same column as that from which the first correction was taken. Separate corrections are given in the lower part for lower (L) and upper (U) limbs. All corrections are to be **added** to apparent altitude, *but 30′ is to be subtracted from the altitude of the upper limb.*

For corrections for pressure and temperature see page A4.

For bubble sextant observations ignore dip, take the mean of upper and lower limb corrections and subtract 15′ from the altitude.

App. Alt. = Apparent altitude = Sextant altitude corrected for index error and dip.

XXXIV

Figure 19-14A. *Altitude Correction Tables 0°–35°—Moon, Nautical Almanac.*

ALTITUDE CORRECTION TABLES 35°-90°—MOON

App. Alt.	35°-39° Corrⁿ	40°-44° Corrⁿ	45°-49° Corrⁿ	50°-54° Corrⁿ	55°-59 Corrⁿ	60°-64 Corrⁿ	65°-69° Corrⁿ	70°-74° Corrⁿ	75°-79° Corrⁿ	80°-84° Corrⁿ	85°-89° Corrⁿ	App. Alt.
00	35 56·5	40 53·7	45 50·5	50 46·9	55 43·1	60 38·9	65 34·6	70 30·1	75 25·3	80 20·5	85 15·6	00
10	56·4	53·6	50·4	46·8	42·9	38·8	34·4	29·9	25·2	20·4	15·5	10
20	56·3	53·5	50·2	46·7	42·8	38·7	34·3	29·7	25·0	20·2	15·3	20
30	56·2	53·4	50·1	46·5	42·7	38·5	34·1	29·6	24·9	20·0	15·1	30
40	56·2	53·3	50·0	46·4	42·5	38·4	34·0	29·4	24·7	19·9	15·0	40
50	56·1	53·2	49·9	46·3	42·4	38·2	33·8	29·3	24·5	19·7	14·8	50
00	36 56·0	41 53·1	46 49·8	51 46·2	56 42·3	61 38·1	66 33·7	71 29·1	76 24·4	81 19·6	86 14·6	00
10	55·9	53·0	49·7	46·0	42·1	37·9	33·5	29·0	24·2	19·4	14·5	10
20	55·8	52·8	49·5	45·9	42·0	37·8	33·4	28·8	24·1	19·2	14·3	20
30	55·7	52·7	49·4	45·8	41·8	37·7	33·2	28·7	23·9	19·1	14·1	30
40	55·6	52·6	49·3	45·7	41·7	37·5	33·1	28·5	23·8	18·9	14·0	40
50	55·5	52·5	49·2	45·5	41·6	37·4	32·9	28·3	23·6	18·7	13·8	50
00	37 55·4	42 52·4	47 49·1	52 45·4	57 41·4	62 37·2	67 32·8	72 28·2	77 23·4	82 18·6	87 13·7	00
10	55·3	52·3	49·0	45·3	41·3	37·1	32·6	28·0	23·3	18·4	13·5	10
20	55·2	52·2	48·8	45·2	41·2	36·9	32·5	27·9	23·1	18·2	13·3	20
30	55·1	52·1	48·7	45·0	41·0	36·8	32·3	27·7	22·9	18·1	13·2	30
40	55·0	52·0	48·6	44·9	40·9	36·6	32·2	27·6	22·8	17·9	13·0	40
50	55·0	51·9	48·5	44·8	40·8	36·5	32·0	27·4	22·6	17·8	12·8	50
00	38 54·9	43 51·8	48 48·4	53 44·6	58 40·6	63 36·4	68 31·9	73 27·2	78 22·5	83 17·6	88 12·7	00
10	54·8	51·7	48·2	44·5	40·5	36·2	31·7	27·1	22·3	17·4	12·5	10
20	54·7	51·6	48·1	44·4	40·3	36·1	31·6	26·9	22·1	17·3	12·3	20
30	54·6	51·5	48·0	44·2	40·2	35·9	31·4	26·8	22·0	17·1	12·2	30
40	54·5	51·4	47·9	44·1	40·1	35·8	31·3	26·6	21·8	16·9	12·0	40
50	54·4	51·2	47·8	44·0	39·9	35·6	31·1	26·5	21·7	16·8	11·8	50
00	39 54·3	44 51·1	49 47·6	54 43·9	59 39·8	64 35·5	69 31·0	74 26·3	79 21·5	84 16·6	89 11·7	00
10	54·2	51·0	47·5	43·7	39·6	35·3	30·8	26·1	21·3	16·5	11·5	10
20	54·1	50·9	47·4	43·6	39·5	35·2	30·7	26·0	21·2	16·3	11·4	20
30	54·0	50·8	47·3	43·5	39·4	35·0	30·5	25·8	21·0	16·1	11·2	30
40	53·9	50·7	47·2	43·3	39·2	34·9	30·4	25·7	20·9	16·0	11·0	40
50	53·8	50·6	47·0	43·2	39·1	34·7	30·2	25·5	20·7	15·8	10·9	50

H.P.	L U	L U	L U	L U	L U	L U	L U	L U	L U	L U	L U	H.P.
54·0	1·1 1·7	1·3 1·9	1·5 2·1	1·7 2·4	2·0 2·6	2·3 2·9	2·6 3·2	2·9 3·5	3·2 3·8	3·5 4·1	3·8 4·5	54·0
54·3	1·4 1·8	1·6 2·0	1·8 2·2	2·0 2·5	2·3 2·7	2·5 3·0	2·8 3·2	3·0 3·5	3·3 3·8	3·6 4·1	3·9 4·4	54·3
54·6	1·7 2·0	1·9 2·2	2·1 2·4	2·3 2·6	2·5 2·8	2·7 3·0	3·0 3·3	3·2 3·5	3·5 3·8	3·7 4·1	4·0 4·3	54·6
54·9	2·0 2·2	2·2 2·3	2·3 2·5	2·5 2·7	2·7 2·9	2·9 3·1	3·2 3·3	3·4 3·5	3·6 3·8	3·9 4·0	4·1 4·3	54·9
55·2	2·3 2·3	2·5 2·4	2·6 2·6	2·8 2·8	3·0 2·9	3·2 3·1	3·4 3·3	3·6 3·5	3·8 3·7	4·0 4·0	4·2 4·2	55·2
55·5	2·7 2·5	2·8 2·6	2·9 2·7	3·1 2·9	3·2 3·0	3·4 3·2	3·6 3·4	3·7 3·5	3·9 3·7	4·1 3·9	4·3 4·1	55·5
55·8	3·0 2·6	3·1 2·7	3·2 2·8	3·3 3·0	3·5 3·1	3·6 3·3	3·8 3·4	3·9 3·6	4·1 3·7	4·2 3·9	4·4 4·0	55·8
56·1	3·3 2·8	3·4 2·9	3·5 3·0	3·6 3·1	3·7 3·2	3·8 3·3	4·0 3·4	4·1 3·6	4·2 3·7	4·4 3·8	4·5 4·0	56·1
56·4	3·6 2·9	3·7 3·0	3·8 3·1	3·9 3·2	3·9 3·3	4·0 3·4	4·1 3·5	4·3 3·6	4·4 3·7	4·5 3·8	4·6 3·9	56·4
56·7	3·9 3·1	4·0 3·1	4·1 3·2	4·1 3·3	4·2 3·3	4·3 3·4	4·3 3·5	4·4 3·6	4·5 3·7	4·6 3·8	4·7 3·8	56·7
57·0	4·3 3·2	4·3 3·3	4·3 3·3	4·4 3·4	4·4 3·4	4·5 3·5	4·5 3·5	4·6 3·6	4·7 3·6	4·7 3·7	4·8 3·8	57·0
57·3	4·6 3·4	4·6 3·4	4·6 3·4	4·6 3·5	4·7 3·5	4·7 3·5	4·7 3·6	4·8 3·6	4·8 3·6	4·8 3·7	4·9 3·7	57·3
57·6	4·9 3·6	4·9 3·6	4·9 3·6	4·9 3·6	4·9 3·6	4·9 3·6	4·9 3·6	4·9 3·6	5·0 3·6	5·0 3·6	5·0 3·6	57·6
57·9	5·2 3·7	5·2 3·7	5·2 3·7	5·2 3·7	5·2 3·7	5·1 3·6	5·1 3·6	5·1 3·6	5·1 3·6	5·1 3·6	5·1 3·6	57·9
58·2	5·5 3·9	5·5 3·8	5·5 3·8	5·4 3·8	5·4 3·7	5·4 3·7	5·3 3·7	5·3 3·6	5·3 3·6	5·2 3·5	5·2 3·5	58·2
58·5	5·9 4·0	5·8 4·0	5·8 3·9	5·7 3·9	5·6 3·8	5·6 3·8	5·5 3·7	5·5 3·6	5·4 3·6	5·3 3·5	5·3 3·4	58·5
58·8	6·2 4·2	6·1 4·1	6·0 4·1	6·0 4·0	5·9 3·9	5·8 3·8	5·7 3·7	5·6 3·6	5·5 3·5	5·4 3·5	5·3 3·4	58·8
59·1	6·5 4·3	6·4 4·3	6·3 4·2	6·2 4·1	6·1 4·0	6·0 3·9	5·9 3·8	5·8 3·6	5·7 3·5	5·6 3·4	5·4 3·3	59·1
59·4	6·8 4·5	6·7 4·4	6·6 4·3	6·5 4·2	6·4 4·1	6·2 3·9	6·1 3·8	6·0 3·7	5·8 3·5	5·7 3·4	5·5 3·2	59·4
59·7	7·1 4·6	7·0 4·5	6·9 4·4	6·8 4·3	6·6 4·1	6·5 4·0	6·3 3·8	6·2 3·7	6·0 3·5	5·8 3·3	5·6 3·2	59·7
60·0	7·5 4·8	7·3 4·7	7·2 4·5	7·0 4·4	6·9 4·2	6·7 4·0	6·5 3·9	6·3 3·7	6·1 3·5	5·9 3·3	5·7 3·1	60·0
60·3	7·8 5·0	7·6 4·8	7·5 4·7	7·3 4·5	7·1 4·3	6·9 4·1	6·7 3·9	6·5 3·7	6·3 3·5	6·0 3·2	5·8 3·0	60·3
60·6	8·1 5·1	7·9 5·0	7·7 4·8	7·6 4·6	7·3 4·4	7·1 4·2	6·9 3·9	6·7 3·7	6·4 3·4	6·2 3·2	5·9 2·9	60·6
60·9	8·4 5·3	8·2 5·1	8·0 4·9	7·8 4·7	7·6 4·5	7·3 4·2	7·1 4·0	6·8 3·7	6·6 3·4	6·3 3·2	6·0 2·9	60·9
61·2	8·7 5·4	8·5 5·2	8·3 5·0	8·1 4·8	7·8 4·5	7·6 4·3	7·3 4·0	7·0 3·7	6·7 3·4	6·4 3·1	6·1 2·8	61·2
61·5	9·1 5·6	8·8 5·4	8·6 5·1	8·3 4·9	8·1 4·6	7·8 4·3	7·5 4·0	7·2 3·7	6·9 3·4	6·5 3·1	6·2 2·7	61·5

Figure 19-14B. *Altitude Correction Tables 35°–90°—Moon*, Nautical Almanac.

To illustrate the use of the altitude correction tables of the *Nautical Almanac* to obtain the correction for an observation of a star, suppose that the star Dubhe was observed with a sextant having an IC of +1.2′, and its hs was recorded as 24° 37.7′. The observer's height of eye is 37.6 feet. To find the apparent altitude and then the altitude correction, the foregoing information is first entered on a sight reduction form, the applicable portion of which is reproduced below.

Body	DUBHE
IC	+1.2′
D (Ht 37.6′)	-5.9′
Sum	-4.7′
hs	24° 37.7′
ha	24° 33.0′

The apparent altitude ha is found by applying the sum of the IC and dip corrections, −4.7′, to the hs. Entering the Altitude Correction Table of the *Nautical Almanac* shown in Figure 19-13A with an ha of 24° 33.0′, a correction of −2.1 is obtained. Applying this correction to the ha, the form is completed as follows:

Body	DUBHE
IC	+1.2′
D (Ht 37.6′)	-5.9′
Sum	-4.7′
hs	24° 37.7′
ha	24° 33.0′
Alt Corr	-2.1′
Add'l Corr Moon HP/Corr	
Ho	24° 30.9′

If, as is usually the case for a star observed under normal atmospheric conditions, no "additional" correction applies, the observed altitude Ho is obtained by subtracting the altitude correction from ha as shown above.

The altitude correction tables of the *Nautical Almanac* are based on the assumption that near normal atmospheric conditions of temperature and barometric pressure prevail; the assumed temperature is 50°F (10°C) and the assumed pressure is 29.83 inches (1,010 millibars) of mercury. If actual atmospheric conditions differ markedly from these values, the atmospheric density and therefore its refractive characteristics would be affected to some degree. Under these conditions, it is necessary to apply an additional refraction correction to the apparent altitude, especially when the ha is 10° or less. A combined correction table for nonstandard air temperature and nonstandard atmospheric pressure is given on page A4 of the *Nautical Almanac,* reproduced in Figure 19-15 on page 354. To use the table, the top half is first entered, using as a vertical argument the temperature and as a horizontal argument the pressure. The point at which imaginary lines from these arguments cross locates a zone letter. Using as arguments this zone letter and the apparent altitude, a correction is then found; interpolation to the nearest tenth is necessary for apparent altitudes between tabulated values. The resulting correction is entered in the "Add'l Corr" space on the sight form, and applied to the ha along with the altitude correction for standard conditions to obtain the Ho.

In practice, corrections for unusual temperatures and barometric pressure conditions are generally not applied to apparent altitudes greater than 10°, as the amount of the correction for higher altitudes is so small as to be considered insignificant for most applications. Moreover, since atmospheric refraction of light from low-altitude bodies can vary unpredictably even in standard atmospheric conditions, and more so in nonstandard conditions, most navigators will normally not observe bodies with altitudes lower than 10°, unless there are no alternatives. Thus, neither Table A3 nor the additional corrections table, Table A4, are used very often in practical navigation.

Adjustments for Equivalent Reading for the Center of the Body

As was mentioned earlier in this chapter, the observed altitude Ho of a celestial body is defined as the angle that would be formed at the center of the earth between the observer's celestial horizon and the line of sight to the center of the body. Since a star is so far distant from earth, it always appears as a point source of light, having no measurable diameter. On the other hand, the sun, the moon, and at times the planets Venus and Mars do have significant diameters as viewed from earth. Although sextant observations could be made to the approximate centers of the sun and moon, it is the preferred procedure to bring either their upper or lower limbs tangent to the sea horizon, for increased accuracy. When this is done, there are three types of corrections that must be made to arrive at an equivalent reading for the center of the body: semidiameter, augmentation, and phase.

Semidiameter, abbreviated SD, is perhaps the most obvious of these three conditions. As can be seen from Figure 19-16, the apparent altitude as measured by the sextant must be decreased by the amount of one-half the diameter of the body as measured by the sextant when the upper limb is observed, and increased by this amount if the lower limb is observed. The semidiameter of the sun varies from a little less than 15.8' early in July, when the sun is at its greatest distance, to about 16.3' early in January, when the earth is nearest the sun. The moon also varies in apparent size as it moves in its orbit around the earth; its semidiameter ranges from 14.7' to 16.8' over the period of each

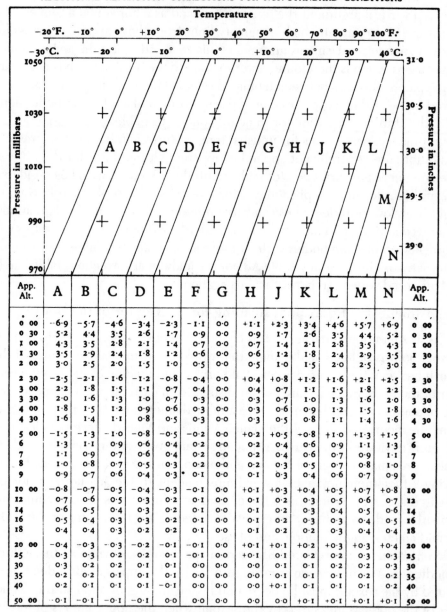

App. Alt.	A	B	C	D	E	F	G	H	J	K	L	M	N	App. Alt.
0 00	−6·9	−5·7	−4·6	−3·4	−2·3	−1·1	0·0	+1·1	+2·3	+3·4	+4·6	+5·7	+6·9	0 00
0 30	5·2	4·4	3·5	2·6	1·7	0·9	0·0	0·9	1·7	2·6	3·5	4·4	5·2	0 30
1 00	4·3	3·5	2·8	2·1	1·4	0·7	0·0	0·7	1·4	2·1	2·8	3·5	4·3	1 00
1 30	3·5	2·9	2·4	1·8	1·2	0·6	0·0	0·6	1·2	1·8	2·4	2·9	3·5	1 30
2 00	3·0	2·5	2·0	1·5	1·0	0·5	0·0	0·5	1·0	1·5	2·0	2·5	3·0	2 00
2 30	−2·5	−2·1	−1·6	−1·2	−0·8	−0·4	0·0	+0·4	+0·8	+1·2	+1·6	+2·1	+2·5	2 30
3 00	2·2	1·8	1·5	1·1	0·7	0·4	0·0	0·4	0·7	1·1	1·5	1·8	2·2	3 00
3 30	2·0	1·6	1·3	1·0	0·7	0·3	0·0	0·3	0·7	1·0	1·3	1·6	2·0	3 30
4 00	1·8	1·5	1·2	0·9	0·6	0·3	0·0	0·3	0·6	0·9	1·2	1·5	1·8	4 00
4 30	1·6	1·4	1·1	0·8	0·5	0·3	0·0	0·3	0·5	0·8	1·1	1·4	1·6	4 30
5 00	−1·5	−1·3	−1·0	−0·8	−0·5	−0·2	0·0	+0·2	+0·5	+0·8	+1·0	+1·3	+1·5	5 00
6	1·3	1·1	0·9	0·6	0·4	0·2	0·0	0·2	0·4	0·6	0·9	1·1	1·3	6
7	1·1	0·9	0·7	0·6	0·4	0·2	0·0	0·2	0·4	0·6	0·7	0·9	1·1	7
8	1·0	0·8	0·7	0·5	0·3	0·2	0·0	0·2	0·3	0·5	0·7	0·8	1·0	8
9	0·9	0·7	0·6	0·4	0·3	0·1	0·0	0·1	0·3	0·4	0·6	0·7	0·9	9
10 00	−0·8	−0·7	−0·5	−0·4	−0·3	−0·1	0·0	+0·1	+0·3	+0·4	+0·5	+0·7	+0·8	10 00
12	0·7	0·6	0·5	0·3	0·2	0·1	0·0	0·1	0·2	0·3	0·5	0·6	0·7	12
14	0·6	0·5	0·4	0·3	0·2	0·1	0·0	0·1	0·2	0·3	0·4	0·5	0·6	14
16	0·5	0·4	0·3	0·3	0·2	0·1	0·0	0·1	0·2	0·3	0·3	0·4	0·5	16
18	0·4	0·4	0·3	0·2	0·2	0·1	0·0	0·1	0·2	0·2	0·3	0·4	0·4	18
20 00	−0·4	−0·3	−0·3	−0·2	−0·1	−0·1	0·0	+0·1	+0·1	+0·2	+0·3	+0·3	+0·4	20 00
25	0·3	0·3	0·2	0·2	0·1	−0·1	0·0	+0·1	0·1	0·2	0·2	0·3	0·3	25
30	0·3	0·2	0·2	0·1	0·1	0·0	0·0	0·0	0·1	0·1	0·2	0·2	0·3	30
35	0·2	0·2	0·1	0·1	0·1	0·0	0·0	0·0	0·1	0·1	0·1	0·2	0·2	35
40	0·2	0·1	0·1	0·1	−0·1	0·0	0·0	0·0	+0·1	0·1	0·1	0·1	0·2	40
50 00	··0·1	−0·1	−0·1	−0·1	0·0	0·0	0·0	0·0	0·0	+0·1	+0·1	+0·1	+0·1	50 00

The graph is entered with arguments temperature and pressure to find a zone letter; using as arguments this zone letter and apparent altitude (sextant altitude corrected for dip), a correction is taken from the table. This correction is to be applied to the sextant altitude in addition to the corrections for standard conditions (for the Sun, planets and stars from the inside front cover and for the Moon from the inside back cover).

242-578 () · 69

Figure 19-15. *Additional Corrections table, p. A4, Nautical Almanac.*

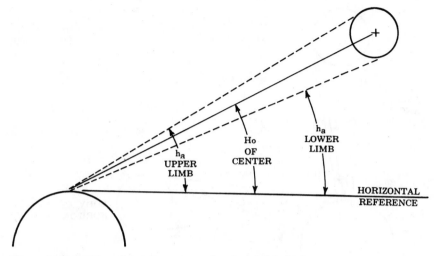

Figure 19-16. *Effect of semidiameter on the altitude of a body.*

month. Although the navigational planets do have a measurable semidiameter at certain times of the year, it is customary to observe their centers at these times, thereby making any semidiameter correction for the planets unnecessary.

Although a correction for the semidiameter of the sun and moon could be made by using daily semidiameter figures for these bodies, which are given at the bottom of each daily page of the *Nautical Almanac,* the usual procedure is simply to use the *Nautical Almanac* Altitude Correction Tables for the sun and the moon. In the sun tables located inside the front cover, the semidiameter correction and the parallax correction to be discussed below are combined with the refraction correction to produce a single "altitude" correction to ha. In the moon tables, located inside the back cover, the semidiameter correction is combined with the augmentation correction discussed below as well as the re-fraction correction; a separate correction for parallax is also extracted. The sun Altitude Correction Tables are shown in Figures 19-13A and 19-13B. The table of Figure 19-13A, located on the left-hand side of the inside front cover of the *Almanac,* is, like the stars and planets table, a critical value table for ha be-tween 10° and 90°, while the table on the right-hand side pictured in Figure 19-13B for ha between 0° and 10° requires interpolation. In each table, there are two columns of corrections; one is for upper or lower limb observations made from October through March, and the other is for observations made from April through September. The moon Altitude Correction Tables are shown in Figures 19-14A and 19-14B; their use will be described in the fol-lowing section.

The increase in apparent size of the sun and moon as a result of increase in the apparent altitude of these bodies is called *augmentation.* The augmentation effect is illustrated in Figure 19-17.

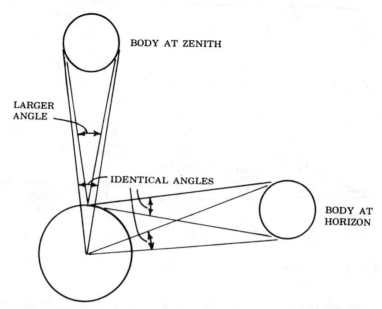

Figure 19-17. *The augmentation effect.*

If the celestial body is near the visible horizon of the observer on the earth's surface, its distance is about the same as it would be as viewed from the center of the earth. If the body is near the zenith, however, its distance is decreased by the radius of the earth. Hence, the body appears larger than it should when it is near the zenith of the observer, assuming its mean distance from earth is always the same. The augmentation correction for the sun is so small (a maximum of ⅟₂₄ of 1″ of arc) that it can be ignored, while the augmentation correction for the moon as mentioned above is included in the moon Altitude Correction Tables on the inside back cover of the *Nautical Almanac* (Figures 19-14A and 19-14B).

A *phase* correction is necessary to compensate for the fact that the actual centers of the moon and sometimes the inferior planet Venus may differ from the apparent centers because of the phase of the body. A phase correction for the moon is unnecessary if its upper or lower limb is observed. A phase correction for Venus is included in the tabulations for its GHA and declination in the daily pages of the *Nautical Almanac*. The superior navigational planets Mars, Jupiter, and Saturn have no phases, so no phase corrections are necessary for them.

Adjustment to Equivalent Reading at the Center of the Earth

As was the case with semidiameter, the need to adjust the apparent altitude of a celestial body located within the solar system for the radius of the earth should be readily apparent. The difference in the apparent altitude of a body

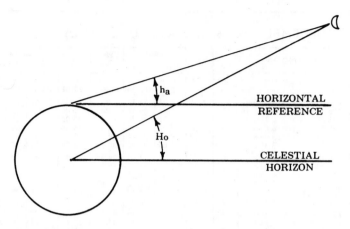

Figure 19-18. *The effect of parallax.*

within the solar system as viewed from the surface of the earth and from its center is called *parallax*. The effects of parallax are greatest when the body is near the visible horizon, and least when the body is near the observer's zenith. The maximum value of parallax, occurring when a body is on the visible horizon, is called *horizontal parallax*, abbreviated *HP.*

In Figure 19-18, the angle formed at the center of the earth between the celestial horizon and the line of sight to the body, Ho, is greater than the corresponding angle ha formed at the earth's surface between the horizontal reference plane parallel to the celestial horizon and the line of sight to the body, because of the radius of the earth. The nearer the body is to the earth, the greater the difference in the two angles becomes. Hence, it follows that the effect of parallax for the moon is most pronounced, followed by the parallax of the sun and finally that of the planets and stars. A star is so far distant that incoming light rays from it are virtually parallel at all points between the earth's surface and center, thus rendering the effect of parallax negligible for a stellar observation. The effect for the moon, on the other hand, is so great that the values of horizontal parallax are tabulated for each hour in the daily pages of the *Nautical Almanac,* for use as described below in the altitude correction tables of the moon. The parallax correction for the sun is included in the Altitude Correction Tables for the sun in the *Almanac* (see Figure 19-13A), and this correction forms the basis for the additional corrections for Venus and Mars in the same table.

When using the Altitude Correction Tables for the moon in the *Nautical Almanac* (Figures 19-14A and B), the apparent altitude ha of either the upper or lower limb of the moon is the entering argument. The left-hand table of Figure 19-14A is used for values of ha between 0° and 35°, and the right-hand table of Figure 19-14B is for values between 35° and 90°. The total correction

G.M.T.	SUN G.H.A.	SUN Dec.	MOON G.H.A.	v	MOON Dec.	d	H.P.	Lat.	Twilight Naut.	Twilight Civil	Sun-rise	Moonrise 15	16	17	18
d h	° '	° '	° '	'	° '	'	'	°	h m	h m	h m	h m	h m	h m	h m
15 00	181 16.4	S23 15.0	292 09.8	10.9	N 6 43.8	13.7	59.1	N 72	08 21	10 49	■	23 01	25 03	01 03	03 07
01	196 16.1	15.2	306 39.7	10.9	6 30.1	13.7	59.1	N 70	08 01	09 49	■	23 03	24 56	00 56	02 50
02	211 15.8	15.3	321 09.6	11.0	6 16.4	13.8	59.0	68	07 46	09 14	■	23 04	24 51	00 51	02 36
03	226 15.5 ··	15.4	335 39.6	11.1	6 02.6	13.8	59.0	66	07 33	08 49	10 29	23 05	24 46	00 46	02 26
04	241 15.2	15.6	350 09.7	11.2	5 48.8	13.8	59.0	64	07 22	08 30	09 47	23 06	24 42	00 42	02 17
05	256 14.9	15.7	4 39.9	11.2	5 35.0	13.8	58.9	62	07 12	08 14	09 19	23 07	24 39	00 39	02 09
06	271 14.6	S23 15.8	19 10.1	11.3	N 5 21.2	13.8	58.9	60	07 04	08 00	08 58	23 08	24 36	00 36	02 02
07	286 14.2	16.0	33 40.4	11.3	5 07.4	13.8	58.9	N 58	06 56	07 49	08 41	23 09	24 34	00 34	01 57
S 08	301 13.9	16.1	48 10.7	11.4	4 53.6	13.9	58.8	56	06 50	07 39	08 26	23 10	24 32	00 32	01 51
A 09	316 13.6 ··	16.2	62 41.1	11.5	4 39.7	13.8	58.8	54	06 44	07 30	08 13	23 10	24 30	00 30	01 47
T 10	331 13.3	16.4	77 11.6	11.5	4 25.9	13.9	58.8	52	06 38	07 22	08 02	23 11	24 28	00 28	01 43
U 11	346 13.0	16.5	91 42.1	11.6	4 12.0	13.8	58.7	50	06 33	07 14	07 52	23 11	24 26	00 26	01 39
R 12	1 12.7	S23 16.6	106 12.7	11.7	N 3 58.2	13.9	58.7	45	06 21	06 58	07 32	23 12	24 22	00 22	01 31
D 13	16 12.4	16.8	120 43.4	11.7	3 44.3	13.8	58.6	N 40	06 11	06 45	07 15	23 13	24 20	00 20	01 24
A 14	31 12.1	16.9	135 14.1	11.7	3 30.5	13.8	58.6	35	06 01	06 33	07 01	23 14	24 17	00 17	01 19
Y 15	46 11.8 ··	17.0	149 44.8	11.8	3 16.6	13.8	58.6	30	05 53	06 22	06 49	23 14	24 15	00 15	01 14
16	61 11.5	17.1	164 15.6	11.9	3 02.8	13.9	58.5	20	05 36	06 04	06 28	23 16	24 11	00 11	01 05
17	76 11.2	17.2	178 46.5	11.9	2 48.9	13.9	58.5	N 10	05 20	05 46	06 09	23 17	24 08	00 08	00 57
18	91 10.9	S23 17.4	193 17.4	12.0	N 2 35.0	13.8	58.5	0	05 03	05 29	05 52	23 18	24 04	00 04	00 50
19	106 10.6	17.5	207 48.4	12.0	2 21.2	13.8	58.4	S 10	04 44	05 11	05 34	23 19	24 01	00 01	00 44
20	121 10.3	17.6	222 19.4	12.1	2 07.4	13.9	58.4	20	04 22	04 51	05 16	23 20	23 58	24 36	00 36
								30	03 53	04 26	04 54	23 21	23 54	24 28	00 28

Figure 19-19. *Excerpt from right-hand daily page,* Nautical Almanac, *15 December.*

is in two parts, the first for the effects of refraction, semidiameter, and augmentation, and the second for the effects of parallax. The first correction is obtained by entering the upper half of the appropriate table in the column containing values of ha; six values of the correction are given for each degree of the apparent altitude. The correction should be interpolated to the nearest tenth if an entering argument is between the tabulated values. To obtain the second correction, the column containing the entering ha is followed down to the lower half of the table. Here, two corrections are located in each column, one for upper limb observations (U) and the other for lower limb (L). The horizontal entering argument used in the lower half of the moon table is the horizontal parallax (HPO) of the moon at the GMT of the observation, obtained from the daily pages of the *Almanac;* the HP extracted from the *Almanac* is the value tabulated for the whole hour of GMT closest to the GMT of the observation. Interpolation is performed in the lower table if the actual value of HP extracted from the *Almanac* lies between two tabulated values. After the HP correction has been obtained, it is added to the altitude correction obtained from the top half of the table. Finally, if the upper limb of the moon was observed, 30' is then subtracted from this combined correction to yield the total altitude and parallax correction for the moon.

As an example of finding the Ho of the moon, suppose that the sextant altitude of the lower limb of the moon were obtained at GMT 02-45-04 on 15 December and found to be 49° 45.6'. Height of eye is 16 feet and IC is +1.5'. Using the *Nautical Almanac* and the sight form, an ha of 49° 43.2' is computed. To enter the altitude correction tables for the moon shown in Figure 19-14B, the value of the HP of the moon for the time and date of the observation must be found. This information is obtained from the daily pages of the *Almanac*, a portion of which appears in Figure 19-19 for the day in question.

The HP of the moon listed for GMT 0300 on 15 December, the value closest to the GMT of the observation, is 59.0. Entering the moon altitude table of Figure 19-14B with ha, the 49° portion locates the observer in the third column. The tabulated correction 49° 40.0' is 47.2'; and for 49° 50.0', 47.0'. Interpolation between these values for 49° 43.2' yields an altitude correction of 47.1'. Next, the third column is followed down into the lower half of the table. Since a lower limb was shot, the figures under the column heading "L" will be used to obtain the parallax correction. Tabulated values for an HP of 58.8 is 6.0'; and for an HP of 59.1, 6.3'. Interpolation between these values for an HP of 59.0 results in an HP correction of 6.2'. Since the lower limb was observed, it remains only to enter these values in a sight form, and apply them to ha to obtain the observed altitude Ho:

Body	MOON (LL)
GMT	
IC	+1.5'
D (Ht 16')	-3.9'
Sum	-2.4'
hs	49° 45.6'
ha	49° 43.2'
Alt Corr	+47.1'
Add'l Corr Moon HP/Corr	59.0'/ +6.2'
Ho	50° 36.5'

The augmentation effect correction for lunar observations discussed in the preceding section can also be considered an adjustment of the apparent altitude at the earth's surface to obtain the desired observed altitude at the center of the earth.

As an example of the application of the additional correction parallax tables for a planet, suppose that the sextant altitude of Venus was observed on 31 December and recorded as 15° 13.5'. Height of eye is 15 feet and IC is −1.3'. Entering this information on a sight form, ha is first computed:

Body	VENUS
IC	-1.3'
D (Ht 15')	-3.8'
Sum	-5.1'
hs	15° 13.5'
ha	15° 08.4'

The entering arguments for the additional correction tables for Venus and Mars are the GMT date of the observation and the apparent altitude ha of the planet. Using the ha for this example, 15° 08.4′, the altitude correction tables are entered once for the basic altitude correction, –3.5′, and then again for the additional correction for Venus, +0.6 in this case.

If a correction for nonstandard temperature and barometric pressure conditions applied in this example, its value would also be entered in the "Add'l Corr" space on the sight form, along with the additional parallax correction. These two additional corrections together with the basic altitude correction would then be added algebraically to ha to form the observed altitude Ho:

Body	VENUS
IC	-1.3′
D (Ht 15′)	-3.8′
Sum	-5.1′
hs	15° 13.5′
ha	15° 08.4′
Alt Corr	-3.5′
Add'l Corr Moon HP/Corr	+.6′
Ho	15° 05.5′

Sextant Altitude Corrections Using the *Air Almanac*

As was stated at the beginning of the section on sextant altitude corrections, the *Air Almanac* can also be used to obtain corrections for each of the categories of possible error discussed in the preceding sections, but the *Nautical Almanac* or a calculator is generally preferred in marine navigation because of the increased precision that they yield. In the *Air Almanac* the refraction correction for all bodies is extracted from the same refraction table, using ha as the entering argument. When applicable, the effects of semidiameter, augmentation, phase, and parallax must be separately reckoned using data in the daily pages, and combined with the extracted refraction correction to form the total correction to ha for each body observed. An additional adjustment to the resulting line of position necessitated by the Coriolis effect on a fast-moving aircraft is also required in air navigation; tables for this correction are also contained in the *Air Almanac*. The use of the *Air Almanac* in determining sextant altitude corrections will be explained in chapter 21.

SUMMARY

This chapter has discussed the marine sextant and its role in the practice of celestial navigation at sea. While there are many different models of sextants available, all are designed to measure the angle between a celestial body and the visible horizon with great precision. The sextant must be handled with skill, which develops only with experience, in order to produce observed sextant altitudes that yield highly accurate celestial LOPs.

After the sextant altitude of a celestial body relative to the visible horizon has been obtained, certain corrections must always be applied to account for various errors inherent in any sextant altitude observation. Certain adjustments must then be made to the resulting apparent altitude to obtain the observed altitude relative to the celestial horizon, upon which the solution of the celestial or navigational triangle is based. These corrections and adjustments are tabulated in several tables in both the *Nautical Almanac* and the *Air Almanac*; they can also be computed with an electronic calculator using formulas given in the *Nautical Almanac*, or be applied automatically by computer programs designed for celestial sight reduction. The corrections and adjustments applying to each of the bodies normally observed in celestial navigation when the *Nautical Almanac* is used are summarized below:

Corrections for *IC* and *Dip* must be applied to all sextant altitude observations to obtain the apparent altitude; the apparent altitude is then used as an entering argument in the *Nautical Almanac* to find the appropriate corrections and adjustments for the body:

Venus and *Mars*—Corrections must be made for refraction and parallax; they are obtained from the Altitude Correction Table A2 of the *Nautical Almanac*.

Other *planets* and *stars*—A single correction must be made for refraction; it is obtained from Table A2 of the *Nautical Almanac*.

The *sun*—Corrections must be made for refraction, parallax, and semidiameter. All are combined in Table A2 of the *Nautical Almanac*.

The *moon*—Corrections must be made for refraction parallax, semidiameter, and augmentation. Two combined corrections for all of these effects are obtained from the Moon Tables of the *Nautical Almanac*.

If the apparent altitude of the planets, stars, or the sun is less than 10°, Table A3 instead of A2 of the *Nautical Almanac* is used. If this low-altitude observation was made under nonstandard temperature and barometric pressure conditions, an additional correction from Table A4 of the *Nautical Almanac* is applied to the apparent altitude.

In the following chapters, the procedures used to convert the observed altitude into a celestial line of position, and the combination of several such LOPs to form a celestial fix or running fix, will be presented. Chapter 20 explains the use of the *Nautical Almanac* in conjunction with the *Sight Reduction Tables for Marine Navigation* to obtain the quantities required by the altitude-intercept method; chapter 21 demonstrates the use of the *Air Almanac* with the *Sight Reduction Tables for Air Navigation* to determine the same quantities; and chapter 22 gives the procedures for using the electronic calculator, the computer, or a set of concise sight reduction tables in the *Nautical Almanac* as an alternative to these. Chapter 23 sets forth the procedures for plotting the celestial fix and running fix. Regardless of the methods employed, the basic techniques of sextant observation and the application of sextant altitude corrections described herein remain essentially unchanged.

THE COMPLETE SOLUTION BY THE *NAUTICAL ALMANAC* AND *TABLES NO. 229*

20

In chapter 17, the technique of plotting the celestial LOP by the altitude-intercept method was demonstrated. It was pointed out that the method is based on the solution of the navigational triangle for the computed altitude and true azimuth of the body from an assumed position of the observer.

The process of deriving the information needed to plot a celestial line of position from an observation of a selected celestial body is called *sight reduction*. A complete set of calculations performed during the sight reduction process is referred to as the *complete solution* for the body observed. The *Sight Reduction Tables for Marine Navigation, No. 229,* yield a solution to the navigational triangle in the form of a computed altitude Hc and an azimuth angle Z for an observed celestial body from an assumed position at a given time of observation; the azimuth angle Z is then immediately converted to a true azimuth Zn to plot the celestial LOP by the altitude-intercept method. Entering arguments for *Tables No. 229* are three quantities from which two legs and an included vertex of the navigational triangle are derived—the assumed latitude (colatitude), the exact declination of the GP of the body (polar distance), and the local hour angle (meridian angle *t*). Thus, *Tables No. 229* are in reality a set of precalculated solutions for the third leg (coaltitude) and an interior angle (azimuth angle) of the navigational triangle, given the other two sides and the included angle between them.

The *Nautical Almanac* (or the *Air Almanac*) is used during sight reduction to provide a set of coordinates for the GP of the observed celestial body at the time of the observation—the exact declination and Greenwich hour angle (GHA). These coordinates, in combination with the terrestrial coordinates of the assumed position, form the three entering arguments for *Tables No. 229*: integral (whole) degrees of declination, local hour angle (from the GHA and assumed longitude), and assumed latitude.

In this chapter, the use of the *Nautical Almanac* in conjunction with the *Sight Reduction Tables for Marine Navigation No. 229,* to obtain the complete solution of a celestial observation will be demonstrated. The first few sections will deal with the layout of the *Almanac* and examples of how to use it to determine the GHA and declination for the bodies observed in celestial navigation. The determination of the local hour angle will then be explained, followed by a discussion of the layout and use of *Tables No. 229*. In chapter 21, the use of the *Air Almanac* with the *Sight Reduction Tables for Air Navigation, No. 249*, will be examined. While it is possible to use either almanac with either set of sight reduction tables, experience has shown that the introduction of these publications to the student navigator is usually best accomplished by presenting them in this manner.

THE *NAUTICAL ALMANAC*

As has been mentioned, the *Nautical Almanac* is used during sight reduction to obtain the exact declination and GHA of a celestial body at the moment of its observation, for further use with the assumed position as entering arguments for the sight reduction tables. The *Nautical Almanac* is published annually in a single volume by the Naval Observatory in the United States, and by Her Majesty's Stationery Office in England. The data in each edition of the almanac are compiled for the sun, moon, navigational planets, and stars for the entire year for which the almanac is printed. The data pertaining to the sun and navigational stars may be used for the succeeding year as well, by following instructions given for this purpose in the "Explanation" section.

In the white daily pages comprising the body of the *Nautical Almanac*, the values of GHA and declination for each hour of Greenwich Mean Time (GMT) are given for a 3-day period for Aries and the navigational planets on the left-hand pages, and for the sun and the moon on the right-hand pages. The left-hand pages also contain a listing of the sidereal hour angle (SHA) and declination of each of 57 so-called *navigational stars*, selected primarily on the basis of their magnitude. The listed values of SHA and declination change slightly from one 3-day period to the next because of the combined effects of precession and nutation, described in chapter 16, and *aberration*. This latter effect is caused by bending of the incoming light rays from the stars as a result of the earth's orbital velocity; the effect of aberration is cyclical, varying as the orbital velocity of the earth varies over the period of a year. Various tables used for the prediction of certain rising and setting phenomena appear on the right-hand pages in addition to the sun and moon data. The equation of time discussed in chapter 18, and the times of the meridian passage and phase of the moon appear at the bottom of the right-hand pages as well. The use of these tables will be covered later in this text. A typical pair of daily pages appear in Figures 20-1A and 20-1B.

G.M.T.	ARIES G.H.A.	VENUS −44 G.H.A.	Dec.	MARS +1·8 G.H.A.	Dec.	JUPITER −1·3 G.H.A.	Dec.	SATURN +0·1 G.H.A.	Dec.	STARS Name	S.H.A.	Dec.
15 00	83 14·2	221 28·5	S13 14·2	229 56·4	S12 25·1	211 11·1	S17 58·5	38 16·8	N14 29·8	Acamar	315 42·1	S 40 25·2
01	98 16·7	236 29·7	14·3	244 57·4	25·7	226 13·0	58·7	53 19·4	29·7	Achernar	335 49·9	S 57 23·2
02	113 19·1	251 30·9	14·4	259 58·3	26·2	241 15·0	58·8	68 22·0	29·7	Acrux	173 45·6	S 62 56·1
03	128 21·6	266 32·1 · ·	14·5	274 59·2 · ·	26·8	256 16·9 · · ·	58·9	83 24·6 · ·	29·7	Adhara	255 37·3	S 28 55·7
04	143 24·1	281 33·3	14·6	290 00·2	27·3	271 18·8	59·0	98 27·2	29·6	Aldebaran	291 25·7	N 16 27·3
05	158 26·5	296 34·5	14·7	305 01·1	27·8	286 20·8	59·2	113 29·8	29·6			
06	173 29·0	311 35·7	S13 14·8	320 02·0	S12 28·4	301 22·7	S17 59·3	128 32·4	N14 29·6	Alioth	166 48·6	N 56 06·7
T 07	188 31·5	326 36·8	14·9	335 03·0	28·9	316 24·7	59·4	143 35·0	29·5	Alkaid	153 24·1	N 49 27·1
U 08	203 33·9	341 38·0	15·0	350 03·9	29·5	331 26·6	59·5	158 37·6	29·5	Al Na'ir	28 23·8	S 47 06·4
E 09	218 36·4	356 39·2 · ·	15·1	5 04·9 · ·	30·0	346 28·5 · ·	59·7	173 40·2 · ·	29·5	Alnilam	276 18·4	S 1 13·0
S 10	233 38·9	11 40·4	15·2	20 05·8	30·5	1 30·5	59·8	188 42·8	29·4	Alphard	218 27·3	S 8 31·8
D 11	248 41·3	26 41·6	15·4	35 06·7	31·1	16 32·4	17 59·9	203 45·4	29·4			
A 12	263 43·8	41 42·7	S13 15·5	50 07·7	S12 31·6	31 34·3	S18 00·0	218 48·0	N14 29·4	Alphecca	126 38·3	N 26 48·5
Y 13	278 46·3	56 43·9	15·6	65 08·6	32·1	46 36·3	00·2	233 50·7	29·4	Alpheratz	358 16·7	N 28 56·1
14	293 48·7	71 45·1	15·7	80 09·5	32·7	61 38·2	00·3	248 53·3	29·3	Altair	62 39·7	N 8 47·4
15	308 51·2	86 46·2 · ·	15·8	95 10·5 · ·	33·2	76 40·2 · ·	00·4	263 55·9 · ·	29·3	Ankaa	353 46·9	S 42 28·0
16	323 53·6	101 47·4	16·0	110 11·4	33·8	91 42·1	00·5	278 58·5	29·3	Antares	113 05·9	S 26 22·2
17	338 56·1	116 48·5	16·1	125 12·3	34·3	106 44·0	00·7	294 01·1	29·2			
18	353 58·6	131 49·7	S13 16·2	140 13·3	S12 34·8	121 46·0	S18 00·8	309 03·7	N14 29·2	Arcturus	146 25·0	N 19 19·8
19	9 01·0	146 50·8	16·3	155 14·2	35·4	136 47·9	00·9	324 06·3	29·2	Atria	108 37·1	S 68 58·7
20	24 03·5	161 52·0	16·4	170 15·1	35·9	151 49·8	01·0	339 08·9	29·1	Avior	234 30·8	S 59 24·7
21	39 06·0	176 53·1 · ·	16·6	185 16·1 · ·	36·5	166 51·8 · ·	01·2	354 11·5 · ·	29·1	Bellatrix	279 05·9	N 6 19·6
22	54 08·4	191 54·3	16·7	200 17·0	37·0	181 53·7	01·3	9 14·1	29·1	Betelgeuse	271 35·5	N 7 24·3
23	69 10·9	206 55·4	16·8	215 17·9	37·5	196 55·7	01·4	24 16·7	29·0			
16 00	84 13·4	221 56·6	S13 17·0	230 18·9	S12 38·1	211 57·6	S18 01·5	39 19·3	N14 29·0	Canopus	264 09·8	S 52 40·6
01	99 15·8	236 57·7	17·1	245 19·8	38·6	226 59·5	01·7	54 21·9	29·0	Capella	281 21·2	N 45 58·4
02	114 18·3	251 58·8	17·2	260 20·7	39·1	242 01·5	01·8	69 24·5	29·0	Deneb	49 53·7	N 45 10·7
03	129 20·8	267 00·0 · ·	17·4	275 21·7 · ·	39·7	257 03·4 · ·	01·9	84 27·1 · ·	28·9	Denebola	183 06·2	N 14 43·9
04	144 23·2	282 01·1	17·5	290 22·6	40·2	272 05·4	02·0	99 29·7	28·9	Diphda	349 27·8	S 18 08·8
05	159 25·7	297 02·2	17·6	305 23·5	40·8	287 07·3	02·2	114 32·4	28·9			
06	174 28·1	312 03·4	S13 17·8	320 24·5	S12 41·3	302 09·2	S18 02·3	129 35·0	N14 28·8	Dubhe	194 30·3	N 61 54·2
W 07	189 30·6	327 04·5	17·9	335 25·4	41·8	317 11·2	02·4	144 37·6	28·8	Elnath	278 52·6	N 28 35·2
E 08	204 33·1	342 05·6	18·1	350 26·3	42·4	332 13·1	02·5	159 40·2	28·8	Eltanin	91 01·5	N 51 29·4
D 09	219 35·5	357 06·7 · ·	18·2	5 27·3 · ·	42·9	347 15·0 · ·	02·7	174 42·8 · ·	28·7	Enif	34 18·7	N 9 44·5
N 10	234 38·0	12 07·8	18·4	20 28·2	43·4	2 17·0	02·8	189 45·4	28·7	Fomalhaut	15 59·1	S 29 46·7
E 11	249 40·5	27 09·0	18·5	35 29·1	44·0	17 18·9	02·9	204 48·0	28·7			
S 12	264 42·9	42 10·1	S13 18·6	50 30·1	S12 44·5	32 20·9	S18 03·0	219 50·6	N14 28·7	Gacrux	172 37·0	S 56 56·8
D 13	279 45·4	57 11·2	18·8	65 31·0	45·0	47 22·8	03·2	234 53·2	28·6	Gienah	176 25·3	S 17 22·8
A 14	294 47·9	72 12·3	18·9	80 31·9	45·6	62 24·7	03·3	249 55·8	28·6	Hadar	149 34·0	S 60 13·9
Y 15	309 50·3	87 13·4 · ·	19·1	95 32·9 · ·	46·1	77 26·7 · ·	03·4	264 58·4 · ·	28·6	Hamal	328 36·7	N 23 19·8
16	324 52·8	102 14·5	19·2	110 33·8	46·6	92 28·6	03·5	280 01·0	28·5	Kaus Aust.	84 26·6	S 34 24·1
17	339 55·3	117 15·6	19·4	125 34·7	47·2	107 30·6	03·7	295 03·6	28·5			
18	354 57·7	132 16·7	S13 19·5	140 35·7	S12 47·7	122 32·5	S18 03·8	310 06·2	N14 28·5	Kochab	137 19·1	N 74 16·1
19	10 00·2	147 17·8	19·7	155 36·6	48·3	137 34·4	03·9	325 08·8	28·4	Markab	14 10·3	N 15 03·1
20	25 02·6	162 18·8	19·9	170 37·5	48·8	152 36·4	04·0	340 11·4	28·4	Menkar	314 48·2	N 3 58·8
21	40 05·1	177 19·9 · ·	20·0	185 38·5 · ·	49·3	167 38·3 · ·	04·2	355 14·0 · ·	28·4	Menkent	148 45·7	S 36 13·6
22	55 07·6	192 21·0	20·2	200 39·4	49·9	182 40·3	04·3	10 16·6	28·4	Miaplacidus	221 46·3	S 69 35·5
23	70 10·0	207 22·1	20·3	215 40·3	50·4	197 42·2	04·4	25 19·2	28·3			
17 00	85 12·5	222 23·2	S13 20·5	230 41·3	S12 50·9	212 44·1	S18 04·5	40 21·8	N14 28·3	Mirfak	309 25·9	N 49 45·8
01	100 15·0	237 24·3	20·7	245 42·2	51·5	227 46·1	04·7	55 24·4	28·3	Nunki	76 38·2	S 26 20·2
02	115 17·4	252 25·3	20·8	260 43·1	52·0	242 48·0	04·8	70 27·0	28·2	Peacock	54 09·7	S 56 50·0
03	130 19·9	267 26·4 · ·	21·0	275 44·1 · ·	52·5	257 50·0 · ·	04·9	85 29·6 · ·	28·2	Pollux	244 06·4	N 28 05·9
04	145 22·4	282 27·5	21·2	290 45·0	53·1	272 51·9	05·0	100 32·2	28·2	Procyon	245 32·8	N 5 18·1
05	160 24·8	297 28·5	21·3	305 45·9	53·6	287 53·8	05·2	115 34·8	28·2			
06	175 27·3	312 29·6	S13 21·5	320 46·9	S12 54·1	302 55·8	S18 05·3	130 37·4	N14 28·1	Rasalhague	96 36·5	N 12 34·7
T 07	190 29·7	327 30·7	21·7	335 47·8	54·7	317 57·7	05·4	145 40·0	28·1	Regulus	208 17·3	N 12 06·5
H 08	205 32·2	342 31·7	21·8	350 48·7	55·2	332 59·7	05·5	160 42·6	28·1	Rigel	281 42·4	S 8 13·9
U 09	220 34·7	357 32·8 · ·	22·0	5 49·6 · ·	55·7	348 01·6 · ·	05·7	175 45·2 · ·	28·0	Rigil Kent.	140 36·2	S 60 42·8
R 10	235 37·1	12 33·8	22·2	20 50·6	56·3	3 03·5	05·8	190 47·8	28·0	Sabik	102 49·6	S 15 41·5
S 11	250 39·6	27 34·9	22·4	35 51·5	56·9	18 05·5	05·9	205 50·4	28·0			
D 12	265 42·1	42 36·0	S13 22·5	50 52·4	S12 57·3	33 07·4	S18 06·0	220 53·0	N14 27·9	Schedar	350 17·1	N 56 23·1
A 13	280 44·5	57 37·0	22·7	65 53·4	57·8	48 09·4	06·2	235 55·6	27·9	Shaula	97 05·7	S 37 05·2
Y 14	295 47·0	72 38·0	22·9	80 54·3	58·4	63 11·3	06·3	250 58·2	27·9	Sirius	259 01·6	S 16 40·4
15	310 49·5	87 39·1 · ·	23·1	95 55·2 · ·	58·9	78 13·2 · ·	06·4	266 00·8 · ·	27·9	Spica	159 05·1	S 11 00·6
16	325 51·9	102 40·1	23·3	110 56·2	12 59·4	93 15·2	06·5	281 03·4	27·8	Suhail	223 15·8	S 43 18·6
17	340 54·4	117 41·2	23·4	125 57·1	13 00·0	108 17·1	06·6	296 06·0	27·8			
18	355 56·9	132 42·2	S13 23·6	140 58·0	S13 00·5	123 19·1	S18 06·8	311 08·6	N14 27·8	Vega	81 01·0	N 38 45·3
19	10 59·3	147 43·2	23·8	155 58·9	01·0	138 21·0	06·9	326 11·2	27·7	Zuben'ubi	137 43·1	S 15 55·4
20	26 01·8	162 44·3	24·0	170 59·9	01·6	153 22·9	07·0	341 13·8	27·7		S.H.A.	Mer. Pass.
21	41 04·2	177 45·3 · ·	24·2	186 00·8 · ·	02·1	168 24·9 · ·	07·1	356 16·4 · ·	27·7		° ′	h m
22	56 06·7	192 46·3	24·4	201 01·7	02·6	183 26·8	07·3	11 19·0	27·7	Venus	137 43·2	9 12
23	71 09·2	207 47·4	24·6	216 02·7	03·2	198 28·8	07·4	26 21·6	27·6	Mars	146 05·5	8 38
Mer. Pass. 18 20·1		v 1·1	d 0·1	v 0·9	d 0·5	v 1·9	d 0·1	v 2·6	d 0·0	Jupiter	127 44·2	9 51
										Saturn	315 06·0	21 19

Figure 20-1A. *Sample left-hand daily page,* Nautical Almanac, *15, 16, 17 December.*

SUN and MOON

G.M.T.	SUN G.H.A.	SUN Dec.	MOON G.H.A.	v	MOON Dec.	d	H.P.
15 00	181 17.8	S 23 14.3	333 25.4	9.4	N25 19.4	6.5	55.8
01	196 17.5	14.4	347 53.8	9.4	25 12.9	6.7	55.7
02	211 17.2	14.5	2 22.2	9.5	25 06.2	6.7	55.7
03	226 16.9	.. 14.7	16 50.7	9.7	24 59.5	6.9	55.7
04	241 16.7	14.8	31 19.4	9.7	24 52.6	7.0	55.7
05	256 16.4	15.0	45 48.1	9.7	24 45.6	7.1	55.6
06	271 16.1	S 23 15.1	60 16.9	9.9	N24 38.5	7.2	55.6
T 07	286 15.8	15.2	74 45.8	10.0	24 31.3	7.3	55.6
U 08	301 15.5	15.4	89 14.8	10.1	24 24.0	7.5	55.6
E 09	316 15.2	.. 15.5	103 43.9	10.2	24 16.5	7.6	55.6
S 10	331 14.9	15.6	118 13.1	10.3	24 08.9	7.7	55.5
D 11	346 14.6	15.8	132 42.4	10.4	24 01.2	7.8	55.5
A 12	1 14.3	S 23 15.9	147 11.8	10.5	N23 53.4	7.9	55.5
Y 13	16 14.0	16.0	161 41.3	10.5	23 45.5	8.0	55.5
14	31 13.7	16.2	176 10.8	10.7	23 37.5	8.1	55.4
15	46 13.4	.. 16.3	190 40.5	10.8	23 29.4	8.3	55.4
16	61 13.1	16.4	205 10.3	10.9	23 21.1	8.3	55.4
17	76 12.8	16.6	219 40.2	10.9	23 12.8	8.5	55.4
18	91 12.5	S 23 16.7	234 10.1	11.1	N23 04.3	8.5	55.4
19	106 12.1	16.8	248 40.2	11.1	22 55.8	8.7	55.3
20	121 11.8	16.9	263 10.3	11.3	22 47.1	8.7	55.3
21	136 11.5	.. 17.1	277 40.6	11.3	22 38.4	8.9	55.3
22	151 11.2	17.2	292 10.9	11.5	22 29.5	8.9	55.3
23	166 10.9	17.3	306 41.4	11.5	22 20.6	9.1	55.2
16 00	181 10.6	S 23 17.4	321 11.9	11.7	N22 11.5	9.1	55.2
01	196 10.3	17.6	335 42.6	11.7	22 02.4	9.3	55.2
02	211 10.0	17.7	350 13.3	11.8	21 53.1	9.3	55.2
03	226 09.7	.. 17.8	4 44.1	11.9	21 43.8	9.4	55.2
04	241 09.4	17.9	19 15.0	12.1	21 34.4	9.6	55.1
05	256 09.1	18.0	33 46.1	12.1	21 24.8	9.6	55.1
06	271 08.8	S 23 18.2	48 17.2	12.2	N21 15.2	9.7	55.1
W 07	286 08.5	18.3	62 48.4	12.3	21 05.5	9.8	55.1
E 08	301 08.2	18.4	77 19.7	12.4	20 55.7	9.8	55.1
D 09	316 07.9	.. 18.5	91 51.1	12.5	20 45.9	10.0	55.0
N 10	331 07.6	18.6	106 22.6	12.5	20 35.9	10.0	55.0
E 11	346 07.3	18.7	120 54.1	12.7	20 25.9	10.2	55.0
S 12	1 07.0	S 23 18.9	135 25.8	12.8	N20 15.7	10.2	55.0
D 13	16 06.7	19.0	149 57.6	12.8	20 05.5	10.3	55.0
A 14	31 06.4	19.1	164 29.4	12.9	19 55.2	10.3	55.0
Y 15	46 06.1	.. 19.2	179 01.3	13.1	19 44.9	10.5	54.9
16	61 05.8	19.3	193 33.4	13.1	19 34.4	10.5	54.9
17	76 05.5	19.4	208 05.5	13.2	19 23.9	10.6	54.9
18	91 05.2	S 23 19.5	222 37.7	13.3	N19 13.3	10.7	54.9
19	106 04.9	19.6	237 10.0	13.3	19 02.6	10.7	54.9
20	121 04.6	19.7	251 42.4	13.4	18 51.9	10.8	54.8
21	136 04.3	.. 19.8	266 14.8	13.6	18 41.1	10.9	54.8
22	151 04.0	19.9	280 47.4	13.6	18 30.2	11.0	54.8
23	166 03.7	20.0	295 20.0	13.7	18 19.2	11.0	54.8
17 00	181 03.4	S 23 20.2	309 52.7	13.8	N18 08.2	11.1	54.8
01	196 03.1	20.3	324 25.5	13.9	17 57.1	11.2	54.8
02	211 02.8	20.4	338 58.4	14.0	17 45.9	11.2	54.7
03	226 02.5	.. 20.5	353 31.4	14.0	17 34.7	11.3	54.7
04	241 02.2	20.6	8 04.4	14.2	17 23.4	11.4	54.7
05	256 01.9	20.7	22 37.6	14.2	17 12.0	11.4	54.7
06	271 01.6	S 23 20.8	37 10.8	14.3	N17 00.6	11.5	54.7
T 07	286 01.2	20.9	51 44.1	14.3	16 49.1	11.5	54.7
H 08	301 00.9	21.0	66 17.4	14.5	16 37.6	11.7	54.7
U 09	316 00.6	.. 21.0	80 50.9	14.5	16 25.9	11.6	54.7
R 10	331 00.3	21.1	95 24.4	14.6	16 14.3	11.8	54.6
11	346 00.0	21.2	109 58.0	14.7	16 02.5	11.7	54.6
S 12	0 59.7	S 23 21.3	124 31.7	14.7	N15 50.8	11.9	54.6
D 13	15 59.4	21.4	139 05.4	14.8	15 38.9	11.9	54.6
A 14	30 59.1	21.5	153 39.2	14.9	15 27.0	11.9	54.6
Y 15	45 58.8	.. 21.6	168 13.1	15.0	15 15.1	12.0	54.6
16	60 58.5	21.7	182 47.1	15.0	15 03.1	12.1	54.5
17	75 58.2	21.8	197 21.1	15.1	14 51.0	12.1	54.5
18	90 57.9	S 23 21.9	211 55.2	15.1	N14 38.9	12.2	54.5
19	105 57.6	22.0	226 29.3	15.3	14 26.7	12.2	54.5
20	120 57.3	22.1	241 03.6	15.3	14 14.5	12.2	54.5
21	135 57.0	.. 22.1	255 37.9	15.3	14 02.3	12.4	54.5
22	150 56.7	22.2	270 12.2	15.5	13 49.9	12.5	54.5
23	165 56.4	22.3	284 46.7	15.4	13 37.6	12.4	54.5
	S.D. 16.3	d 0.1	S.D. 15.1		15.0		14.9

Twilight, Sunrise and Moonrise

Lat.	Naut.	Civil	Sun-rise	Moonrise 15	16	17	18
N 72	08 20	10 48	■	□	□	18 22	20 42
N 70	08 01	09 48	■	□	16 01	18 55	20 56
68	07 45	09 14	■	□	17 08	19 18	21 08
66	07 32	08 49	10 28	15 14	17 44	19 36	21 17
64	07 21	08 29	09 47	16 14	18 09	19 51	21 25
62	07 12	08 13	09 19	16 47	18 29	20 03	21 31
60	07 03	08 00	08 58	17 11	18 45	20 13	21 37
N 58	06 56	07 48	08 40	17 31	18 58	20 22	21 42
56	06 49	07 38	08 26	17 47	19 10	20 30	21 47
54	06 43	07 29	08 13	18 01	19 20	20 36	21 51
52	06 38	07 21	08 02	18 13	19 29	20 43	21 54
50	06 32	07 14	07 52	18 23	19 37	20 48	21 58
45	06 21	06 58	07 32	18 45	19 54	21 00	22 05
N 40	06 10	06 44	07 15	19 03	20 07	21 10	22 10
35	06 01	06 33	07 01	19 18	20 19	21 18	22 16
30	05 52	06 22	06 49	19 31	20 29	21 26	22 20
20	05 36	06 04	06 28	19 52	20 47	21 38	22 28
N 10	05 20	05 46	06 09	20 11	21 02	21 49	22 34
0	05 03	05 29	05 52	20 29	21 16	22 00	22 41
S 10	04 44	05 11	05 34	20 46	21 30	22 10	22 47
20	04 22	04 51	05 15	21 05	21 45	22 21	22 53
30	03 53	04 26	04 53	21 26	22 02	22 33	23 01
35	03 34	04 11	04 41	21 39	22 12	22 40	23 05
40	03 11	03 53	04 26	21 53	22 23	22 48	23 10
45	02 41	03 30	04 08	22 10	22 36	22 58	23 16
S 50	01 56	03 01	03 45	22 31	22 52	23 09	23 22
52	01 28	02 46	03 34	22 41	23 00	23 14	23 26
54	00 44	02 28	03 22	22 52	23 08	23 20	23 29
56	////	02 06	03 08	23 04	23 17	23 26	23 33
58	////	01 37	02 52	23 19	23 28	23 33	23 37
S 60	////	00 49	02 31	23 36	23 40	23 41	23 41

Sunset, Twilight and Moonset

Lat.	Sun-set	Civil	Naut.	Moonset 15	16	17	18
N 72	■	13 02	15 30	□	□	13 37	12 45
N 70	■	14 02	15 49	□	14 25	13 03	12 28
68	■	14 36	16 05	□	13 16	12 38	12 15
66	13 22	15 01	16 18	13 29	12 40	12 19	12 04
64	14 04	15 21	16 29	12 29	12 13	12 03	11 55
62	14 31	15 37	16 39	11 55	11 53	11 50	11 47
60	14 53	15 50	16 47	11 30	11 36	11 39	11 40
N 58	15 10	16 02	16 54	11 10	11 22	11 29	11 34
56	15 25	16 12	17 01	10 53	11 10	11 20	11 28
54	15 38	16 21	17 07	10 39	10 59	11 13	11 23
52	15 49	16 29	17 13	10 27	10 49	11 06	11 19
50	15 58	16 37	17 18	10 16	10 41	11 00	11 15
45	16 19	16 54	17 30	09 53	10 23	10 46	11 06
N 40	16 36	17 06	17 41	09 34	10 08	10 35	10 58
35	16 50	17 18	17 50	09 19	09 55	10 25	10 52
30	17 02	17 29	17 59	09 05	09 44	10 17	10 46
20	17 23	17 48	18 15	08 42	09 25	10 02	10 36
N 10	17 42	18 05	18 31	08 22	09 08	09 48	10 28
0	17 59	18 22	18 48	08 03	08 52	09 37	10 20
S 10	18 17	18 40	19 07	07 44	08 36	09 25	10 11
20	18 36	19 00	19 29	07 23	08 19	09 12	10 02
30	18 58	19 25	19 58	06 59	07 59	08 57	09 52
35	19 11	19 40	20 17	06 45	07 47	08 48	09 46
40	19 26	19 58	20 40	06 29	07 34	08 38	09 39
45	19 44	20 21	21 10	06 09	07 18	08 26	09 31
S 50	20 06	20 50	21 56	05 45	06 58	08 11	09 22
52	20 17	21 06	22 23	05 33	06 49	08 04	09 17
54	20 29	21 23	23 08	05 19	06 39	07 57	09 12
56	20 43	21 45	////	05 03	06 26	07 48	09 07
58	21 00	22 15	////	04 44	06 13	07 39	09 01
S 60	21 20	23 04	////	04 21	05 56	07 27	08 54

SUN and MOON daily

Day	SUN Eqn. of Time 00h	12h	Mer. Pass.	MOON Mer. Pass. Upper	Lower	Age	Phase
15	05 12	04 58	11 55	01 50	14 16	17	◐
16	04 43	04 29	11 56	02 40	15 04	18	
17	04 14	04 00	11 56	03 27	15 49	19	

Figure 20-1B. *Sample right-hand daily page,* Nautical Almanac, 15, 16, 17 December.

The extraction of the GHA and declination from the *Nautical Almanac* varies somewhat for each of the four types of bodies observed in celestial navigation. These variations are best illustrated by means of examples. Consequently, the following four sections of this chapter will each present an example of the use of the *Nautical Almanac* to find the GHA and declination of one of the four types of bodies normally observed—a star, a planet, the sun, and the moon. In each of the following examples, the observed altitude Ho of the body will be considered already determined by the methods discussed in the last chapter.

Because of the many additions, subtractions, and interpolations necessary during manual sight reduction, it is customary to use a standardized form called a *sight reduction form* on which to record the various calculations. Many of the forms developed by the U.S. Navy for sight reduction and other celestial applications will be utilized for the examples presented in this and succeeding chapters of this book. The use of the top portion of the form for sight reductions using the *Nautical Almanac* in conjunction with *Tables No. 229* has already been presented in chapter 19, in connection with the determination of the observed altitude Ho; the use of the remaining sections of the form will be demonstrated here.

Determining the GHA and Declination of a Star

As an example of the use of the *Nautical Almanac* to obtain the GHA and declination of a star for a given time, suppose that the star Canopus was observed and was found to have an Ho of 50° 39.4′ at a GMT of 17-12-09 on 16 December. The ship's DR position at the time of the observation was L 34° 19.0′S. λ 163° 05.7′E. As previously mentioned, the Greenwich Mean Time and date of celestial observations are the primary entering arguments for the *Nautical Almanac,* since these form the basis by which all hourly values of coordinates in the daily pages are tabulated. Since Canopus is a star, its GHA is not separately tabulated, but rather it must be computed by the use of the formula

$$\text{GHA}\star = \text{GHA}\Upsilon + \text{SHA}\star$$

previously introduced in chapter 16.

To find the GHAϒ for the time of the observation, its tabulated value for the whole hour of GMT immediately preceding the observation is first obtained and recorded on the sight form as the GHA (h). From Figure 20-1A, the tabulation for 1700 GMT on 15 December is 338° 56.1′. Next, the amount of the additional movement of Aries during the remaining 12 minutes 9 seconds until 07-12-09 GMT must be determined and added to the tabulated GHA. This additional movement of Aries is referred to as the increment, abbreviated "Incre" on the form. A set of increments and corrections tables designed for obtaining this and certain other related increments and corrections is printed on

12	SUN PLANETS	ARIES	MOON	v or Corrⁿ d	v or Corrⁿ d	v or Corrⁿ d
s	° ′	° ′	° ′	′ ′	′ ′	′ ′
00	3 00·0	3 00·5	2 51·8	0·0 0·0	6·0 1·3	12·0 2·5
01	3 00·3	3 00·7	2 52·0	0·1 0·0	6·1 1·3	12·1 2·5
02	3 00·5	3 01·0	2 52·3	0·2 0·0	6·2 1·3	12·2 2·5
03	3 00·8	3 01·2	2 52·5	0·3 0·1	6·3 1·3	12·3 2·6
04	3 01·0	3 01·5	2 52·8	0·4 0·1	6·4 1·3	12·4 2·6
05	3 01·3	3 01·7	2 53·0	0·5 0·1	6·5 1·4	12·5 2·6
06	3 01·5	3 02·0	2 53·2	0·6 0·1	6·6 1·4	12·6 2·6
07	3 01·8	3 02·2	2 53·5	0·7 0·1	6·7 1·4	12·7 2·6
08	3 02·0	3 02·5	2 53·7	0·8 0·2	6·8 1·4	12·8 2·7
09	3 02·3	3 02·7	2 53·9	0·9 0·2	6·9 1·4	12·9 2·7
10	3 02·5	3 03·0	2 54·2	1·0 0·2	7·0 1·5	13·0 2·7
11	3 02·8	3 03·3	2 54·4	1·1 0·2	7·1 1·5	13·1 2·7
12	3 03·0	3 03·5	2 54·7	1·2 0·3	7·2 1·5	13·2 2·8
13	3 03·3	3 03·8	2 54·9	1·3 0·3	7·3 1·5	13·3 2·8
14	3 03·5	3 04·0	2 55·1	1·4 0·3	7·4 1·5	13·4 2·8
15	3 03·8	3 04·3	2 55·4	1·5 0·3	7·5 1·6	13·5 2·8
16	3 04·0	3 04·5	2 55·6	1·6 0·3	7·6 1·6	13·6 2·8
17	3 04·3	3 04·8	2 55·9	1·7 0·4	7·7 1·6	13·7 2·9
18	3 04·5	3 05·0	2 56·1	1·8 0·4	7·8 1·6	13·8 2·9
19	3 04·8	3 05·3	2 56·3	1·9 0·4	7·9 1·6	13·9 2·9
20	3 05·0	3 05·5	2 56·6	2·0 0·4	8·0 1·7	14·0 2·9
21	3 05·3	3 05·8	2 56·8	2·1 0·4	8·1 1·7	14·1 2·9
22	3 05·5	3 06·0	2 57·0	2·2 0·5	8·2 1·7	14·2 3·0
23	3 05·8	3 06·3	2 57·3	2·3 0·5	8·3 1·7	14·3 3·0
24	3 06·0	3 06·5	2 57·5	2·4 0·5	8·4 1·8	14·4 3·0
25	3 06·3	3 06·8	2 57·8	2·5 0·5	8·5 1·8	14·5 3·0
26	3 06·5	3 07·0	2 58·0	2·6 0·5	8·6 1·8	14·6 3·0
27	3 06·8	3 07·3	2 58·2	2·7 0·6	8·7 1·8	14·7 3·1
28	3 07·0	3 07·5	2 58·5	2·8 0·6	8·8 1·8	14·8 3·1
29	3 07·3	3 07·8	2 58·7	2·9 0·6	8·9 1·9	14·9 3·1
30	3 07·5	3 08·0	2 59·0	3·0 0·6	9·0 1·9	15·0 3·1
31	3 07·8	3 08·3	2 59·2	3·1 0·6	9·1 1·9	15·1 3·1
32	3 08·0	3 08·5	2 59·4	3·2 0·7	9·2 1·9	15·2 3·2
33	3 08·3	3 08·8	2 59·7	3·3 0·7	9·5 1·9	15·3 3·2
34	3 08·5	3 09·0	2 59·9	3·4 0·7	9·4 2·0	15·4 3·2
35	3 08·8	3 09·3	3 00·2	3·5 0·7	9·5 2·0	15·5 3·2
36	3 09·0	3 09·5	3 00·4	3·6 0·8	9·6 2·0	15·6 3·3
37	3 09·3	3 09·8	3 00·6	3·7 0·8	9·7 2·0	15·7 3·3
38	3 09·5	3 10·0	3 00·9	3·8 0·8	9·8 2·0	15·8 3·3
39	3 09·8	3 10·3	3 01·1	3·9 0·8	9·9 2·1	15·9 3·3
40	3 10·0	3 10·5	3 01·3	4·0 0·8	10·0 2·1	16·0 3·3
41	3 10·3	3 10·8	3 01·6	4·1 0·9	10·1 2·1	16·1 3·4
42	3 10·5	3 11·0	3 01·8	4·2 0·9	10·2 2·1	16·2 3·4
43	3 10·8	3 11·3	3 02·1	4·3 0·9	10·3 2·1	16·3 3·4
44	3 11·0	3 11·5	3 02·3	4·4 0·9	10·4 2·2	16·4 3·4
45	3 11·3	3 11·8	3 02·5	4·5 0·9	10·5 2·2	16·5 3·4
46	3 11·5	3 12·0	3 02·8	4·6 1·0	10·6 2·2	16·6 3·5
47	3 11·8	3 12·3	3 03·0	4·7 1·0	10·7 2·2	16·7 3·5
48	3 12·0	3 12·5	3 03·3	4·8 1·0	10·8 2·3	16·8 3·5
49	3 12·3	3 12·8	3 03·5	4·9 1·0	10·9 2·3	16·9 3·5
50	3 12·5	3 13·0	3 03·7	5·0 1·0	11·0 2·3	17·0 3·5
51	3 12·8	3 13·3	3 04·0	5·1 1·1	11·1 2·3	17·1 3·6
52	3 13·0	3 13·5	3 04·2	5·2 1·1	11·2 2·3	17·2 3·6
53	3 13·3	3 13·8	3 04·4	5·3 1·1	11·3 2·4	17·3 3·6
54	3 13·5	3 14·0	3 04·7	5·4 1·1	11·4 2·4	17·4 3·6
55	3 13·8	3 14·3	3 04·9	5·5 1·1	11·5 2·4	17·5 3·6
56	3 14·0	3 14·5	3 05·2	5·6 1·2	11·6 2·4	17·6 3·7
57	3 14·3	3 14·8	3 05·4	5·7 1·2	11·7 2·4	17·7 3·7
58	3 14·5	3 15·0	3 05·6	5·8 1·2	11·8 2·5	17·8 3·7
59	3 14·8	3 15·3	3 05·9	5·9 1·2	11·9 2·5	17·9 3·7
60	3 15·0	3 15·5	3 06·1	6·0 1·3	12·0 2·5	18·0 3·8

Figure 20-2. *Increments and Corrections, 12 minutes*, Nautical Almanac.

yellow pages ii through xxxi located in the back of the *Almanac*. There is a separate table for each additional minute of GMT, and for each minute, the various increments and corrections corresponding with each of the 60 seconds comprising the minute are listed in vertical columns. Figure 20-2 pictures the table "Increments and Corrections" for 12 minutes. From this table the GHA Aries increment for 12 minutes 9 seconds is found to be 3° 02.7'; this figure is inserted on the form.

The SHA of Canopus listed on the daily page shown in Figure 20-1A is 264° 09.8'. After this value has been entered on the form, the GHA of Canopus is computed:

$$\text{GHA Canopus} = 338° \; 56.1' + \underbrace{3° \; 02.7'}_{\text{GHA}\Upsilon} + \underbrace{264° \; 09.8'}_{\text{SHA}\star} = 606° \; 08.6'$$

Since this result is greater than 360°, 360° is subtracted to yield the equivalent GHA of Canopus of 246° 08.6'.

If the navigator desires, with the GHA computed, the hour circle of Canopus can now be located relative to the Greenwich meridian (G) on a time diagram, for later use in determining the LHA.

To complete the use of the *Almanac* for Canopus, the declination of Canopus is obtained from its listing alongside the SHA on the daily page of Figure 20-1A and entered on the form as the "Table Declination (Tab Dec.)." The form now appears as in Figure 20-3, page 370. After the LHA has been computed by methods to be discussed shortly, the form will be ready for use with *Sight Reduction Tables No. 229*.

Determining the GHA and Declination of a Planet

The use of the *Nautical Almanac* to find the GHA and declination of a navigational planet differs somewhat from the procedure just described for a star, because both the GHA and declination for the navigational planets are tabulated in the *Almanac* for each whole hour of GMT. These tabulations are necessitated by the fact that the planets, being relatively close to earth in the solar system, seem to move across the unchanging patterns of the stars from one hour to the next.

Suppose that the planet Venus was observed at GMT 17-04-12 on the same occasion as the star Canopus, and its observed altitude was determined to be 16° 38.6'. The DR position at the time of the sight was L 34° 17.0'S, λ 163° 09.1'E.

The *Nautical Almanac* is first entered to compute the GHA of Venus for the time of the observation. The tabulated value of the GHA of Venus for the whole

Body	CANOPUS
GMT	17-12-09 Dec16

Ho	50° 39.4'
GHA (h)	338° 56.1'
Incre. (m/s)	3° 02.7'
v/v Corr SHA	264° 09.8'
Total GHA	606° 08.6'
±360°	246° 08.6'
aλ (+E,-W)	
LHA	
Tab Dec	S 52° 40.6'

Figure 20-3. *Canopus sight reduction form, GHA and declination determined.*

hour of GMT immediately preceding the observation is recorded on the form. In addition, the v value at the bottom of the column containing the Venus tabulations is noted and recorded on the sight form in the space provided. This v value represents the average irregularity in the rate of increase of the GHA of Venus from one hour to the next over the 3-day period covered by the daily page, as a result of the motion of the planet in its orbit. Because the rate of increase of the GHA is not constant, a v correction derived from the v value must be applied to the tabulated GHA of the planet, in addition to the GHA increment. The v correction, like the GHA increment, is taken from the appropriate "Increments and Corrections" table in the back of the *Almanac.*

It should be mentioned here that all bodies in the solar system are characterized by this irregular rate of increase of GHA. Thus, a v correction is applied to the GHA increments of all the navigational planets and the moon; the v correction for the sun is so small that it is ignored when using the *Nautical Almanac.* The v corrections for Mars, Jupiter, Saturn, and the moon are always positive; at certain times of the year, the Venus v correction can be negative.

Returning to the example at hand, a GHA increment for 4 minutes 12 seconds is found from the 4-minute "Increments and Corrections" table and entered on the sight form; a portion of the 4-minute table is shown in Figure 20-4. To find the v correction, the columns in the appropriate table with the headings v or d $Corr^n$ are used (the d correction will be explained below). The left side of each column contains v or d values, and the right side, the corresponding v or d correction. In this case, a v correction of +0.1' corresponding to a v value of +1.1 is obtained and entered on the form as the "v Corr^n." The tabulated GHA, GHA increment, and v correction are all summed to yield the

$\overset{\text{m}}{4}$	SUN PLANETS	ARIES	MOON	v or Corrⁿ d		v or Corrⁿ d		v or Corrⁿ d	
s	° ′	° ′	° ′	′	′	′	′	′	′
00	1 00·0	1 00·2	0 57·3	0·0	0·0	6·0	0·5	12·0	0·9
01	1 00·3	1 00·4	0 57·5	0·1	0·0	6·1	0·5	12·1	0·9
02	1 00·5	1 00·7	0 57·7	0·2	0·0	6·2	0·5	12·2	0·9
03	1 00·8	1 00·9	0 58·0	0·3	0·0	6·3	0·5	12·3	0·9
04	1 01·0	1 01·2	0 58·2	0·4	0·0	6·4	0·5	12·4	0·9
05	1 01·3	1 01·4	0 58·5	0·5	0·0	6·5	0·5	12·5	0·9
06	1 01·5	1 01·7	0 58·7	0·6	0·0	6·6	0·5	12·6	0·9
07	1 01·8	1 01·9	0 58·9	0·7	0·1	6·7	0·5	12·7	1·0
08	1 02·0	1 02·2	0 59·2	0·8	0·1	6·8	0·5	12·8	1·0
09	1 02·3	1 02·4	0 59·4	0·9	0·1	6·9	0·5	12·9	1·0
10	1 02·5	1 02·7	0 59·7	1·0	0·1	7·0	0·5	13·0	1·0
11	1 02·8	1 02·9	0 59·9	1·1	0·1	7·1	0·5	13·1	1·0
12	1 03·0	1 03·2	1 00·1	1·2	0·1	7·2	0·5	13·2	1·0
13	1 03·3	1 03·4	1 00·4	1·3	0·1	7·3	0·5	13·3	1·0
14	1 03·5	1 03·7	1 00·6	1·4	0·1	7·4	0·6	13·4	1·0
15	1 03·8	1 03·9	1 00·8	1·5	0·1	7·5	0·6	13·5	1·0
16	1 04·0	1 04·2	1 01·1	1·6	0·1	7·6	0·6	13·6	1·0
17	1 04·3	1 04·4	1 01·3	1·7	0·1	7·7	0·6	13·7	1·0
18	1 04·5	1 04·7	1 01·6	1·8	0·1	7·8	0·6	13·8	1·0
19	1 04·8	1 04·9	1 01·8	1·9	0·1	7·9	0·6	13·9	1·0
20	1 05·0	1 05·2	1 0		0·2	8·0	0·6	14	
						8·1			

Figure 20-4. *Portion of 4-minute Increments and Corrections table,* Nautical Almanac.

GHA of Venus of 117° 51.6′. If desired, the hour circle of Venus can now be located on a time diagram, for later use in determining the LHA.

To find the declination of Venus for the time of the observation, the tabulated declination for the whole hour of GMT preceding the sight is first extracted and entered on the form as the tabulated declination (Tab Dec). Just as the GHA of a body in the solar system is continually changing from one hour to the next, the declination of these bodies is also changing. In contrast to the irregular rate of change of GHA, however, the rate of change in declination is nearly constant. The *d* value appearing at the bottom of the Venus column represents the average hourly rate of change of its declination over the 3-day period covered by the daily page. A *d* correction (actually analogous to the GHA increment) derived from this *d* value is found from the appropriate "Increments and Corrections" table and applied to the tabulated declination to form the true declination of Venus at the time of the observation. The sign of the *d* value and hence the sign of the *d* correction is positive if the trend of the tabulated declination is increasing. If the tabulated declination is decreasing, the sign is negative. In this case, a *d* correction of 0.0 is obtained from the 4-minute table partially shown in Figure 20-4, using a *d* value of 0.1 as entering argument in the "*v* or *d* correction" column. Thus, the true declination of Venus at GMT 17-04-12 is S13° 16.1′.

The use of the *Nautical Almanac* for Venus is now complete; the form at this point is shown in Figure 20-5.

Body	VENUS
GMT	17-04-12 Dec16

Ho	16° 38.6'
GHA (h)	116° 48.5'
Incre. (m/s)	1° 03.0'
v/v Corr SHA	1.1'/ 0.1'
Total GHA	117° 51.6'
±360°	
aλ (+E,-W)	
LHA	
Tab Dec	S 13° 16.1'
d#/d Corr	0.1'/ 0
True Dec	S 13° 16.1'

Figure 20-5. *Venus sight reduction form, GHA and declination determined.*

Determining the GHA and Declination of the Moon

The use of the *Nautical Almanac* to find the GHA and declination of the moon is virtually identical to the procedure just described for the planets, with the exception of the *v* and *d* corrections. Because the moon is much closer to the earth than the planets, the rate of irregular change of its GHA and the rate of change of its declination are both so rapid as to necessitate the tabulation of *v* and *d* values for each hour of GMT, rather than to average values for the 3-day period. After recording the tabulated GHA and declination for the whole hour of GMT immediately preceding the time of the sight, the *v* and *d* values for this same time are also recorded. The GHA increment, *v* correction, and *d* correction are then determined as before, using the appropriate "Increments and Corrections" page. The sign of the *v* value, and hence the *v* correction for the moon, is always positive; the sign of the *d* value and the resulting *d* correction is determined by inspection of the trend of increase or decrease in the tabulated declination near the time of the observation. A partially completed sight form for an upper-limb observation of the moon appears in Figure 20-6.

Determining the GHA and Declination of the Sun

To complete the examples of the use of the *Nautical Almanac* during sight reduction to determine the GHA and declination of the various types of celestial bodies normally observed, the following example for a lower-limb observation of the sun is included. The computation is essentially the same as that for a planet, with the difference being that no *v* correction is applied to the sum of

Body	MOON (UL)
GMT	16-58-57 Dec16

Ho	28° 35.4′
GHA (h)	205° 10.3′
Incre. (m/s)	14° 04.0′
v/v Corr SHA	10.9′/ 10.6′
Total GHA	219° 24.9′
±360°	
aλ (+E,-W)	
LHA	
Tab Dec	N 23° 21.1′
d#/d Corr	-8.3′/ -8.1′
True Dec	N 23° 13.0′

Figure 20-6. Portion of a moon (UL) sight reduction form.

the tabulated GHA and the GHA increment; the irregularity in the hourly rate of change of the GHA of the sun is so small as to be disregarded for most practical purposes. The sign of the d correction is determined by inspection of the trend of the hourly tabulated declination values. A partially completed sun-sight form appears in Figure 20-7.

Body	SUN (LL)
GMT	12-12-06 Dec16

Ho	10° 16.8′
GHA (h)	4° 08.5′
Incre. (m/s)	3° 01.5′
v/v Corr SHA	
Total GHA	4° 08.5′
±360°	
aλ (+E,-W)	
LHA	
Tab Dec	S 23° 18.9′
d#/d Corr	0.1′/ 0
True Dec	S 23° 18.9′

Figure 20-7. Portion of a sun (LL) sight reduction form.

DETERMINING THE LOCAL HOUR ANGLE

In the preceding four sections, examples of the use of the *Nautical Almanac* to find the GHA and declination of each of the four types of bodies observed in the practice of celestial navigation were given. The declination so determined forms one of the three entering arguments for *Sight Reduction Tables No. 229*. In this section, the methods by which the GHA is combined with the assumed longitude to produce a second entering argument, LHA, will be set forth.

In chapter 16, the local hour angle (LHA) of a celestial body is defined as the hour angle measured from the observer's celestial meridian to the hour circle of the body. It is always measured in a westerly direction, and increases from 0° through 360° as the earth rotates beneath the celestial sphere. For the purpose of sight reduction, LHA is measured from the meridian of the assumed position. Since the assumed longitude is defined as the horizontal angle between the Greenwich meridian and the location of the AP of the observer, the following relationship exists between the local hour angle (LHA), the GHA of the body, and the assumed longitude (aλ):

$$\text{LHA} = \text{GHA} \begin{array}{l} + \text{ a}\lambda \text{ (E)} \\ - \text{ a}\lambda \text{ (W)} \end{array}$$

This relationship is depicted in Figure 20-8 on page 375. Occasionally, when the assumed longitude is west its value may be greater than the GHA of the body; this would result in a negative value for the LHA. In these cases, 360° is added to the GHA prior to subtracting the westerly assumed longitude. An equivalent positive LHA results.

Since the entering argument for *Tables No. 229* must be an integral LHA, the assumed longitude is chosen such that when it is combined with the GHA determined from the *Nautical* (or *Air*) *Almanac*, an integral LHA results. Thus, if the observer is located in east longitude at the time of the observation, the minutes of the assumed longitude are selected in such a manner that when they are added to the minutes of the GHA a whole degree results. If the observer is in west longitude, the minutes of the assumed longitude chosen should be equal to the minutes of the GHA, so that when the assumed longitude is subtracted from the GHA an integral LHA results. In order that the intercept distance from the AP of the observer to the LOP produced from an observation will not be excessive, the following rule has been developed:

> The assumed longitude chosen should not be more than 30 minutes of longitude from the DR longitude for the time of the observation.

As examples of the determination of local hour angles for use as entering arguments in *Sight Reduction Tables No. 229*, the integral LHA for each of the

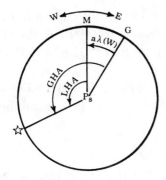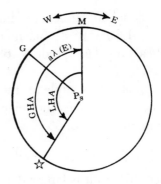

Figure 20-8. *Two possible relationships between a star and the Greenwich and local meridians.*

four examples of the use of the *Nautical Almanac* given previously will be determined, as outlined in the following paragraphs.

In the first example, the GHA of the star Canopus at the time of the observation was determined to be 246° 08.6′; the DR longitude for this time was λ 163° 05.7′E. To compute the exact integral value of the LHA from an assumed position close to the DR position at the time of the observation, the formula

$$LHA = GHA + a\lambda(E)$$

applies, since the DR longitude is east. Since the GHA and assumed longitude are to be added to form the LHA, it follows that to obtain an integral LHA the minutes of the selected assumed longitude should be chosen such that when they are added to the minutes of the GHA the result will be 60 minutes or one degree. The minutes (m) of the assumed longitude must therefore satisfy the equation

$$60 = 08.6 + m$$

Hence, the minutes chosen must be 51.4. Now, care must be taken to choose the degrees of the assumed longitude so that the result will be within 30 minutes of the DR longitude. An assumed longitude 163° 51.4′ would exceed this limit, so the value 162° 51.4′ is selected. Adding this value to the computed GHA yields the integral LHA 409°:

$$246° 08.6′ + 162° 51.4′ = 409°$$

Since this result is greater than 360°, 360° is subtracted to form the equivalent LHA 49°.

In the second example, the GHA of the planet Venus was found to be 117° 51.6′, and the DR longitude was 163° 09.1′E at the time of the observation. The formula LHA = GHA + aλ(E) again applies. To obtain an integral LHA, the minutes of the GHA when added to the minutes of the assumed longitude must equal 60. Hence, the minutes of the assumed longitude chosen must be (60' – 51.6') or 08.4'. Inasmuch as the DR longitude is 163° 09.1'E, the complete assumed longitude is 163° 08.4'. The LHA is then 117° 51.6' + 163° 08.4' or 281°.

In the third example for the moon, the GHA was 219° 24.9' and the DR longitude 163° 11.7'E. The selected assumed longitude is 163° 35.1', resulting in an integral LHA of 383°:

$$219° 24.9' + 163° 35.1' = 383°$$

Since the result is larger than 360°, the equivalent LHA of 23° is formed by subtraction of 360°.

In the sun example, the GHA was 4° 08.5' and the DR longitude was 64° 48.0'W. In this situation the formula

$$LHA = GHA - a\lambda(W)$$

applies, and the minutes of the GHA and assumed longitude must be identical to produce an integral LHA. Thus, the minutes of the assumed longitude must be 08.5'. Considering the DR position, care must again be taken to select the proper degrees of the assumed longitude. In accordance with the "30-minute" rule, the assumed longitude chosen is 65° 08.5'. Had a value of 64° 08.5' been chosen, the assumed longitude would have been more than 30 minutes from the DR longitude (48.0' – 08.5' = 39.5'). Since the GHA is smaller than the assumed longitude, 360° is added to the GHA, and the resulting LHA is given by the following expression:

$$(360° + 4° 08.5') - 65° 08.5' = 299°$$

Certain of the LHAs calculated as examples above will be used as entering arguments for *Tables No. 229* later in this chapter.

DETERMINING THE ASSUMED LATITUDE

The remaining entering argument for *Tables No. 229* is an integral value of assumed latitude. The assumed latitude, similar to the assumed longitude, should not be more than 30 minutes of latitude from the DR latitude. The following rule is used for determining the assumed latitude:

> The assumed latitude chosen should be the closest whole degree of latitude to the DR latitude for the time of the observation.

Thus, the assumed latitudes for the Canopus, Venus, and moon observations used as examples earlier are all 34°S, as the DR latitudes for all three of these observations varied between 34° 15.5'S and 34° 19.0'S. The assumed latitude for the sun observation, based on the DR latitude 31° 08.1'N, is 31°N.

THE *SIGHT REDUCTION TABLES FOR MARINE NAVIGATION, NO. 229*

The remainder of this chapter will be concerned with the use of the *Sight Reduction Tables for Marine Navigation, No. 229,* to obtain the complete solution for a celestial observation. As was mentioned earlier, *Tables No. 229* are, in effect, a set of precalculated solutions for the computed altitude Hc and the azimuth angle Z of the navigational triangle, given two other sides and the included angle between them. The tables themselves are divided into six volumes, each covering a basic 15° band of latitude. A 1° overlap occurs between the volumes, so that volume 1 covers latitudes between 0° and 15°, volume 2 covers 15° to 30°, volume 3 covers 30° to 45°, and so on. Entering arguments for the tables are integral (whole number) degrees of local hour angle (LHA), assumed latitude, and declination. Values of Hc and Z are tabulated for each whole degree of each of the entering arguments, and interpolation tables are included inside the front and back covers of each volume for interpolating both Hc and Z for the exact declination. No interpolation for LHA or assumed latitude is necessary since, as has been explained, the assumed position from which the intercept distance is laid down is selected to yield an integral LHA and assumed latitude.

Each volume of *Tables No. 229* contains two sets of tabulations for all integral degrees of LHA between 0° and 360°. One set, comprising the front half of the volume, is for the first 8° of latitude covered by that volume, and the other set, comprising the second half of the volume, is for the remaining 8°. The values of the local hour angle, considered to be the primary entering argument within each of the 8° latitude "zones" thus established, are prominently displayed both at the top and bottom of each page. The 8° of latitude within each zone form the horizontal argument for each LHA page, while the vertical argument is declination. The tabulations are arranged in columns, with one column for each of the 8° of latitude covered.

Instructions printed across the top and bottom of each page indicate whether the tabulations on that page are for assumed latitudes on the same side of the celestial equator as the declination (latitude *same* name as declination) or for assumed latitudes on the opposite side of the equator (latitude *contrary* name to declination). If same name and contrary name tabulations appear on the same page, they are separated in each column by a horizontal line that forms a configuration across the page resembling the profile of a staircase. Each horizontal line indicates the degree of declination for that particular LHA and assumed latitude combination in which the visible sea horizon occurs.

Figure 20-9A on page 378 depicts a left-hand page of volume 3 of *Tables No. 229* for LHAs of 49° and 311° for the 30° to 37° latitude zone with latitude the same name as declination. Figure 20-9B shows a right-hand page for the same

49°, 311° L.H.A. LATITUDE SAME NAME AS DECLINATION

N. Lat. { L.H.A. greater than 180° .....Zn=Z ; L.H.A. less than 180° ....Zn=360°−Z }

Dec.	30° Hc	30° d	30° Z	31° Hc	31° d	31° Z	32° Hc	32° d	32° Z	33° Hc	33° d	33° Z	34° Hc	34° d	34° Z	35° Hc	35° d	35° Z	36° Hc	36° d	36° Z	37° Hc	37° d	37° Z	Dec.
°	° ′	′	°	° ′	′	°	° ′	′	°	° ′	′	°	° ′	′	°	° ′	′	°	° ′	′	°	° ′	′	°	°
0	34 37.3	+36.3	113.5	34 13.1	+37.2	114.1	33 48.3	+38.1	114.7	33 22.9	+38.9	115.3	32 57.0	+39.8	115.9	32 30.5	+40.6	116.5	32 03.4	+41.5	117.1	31 35.9	+42.2	117.6	0
1	35 13.6	35.7	112.5	34 50.3	36.7	113.2	34 26.4	37.6	113.8	34 01.8	38.6	114.4	33 36.8	39.4	115.0	33 11.1	40.2	115.6	32 44.9	41.0	116.2	32 18.1	41.9	116.8	1
2	35 49.3	35.3	111.5	35 27.0	36.2	112.2	35 04.0	37.2	112.9	34 40.4	38.1	113.5	34 16.2	38.9	114.1	33 51.3	39.9	114.7	33 25.9	40.8	115.3	33 00.0	41.5	115.9	2
3	36 24.6	34.7	110.5	36 03.2	35.7	111.2	35 41.2	36.7	111.9	35 18.5	37.6	112.5	34 55.1	38.6	113.2	34 31.2	39.5	113.8	34 06.7	40.3	114.5	33 41.5	41.2	115.1	3
4	36 59.3	34.2	109.5	36 38.9	35.3	110.2	36 17.9	36.2	110.9	35 56.1	37.2	111.6	35 33.7	38.1	112.3	35 10.7	39.0	112.9	34 47.0	39.9	113.6	34 22.7	40.8	114.2	4
5	37 33.5	+33.7	108.5	37 14.2	+34.7	109.2	36 54.1	+35.7	109.9	36 33.3	+36.7	110.6	36 11.8	+37.7	111.3	35 49.7	+38.6	112.0	35 26.9	+39.5	112.6	35 03.5	+40.4	113.3	5
6	38 07.2	33.1	107.4	37 48.9	34.1	108.2	37 29.8	35.2	108.9	37 10.0	36.2	109.6	36 49.5	37.2	110.3	36 28.3	38.1	111.0	36 06.4	39.1	111.7	35 43.9	40.0	112.4	6
7	38 40.3	32.6	106.4	38 23.0	33.6	107.1	38 05.0	34.6	107.9	37 46.2	35.7	108.6	37 26.7	36.6	109.4	37 06.4	37.7	110.1	36 45.5	38.6	110.8	36 23.9	39.5	111.5	7
8	39 12.9	31.9	105.3	38 56.6	33.0	106.1	38 39.6	34.1	106.8	38 21.9	35.1	107.6	38 03.3	36.2	108.4	37 44.1	37.2	109.1	37 24.1	38.2	109.8	37 03.4	39.1	110.5	8
9	39 44.8	31.2	104.2	39 29.6	32.5	105.0	39 13.7	33.5	105.8	38 57.0	34.6	106.6	38 39.5	35.6	107.3	38 21.3	36.6	108.1	38 02.3	37.6	108.8	37 42.5	38.6	109.6	9
10	40 16.0	+30.7	103.1	40 02.1	+31.7	103.9	39 47.2	+32.9	104.7	39 31.6	+34.0	105.5	39 15.1	+35.1	106.3	38 57.9	+36.1	107.1	38 39.9	+37.1	107.8	38 21.1	+38.2	108.6	10
11	40 46.7	29.9	101.9	40 33.8	31.2	102.8	40 20.1	32.3	103.6	40 05.6	33.4	104.4	39 50.2	34.5	105.2	39 34.0	35.6	106.0	39 17.0	36.7	106.8	38 59.3	37.6	107.6	11
12	41 16.6	29.3	100.8	41 05.0	30.4	101.7	40 52.4	31.6	102.5	40 39.0	32.7	103.3	40 24.7	33.9	104.2	40 09.6	35.0	105.0	39 53.7	36.0	105.8	39 36.9	37.1	106.6	12
13	41 45.9	28.5	99.6	41 35.4	29.8	100.5	41 24.0	31.0	101.4	41 11.7	32.2	102.2	40 58.6	33.2	103.1	40 44.6	34.4	103.9	40 29.7	35.5	104.8	40 14.0	36.5	105.6	13
14	42 14.4	27.9	98.4	42 05.2	29.0	99.3	41 55.0	30.2	100.2	41 43.9	31.4	101.1	41 31.8	32.7	102.0	41 19.0	33.7	102.8	41 05.2	34.9	103.7	40 50.5	36.0	104.5	14
15	42 42.3	+27.0	97.2	42 34.2	+28.3	98.2	42 25.2	+29.6	99.1	42 15.3	+30.8	100.0	42 04.5	+31.9	100.9	41 52.7	+33.1	101.7	41 40.1	+34.2	102.6	41 26.5	+35.4	103.5	15
16	43 09.3	26.2	96.0	43 02.5	27.5	97.0	42 54.8	28.7	97.9	42 46.1	30.0	98.8	42 36.4	31.2	99.7	42 25.8	32.5	100.6	42 14.3	33.6	101.5	42 01.9	34.7	102.4	16
17	43 35.5	25.5	94.8	43 30.0	26.8	95.7	43 23.5	28.1	96.7	43 16.1	29.3	97.6	43 07.6	30.6	98.5	42 58.3	31.7	99.5	42 47.9	32.9	100.4	42 36.6	34.1	101.3	17
18	44 01.0	24.6	93.5	43 56.8	25.9	94.5	43 51.6	27.2	95.5	43 45.4	28.5	96.4	43 38.2	29.8	97.4	43 30.0	31.0	98.3	43 20.8	32.3	99.2	43 10.7	33.5	100.2	18
19	44 25.6	23.7	92.3	44 22.7	25.1	93.2	44 18.8	26.4	94.2	44 13.9	27.7	95.2	44 08.0	29.0	96.2	44 01.0	30.3	97.1	43 53.1	31.5	98.1	43 44.2	32.7	99.0	19
20	44 49.3	+22.9	91.0	44 47.8	+24.2	92.0	44 45.2	+25.6	93.0	44 41.6	+26.9	93.9	44 37.0	+28.2	94.9	44 31.3	+29.5	95.9	44 24.6	+30.8	96.9	44 16.9	+32.0	97.9	20
21	45 12.2	21.9	89.7	45 12.0	23.4	90.7	45 10.8	24.7	91.7	45 08.5	26.1	92.7	45 05.2	27.4	93.7	45 00.8	28.7	94.7	44 55.4	30.0	95.7	44 48.9	31.3	96.7	21
22	45 34.1	21.1	88.3	45 35.4	22.4	89.4	45 35.5	23.8	90.4	45 34.6	25.1	91.4	45 32.6	26.5	92.4	45 29.5	27.9	93.4	45 25.4	29.1	94.5	45 20.2	30.5	95.5	22
23	45 55.2	20.0	87.0	45 57.8	21.5	88.0	45 59.3	22.9	89.1	45 59.7	24.3	90.1	45 59.1	25.6	91.1	45 57.4	27.0	92.2	45 54.5	28.4	93.2	45 50.7	29.6	94.2	23
24	46 15.2	19.1	85.6	46 19.3	20.5	86.7	46 22.2	21.9	87.7	46 24.0	23.4	88.8	46 24.7	24.8	89.8	46 24.4	26.1	90.9	46 22.9	27.5	91.9	46 20.3	28.9	93.0	24
25	46 34.3	+18.1	84.3	46 39.8	+19.5	85.3	46 44.1	+21.0	86.4	46 47.4	+22.4	87.4	46 49.5	+23.8	88.5	46 50.5	+25.2	89.6	46 50.4	+26.6	90.6	46 49.2	+27.9	91.7	25
26	46 52.4	17.1	82.9	46 59.3	18.5	83.9	47 05.1	20.0	85.0	47 09.8	21.4	86.1	47 13.3	22.8	87.2	47 15.7	24.3	88.2	47 17.0	25.7	89.3	47 17.1	27.1	90.4	26
27	47 09.5	16.0	81.5	47 17.8	17.5	82.5	47 25.1	18.9	83.6	47 31.2	20.4	84.7	47 36.1	21.9	85.8	47 40.0	23.3	86.9	47 42.7	24.7	88.0	47 44.2	26.2	89.1	27
28	47 25.5	15.0	80.0	47 35.3	16.5	81.1	47 44.0	17.9	82.2	47 51.6	19.4	83.3	47 58.0	20.9	84.4	48 03.3	22.3	85.5	48 07.4	23.8	86.6	48 10.4	25.1	87.7	28
29	47 40.5	13.9	78.6	47 51.8	15.3	79.7	48 01.9	16.9	80.8	48 11.0	18.3	81.9	48 18.9	19.8	83.0	48 25.6	21.3	84.1	48 31.2	22.7	85.2	48 35.5	24.3	86.4	29
30	47 54.4	+12.7	77.2	48 07.1	+14.3	78.2	48 18.8	+15.7	79.3	48 29.3	+17.3	80.5	48 38.7	+18.7	81.6	48 46.9	+20.2	82.7	48 53.9	+21.7	83.8	48 59.8	+23.1	85.0	30
31	48 07.1	11.7	75.7	48 21.4	13.1	76.8	48 34.5	14.7	77.9	48 46.6	16.1	79.0	48 57.4	17.7	80.1	49 07.1	19.2	81.3	49 15.6	20.7	82.4	49 22.9	22.2	83.6	31
32	48 18.8	10.5	74.2	48 34.5	12.1	75.3	48 49.2	13.5	76.4	49 02.7	15.0	77.5	49 15.1	16.5	78.7	49 26.3	18.0	79.8	49 36.3	19.5	81.0	49 45.1	21.1	82.1	32
33	48 29.3	9.4	72.7	48 46.6	10.8	73.8	49 02.7	12.4	74.9	49 17.7	13.9	76.1	49 31.6	15.4	77.2	49 44.3	16.9	78.3	49 55.8	18.5	79.5	50 06.2	19.9	80.7	33
34	48 38.7	8.2	71.3	48 57.4	9.7	72.3	49 15.1	11.2	73.4	49 31.6	12.7	74.6	49 47.0	14.2	75.7	50 01.2	15.8	76.9	50 14.3	17.3	78.0	50 26.1	18.8	79.2	34
35	48 46.9	+7.0	69.8	49 07.1	+8.5	70.8	49 26.3	+10.0	71.9	49 44.3	+11.5	73.1	50 01.2	+13.1	74.2	50 17.0	+14.6	75.4	50 31.6	+16.1	76.5	50 44.9	+17.7	77.7	35
36	48 53.9	5.9	68.2	49 15.6	7.3	69.3	49 36.3	8.8	70.4	49 55.8	10.4	71.5	50 14.3	11.8	72.7	50 31.6	13.3	73.8	50 47.7	14.9	75.0	51 02.6	16.5	76.2	36
37	48 59.8	4.6	66.7	49 22.9	6.2	67.8	49 45.1	7.6	68.9	50 06.2	9.1	70.0	50 26.1	16.6	71.1	50 44.9	12.2	72.3	51 02.6	13.7	73.5	51 19.1	15.2	74.7	37
38	49 04.4	3.5	65.2	49 29.1	4.9	66.3	49 52.7	6.4	67.4	50 15.3	7.8	68.5	50 36.7	9.4	69.6	50 57.1	10.9	70.7	51 16.3	12.5	71.9	51 34.3	14.1	73.1	38
39	49 07.9	2.2	63.7	49 34.0	3.6	64.7	49 59.1	5.1	65.8	50 23.1	6.6	66.9	50 46.1	8.1	68.0	51 08.0	9.7	69.2	51 28.8	11.2	70.3	51 48.4	12.7	71.5	39
40	49 10.1	+1.0	62.2	49 37.6	+2.5	63.2	50 04.2	+3.9	64.3	50 29.7	+5.4	65.3	50 54.2	+6.9	66.5	51 17.7	+8.3	67.6	51 40.0	+9.9	68.8	52 01.1	+11.5	70.0	40
41	49 11.1	−0.2	60.6	49 40.1	+1.2	61.6	50 08.1	2.6	62.7	50 35.1	4.1	63.8	51 01.1	5.6	64.9	51 26.0	7.1	66.0	51 49.9	8.6	67.2	52 12.6	10.1	68.4	41
42	49 10.9	1.4	59.1	49 41.3	0.0	60.1	50 10.7	1.4	61.1	50 39.2	2.8	62.2	51 06.7	4.3	63.3	51 33.1	5.8	64.4	51 58.5	7.3	65.6	52 22.7	8.9	66.7	42
43	49 09.5	2.6	57.6	49 41.3	−1.3	58.6	50 12.1	+0.2	59.6	50 42.0	1.6	60.6	51 11.0	3.0	61.7	51 38.9	4.5	62.8	52 05.8	5.9	64.0	52 31.6	7.5	65.1	43
44	49 06.9	3.8	56.0	49 40.0	2.5	57.0	50 12.3	−1.2	58.0	50 43.6	+0.2	59.0	51 14.0	1.6	60.1	51 43.4	3.1	61.2	52 11.7	4.7	62.3	52 39.1	6.1	63.5	44
45	49 03.1	−5.0	54.5	49 37.5	3.7	55.5	50 11.1	−2.3	56.5	50 43.8	−1.0	57.5	51 15.6	+0.4	58.5	51 46.5	+1.8	59.6	52 16.4	+3.2	60.7	52 45.2	+4.8	61.9	45
46	48 58.1	6.2	53.0	49 33.8	4.9	53.9	50 08.8	3.7	54.9	50 42.8	2.3	55.9	51 16.0	−0.9	56.9	51 48.3	+0.5	58.0	52 19.6	2.0	59.1	52 50.0	3.4	60.2	46
47	48 51.9	7.4	51.5	49 28.9	6.1	52.4	50 05.1	4.9	53.3	50 40.5	3.5	54.3	51 15.1	2.2	55.3	51 48.8	0.8	56.4	52 21.6	+0.5	57.4	52 53.4	2.0	58.5	47
48	48 44.5	8.5	50.0	49 22.8	7.4	50.9	50 00.2	6.1	51.8	50 37.0	4.9	52.7	51 12.9	3.5	53.7	51 48.0	2.2	54.7	52 22.1	−0.8	55.8	52 55.4	+0.6	56.9	48
49	48 36.0	9.8	48.5	49 15.4	8.6	49.3	49 54.1	7.3	50.2	50 32.1	6.1	51.2	51 09.4	4.9	52.1	51 45.8	3.5	53.1	52 21.3	2.1	54.2	52 56.0	−0.7	55.2	49
50	48 26.2	−10.8	47.0	49 06.8	−9.7	47.8	49 46.8	−8.6	48.7	50 26.0	−7.3	49.6	51 04.5	−6.1	50.5	51 42.3	−4.8	51.5	52 19.2	−3.5	52.5	52 55.3	−2.2	53.6	50
51	48 15.4	12.0	45.5	48 57.1	10.9	46.3	49 38.2	9.7	47.2	50 18.7	8.6	48.0	50 58.4	7.3	49.0	51 37.5	6.2	49.9	52 15.7	4.8	50.9	52 53.1	3.5	51.9	51
52	48 03.4	13.1	44.0	48 46.2	12.0	44.8	49 28.5	11.0	45.6	50 10.1	9.8	46.5	50 51.1	8.7	47.4	51 31.3	7.4	48.3	52 10.9	6.2	49.3	52 49.6	4.9	50.3	52
53	47 50.3	14.2	42.6	48 34.2	13.2	43.3	49 17.5	12.1	44.1	50 00.3	11.0	45.0	50 42.4	9.9	45.8	51 23.9	8.7	46.7	52 04.7	7.5	47.6	52 44.7	6.2	48.6	53
54	47 36.1	15.3	41.1	48 21.0	14.3	41.9	49 05.4	13.3	42.6	49 49.3	12.3	43.4	50 32.5	11.1	44.3	51 15.2	10.0	45.1	51 57.2	8.9	46.0	52 38.5	7.6	47.0	54
55	47 20.8	−16.4	39.7	48 06.7	−15.4	40.4	48 52.1	14.4	41.2	49 37.0	13.4	41.9	50 21.4	12.3	42.7	51 05.2	11.3	43.6	51 48.3	10.1	44.4	52 30.9	9.0	45.3	55
56	47 04.4	17.3	38.3	47 51.3	16.5	39.0	48 37.7	15.5	39.7	49 23.6	14.5	40.4	50 09.1	13.6	41.2	50 53.9	12.5	42.0	51 38.2	11.4	42.8	52 21.9	10.3	43.7	56
57	46 47.1	18.4	36.9	47 34.8	17.5	37.5	48 22.2	16.6	38.2	49 09.1	15.7	38.9	49 55.5	14.7	39.7	50 41.4	13.7	40.5	51 26.8	12.6	41.3	52 11.6	11.5	42.1	57
58	46 28.7	19.4	35.5	47 17.3	18.5	36.1	48 05.6	17.7	36.8	48 53.4	16.8	37.5	49 40.8	15.9	38.2	50 27.7	14.9	38.9	51 14.2	13.9	39.7	52 00.1	12.9	40.5	58
59	46 09.3	20.4	34.1	46 58.8	19.6	34.7	47 47.9	18.8	35.4	48 36.6	17.9	36.0	49 24.9	17.0	36.7	50 12.8	16.0	37.4	51 00.3	15.2	38.1	51 47.2	14.1	38.9	59
60	45 48.9	−21.3	32.8	46 39.2	−20.6	33.3	47 29.1	19.7	33.9	48 18.7	−18.9	34.6	49 07.9	−18.1	35.2	49 56.8	−17.3	35.9	50 45.1	16.3	36.6	51 33.1	15.4	37.4	60
61	45 27.6	22.2	31.4	46 18.6	21.5	32.0	47 09.4	20.8	32.6	47 59.8	20.0	33.1	48 49.8	19.1	33.8	49 39.5	18.3	34.4	50 28.8	17.5	35.1	51 17.7	16.6	35.8	61
62	45 05.4	23.2	30.1	45 57.1	22.4	30.6	46 48.6	21.7	31.2	47 39.8	21.0	31.7	48 30.7	20.3	32.3	49 21.2	19.5	33.0	50 11.3	18.6	33.6	51 01.1	17.7	34.3	62
63	44 42.2	24.0	28.8	45 34.7	23.4	29.3	46 26.9	22.7	29.8	47 18.8	22.0	30.4	48 10.4	21.3	30.9	49 01.7	20.5	31.5	49 52.7	19.7	32.1	50 43.4	19.0	32.8	63
64	44 18.2	24.8	27.5	45 11.3	24.2	28.0	46 04.2	23.6	28.5	46 56.8	23.0	29.0	47 49.1	22.2	29.5	48 41.2	21.6	30.1	49 33.0	20.8	30.7	50 24.4	20.0	31.3	64
65	43 53.4	−25.7	26.3	44 47.1	−25.1	26.7	45 40.6	−24.5	27.2	46 33.8	−23.9	27.6	47 26.9	−23.3	28.1	48 19.6	−22.5	28.7	49 12.2	−21.9	29.2	50 04.4	−21.2	29.8	65
66	43 27.7	26.5	25.0	44 22.0	26.0	25.4	45 16.1	25.4	25.9	46 09.9	24.8	26.3	47 03.6	24.2	26.8	47 57.1	23.6	27.3	48 50.3	22.9	27.8	49 43.2	22.2	28.3	66
67	43 01.2	27.3	23.8	43 56.0	26.8	24.2	44 50.7	26.3	24.6	45 45.1	75.7	25.0	46 39.4	25.1	25.4	47 33.5	24.5	25.9	48 27.4	24.0	26.4	49 21.0	23.3	26.9	67
68	42 33.9	28.0	22.6	43 29.2	27.5	22.9	44 24.4	27.0	23.3	45 19.4	26.5	23.7	46 14.3	26.0	24.1	47 09.0	25.5	24.6	48 03.4	24.9	25.0	48 57.7	24.3	25.5	68
69	42 05.9	28.8	21.4	43 01.7	28.3	21.7	43 57.4	27.9	22.1	44 52.9	27.4	22.4	45 48.3	26.9	22.8	46 43.5	26.4	23.2	47 38.5	25.8	23.7	48 33.4	25.3	24.1	69
70	41 37.1	−29.5	20.2	42 33.4	−29.1	20.5	43 29.5	−28.6	20.8	44 25.5	−28.2	21.2	45 21.4	−27.7	21.6	46 17.1	−27.2	21.9	47 12.7	26.7	22.3	48 08.1	−26.2	22.8	70
71	41 07.6	30.1	19.0	42 04.3	29.8	19.3	43 00.9	29.4	19.6	43 57.3	28.9	20.0	44 53.7	28.6	20.3	45 49.9	28.1	20.6	46 46.0	27.7	21.0	47 41.9	27.2	21.4	71
72	40 37.5	30.9	17.9	41 34.5	30.4	18.2	42 31.5	30.1	18.4	43 28.4	29.8	18.7	44 25.1	29.3	19.1	45 21.8	28.9	19.4	46 18.3	28.5	19.7	47 14.7	28.0	20.1	72
73	40 06.6	31.4	16.8	41 04.1	31.2	17.0	42 01.4	30.8	17.3	42 58.6	30.4	17.6	43 55.8	30.1	17.8	44 52.9	29.7	18.1	45 49.8	29.3	18.5	46 46.7	28.9	18.8	73
74	39 35.2	32.1	15.7	40 32.9	31.8	15.9	41 30.6	31.5	16.1	42 28.2	31.2	16.4	43 25.7	30.8	16.6	44 23.2	30.5	16.9	45 20.5	30.1	17.2	46 17.8	29.8	17.5	74
75	39 03.1	−32.7	14.6	40 01.1	−32.4	14.8	40 59.1	−32.1	15.0	41 57.0	−31.8	15.2	42 54.9	−31.5	15.5	43 52.7	31.2	15.7	44 50.4	−30.9	16.0	45 48.0	−30.5	16.3	75
76	38 30.4	33.3	13.5	39 28.7	33.0	13.7	40 27.0	32.8	13.9	41 25.2	32.5	14.1	42 23.4	32.3	14.3	43 21.5	32.0	14.5	44 19.5	31.6	14.8	45 17.5	31.3	15.0	76
77	37 57.1	33.9	12.4	38 55.7	33.7	12.6	39 54.2	33.4	12.8	40 52.7	33.1	13.0	41 51.1	32.9	13.2	42 49.5	32.6	13.4	43 47.9	32.4	13.6	44 46.2	32.1	13.8	77
78	37 23.2	34.3	11.4	38 22.0	34.1	11.5	39 20.8	33.9	11.7	40 19.6	33.8	11.9	41 18.2	33.5	12.1	42 16.9	33.3	12.2	43 15.5	33.0	12.4	44 14.1	32.8	12.7	78
79	36 48.9	34.9	10.4	37 47.9	34.8	10.5	38 46.9	34.6	10.6	39 45.8	34.3	10.8	40 44.7	34.1	11.0	41 43.6	33.9	11.1	42 42.5	33.7	11.3	43 41.3	33.5	11.5	79
80	36 14.0	−35.5	9.4	37 13.1	−35.2	9.5	38 12.3	−35.1	9.6	39 11.5	−34.9	9.7	40 10.6	−34.7	9.9	41 09.7	34.6	10.0	42 08.8	34.4	10.2	43 07.8	34.2	10.3	80
81	35 38.5	35.9	8.4	36 37.9	35.8	8.5	37 37.2	35.6	8.6	38 36.6	35.5	8.7	39 35.9	35.3	8.8	40 35.1	35.1	8.9	41 34.4	35.0	9.1	42 33.6	34.8	9.2	81
82	35 02.6	36.3	7.4	36 02.1	36.2	7.5	37 01.6	36.1	7.6	38 01.1	36.0	7.7	39 00.6	35.9	7.8	40 00.0	35.7	7.9	40 59.4	35.5	8.0	41 58.8	35.4	8.1	82
83	34 26.3	36.8	6.4	35 25.9	36.7	6.5	36 25.5	36.6	6.6	37 25.1	36.5	6.7	38 24.7	36.4	6.7	39 24.3	36.3	6.8	40 23.9	36.2	6.9	41 23.4	36.0	7.0	83
84	33 49.5	37.3	5.4	34 49.2	37.2	5.5	35 48.9	37.1	5.6	36 48.6	37.0	5.7	37 48.3	36.9	5.7	38 48.0	36.8	5.8	39 47.7	36.7	5.9	40 47.4	36.6	6.0	84
85	33 12.2	−37.7	4.5	34 12.0	−37.6	4.6	35 11.8	−37.5	4.6	36 11.6	−37.4	4.7	37 11.4	−37.3	4.7	38 11.2	−37.3	4.8	39 11.0	−37.2	4.9	40 10.8	37.1	4.9	85
86	32 34.5	38.0	3.6	33 34.4	38.0	3.6	34 34.3	38.0	3.7	35 34.2	37.9	3.7	36 34.1	37.9	3.8	37 33.9	37.8	3.8	38 33.8	37.7	3.9	39 33.7	37.7	3.9	86
87	31 56.5	38.5	2.7	32 56.4	38.4	2.7	33 56.3	38.3	2.7	34 56.3	38.4	2.8	35 56.2	38.3	2.8	36 56.1	38.2	2.8	37 56.1	38.3	2.9	38 56.0	38.2	2.9	87
88	31 18.0	38.8	1.8	32 18.0	38.8	1.8	33 18.0	38.8	1.8	34 17.9	38.7	1.8	35 17.9	38.7	1.8	36 17.9	38.7	1.9	37 17.8	38.7	1.9	38 17.8	38.7	1.9	88
89	30 39.2	39.2	0.9	31 39.2	39.2	0.9	32 39.2	39.2	0.9	33 39.2	39.2	0.9	34 39.2	39.2	0.9	35 39.2	39.2	0.9	36 39.1	39.1	0.9	37 39.1	39.1	1.0	89
90	30 00.0	39.5	0.0	31 00.0	−39.5	0.0	32 00.0	−39.5	0.0	33 00.0	−39.6	0.0	34 00.0	39.6	0.0	35 00.0	−39.6	0.0	36 00.0	39.6	0.0	37 00.0	−39.6	0.0	90
	30°			31°			32°			33°			34°			35°			36°			37°			

49°, 311° L.H.A. LATITUDE SAME NAME AS DECLINATION

Figure 20-9A. *Same-name page,* Tables No. 229.

LATITUDE CONTRARY NAME TO DECLINATION — L.H.A. 61°, 299°

Dec.	30° Hc	30° d	30° Z	31° Hc	31° d	31° Z	32° Hc	32° d	32° Z	33° Hc	33° d	33° Z
°	° ′	′	°	° ′	′	°	° ′	′	°	° ′	′	°
0	24 49.5	−33.2	105.5	24 33.3	−34.1	105.9	24 16.6	−35.0	106.4	23 59.5	−35.9	106.8
1	24 16.3	33.5	106.4	23 59.2	34.5	106.8	23 41.6	35.4	107.3	23 23.6	36.3	107.7
2	23 42.8	33.9	107.3	23 24.7	34.8	107.7	23 06.2	35.6	108.1	22 47.3	36.4	108.5
3	23 08.9	34.2	108.2	22 49.9	35.0	108.6	22 30.6	35.9	109.0	22 10.9	36.8	109.4
4	22 34.7	34.4	109.1	22 14.9	35.3	109.5	21 54.7	36.2	109.9	21 34.1	37.0	110.3
5	22 00.3	−34.8	110.0	21 39.6	−35.6	110.4	21 18.5	−36.4	110.7	20 57.1	−37.3	111.1
6	21 25.5	35.0	110.9	21 04.0	35.9	111.2	20 42.1	36.7	111.6	20 19.8	37.5	111.9
7	20 50.5	35.3	111.7	20 28.1	36.1	112.1	20 05.4	36.9	112.4	19 42.3	37.7	112.8
8	20 15.2	35.5	112.6	19 52.0	36.3	112.9	19 28.5	37.2	113.3	19 04.6	37.9	113.6
9	19 39.7	35.7	113.5	19 15.7	36.6	113.8	18 51.3	37.3	114.1	18 26.7	38.2	114.4
10	19 04.0	−36.1	114.3	18 39.1	−36.8	114.6	18 14.0	−37.6	114.9	17 48.5	−38.3	115.2
11	18 27.9	36.2	115.2	18 02.3	37.0	115.5	17 36.4	37.8	115.7	17 10.2	38.6	116.0
12	17 51.7	36.4	116.0	17 25.3	37.2	116.3	16 58.6	38.0	116.6	16 31.6	38.7	116.8
13	17 15.3	36.7	116.8	16 48.1	37.4	117.1	16 20.6	38.1	117.4	15 52.9	38.9	117.6
14	16 38.6	36.8	117.7	16 10.7	37.6	117.9	15 42.5	38.4	118.2	15 14.0	39.0	118.4
15	16 01.8	−37.0	118.5	15 33.1	−37.8	118.7	15 04.1	−38.5	119.0	14 35.0	−39.3	119.2
16	15 24.8	37.2	119.3	14 55.3	37.9	119.5	14 25.6	38.6	119.8	13 55.7	39.3	120.0
17	14 47.6	37.4	120.1	14 17.4	38.1	120.3	13 47.0	38.8	120.5	13 16.4	39.5	120.8
18	14 10.2	37.5	120.9	13 39.3	38.3	121.1	13 08.2	39.0	121.3	12 36.9	39.7	121.5
19	13 32.7	37.7	121.7	13 01.0	38.4	121.9	12 29.2	39.1	122.1	11 57.2	39.7	122.3
20	12 55.0	−37.9	122.5	12 22.6	−38.5	122.7	11 50.1	−39.2	122.9	11 17.5	−39.9	123.1
21	12 17.1	37.9	123.3	11 44.1	38.6	123.5	11 10.9	39.3	123.7	10 37.6	40.0	123.8
22	11 39.2	38.1	124.1	11 05.5	38.8	124.3	10 31.6	39.4	124.4	9 57.6	40.1	124.6
23	11 01.1	38.2	124.9	10 26.7	38.9	125.0	9 52.2	39.6	125.2	9 17.5	40.2	125.3
24	10 22.9	38.4	125.7	9 47.8	39.0	125.8	9 12.6	39.6	126.0	8 37.3	40.2	126.1
25	9 44.5	−38.4	126.5	9 08.8	−39.1	126.6	8 33.0	−39.7	126.7	7 57.1	−40.4	126.8
26	9 06.1	38.6	127.2	8 29.7	39.2	127.4	7 53.3	39.8	127.5	7 16.7	40.4	127.6
27	8 27.5	38.6	128.0	7 50.5	39.2	128.1	7 13.5	39.9	128.2	6 36.3	40.5	128.3
28	7 48.9	38.7	128.8	7 11.3	39.4	128.9	6 33.6	40.0	129.0	5 55.8	40.6	129.1
29	7 10.2	38.8	129.6	6 31.9	39.4	129.6	5 53.6	40.0	129.7	5 15.2	40.6	129.8
30	6 31.4	−38.9	130.3	5 52.5	−39.4	130.4	5 13.6	−40.0	130.5	4 34.6	−40.6	130.5
31	5 52.5	38.9	131.1	5 13.1	39.5	131.2	4 33.6	40.2	131.2	3 54.0	40.7	131.3
32	5 13.6	39.0	131.9	4 33.6	39.6	131.9	3 53.4	40.1	132.0	3 13.3	40.7	132.0
33	4 34.6	39.0	132.6	3 54.0	39.6	132.7	3 13.3	40.2	132.7	2 32.6	40.8	132.8
34	3 55.6	39.1	133.4	3 14.4	39.7	133.4	2 33.1	40.2	133.5	1 51.8	40.7	133.5
35	3 16.5	−39.1	134.1	2 34.7	−39.6	134.2	1 52.9	−40.2	134.2	1 11.1	−40.8	134.2
36	2 37.4	39.1	134.9	1 55.1	39.7	134.9	1 12.7	40.3	134.9	0 30.3	−40.8	135.0
37	1 58.3	39.2	135.7	1 15.4	39.7	135.7	0 32.4	−40.2	135.7	0 10.5	+40.8	44.3
38	1 19.1	39.1	136.4	0 35.7	−39.7	136.4	0 07.8	+40.2	43.6	0 51.3	40.7	43.6
39	0 40.0	39.2	137.2	0 04.0	+39.7	42.8	0 48.0	40.3	42.8	1 32.0	40.8	42.8
40	0 00.8	−39.2	137.9	0 43.7	+39.7	42.1	1 28.3	+40.2	42.1	2 12.8	+40.7	42.1
41	0 38.4	+39.1	41.3	1 23.4	39.7	41.3	2 08.5	40.2	41.3	2 53.5	40.7	41.4
42	1 17.5	39.2	40.6	2 03.1	39.7	40.6	2 48.7	40.1	40.6	3 34.2	40.7	40.6
43	1 56.7	39.1	39.8	2 42.8	39.6	39.8	3 28.8	40.2	39.9	4 14.9	40.6	39.9
44	2 35.8	39.1	39.0	3 22.4	39.6	39.1	4 09.0	40.1	39.1	4 55.5	40.6	39.2
45	3 14.9	+39.1	38.3	4 02.0	+39.6	38.3	4 49.1	+40.0	38.4	5 36.1	+40.5	38.4
46	3 54.0	39.0	37.5	4 41.6	39.5	37.6	5 29.1	40.0	37.6	6 16.6	40.5	37.7
47	4 33.0	39.0	36.8	5 21.1	39.4	36.8	6 09.1	39.9	36.9	6 57.1	40.4	36.9
48	5 12.0	38.9	36.0	6 00.5	39.4	36.0	6 49.0	39.9	36.1	7 37.5	40.3	36.2
49	5 50.9	38.9	35.2	6 39.9	39.3	35.3	7 28.9	39.8	35.4	8 17.8	40.2	35.4
50	6 29.8	+38.8	34.5	7 19.2	+39.3	34.5	8 08.7	+39.6	34.6	8 58.0	+40.1	34.7
51	7 08.6	38.7	33.7	7 58.5	39.1	33.8	8 48.3	39.6	33.8	9 38.1	40.1	33.9
52	7 47.3	38.6	32.9	8 37.6	39.1	33.0	9 27.9	39.5	33.1	10 18.2	39.9	33.2
53	8 25.9	38.6	32.1	9 16.7	39.0	32.2	10 07.4	39.4	32.3	10 58.1	39.8	32.4
54	9 04.5	38.4	31.4	9 55.7	38.9	31.5	10 46.8	39.3	31.6	11 37.9	39.7	31.7
55	9 42.9	+38.4	30.6	10 34.6	+38.7	30.7	11 26.1	+39.2	30.8	12 17.6	+39.6	30.9
56	10 21.3	38.2	29.8	11 13.3	38.6	29.9	12 05.3	39.0	30.0	12 57.2	39.4	30.1
57	10 59.5	38.1	29.0	11 51.9	38.5	29.1	12 44.3	38.9	29.2	13 36.6	39.3	29.3
58	11 37.6	38.0	28.2	12 30.4	38.4	28.3	13 23.2	38.8	28.5	14 15.9	39.2	28.6
59	12 15.6	37.8	27.5	13 08.8	38.2	27.6	14 02.0	38.6	27.7	14 55.1	39.0	27.8
60	12 53.4	+37.7	26.7	13 47.0	+38.1	26.8	14 40.6	+38.4	26.9	15 34.1	+38.8	27.0
61	13 31.1	37.6	25.9	14 25.1	37.9	26.0	15 19.0	38.3	26.1	16 12.9	38.6	26.2
62	14 08.7	37.3	25.1	15 03.0	37.7	25.2	15 57.3	38.0	25.3	16 51.5	38.4	25.4
63	14 46.0	37.2	24.2	15 40.7	37.6	24.4	16 35.3	38.0	24.5	17 29.9	38.3	24.6
64	15 23.2	37.1	23.4	16 18.3	37.3	23.5	17 13.3	37.7	23.7	18 08.2	38.0	23.8
65	16 00.3	+36.8	22.6	16 55.6	+37.2	22.7	17 51.0	+37.5	22.9	18 46.2	+37.9	23.0
66	16 37.1	36.7	21.8	17 32.8	37.0	21.9	18 28.5	37.2	22.0	19 24.1	37.6	22.2
67	17 13.8	36.4	21.0	18 09.8	36.7	21.1	19 05.7	37.1	21.2	20 01.7	37.3	21.3
68	17 50.2	36.3	20.1	18 46.5	36.6	20.2	19 42.8	36.8	20.4	20 39.0	37.2	20.5
69	18 26.5	36.0	19.3	19 23.1	36.3	19.4	20 19.6	36.6	19.5	21 16.2	36.8	19.7
70	19 02.5	+35.7	18.4	19 59.4	+36.0	18.6	20 56.2	+36.4	18.7	21 53.0	+36.7	18.8
71	19 38.2	35.6	17.6	20 35.4	35.8	17.7	21 32.6	36.0	17.8	22 29.7	36.3	18.0
72	20 13.8	35.3	16.7	21 11.2	35.6	16.8	22 08.6	35.8	17.0	23 06.0	36.1	17.1
73	20 49.1	35.0	15.9	21 46.8	35.2	16.0	22 44.4	35.5	16.1	23 42.1	35.7	16.2
74	21 24.1	34.7	15.0	22 22.0	35.0	15.1	23 19.9	35.3	15.2	24 17.8	35.5	15.3
75	21 58.8	+34.5	14.1	22 57.0	+34.7	14.2	23 55.2	+34.9	14.3	24 53.3	+35.1	14.5
76	22 33.3	34.2	13.2	23 31.7	34.4	13.3	24 30.1	34.6	13.4	25 28.4	34.8	13.6
77	23 07.5	33.9	12.4	24 06.1	34.1	12.4	25 04.7	34.2	12.5	26 03.2	34.5	12.7
78	23 41.4	33.5	11.5	24 40.2	33.7	11.5	25 38.9	34.0	11.6	26 37.7	34.1	11.7
79	24 14.9	33.3	10.5	25 13.9	33.4	10.6	26 12.9	33.5	10.7	27 11.8	33.8	10.8
80	24 48.2	+32.9	9.6	25 47.3	+33.1	9.7	26 46.4	+33.3	9.8	27 45.6	+33.3	9.9
81	25 21.1	32.5	8.7	26 20.4	32.7	8.8	27 19.7	32.8	8.9	28 18.9	33.0	8.9
82	25 53.6	32.2	7.8	26 53.1	32.3	7.8	27 52.5	32.4	7.9	28 51.9	32.6	8.0
83	26 25.8	31.8	6.8	27 25.4	31.9	6.9	[illegible]	32.1	7.0	29 24.5	32.1	7.0
84	26 57.6	31.5	5.9	27 57.3	31.5	5.9	28 [illegible]	31.6	6.0	29 56.6	31.8	6.1
85	27 29.1	+31.0	4.9	28 28.8	+31.1	5.0	29 [illegible]	+31.2	5.0	30 28.4	+31.2	5.1
86	28 00.1	30.6	4.0	28 59.9	30.7	4.0	29 59.8	30.7	4.0	30 59.6	30.9	4.1
87	28 30.7	30.2	3.0	29 30.6	30.3	3.0	30 30.5	30.3	3.0	31 30.5	30.5	3.1
88	29 00.9	29.8	2.0	30 00.9	[illegible]	2.0	31 00.8	29.9	2.0	32 00.8	29.9	2.1
89	29 30.7	29.3	1.0	30 30.7	29.3	1.0	31 30.7	29.3	1.0	32 30.7	29.3	1.0
90	30 00.0	+28.9	0.0	31 00.0	+28.8	0.0	32 00.0	+28.8	0.0	33 00.0	+28.8	0.0

Dec.	34° Hc	34° d	34° Z	35° Hc	35° d	35° Z	36° Hc	36° d	36° Z	37° Hc	37° d	37° Z
°	° ′	′	°	° ′	′	°	° ′	′	°	° ′	′	°
0	23 41.9	−36.8	107.2	23 23.9	−37.6	107.6	23 05.6	−38.5	108.0	22 46.8	−39.3	108.4
1	23 05.1	37.0	108.1	22 46.3	37.9	108.5	22 27.1	38.7	108.9	22 07.5	39.5	109.3
2	22 28.1	37.4	108.9	22 08.4	38.1	109.3	21 48.4	39.0	109.7	21 28.0	39.8	110.1
3	21 50.7	37.5	109.8	21 30.3	38.4	110.2	21 09.4	39.2	110.5	20 48.2	40.0	110.9
4	21 13.2	37.9	110.6	20 51.9	38.7	111.0	20 30.2	39.4	111.3	20 08.2	40.2	111.7
5	20 35.3	−38.0	111.4	20 13.2	−38.8	111.8	19 50.8	−39.7	112.1	19 28.0	−40.4	112.5
6	19 57.3	38.3	112.3	19 34.4	39.1	112.6	19 11.1	39.8	112.9	18 47.6	40.6	113.2
7	19 19.0	38.5	113.1	18 55.3	39.3	113.4	18 31.3	40.0	113.7	18 07.0	40.8	114.0
8	18 40.5	38.8	113.9	18 16.0	39.5	114.2	17 51.3	40.3	114.5	17 26.2	40.9	114.8
9	18 01.7	38.9	114.7	17 36.5	39.6	115.0	17 11.0	40.4	115.3	16 45.3	41.1	115.6
10	17 22.8	−39.1	115.5	16 56.9	−39.9	115.8	16 30.6	−40.5	116.1	16 04.2	−41.3	116.3
11	16 43.7	39.3	116.3	16 17.0	40.0	116.6	15 50.1	40.8	116.8	15 22.9	41.5	117.1
12	16 04.4	39.4	117.1	15 37.0	40.2	117.3	15 09.3	40.8	117.6	14 41.4	41.5	117.8
13	15 25.0	39.6	117.9	14 56.8	40.3	118.1	14 28.5	41.1	118.3	13 59.9	41.7	118.6
14	14 45.4	39.8	118.6	14 16.5	40.5	118.9	13 47.4	41.1	119.1	13 18.2	41.9	119.3
15	14 05.6	−39.9	119.4	13 36.0	−40.6	119.6	13 06.3	−41.3	119.8	12 36.3	−41.9	120.0
16	13 25.7	40.1	120.2	12 55.4	40.7	120.4	12 25.0	41.4	120.6	11 54.4	42.1	120.8
17	12 45.6	40.2	121.0	12 14.7	40.9	121.1	11 43.6	41.6	121.3	11 12.3	42.2	121.5
18	12 05.4	40.3	121.7	11 33.8	41.0	121.9	11 02.0	41.6	122.1	10 30.1	42.3	122.2
19	11 25.1	40.4	122.5	10 52.8	41.1	122.6	10 20.4	41.7	122.8	9 47.8	42.3	122.9
20	10 44.7	−40.6	123.2	10 11.7	−41.2	123.4	9 38.7	−41.9	123.5	9 05.5	−42.5	123.7
21	10 04.1	40.6	124.0	9 30.5	41.2	124.1	8 56.8	41.9	124.3	8 23.0	42.5	124.4
22	9 23.5	40.7	124.7	8 49.3	41.4	124.9	8 14.9	42.0	125.0	7 40.5	42.6	125.1
23	8 42.8	40.9	125.5	8 07.9	41.4	125.6	7 32.9	42.0	125.7	6 57.9	42.7	125.8
24	8 01.9	40.9	126.2	7 26.5	41.6	126.3	6 50.9	42.1	126.4	6 15.2	42.7	126.5
25	7 21.0	−40.9	126.9	6 44.9	−41.5	127.0	6 08.8	−42.2	127.1	5 32.5	−42.8	127.2
26	6 40.1	41.1	127.7	6 03.4	41.7	127.8	5 26.6	42.2	127.8	4 49.7	42.8	127.9
27	5 59.0	41.1	128.4	5 21.7	41.7	128.5	4 44.4	42.3	128.6	4 06.9	42.8	128.6
28	5 17.9	41.1	129.1	4 40.0	41.7	129.2	4 02.1	42.3	129.3	3 24.1	42.9	129.3
29	4 36.8	41.2	129.9	3 58.3	41.8	129.9	3 19.8	42.4	130.0	2 41.2	42.9	130.0
30	3 55.6	−41.2	130.6	3 16.5	−41.8	130.7	2 37.4	−42.3	130.7	1 58.3	−42.9	130.7
31	3 14.4	41.3	131.3	2 34.7	41.8	131.4	1 55.1	42.4	131.4	1 15.4	43.0	131.4
32	2 33.1	41.3	132.1	1 52.9	41.8	132.1	1 12.7	42.4	132.1	0 32.4	−42.9	132.1
33	1 51.8	41.3	132.8	1 11.1	41.9	132.8	0 30.3	−42.4	132.8	0 10.5	+42.9	47.2
34	1 10.5	41.3	133.5	0 29.2	−41.8	133.5	0 12.1	+42.4	46.5	0 53.4	42.9	46.5
35	0 29.2	−41.3	134.2	0 12.6	+41.9	45.8	0 54.5	+42.4	45.8	1 36.3	+43.0	45.8
36	0 12.1	+41.3	45.0	0 54.5	41.8	45.0	1 36.9	42.4	45.1	2 19.3	42.9	45.1
37	0 53.4	41.3	44.3	1 36.3	41.9	44.3	2 19.3	42.3	44.4	3 02.2	42.8	44.4
38	1 34.7	41.3	43.6	2 18.2	41.8	43.6	3 01.6	42.3	43.6	3 45.0	42.8	43.7
39	2 16.0	41.3	42.9	3 00.0	41.8	42.9	3 43.9	42.3	42.9	4 27.8	42.8	43.0
40	2 57.3	+41.2	42.1	3 41.8	+41.7	42.2	4 26.2	+42.3	42.2	5 10.6	+42.8	42.3
41	3 38.5	41.2	41.4	4 23.5	41.7	41.5	5 08.5	42.2	41.5	5 53.4	42.7	41.6
42	4 19.7	41.2	40.7	5 05.2	41.7	40.7	5 50.7	42.1	40.8	6 36.1	42.6	40.9
43	5 00.9	41.1	39.9	5 46.9	41.6	40.0	6 32.8	42.1	40.1	7 18.7	42.6	40.2
44	5 42.0	41.1	39.2	6 28.5	41.5	39.3	7 14.9	42.0	39.4	8 01.3	42.5	39.4
45	6 23.1	+41.0	38.5	7 10.0	+41.5	38.6	7 56.9	+42.0	38.6	8 43.8	+42.4	38.7
46	7 04.1	40.9	37.7	7 51.5	41.4	37.8	8 38.9	41.8	37.9	9 26.2	42.3	38.0
47	7 45.0	40.9	37.0	8 32.9	41.3	37.1	9 20.7	41.8	37.2	10 08.5	42.2	37.3
48	8 25.9	40.7	36.3	9 14.2	41.2	36.4	10 02.5	41.7	36.5	10 50.7	42.1	36.6
49	9 06.6	40.7	35.5	9 55.4	41.2	35.6	10 44.2	41.5	35.7	11 32.8	42.1	35.8
50	9 47.3	+40.6	34.8	10 36.6	+41.0	34.9	11 25.7	+41.5	35.0	12 14.9	+41.9	35.1
51	10 27.9	40.5	34.0	11 17.6	40.9	34.1	12 07.2	41.4	34.3	12 56.8	41.7	34.4
52	11 08.4	40.3	33.3	11 58.5	40.8	33.4	12 48.6	41.2	33.5	13 38.5	41.7	33.6
53	11 48.7	40.3	32.5	12 39.3	40.7	32.6	13 29.8	41.1	32.8	14 20.2	41.5	32.9
54	12 29.0	40.1	31.8	13 20.0	40.5	31.9	14 10.9	40.9	32.0	15 01.7	41.4	32.2
55	13 09.1	+40.0	31.0	14 00.5	+40.4	31.1	14 51.8	+40.8	31.3	15 43.1	+41.2	31.4
56	13 49.1	39.8	30.2	14 40.9	40.2	30.4	15 32.6	40.7	30.5	16 24.3	41.0	30.7
57	14 28.9	39.7	29.5	15 21.1	40.1	29.6	16 13.3	40.4	29.7	17 05.3	40.9	29.9
58	15 08.6	39.5	28.7	16 01.2	39.9	28.8	16 53.7	40.3	29.0	17 46.2	40.7	29.1
59	15 48.1	39.4	27.9	16 41.1	39.7	28.1	17 34.0	40.1	28.2	18 26.9	40.5	28.4
60	16 27.5	+39.2	27.1	17 20.8	+39.6	27.3	18 14.1	+40.0	27.4	19 07.4	+40.3	27.6
61	17 06.7	39.0	26.3	18 00.4	39.4	26.5	18 54.1	39.7	26.6	19 47.7	40.1	26.8
62	17 45.7	38.8	25.5	18 39.8	39.1	25.7	19 33.8	39.5	25.8	20 27.8	39.8	26.0
63	18 24.5	38.5	24.7	19 18.9	39.0	24.9	20 13.3	39.3	25.0	21 07.6	39.7	25.2
64	19 03.0	38.4	23.9	19 57.9	38.7	24.1	20 52.6	39.1	24.2	21 47.3	39.4	24.4
65	19 41.4	+38.2	23.1	20 36.6	+38.5	23.3	21 31.7	+38.8	23.4	22 26.7	+39.2	23.6
66	20 19.6	37.9	22.3	21 15.1	38.2	22.4	22 10.5	38.6	22.6	23 05.9	38.9	22.8
67	20 57.5	37.7	21.5	21 53.3	38.0	21.6	22 49.1	38.3	21.8	23 44.8	38.6	21.9
68	21 35.2	37.4	20.6	22 31.3	37.8	20.8	23 27.4	38.1	20.9	24 23.4	38.4	21.1
69	22 12.6	37.2	19.8	23 09.1	37.4	19.9	24 05.5	37.7	20.1	25 01.8	38.1	20.2
70	22 49.8	+36.9	18.9	23 46.5	+37.2	19.1	24 43.2	+37.5	19.2	25 39.9	+37.7	19.4
71	23 26.7	36.6	18.1	24 23.7	36.9	18.2	25 20.7	37.2	18.4	26 17.6	37.5	18.5
72	24 03.3	36.4	17.2	25 00.6	36.6	17.4	25 57.9	36.8	17.5	26 55.1	37.1	17.6
73	24 39.7	36.0	16.3	25 37.2	36.3	16.5	26 34.7	36.6	16.6	27 32.2	36.8	16.8
74	25 15.7	35.7	15.5	26 13.5	35.9	15.6	27 11.3	36.1	15.7	28 09.0	36.4	15.9
75	25 51.4	+35.3	14.6	26 49.4	+35.6	14.7	27 47.4	+35.9	14.8	28 45.4	+36.1	15.0
76	26 26.7	35.1	13.7	27 25.0	35.3	13.8	28 23.3	35.4	13.9	29 21.5	35.7	14.0
77	27 01.8	34.6	12.8	28 00.3	34.8	12.9	28 58.7	35.1	13.0	29 57.2	35.3	13.1
78	27 36.4	34.3	11.8	28 35.1	34.5	12.0	29 33.8	34.7	12.1	30 32.5	34.9	12.2
79	28 10.7	34.0	10.9	29 09.6	34.2	11.0	30 08.5	34.3	11.1	31 07.4	34.5	11.2
80	28 44.7	+33.5	10.0	29 43.8	+33.7	10.1	30 42.8	+33.9	10.2	31 41.9	+34.0	10.3
81	29 18.2	33.1	9.0	30 17.5	33.2	9.1	31 16.7	33.4	9.2	32 15.9	33.6	9.3
82	29 51.3	32.7	8.1	30 50.7	32.9	8.2	31 50.1	33.0	8.2	32 49.5	33.1	8.3
83	30 24.0	32.3	7.1	31 23.6	32.4	7.2	32 23.1	32.5	7.3	33 22.6	32.6	7.3
84	30 56.3	31.8	6.1	31 56.0	31.9	6.2	32 55.6	32.0	6.3	33 54.8	32.2	6.3
85	31 28.1	+31.4	5.1	32 27.9	+31.4	5.2	33 27.6	+31.6	5.2	34 27.4	+31.6	5.3
86	31 59.5	30.9	4.1	32 59.3	31.0	4.2	33 59.2	31.0	4.2	34 59.0	31.1	4.3
87	32 30.4	30.4	3.1	33 30.3	30.4	3.1	34 30.2	30.5	3.2	35 30.1	30.5	3.2
88	33 00.8	29.8	2.1	34 00.7	29.9	2.1	35 00.7	29.9	2.1	36 00.6	30.0	2.2
89	33 30.6	29.4	1.0	34 30.6	29.4	1.1	35 30.6	29.4	1.1	36 30.6	29.4	1.1
90	34 00.0	+28.8	0.0	35 00.0	+28.8	0.0	36 00.0	+28.8	0.0	37 00.0	+28.8	0.0

S. Lat. { L.H.A. greater than 180°......Zn = 180° − Z; L.H.A. less than 180°............Zn = 180° + Z }

LATITUDE SAME NAME AS DECLINATION — L.H.A. 119°, 241°

Figure 20-9B. *Contrary-name page,* Tables No. 229.

latitude zone for LHAs of 61°, 299°, 119° and 241°; the top of the page is used if the latitude and the declination are of contrary name, and the bottom of the page is for use if the latitude and declination are of the same name.

As was the case with the *Nautical Almanac,* the use of the *Sight Reduction Tables No. 229* is best explained by means of examples. Because the procedures for the use of the tables during sight reduction are the same for any type of celestial body observed, only two of the four sample problems introduced earlier in this chapter will be completed here—Canopus and the sun. The complete solutions for the planet Venus and the moon will appear in chapter 23 in connection with the plot of a celestial fix.

Completing the Sight Reduction by *Tables No. 229*—First Example

As the first example of the completion of the sight reduction process by use of *Sight Reduction Tables No. 229,* the Canopus form shown in Figure 20-3 will be completed. The entering arguments, determined in previous sections of this chapter, are reproduced below:

LHA	49°
True Dec	S 52° 40.6'
a LAT	34° S Same

The first step in entering *Tables No. 229* is to select the proper volume and page number. Since the assumed latitude in this example is 34°S, volume 3 containing tabulations for latitudes from 30° to 45° is selected. The volume is opened to the page containing tabulations for LHA of 49° with latitude the same name as declination, in the 30° to 37° latitude zone. This is the page appearing in Figure 20-9A.

The first quantity extracted from the table is the tabulated value of computed altitude Hc in the column of figures corresponding to the assumed latitude for the nearest whole degree of declination less than the exact declination. In this example an Hc of 50° 51.1' is extracted, using as a horizontal argument 34° of latitude, and as a vertical argument 52° of declination. In practice, to save time the azimuth angle Z would also be recorded at this time for future use, but to simplify this initial explanation, this will be omitted here.

Next, the exact value of Hc corresponding to the exact declination of the body to the time of observation must be determined by interpolation. To simplify this interpolation process, a set of interpolation tables is included inside the front and back cover of each volume.

The entering arguments for the interpolation tables are the declination increment (Dec Inc)—the remaining minutes and tenths of the exact declination—and the altitude difference d between the two tabulated Hc's bracketing the exact declination. The value of d between successive tabulated Hc's has been precalculated and appears in the center of each column of tabulations. Both the declination increment and altitude difference d are recorded in the appropriate spaces on the sight form. In this case, the declination increment is 40.6′, and the altitude difference d is –8.7′. If the Hc decreases in value with increasing declination, as is the case in this instance, the sign of the altitude difference d is negative; this is indicated in the tables by the placement of a minus sign adjacent to the initial negative value and every fifth value thereafter. If the Hc increases with increasing declination, the sign of the altitude difference d is positive, indicated by a plus sign.

In almost all cases, two increments are extracted from the interpolation table—one for the tens of minutes of the altitude difference d and the other for the remaining units and tenths. Adding the two parts together yields the total interpolation correction (Total Corr), which in turn is added algebraically to the tabulated Hc to obtain the final computed altitude. In about 1 percent of all cases, a third increment called a "double second difference" correction (DS Corr) must also be found. Occasions for which this is necessitated are indicated in the tables by the d value being printed in italic type followed by a dot.

The interpolation tables inside the front cover are used for declination increments in the range 0.0′ to 31.9′, and those inside the back cover for the range 28.0′ to 59.9′. Since the declination increment in this example is 40.6′, the tables inside the back cover are used. An extract from the appropriate table appears in Figure 20-10.

The interpolation table is entered first for the "tens" increment and then for the "units" increment of the interpolation correction. For the tens increment

Dec. Inc.	Tens 10′	20′	30′	40′	50′	Decimals	Units 0′	1′	2′	3′	4′	5′	6′	7′	8′	9′
36.0	6.0	12.0	18.0	24.0	30.0	.0	0.0	0.6	1.2	1.8	2.4	3.0	3.6	4.3	4.9	5.5
			18.0	24.0	30.1	.1									4.9	5.5
39.9	6.7															
40.0	6.6	13.3	20.0	26.6	33.3	.0	0.0	0.7	1.3	2.0	2.7	3.4	4.0	4.7	5.4	6.1
40.1	6.7	13.3	20.0	26.7	33.4	.1	0.1	0.7	1.4	2.1	2.8	3.4	4.1	4.8	5.5	6.1
40.2	6.7	13.4	20.1	26.8	33.5	.2	0.1	0.8	1.5	2.2	2.8	3.5	4.2	4.9	5.5	6.2
40.3	6.7	13.4	20.1	26.9	33.6	.3	0.2	0.9	1.6	2.2	2.9	3.6	4.3	4.9	5.6	6.3
40.4	6.7	13.5	20.2	26.9	33.7	.4	0.3	0.9	1.6	2.3	3.0	3.6	4.3	5.0	5.7	6.3
40.5	6.8	13.5	20.3	27.0	33.8	.5	0.3	1.0	1.7	2.4	3.0	3.7	4.4	5.1	5.7	6.4
40.6	6.8	13.5	20.3	27.1	33.8	.6	0.4	1.1	1.8	2.4	3.1	3.8	4.5	5.1	5.8	6.5
40.7	6.8	13.6	20.4	27.2	33.9	.7	0.5	1.1	1.8	2.5	3.2	3.8	4.5	5.2	5.9	6.5
40.8	6.8	13.6	20.4	27.2	34.0	.8	0.5	1.2	1.9	2.6	3.2	3.9	4.6	5.3	5.9	6.6
40.9	6.9	13.7	20.5	27.3	34.1	.9	0.6	1.3	2.0	2.6	3.3	4.0	4.7	5.3	6.0	6.7
		13.6	20.5	27.3	34.1	.0							4.1	4.8	5.5	6.2
															5.6	6.3

Double Second Diff.	Corr.
0.8	
	0.1
2.5	
	0.2
0.9	
	0.1
2.8	
	0.2
4.6	
	0.3
6.5	
	0.4
8.3	
	0.5
10.2	
	0.6
12.0	
	0.7
13.9	
	0.8
15.7	
	0.9
17.6	
	1.0
19.4	
	1.1
21.3	
	1.2

Figure 20-10. *Extract from interpolation tables,* Tables No. 229.

the table is entered directly across from the declination increment, 40.6′ in this case, and the increment beneath the appropriate number of tens of altitude difference is recorded. Here, no tens increment is extracted, since the altitude difference d is 08.7′. To find the units increment, the appropriate units column (in this case the one headed 8′) is followed down the page in the group of tabulations most nearly opposite the declination increase until the appropriate decimal fraction is reached (0.7 in this case). For this example, the units increment is −5.9; it is negative, since the altitude difference d is negative.

Had it been necessary to find a double second difference correction, the difference between the two d values directly above and directly below the d value corresponding to the integral portion of the actual declination is mentally computed. Using this "double second difference" as an entering argument, the right-hand side of the interpolation table is used to find the correction. There are several complete DS interpolation sections on each page; the section used is the one most nearly opposite the original declination increment. If a DS correction is necessary, it is entered on the form and always added to the total of the tens and units increments to form the total interpolation correction.

After extracting, recording, and summing the two increments to form the total interpolation correction, −5.9 in this case, the correction is added to the tabulated Hc recorded earlier to obtain the final computed altitude of 50° 45.2′. To find the intercept distance a, this value is compared with the observed altitude Ho 50° 39.4′ determined earlier. Since Ho is less than Hc in this case, the intercept distance, 5.8 miles, is labeled "Away" (A) from the direction of the GP of the body.

To find the true azimuth of the GP from the AP, the azimuth angle Z must be determined. To compute the value for Z, it is necessary to interpolate between the values of Z tabulated in *Tables No. 229* for the whole degrees of declination bracketing the true declination in the same latitude column used previously. Inasmuch as the difference between successive tabulated azimuth angles is normally small, interpolation is usually done mentally. The interpolation tables can be used for this interpolation, however, by substituting the difference between the two tabulated azimuth angles as a difference d. In this example, the azimuth angle tabulated for declination 52° is 47.4° and the angle for declination 53° is 45.8°. Hence, the value of the azimuth angle for a true

declination 52° 40.6′ is 46.3° $\left(47.4' - 45.8' = (-)1.6°; \dfrac{40.6'}{60'} \times (-)1.6° = \right.$
$\left. (-)1.1°; 47.4° (-)1.1° = 46.3° \right)$. Since the assumed latitude lies in the southern hemisphere and the body lies west of the observer (LHA is less than 180°), the prefix S and the suffix W are applied: S46.3°W.

The final step in completing the sight reduction form is to convert the azimuth angle just computed to the true azimuth Zn of the body from the AP of the observer. One method of accomplishing this by sketching the angle was presented in chapter 16. The easiest method, however, is simply to use the conversion formulas printed on each page of *Tables No. 229*:

N. Lat. $\begin{cases} \text{LHA greater than } 180° \ldots\ldots\ldots \text{Zn} = \text{Z} \\ \text{LHA less than } 180° \ldots\ldots\ldots\ldots \text{Zn} = 360° - \text{Z} \end{cases}$

S. Lat. $\begin{cases} \text{LHA greater than } 180° \ldots\ldots\ldots \text{Zn} = 180° - \text{Z} \\ \text{LHA less than } 180° \ldots\ldots\ldots\ldots \text{Zn} = 180° + \text{Z} \end{cases}$

By either method the azimuth angle S46° .3'W is converted to a true azimuth 226.3°T. The completed sight reduction for Canopus appears in Figure 20-11.

Body	CANOPUS
GMT	17-12-09 Dec16
Ho	50° 39.4'
GHA (h)	338° 56.1'
Incre. (m/s)	3° 02.7'
v/v Corr SHA	264° 09.8'
Total GHA	606° 08.6'
±360°	246° 08.6'
aλ (+E,-W)	162° 51.4'E
LHA	49°
Tab Dec	
d#/d Corr	
True Dec	S 52° 40.6'
a LAT Same Contrary	34°S Same
Dec Inc/d	40.6/-8.7
Tens/DSD	0
Units/DSD Corr	-5.9'
Total Corr	-5.9'
Hc (Tab)	50° 51.1'
Hc (Comp)	50° 45.2'
Ho	50° 39.4'
a	5.8'A
Z	S 46.3° W
Zn	226.3°T

Figure 20-11. *Complete solution for the star Canopus by* Tables No. 229.

In chapter 23 the plot of the resulting Canopus LOP will be combined with the LOPs resulting from the Venus and moon sight reductions partially completed earlier in this chapter to form a celestial fix.

Completing the Sight Reduction by *Tables No. 229*—Second Example

As the second example of the use of *Tables No. 229*, the sun sight reduction begun as the last example of the use of the *Nautical Almanac* will be completed here. The entering arguments for the tables for this example follow:

LHA	299°
True Dec	S 23° 18.9′
a LAT	31° N Cont

Opening the tables to the LHA 299° page shown in Figure 20-9B, the tabulated Hc corresponding to declination 23° and latitude 31° is first recorded, along with the altitude difference d; they are 10° 26.7′ and −38.9, respectively. To eliminate the necessity to refer back to the LHA page, the azimuth angle Z

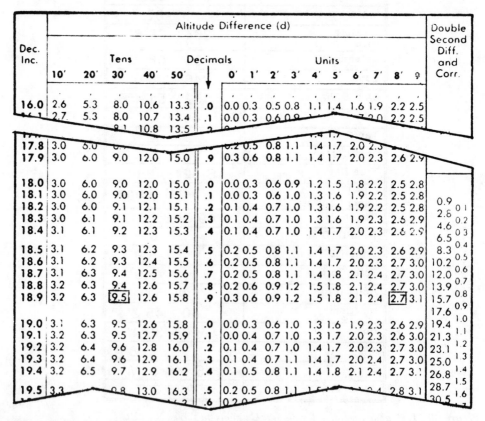

Figure 20-12. *Extract from interpolation tables,* Tables No. 229.

Body	SUN (LL)
GMT	12-12-06 Dec16
Ho	10° 16.8′
GHA (h)	4° 08.5′
Incre. (m/s)	3° 01.5′
v/v Corr SHA	
Total GHA	4° 08.5′
±360°	364° 08.5′
aλ (+E,-W)	65° 08.5′
LHA	299°
Tab Dec	S 23° 18.9′
d#/d Corr	0.1′/ 0
True Dec	S 23° 18.9′
a LAT Same Contrary	31° N Cont
Dec Inc/d	18.9′/-38.9′
Tens/DSD	-9.5′
Units/DSD Corr	-2.7
Total Corr	-12.2′
Hc (Tab)	10° 26.7′
Hc (Comp)	10° 14.5′
Ho	10° 16.8′
a	2.3′T
Z	N 125.2° E
Zn	125.2° T

Figure 20-13. *Complete solution for the sun (LL) by* Tables No. 229.

values corresponding to declinations 23° and 24° are also recorded on a sheet of scratch paper; they are 125.0° and 125.8°, respectively.

Turning to the interpolation tables inside the front cover, the appropriate portion of which is shown in Figure 20-12, the correction to the tabulated Hc is first found. The tens increment is 9.5′, and the units increment is 2.7′; these values are boxed in the figure. No double second difference correction applies, so the total correction to the tabulated Hc is the sum of the tens and units increments, or 12.2′. Applying this correction yields the computed altitude 10° 14.5′.

Comparing this Hc with the Ho 10° 16.8′ results in an intercept distance of 2.3 miles; since Ho is greater than Hc, the intercept is labeled "Toward" (T) the GP of the body.

Referring to the azimuth angles noted earlier, the difference between the two is 0.8°. Entering the interpolation tables once again, using this figure as a difference d, a correction to the tabulated angle of 0.2° results. Hence, the azimuth angle is N125.2°E (east, because the LHA is greater than 180°). Conversion of this azimuth angle yields a true azimuth of 125.2°T.

The completed sun-sight reduction form appears in Figure 20-13. In chapter 23, the resulting LOP will be combined with a second LOP obtained from a later sun observation to form a running fix.

SUMMARY

In this chapter, the reduction of the observation of a celestial body by the use of the *Nautical Almanac* in conjunction with the *Sight Reduction Tables for Marine Navigation, No. 229*, was demonstrated by means of several examples. Use of these publications during sight reduction is rather tedious, but has the positive advantage that a high degree of accuracy is attainable in the resulting celestial LOP. In the following chapter, the use of the *Air Almanac* with the *Sight Reduction Tables for Air Navigation* is demonstrated for several of the same observations presented as examples in this chapter. The increased speed of computation using these latter publications will become obvious, but it should be borne in mind that the precision of the resulting Hc and Zn is less than that achieved by use of the *Nautical Almanac* and *Tables No. 229*.

THE COMPLETE SOLUTION BY THE *AIR ALMANAC* AND TABLES NO. 249

21

In the last chapter the complete solution for a celestial line of position by the altitude-intercept method using the *Nautical Almanac* in conjunction with the *Sight Reduction Tables for Marine Navigation, No. 229*, was discussed. In this chapter an alternative method for the complete solution will be examined, using the *Air Almanac* with the *Sight Reduction Tables for Air Navigation, No. 249*. Although either of these two almanacs could be used with either of these two sight reduction tables, in practice they are usually employed as the titles of the chapters indicate; the *Nautical Almanac* and *Tables No. 229* are generally preferred in marine navigation in order to take advantage of the finer precision inherent in this combination. The greater ease of obtaining the complete solution by the *Air Almanac* and *Tables No. 249*, however, has resulted in their adoption by many marine navigators, particularly when operating in rough weather on the high seas, or in small vessels in which highly precise celestial observations are very difficult to make even under optimal weather conditions.

THE *AIR ALMANAC*

The *Air Almanac* is published annually by the U.S. Naval Observatory. It is bound with plastic rings to allow the pages to be removed if desired for more convenient use.

The contents of the *Air Almanac* are basically the same as the *Nautical Almanac*, the primary differences being in the arrangement and precision of the data. The *Air Almanac* is designed for speed and ease of computation, at the cost of some small degree of precision. As an example, when the *Air Almanac* is to be used to obtain sextant altitude corrections to obtain Ho, the hs measured by the sextant is normally rounded off to the nearest whole minute. All corrections to be applied to hs are tabulated to the nearest minute in various critical type tables, with the exception of the refraction correction; at very low altitudes the refraction correction is taken to the nearest 2 or 5 minutes. Fig-

CORRECTIONS TO BE APPLIED TO MARINE SEXTANT ALTITUDES

MARINE SEXTANT ERROR Sextant Number Index Error	CORRECTIONS In addition to sextant error and dip, corrections are to be applied for: Refraction Semi-diameter (for the Sun and Moon) Parallax (for the Moon) Dome refraction (if applicable)	CORRECTION FOR DIP OF THE HORIZON To be subtracted from sextant altitude								
		Ht.	Dip	Ht.	Dip	Ht.	Dip	Ht.	Dip	Ht.	Dip
		Ft.	,	Ft.	,	Ft.	,	Ft	,	Ft.	
		0		114		437		968		1 707	
		2	1	137	11	481	21	1 033	31	1 792	41
		6	2	162	12	527	22	1 099	32	1 880	42
		12	3	189	13	575	23	1 168	33	1 970	43
		21	4	218	14	625	24	1 239	34	2 061	44
		31	5	250	15	677	25	1 311	35	2 155	45
		43	6	283	16	731	26	1 386	36	2 251	46
		58	7	318	17	787	27	1 463	37	2 349	47
		75	8	356	18	845	28	1 543	38	2 449	48
		93	9	395	19	906	29	1 624	39	2 551	49
		114	10	437	20	968	30	1 707	40	2 655	50

Figure 21-1A. *Dip correction table,* Air Almanac.

ure 21-1A shows the dip table for the *Air Almanac*, which is located on the outside back cover, and Figure 21-1B depicts the refraction correction table on the inside back cover.

The sextant altitude hs corrected for IC is the entering argument for both tables. For surface navigation, the left-hand column in the refraction table headed 0 (for 0 altitude) is used. The refraction correction R_0 corresponding to the range of sextant altitude containing the given hs is the value extracted; it is always subtracted from hs. Under abnormal temperature conditions a multiplication factor f is obtained from the lower part of the table and applied to the basic R_0 correction by means of the nomogram on the right side. (Note the instruction in the lower right corner to use $R = R_0$ where the R_0 correction is less than 10'.)

When a sun or moon observation is made, semidiameter (SD) and parallax corrections are made using data on the daily pages. The semidiameter of the moon is always given to the nearest minute, while the semidiameter of the sun is given to the nearest tenth of a minute. The SDs are either added to or subtracted from hs, depending on whether a lower- or upper-limb observation was made. The value of parallax for the moon is found from a critical value table, using the hs corrected for IC, dip, refraction, and SD as entering argument; it is always added to the corrected hs.

The daily pages of the *Air Almanac* contain much of the same data found on the corresponding daily pages of the *Nautical Almanac*, the main difference being in the frequency of tabulations. Whereas in the *Nautical Almanac* tabulated values of GHA and declination appear for every whole hour of GMT, in the *Air Almanac* the tabulations are for every 10 minutes of GMT. The increased frequency of tabulations eliminates the necessity for any v or d corrections to the GHA and declination. In fact, because the exact calculated value of

CORRECTIONS TO BE APPLIED TO SEXTANT ALTITUDE

REFRACTION

To be subtracted from sextant altitude (referred to as observed altitude in A.P. 3270)

R_0	Height above sea level in units of 1 000 ft.												R_2	$R = R_2 \times f$
	0	5	10	15	20	25	30	35	40	45	50	55		0·9 1·0 1·1 1·2
						Sextant Altitude								R
	′	′	′	′	′	′	′	′	′	′	′	′	′	′ ′ ′ ′
0	90	90	90	90	90	90	90	90	90	90	90	90	0	0 0 0 0
1	63	59	55	51	46	41	36	31	26	20	17	13		1 1 1 1
2	33	29	26	22	19	16	14	11	9	7	6	4	1	1 1 1 1
3	21	19	16	14	12	10	8	7	5	4	2 40	1 40	2	2 2 2 2
4	16	14	12	10	8	7	6	5	3 10	2 20	1 30	0 40	3	3 3 3 4
5	12	11	9	8	7	5	4 00	3 10	2 10	1 30	0 39	+0 05	4	4 4 4 5
6	10	9	7	5 50	4 50	3 50	3 10	2 20	1 30	0 49	+0 11	−0 19	5	5 5 5 6
7	8 10	6 50	5 50	4 50	4 00	3 00	2 20	1 50	1 10	0 24	−0 11	−0 38	6	5 6 7 7
8	6 50	5 50	5 00	4 00	3 10	2 30	1 50	1 20	0 38	+0 04	−0 28	−0 54	7	6 7 8 8
9	6 00	5 10	4 10	3 20	2 40	2 00	1 30	1 00	0 19	−0 13	−0 42	−1 08	8	7 8 9 10
10	5 20	4 30	3 40	2 50	2 10	1 40	1 10	0 35	+0 03	−0 27	−0 53	−1 18	9	8 9 10 11
12	4 30	3 40	2 50	2 20	1 40	1 10	0 37	+0 11	−0 16	−0 43	−1 08	−1 31	10	9 10 11 12
14	3 30	2 50	2 10	1 40	1 10	0 34	+0 09	−0 14	−0 37	−1 00	−1 23	−1 44	12	11 12 13 14
16	2 50	2 10	1 40	1 10	0 37	+0 10	−0 13	−0 34	−0 53	−1 14	−1 35	−1 56	14	13 14 15 17
18	2 20	1 40	1 20	0 43	+0 15	−0 08	−0 31	−0 52	−1 08	−1 27	−1 46	−2 05	16	14 16 18 19
20	1 50	1 20	0 49	+0 23	−0 02	−0 26	−0 46	−1 06	−1 22	−1 39	−1 57	−2 14	18	16 18 20 22
25	1 12	0 44	+0 19	−0 06	−0 28	−0 48	−1 09	−1 27	−1 42	−1 58	−2 14	−2 30	20	18 20 22 24
30	0 34	+0 10	−0 13	−0 36	−0 55	−1 14	−1 32	−1 51	−2 06	−2 21	−2 34	−2 49	25	22 25 28 30
35	+0 06	−0 16	−0 37	−0 59	−1 17	−1 33	−1 51	−2 07	−2 23	−2 37	−2 51	−3 04	30	27 30 33 36
40	−0 18	−0 37	−0 58	−1 16	−1 34	−1 49	−2 06	−2 22	−2 35	−2 49	−3 03	−3 16	35	31 35 38 42
45		−0 53	−1 14	−1 31	−1 47	−2 03	−2 18	−2 33	−2 47	−2 59	−3 13	−3 25	40	36 40 44 48
50		−1 10	−1 28	−1 44	−1 59	−2 15	−2 28	−2 43	−2 56	−3 08	−3 22	−3 33	45	40 45 50 54
55			−1 40	−1 53	−2 09	−2 24	−2 38	−2 52	−3 04	−3 17	−3 29	−3 41	50	45 50 55 60
60				−2 03	−2 18	−2 33	−2 46	−3 01	−3 12	−3 25	−3 37	−3 48	55	49 55 60 66
							−2 53	−3 07	−3 19	−3 31	−3 42	−3 53	60	54 60 66 72

f	0	5	10	15	20	25	30	35	40	45	50	55	f	0·9 1·0 1·1 1·2
						Temperature in °C.								f
0·9	+47	+36	+27	+18	+10	+ 3	− 5	−13					0·9	Where R_2 is
1·0	+26	+16	+ 6	− 4	−13	−22	−31	−40	For these heights no				1·0	less than 10′
1·1	+ 5	− 5	−15	−25	−36	−46	−57	−68	temperature correction				1·1	or the height
1·2	−16	−25	−36	−46	−58	−71	−83	−95	is necessary, so use				1·2	greater than
	−37	−45	−56	−67	−81	−95			$R = R_0$					35 000 ft. use
														$R = R_2$

Choose the column appropriate to height, in units of 1 000 ft., and find the range of altitude in which the sextant altitude lies; the corresponding value of R_0 is the refraction, to be subtracted from sextant altitude, unless conditions are extreme. In that case find f from the lower table, with critical argument temperature. Use the table on the right to form the refraction, $R = R_0 \times f$.

Figure 21-1B. *Refraction correction table,* Air Almanac.

the GHA of the sun correct to the nearest tenth is printed for every 10 minutes of GMT, it is possible to obtain a more precise value of GHA for the sun than when using the *Nautical Almanac,* wherein the v correction for the sun is ignored. The GHA of Aries is also given to the nearest tenth, but the GHAs and declinations of the navigational planets and the moon are given only to the nearest minute.

Each calendar day is covered by a single leaf in the *Air Almanac.* The front side of each leaf, or right-hand pages as viewed in the almanac, contain tabulations for every 10 minutes of GMT from 0 hours 0 minutes to 11 hours 50 minutes. The reverse sides—the left-hand pages—are for times from 12 hours 0 minutes to 23 hours 50 minutes.

A typical set of daily pages from an *Air Almanac* for 15 December of the same year in which the examples in chapter 20 are set is shown in Figures 21-2A and 21-2B. These may be compared with the corresponding daily pages from the *Nautical Almanac* shown in Figures 20-1A and 20-1B on pages 365 and 366.

The average values of SHA and declination for the year of publication for the fifty-seven navigational stars are tabulated on the inside front cover, along with a table for interpolation of GHA for the sun, Aries, the planets, and the moon for time increments between the 10-minute tabulations in the daily pages. Asterisks (*) by the star numbers denote those that appear in *Tables No. 249* Volume 1, while daggers (†) after their names indicate those with declinations suitable for use with volumes 2 and 3. Both of these tables appear in Figure 21-3 on page 393.

Monthly tabulations of SHA and declination for the fifty-seven navigational stars precise to the nearest tenth of a minute for applications requiring more precision are included in the back of each *Air Almanac* volume. Tables for finding the times of civil twilight, sunrise, and sunset, and for interpolating the times of moonrise and moonset are also included in the back of each volume, along with sky diagrams for various times of the year.

Use of the *Air Almanac* during Sight Reduction

The *Air Almanac*, like the *Nautical Almanac*, is used during sight reduction to obtain various celestial coordinates from which certain of the entering arguments for the sight reduction tables to be used are derived. *Tables No. 249* are made up of three volumes. Volume 1 uses integral values for the local hour angle of Aries (LHA♈) and the assumed latitude of the observer as entering arguments, while the other two volumes use the same arguments as *Tables No. 229*—integral degrees of declination and LHA of the body, and the integral assumed latitude of the observer. When volume 1 of *Tables No. 249* is to be used, the *Air Almanac* furnishes the GHA♈ from which LHA♈ is produced; for volumes 2 and 3, it provides the declination and GHA of the body, from which the LHA is derived.

There are a number of sight reduction forms that have been developed for use with the *Air Almanac* and *Tables No. 249*. Two forms developed by the U.S. Navy for use with the *Air Almanac* and either volume 1 or volumes 2 or 3 of *Tables No. 249* will be used as the media on which the sample complete solutions demonstrated in the remainder of this chapter will be worked.

As was the case earlier when the role of the *Nautical Almanac* during sight reduction was examined, the use of the *Air Almanac* during sight reduction is best explained by means of examples. For purposes of comparison, the examples below are two of the same observations discussed in chapter 20—the star Canopus and the upper limb of the moon. Because no *v* or *d* correction is nec-

GREENWICH A. M. DECEMBER 15 (TUESDAY)

GMT	SUN GHA	SUN Dec.	ARIES GHA ♈	VENUS −4.4 GHA	VENUS Dec.	JUPITER −1.3 GHA	JUPITER Dec.	SATURN 0.1 GHA	SATURN Dec.	MOON GHA	MOON Dec.
h m	° ′	° ′	° ′	° ′	° ′	° ′	° ′	° ′	° ′	° ′	° ′
00 00	181 18.0	S23 14.3	83 14.2	221 29	S13 14	211 11	S17 59	38 17	N14 30	333 25	N25 19
10	183 47.9	14.3	85 44.6	223 59		213 41		40 47		335 50	18
20	186 17.9	14.3	88 15.0	226 29		216 12		43 18		338 15	17
30	188 47.8 ·	14.3	90 45.5	228 59 ·		218 42 ·		45 48 ·		340 40 ·	16
40	191 17.8	14.4	93 15.9	231 29		221 12		48 19		343 04	15
50	193 47.7	14.4	95 46.3	234 00		223 43		50 49		345 29	13
01 00	196 17.7	S23 14.4	98 16.7	236 30	S13 14	226 13	S17 59	53 19	N14 30	347 54	N25 12
10	198 47.6	14.4	100 47.1	239 00		228 43		55 50		350 19	11
20	201 17.6	14.5	103 17.5	241 30		231 14		58 20		352 43	10
30	203 47.5 ·	14.5	105 47.9	244 00 ·		233 44 ·		60 51 ·		355 08 ·	09
40	206 17.5	14.5	108 18.3	246 31		236 14		63 21		357 33	08
50	208 47.4	14.5	110 48.7	249 01		238 45		65 52		359 57	07
02 00	211 17.4	S23 14.5	113 19.1	251 31	S13 14	241 15	S17 59	68 22	N14 30	2 22	N25 06
10	213 47.4	14.6	115 49.6	254 01		243 45		70 52		4 47	05
20	216 17.3	14.6	118 20.0	256 31		246 16		73 23		7 12	03
30	218 47.3 ·	14.6	120 50.4	259 02 ·		248 46 ·		75 53 ·		9 36 ·	02
40	221 17.2	14.6	123 20.8	261 32		251 16		78 24		12 01	01
50	223 47.2	14.7	125 51.2	264 02		253 47		80 54		14 26	25 00
03 00	226 17.1	S23 14.7	128 21.6	266 32	S13 15	256 17	S17 59	83 25	N14 30	16 51	N24 59
10	228 47.1	14.7	130 52.0	269 02		258 47		85 55		19 15	58
20	231 17.0	14.7	133 22.4	271 33		261 18		88 25		21 40	57
30	233 47.0 ·	14.8	135 52.8	274 03 ·		263 48 ·		90 56 ·		24 05 ·	55
40	236 16.9	14.8	138 23.3	276 33		266 18		93 26		26 30	54
50	238 46.9	14.8	140 53.7	279 03		268 48		95 57		28 55	53
04 00	241 16.8	S23 14.8	143 24.1	281 33	S13 15	271 19	S17 59	98 27	N14 30	31 19	N24 52
10	243 46.8	14.8	145 54.5	284 04		273 49		100 58		33 44	51
20	246 16.7	14.9	148 24.9	286 34		276 19		103 28		36 09	50
30	248 46.7 ·	14.9	150 55.3	289 04 ·		278 50 ·		105 59 ·		38 34 ·	49
40	251 16.6	14.9	153 25.7	291 34		281 20		108 29		40 59	47
50	253 46.6	14.9	155 56.1	294 04		283 50		110 59		43 23	46
05 00	256 16.5	S23 15.0	158 26.5	296 35	S13 15	286 21	S17 59	113 30	N14 30	45 48	N24 45
10	258 46.5	15.0	160 57.0	299 05		288 51		116 00		48 13	44
20	261 16.4	15.0	163 27.4	301 35		291 21		118 31		50 38	43
30	263 46.4 ·	15.0	165 57.8	304 05 ·		293 52 ·		121 01 ·		53 03 ·	41
40	266 16.3	15.1	168 28.2	306 35		296 22		123 32		55 27	40
50	268 46.3	15.1	170 58.6	309 06		298 52		126 02		57 52	39
06 00	271 16.2	S23 15.1	173 29.0	311 36	S13 15	301 23	S17 59	128 32	N14 30	60 17	N24 38
10	273 46.2	15.1	175 59.4	314 06		303 53		131 03		62 42	37
20	276 16.1	15.1	178 29.8	316 36		306 23		133 33		65 07	36
30	278 46.1 ·	15.2	181 00.2	319 06 ·		308 54 ·		136 04 ·		67 31 ·	34
40	281 16.0	15.2	183 30.6	321 36		311 24		138 34		69 56	33
50	283 46.0	15.2	186 01.1	324 07		313 54		141 05		72 21	32
07 00	286 15.9	S23 15.2	188 31.5	326 37	S13 15	316 25	S17 59	143 35	N14 30	74 46	N24 31
10	288 45.9	15.3	191 01.9	329 07		318 55		146 05		77 11	29
20	291 15.8	15.3	193 32.3	331 37		321 25		148 36		79 35	28
30	293 45.8 ·	15.3	196 02.7	334 07 ·		323 56 ·		151 06 ·		82 00 ·	27
40	296 15.7	15.3	198 33.1	336 38		326 26		153 37		84 25	26
50	298 45.7	15.4	201 03.5	339 08		328 56		156 07		86 50	25
08 00	301 15.6	S23 15.4	203 33.9	341 38	S13 15	331 27	S18 00	158 38	N14 30	89 15	N24 23
10	303 45.6	15.4	206 04.3	344 08		333 57		161 08		91 40	22
20	306 15.5	15.4	208 34.8	346 38		336 27		163 38		94 05	21
30	308 45.5 ·	15.4	211 05.2	349 09 ·		338 58 ·		166 09 ·		96 29 ·	20
40	311 15.4	15.5	213 35.6	351 39		341 28		168 39		98 54	18
50	313 45.4	15.5	216 06.0	354 09		343 58		171 10		101 19	17
09 00	316 15.3	S23 15.5	218 36.4	356 39	S13 15	346 29	S18 00	173 40	N14 29	103 44	N24 16
10	318 45.3	15.5	221 06.8	359 09		348 59		176 11		106 09	15
20	321 15.2	15.6	223 37.2	1 40		351 29		178 41		108 34	13
30	323 45.2 ·	15.6	226 07.6	4 10 ·		354 00 ·		181 12 ·		110 59 ·	12
40	326 15.1	15.6	228 38.0	6 40		356 30		183 42		113 23	11
50	328 45.1	15.6	231 08.5	9 10		359 00		186 12		115 48	10
10 00	331 15.0	S23 15.6	233 38.9	11 40	S13 15	1 31	S18 00	188 43	N14 29	118 13	N24 08
10	333 45.0	15.7	236 09.3	14 11		4 01		191 13		120 38	07
20	336 14.9	15.7	238 39.7	16 41		6 31		193 44		123 03	06
30	338 44.9 ·	15.7	241 10.1	19 11 ·		9 01 ·		196 14 ·		125 28 ·	04
40	341 14.8	15.7	243 40.5	21 41		11 32		198 45		127 53	03
50	343 44.8	15.8	246 10.9	24 11		14 02		201 15		130 18	02
11 00	346 14.7	S23 15.8	248 41.3	26 42	S13 15	16 32	S18 00	203 45	N14 29	132 42	N24 01
10	348 44.7	15.8	251 11.7	29 12		19 03		206 16		135 07	23 59
20	351 14.6	15.8	253 42.1	31 42		21 33		208 46		137 32	58
30	353 44.6 ·	15.8	256 12.6	34 12 ·		24 03 ·		211 17 ·		139 57 ·	57
40	356 14.5	15.9	258 43.0	36 42		26 34		213 47		142 22	55
50	358 44.5	15.9	261 13.4	39 13		29 04		216 18		144 47	54

Moonrise table:

Lat.	Moonrise	Diff.
N °	h m	m
72	□	*
70	□	*
68	□	*
66	15 14	*
64	16 14	*
62	16 47	53
60	17 11	47
58	17 31	44
56	17 47	41
54	18 01	39
52	18 13	38
50	18 23	37
45	18 45	34
40	19 03	32
35	19 18	31
30	19 31	30
20	19 52	28
10	20 11	26
0	20 29	25
10	20 46	23
20	21 05	22
30	21 26	20
35	21 39	18
40	21 53	17
45	22 10	15
50	22 31	13
52	22 41	12
54	22 52	10
56	23 04	08
58	23 19	06
60	23 36	03
S		

Moon's P. in A.

Alt. °	Corr. +	Alt. °	Corr. +
0	56	55	31
3	55	56	30
11	54	57	29
15	53	59	28
19	52	60	27
22	51	61	26
27	50	62	25
29	49	63	24
31	48	65	23
33	47	66	22
35	46	67	21
38	45	68	20
40	44	69	19
41	43	70	18
43	42	71	17
44	41	72	16
46	40	73	15
47	39	74	14
48	38	75	13
50	37	77	12
51	36	78	11
52	35	79	10
54	34	80	
55	33		
56	32		
	31		

Sun SD 16.3

Moon SD 15′

Age 16d

Figure 21-2A. *Sample daily page,* Air Almanac, *Greenwich A.M., 15 December.*

GREENWICH P. M. DECEMBER 15 (TUESDAY)

GMT	☉ SUN GHA	Dec.	ARIES GHA ♈	VENUS −4.4 GHA	Dec.	JUPITER −1.3 GHA	Dec.	SATURN 0.1 GHA	Dec.	☽ MOON GHA	Dec.
h m	° '	° '	° '	° '	° '	° '	° '	° '	° '	° '	° '
12 00	1 14.4	S23 15.9	263 43.8	41 43	S13 16	31 34	S18 00	218 48	N14 29	147 12	N23 53
10	3 44.4	15.9	266 14.2	44 13		34 05		221 18		149 37	51
20	6 14.3	16.0	268 44.6	46 43		36 35		223 49		152 02	50
30	8 44.3 ·	16.0	271 15.0	49 13 ·		39 05 ·		226 19 ·		154 27 ·	49
40	11 14.2	16.0	273 45.4	51 44		41 36		228 50		156 52	47
50	13 44.2	16.0	276 15.8	54 14		44 06		231 20		159 16	46
13 00	16 14.1	S23 16.0	278 46.3	56 44	S13 16	46 36	S18 00	233 51	N14 29	161 41	N23 45
10	18 44.1	16.1	281 16.7	59 14		49 07		236 21		164 06	44
20	21 14.0	16.1	283 47.1	61 44		51 37		238 52		166 31	42
30	23 44.0 ·	16.1	286 17.5	64 15 ·		54 07 ·		241 22 ·		168 56 ·	41
40	26 13.9	16.1	288 47.9	66 45		56 38		243 52		171 21	40
50	28 43.9	16.2	291 18.3	69 15		59 08		246 23		173 46	38
14 00	31 13.8	S23 16.2	293 48.7	71 45	S13 16	61 38	S18 00	248 53	N14 29	176 11	N23 37
10	33 43.8	16.2	296 19.1	74 15		64 09		251 24		178 36	35
20	36 13.7	16.2	298 49.5	76 45		66 39		253 54		181 01	34
30	38 43.7 ·	16.3	301 20.0	79 16 ·		69 09 ·		256 25 ·		183 26 ·	33
40	41 13.6	16.3	303 50.4	81 46		71 40		258 55		185 51	31
50	43 43.6	16.3	306 20.8	84 16		74 10		261 25		188 16	30
15 00	46 13.5	S23 16.3	308 51.2	86 46	S13 16	76 40	S18 00	263 56	N14 29	190 41	N23 29
10	48 43.5	16.3	311 21.6	89 16		79 11		266 26		193 06	27
20	51 13.4	16.3	313 52.0	91 47		81 41		268 57		195 31	26
30	53 43.4 ·	16.4	316 22.4	94 17 ·		84 11 ·		271 27 ·		197 55 ·	25
40	56 13.3	16.4	318 52.8	96 47		86 41		273 58		200 20	23
50	58 43.3	16.4	321 23.2	99 17		89 12		276 28		202 45	22
16 00	61 13.2	S23 16.4	323 53.6	101 47	S13 16	91 42	S18 01	278 59	N14 29	205 10	N23 20
10	63 43.2	16.5	326 24.1	104 18		94 12		281 29		207 35	19
20	66 13.1	16.5	328 54.5	106 48		96 43		283 59		210 00	18
30	68 43.1 ·	16.5	331 24.9	109 18 ·		99 13 ·		286 30 ·		212 25 ·	16
40	71 13.0	16.5	333 55.3	111 48		101 43		289 00		214 50	15
50	73 43.0	16.5	336 25.7	114 18		104 14		291 31		217 15	14
17 00	76 12.9	S23 16.6	338 56.1	116 49	S13 16	106 44	S18 01	294 01	N14 29	219 40	N23 12
10	78 42.9	16.6	341 26.5	119 19		109 14		296 32		222 05	11
20	81 12.8	16.6	343 56.9	121 49		111 45		299 02		224 30	09
30	83 42.8 ·	16.6	346 27.3	124 19 ·		114 15 ·		301 32 ·		226 55 ·	08
40	86 12.7	16.6	348 57.8	126 49		116 45		304 03		229 20	06
50	88 42.7	16.7	351 28.2	129 20		119 16		306 33		231 45	05
18 00	91 12.6	S23 16.7	353 58.6	131 50	S13 16	121 46	S18 01	309 04	N14 29	234 10	N23 04
10	93 42.6	16.7	356 29.0	134 20		124 16		311 34		236 35	02
20	96 12.5	16.7	358 59.4	136 50		126 47		314 05		239 00	23 01
30	98 42.5 ·	16.8	1 29.8	139 20 ·		129 17 ·		316 35 ·		241 25	22 59
40	101 12.4	16.8	4 00.2	141 50		131 47		319 05		243 50	58
50	103 42.4	16.8	6 30.6	144 21		134 18		321 36		246 15	57
19 00	106 12.3	S23 16.8	9 01.0	146 51	S13 16	136 48	S18 01	324 06	N14 29	248 40	N22 55
10	108 42.3	16.8	11 31.5	149 21		139 18		326 37		251 05	54
20	111 12.2	16.9	14 01.9	151 51		141 49		329 07		253 30	52
30	113 42.2 ·	16.9	16 32.3	154 21 ·		144 19 ·		331 38 ·		255 55 ·	51
40	116 12.1	16.9	19 02.7	156 52		146 49		334 08		258 20	49
50	118 42.1	16.9	21 33.1	159 22		149 19		336 38		260 45	48
20 00	121 12.0	S23 16.9	24 03.5	161 52	S13 17	151 50	S18 01	339 09	N14 29	263 10	N22 46
10	123 42.0	17.0	26 33.9	164 22		154 20		341 39		265 35	45
20	126 11.9	17.0	29 04.3	166 52		156 50		344 10		268 01	43
30	128 41.9 ·	17.0	31 34.7	169 23 ·		159 21 ·		346 40 ·		270 26 ·	42
40	131 11.8	17.0	34 05.1	171 53		161 51		349 11		272 51	41
50	133 41.8	17.0	36 35.6	174 23		164 21		351 41		275 16	39
21 00	136 11.7	S23 17.1	39 06.0	176 53	S13 17	166 52	S18 01	354 12	N14 29	277 41	N22 38
10	138 41.7	17.1	41 36.4	179 23		169 22		356 42		280 06	36
20	141 11.6	17.1	44 06.8	181 54		171 52		359 12		282 31	35
30	143 41.6 ·	17.1	46 37.2	184 24 ·		174 23 ·		1 43 ·		284 56 ·	33
40	146 11.5	17.2	49 07.6	186 54		176 53		4 13		287 21	32
50	148 41.5	17.2	51 38.0	189 24		179 23		6 44		289 46	30
22 00	151 11.4	S23 17.2	54 08.4	191 54	S13 17	181 54	S18 01	9 14	N14 29	292 11	N22 29
10	153 41.4	17.2	56 38.8	194 24		184 24		11 45		294 36	27
20	156 11.3	17.2	59 09.3	196 55		186 54		14 15		297 01	26
30	158 41.2 ·	17.3	61 39.7	199 25 ·		189 25 ·		16 45 ·		299 26 ·	24
40	161 11.2	17.3	64 10.1	201 55		191 55		19 16		301 51	23
50	163 41.1	17.3	66 40.5	204 25		194 25		21 46		304 16	21
23 00	166 11.1	S23 17.3	69 10.9	206 55	S13 17	196 56	S18 01	24 17	N14 29	306 42	N22 20
10	168 41.0	17.3	71 41.3	209 26		199 26		26 47		309 07	18
20	171 11.0	17.4	74 11.7	211 56		201 56		29 18		311 32	17
30	173 40.9 ·	17.4	76 42.1	214 26 ·		204 27 ·		31 48 ·		313 57 ·	15
40	176 10.9	17.4	79 12.5	216 56		206 57		34 18		316 22	14
50	178 40.8	17.4	81 43.0	219 26		209 27		36 49		318 47	12

Moonset

Lat.	Moonset	Diff.
N °	h m	m
72	· □	*
70	□	*
68	□	*
66	13 29	*
64	12 29	*
62	11 55	−01
60	11 30	+05
58	11 10	08
56	10 53	11
54	10 39	12
52	10 27	14
50	10 16	15
45	09 53	17
40	09 34	19
35	09 19	20
30	09 05	21
20	08 42	23
10	08 22	24
0	08 03	26
10	07 44	27
20	07 23	29
30	06 59	30
35	06 45	31
40	06 29	33
45	06 09	34
50	05 45	36
52	05 33	37
54	05 19	39
56	05 03	41
58	04 44	43
60	04 21	47
S		

Moon's P. in A.

Alt. °	Corr. +	Alt. °	Corr. +
0	55	56	·30
10	55	57	29
14	54	59	28
18	53	60	27
24	52	61	26
26	51	62	25
28	50	63	24
30	49	64	23
32	48	66	22
34	47	67	21
36	46	68	20
38	45	69	19
39	44	70	18
41	43	71	17
42	42	72	16
44	41	73	15
45	40	74	14
47	39	75	13
48	38	76	12
50	37	78	11
51	36	79	10
52	35	80	
54	34		
55	33		
57	32		
	31		
	30		

Sun SD 16.3
Moon SD 15'
Age 17d

Figure 21-2B. Sample daily page, Air Almanac, Greenwich P.M., 15 December.

INTERPOLATION OF G.H.A.

Increment to be added for intervals of G.M.T. to G.H.A. of:
Sun, Aries (♈) and planets; Moon

No.	Name	Mag.	S.H.A.	Dec.
7*	*Acamar*	3·1	315 42	S.40 25
5*	*Achernar*	0·6	335 50	S.57 23
30*	*Acrux*	1·1	173 46	S.62 56
19	*Adhara* †	1·6	255 38	S.28 56
10*	*Aldebaran* †	1·1	291 26	N.16 27
32*	*Alioth*	1·7	166 49	N.56 07
34*	*Alkaid*	1·9	153 24	N.49 27
55	*Al Na'ir*	2·2	28 24	S.47 06
15*	*Alnilam* †	1·8	276 19	S. 1 13
25*	*Alphard* †	2·2	218 28	S. 8 32
41*	*Alphecca* †	2·3	126 38	N.26 49
1*	*Alpheratz* †	2·2	358 17	N.28 56
51*	*Altair* †	0·9	62 40	N. 8 47
2	*Ankaa*	2·4	353 47	S.42 28
42*	*Antares* †	1·2	113 06	S.26 22
37*	*Arcturus* †	0·2	146 25	N.19 20
43	*Atria*	1·9	108 37	S.68 59
22	*Avior*	1·7	234 31	S.59 25
13	*Bellatrix* †	1·7	279 06	N. 6 20
16*	*Betelgeuse* †	0·1–1·2	271 36	N. 7 24
17*	*Canopus*	−0·9	264 10	S.52 40
12*	*Capella*	0·2	281 22	N.45 58
53*	*Deneb*	1·3	49 53	N.45 11
28*	*Denebola* †	2·2	183 06	N.14 44
4*	*Diphda* †	2·2	349 28	S.18 09
27*	*Dubhe*	2·0	194 31	N.61 54
14	*Elnath* †	1·8	278 53	N.28 35
47	*Eltanin*	2·4	91 01	N.51 30
54*	*Enif*	2·5	34 19	N. 9 45
56*	*Fomalhaut* †	1·3	15 59	S.29 47
31	*Gacrux*	1·6	172 37	S.56 57
29*	*Gienah* †	2·8	176 26	S.17 23
35	*Hadar*	0·9	149 34	S.60 14
6*	*Hamal*	2·2	328 37	N.23 20
48	*Kaus Aust.*	2·0	84 26	S.34 24
40*	*Kochab*	2·2	137 19	N.74 16
57	*Markab* †	2·6	14 10	N.15 03
8*	*Menkar* †	2·8	314 48	N. 3 59
36	*Menkent*	2·3	148 46	S.36 14
24*	*Miaplacidus*	1·8	221 47	S.69 35
9*	*Mirfak*	1·9	309 26	N.49 46
50*	*Nunki* †	2·1	76 38	S.26 20
52*	*Peacock*	2·1	54 09	S.56 50
21*	*Pollux* †	1·2	244 07	N.28 06
20*	*Procyon* †	0·5	245 33	N. 5 18
46*	*Rasalhague* †	2·1	96 36	N.12 35
26*	*Regulus* †	1·3	208 18	N.12 07
11*	*Rigel* †	0·3	281 43	S. 8 14
38*	*Rigil Kent.*	0·1	140 36	S.60 43
44	*Sabik* †	2·6	102 50	S.15 41
3*	*Schedar*	2·5	350 17	N.56 23
45*	*Shaula*	1·7	97 06	S.37 05
18*	*Sirius* †	−1·6	259 02	S.16 40
33*	*Spica* †	1·2	159 05	S.11 01
23*	*Suhail*	2·2	223 16	S.43 19
49*	*Vega*	0·1	81 01	N.38 45
39	*Zuben'ubi* †	2·9	137 41	S.15 55

SUN, etc.	MOON	SUN, etc.	MOON	SUN, etc.	MOON
m s ° ′	m s ° ′	m s ° ′	m s ° ′	m s ° ′	m s ° ′
00 00 00 00	0 00 00 00	03 17 00 50	03 25	06 37 1 40	06 52
01 0 01	00 02	21 0 51	03 29	41 1 41	06 56
05 0 02	00 06	25 0 52	03 33	45 1 42	07 00
09 0 03	00 10	29 0 53	03 37	49 1 43	07 04
13 0 04	00 14	33 0 54	03 41	53 1 44	07 08
17 0 05	00 18	37 0 55	03 45	06 57 1 45	07 13
21 0 06	00 22	41 0 56	03 49	07 01 1 46	07 17
25 0 07	00 26	45 0 57	03 54	05 1 47	07 21
29 0 08	00 31	49 0 58	03 58	09 1 48	07 25
33 0 09	00 35	53 0 59	04 02	13 1 49	07 29
37 0 10	00 39	03 57 1 00	04 06	17 1 50	07 33
41 0 11	00 43	04 01 1 01	04 10	21 1 51	07 37
45 0 12	00 47	05 1 02	04 14	25 1 52	07 42
49 0 13	00 51	09 1 03	04 19	29 1 53	07 46
53 0 14	00 55	13 1 04	04 23	33 1 54	07 50
00 57 0 15	01 00	17 1 05	04 27	37 1 55	07 54
01 01 0 16	01 04	21 1 06	04 31	41 1 56	07 58
05 0 17	01 08	25 1 07	04 35	45 1 57	08 02
09 0 18	01 12	29 1 08	04 39	49 1 58	08 06
13 0 19	01 16	33 1 09	04 43	53 1 59	08 11
17 0 20	01 20	37 1 10	04 48	07 57 2 00	08 15
21 0 21	01 24	41 1 11	04 52	08 01 2 01	08 19
25 0 22	01 29	45 1 12	04 56	05 2 02	08 23
29 0 23	01 33	49 1 13	05 00	09 2 03	08 27
33 0 24	01 37	53 1 14	05 04	13 2 04	08 31
37 0 25	01 41	04 57 1 15	05 08	17 2 05	08 35
41 0 26	01 45	05 01 1 16	05 12	21 2 06	08 40
45 0 27	01 49	05 1 17	05 17	25 2 07	08 44
49 0 28	01 53	09 1 18	05 21	29 2 08	08 48
53 0 29	01 58	13 1 19	05 25	33 2 09	08 52
01 57 0 30	02 02	17 1 20	05 29	37 2 10	08 56
02 01 0 31	02 06	21 1 21	05 33	41 2 11	09 00
05 0 32	02 10	25 1 22	05 37	45 2 12	09 04
09 0 33	02 14	29 1 23	05 41	49 2 13	09 09
13 0 34	02 18	33 1 24	05 46	53 2 14	09 13
17 0 35	02 22	37 1 25	05 50	08 57 2 15	09 17
21 0 36	02 27	41 1 26	05 54	09 01 2 16	09 21
25 0 37	02 31	45 1 27	05 58	05 2 17	09 25
29 0 38	02 35	49 1 28	06 02	09 2 18	09 29
33 0 39	02 39	53 1 29	06 06	13 2 19	09 33
37 0 40	02 43	05 57 1 30	06 10	17 2 20	09 38
41 0 41	02 47	06 01 1 31	06 15	21 2 21	09 42
45 0 42	02 51	05 1 32	06 19	25 2 22	09 46
49 0 43	02 56	09 1 33	06 23	29 2 23	09 50
53 0 44	03 00	13 1 34	06 27	33 2 24	09 54
02 57 0 45	03 04	17 1 35	06 31	37 2 25	09 58
03 01 0 46	03 08	21 1 36	06 35	41 2 26	10 00
05 0 47	03 12	25 1 37	06 39	45 2 27	
09 0 48	03 16	29 1 38	06 44	49 2 28	
13 0 49	03 20	33 1 39	06 48	53 2 29	
17 0 50	03 25	37 1 40	06 52	09 57 2 30	
03 21	03 29	06 41	06 56	10 00	

* Stars used in H.O. 249 (A.P. 3270) Vol. 1.
† Stars that may be used with Vols. 2 and 3.

***Figure 21-3.** Inside front cover,* Air Almanac.

essary when the *Air Almanac* is used, the procedures for finding the GHA and declination of a planet or the sun are virtually identical to the procedures for finding these same coordinates for a star or the moon. Separate examples of the complete solutions of the planets and the sun, therefore, will not be presented in this chapter.

Finding the Complete Solution of a Star by the *Air Almanac*

As the first example of the complete solution for a star by the use of the *Air Almanac,* let us consider once again the Canopus sight used as an example in chapter 20. The star Canopus was observed at GMT 17-12-09 on 15 December and found to have a sextant altitude hs of 50° 46′, rounded off to the nearest whole minute. The DR position at the time of the observation was L 34° 19.0′S, λ 163° 05.7′E, and the index error was 0.3′ off the arc.

The first step in the complete solution using the *Air Almanac* is to find the observed altitude Ho by application of the corrections for IC, dip, and refraction to the hs. The index correction in this case is considered insignificant, the dip correction from the dip table in Figure 21-1A is 7′, and the refraction correction R_0 from the refraction table of Figure 21-1B is –1′. Entering these values on the form and adding them algebraically yields the total correction to hs of –8′. Applying this correction to hs, an Ho of 50° 38′ results; note that this result differs by 1.4′ from the Ho as computed by the *Nautical Almanac* (see Figure 20-4). At this point, the Canopus form appears as shown in Figure 21-4A.

Inasmuch as volume 1 of *Tables No. 249* can be used to find the computed altitude Hc and the true azimuth Zn of Canopus in this case, the only coordinate of interest that must be obtained from the *Air Almanac* is the GHA♈ for the time of observation, for later conversion to an integral LHA♈, one of the two entering arguments for volume 1; the other argument is the star name. The tabulated value of GHA♈ for GMT 17-10-00 on 15 December from the daily page reproduced in Figure 21-2B is 341° 26.5′; this figure is entered on the form opposite "GHA♈." Next, the increment to GHA♈ corresponding to the remaining 2 minutes 9 seconds of GMT is found from the GHA interpola-

Body	CANOPUS
GMT	17-12-09 Dec15
IC	+.3′
D (Ht 44′)	-7′
R	-1′
Total Corr	-8′
hs	50° 46′
Ho	50° 38′

Figure 21-4A. *Canopus* Air Almanac *form, Ho computed.*

tion table on the inside front cover, pictured in Figure 21-3. It is +0° 32′. Note that 02-09 is an exact tabulated argument; as is the case in all critical value tables of this type, the correction corresponding to the preceding interval is always extracted when entering arguments are identical to tabulated arguments. Adding the GHAϓ increment to the tabulated GHAϓ yields a total GHAϓ for GMT 17-12-09 of 341° 58.5′.

If one of the 57 navigational stars other than those listed in volume 1 of *Tables No. 249* had been observed, it would be necessary to determine the LHA and declination of the star for later use with either volume 2 or 3. In this case, the *Air Almanac,* like the *Nautical Almanac,* would be used to find the GHA and declination of the star. The GHA of the star is found by the formula

$$\text{GHA} \star = \text{GHA} ϓ + \text{SHA} \star$$

and the local hour angle by the relationship

$$\text{LHA} \star = \text{GHA} \star \begin{array}{l} + \ a\lambda \ (\text{E}) \\ - \ a\lambda \ (\text{W}) \end{array}$$

The sidereal hour angle (SHA) and declination of the star are found from one of the two tables included in the *Air Almanac* for this purpose, located inside the front cover and following the daily pages.

Since an integral LHAϓ is required as an entering argument for volume 1 of *Tables No. 249,* and the DR longitude is east, an assumed longitude (*a*λ) is chosen such that when it is added to the total GHAϓ a whole degree of LHAϓ will result. As before, the longitude selected should also lie within 30′ of the DR longitude. Thus, the appropriate assumed longitude for this example is 163° 01.5E. Adding this value to GHAϓ and subtracting 360° results in an LHAϓ of 145°. The form now appears as shown in Figure 21-4B, ready for use with *Tables No. 249.*

Determining the Ho, GHA, Declination, and LHA of the Moon

When the complete solution for any body other than one of the seven navigational stars featured in volume 1 of *Tables No. 249* is to be found, the *Air Almanac* is used to obtain the GHA and declination of the body at the time of observation for later use with volume 2 or 3. In the following example, the upper limb of the moon was observed on the same occasion as the star Canopus of the example above. The GMT of the observation was 16-58-57 and its sextant altitude hs to the nearest minute was 28° 10′. The DR position at the time of the sight was L 34° 15.5′S, λ 163° 11.7′E.

As in the previous example, the first step in finding the complete solution is to obtain the observed altitude Ho. Since the moon was observed, additional corrections for semidiameter and parallax must be applied to hs, along with the usual IC, dip, and refraction corrections. The IC correction is 0, the dip cor-

Body	CANOPUS
GMT	17-12-09 Dec15
IC	+.3'
D (Ht 44')	-7'
R	-1'
Total Corr	-8'
hs	50° 46'
Ho	50° 38'
GHA γ (h, 10m)	341° 26.5'
Incre. (m/s)	0° 32.0'
Total GHA γ	341° 58.5'
±360°	
aλ (+E,-W)	163° 01.5' E
LHA γ	145°
a LAT	34 S

Figure 21-4B. *Canopus* Air Almanac *form, ready for* Tables No. 249.

rection from the dip table is –7′, and the refraction correction R_0 from the refraction table is –2′. The semidiameter of the moon on this date is found from the daily page (Figure 21-2B) as 15′; since the upper limb of the moon was observed, the sign of this SD correction is negative. Finally, the parallax correction for the moon is found from the "P in A" table on the appropriate side of the daily page. The entering argument for this table is the sextant altitude hs corrected for IC, dip, refraction, and semidiameter: in this case, a value of +49′ is extracted, corresponding to a corrected hs of 27° 46′. Applying this "P in A" correction results in an Ho of 28° 35′, a value that differs only 0.4′ from the corresponding Ho found by the use of the *Nautical Almanac*. The partially completed form appears in Figure 21-5A.

To find the GHA and declination of the moon, the values for these quantities tabulated for the 10 minutes of GMT immediately preceding the time of observation are extracted. The tabulated GHA of the moon for GMT 16-50-00 on 15 December is 217° 15′, and the tabulated declination is N23° 14′. A GHA increment of 2° 10′ is obtained from the GHA interpolation table for the moon for the additional 8 minutes and 57 seconds of time. This value is added to the tabulated GHA to obtain the total GHA for 16-58-57 of 219° 25′. No correction to the tabulated declination is necessary.

Finally, the LHA of the moon to the nearest whole degree is computed in the usual manner by applying a suitably chosen assumed longitude to the total GHA:

Body	MOON (UL)
GMT	16-58-57 Dec15
IC	+.3'
D (Ht 44')	-7'
R	-2'
SD	-15'
P. in A. (Moon)	+49'
Total Corr	+35'
hs	28° 10'
Ho	28° 35'

Figure 21-5A. Moon Air Almanac *form, Ho computed.*

$$219°\ 25' + 163°\ 35' = 383°$$

Subtracting 360° yields an LHA of the moon at the time of observation of 23°. At this point the form is ready for use with *Tables No. 249*, and it appears as shown in Figure 21-5B.

Body	MOON (UL)
GMT	16-58-57 Dec15
IC	+.3'
D (Ht 44')	-7'
R	-2'
SD	-15'
P. in A. (Moon)	+49'
Total Corr	+35'
hs	28° 10'
Ho	28° 35'
GHA (h, 10m)	217° 15'
Incre. (m/s)	2° 10'
Total GHA	219° 25'
±360°	
aλ (+E,-W)	163° 35' E
LHA	23°
Tab Dec	N 23° 14'
a LAT Same Contrary	34° S Cont

Figure 21-5B. Moon Air Almanac *form, ready for* Tables No. 249.

Determining the Ho, GHA, Declination, and LHA of the Sun and Planets

The procedure for determining the Ho, GHA, declination, and LHA of the sun is basically the same as in the example just discussed for the moon; the only difference is that no parallax correction to the sextant altitude hs is necessary for the sun. The procedure for the navigational planets is identical to the procedure for a star not listed in volume 1 of *Tables No. 249*. The IC, dip, and refraction corrections are first found and applied to the hs to obtain Ho, then the daily pages and the GHA interpolation tables are used to obtain the values of GHA and declination. An integral value of LHA is determined by choosing a suitable assumed longitude and applying it to the total GHA in the normal manner.

SIGHT REDUCTION TABLES FOR AIR NAVIGATION, NO. 249

After either the LHAϓ or the LHA and declination of the body have been determined by means of the *Air Almanac*, the sight reduction form is ready for use with the *Sight Reduction Tables for Air Navigation, No. 249*. As the name indicates, these tables are designed primarily for the use of air navigators, but like the *Air Almanac*, they have found considerable favor with surface navigators in cases where their speed and ease of use offset the less precise computations they contain.

Tables No. 249 are published in three volumes. Like *Tables No. 229*, they are inspection tables designed for use with an assumed position, but they differ in the degree of precision. Whereas in *Tables No. 229* all computations are made to the nearest tenth of a minute, in *Tables No. 249* computed altitudes are stated only to the nearest whole minute. There are other differences as well. Volume 1 is specifically designed to reduce the time and effort required to obtain a celestial fix to a minimum, by listing calculated computed altitudes and true azimuths for a given set of seven navigational stars selected on a worldwide basis. Volumes 2 and 3 are similar to *Tables No. 229* in concept. They provide a computed Hc and azimuth angle Z for all bodies having a declination less than 30° north or south.

Volume 1 of *Tables No. 249* is arranged for entering with a whole degree of assumed latitude, a whole degree of local hour angle of Aries, and the appropriate star name. All integral degrees of latitude from 89° north to 89° south are included. For each degree of latitude and of LHAϓ, seven selected stars are tabulated; they are selected chiefly for good distribution in azimuth, for their magnitude and altitude, and for continuity both in latitude and hour angle. Each combination of seven stars is tabulated for 15° of LHAϓ, and the three with the best distribution in azimuth and altitude for fixing purposes are marked with asterisks. First-magnitude star names are capitalized, and second-

and third-magnitude stars appear in lowercase letters. Inasmuch as the coordinates of the selected stars change slightly from year to year primarily as a result of precession and nutation of the earth's axis, volume 1 is recomputed and reissued once every 5 years. The year for which volume 1 is issued is called an *epoch* year (not necessarily the year for which tabulations in volume 1 are computed). This procedure causes no difficulties, however, because a precession and nutation table is included in the back of each issue for adjusting LOP and fix positions obtained using the tabulated figures for any year in the period 4 years before and 4 years after the epoch year.

In practice, when volume 1 is to be used for sight reduction purposes, the navigator first determines approximate values of LHA♈ and the assumed latitude based upon the ship's projected DR position for the probable time of observation. Volume 1 is then entered using these coordinates as arguments to find which seven stars are tabulated for coordinates as arguments to find which seven stars are tabulated for that approximate position and time. Thus, the navigator can determine beforehand exactly which stars should be observed to obtain a fix using volume 1 of *Tables No. 249* and, of the seven, which three will provide the best fix. Furthermore, because the calculated Hc and true azimuth Zn are given for each star, he or she knows precisely where to look in the heavens for each of the stars to be observed. Hence, volume 1 can be considered to be a kind of *starfinder* for selected stars, in addition to its prime function as a sight reduction table.

For every degree of latitude between 89° north and 89° south, there are two pages in volume 1. Left-hand pages contain tabulated Hc's and Zn's for seven selected stars for LHAs between 0° and 179°, while right-hand pages contain tabulations for LHAs between 180° and 359°. The appropriate volume 1 page for latitude 34° south and LHAs from 0° to 179° for the epoch of the example problems of this chapter is reproduced in Figure 21-6A.

As mentioned previously, volumes 2 and 3 of *Tables No. 249* are similar in format to *Tables No. 229*, with the main differences being that the tabulated Hc and Z values are precise only to the nearest whole minute, and the range of declination covered is from N30° to S30°. Hence, complete data for the reduction of sights of the sun, moon, navigational planets, and most but not all navigational stars are provided. Volume 2 contains tabulations for assumed latitudes from 0° to 39°, and volume 3 is for assumed latitudes from 40° to 89°. The entering arguments are a whole degree of assumed latitude, a whole degree of declination of same or contrary name to the latitude, and a whole degree of LHA. For each set of entering arguments, the tables contain a computed altitude under the heading "Hc," a computed altitude difference d in minutes between the tabulated Hc and the Hc for a declination one degree higher, and an azimuth angle under the heading "Z." A typical page from volume 2 appears in Figure 21-6B.

LAT 34°S

LHA ♈	Hc	Zn	Hc	Zn	Hc	Zn	Hc	Zn	Hc	Zn	Hc	Zn	Hc	Zn
	*Alpheratz		Hamal		*RIGEL		CANOPUS		ACHERNAR		*Peacock		Enif	
0	27 03	002	25 23	032	14 17	090	23 11	139	61 29	152	47 11	221	35 28	317
1	27 04	001	25 49	031	15 07	090	23 44	139	61 52	153	46 38	221	34 54	316
2	27 04	000	26 14	030	15 57	089	24 17	138	62 14	154	46 06	221	34 19	315
3	27 04	359	26 39	029	16 47	089	24 50	138	62 35	155	45 33	221	33 44	314
4	27 02	358	27 03	028	17 36	088	25 23	138	62 56	156	45 01	221	33 08	313
5	27 00	357	27 26	027	18 26	088	25 57	138	63 16	157	44 28	221	32 31	312
6	26 57	356	27 48	026	19 16	087	26 30	137	63 35	158	43 55	221	31 54	311
7	26 53	355	28 10	025	20 05	086	27 04	137	63 54	159	43 22	221	31 16	310
8	26 48	354	28 31	024	20 55	086	27 38	137	64 11	160	42 50	221	30 38	310
9	26 42	353	28 51	024	21 45	085	28 12	137	64 28	161	42 17	221	29 59	309
10	26 35	352	29 11	023	22 34	085	28 46	136	64 44	162	41 44	221	29 20	308
11	26 28	351	29 30	022	23 24	084	29 21	136	64 58	163	41 11	221	28 41	307
12	26 20	350	29 47	021	24 13	083	29 55	136	65 12	164	40 38	221	28 01	306
13	26 10	349	30 05	020	25 03	083	30 30	136	65 25	165	40 05	221	27 21	305
14	26 01	348	30 21	019	25 52	082	31 05	136	65 37	167	39 33	221	26 40	305
	*Hamal		ALDEBARAN		RIGEL		*CANOPUS		ACHERNAR		*FOMALHAUT		Alpheratz	
15	30 36	017	18 18	054	26 41	082	31 40	135	65 48	168	63 27	271	25 50	347
16	30 51	016	18 58	054	27 30	081	32 15	135	65 58	169	62 37	270	25 38	346
17	31 04	015	19 38	053	28 19	080	32 50	135	66 07	170	61 47	269	25 26	345
18	31 17	014	20 18	052	29 08	080	33 25	135	66 15	172	60 58	269	25 13	344
19	31 29	013	20 57	051	29 57	079	34 00	135	66 21	173	60 08	268	24 59	343
20	31 40	012	21 35	051	30 46	078	34 36	134	66 27	174	59 18	268	24 44	342
21	31 50	011	22 14	050	31 35	078	35 11	134	66 31	176	58 28	267	24 29	341
22	31 59	010	22 52	049	32 23	077	35 47	134	66 34	177	57 39	267	24 13	341
23	32 08	009	23 29	048	33 12	076	36 23	134	66 36	178	56 49	266	23 56	340
24	32 15	008	24 06	047	34 00	076	36 59	134	66 37	180	55 59	266	23 38	339
25	32 22	007	24 42	047	34 48	075	37 35	134	66 36	181	55 10	265	23 20	338
26	32 27	006	25 18	046	35 36	074	38 11	134	66 35	183	54 20	265	23 01	337
27	32 32	005	25 54	045	36 24	074	38 47	134	66 32	184	53 31	265	22 41	336
28	32 35	004	26 28	044	37 12	073	39 23	133	66 28	185	52 41	264	22 20	335
29	32 38	003	27 03	043	37 59	072	39 59	133	66 23	187	51 52	264	21 59	334
	Hamal		ALDEBARAN		*SIRIUS		CANOPUS		*ACHERNAR		FOMALHAUT		*Alpheratz	
30	32 40	001	27 37	042	24 48	094	40 35	133	66 17	188	51 02	263	21 37	333
31	32 41	000	28 10	042	25 38	093	41 11	133	66 09	189	50 13	263	21 15	333
32	32 41	359	28 43	041	26 28	093	41 48	133	66 01	190	49 24	262	20 52	332
33	32 39	358	29 15	040	27 17	092	42 24	133	65 51	192	48 34	262	20 28	331
34	32 37	357	29 46	039	28 07	092	43 00	133	65 41	193	47 45	261	20 03	330
35	32 34	356	30 17	038	28 57	091	43 37	133	65 29	194	46 56	261	19 38	329
36	32 31	355	30 47	037	29 47	091	44 13	133	65 16	195	46 07	261	19 13	329
37	32 26	354	31 17	036	30 36	090	44 49	133	65 03	197	45 18	260	18 46	328
38	32 20	353	31 46	035	31 26	090	45 26	133	64 48	198	44 29	260	18 19	327
39	32 13	352	32 14	034	32 16	089	46 02	133	64 33	199	43 40	259	17 52	326
40	32 06	351	32 41	033	33 06	088	46 38	133	64 16	200	42 51	259	17 24	325
41	31 57	350	33 08	032	33 55	088	47 15	133	63 59	201	42 02	259	16 55	325
42	31 48	349	33 34	031	34 45	087	47 51	133	63 41	202	41 14	258	16 26	324
43	31 37	347	33 59	030	35 35	087	48 27	133	63 22	203	40 25	258	15 57	323
44	31 26	346	34 24	029	36 24	086	49 04	133	63 02	204	39 36	257	15 27	322
	*ALDEBARAN		BETELGEUSE		SIRIUS		*CANOPUS		ACHERNAR		*FOMALHAUT		Hamal	
45	34 47	028	31 42	053	37 14	086	49 40	133	62 42	205	38 48	257	31 14	345
46	35 10	027	32 21	052	38 03	085	50 16	134	62 21	206	37 59	257	31 01	344
47	35 32	026	33 01	051	38 53	084	50 52	134	61 59	206	37 11	256	30 47	343
48	35 53	025	33 39	051	39 42	084	51 28	134	61 36	207	36 23	256	30 32	342
49	36 13	023	34 17	050	40 32	083	52 03	134	61 13	208	35 35	255	30 17	341
50	36 33	022	34 55	049	41 21	082	52 39	134	60 49	209	34 47	255	30 00	340
51	36 51	021	35 32	048	42 10	082	53 15	134	60 25	210	33 59	254	29 43	339
52	37 09	020	36 09	047	43 00	081	53 50	135	60 00	210	33 11	254	29 25	338
53	37 25	019	36 44	046	43 49	080	54 26	135	59 35	211	32 23	254	29 06	337
54	37 41	018	37 20	045	44 38	080	55 01	135	59 09	212	31 35	253	28 46	336
55	37 55	017	37 55	044	45 27	079	55 36	135	58 43	212	30 48	253	28 26	335
56	38 09	015	38 29	043	46 15	078	56 10	136	58 16	213	30 00	252	28 05	334
57	38 22	014	39 02	042	47 04	077	56 45	136	57 49	213	29 13	252	27 43	333
58	38 33	013	39 35	041	47 52	077	57 19	136	57 22	214	28 26	252	27 20	332
59	38 44	012	40 07	040	48 41	076	57 51	137	56 54	214	27 39	251	26 57	331
	ALDEBARAN		*BETELGEUSE		SIRIUS		*CANOPUS		ACHERNAR		*FOMALHAUT		Hamal	
60	38 54	011	40 38	038	49 29	075	58 27	137	56 26	215	26 52	251	26 32	331
61	39 02	009	41 09	037	50 17	074	59 01	138	55 57	215	26 05	250	26 08	330
62	39 10	008	41 38	036	51 05	073	59 34	138	55 28	216	25 18	250	25 42	329
63	39 16	007	42 07	035	51 52	073	60 07	139	54 59	216	24 31	249	25 16	328
64	39 22	006	42 35	034	52 40	072	60 40	139	54 30	216	23 45	249	24 49	327
65	39 26	004	43 03	033	53 27	071	61 12	140	54 00	217	22 58	249	24 22	326
66	39 29	003	43 29	031	54 14	070	61 44	140	53 30	217	22 12	248	23 54	325
67	39 32	002	43 54	030	55 00	069	62 16	141	53 00	218	21 26	248	23 25	324
68	39 33	001	44 19	029	55 46	068	62 47	142	52 29	218	20 40	247	22 56	323
69	39 33	359	44 42	028	56 32	067	63 17	143	51 59	218	19 54	247	22 26	323
70	39 32	358	45 05	026	57 18	066	63 47	143	51 28	218	19 09	246	21 55	322
71	39 30	357	45 26	025	58 03	065	64 17	144	50 57	219	18 23	246	21 24	321
72	39 27	356	45 47	024	58 48	064	64 45	145	50 26	219	17 38	246	20 53	320
73	39 22	354	46 06	022	59 32	062	65 14	146	49 55	219	16 53	245	20 21	319
74	39 17	353	46 25	021	60 16	061	65 41	147	49 23	219	16 07	245	19 48	319
	BETELGEUSE		*PROCYON		Suhail		*ACRUX		ACHERNAR		*Diphda		ALDEBARAN	
75	46 42	020	35 52	051	42 01	120	21 11	153	48 52	219	30 54	268	39 11	352
76	46 58	018	36 31	050	42 44	120	21 34	153	48 20	220	30 05	268	39 03	351
77	47 13	017	37 09	049	43 27	120	21 57	152	47 48	220	29 15	267	38 55	350
78	47 27	015	37 46	048	44 10	120	22 20	152	47 16	220	28 25	266	38 45	348
79	47 39	014	38 23	047	44 53	120	22 43	152	46 44	220	27 36	266	38 35	347
80	47 50	012	38 59	046	45 36	120	23 07	152	46 12	220	26 46	265	38 23	346
81	48 01	011	39 35	045	46 20	119	23 30	151	45 40	220	25 56	265	38 10	345
82	48 09	010	40 10	044	47 03	119	23 54	151	45 08	220	25 07	264	37 57	344
83	48 17	008	40 45	043	47 47	119	24 18	151	44 36	220	24 17	264	37 42	342
84	48 23	007	41 18	042	48 30	119	24 43	151	44 04	220	23 28	263	37 27	341
85	48 28	005	41 51	041	49 13	119	25 07	150	43 31	220	22 39	263	37 10	340
86	48 32	004	42 24	040	49 57	119	25 32	150	42 59	221	21 49	262	36 53	339
87	48 35	002	42 55	039	50 41	119	25 56	150	42 27	221	21 00	262	36 35	338
88	48 36	001	43 26	038	51 24	119	26 21	150	41 55	221	20 11	261	36 15	337
89	48 36	359	43 56	036	52 08	119	26 46	150	41 22	221	19 22	261	35 55	336

LHA ♈	Hc	Zn	Hc	Zn	Hc	Zn	Hc	Zn	Hc	Zn	Hc	Zn	Hc	Zn
	PROCYON		REGULUS		*Suhail		ACRUX		*ACHERNAR		RIGEL		*BETELGEUSE	
90	44 25	035	15 29	063	52 51	119	27 12	149	40 50	221	62 04	335	48 34	358
91	44 53	034	16 13	063	53 35	119	27 37	149	40 18	221	61 42	333	48 31	356
92	45 21	033	16 58	062	54 19	119	28 03	149	39 45	220	61 18	331	48 27	355
93	45 47	031	17 41	061	55 02	119	28 28	149	39 13	220	60 53	329	48 22	353
94	46 13	030	18 25	061	55 46	119	28 54	149	38 41	220	60 27	327	48 15	352
95	46 37	029	19 08	060	56 30	119	29 20	149	38 08	220	59 59	325	48 08	350
96	47 01	027	19 51	059	57 13	119	29 46	148	37 36	220	59 30	324	47 58	349
97	47 23	026	20 33	058	57 57	119	30 12	148	37 04	220	59 00	322	47 48	347
98	47 44	025	21 16	058	58 40	119	30 38	148	36 32	220	58 29	320	47 36	346
99	48 05	023	21 57	057	59 24	119	31 05	148	36 00	220	57 56	319	47 24	344
100	48 24	022	22 39	056	60 07	119	31 31	148	35 28	220	57 23	317	47 10	343
101	48 42	021	23 20	055	60 51	119	31 58	148	34 56	220	56 49	316	46 54	342
102	48 59	019	24 01	055	61 34	119	32 24	148	34 24	220	56 14	314	46 38	340
103	49 14	018	24 41	054	62 17	120	32 51	147	33 53	220	55 37	313	46 21	339
104	49 29	016	25 21	053	63 01	120	33 18	147	33 21	219	55 01	311	46 02	337
	PROCYON		REGULUS		*Gienah		ACRUX		*ACHERNAR		RIGEL		*BETELGEUSE	
105	49 42	015	26 01	052	18 54	099	33 45	147	32 50	219	54 23	310	45 42	336
106	49 54	013	26 40	052	19 43	098	34 12	147	32 18	219	53 45	309	45 22	335
107	50 04	012	27 19	051	20 32	098	34 39	147	31 47	219	53 06	308	45 00	333
108	50 14	010	27 57	050	21 22	097	35 06	147	31 16	219	52 26	306	44 37	332
109	50 22	008	28 35	049	22 11	097	35 33	147	30 44	219	51 45	305	44 13	331
110	50 28	007	29 12	048	23 00	096	36 00	147	30 13	218	51 05	304	43 49	330
111	50 34	005	29 49	047	23 50	096	36 27	147	29 42	218	50 23	303	43 23	328
112	50 38	004	30 25	046	24 39	095	36 54	147	29 12	218	49 41	302	42 56	327
113	50 40	002	31 01	045	25 29	094	37 22	147	28 41	218	48 59	301	42 29	326
114	50 42	001	31 36	045	26 19	094	37 49	147	28 11	218	48 16	300	42 01	325
115	50 42	359	32 11	044	27 08	093	38 16	147	27 40	217	47 32	299	41 32	324
116	50 40	358	32 45	043	27 58	093	38 43	147	27 10	217	46 48	298	41 02	322
117	50 37	356	33 18	042	28 48	092	39 11	147	26 40	217	46 04	297	40 31	321
118	50 33	354	33 51	041	29 37	092	39 38	147	26 10	217	45 20	296	40 00	320
119	50 28	353	34 23	040	30 27	091	40 05	147	25 40	217	44 35	295	39 27	319
	REGULUS		*SPICA		ACRUX		*CANOPUS		RIGEL		BETELGEUSE		*PROCYON	
120	34 55	039	13 37	094	40 33	147	64 35	215	43 49	294	38 55	318	50 21	351
121	35 26	038	14 27	094	41 00	147	64 06	216	43 04	293	38 21	317	50 13	350
122	35 56	037	15 16	093	41 27	147	63 36	217	42 18	292	37 47	316	50 03	348
123	36 25	036	16 06	093	41 54	147	63 06	218	41 32	292	37 12	315	49 52	347
124	36 54	035	16 56	092	42 21	147	62 36	218	40 45	291	36 36	314	49 40	345
125	37 22	034	17 45	091	42 48	147	62 04	219	39 59	290	36 00	313	49 27	344
126	37 49	032	18 35	091	43 15	147	61 33	220	39 12	289	35 24	312	49 12	342
127	38 15	031	19 25	090	43 42	147	61 01	220	38 25	288	34 46	311	48 56	341
128	38 40	030	20 15	090	44 09	147	60 28	221	37 38	288	34 09	310	48 39	339
129	39 05	029	21 04	089	44 36	147	59 56	221	36 50	287	33 30	309	48 21	338
130	39 29	028	21 54	089	45 03	148	59 22	222	36 02	286	32 52	308	48 02	336
131	39 52	027	22 44	088	45 29	148	58 49	222	35 14	285	32 12	307	47 42	335
132	40 14	026	23 33	088	45 56	148	58 15	223	34 26	285	31 33	307	47 20	334
133	40 35	024	24 23	087	46 22	148	57 41	223	33 38	284	30 53	306	46 57	332
134	40 55	023	25 13	086	46 48	148	57 07	224	32 50	283	30 12	305	46 34	331
	*REGULUS		SPICA		*ACRUX		CANOPUS		*RIGEL		BETELGEUSE		PROCYON	
135	41 14	022	26 02	086	47 14	148	56 33	224	32 01	283	29 31	304	46 09	330
136	41 32	021	26 52	085	47 40	149	55 58	224	31 13	282	28 50	303	45 44	328
137	41 49	019	27 42	085	48 06	149	55 23	225	30 24	281	28 08	302	45 17	327
138	42 05	018	28 31	084	48 32	149	54 48	225	29 35	281	27 26	302	44 50	326
139	42 20	017	29 21	083	48 57	149	54 13	225	28 46	280	26 43	301	44 21	325
140	42 34	016	30 10	083	49 22	150	53 38	225	27 57	279	26 00	300	43 52	323
141	42 47	014	30 59	082	49 47	150	53 02	226	27 08	279	25 17	299	43 22	322
142	42 58	013	31 48	082	50 12	150	52 26	226	26 19	278	24 34	299	42 51	321
143	43 09	012	32 38	081	50 36	151	51 51	226	25 30	277	23 50	298	42 19	320
144	43 18	010	33 27	080	51 01	151	51 15	226	24 40	277	23 06	297	41 47	319
145	43 27	009	34 16	080	51 25	151	50 39	226	23 51	276	22 21	296	41 14	318
146	43 34	008	35 05	079	51 49	152	50 03	226	23 01	276	21 37	296	40 40	317
147	43 40	006	35 53	078	52 12	152	49 27	227	22 12	275	20 52	295	40 05	316
148	43 45	005	36 42	078	52 35	152	48 51	227	21 22	274	20 07	294	39 30	314
149	43 49	004	37 30	077	52 58	153	48 14	227	20 33	274	19 21	294	38 55	313
	REGULUS		*Denebola		SPICA		*ACRUX		CANOPUS		*SIRIUS		PROCYON	
150	43 51	002	34 57	032	38 19	076	53 21	153	47 38	227	42 55	279	38 18	312
151	43 53	001	35 23	031	39 07	075	53 43	154	47 02	227	42 06	278	37 41	311
152	43 53	000	35 48	030	39 55	075	54 05	154	46 25	227	41 17	278	37 04	310
153	43 52	358	36 13	029	40 43	074	54 26	155	45 49	227	40 28	277	36 25	310
154	43 50	357	36 37	028	41 31	073	54 48	155	45 13	227	39 38	276	35 47	309
155	43 47	356	37 00	027	42 18	072	55 08	156	44 36	227	38 49	276	35 08	308
156	43 42	354	37 22	026	43 05	072	55 29	156	44 00	227	37 59	275	34 28	307
157	43 37	353	37 43	025	43 53	071	55 48	157	43 24	227	37 10	274	33 48	306
158	43 30	351	38 03	023	44 39	070	56 08	157	42 47	227	36 20	274	33 07	305
159	43 22	350	38 22	022	45 26	069	56 27	158	42 11	227	35 30	273	32 26	304
160	43 13	349	38 41	021	46 12	068	56 45	158	41 35	227	34 41	273	31 45	303
161	43 03	348	38 58	020	46 58	067	57 03	159	40 58	227	33 51	272	31 03	302
162	42 51	346	39 14	019	47 44	066	57 21	160	40 22	227	33 01	271	30 21	302
163	42 39	345	39 30	017	48 30	065	57 38	160	39 46	227	32 12	271	29 39	301
164	42 25	344	39 44	016	49 15	065	57 54	161	39 10	227	31 22	270	28 56	300
	ARCTURUS		*ANTARES		ACRUX		*CANOPUS		SIRIUS		PROCYON		*REGULUS	
165	19 25	049	20 41	109	58 10	162	38 34	226	30 32	270	28 12	299	42 11	342
166	20 02	048	21 28	108	58 25	163	37 58	226	29 42	269	27 29	298	41 55	341
167	20 39	047	22 15	108	58 40	163	37 22	226	28 53	269	26 45	298	41 38	340
168	21 15	046	23 03	107	58 54	164	36 46	226	28 03	268	26 01	297	41 21	339
169	21 51	046	23 50	107	59 07	165	36 10	226	27 13	268	25 16	296	41 02	337
170	22 26	045	24 38	106	59 20	166	35 34	226	26 23	267	24 31	295	40 42	336
171	23 01	044	25 26	106	59 32	166	34 59	226	25 34	267	23 46	295	40 22	335
172	23 35	043	26 14	105	59 43	167	34 23	226	24 44	266	23 01	294	40 00	334
173	24 09	042	27 02	105	59 54	168	33 48	225	23 55	265	22 16	293	39 38	333
174	24 42	041	27 50	104	60 04	169	33 12	225	23 05	265	21 30	293	39 14	331
175	25 14	041	28 38	104	60 13	170	32 37	225	22 15	264	20 44	292	38 50	330
176	25 46	040	29 26	104	60 22	171	32 02	225	21 26	264	19 58	291	38 25	329
177	26 18	039	30 15	103	60 30	171	31 27	225	20 37	263	19 11	291	37 59	328
178	26 49	038	31 03	103	60 37	172	30 52	224	19 47	263	18 24	290	37 32	327
179	27 19	037	31 52	102	60 43	173	30 17	224	18 58	262	17 38	289	37 05	326

Figure 21-6A. *Typical left-hand page, Lat 34°S, LHAs 0°–179°, volume 1,* Tables No. 249.

N. Lat. {LHA greater than 180°……… Zn=Z; LHA less than 180°……… Zn=360−Z}

DECLINATION (15°-29°) CONTRARY NAME TO LATITUDE

LHA	15° Hc	d	Z	16° Hc	d	Z	17° Hc	d	Z	18° Hc	d	Z	19° Hc	d	Z
	° ′	′	°	° ′	′	°	° ′	′	°	° ′	′	°	° ′	′	°
69	08 11	38	114	07 33	37	115	06 56	38	116	06 18	38	117	05 40	37	118
68	08 56	38	115	08 18	38	116	07 40	38	117	07 02	38	117	06 24	38	118
67	09 41	38	116	09 03	38	116	08 25	38	117	07 47	39	118	07 08	38	119
66	10 26	39	116	09 47	38	117	09 09	39	118	08 30	38	119	07 52	39	119
65	11 10	−38	117	10 32	−39	118	09 53	−39	118	09 14	−39	119	08 35	−39	120
64	11 54	39	118	11 15	39	118	10 36	39	119	09 57	39	120	09 18	39	121
63	12 38	39	118	11 59	39	119	11 20	40	120	10 40	39	120	10 01	40	121
62	13 22	39	119	12 43	40	120	12 03	40	120	11 23	40	121	10 43	40	122
61	14 06	40	119	13 26	40	120	12 46	41	121	12 05	40	122	11 25	40	123
60	14 49	−40	120	14 09	−41	121	13 28	−40	122	12 48	−41	122	12 07	−41	123
59	15 32	41	121	14 51	41	122	14 10	41	122	13 29	41	123	12 48	41	124
58	16 14	41	121	15 33	41	122	14 52	41	123	14 11	41	124	13 30	42	125
57	16 57	42	122	16 15	41	123	15 34	42	124	14 52	41	124	14 11	42	125
56	17 39	42	123	16 57	42	124	16 15	42	124	15 33	42	125	14 51	42	126
55	18 20	−42	124	17 38	−42	124	16 56	−42	125	16 14	−43	126	15 31	−42	127
54	19 01	42	124	18 19	42	125	17 37	43	126	16 54	43	127	16 11	43	127
53	19 42	42	125	19 00	43	126	18 17	43	127	17 34	44	127	16 50	43	128
52	20 23	43	126	19 40	43	126	18 57	44	127	18 13	44	128	17 29	43	129
51	21 03	43	127	20 20	44	128	19 36	44	128	18 52	44	129	18 08	44	129
50	21 43	−44	127	20 59	−44	128	20 15	−44	129	19 31	−45	129	18 46	−44	130
49	22 22	44	128	21 38	44	129	20 54	45	129	20 09	45	130	19 24	45	131
48	23 01	44	129	22 17	45	130	21 32	45	130	20 47	45	131	20 02	46	132
47	23 40	45	130	22 55	45	130	22 10	46	131	21 24	45	132	20 39	46	132
46	24 18	45	130	23 33	46	131	22 47	46	132	22 01	46	132	21 15	46	133
45	24 56	−46	131	24 10	−46	132	23 24	−46	133	22 38	−47	133	21 51	−46	134
44	25 33	46	132	24 47	47	133	24 00	46	133	23 14	47	134	22 27	47	135
43	26 10	47	133	25 23	47	134	24 36	47	134	23 49	47	135	23 02	48	136
42	26 46	47	134	25 59	47	134	25 12	48	135	24 24	47	136	23 37	48	136
41	27 22	48	135	26 34	47	135	25 47	48	136	24 59	48	137	24 11	49	137
40	27 57	−48	135	27 09	−48	136	26 21	−48	137	25 33	−49	137	24 44	−49	138
39	28 32	49	136	27 43	48	137	26 55	49	138	26 06	49	138	25 17	49	139
38	29 06	49	137	28 17	49	138	27 28	49	138	26 39	49	139	25 50	50	140
37	29 39	49	138	28 50	49	139	28 01	50	139	27 11	49	140	26 22	50	141
36	30 12	49	139	29 23	50	140	28 33	50	140	27 43	50	141	26 53	51	142
35	30 45	−50	140	29 55	−50	141	29 05	−51	141	28 14	−51	142	27 23	−50	142
34	31 17	51	141	30 26	51	141	29 35	50	142	28 45	51	143	27 54	52	143
33	31 48	51	142	30 57	51	142	30 06	52	143	29 14	51	144	28 23	52	144
32	32 18	51	143	31 27	52	143	30 35	51	144	29 44	52	145	28 52	52	145
31	32 48	52	144	31 56	52	144	31 04	52	145	30 12	52	146	29 20	52	146
30	33 17	−52	145	32 25	−53	145	31 32	−52	146	30 40	−53	146	29 47	−52	147
29	33 45	52	146	32 53	53	146	32 00	53	147	31 07	53	147	30 14	53	148
28	34 13	53	147	33 20	53	148	32 27	53	148	31 34	54	148	30 40	53	149
27	34 40	53	148	33 47	54	148	32 53	54	149	31 59	54	149	31 05	54	150
26	35 06	54	149	34 12	54	149	33 18	54	150	32 24	54	150	31 30	54	151
25	35 31	−54	150	34 37	−54	150	33 43	−55	151	32 48	−54	151	31 54	−55	152
24	35 56	55	151	35 01	54	152	34 07	55	152	33 12	55	153	32 17	55	153
23	36 20	55	152	35 25	55	153	34 30	56	153	33 34	55	154	32 39	55	154
22	36 43	56	153	35 47	55	154	34 52	56	154	33 56	56	155	33 00	55	155
21	37 05	56	154	36 09	56	155	35 13	56	155	34 17	56	156	33 21	56	156
20	37 26	−56	155	36 30	−57	156	35 33	−56	156	34 37	−56	157	33 41	−57	157
19	37 46	56	157	36 50	57	157	35 53	57	157	34 56	56	158	34 00	57	158
18	38 05	56	158	37 09	57	158	36 12	57	159	35 15	57	159	34 18	57	159
17	38 24	57	159	37 27	58	159	36 29	57	160	35 32	57	160	34 35	57	160
16	38 41	57	160	37 44	58	160	36 46	57	161	35 49	58	161	34 51	58	162
15	38 58	−58	161	38 00	−58	162	37 02	−58	162	36 04	−58	162	35 06	−57	163
14	39 13	58	162	38 15	58	163	37 17	58	163	36 19	58	163	35 21	58	164
13	39 28	59	164	38 29	58	164	37 31	58	164	36 33	59	165	35 34	58	165
12	39 41	58	165	38 43	59	165	37 44	58	165	36 46	59	166	35 47	59	166
11	39 54	59	166	38 55	59	166	37 56	59	167	36 57	59	167	35 58	59	167
10	40 05	−59	167	39 06	−59	168	38 07	−59	168	37 08	−59	168	36 09	−59	168
9	40 15	59	169	39 16	59	169	38 17	59	169	37 18	59	169	36 19	60	169
8	40 25	60	170	39 25	59	170	38 26	59	170	37 27	60	170	36 27	59	171
7	40 33	60	171	39 33	59	171	38 34	60	171	37 34	59	172	36 35	60	172
6	40 40	59	172	39 41	60	173	38 41	60	173	37 41	59	173	36 42	60	173
5	40 46	−60	174	39 46	−59	174	38 47	−60	174	37 47	−60	174	36 47	−60	174
4	40 51	60	175	39 51	59	175	38 52	60	175	37 52	60	175	36 52	60	175
3	40 55	60	176	39 55	60	176	38 55	60	176	37 55	60	176	36 55	59	177
2	40 58	60	177	39 58	60	178	38 58	60	178	37 58	60	178	36 58	60	178
1	40 59	59	179	40 00	60	179	39 00	60	179	38 00	60	179	37 00	60	179
0	41 00	−60	180	40 00	−60	180	39 00	−60	180	38 00	−60	180	37 00	−60	180

LHA	20° Hc	d	Z	21° Hc	d	Z	22° Hc	d	Z	23° Hc	d	Z	24° Hc	d	Z
	° ′	′	°	° ′	′	°	° ′	′	°	° ′	′	°	° ′	′	°
69	05 03	38	118	04 25	38	119	03 47	38	120	03 09	38	121	02 31	38	121
68	05 46	38	119	05 08	38	120	04 30	38	120	03 52	38	121	03 14	39	122
67	06 30	39	120	05 51	38	120	05 13	39	121	04 34	38	122	03 56	39	123
66	07 13	39	120	06 34	39	121	05 55	39	122	05 16	39	122	04 37	39	123
65	07 56	−39	121	07 17	−39	122	06 38	−40	122	05 58	−39	123	05 19	−39	124
64	08 39	40	121	07 59	39	122	07 20	40	123	06 40	40	124	06 00	40	124
63	09 21	40	122	08 41	40	123	08 01	40	124	07 21	40	124	06 41	40	125
62	10 03	40	123	09 23	40	123	08 43	41	124	08 02	40	125	07 22	41	126
61	10 45	41	123	10 04	40	124	09 24	41	125	08 43	41	126	08 02	41	127
60	11 26	−41	124	10 45	−41	125	10 04	−41	125	09 23	−41	126	08 42	−41	127
59	12 07	41	125	11 26	41	125	10 45	42	126	10 03	41	127	09 22	42	128
58	12 48	41	125	12 07	42	126	11 25	42	127	10 43	42	128	10 01	42	128
57	13 29	42	126	12 47	43	127	12 04	42	127	11 22	42	128	10 40	43	129
56	14 09	43	127	13 26	42	127	12 44	43	128	12 01	43	129	11 18	42	129
55	14 49	−43	127	14 06	−43	128	13 23	−43	129	12 40	−43	129	11 57	−44	130
54	15 28	43	128	14 45	43	129	14 02	44	129	13 18	43	130	12 35	44	131
53	16 07	44	129	15 23	43	129	14 40	44	130	13 56	44	131	13 12	44	132
52	16 46	44	129	16 02	44	130	15 18	45	131	14 33	44	132	13 49	44	132
51	17 24	44	130	16 40	45	131	15 55	44	132	15 11	45	132	14 26	45	133
50	18 02	−45	131	17 17	−45	132	16 32	−45	132	15 47	−45	133	15 02	−45	134
49	18 39	45	132	17 54	45	132	17 09	46	133	16 23	45	134	15 38	46	134
48	19 16	45	132	18 31	46	133	17 45	46	134	16 59	46	134	16 13	46	135
47	19 53	46	133	19 07	46	134	18 21	46	134	17 35	47	135	16 48	46	136
46	20 29	46	134	19 43	47	135	18 56	47	135	18 09	46	136	17 23	47	137
45	21 05	−47	135	20 18	−47	135	19 31	−47	136	18 44	−47	137	17 57	−48	137
44	21 40	47	135	20 53	48	136	20 05	47	137	19 18	48	137	18 30	47	138
43	22 14	47	136	21 27	48	137	20 39	48	138	19 51	48	138	19 03	48	139
42	22 49	48	137	22 01	48	138	21 13	49	138	20 24	48	139	19 36	49	140
41	23 22	48	138	22 34	49	139	21 45	48	139	20 57	49	140	20 08	49	140
40	23 55	−48	139	23 07	−49	139	22 18	−49	140	21 29	−50	141	20 39	−49	141
39	24 28	49	140	23 39	50	140	22 49	49	141	22 00	50	141	21 10	50	142
38	25 00	50	140	24 10	49	141	23 21	50	142	22 31	50	142	21 41	51	143
37	25 32	50	141	24 42	51	142	23 51	50	142	23 01	51	143	22 10	50	144
36	26 02	50	142	25 12	51	143	24 21	50	143	23 31	51	144	22 40	51	144
35	26 33	−51	143	25 42	−51	144	24 51	−51	144	24 00	−52	145	23 08	−51	145
34	27 02	51	144	26 11	51	144	25 20	52	145	24 28	52	146	23 36	51	146
33	27 31	51	145	26 40	52	145	25 48	52	146	24 56	52	147	24 04	52	147
32	28 00	52	146	27 08	53	146	26 15	52	147	25 23	52	147	24 31	53	148
31	28 28	53	147	27 35	53	147	26 42	52	148	25 50	53	148	24 57	53	149
30	28 55	−53	148	28 02	−53	148	27 09	−54	149	26 15	−53	149	25 22	−53	150
29	29 21	53	149	28 28	54	149	27 34	53	150	26 41	54	150	25 47	54	151
28	29 47	54	150	28 53	54	150	27 59	54	151	27 05	54	151	26 11	54	151
27	30 11	54	150	29 17	54	151	28 23	54	151	27 29	54	152	26 35	55	152
26	30 36	55	151	29 41	54	152	28 47	55	152	27 52	55	153	26 57	55	153
25	30 59	−55	152	30 04	−55	153	29 09	−55	153	28 14	−55	154	27 19	−55	154
24	31 22	55	153	30 27	56	154	29 31	55	154	28 36	56	155	27 40	55	155
23	31 44	56	154	30 48	56	155	29 52	55	155	28 57	56	156	28 01	56	156
22	32 05	56	156	31 09	56	156	30 13	56	156	29 17	56	157	28 21	56	157
21	32 25	56	157	31 29	57	157	30 32	56	157	29 36	56	158	28 40	57	158
20	32 44	−56	158	31 48	−57	158	30 51	−56	158	29 55	−57	159	28 58	−57	159
19	33 03	57	159	32 06	57	159	31 09	57	159	30 12	57	160	29 15	57	160
18	33 21	57	160	32 24	58	160	31 26	57	160	30 29	57	161	29 32	58	161
17	33 38	58	161	32 40	57	161	31 43	58	161	30 45	58	162	29 47	57	162
16	33 53	57	162	32 56	58	162	31 58	58	163	31 00	58	163	30 02	58	163
15	34 09	−58	163	33 11	−58	163	32 13	−59	164	31 14	−58	164	30 16	−58	164
14	34 23	59	164	33 24	58	164	32 26	58	165	31 28	58	165	30 30	59	165
13	34 36	59	165	33 37	58	165	32 39	59	166	31 40	58	166	30 42	59	166
12	34 48	58	166	33 50	59	167	32 51	59	167	31 52	59	167	30 53	58	167
11	35 00	59	167	34 01	59	168	33 02	59	168	32 03	59	168	31 04	59	168
10	35 10	−59	169	34 11	−59	169	33 12	−59	169	32 13	−59	169	31 14	−60	169
9	35 19	59	170	34 20	59	170	33 21	59	170	32 22	60	170	31 22	59	170
8	35 28	59	171	34 29	60	171	33 29	59	171	32 30	59	171	31 31	60	171
7	35 35	59	172	34 36	60	172	33 36	59	172	32 37	60	172	31 37	59	[illegible]
6	35 42	60	173	34 42	59	173	33 43	60	173	32 43	60	173	31 43	59	[illegible]
5	35 47	−59	174	34 48	−60	174	33 48	−60	174	32 48	−60	174	31 48	−59	[illegible]
4	35 52	60	175	34 52	60	175	33 52	60	176	32 52	59	176	31 53	60	176
3	35 56	60	177	34 56	60	177	33 56	60	177	32 56	60	177	31 56	60	177
2	35 58	60	178	34 58	60	178	33 58	60	178	32 58	60	178	31 58	60	178
1	36 00	60	179	35 00	60	179	34 00	60	179	33 00	60	179	32 00	60	179
0	36 00	−60	180	35 00	−60	180	34 00	−60	180	33 00	−60	180	32 00	−60	180

LHA	25° Hc	d	Z	26° Hc	d	Z	27° Hc	d	Z	28° Hc	d	Z	29° Hc	d	Z	LHA
	° ′	′	°	° ′	′	°	° ′	′	°	° ′	′	°	° ′	′	°	
69	01 53	38	122	01 15	38	123	00 37	38	124	−0 01	38	124	−0 39	38	125	291
68	02 35	38	123	01 57	38	124	01 19	39	124	00 40	38	125	00 02	39	126	292
67	03 17	39	123	02 38	38	124	02 00	39	125	01 21	39	126	00 42	39	126	293
66	03 58	39	124	03 19	39	125	02 40	39	125	02 01	39	126	01 22	39	127	294
65	04 40	−40	125	04 00	−39	125	03 21	−40	126	02 41	−40	127	02 01	−39	128	295
64	05 20	39	125	04 41	40	126	04 01	40	127	03 21	40	127	02 41	40	128	296
63	06 01	40	126	05 21	41	127	04 40	40	127	04 00	40	128	03 20	41	129	297
62	06 41	40	126	06 01	41	127	05 20	41	128	04 39	41	129	03 58	41	129	298
61	07 21	41	127	06 40	41	128	05 59	41	128	05 18	41	129	04 37	42	130	299
60	08 01	−42	128	07 19	−41	128	06 38	−42	129	05 56	−41	130	05 15	−42	131	300
59	08 40	42	128	07 58	42	129	07 16	42	130	06 34	41	130	05 53	42	131	301
58	09 19	42	129	08 37	43	130	07 54	42	130	07 12	42	131	06 30	43	132	302
57	09 57	42	130	09 15	43	130	08 32	42	131	07 50	43	132	07 07	43	132	303
56	10 36	43	130	09 53	43	131	09 10	43	132	08 27	44	132	07 43	43	133	304
55	11 13	−43	131	10 30	−43	132	09 47	−44	132	09 03	−43	133	08 20	−44	134	305
54	11 51	44	132	11 07	44	132	10 23	44	133	09 39	44	134	08 55	44	134	306
53	12 28	44	132	11 44	44	133	11 00	45	134	10 15	44	134	09 31	45	135	307
52	13 05	45	133	12 20	45	134	11 35	44	134	10 51	45	135	10 06	45	136	308
51	13 41	45	134	12 56	45	134	12 11	45	135	11 26	46	136	10 40	45	136	309
50	14 17	−46	134	13 31	−45	135	12 46	−46	136	12 00	−45	136	11 15	−46	137	310
49	14 52	46	135	14 06	45	136	13 21	46	136	12 35	47	137	11 48	46	138	311
48	15 27	46	136	14 41	46	136	13 55	47	137	13 08	46	138	12 22	47	138	312
47	16 02	47	136	15 15	47	137	14 28	46	138	13 42	47	138	12 55	47	139	313
46	16 36	47	137	15 49	47	138	15 02	48	139	14 14	47	139	13 27	47	140	314
45	17 09	−47	138	16 22	−48	139	15 34	−47	139	14 47	−48	140	13 59	−48	140	315
44	17 43	48	139	16 55	48	139	16 07	48	140	15 19	49	141	14 30	48	141	316
43	18 15	48	139	17 27	49	140	16 38	48	141	15 50	49	141	15 01	48	142	317
42	18 47	48	140	17 59	49	141	17 10	49	141	16 21	49	142	15 32	49	143	318
41	19 19	49	141	18 30	49	142	17 41	50	142	16 51	49	143	16 02	50	143	319
40	19 50	−50	142	19 00	−49	142	18 11	−50	143	17 21	−50	144	16 31	−50	144	320
39	20 20	49	143	19 31	50	143	18 41	51	144	17 50	50	144	17 00	50	145	321
38	20 50	50	143	20 00	50	144	19 10	51	145	18 19	50	145	17 29	51	146	322
37	21 20	51	144	20 29	51	145	19 38	51	145	18 47	51	146	17 56	51	146	323
36	21 49	51	145	20 58	52	146	20 06	51	146	19 15	51	147	18 24	52	147	324
35	22 17	−52	146	21 25	−51	146	20 34	−52	147	19 42	−52	148	18 50	−52	148	325
34	22 45	52	147	21 53	52	147	21 01	53	148	20 08	52	148	19 16	52	149	326
33	23 12	53	148	22 19	52	148	21 27	53	149	20 34	52	149	19 42	53	150	327
32	23 38	53	148	22 45	53	149	21 52	52	149	21 00	53	150	20 07	53	150	328
31	24 04	53	149	23 11	54	150	22 17	53	150	21 24	53	151	20 31	54	151	329
30	24 29	−54	150	23 35	−53	151	22 42	−54	151	21 48	−54	152	20 54	−53	152	330
29	24 53	54	151	23 59	53	152	23 06	54	152	22 12	55	153	21 17	54	153	331
28	25 17	54	152	24 23	54	152	23 29	55	153	22 34	54	153	21 40	55	154	332
27	25 40	54	153	24 46	55	153	23 51	55	154	22 56	55	154	22 01	55	155	333
26	26 02	54	154	25 08	55	154	24 13	56	155	23 17	55	155	22 22	55	156	334
25	26 24	−55	155	25 29	−56	155	24 33	−55	156	23 38	−55	156	22 43	−56	156	335
24	26 45	56	156	25 49	55	156	24 54	56	157	23 58	56	157	23 02	56	157	336
23	27 05	56	157	26 09	56	157	25 13	56	157	24 17	56	158	23 21	56	158	337
22	27 25	57	158	26 28	56	158	25 32	56	158	24 36	57	159	23 39	56	159	338
21	27 43	56	159	26 47	57	159	25 50	57	159	24 53	56	160	23 57	57	160	339
20	28 01	−57	159	27 04	−57	160	26 07	−57	160	25 10	−57	161	24 13	−57	161	340
19	28 18	57	160	27 21	57	161	26 24	57	161	25 27	58	161	24 29	57	162	341
18	28 34	57	161	27 37	57	162	26 40	58	162	25 42	58	162	24 44	57	163	342
17	28 50	58	162	27 52	58	163	26 54	57	163	25 57	58	163	24 59	58	164	343
16	29 04	57	163	28 07	58	164	27 09	58	164	26 11	58	164	25 13	59	165	344
15	29 18	−58	164	28 20	−58	165	27 22	−58	165	26 24	−59	165	25 25	−58	166	345
14	29 31	58	165	28 33	59	166	27 34	58	166	26 36	59	166	25 37	58	166	346
13	29 43	58	166	28 45	59	167	27 46	59	167	26 47	58	167	25 49	59	167	347
12	29 55	59	167	28 56	59	168	27 57	59	168	26 58	59	168	25 59	59	168	348
11	30 05	59	169	29 06	59	169	28 07	59	169	27 08	59	169	26 09	59	169	349
10	30 14	−59	170	29 15	−59	170	28 16	−59	170	27 17	−59	170	26 18	−60	170	350
9	30 23	59	171	29 24	60	171	28 24	59	171	27 25	59	171	26 26	60	171	351
8	30 31	60	172	29 31	59	172	28 32	60	172	27 32	59	172	26 33	60	172	352
7	30 38	60	173	29 38	60	173	28 38	59	173	27 39	60	173	26 39	59	173	353
6	30 44	60	174	29 44	60	174	28 44	60	174	27 44	59	174	26 45	60	174	354
5	30 49	−60	175	29 49	−60	175	28 49	−60	175	27 49	−60	175	26 49	−59	175	355
4	30 53	60	176	29 53	60	176	28 53	60	176	27 53	60	176	26 53	60	176	356
3	30 56	60	177	29 56	60	177	28 56	60	177	27 56	60	177	26 56	60	177	357
2	30 58	60	178	29 58	60	178	28 58	60	178	27 58	60	178	26 58	60	178	358
1	31 00	60	179	30 00	60	179	29 00	60	179	28 00	60	179	27 00	60	179	359
0	31 00	−60	180	30 00	−60	180	29 00	−60	180	28 00	−60	180	27 00	−60	180	360

S. Lat. {LHA greater than 180°……… Zn=180−Z; LHA less than 180°……… Zn=180+Z}

DECLINATION (15°-29°) CONTRARY NAME TO LATITUDE

LAT 34°

Figure 21-6B. Lat 34° page, volume 2, Tables No. 249.

TABLE 4.—Correction to Tabulated Altitude

d / ′	1 2 3	4 5 6	7 8 9	10 11 12	13 14 15	16 17 18	19 20 21	22 23 24	25 26 27	28 29 30	31 32 33	8	49 50 51	52 53 54	55 56 57	58 59 60	d / ′
0	0 0 0	0 0 0	0 0 0	0 0 0	0 0 0	0 0 0	0 0 0	0 0 0	0 0 0	0 0 0	0 0 0	0	0 0 0	0 0 0	0 0 0	0 0 0	0
1	0 0 0	0 0 0	0 0 0	0 0 0	0 0 0	0 0 0	0 0 0	0 0 0	0 0 0	0 0 0	1 1	1	1 1 1	1 1 1	1 1 1	1 1 1	1
2	0 0 0	0 0 0	0 0 0	0 0 0	0 0 0	1 1 1	1 1 1	1 1 1	1 1 1	1 1 1	1 1	2	2 2 2	2 2 2	2 2 2	2 2 2	2
3	0 0 0	0 0 0	0 0 0	0 1 1	1 1 1	1 1 1	1 1 1	1 1 1	1 1 1	1 1 2	2 2 2	2	2 2 3	3 3 3	3 3 3	3 3 3	3
4	0 0 0	0 0 0	0 1 1	1 1 1	1 1 1	1 1 1	1 1 1	1 2 2	2 2 2	2 2 2	2 2	3 3	3 3 3	3 4 4	4 4 4	4 4 4	4
5	0 0 0	0 0 0	1 1 1	1 1 1	1 1 1	1 1 2	2 2 2	2 2 2	2 2 2	2 2 2	3 3	4 4	4 4 4	4 4 4	5 5 5	5 5 5	5
6	0 0 0	0 0 1	1 1 1	1 1 1	1 1 2	2 2 2	2 2 2	2 2 2	2 3 3	3 3 3	3	5 5	5 5 5	5 5 5	6 6 6	6 6 6	6
7	0 0 0	0 1 1	1 1 1	1 1 1	2 2 2	2 2 2	2 2 2	3 3 3	3 3 3	3 3 4	4	5 6	6 6 6	6 6 6	6 7 7	7 7 7	7
8	0 0 0	1 1 1	1 1 1	1 1 2	2 2 2	2 2 2	3 3 3	3 3 3	3 3 4	4 4 4	4	6 6	7 7 7	7 7 7	7 7 8	8 8 8	8
9	0 0 0	1 1 1	1 1 1	2 2 2	2 2 2	2 3 3	3 3 3	3 3 4	4 4 4	4 4 4	5	7 7	7 8 8	8 8 8	8 8 9	9 9 9	9
10	0 0 0	1 1 1	1 1 2	2 2 2	2 2 2	3 3 3	3 3 4	4 4 4	4 4 4	5 5 5	5	8 8 8	8 8 8	9 9 9	9 9 10	10 10 10	10
11	0 0 1	1 1 1	1 1 2	2 2 2	2 3 3	3 3 3	3 4 4	4 4 4	5 5 5	5 5 6	6	8 9 9	9 9 9	10 10 10	10 10 10	11 11 11	11
12	0 0 1	1 1 1	1 2 2	2 2 2	3 3 3	3 3 4	4 4 4	4 5 5	5 5 5	6 6 6		9 9 10	10 10 10	10 11 11	11 11 11	12 12 12	12
13	0 0 1	1 1 1	2 2 2	2 2 3	3 3 3	3 4 4	4 4 5	5 5 5	5 6 6	6 6 6		10 10 10	11 11 11	11 11 12	12 12 12	13 13 13	13
14	0 0 1	1 1 1	2 2 2	2 3 3	3 3 4	4 4 4	4 5 5	5 5 6	6 6 6	7 7 7		11 11 11	11 12 12	12 12 13	13 13 13	14 14 14	14
15	0 0 1	1 1 2	2 2 2	2 3 3	3 4 4	4 4 4	5 5 5	6 6 6	6 6 7	7 7 8		12 12 12	12 12 13	13 13 14	14 14 14	14 15 15	15
16	0 1 1	1 1 2	2 2 2	3 3 3	3 4 4	4 5 5	5 5 6	6 6 6	7 7 7	7 8 8		12 13 13	13 13 14	14 14 14	15 15 15	15 16 16	16
17	0 1 1	1 1 2	2 2 3	3 3 3	4 4 4	5 5 5	5 6 6	6 7 7	7 7 8	8 8 8		13 13 14	14 14 14	15 15 15	16 16 16	16 17 17	17
18	0 1 1	1 2 2	2 2 3	3 3 4	4 4 4	5 5 5	6 6 6	7 7 7	8 8 8	8 9 9		14 14 14	15 15 15	16 16 16	16 17 17	17 18 18	18
19	0 1 1	1 2 2	2 3 3	3 3 4	4 4 5	5 5 6	6 6 7	7 7 8	8 8 9	9 9 1		15 15 15	16 16 16	16 17 17	17 18 18	18 19 19	19
20	0 1 1	1 2 2	2 3 3	3 4 4	4 5 5	5 6 6	6 7 7	7 8 8	8 9 9	9 10 1		15 16 16	16 17 17	17 18 18	18 19 19	19 20 20	20
21	0 1 1	1 2 2	2 3 3	4 4 4	5 5 5	6 6 6	7 7 7	8 8 8	9 9 9	10 10		16 16 17	17 18 18	18 19 19	19 20 20	20 21 21	21
22	0 1 1	1 2 2	3 3 3	4 4 4	5 5 6	6 6 7	7 7 8	8 8 9	9 10 10	10 11		17 17 18	18 18 19	19 19 20	20 21 21	21 22 22	22
23	0 1 1	2 2 2	3 3 3	4 4 5	5 5 6	6 7 7	7 8 8	8 9 9	10 10 10	11 11 1		18 18 18	19 19 20	20 20 21	21 21 22	22 23 23	23
24	0 1 1	2 2 2	3 3 4	4 4 5	5 6 6	6 7 7	8 8 8	9 9 10	10 10 11	11 12 1		18 19 19	20 20 20	21 21 22	22 22 23	23 24 24	24
25	0 1 1	2 2 2	3 3 4	4 5 5	5 6 6	7 7 8	8 8 9	9 10 10	10 11 11	12 12 12		19 20 20	20 21 21	22 22 22	23 23 24	24 25 25	25
26	0 1 1	2 2 3	3 3 4	4 5 5	6 6 6	7 7 8	8 9 9	10 10 10	11 11 12	12 13 13		20 20 21	21 22 22	23 23 23	24 24 25	25 26 26	26
43	1	2 2 3	3 4 4	4 5 5	6 6 7	7 8 8	9 9 9	10 10 11	11 12 12	13 13 14		21 21 22	22 22 23	23 24 24	25 25 26	26 27 27	27
44	1 1 2	[illegible]	3 4 4	5 5 6	6 7	7 8 8	9 9 10	10 11 11	12 12 13	13 14 14		21 22 22	23 23 24	24 25 25	26	28 28	28
45	1 2 2	3 4 4	[illegible]	[illegible]	10 10 11	12 13 14	[illegible]	11 11 12	12 13 13	22 22	2	34 35	24 24 25	25 26	41 42 43	44 44 45	[illegible]
46	1 2 2	3 4 5	5 6 7	8 8 9	10 11 12	12 13 14	15 15 16	17	20 21	21 22 23	24	35 36 37	38	41 41	42 43 44	44 45 46	46
47	1 2 2	3 4 5	5 6 7	8 9 9	10 11 12	13 13 14	15 16 16	17 18 19	20 20 21	22 23 24	24	36 37 38	38 39 40	41 42 42	43 44 45	45 46 47	47
48	1 2 2	3 4 5	6 6 7	8 9 10	10 11 12	13 14 14	15 16 17	18 18 19	20 21 22	22 23 24	25	37 38 38	39 40 41	42 42 43	44 45 46	46 47 48	48
49	1 2 2	3 4 5	6 7 7	8 9 10	11 11 12	13 14 15	16 16 17	18 19 20	20 21 22	23 24 24	25	38 38 39	40 41 42	42 43 44	45 46 47	47 48 49	49
50	1 2 2	3 4 5	6 7 8	8 9 10	11 12 12	13 14 15	16 17 18	18 19 20	21 22 22	23 24 25	26	39 40	41 42 42	43 44 45	46 47 48	48 49 50	50
51	1 2 3	3 4 5	6 7 8	8 9 10	11 12 13	14 14 15	16 17 18	19 20 20	21 22 23	24 25 26	26	40 41	42 42 43	44 45 46	47 48 48	49 50 51	51
52	1 2 3	3 4 5	6 7 8	9 10 10	11 12 13	14 15 16	16 17 18	19 20 21	22 23 23	24 25 26	27	41 42	42 43 44	45 46 47	48 49 49	50 51 52	52
53	1 2 3	4 4 5	6 7 8	9 10 11	11 12 13	14 15 16	17 18 19	19 20 21	22 23 24	25 26 26	27	42 42	43 44 45	46 47 48	49 49 50	51 52 53	53
54	1 2 3	4 4 5	6 7 8	9 10 11	12 13 14	14 15 16	17 18 19	20 21 22	22 23 24	25 26 27	28	42 43	44 45 46	47 48 49	50 50 51	52 53 54	54
55	1 2 3	4 5 6	6 7 8	9 10 11	12 13 14	15 16 16	17 18 19	20 21 22	23 24 25	26 27 28	28	43 44	45 46 47	48 49 50	50 51 52	53 54 55	55
56	1 2 3	4 5 6	7 7 8	9 10 11	12 13 14	15 16 17	18 19 20	21 21 22	23 24 25	26 27 28	29	44 45	46 47 48	49 49 50	51 52 53	54 55 56	56
57	1 2 3	4 5 6	7 8 9	10 10 11	12 13 14	15 16 17	18 19 20	21 22 23	24 25 26	27 28 28	29	45 46	47 48 48	49 50 51	52 53 54	55 56 57	57
58	1 2 3	4 5 6	7 8 9	10 11 12	13 14 14	15 16 17	18 19 20	21 22 23	24 25 26	27 28 29	30	45 46	47 48 49	50 51 52	53 54 55	56 57 58	58
59	1 2 3	4 5 6	7 8 9	10 11 12	13 14 15	16 17 18	19 20 21	22 23 24	25 26 27	28 29 30	30	46 47	48 49 50	51 52 53	54 55 56	57 58 59	59

Figure 21-6C. *Portion of Table 4, volume 2,* Tables No. 249.

The tables are entered in the column containing tabulations for the whole degree of declination less than the true declination. The tabulated Hc in this column directly across from the appropriate LHA is recorded, as well as the d value and the azimuth angle Z. The Hc thus extracted must then be adjusted to the proper value corresponding to the true declination. To do this, a "declination correction" is found mathematically by multiplying the tabulated d value by the additional minutes of the true declination divided by 60. For the sake of convenience, a table containing precomputed products for this multiplication is included in the back of volumes 2 and 3 as Table 4; a portion of this table appears in Figure 21-6C. The table uses as entering arguments the d value and minutes of the true declination. After the correction is found, it is applied to the tabulated altitude according to the sign of the d value in the main table.

After the Hc corresponding to the true declination is obtained, the azimuth angle Z is then converted to true azimuth Zn by the rules given on each page, which are identical to the rules given in *Tables No. 229* for the same purpose.

In the following two sections of this chapter, the examples begun earlier in the chapter will be completed using volumes 1 and 2 of *Tables No. 249* as ap-

Body	CANOPUS
GMT	17-12-09 Dec15
IC	+.3'
D (Ht 44')	-7'
R	-1'
Total Corr	-8'
hs	50° 46'
Ho	50° 38'
GHA γ (h, 10m)	341° 26.5'
Incre. (m/s)	0° 32.0'
Total GHA γ	341° 58.5'
±360°	
aλ (+E,-W)	163° 01.5' E
LHA γ	145°
a LAT	34 S
Hc	50° 39'
Ho	50° 38'
a	1'A
Zn	226° T
P and N Corr	N/A

Figure 21-7. *Completed Canopus sight form by* Tables No. 249.

propriate. The convenience afforded by volume 1 of *Tables No. 249* in comparison to *Tables No. 229* or volumes 2 and 3 of *Tables No. 249* will soon become apparent to the inexperienced navigator in the course of reading over these two examples.

Completing the Sight Reduction by *Tables No. 249*—First Example

As the first example of the completion of the sight reduction process by *Tables No. 249*, let us return to the Canopus form shown in Figure 21-4B. Since the star Canopus is one of the seven stars tabulated for 34° south latitude (the assumed latitude) and LHAϒ of 145°, volume 1 can be used for direct extraction of the Hc and Zn. From Figure 21-6A, the tabulated Hc for Canopus is 50° 39', and the tabulated Zn is 226°T. It remains only to compare this Hc with the Ho to find the intercept distance a. Since the Ho is 50° 38' and the Hc is 50° 39', the intercept distance is 1 mile away (A) from the assumed position at L 34°S, λ 163° 01.5'E. The completed form appears in Figure 21-7.

Note that when volume 1 of *Tables No. 249* is used, no interpolations of any kind are required. After the LHAϒ and assumed latitude have been determined, a simple table look-up is all that is necessary to find Hc and Zn. The LHA♈ can be found through the use of either the *Nautical* or *Air Almanac*, and the assumed latitude is simply the whole degree of latitude nearest the DR position of the observer at the time of the sight.

Completing the Sight Reduction by *Tables No. 249*— Second Example

As a second and perhaps more complicated example of the use of *Tables No. 249* in sight reduction, let us complete the solution for the moon sight of Figure 21-5B. From the form, the LHA as determined by the *Air Almanac* was 23°, and the declination is N23° 14′. The remaining entering argument for volume 2 of *Tables No. 249*, the assumed latitude, is 34° south.

The assumed latitude is the primary entering argument for both volumes 2 and 3 of *Tables No. 249*. Once the proper latitude pages have been located, the page of interest is selected by referring to the declination. In this case, the latitude 34° page that contains tabulations for declinations of 15°–29° contrary to the name of the latitude and LHAs from 0° to 69° is the page chosen; this page appears in Figure 21-6B. The tabulated values of Hc and Z selected are those contained in the 23° declination column directly opposite the vertical 23° LHA argument. Hc is 28° 57′, and Z is 156°. To find the Hc corresponding with the true declination 23° 14′, the difference d between the aforementioned Hc and the tabulation for the next higher degree of declination is recorded for use with Table 4. In this case the d value is –56′. Using this value as a vertical argument and the 14′ additional declination as a horizontal argument, the declination correction as found from Table 4 of Figure 21-6C is –13′. Thus, the computed Hc corresponding with true declination 23° 14′ is given by the following expression:

$$28° \ 57′ - 13′ = 28° \ 44′$$

No interpolation is necessary for the azimuth angle Z.

To complete the solution, Z is converted to Zn by the rule for south latitudes printed on each page:

S. Lat. $\begin{cases} \text{LHA greater than } 180° \ \ldots\ldots\ldots \ \text{Zn} = 180 - \text{Z} \\ \text{LHA less than } 180° \ \ldots\ldots\ldots \ \text{Zn} = 180 + \text{Z} \end{cases}$

Thus, Zn = 180° + 156° = 336°T. Comparison of Hc with the Ho determined previously, 28° 35′, yields an intercept distance a of 9 miles "away" from the GP of the moon—away, since Hc is greater than Ho. The completed moon sight reduction form appears in Figure 21-8.

Body	MOON (UL)
GMT	16-58-57 Dec15
IC	+.3'
D (Ht 44')	-7'
R	-2'
SD	-15'
P. in A. (Moon)	+49'
Total Corr	+35'
hs	28° 10'
Ho	28° 35'
GHA (h, 10m)	217° 15'
Incre. (m/s)	2° 10'
Total GHA	219° 25'
±360°	
aλ (+E,-W)	163° 35' E
LHA	23°
Tab Dec	N 23° 14'
a LAT Same Contrary	34° S Cont
Dec Inc/d	14'/-56'
Dec Corr	-13'
Hc (Tab)	28° 57'
Hc (Comp)	28° 44'
Ho	28° 35'
a	9'A
Z	S 156° W
Zn	336° T

Figure 21-8. Completed moon sight form by Tables No. 249.

PLOTTING THE CELESTIAL LOP DETERMINED BY THE *AIR ALMANAC* AND *TABLES NO. 249*

In surface navigation, the celestial LOP as determined by the use of the *Air Almanac* and *Tables No. 249* is plotted just as is done for computations based on the *Nautical Almanac* and *Tables No. 229*. The assumed position corresponding to the assumed latitude and longitude chosen for the sight solution is first plotted, then the intercept distance a is laid down along the true azimuth line and the LOP is plotted relative to it. Several LOPs may be combined to form a fix or running fix, or an estimated position may be located on a single

Body	MOON (UL)	VENUS
GMT	16-58-57 Dec15	17-04-12 Dec15
IC	+.3'	+.3'
D (Ht 44')	-7'	-7'
R	-2'	-3'
SD	-15'	
P. in A. (Moon)	+49'	
Total Corr	+35'	-10'
hs	28° 10'	16° 47'
Ho	28° 35'	16° 37'
GHA (h, 10m)	217° 15'	117° 52'
Incre. (m/s)	2° 10'	1° 03'
SHA (star)		
Total GHA	219° 25'	117° 52'
±360°		
aλ (+E,-W)	163° 35' E	163° 08' E
LHA	23°	281°
Tab Dec	N 23° 14'	S 13° 16'
a LAT Same Contrary	34° S Cont	34° S Same
Dec Inc/d	14'/-56'	16'/+32'
Dec Corr	-13'	+9'
Hc (Tab)	28° 57'	16° 15'
Hc (Comp)	28° 44'	16° 24'
Ho	28° 35'	16° 37'
a	9'A	13'T
Z	S 156° W	S 84.7° E
Zn	336° T	095.3° T

Body	CANOPUS
GMT	17-12-09 Dec15
IC	+.3'
D (Ht 44')	-7'
R	-1'
Total Corr	-8'
hs	50° 46'
Ho	50° 38'
GHA γ (h, 10m)	341° 26.5'
Incre. (m/s)	0° 32.0'
Total GHA γ	341° 58.5'
±360°	
aλ (+E,-W)	163° 01.5' E
LHA γ	145°
a LAT	34 S
Hc	50° 39'
Ho	50° 38'
a	1'A
Zn	226° T
P and N Corr	N/A

Figure 21-9. *Complete solutions for the moon, Venus, and Canopus by the* Air Almanac *and* Tables No. 249.

LOP, by methods to be discussed in the following chapter. Figure 21-9 pictures the completed sight forms for Canopus and the moon discussed in the preceding sections, as well as a complete solution by the *Air Almanac* and *Tables No. 249* for the Venus sight of chapter 20.

When volume 1 of *Tables No. 249* is used for sight reduction in a year other than the designated Epoch year, it will frequently be necessary to correct the resulting LOP for the effects of precession and nutation of the earth. These corrections are tabulated in Table 5, located in the back of volume 1, for each year in which they apply during the period 4 years before and 4 years after the Epoch year. A portion of Table 5 containing corrections for 1999 from the Epoch 1995 edition of *Tables No. 249* is reproduced in Figure 21-10. Entering arguments for the table are the nearest tabulated values of LHAγ and latitude. No interpolation is required. If, for instance, the year in which the observations in this chapter were made were 1999, and the Epoch 1995 edition of volume 1 had been used to obtain Hc and Zn, the Canopus LOP would have to be shifted a distance of 2 miles in the direction 120°T.

If volume 1 is exclusively used for all LOPs of a celestial fix, any required precession and nutation correction should be applied to the fix position itself rather than the individual LOPs.

| | North latitudes | | | | | | | 1999 | South latitudes | | | | | | | |
LHA γ	N 89°	N 80°	N 70°	N 60°	N 50°	N 40°	N 20°	0°	S 20°	S 40°	S 50°	S 60°	S 70°	S 80°	S 89°	LHA γ
0	I 010	I 030	2 040	2 050	3 060	3 060	3 070	4 070	3 060	3 060	2 050	2 050	2 040	I 020	I 000	0
30	I 040	I 050	2 060	3 060	3 070	3 070	4 070	3 070	3 070	3 060	2 050	2 040	I 020	I 350	I 330	30
60	I 070	I 070	2 080	3 080	3 080	3 080	4 080	3 080	3 080	2 070	I 060	I 040	I 350	I 310	I 300	60
90	I 100	2 090	2 090	3 090	3 090	3 090	4 090	3 090	3 090	2 090	I 100	0 —	0 —	I 260	I 270	90
120	I 120	2 120	2 110	3 110	3 110	3 100	4 100	3 100	3 110	2 110	I 120	I 140	I 180	I 220	I 230	120
150	I 150	2 140	2 130	3 120	3 120	3 110	4 110	4 110	3 110	2 120	2 130	2 140	I 160	I 180	I 200	150
180	I 180	I 160	2 140	2 130	2 130	3 120	3 120	4 110	3 110	3 120	3 120	2 130	2 140	2 150	I 170	180
210	I 210	I 190	2 160	2 140	2 130	2 120	3 110	3 110	4 110	3 110	3 110	3 120	2 120	2 130	I 140	210
240	I 240	I 230	I 190	I 140	I 120	2 110	3 100	3 100	4 100	3 100	3 100	3 100	2 100	2 110	I 110	240
270	I 270	I 280	0 —	0 —	I 080	2 090	3 090	3 090	4 090	3 090	3 090	3 090	2 090	2 090	I 080	270
300	I 310	I 320	I 000	I 040	I 060	2 070	3 070	3 080	4 080	3 080	3 070	3 070	2 070	2 060	I 060	300
330	I 340	I 000	I 020	2 040	2 050	2 060	3 070	4 070	4 070	3 070	3 060	3 060	2 050	2 040	I 030	330
360	I 010	2 030	2 040	2 050	3 060	3 060	3 070	4 070	3 060	3 060	2 050	2 050	2 040	I 020	I 000	360

Figure 21-10. 1999 Corrections for Precession and Nutation, Table 5, Epoch 1995 edition of volume 1, Tables No. 249.

THE USE OF *TABLES NO. 249* AS A STARFINDER

As was mentioned earlier in this chapter, volume 1 of *Tables No. 249* may be used as a *starfinder* for precalculation of the approximate sextant altitudes and true azimuths of the optimum seven navigational stars for any particular location and time. To use volume 1 in this manner, an LHAϒ and assumed latitude are chosen based on the projected DR track and predicted time of observation. The tables are then opened to the appropriate latitude page, and the names and coordinates of the stars tabulated for the precomputed LHAϒ are extracted and recorded. A sample list of seven stars preselected for observation when the assumed latitude is 34° south and the LHAϒ is 145° (see Figure 21-6A, page 400) appears below.

	Regulus	Spica	Acrux	Canopus	Rigel	Betelgeuse	Procyon
Hc's	43.5°	34.5°	51.5°	50.5°	24.0°	22.5°	41°
Zn's	009°	080°	151°	226°	276°	296°	318°

Chapter 24 discusses a more sophisticated starfinder called the Rude Starfinder, by means of which the coordinates of all possible navigational stars and planets observable from a given projected position may be determined. Volume 1 of *Tables No. 249*, however, is very convenient to use in this mode if a simple 3- or 4-star fix is all that is desired.

SUMMARY

This chapter has examined the altitude-intercept method of the complete solution for one or more celestial LOPs using the *Air Almanac* in conjunction with *Sight Reduction Tables No. 249*. The relative ease of the sight solution using volume 1 of *Tables No. 249* in comparison to volumes 2 and 3 and *Tables No. 229* was stressed, and the secondary use of volume 1 as a starfinder was discussed.

It should be emphasized that the convenience of the *Air Almanac* and *Tables No. 249* is gained at the expense of some precision in the resulting celestial LOP. The precision lost is roughly comparable to the observation errors experienced on a medium-sized ship in rough weather. Until they have gained some proficiency with the sextant, inexperienced navigators should take advantage of the greater precision afforded by the *Nautical Almanac* used in combination with *Tables No. 229*, or values computed with an electronic calculator, to offset any possible observation errors. Later on, the navigator can then switch to the less precise though more rapid *Air Almanac* and *Tables No. 249* in situations that do not demand high precision.

ALTERNATIVE METHODS OF SIGHT REDUCTION AND NAVIGATIONAL COMPUTATIONS

22

The rapid proliferation of the low-cost handheld *electronic calculator* during the last two decades has certainly been one of the most remarkable technological developments of this century. The incorporation of trigonometric functions in most calculators, and the programmable features of many, have resulted in a great deal of renewed interest in sight reduction of celestial observations by formulas derived from the old cosine-haversine trigonometric formulas, as an alternative to the solution by use of sight reduction tables. Any calculator capable of handling trigonometric functions can be used in this endeavor. Programmable models, however, especially those with the capability of storing user-oriented programs on magnetic storage devices, greatly facilitate the calculator solution of the sight reduction problem, as well as a great many other common navigational computations.

Many of the more sophisticated calculators now come equipped with various "applications programs" in the field of navigation. These programs, written by the manufacturers, can be fed into the calculator memory from one of several kinds of external magnetic storage media. After reading the program into the calculator, the operator then keys in the values of the appropriate variables called for by the particular program, pushes the "execute" key, and receives the answer. The two largest calculator manufacturers in the United States, Hewlett-Packard and Texas Instruments, and several other manufacturers offer several models of programmable calculators that come with a supply of navigational applications programs.

In addition to the general purpose programmable calculators that can be set up to solve various specific navigational problems, there are also available several makes of so-called *navigation calculators* that feature permanently built-in navigation programs. The more sophisticated of these are actually microcomputers, with perpetual almanacs, sight reduction algorithms, and other navigational routines permanently resident in their internal memories. They are capable of calculating a single LOP, or a fix from two LOPs, from sights of the sun, moon, or navigational planets and stars, simply by entering into the calculator the sextant IC, the height of eye, the sextant angle, the GMT and the DR position for that time, and an identifying number assigned to the body. The less-sophisticated models will perform the same calculations but require, in addition to the above variables, some appropriate almanac data such as the SHA of the body observed.

Another electronic tool that has enjoyed exponential growth in both quantity and capability over the last few years is the personal *electronic computer,* or PC. While their computational speeds, internal memory, and ease of use have all increased dramatically, their size and price have both decreased apace, to the point where they are now no more expensive than a moderately capable electronic navigation receiver was in the early 1980s. Consequently, today computers are seeing ever-increasing use for navigational purposes on all types and sizes of surface vessels from naval warships and merchant steamships to pleasure craft.

Paralleling the adoption of the computer for navigational purposes in Navy ships has been the development of the capability to integrate and disseminate navigational information throughout the ship via interconnected computer systems to all watchstanding and weapons control stations where it is needed. This capability, called *navigation sensor system interface* (NAVSSI), is currently scheduled for installation in all aircraft carriers, cruisers, destroyers, and amphibious assault ships by the year 2000. Similar systems have already been in use in nuclear submarines for some time.

When high-quality electronic navigation charts (ENC) of the type discussed in chapter 4 become available, NAVSSI systems will be upgraded to display the appropriate DNC as background on navigational displays on the bridge, charthouse, and in CIC, eventually eliminating the need for paper charts. Moreover, once they have been further upgraded to accept input from other sensor systems, such as radar and the fathometer, NAVSSI systems will also eliminate the need for manual logging of much of the data now required to be maintained in such records as the *Ship's Deck Log, Bearing Book, Ship's Position Log,* and *Navigation Workbook.*

The remainder of this chapter will cover the basic formulas used for celestial sight reduction by electronic calculator, and present an overview of some of the more capable navigation application programs now available not only for sight reduction but also for many other navigational computations on the personal computer. For those traditionalists who would still prefer to take pencil in hand to reduce their celestial sights, but are willing to try something a bit newer than the multivolumed DMA sight-reduction tables described earlier in this text, an explanation of the *Nautical Almanac Concise Sight Reduction Tables* is also included.

SOME BASIC FORMULAS FOR CALCULATOR NAVIGATION

Beginning with the 1989 edition, the *Nautical Almanac* contains a section on sight reduction procedures for the use of the handheld electronic calculator. To reduce a celestial sight by means of the calculator, first the observed altitude is derived from the sextant altitude as explained in chapters 19 and 21, except that residual minutes of arc must normally be expressed as decimal fractions of a degree, since most calculators cannot handle arcs expressed in degrees and minutes. Then, the computed altitude Hc and the true azimuth Zn are calculated, using the following steps (reproduced from the *Nautical Almanac*); be sure to use the sign conventions for *Lat* and *Dec* listed in note 4 below:

Step 1. Calculate the local hour angle

$$LHA = GHA + Long$$

Step 2. Calculate the altitude from

$$H_c = \sin^{-1}(\sin Lat \sin Dec + \cos Lat \cos Dec \cos LHA)$$

where $\sin^{-1}$ is the inverse function of sine.

Step 3. Calculate *X*, *Y*, and *A* from

$$X = \cos Lat \sin Dec - \sin Lat \cos Dec \cos LHA$$

$$Y = -\cos Dec \sin LHA$$

$$A = \tan^{-1}(Y/X)$$

where $\tan^{-1}$ is the inverse function of tangent.

Step 4. Determine the azimuth Z:

If $X < 0$ then $Z = A + 180°$

If $X > 0$ and $Y < 0$ then $Z = A + 360°$

Otherwise $Z = A$

Notes

1. In step 1 it is not essential to add or subtract multiples of 360° to *LHA* to set it in the range 0° to 360°.
2. In step 3 to avoid division by zero at the east and west cardinal points, which will produce an overflow in the calculator, add a very small number to the denominator *X* before the division; for example add $1.0 \times 10^{-10} \equiv 1.0E - 10$.

3. For step 4, many calculators have the facility to convert directly from rectangular coordinates (x, y) to polar coordinates (r, θ). Set $x = X$, $y = Y$ and convert to polar coordinates. Then $Z = \theta$.

4. *Sign conventions.* For declination (*Dec*) and latitude (*Lat*), north is positive (+) and south is negative (–). For longitude (*Long*) east is positive (+) and west is negative (–).

Once calculated, Hc can be compared to Ho to find the intercept distance a to the nearest tenth of a mile, and the LOP can be plotted using the intercept method.

If desired, the calculator can also be used to find the various corrections applicable to the sextant altitude hs to obtain Ho:

Step 1. Calculate dip

$$D = 0.0293° \sqrt{h}$$

where h is the height of eye above the horizon (i.e., above sea level) in meters. (Note: Dip thus calculated is in decimal fractions of degrees.)

Step 2. Calculate apparent altitude

$$Ha = hs + I - D$$

where I is the sextant index error.

Step 3. Calculate refraction (R) at a standard temperature of 10° Celsius (C) and pressure of 1010 millibars (mb)

$$R_0 = 0.0167°/\tan[Ha + 7.31/(Ha + 4.4)]$$

If the temperature T °C and pressure P mb are known, calculate the refraction from

$$R = f R_0 \qquad \text{where} \qquad f = 0.28P/T + 273)$$

otherwise set $\qquad R = R_0$

Step 4. Calculate the parallax in altitude (PA) from the horizontal parallax (HP) and the apparent altitude (Ha) for the sun, moon, Venus, and Mars as follows:

$$PA = HP \cos Ha$$

For the sun $HP = 0.0024°$. This correction is very small and could be ignored. For the moon HP is taken for the nearest hour from the main tabular page of the *Nautical Almanac* and converted to degrees.

For Venus and Mars, the *HP* is taken from a critical table in the instructions section of the *Nautical Almanac* and converted to degrees.

For the navigational stars and the remaining planets, Jupiter and Saturn, set *PA* = 0.

If an error of 0.2′ is significant the expression for the parallax in altitude for the Moon should include a small correction *OB* for the oblateness of the Earth as follows:

$$PA = HP \cos Ha + OB$$

where $OB = -0.0032° \sin^2 Lat \cos Ha + 0.0032° \sin (2Lat) \cos Z \sin Ha$

At midlatitudes and for altitudes of the moon below 60°, a simple approximation to *OB* is

$$OB = -0.0017° \cos Ha$$

Step 5. Calculate the semidiameter for the Sun and Moon as follows:

Sun: *S* is taken from the main tabular page of the *Nautical Almanac* and converted to degrees.

Moon: $S = 0.2724° HP$ where *HP* is taken for the nearest hour from the main tabular page and converted to degrees.

Step 4. Calculate the observed altitude

$$Ho = Ha - R + PA \pm S$$

where the plus sign is used if the lower limb of the sun or moon was observed and the minus sign if the upper limb was observed.

A standardized sight reduction form developed by the U.S. Navy NROTC program for use in sight reduction by electronic calculator is included among the various types of sight reduction forms reproduced in appendix D; see Figure D-4. For this form the various corrections to be applied to the sextant altitude can either be calculated as per the steps given above, or obtained from the appropriate tables of the *Nautical* or *Air Almanac*. The Tab GHA and declination data can either be obtained from one of the printed almanacs, or if the navigator is so equipped, from almanac data resident in various PC or calculator navigation applications programs now available.

In the voyage planning chapter of part 1, the methods of determining rhumb line approximations to great-circle routes graphically by use of a gnomonic and Mercator chart projection were presented. Rhumb line approximations to the great circle may also be calculated by use of *Tables No. 229* using the procedure described below.

Enter 229 with:

(1) Latitude of the point of departure as *Lat.*
(2) Latitude of the destination as *Declination.*
(3) Difference in longitude as *LHA.*

If the respondent values of Hc and Z correspond to those of a celestial body *above* the celestial horizon, distance (D) = 90 – Hc. Initial course angle (C) = Z.

If the respondent values correspond to those of a celestial body *below* the celestial horizon (i.e., when the C-S line is crossed), D = 90 + Hc and C = 180 – Z.

Course angle (C) is assigned a prefix N or S depending upon the latitude of the point of departure. C receives the suffix E or W of the general direction of movement. True course is then determined by applying these prefixes/suffixes. For example:

$$\begin{aligned}
\text{N60E} &= 000° + 60°\text{E} = 060°\text{T} \\
\text{N60W} &= 000° - 60°\text{W} = 300°\text{T} \\
\text{S60E} &= 180° - 60°\text{E} = 120°\text{T} \\
\text{S60W} &= 180° + 60°\text{W} = 240°\text{T}
\end{aligned}$$

As in sight reduction, the calculator may be used in lieu of the tables by use of the following formulas:

$$C = \tan^{-1}\left[\frac{\sin \text{DLo}}{(\cos L_1 \tan L_2) - (\sin L_1 \cos \text{DLo})}\right]$$

$$D = \cos^{-1}[(\sin L_1 \sin L_2) + (\cos L_1 \cos L_2 \cos \text{DLo})]$$

Where: L_1 is the latitude of the point of departure and L_2 is the latitude of the destination. DLo is the difference between the longitude of the point of departure and the destination.

NOTE: When the latitude of the destination is named (i.e., north or south) contrary to that of the point of departure, it is treated as a negative quantity.

If C is computed as a negative quantity, add 180°C.
If DLo > 180°, enter it as a negative quantity.

PC NAVIGATION APPLICATION SOFTWARE

As indicated in the introduction to this chapter, the personal computer (PC) is being used by increasing numbers of navigators of all types of naval, commercial, and private vessels to both automate the process of celestial sight reduction and perform many other types of navigational calculations previously done manually with the help of a number of different reference publications required to be carried on board for the purpose.

In regard to sight reduction, the ability of the computer to quickly sequence through the different formulas presented herein for sight reduction by electronic calculator, or other more sophisticated algorithms, in effect replaces the need for some 20 pounds of sight reduction tables such as *DMAHTC Tables 229* or *249*. The addition of almanac and tide and current data either on external accessible media such as floppy disks or CD-ROMs or resident in internal memory (RAM) can replace traditional almanacs such as the *Nautical* and *Air Almanacs*, as well as traditional reference publications such as the *Tide* and *Current Tables*.

The following is a partial survey, by no means complete, of some of the more capable PC navigation software that has been developed to date. Most of it is available to the general public at prices under a few hundred dollars.

STELLA (System to Estimate Latitude and Longitude Astronomically)

Designed by the U.S. Naval Observatory in response to a U.S. Navy requirement, *STELLA* is a recently developed computer program intended for official use only by the U.S. armed services; it is not warranted for civil use. Although *STELLA* can be run with Microsoft Windows in PCs, it is designed to be used under a DOS operating system and must be installed in DOS mode.

STELLA incorporates almanac algorithms for computation of celestial ephemeris data for any year between 1970 and 2000. It can plot DR position and intended tracks and waypoints; compute rising/setting/transit phenomena; compute gyro error; and do sight reduction and star-sight planning.

Interactive Computer Ephemeris (ICE)

A computerized form of the *Nautical Almanac* that has been incorporated by the U.S. Naval Observatory onto floppy disks, the *ICE* can compute ephemeris data for 1,536 visible stars, the sun, the moon, and the four navigational planets, for a period of 250 years. It is available to private users.

MacNavigator™ (for Macintosh)

MacNavigator, written by Stephen Tripp and distributed by Celestaire, Inc., of Wichita, Kansas, is a fully interactive integrated navigation program for

Macintosh PCs. It will compute waypoints for voyage planning on either rhumb line or great-circle courses; compute DR positions, running fixes, and ETAs; and do sight reduction with a perpetual built-in almanac. The sight reduction subroutine features data entry as blanks to be filled in on a sight form, for possible comparison with a manual form. The program plots up to ten lines of position for respective observations, advances or retards them as desired, and can draw a confidence ellipse for the resulting fix.

PC Navigator

PC Navigator, written by Barney Croger and distributed by Celestaire, Inc., is an integrated assembly of routines incorporating positioning by celestial LOPs, piloting, and DR. It does voyage planning by both great circle and rhumb line. Other features include sight reduction, almanac data for all bodies to the year 2100, running fix computation, computation of current set and drift, star identification, and refraction corrections.

CAP'N (Computerized American Practical Navigator)

The *Cap'n* by Nautical Technologies, Ltd., of Bangor, Maine, is one of the more comprehensive integrated navigation programs currently available. It enables a PC to do almost every manual navigation task described in this text. Among its celestial capabilities are graphic plotting of all LOPs, DRs, and fixes, with Hc and Zn displayed for all bodies; computation of times of twilight, rising and setting and meridian transit for the sun, moon, and navigational planets; and display of the equation of time for LAN.

Other capabilities include (1) voyage planning, wherein it will compute the course, distance, and ETAs between waypoints; (2) log-keeping, using a choice of manual DR positions, computerized DR positions, celestial fixes, or GPS (with an optional GPS interface); (3) tide and current predictions for any subordinate station in North, South, or Central America, the Caribbean, Bermuda, and Hawaii (and Australia and New Zealand as an add-on option) for 300 years, in the form of either sine-wave line graphs or bar charts; (4) retrieval of USCG and DMA *Light List* data; and (5) retrieval of *World Port Index* data for some 6,500 ports and *Coastal Pilot* and *Sailing Directions* data for coastal areas throughout the world.

An optional CAP'N Electronic Charting System available as an add-on to the basic *Cap'n* software allows operation with all standard BSB/NOAA electronic charts. The *Cap'n* has been adopted for use at the U.S. Naval Academy, as well as on many ships of the U.S. Navy.

Tides and Currents™ and Chart View™

Tides and Currents™ and *Chart View*™ are two Windows-95–compatible applications programs designed by Nautical Software, Inc., of Beaverton, Oregon, for tide and current predictions and voyage planning. The former depicts tide and current predictions in both color graphic and spreadsheet format for selected days, weeks, or months for any time between the years 1901 and 2100. Coverage is available for either an east region comprising the east and Gulf coasts of North and Central America and the Bahamas, or a west region comprising the west coasts of North and Central America plus Hawaii and Alaska. *Chart View*™ is a voyage-planning program that can project all manner of planning data onto raster-scanned electronic charts such as those produced by BSB/NOAA. It can also be used in conjunction with *Tides and Currents*™ to project tide and current data onto the electronic planning chart presentation. It features different zoom levels plus seamless scrolling onto adjacent charts.

As stated in the introduction to this section, the foregoing is just a sample of the kinds of PC software now available for use in surface navigation. Further advances and interface capabilities will undoubtedly continue to be developed, which will enable even the smallest seagoing vessel to have many of the same capabilities as NAVSSI-equipped Navy ships in the not too distant future.

NAUTICAL ALMANAC CONCISE SIGHT REDUCTION TABLES

Another innovation begun with the 1989 edition of the *Nautical Almanac* is the inclusion of *Concise Sight Reduction Tables* (hereafter referred to as the *Concise Tables*), developed in the early 1980s by Dr. Paul Janiczek of the U.S. Naval Observatory in Washington, D.C. Covering only some 30 pages in contrast to the multivolumed sight reduction *Tables No. 229 and 249*, the compact arrangement results from dividing the navigational triangle into two right spherical triangles, as shown in Figure 22-1.

The disadvantages of the compact format are that the tables have to be entered twice to solve each resulting triangle, and the Hc obtained is precise only to the nearest minute of arc, with errors of up to 2 minutes of arc possible. True azimuths found by use of the *Concise Tables* are precise only to the nearest whole degree. And, the use of the tables to find Hc and Zn is somewhat more complicated than *Tables No. 229 or 249*. Despite these shortcomings, their small size and inclusion within the *Nautical Almanac* make the *Concise Tables* very convenient for many navigators, especially those of smaller, more unstable craft unable to take advantage of the increased precision offered by the larger sight reduction tables or navigators not having access to a trigonometric calculator or computer.

Completing the Sight Reduction by the *Concise Tables*

As mentioned above, in order to obtain the computed altitude Hc and true azimuth Zn by use of the *Concise Tables*, they must be entered twice, to solve

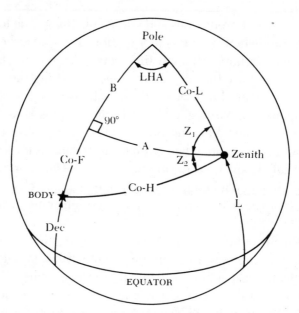

Figure 22-1. *The* Nautical Almanac Concise Sight Reduction Tables *are based upon division of the navigational triangle into two right spherical triangles. Labels for the sides coincide with the entering arguments for the Concise Tables.*

both sides of the split navigational triangle shown in Figure 22-1. The first entering arguments are integral values of LHA and the assumed latitude, both of which are computed just as is done for using *Tables No. 229* and *249*. Each pair of pages in the *Concise Tables* covers 6° of assumed latitudes across the top margins of the tables; left-hand pages are used for LHAs between 0° and 45° or 135° and 180° (left margins of the tables) and 180° to 225° or 315° to 360° (right margins), while right-hand pages are for LHAs in the remaining intervals. See Figures 22-2A and B for sample page-layouts. With the assumed latitude as a vertical argument and the LHA as a horizontal argument, the intermediate angular quantities A, B, and Z_1 are obtained and recorded to the nearest minute from the appropriate column (see Figures 22-2A and B). Then, an intermediate angle F is formed by taking the algebraic sum of the declination of the body (found in the usual manner from the daily pages of the *Nautical Almanac*) and angle B; for purposes of this addition, the declination is negative if the name of the declination (N or S) is contrary to latitude.

Next, the tables are entered a second time, this time using as entering arguments the A-angle in place of latitude as the vertical argument across the top of the tables, and the F-angle in place of LHA as the horizontal argument on the sides of the tables (only the whole degree part of these angles is used as the entering arguments). With these arguments, the appropriate values of the angular quantities H, P, and Z_2 are obtained and recorded to the nearest minute.

As the final step, Hc is calculated from the H-angle by adding to it two cor-

SIGHT REDUCTION TABLE

B: (−) for 90° < LHA < 270°
Dec: (−) for Lat. contrary name

Z₁: same sign as B
Z₂: (−) for F > 90°

Lat./A LHA/F	30° A/H	30° B/P	30° Z₁/Z₂	31° A/H	31° B/P	31° Z₁/Z₂	32° A/H	32° B/P	32° Z₁/Z₂	33° A/H	33° B/P	33° Z₁/Z₂	34° A/H	34° B/P	34° Z₁/Z₂	35° A/H	35° B/P	35° Z₁/Z₂	Lat./A LHA
0	0 00	60 00	90.0	0 00	59 00	90.0	0 00	58 00	90.0	0 00	57 00	90.0	0 00	56 00	90.0	0 00	55 00	90.0	180
1	0 52	60 00	89.5	0 51	59 00	89.5	0 51	58 00	89.5	0 50	57 00	89.5	0 50	56 00	89.4	0 49	55 00	89.4	179
2	1 44	59 59	89.0	1 43	59 00	89.0	1 42	58 00	88.9	1 41	57 00	88.9	1 39	55 59	88.9	1 38	54 59	88.9	178
3	2 36	59 58	88.5	2 34	58 59	88.5	2 33	57 58	88.4	2 31	56 58	88.4	2 29	55 58	88.3	2 27	54 58	88.3	177
4	3 28	59 56	88.0	3 26	58 56	87.9	3 23	57 56	87.9	3 21	56 56	87.8	3 19	55 56	87.8	3 17	54 56	87.7	176
5	4 20	59 54	87.5	4 17	58 54	87.4	4 14	57 54	87.3	4 12	56 54	87.3	4 09	55 54	87.2	4 06	54 54	87.1	175
6	5 12	59 52	87.0	5 08	58 52	86.9	5 05	57 52	86.8	5 02	56 50	86.7	4 58	55 51	86.6	4 55	54 51	86.6	174
7	6 04	59 48	86.5	6 00	58 48	86.4	5 56	57 48	86.3	5 52	56 48	86.2	5 48	55 48	86.1	5 44	54 48	86.0	173
8	6 55	59 45	86.0	6 51	58 45	85.9	6 47	57 45	85.7	6 42	56 45	85.6	6 38	55 44	85.5	6 33	54 44	85.4	172
9	7 47	59 42	85.5	7 42	58 41	85.3	7 37	57 41	85.2	7 32	56 40	85.1	7 27	55 40	84.9	7 22	54 40	84.8	171
10	8 39	59 37	85.0	8 34	58 37	84.8	8 28	57 36	84.6	8 22	56 36	84.5	8 17	55 36	84.4	8 11	54 34	84.2	170
11	9 31	59 32	84.4	9 25	58 32	84.3	9 19	57 31	84.1	9 13	56 31	84.0	9 06	55 30	83.8	9 00	54 30	83.6	169
12	10 22	59 27	83.9	10 16	58 26	83.8	10 09	57 26	83.6	10 03	56 25	83.4	9 56	55 25	83.2	9 48	54 24	83.0	168
13	11 14	59 21	83.4	11 07	58 20	83.2	11 00	57 20	83.0	10 52	56 19	82.8	10 45	55 18	82.6	10 37	54 18	82.5	167
14	12 06	59 15	82.9	11 58	58 14	82.7	11 50	57 13	82.5	11 42	56 12	82.3	11 34	55 12	82.1	11 26	54 11	81.9	166
15	12 57	59 08	82.4	12 49	58 07	82.1	12 41	57 06	81.9	12 32	56 05	81.7	12 23	55 04	81.5	12 14	54 04	81.3	165
16	13 49	59 01	81.8	13 40	57 59	81.6	13 31	56 58	81.4	13 22	55 57	81.1	13 13	54 57	80.9	13 03	53 56	80.7	164
17	14 40	58 53	81.3	14 31	57 51	81.1	14 21	56 50	80.8	14 12	55 49	80.5	14 02	54 48	80.3	13 51	53 47	80.1	163
18	15 31	58 44	80.8	15 22	57 43	80.5	15 12	56 42	80.2	15 01	55 40	80.0	14 51	54 39	79.7	14 40	53 38	79.4	162
19	16 23	58 35	80.2	16 12	57 34	79.9	16 02	56 33	79.6	15 51	55 31	79.4	15 40	54 30	79.1	15 28	53 29	78.8	161
20	17 14	58 26	79.7	17 03	57 24	79.4	16 52	56 23	79.1	16 40	55 21	78.8	16 28	54 20	78.5	16 16	53 19	78.2	160
21	18 05	58 16	79.1	17 53	57 14	78.8	17 42	56 13	78.5	17 29	55 11	78.2	17 17	54 09	77.9	17 04	53 08	77.6	159
22	18 56	58 05	78.6	18 44	57 03	78.2	18 31	56 01	77.9	18 19	55 00	77.6	18 06	53 58	77.3	17 52	52 56	77.0	158
23	19 47	57 54	78.0	19 34	56 52	77.7	19 21	55 50	77.3	19 08	54 48	77.0	18 54	53 46	76.6	18 40	52 44	76.3	157
24	20 37	57 42	77.4	20 24	56 40	77.1	20 11	55 38	76.7	19 57	54 36	76.4	19 42	53 34	76.0	19 28	52 32	75.7	156
25	21 28	57 30	76.9	21 14	56 27	76.5	21 00	55 25	76.1	20 46	54 23	75.7	20 31	53 21	75.4	20 15	52 19	75.0	155
26	22 19	57 17	76.3	22 04	56 14	75.9	21 49	55 12	75.5	21 34	54 09	75.1	21 19	53 07	74.7	21 03	52 05	74.4	154
27	23 09	57 03	75.7	22 54	56 00	75.3	22 39	54 57	74.9	22 23	53 55	74.5	22 07	52 52	74.1	21 50	51 51	73.7	153
28	23 59	56 49	75.1	23 44	55 46	74.7	23 28	54 43	74.3	23 11	53 40	73.8	22 54	52 37	73.4	22 37	51 35	73.0	152
29	24 50	56 34	74.5	24 33	55 31	74.1	24 17	54 27	73.6	23 59	53 24	73.2	23 42	52 22	72.8	23 24	51 19	72.4	151
30	25 40	56 19	73.9	25 23	55 15	73.4	25 05	54 11	73.0	24 48	53 08	72.5	24 29	52 05	72.1	24 11	51 03	71.7	150
31	26 29	56 02	73.3	26 12	54 58	72.8	25 54	53 54	72.3	25 35	52 51	71.9	25 17	51 48	71.4	24 57	50 45	71.0	149
32	27 19	55 45	72.6	27 01	54 41	72.2	26 42	53 37	71.6	26 23	52 33	71.2	26 04	51 30	70.7	25 44	50 27	70.3	148
33	28 09	55 27	72.0	27 50	54 23	71.5	27 31	53 19	71.0	27 11	52 15	70.5	26 50	51 12	70.0	26 30	50 08	69.6	147
34	28 58	55 09	71.4	28 38	54 04	70.9	28 19	53 00	70.3	27 58	51 56	69.8	27 37	50 52	69.3	27 16	49 49	68.8	146
35	29 47	54 49	70.7	29 27	53 44	70.2	29 06	52 40	69.6	28 45	51 36	69.1	28 24	50 32	68.6	28 01	49 29	68.1	145
36	30 36	54 29	70.0	30 15	53 24	69.5	29 54	52 19	68.9	29 32	51 15	68.4	29 10	50 11	67.9	28 47	49 07	67.4	144
37	31 25	54 08	69.4	31 03	53 03	68.8	30 41	51 58	68.2	30 19	50 53	67.7	29 56	49 49	67.2	29 32	48 45	66.6	143
38	32 13	53 46	68.7	31 51	52 40	68.1	31 28	51 35	67.5	31 05	50 30	66.9	30 41	49 26	66.4	30 17	48 23	65.9	142
39	33 02	53 23	68.0	32 39	52 17	67.4	32 15	51 12	66.8	31 51	50 07	66.2	31 27	49 03	65.6	31 02	47 59	65.1	141
40	33 50	53 00	67.2	33 26	51 53	66.6	33 02	50 48	66.0	32 37	49 43	65.4	32 12	48 38	64.9	31 46	47 34	64.3	140
41	34 37	52 35	66.5	34 13	51 29	65.9	33 48	50 23	65.3	33 23	49 17	64.7	32 57	48 13	64.1	32 30	47 09	63.5	139
42	35 25	52 09	65.8	35 00	51 03	65.1	34 34	49 56	64.5	34 08	48 51	63.9	33 42	47 46	63.3	33 14	46 42	62.7	138
43	36 12	51 43	65.0	35 46	50 36	64.3	35 20	49 29	63.7	34 53	48 24	63.1	34 26	47 19	62.5	33 58	46 15	61.9	137
44	36 59	51 15	64.2	36 33	50 08	63.6	36 06	49 01	62.9	35 38	47 55	62.3	35 10	46 51	61.6	34 41	45 46	61.0	136
45	37 46	50 46	63.4	37 19	49 39	62.7	36 51	48 32	62.1	36 22	47 26	61.4	35 53	46 21	60.8	35 24	45 17	60.2	135

Figure 22-2A. *Sample left-hand page from* Nautical Almanac Concise Sight Reduction Tables.

Lat./A →		30°			31°			32°			33°			34°			35°			Lat./A	
LHA/F		A/H	B/P	Z1/Z2	A/H	B/P	Z1/Z2	A/H	B/P	Z1/Z2	A/H	B/P	Z1/Z2	A/H	B/P	Z1/Z2	A/H	B/P	Z1/Z2	LHA	
45	135	37 46	50 46	63.4	37 19	49 39	62.7	36 51	48 32	62.1	36 22	47 26	61.4	35 53	46 21	60.8	35 24	45 17	60.2	225	315
46	134	38 32	50 16	62.6	38 04	49 08	61.9	37 36	48 02	61.2	37 06	46 56	60.6	36 37	45 51	59.9	36 06	44 46	59.3	226	314
47	133	39 18	49 45	61.8	38 49	48 37	61.1	38 20	47 30	60.4	37 50	46 24	59.7	37 19	45 19	59.1	36 48	44 15	58.4	227	313
48	132	40 04	49 13	61.0	39 34	48 05	60.2	39 04	46 58	59.5	38 33	45 51	58.8	38 02	44 46	58.2	37 30	43 42	57.5	228	312
49	131	40 49	48 39	60.1	40 19	47 31	59.4	39 48	46 24	58.6	39 16	45 18	58.0	38 44	44 12	57.3	38 11	43 08	56.6	229	311
50	130	41 34	48 04	59.2	41 03	46 56	58.5	40 31	45 49	57.7	39 59	44 42	57.0	39 26	43 37	56.3	38 52	42 33	55.6	230	310
51	129	42 18	47 28	58.3	41 46	46 20	57.5	41 14	45 12	56.8	40 41	44 06	56.1	40 07	43 01	55.4	39 32	41 57	54.7	231	309
52	128	43 02	46 50	57.4	42 29	45 42	56.6	41 56	44 34	55.9	41 22	43 28	55.1	40 47	42 23	54.4	40 12	41 19	53.7	232	308
53	127	43 46	46 11	56.4	43 12	45 03	55.6	42 38	43 55	54.9	42 03	42 49	54.1	41 28	41 44	53.4	40 52	40 41	52.7	233	307
54	126	44 29	45 31	55.5	43 54	44 22	54.7	43 19	43 15	53.9	42 44	42 09	53.1	42 07	41 04	52.4	41 30	40 01	51.7	234	306
55	125	45 11	44 49	54.5	44 36	43 40	53.7	44 00	42 33	52.9	43 24	41 27	52.1	42 46	40 23	51.4	42 09	39 19	50.7	235	305
56	124	45 53	44 05	53.5	45 17	42 57	52.6	44 40	41 50	51.8	44 03	40 44	51.1	43 25	39 40	50.3	42 46	38 37	49.6	236	304
57	123	46 35	43 20	52.4	45 58	42 11	51.6	45 20	41 05	50.8	44 42	39 59	50.0	44 03	38 55	49.3	43 24	37 53	48.5	237	303
58	122	47 16	42 33	51.3	46 38	41 25	50.5	45 59	40 18	49.7	45 20	39 13	48.9	44 40	38 09	48.2	44 00	37 07	47.5	238	302
59	121	47 56	41 44	50.2	47 17	40 36	49.4	46 38	39 30	48.6	45 58	38 25	47.8	45 17	37 22	47.1	44 36	36 20	46.3	239	301
60	120	48 35	40 54	49.1	47 56	39 46	48.3	47 16	38 40	47.5	46 35	37 36	46.7	45 54	36 33	45.9	45 11	35 32	45.2	240	300
61	119	49 14	40 01	47.9	48 34	38 54	47.1	47 53	37 48	46.3	47 11	36 45	45.5	46 29	35 42	44.7	45 46	34 42	44.0	241	299
62	118	49 53	39 07	46.8	49 11	38 00	45.9	48 29	36 55	45.1	47 46	35 52	44.3	47 03	34 50	43.6	46 19	33 50	42.8	242	298
63	117	50 30	38 11	45.5	49 48	37 04	44.7	49 05	36 00	43.9	48 21	34 57	43.1	47 37	33 57	42.3	46 53	32 57	41.6	243	297
64	116	51 07	37 13	44.3	50 23	36 07	43.4	49 40	35 03	42.6	48 55	34 01	41.8	48 10	33 01	41.1	47 25	32 03	40.4	244	296
65	115	51 43	36 12	43.0	50 58	35 07	42.2	50 14	34 04	41.3	49 28	33 03	40.6	48 43	32 04	39.8	47 56	31 07	39.1	245	295
66	114	52 18	35 10	41.7	51 33	34 06	40.8	50 47	33 04	40.0	50 01	32 04	39.3	49 14	31 05	38.5	48 27	30 09	37.8	246	294
67	113	52 52	34 05	40.3	52 06	33 01	39.5	51 19	32 01	38.7	50 32	31 02	37.9	49 44	30 05	37.2	48 56	29 10	36.5	247	293
68	112	53 25	32 59	38.9	52 38	31 56	38.1	51 50	30 57	37.3	51 02	29 59	36.6	50 14	29 03	35.8	49 25	28 09	35.2	248	292
69	111	53 57	31 50	37.5	53 09	30 49	36.7	52 21	29 50	35.9	51 32	28 53	35.2	50 43	27 59	34.5	49 53	27 06	33.8	249	291
70	110	54 28	30 39	36.1	53 39	29 39	35.2	52 50	28 42	34.5	52 00	27 46	33.8	51 10	26 53	33.1	50 20	26 02	32.4	250	290
71	109	54 58	29 25	34.6	54 08	28 27	33.8	53 18	27 31	33.0	52 28	26 38	32.3	51 37	25 46	31.6	50 46	24 56	31.0	251	289
72	108	55 27	28 09	33.0	54 37	27 13	32.2	53 46	26 19	31.5	52 54	25 27	30.8	52 03	24 37	30.2	51 10	23 49	29.5	252	288
73	107	55 55	26 51	31.4	55 03	25 57	30.7	54 12	25 04	30.0	53 19	24 14	29.3	52 27	23 26	28.7	51 34	22 40	28.1	253	287
74	106	56 21	25 31	29.8	55 29	24 39	29.1	54 36	23 48	28.4	53 43	23 00	27.8	52 50	22 14	27.1	51 57	21 29	26.6	254	286
75	105	56 46	24 09	28.2	55 53	23 18	27.5	55 00	22 30	26.8	54 06	21 44	26.2	53 12	21 00	25.6	52 18	20 17	25.0	255	285
76	104	57 10	22 44	26.5	56 16	21 56	25.8	55 23	21 10	25.2	54 28	20 26	24.6	53 33	19 44	24.0	52 38	19 04	23.5	256	284
77	103	57 33	21 17	24.8	56 38	20 31	24.1	55 44	19 48	23.5	54 48	19 06	23.0	53 53	18 27	22.4	52 57	17 49	21.9	257	283
78	102	57 54	19 48	23.0	56 59	19 05	22.4	56 03	18 24	21.9	55 07	17 45	21.3	54 11	17 08	20.8	53 15	16 32	20.3	258	282
79	101	58 13	18 17	21.2	57 17	17 37	20.7	56 21	16 59	20.1	55 25	16 22	19.6	54 28	15 48	19.2	53 31	15 15	18.7	259	281
80	100	58 32	16 44	19.4	57 35	16 07	18.9	56 38	15 32	18.4	55 41	14 58	17.9	54 44	14 26	17.5	53 47	13 56	17.1	260	280
81	99	58 48	15 10	17.6	57 51	14 36	17.1	56 53	14 03	16.6	55 56	13 33	16.2	54 58	13 03	15.8	54 00	12 36	15.4	261	279
82	98	59 03	13 33	15.7	58 05	13 02	15.3	57 07	12 33	14.9	56 09	12 06	14.5	55 11	11 40	14.1	54 13	11 14	13.8	262	278
83	97	59 16	11 55	13.8	58 18	11 28	13.4	57 19	11 02	13.0	56 21	10 38	12.7	55 22	10 14	12.4	54 24	9 52	12.1	263	277
84	96	59 28	10 16	11.9	58 29	9 52	11.5	57 30	9 30	11.2	56 31	9 09	10.9	55 32	8 49	10.6	54 33	8 29	10.4	264	276
85	95	59 37	8 35	10.0	58 38	8 15	9.6	57 39	7 56	9.4	56 40	7 39	9.1	55 41	7 22	8.9	54 41	7 06	8.7	265	275
86	94	59 46	6 53	8.0	58 46	6 37	7.7	57 47	6 22	7.5	56 47	6 08	7.3	55 48	5 54	7.1	54 48	5 41	7.0	266	274
87	93	59 52	5 11	6.0	58 52	4 59	5.8	57 52	4 47	5.6	56 53	4 36	5.5	55 53	4 26	5.4	54 53	4 16	5.2	267	273
88	92	59 56	3 28	4.0	58 57	3 19	3.9	57 57	3 12	3.8	56 57	3 05	3.7	55 57	2 58	3.6	54 57	2 51	3.5	268	272
89	91	59 59	1 44	2.0	58 59	1 40	1.9	57 59	1 36	1.9	56 59	1 32	1.8	55 59	1 29	1.8	54 59	1 26	1.7	269	271
90	90	60 00	0 00	0.0	59 00	0 00	0.0	58 00	0 00	0.0	57 00	0 00	0.0	56 00	0 00	0.0	55 00	0 00	0.0	270	270

N. Lat.: for LHA > 180°.... Zn = Z
for LHA < 180°.... Zn = 360° - Z

S. Lat.: for LHA > 180°.... Zn = 180° - Z
for LHA < 180°.... Zn = 180° + Z

Right-side column header relations: Z1 = 180° - Z; Z2 = 180° + Z

Figure 22-2B. *Sample right-hand page from* Nautical Almanac Concise Sight Reduction Tables.

rections from an auxiliary table included for the purpose at the back of the *Concise Tables*; this auxiliary table uses as entering arguments the minutes portion of the F-angle and the P-angle (first correction), and the minutes portion of the A-angle and the Z_2-angle (second correction). The azimuth angle Z is then formed by the algebraic sum of the Z_1 and Z_2 angles (for this purpose Z_2 is negative if F > 90°), and converted to true azimuth Zn by the same rules used for *Tables No. 229* and *249*.

More precise explanations of each step, and more detailed instructions for use of the auxiliary table, are contained in the *Nautical Almanac* in the instructions pages immediately preceding the *Concise Tables*, along with examples. The navigator should refer to these before attempting to reduce any actual sights by use of the *Concise Tables*. The foregoing explanation, however, should be sufficient to give the student a basic understanding of the format and use of the *Concise Tables* as an alternative to *Tables No. 229* and *249* discussed in chapters 20 and 21.

SUMMARY

This chapter has described alternative methods for the solution of the navigational triangle by the handheld electronic calculator, the computer, and by the use of the *Concise Sight Reduction Tables* in the *Nautical Almanac*. The widespread availability of the calculator and the personal computer makes the former methods attractive to many navigators today, while the small size and ease of stowage of the *Concise Tables* makes them a handy alternative to the more traditional but larger *Tables No. 229* and *249*, especially for navigators of small craft.

THE CELESTIAL FIX AND RUNNING FIX

23

The preceding chapters have dealt with the manual solution of the navigational triangle to obtain a celestial LOP by the altitude-intercept method, by use of either the *Nautical Almanac* or *Air Almanac* in conjunction with either the *Sight Reduction Tables for Marine Navigation, No. 229,* or *Air Navigation, No. 249.* In this chapter, the combination of several such LOPs to achieve the ultimate objective of celestial navigation—the celestial fix or running fix—will be addressed.

In celestial navigation, as in piloting, it should be emphasized that neatness and proper labeling standards are essential to accuracy in the navigation plot. It has been shown that the assumed position is based on the DR position at the time of the observation. If the DR position is greatly in error because of inaccuracies in the DR plot, the assumed position chosen may be so far distant from the ship's actual position as to cause the intercept distance to be excessive. This in turn can lead to an error of sizable proportions in the plotted celestial LOP, unless the assumed position is relocated to reduce the intercept distance. In circumstances requiring the ship's position to be speedily determined, the delay thus caused could be unacceptable; hence the importance of an accurate DR plot for celestial navigation.

THE CELESTIAL FIX

In chapter 17 the altitude-intercept method of plotting a single celestial LOP given the assumed position, intercept distance, and true azimuth was demonstrated (see Figure 17-5). It was stated that two or preferably three or more simultaneous celestial LOPs could be crossed to determine the celestial fix position of the observer at the time of the observation.

In practice, sextant observations of the altitudes of celestial bodies are normally not made simultaneously, but rather over an interval of several minutes.

Occasionally, as much as a half hour may elapse between the first and last observations during a round of sextant observations, especially when unfavorable weather conditions prevail. Because the fix is formed by the intersection of two or more *simultaneous* LOPs, it is usually necessary to adjust the positions of all but one of the celestial LOPs resulting from a round of observations, to account for the moving ship's progress during the interval in which the observations were made.

The technique of adjusting the positions of the various LOPs to form a *celestial fix* is similar to that used during piloting to plot a running fix, with one important difference. Whereas in piloting an earlier LOP is always advanced to the time of a later LOP to form the running fix, in celestial navigation LOPs can either be *advanced* or *retired* up to 30 minutes to form the celestial fix. If two celestial LOPs were plotted 20 minutes apart, for example, the first LOP could be advanced 20 minutes to be crossed with the second LOP, or the second LOP could be retired 20 minutes to be crossed with the first LOP. Since the accuracy of a celestial fix is usually considered acceptable if it is within 1 to 2 miles from the true position, advancing or retiring a celestial LOP for up to 30 minutes in this manner does not unduly affect the accuracy of the resultant fix. Advancement of a celestial LOP for periods in excess of a half hour results in a *celestial running fix*, which will be discussed later in this chapter.

Although strictly speaking only two intersecting simultaneous celestial LOPs are required for a celestial fix, in the normal practice of celestial navigation as in piloting at least three or more LOPs having a good spread in azimuth are usually desired. The additional LOPs act as a check on the accuracy of the resulting position.

PLOTTING THE CELESTIAL FIX

The choice of whether to advance or retire celestial LOPs to form the celestial fix mainly depends on the preference of the individual navigator. It should be noted, however, that advancing the LOPs results in the advantage of a later time of fix than would be the case were subsequent LOPs retired to the time of an earlier LOP. For this reason, and also because it has been found that most students grasp the concept of advancing an LOP more readily than retiring one, the former technique will be the one presented herein.

While it would be possible to first plot all the celestial LOPs obtained during a round of observations and then advance certain of them to form the celestial fix, in practice only the APs of the LOPs to be advanced are plotted and advanced. This enhances the neatness of the plot by eliminating unnecessary lines. After the APs have been advanced as necessary, the corresponding LOPs are then plotted relative to them in the usual manner.

Body	MOON (UL)	VENUS	CANOPUS
GMT	16-58-57 Dec16	17-04-12 Dec16	17-12-09 Dec16
IC	+.3'	+.3'	+.3'
D (Ht 44')	-6.4'	-6.4'	-6.4'
Sum	-6.1'	-6.1'	-6.1'
hs	28° 09.6'	16° 47.4'	50° 46.3'
ha	28° 03.5'	16° 41.3'	50° 40.2'
Alt Corr	59.7'	-3.2'	-.8'
Add'l Corr Moon HP/Corr	55.4'/ -27.8	+.5'	
Ho	28° 35.4'	16° 38.6'	50° 39.4'
GHA (h)	205° 10.3'	116° 48.5'	338° 56.1'
Incre. (m/s)	14° 04.0'	1° 03.0'	3° 02.7'
v/v Corr SHA	10.9'/ 10.6'	1.1'/ 0.1'	264° 09.8'
Total GHA	219° 24.9'	117° 51.6'	606° 08.6'
±360°			246° 08.6'
aλ (+E,-W)	163° 35.1'E	163° 08.4'E	162° 51.4'E
LHA	23°	281°	49°
Tab Dec	N 23° 21.1'	S 13° 16.1'	S 52° 40.6'
d#/d Corr	-8.3'/ -8.1'	0.1'/ 0	
True Dec	N 23° 13.0'	S 13° 16.1'	S 52° 40.6'
a LAT Same Contrary	34°S Same	34°S Same	34°S Same
Dec Inc/d	13.0'/-55.8'	16.1'/0.1'	40.6'/-8.7'
Tens/DSD	-10.8'	8.0'	0
Units/DSD Corr	-1.3'	.5'	-5.9'
Total Corr	-12.1'	8.5'	-5.9'
Hc (Tab)	28° 56.7'	16° 15.3'	50° 51.1'
Hc (Comp)	28° 44.6'	16° 23.8'	50° 45.2'
Ho	28° 35.4'	16° 38.6'	50° 39.4'
a	9.2'A	14.8'A	5.8'A
Z	S 155.8° W	S 84.8° E	S 46.3° W
Zn	335.8° T	095.2° T	226.3° T

Figure 23-1. *Complete solutions for the moon, Venus, and Canopus by* Tables No. 229.

As an example of the plot of a celestial fix, consider the three observations used as examples in chapters 20 and 21. The complete solutions by *Tables No. 229* for each of these bodies—the star Canopus, the planet Venus, and the moon—are shown on a single sight reduction form as they would appear in practice in Figure 23-1.

Since the moon and Venus observations were made first, at zone times 0359 (1659 GMT) and 0404 (1704 GMT), respectively (rounded to the nearest minute), their APs will be advanced to the time of the Canopus observation, 0412 (1712 GMT). For purposes of plotting the celestial fix, the times of observation are always rounded to the nearest minute in this manner, with no loss of accuracy resulting. The assumed position for the moon, because it was the first body observed, is designated AP_1; it must be advanced 13 minutes from 0359 to 0412. The assumed position for Venus, AP_2, must be advanced 8 minutes from 0404 to 0412, while the assumed position for Canopus, AP_3, need not be advanced at all. The most convenient method of advancing AP_1 and AP_2 is to use the DR plot constructed earlier to find the DR positions of the ship at the various times of observation, for use during the sight reduction process. It will be assumed here that such a plot has been constructed on a properly labeled SAPS-35, oriented for southern latitudes as shown in Figure 23-2 below.

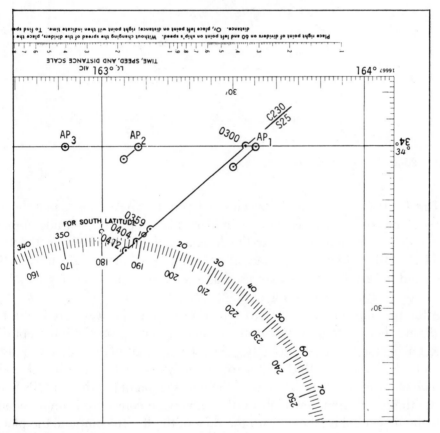

Figure 23-2. *Advancing AP_1 and AP_2 using the DR plot.*

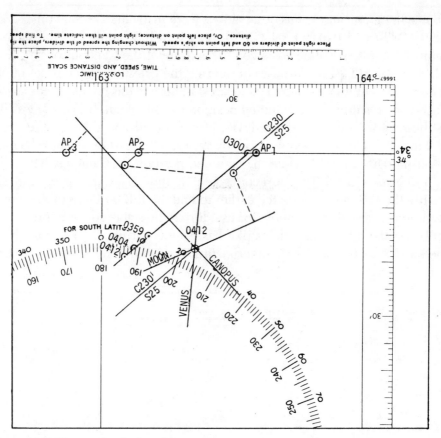

Figure 23-3. *The completed celestial fix.*

The first step in the plot of the celestial fix on the SAPS-35 is to plot the initial positions of all APs based on the assumed latitude and longitude entered on the sight reduction form. Next, the distance and direction between the 0359 and 0412 DR positions are picked off and used to advance AP$_1$. Similarly, the 0404 and 0412 DR positions are used as a reference for advancing AP$_2$. The advanced APs are shown in Figure 23-2.

After the appropriate APs have been advanced, the associated celestial LOPs are then plotted in the normal manner by laying off the intercept distances along the true azimuths either toward or away from the APs as required. Each LOP is labeled with the name of the body upon which it is based, and a ⅛-inch diameter circle is drawn over the intersection of the three LOPs to indicate the fix position. Since all LOPs can now be considered simultaneous, only the fix symbol is labeled with the time of the fix. To complete the plot, a new DR course and speed line is originated from the fix and properly labeled with the ordered course and speed. The completed plot is shown in Figure 23-3 above.

After gaining some proficiency in the techniques of celestial navigation, the navigator will often plot only a single DR position corresponding to the time of one of the observations obtained during a given round of sights, which will then be used to determine the assumed position for all sights reduced on that occasion. In this situation the DR positions corresponding to the time of each sight are not plotted to use as a basis for advancing or retiring the appropriate APs. Instead, the navigator simply makes use of the ship's deck logbook to obtain the ship's ordered course and speed during the observation period, which are then used as a basis for adjusting the appropriate APs.

If a reasonable-sized triangle results rather than a point fix when three celestial LOPs are plotted, or if a polygon is formed by four or more LOPs, the fix position can be assumed to lie approximately at the center of the triangle or polygon if the LOPs have been obtained from celestial bodies well distributed in azimuth. On the other hand, if for some reason such a triangle or polygon is formed by LOPs obtained from bodies all within 180° of azimuth, as might occur when observations are made under cloudy or overcast skies, the fix position may lie *outside* the area enclosed by the LOPs.

In situations in which three or more observations are made of bodies within 180° of azimuth, it is wise to use so-called LOP *bisectors* to determine the fix. In this technique, each angle formed by a pair of LOPs is bisected, with the bisector drawn from the vertex of the angle in the direction of the mean of the azimuths of the two bodies. As an example, consider Figure 23-4, in which LOPs corresponding to three celestial bodies having true azimuths of 224°, 260°, and 340° have been plotted. The bisectors of each of the pairs of lines are computed below:

$$\frac{(224° + 260°)}{2} = 242°$$

$$\frac{(260° + 340°)}{2} = 300°$$

$$\frac{(224° + 340°)}{2} = 282°$$

They appear as blue lines in Figure 23-4. The most probable position for the fix lies at the intersection of the three bisectors rather than within the rather large triangle formed by the LOPs themselves.

If a triangle similar to that appearing in Figure 23-4 results because of a constant observation error, as for example an error caused by a hazy horizon, or a constant math error during sight reduction, a point fix may be obtained by adding or subtracting a uniform correction determined by trial and error to all intercept distances.

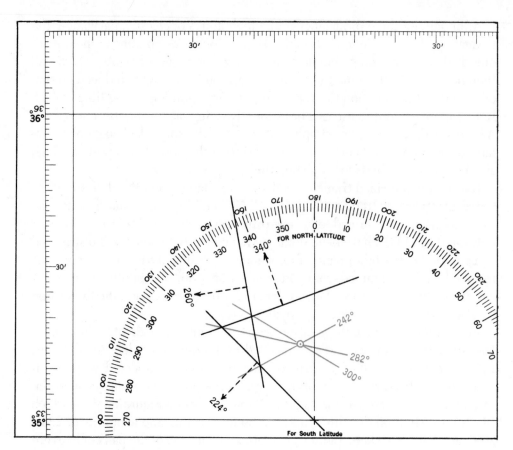

Figure 23-4. *Finding an exterior fix position using bisectors.*

If one LOP plots a considerable distance from all others, misidentification of the celestial body may have occurred, or perhaps a one-time math error during sight reduction. It is for this reason that four or more celestial LOPs are generally preferred in celestial position-fixing, so that those obviously in error can be either corrected or disregarded.

THE ESTIMATED POSITION BY CELESTIAL NAVIGATION

If an appreciable time has elapsed since the determination of the last fix of the ship's position at sea, the cumulative error inherent in the DR plot may build up to the point where the ship's actual position is a good distance away from her DR position for a given time. Under these circumstances, if a single celestial LOP can be obtained and plotted, an *estimated position (EP)* can be determined based on the LOP. If the LOP is accurate, the ship must have been located somewhere on it at the moment of observation. In the absence of a sec-

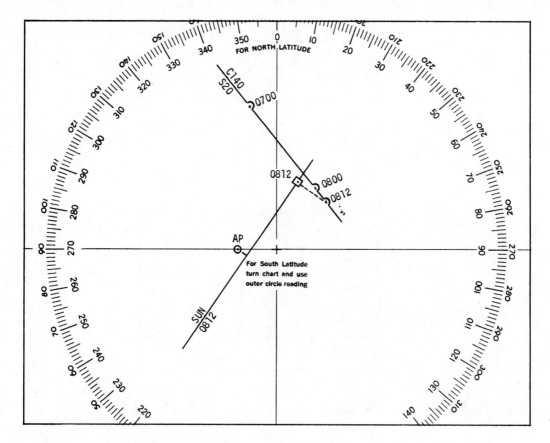

Figure 23-5. *An estimated position based on a single celestial LOP.*

ond intersecting LOP, an estimate of the ship's position on the line can be made by dropping a perpendicular construction line to the single LOP from the ship's DR position at the time the LOP was obtained. The estimated position thus formed represents the closest point on the LOP to the ship's DR position for that time. Such an estimated position based on the 0812 local zone time (1212 GMT) sun line completed as an example in chapter 20 (see Figure 20-13) is illustrated in Figure 23-5. It is labeled with a ⅛-inch square and the time as shown.

After the EP has been plotted, if the ship is well out to sea and there is a high degree of confidence in the single LOP, the navigator might elect to project a DR track from it, so as to obtain more meaningful estimates of the ship's position until either a second LOP is obtained to form a running fix, or two or more simultaneous LOPs are obtained to fix the ship's position. The standard procedure, of course, would be to continue the original DR until either a fix or running fix was obtained.

Body	SUN (LL)	SUN (LL)
GMT	12-12-06 Dec16	15-00-00 Dec16
IC	-.3'	-.3'
D (Ht 44')	-6.4'	-6.4'
Sum	-6.7'	-6.7'
hs	10° 12.4'	35° 13.1'
ha	10° 05.7'	35° 06.4'
Alt Corr	+11.1'	+14.9
Add'l Corr Moon HP/Corr		
Ho	10° 16.8'	35° 21.3'
GHA (h)	4° 08.5'	46° 06.1'
Incre. (m/s)	3° 01.5'	0
v/v Corr SHA		
Total GHA	4° 08.5'	46° 06.1'
±360°	364° 08.5'	406° 06.1'
aλ (+E,-W)	65° 08.5'	64° 06.1
LHA	299°	342°
Tab Dec	S 23° 18.9'	S 23° 19.2'
d#/d Corr	0.1'/ 0	0.1'/ 0
True Dec	S 23° 18.9'	S 23° 19.2'
a LAT Same Contrary	31° N Cont	30° N Cont
Dec Inc/d	18.9'/-38.9'	19.2'/-56.8'
Tens/DSD	-9.5'	-16.0'
Units/DSD Corr	-2.7	-2.2'
Total Corr	-12.2'	-18.2'
Hc (Tab)	10° 26.7'	34° 15.0'
Hc (Comp)	10° 14.5'	34° 56.6'
Ho	10° 16.8'	35° 21.3'
a	2.3'T	34.5'A
Z	N 125.2° E	N 160° E
Zn	125.2° T	160° T

Figure 23-6. *Complete solutions for two sun lines by* Tables No. 229.

THE CELESTIAL RUNNING FIX

The procedure for plotting a running fix involving a celestial line of position at sea is virtually identical to the procedure used during piloting. A celestial LOP may be advanced for any reasonable time interval to be crossed with a subsequent LOP derived from any source, or any earlier LOP may be advanced and crossed with a later celestial LOP to determine a running fix. No arbitrary time limit exists for the advancement of an LOP at sea, because of the less stringent positioning requirements in this environment as opposed to piloting waters. The only criterion that should be applied is the angle of intersection of the advanced and subsequent LOPs.

If a celestial LOP is to be advanced to form a running fix, it is the usual practice to advance the line itself rather than its AP, as is done in plotting a celestial fix. As in piloting, the distance and direction through which the LOP is advanced is based on the DR plot. It is the normal procedure at sea to obtain running fixes using sun LOPs every few hours if possible during the course of the day. The noon position at sea, for example, is normally obtained by advancing a morning sun line to cross the LAN latitude line; a midafternoon running fix is normally plotted by advancing the LAN line to cross an afternoon sun line.

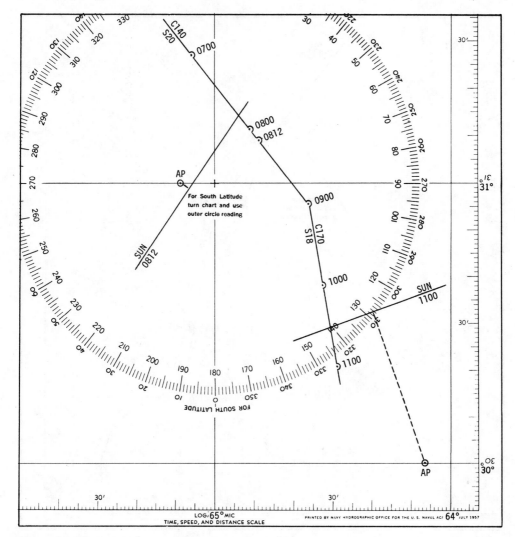

Figure 23-7A. *Two sun lines and the intervening DR plot.*

As an example of a celestial running fix produced from two celestial LOPs, consider the two sun sight solutions shown in Figure 23-6. The 0812 zone time (1212 GMT) sight was presented as an example earlier, and the 1100 zone time (1500 GMT) sight was obtained later in the morning of the same day.

The plot of the first sun line, and an EP based on it, is shown in Figure 23-5. The DR track was continued from the 0812 DR to the time of the second LOP, 1100. After the 1100 sun line is plotted, the SAPS-31 appears as shown in Figure 23-7A above; note the course and speed change at 0900.

To plot the celestial running fix, the 0812 sun line is advanced to the time of the 1100 sun line by reference to the DR plot. For clarity in this example, a

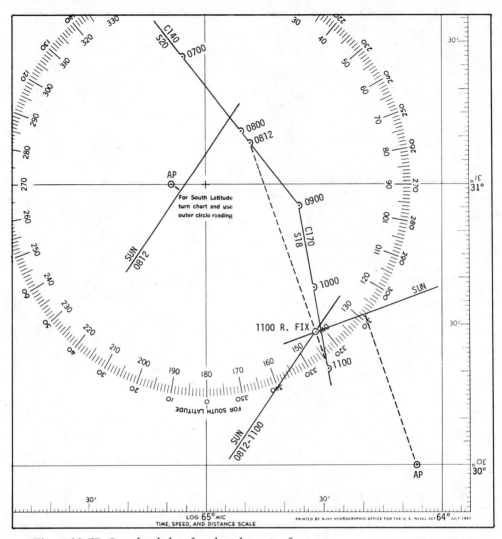

Figure 23-7B. *Completed plot of a celestial running fix.*

construction line has been drawn between the 0812 and 1100 DR positions, showing the distance and direction through which the 0812 line was advanced. The advanced line is so labeled, and the running fix position is symbolized by a ⅛-inch circle placed over the intersection of the two sun lines. After the symbol is labeled with the time, a new DR course and speed line is originated from the running fix position and labeled. The completed plot appears in Figure 23-7B above.

As an alternative to continuing the DR plot from the 0812 DR position, the navigator might have chosen to originate a new DR plot from the 0812 EP, especially if some time had elapsed since the last good fix. On this event, the 0812 sun line is advanced to the time of the subsequent 1100 sun line by reference to

432

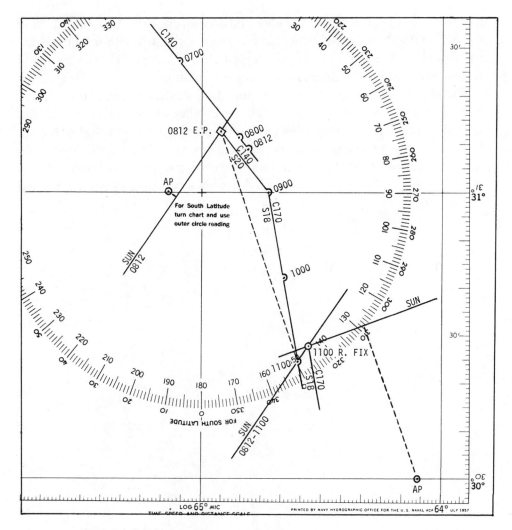

Figure 23-7C. *Plot of a celestial running fix using an EP.*

the distance between the 0812 EP and the 1100 DR. The resulting running fix is shown plotted in Figure 23-7C (above); the fix position so obtained is identical to that produced by the conventional method described above.

SUMMARY

In this chapter, the procedures for plotting the celestial fix and running fix have been presented, together with a discussion of the method used to plot an estimated position at sea based on an observation of a single celestial body. Such a single LOP can be combined with an earlier or later LOP obtained from any other source to form a running fix, with the only restrictions being that the earlier LOP must not be advanced over an excessive time interval, and the two LOPs should form an acceptable angle with one another. The navigator should not hesitate to combine a celestial LOP with an LOP determined from another source to obtain positioning information.

This chapter concludes the presentation of the determination of the celestial fix by the altitude-intercept method. The succeeding chapters on celestial navigation will each deal with special case solutions of the navigational or the celestial triangle for some specific purpose auxiliary to finding the ship's position. The interplay of celestial position-finding with electronically determined fixes will be presented in chapter 32, which describes a typical day's work in navigation at sea.

THE RUDE STARFINDER

24

To solve the navigational triangle for a computed altitude and true azimuth, the navigator must either know beforehand or be able to determine afterward the name of the celestial body observed, so that its GHA and declination can be obtained from either the *Nautical Almanac* or the *Air Almanac*. Several aids are available to the navigator to assist in identifying and locating celestial bodies, among which are star charts, sky diagrams, starfinders, and starfinder computer programs. *Star charts* are representations resembling photographs of the night sky at certain times of the year; *sky diagrams* are drawings of the heavens as they would be seen from certain locations at various times; and *starfinders* are devices intended to furnish the approximate altitude and true azimuth of celestial bodies either before or after navigational observations. Several excellent star charts are sold commercially and may be obtained at most nautical supply stores. In addition, there are many different manuals, textbooks, and almanacs available that contain excellent star charts; among these are *Dutton's Navigation and Piloting*, the *American Practical Navigator* (*Bowditch*), and the *Nautical Almanac*. Each *Air Almanac* contains a set of sky diagrams for various latitudes for the fifteenth day of each of the six months it covers. The use of volume 1 of *No. 249* as a starfinder has already been discussed in chapter 21.

In this chapter, perhaps the most commonly used of all mechanical starfinders, the *Rude Starfinder*, will be examined. This starfinder was originally patented by Captain G. T. Rude, USC&GS, some years ago and was subsequently sold to the old U.S. Navy Hydrographic Office, which improved the device and made it available under the designation 2102-D. Most practicing navigators, however, continue to refer to it as the Rude Starfinder. Currently, it is issued on request to Navy and other government users by the Ship's Parts Control Center, Mechanicsburg, Pennsylvania, and it is produced for sale to commercial and private users by a marine supply firm, Weems and Plath, Inc., of Annapolis, Maryland.

DESCRIPTION OF THE RUDE STARFINDER

The Rude Starfinder, 2102-D, is designed to permit the determination of the approximate apparent altitude and azimuth of any of the 57 selected navigational stars tabulated in the *Nautical* and *Air Almanacs* that appear above the observer's celestial horizon at any given place and time. With some minor manipulations it can also be set up to obtain the positions of the navigational planets, any unlisted stars of interest, and even the sun or moon if desired. The device can also be used to identify an unknown body, given its altitude and true azimuth. The accuracy of the starfinder is generally considered to be about ±3 to 5 degrees in both altitude and true azimuth determinations.

Essentially, the Rude Starfinder consists of the *star base,* an opaque white plastic circular base plate fitted with a peg in the center, and ten circular transparent *templates.*

On one side of the star base the north celestial pole appears at the center, and on the other side, the south celestial pole. All of the 57 selected stars are shown on each side at their positions relative to the appropriate pole in a type of projection called an azimuthal equidistant projection. In this projection, the positions of the stars relative to one another are distorted, but their true declinations and azimuths relative to the pole are correct; hence, the pattern of the stars on the star base does not correspond to their apparent positions as seen in the sky. Each star on the base is labeled, and its magnitude is indicated by its symbol—a large heavy ring indicates first magnitude, an intermediate-sized ring second magnitude, and a small thin ring third magnitude. The celestial equator appears as a solid circle about 4 inches in diameter on each side of the star base, and the periphery of each side is graduated a half-degree of LHA♈.

There are ten templates included for use with the star base. Nine of these are printed with blue ink and are designed for apparent altitude and azimuth determinations, while the tenth, printed in red ink, is intended for the plotting of bodies other than the 57 selected stars on the base plate. There is one blue template for every 10° of latitude between 5° and 85°; one side of each template is for use in north latitudes, the other for south latitudes. Each of these "latitude" templates is printed with a set of oval blue altitude curves at 5° intervals, with the outermost curve representing the observer's celestial horizon, and a second set of radial azimuth curves, also at 5° intervals. The red template is printed with a set of concentric declination circles, one for each 10° of declination, and a set of radial meridian angle lines. In use, the appropriate template is snapped in place on the star base like a record on a phonograph turntable.

Figure 24-1 shows the south celestial pole side of the star base with the red template affixed and the nine blue templates arrayed below.

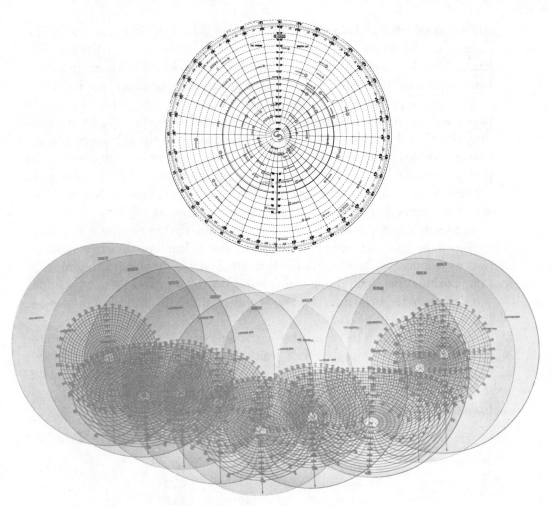

Figure 24-1. *The Rude Starfinder, 2102-D.*

USE OF THE RUDE STARFINDER FOR OBSERVATIONS OF THE 57 SELECTED STARS

Although it can be used for other bodies, the Rude Starfinder is most conveniently suited for determining which of the 57 selected navigational stars will be favorably situated for observation at a given DR position and time. To use the device for this purpose, the LHAƔ must first be determined for the midtime of the period during which the observations are scheduled to be made. In practice, the beginning of civil twilight is generally used for morning sights, and the ending of civil twilight is used for evening sights: both the definition and technique of prediction of times for these phenomena will be discussed in chapter 27. After the LHAƔ for the appropriate time and projected DR position has been computed, the blue template for the latitude closest to the DR latitude is selected and placed over the proper side of the star base plate. The name of the side of the star base used (i.e., N or S) and the name of the template latitude should always be identical. Should the DR position lie exactly between two template latitudes, such as L. 10° 00.0′, either template may be selected.

After the appropriate template has been placed on the proper side of the base plate, the template is rotated until the index arrow (the 0°/180° azimuth line) points to the proper value of LHA𝛶 on the rim of the star base. The apparent altitude ha and true azimuth Zn of all stars of interest located within the perimeter of the blue grid (the outer curve of which represents the observer's celestial horizon) are then recorded to the nearest whole degree. Normally the stars are recorded in the order of increasing true azimuth.

As an example, suppose that the navigator wished to use the Rude Starfinder to record the name, apparent altitude, and true azimuth of all first-magnitude stars with apparent altitudes greater than 10°, suitable for observation from the same position for which the use of volume 1 of *Tables No. 249* as a starfinder was demonstrated in chapter 21. The assumed latitude for this position was 34° south, and the LHA𝛶 was 145°.

Since the latitude is 34° south, the south pole side of the star base is used together with the blue template for latitude 35°S. After aligning the 0°/180° azimuth arrow with an LHA𝛶 of 145°, the device should appear as in Figure 24-2.

The following is a list of the ha's and Zn's of all first-magnitude stars as read from the starfinder in Figure 24-2 to the nearest whole degree:

Star	ha	Zn
* Regulus	42°	008°
* Spica	34°	079°
Hadar	42°	142°
Rigel Kent	38°	143°
* Acrux	53°	150°
Achernar	15°	208°
* Canopus	52°	227°
* Rigel	24°	277°
Sirius	47°	283°
* Betelgeuse	22°	297°
* Procyon	40°	318°
Pollux	21°	332°

Those stars preceded by an asterisk in the preceding list appear also on the list obtained by the use of volume 1 of *Tables No. 249* shown on page 408. As can be seen by a quick comparison of the two lists, the data agree quite closely.

In practice, the navigator should always list more stars than the ones he or she actually expects to observe, as some may be obscured by clouds. Moreover,

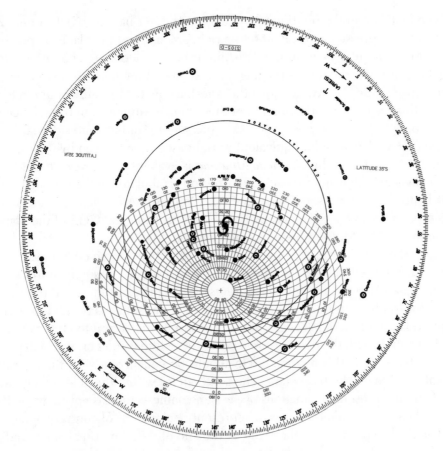

Figure 24-2. *Rude Starfinder set up for LHA♈ 145°, aL 34°S.*

the stars listed should not be confined to those of the first magnitude; all the stars shown on the starfinder are readily visible in clear weather. The most convenient band of apparent altitude for observation purposes lies roughly between 15° and 60°, but it is sometimes preferable to list stars lower or higher than those limits, rather than to have poor distribution in azimuth.

USE OF THE STARFINDER FOR THE NAVIGATIONAL PLANETS

The Rude Starfinder, 2102-D, may be used in a similar manner to find the apparent altitude and true azimuth of the navigational planets that will appear above the celestial horizon at a given position and time. Inasmuch as the planets are continually changing position relative to the "fixed" stars, their positions cannot be permanently printed on the star base along with the 57 selected stars; instead, they must be periodically plotted on the star base by means of the red template. In practice most navigators plot the positions of the planets on the star base about every 10 days to two weeks.

The first step in the plotting of a planet on the star base is to determine the value of an expression equivalent to the right ascension of the planet, expressed in terms of angular measurement. It might be recalled from chapter 16 that right ascension is defined as the angular distance of the hour circle of a celestial body measured eastward along the celestial equator from the hour circle of Aries; this measurement expressed in time units is employed in the science of astronomy to locate bodies on the celestial sphere, rather than sidereal hour angle (SHA). The angular equivalent of right ascension is given by the expression 360° − SHA. If the *Nautical Almanac* is available, the average SHA for each navigational planet for each 3-day period is tabulated at the bottom of the left-hand daily pages (see Figure 20-1A). If only the *Air Almanac* is on hand, the expression 360° − SHA may be found by subtracting the GHA of the planet from the GHAϒ for the time of the observation.

After the value of the right ascension angle has been obtained, the next step is to place the red template on the proper side of the star base; the north sides of the base plate and red template are used for north latitudes, and the south sides for south latitudes. On the red template, a radial line is printed to represent every 10° of meridian angle, and a concentric circle is printed for every 10° of declination, with the median circle being the celestial equator. When in place on the base plate, this median circle should be concurrent with the celestial equator circle on the base plate. The solid circles within the celestial equator circle then represent declinations of the same name as the base plate, while the dashed circles outside the equator represent declinations of contrary name.

The final step is to align the arrow at the end of the 0° meridian angle radial on the template with the calculated value of the right ascension angle on the base plate. After the template has been properly aligned, the body of interest is then plotted on the base plate with a pencil, using the declination scale printed for this purpose alongside the open slot on the 0° meridian angle radial. The declination of the body is obtained from an almanac. If it is of the *same* name as the center of the star base plate, the body is plotted on the side of the celestial equator circle *toward* the center; if the declination is of *contrary* name, the body is plotted on the side of the equator *away from* the center of the base plate.

As an example, let us plot the position of Venus on the south side of the star base plate for the time of the observation used as an example in the three previous chapters. The SHA of Venus for the time of the observation on 15 December can be found from the daily page of the *Nautical Almanac* pictured in Figure 20-1A; it is 137° 43.2′. Hence, the value of the right ascension angle, 360° − SHA, is approximately

$$360° - 137.5° = 222.5°$$

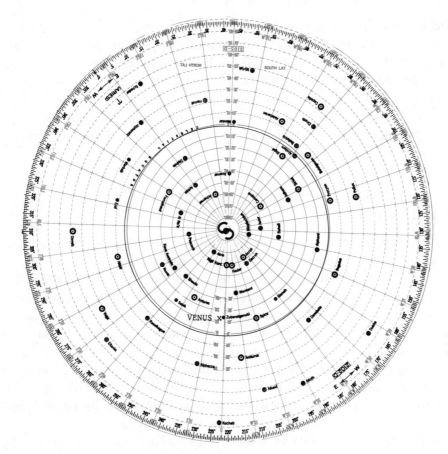

Figure 24-3. *Rude Starfinder set up for plotting the planet Venus.*

The SHA is rounded to the nearest half degree for this computation, in keeping with the precision of the graduations of the star base. Since the approximate position of the observer is in south latitudes at the time at which Venus is to be observed, the south sides of the star base plate and red template are used, and the arrow on the template is aligned with 222.5° on the base plate. The declination of Venus tabulated on the daily page of the almanac for the whole hour of GMT nearest the observation time is S13° 16.1'; this value is also rounded to the nearest half degree, S13.5°, consistent with the precision of the declination graduations of the red template. Because this declination is of the same name as the side of the star base used in this instance, the planet is plotted opposite the innermost 13.5° point on the declination scale of the template; thus, Venus at this time is almost alongside the third-magnitude star Zubenelgenubi. At this point, the starfinder appears as pictured in Figure 24-3.

Finally, the template is removed and the plot of Venus is labeled.

In similar fashion, the remaining three navigational planets could each be plotted on the base plate. After all four have been plotted and labeled, it is a good practice to print the date for which they were plotted somewhere on the base plate, so as to be able to judge when to replot them.

The apparent altitudes and true azimuths of the planets can now be obtained and recorded along with the stars by using the appropriate blue template. With the 35°S template aligned to LHAϓ 145°, the apparent altitude of Venus is read as 18°, and the true azimuth is 097°T.

USE OF THE STARFINDER FOR UNLISTED STARS, THE SUN, OR THE MOON

If the navigator wishes to use the starfinder to obtain the altitude and true azimuth of a star not among the 57 selected navigational stars, the body must first be plotted on the star base plate. To do this, the SHA and declination of the star is first located in an almanac, then the body is plotted in exactly the same manner as described above for a planet. Once it has been located on the star base, an unlisted star may be left in its plotted position indefinitely, as it will not change position relative to the other stars.

Although the location of the sun and moon in the heavens is of course no problem for the navigator, it may on occasion be desirable to use the Rude Starfinder to determine their positions at some future time for tactical planning purposes. Should this become necessary, the body of interest is first plotted on the star base. The GHAϓ and the GHA and declination of the body for the time of interest are extracted from an almanac, and the right ascension angle is found by subtracting the GHA of the body from the GHAϓ. After the body has been plotted using its declination and the red template, the apparent altitude and true azimuth of the body for the time in question are determined by use of the proper blue template.

USE OF THE STARFINDER TO IDENTIFY AN UNKNOWN BODY

Occasionally the navigator may obtain a good sextant altitude of an unknown body not included in the preselected list of stars and planets intended for observation. In such cases, the Rude Starfinder may be used to identify the body, if both its altitude and true azimuth for the time of the sight are recorded. To set up the device for this purpose, both the appropriate blue template and the red template are placed on the proper side of the star base plate, with the red template on top, and all three properly oriented for latitude. The index arrows of both templates are aligned with the proper value of LHAϓ on the base plate. To identify the body, the observed value of the true azimuth and the observed altitude are located on the blue template. If a star appears on the base plate at or near this location, it may be assumed that this was the body observed. If no star appears nearby, however, the red template is used to determine the approximate declination and meridian angle corresponding with the location. Knowing the meridian angle and the ship's longitude, the approximate GHA of the body may be computed, and from that, the approximate SHA. First, the navigator searches the appropriate daily page of the *Nautical* or *Air Almanac* with the GHA and declination to determine if the body is a planet. If it happens that the body is not a planet, the SHA and declination are used with the list of additional stars in the back of the *Nautical Almanac* to identify the body and obtain its exact GHA and declination. Most computer starfinder programs incorporate an identification subroutine that can be used for this purpose.

SUMMARY

This chapter has examined the use of the Rude Starfinder, 2102-D, in obtaining predicted values of the apparent altitude and true azimuth for compilation of a list of selected celestial bodies to be observed at a given location and time. The device is best suited for use with the 57 selected stars tabulated in the *Nautical* and *Air Almanacs*, but if necessary, it can be used to determine the positions of the navigational planets, unlisted stars, and even the sun or moon. It can also be used in conjunction with the *Nautical* or *Air Almanac* to identify an unknown body, if the true azimuth as well as the altitude is recorded at the time of the observation.

THE DETERMINATION OF LATITUDE AT SEA

25

Earlier in this book, the process of determining a celestial fix by the altitude-intercept method of sight reduction was examined in some detail. In this chapter, a special case solution of the celestial triangle to produce a type of LOP called the *latitude line* will be discussed. This LOP may be combined with other LOPs that were obtained simultaneously to form a celestial fix, or with a nonsimultaneous LOP to produce a running fix. If no other LOP is available, the latitude LOP may be used in conjunction with the DR plot to form a highly accurate estimated position.

THE CELESTIAL TRIANGLE AT MERIDIAN TRANSIT OF A BODY

The celestial latitude line of position is a special case solution of the celestial triangle, in which all three vertices of the triangle—the celestial pole, the zenith of the observer, and the position of the body—lie on a single great circle, coincident with a celestial meridian. This condition occurs whenever any celestial body transits the upper or lower branch of the observer's meridian, as shown in Figure 25-1.

Observation of any celestial body at or near the moment of meridian transit produces a highly accurate LOP because its altitude changes most slowly at this time. If a series of observations is commenced shortly before the time of meridian transit and continued until after meridian transit has occurred, a series of nearly identical altitudes results. Reduction of the highest altitude observed in the case of a body at upper transit, or the lowest altitude of a body at lower transit, should therefore yield a highly accurate latitude line in almost every case. The only exception occurs when the navigator's platform is moving rapidly (i.e., in excess of 20 knots or so) in a generally northerly or southerly direction at the time of transit. In such situations, because the vessel is moving either directly toward or away from the GP beneath the transiting body, the

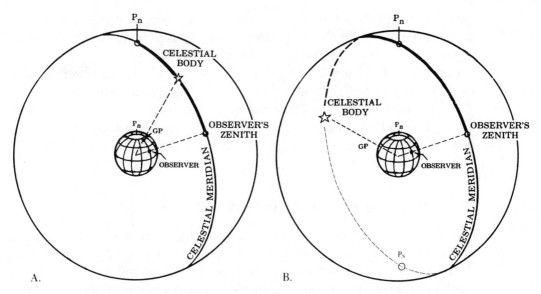

Figure 25-1. *Celestial triangle of a body at (A) upper meridian transit and (B) lower meridian transit.*

vessel motion can cause some distortion in the apparent rate of change of altitude near the time of transit. In these situations, care must be taken to use the altitude recorded when the body bears either due north or south of the observer.

When a very high degree of accuracy in the latitude line is required, many navigators construct a graph such as that shown in Figure 25-2 to assist in picking off the precise altitude at the moment of transit. Such a series of sights is called a time *sight*; the technique can also be used to enhance the precision of observations made at other times, when extreme accuracy is desired. More information on this advanced technique can be obtained from the *American Practical Navigator (Bowditch)*.

In addition to the high accuracy of the resulting LOP, a further advantage of any celestial latitude line obtained by observation of a body at meridian transit is that only the celestial triangle itself is solved, with no requirement for the derivation of a related navigational triangle. No sight reduction tables are re-

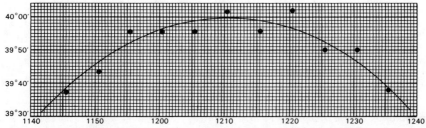

Figure 25-2. *A plot of observed altitudes versus time near meridian transit (upper branch).*

quired to obtain the solution, eliminating the necessity for the selection of an assumed position. Only the declination of the observed body extracted from an almanac for the time of the observation is needed.

In practice, the latitude line is normally obtained by direct observation of one of only two celestial bodies—the pole star Polaris or the sun. Polaris is ideally suited for obtaining a latitude line in north latitudes because it is virtually always at meridian transit, being located approximately at the north celestial pole; the sun is the only celestial body readily observable at transit during daylight hours in mid latitudes.

OBTAINING A LATITUDE LINE BY POLARIS

In the celestial triangle for the star Polaris the position of the star and the north celestial pole P_n are nearly coincident. The sides of the triangle linking the zenith of the observer with the position of the star and with the celestial north pole, 90° – Ho and 90° – Latitude, respectively, are therefore approximately coincident and equal in length, as shown in Figure 25-3A.

If Polaris were in fact exactly coincident with the north celestial pole, the relationship between the two coincident sides of the celestial triangle could be expressed by the following equation:

$$(90° – Ho) = (90° – Latitude)$$

This relationship is clearly indicated in Figure 25-3B, in which a cross section of the celestial sphere bisected along the observer's celestial meridian is shown.

Removing the parentheses and combining terms algebraically, the equation above can be simplified to the following expression:

$$Ho = Latitude$$

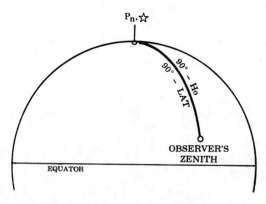

Figure 25-3A. *Celestial triangle for Polaris.*

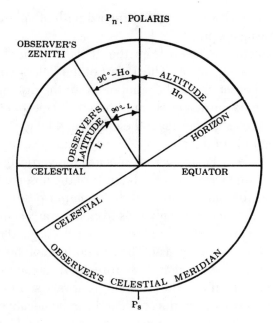

Figure 25-3B. *Celestial triangle for Polaris (side view).*

Thus, the observed altitude Ho of Polaris is approximately equal to the latitude of the observer, a fact of great usefulness to the navigator operating at sea in northern latitudes.

In reality Polaris is *not* exactly coincident with the north celestial pole, but rather it is about three-quarters of a degree off to one side; the exact separation varies somewhat over time, as a result of the effects of the earth's precession and nutation, and the aberration effect. Hence, the star has a small diurnal circle of variable radius about the pole, which must be taken into account in order to obtain a latitude line precise to more than about 1° by Polaris. In the days of sail, this 1° error could be disregarded, but the precision required by modern-day navigation makes it mandatory to take the distance between Polaris and the north celestial pole into account. Tables are provided in both the *Nautical* and *Air Almanacs* for the purpose of correcting the apparent altitude ha of Polaris for this distance at any given moment. In practice, this correction is usually combined with the refraction and temperature/barometric pressure corrections to form a total correction to the apparent altitude ha, even though the Polaris correction is not, strictly speaking, an altitude correction.

For marine navigation, the Polaris correction tables in the *Nautical Almanac* are generally preferred over the corresponding *Air Almanac* tables because of their greater precision. An extract from the Polaris correction tables of a *Nautical Almanac* appears in Figure 25-4. Another part of the tables, not shown in

Figure 25-4, is used for determining the true azimuth of Polaris; it will be discussed in the following chapter.

The Polaris correction is taken from the "Polaris Tables" in three parts, which are designated a_0, a_1, and a_2. The a_0 part of the Polaris correction can be thought of as a compensation for the component of the distance between the position of Polaris in its diurnal circle and the north celestial pole, measured along the observer's celestial meridian; the effect, and hence the a_0 correction, varies with the longitude of the observer. The a_1 part is a compensation for the tilt of the diurnal circle of Polaris with respect to the vertical circle of the observer; the a_1 correction increases as the latitude of the observer increases. The a_2 part compensates for an aberration in the apparent position of Polaris because of bending of the incoming light rays from the star, occurring as a result of the velocity of the earth in its orbit; the orbital velocity, and therefore the a_2 correction, varies with the time of year. The composition of the "Polaris Tables" in the *Nautical Almanac* is such that these three parts are always positive; but after they have been added together, 1° (60′) is always subtracted from their sum. The total Polaris correction thus obtained can be negative at times.

As can be seen from Figure 25-4, there is one column in the tables for every 10° of LHAϓ, and one horizontal section for each of the three parts. The three parts of the Polaris correction are all based on the exact LHAϓ determined from the DR position of the observer at the time of the observation. After the exact value of LHAϓ has been obtained, the 10° LHAϓ column containing this value is entered. The a_0 part of the Polaris correction corresponding to the exact LHAϓ is obtained by mental interpolation from the top section of the column. The a_1 part is extracted from the second section of the column, using as an entering argument the tabulated latitude closest to the DR latitude of the observer. The a_2 part is obtained from the third section of the same column, using the month in which the observation was made as the entering argument. No interpolation is necessary in the a_1 and a_2 sections. An example of the determination of a Polaris correction is given below.

Even though the Polaris sight reduction is uncomplicated, a sight reduction form can be of convenience during the process. Such a form is used as the medium on which the following example Polaris sight is worked.

Suppose that Polaris was observed and found to have a sextant altitude hs of 31° 33.4′ at 01-12-09 GMT (2012 local zone time) 16 December, at which time the DR position of the observer was L 30° 30.5′N, λ 67° 37.2′W. The sextant index correction (IC) is +0.5′, and the height of eye is 44′.

First, the IC and dip corrections are determined, summed, and applied to the sextant altitude to obtain the apparent altitude ha of 31° 27.5′. The refraction correction for this altitude is −1.6′, which when applied to the ha yields an observed altitude of 31° 25.9′. The GHAϓ for the time of the observation is then obtained by combining the tabulated GHAϓ on the daily pages with the

POLARIS (POLE STAR) TABLES
FOR DETERMINING LATITUDE FROM SEXTANT ALTITUDE AND FOR AZIMUTH

L.H.A. ARIES	0°–9°	10°–19°	20°–29°	30°–39°	40°–49°	50°–59°	60°–69°	70°–79°	80°–89°	90°–99°	100°–109°	110°–119°
	a_0	a_0	a_0	a_0	a_0	a_0	a_0	a_0	a_0	a_0	a_0	a_0
0	0 14.2	0 10.2	0 07.6	0 06.7	0 07.4	0 09.6	0 13.4	0 18.6	0 25.0	0 32.4	0 40.7	0 49.5
1	13.7	09.8	07.5	06.7	07.5	09.9	13.9	19.2	25.7	33.2	41.5	50.4
2	13.3	09.5	07.3	06.7	07.7	10.3	14.3	19.8	26.4	34.0	42.4	51.3
3	12.8	09.2	07.2	06.7	07.9	10.6	14.8	20.4	27.1	34.8	43.3	52.2
4	12.4	09.0	07.1	06.8	08.1	11.0	15.3	21.0	27.9	35.7	44.1	53.1
5	0 12.0	0 08.7	0 07.0	0 06.8	0 08.3	0 11.3	0 15.8	0 21.7	0 28.6	0 36.5	0 45.0	0 54.0
6	11.6	08.5	06.9	06.9	08.5	11.7	16.4	22.3	29.4	37.3	45.9	54.9
7	11.2	08.2	06.8	07.0	08.8	12.1	16.9	23.0	30.1	38.1	46.8	55.8
8	10.9	08.0	06.8	07.1	09.1	12.5	17.5	23.6	30.9	39.0	47.7	56.7
9	10.5	07.8	06.7	07.2	09.3	13.0	18.0	24.3	31.7	39.8	48.6	57.6
10	0 10.2	0 07.6	0 06.7	0 07.4	0 09.6	0 13.4	0 18.6	0 25.0	0 32.4	0 40.7	0 49.5	0 58.5

Lat.	a_1	a_1	a_1	a_1	a_1	a_1	a_1	a_1	a_1	a_1	a_1	a_1
0	0.5	0.6	0.6	0.6	0.6	0.5	0.5	0.4	0.3	0.2	0.2	0.1
10	.5	.6	.6	.6	.6	.5	.5	.4	.3	.3	.2	.2
20	.5	.6	.6	.6	.6	.5	.5	.4	.4	.3	.3	.3
30	.6	.6	.6	.6	.6	.6	.5	.5	.4	.4	.4	.4
40	0.6	0.6	0.6	0.6	0.6	0.6	0.6	0.5	0.5	0.5	0.5	0.5
45	.6	.6	.6	.6	.6	.6	.6	.6	.6	.5	.5	.5
50	.6	.6	.6	.6	.6	.6	.6	.6	.6	.6	.6	.6
55	.6	.6	.6	.6	.6	.6	.6	.6	.7	.7	.7	.7
60	.6	.6	.6	.6	.6	.6	.7	.7	.7	.8	.8	.8
62	0.7	0.6	0.6	0.6	0.6	0.6	• 0.7	0.7	0.8	0.8	0.9	0.9
64	.7	.6	.6	.6	.6	.7	.7	.8	.8	.9	0.9	0.9
66	.7	.6	.6	.6	.6	.7	.7	.8	.9	0.9	1.0	1.0
68	0.7	0.6	0.6	0.6	0.6	0.7	0.8	0.8	0.9	1.0	1.1	1.1

Month	a_2	a_2	a_2	a_2	a_2	a_2	a_2	a_2	a_2	a_2	a_2	a_2
Jan.	0.7	0.7	0.7	0.7	0.7	0.7	0.7	0.7	0.7	0.7	0.7	0.7
Feb.	.6	.6	.7	.7	.7	.8	.8	.8	.8	.8	.8	.8
Mar.	.5	.5	.6	.6	.7	.7	.8	.8	.9	.9	.9	0.9
Apr.	0.3	0.4	0.4	0.5	0.6	0.6	0.7	0.8	0.8	0.9	0.9	1.0
May	.2	.3	.3	.4	.4	.5	.6	.6	.7	.8	.8	0.9
June	.2	.2	.2	.3	.3	.4	.4	.5	.5	.6	.7	.8
July	0.2	0.2	0.2	0.2	0.2	0.3	0.3	0.4	0.4	0.5	0.5	0.6
Aug.	.4	.3	.3	.3	.3	.3	.3	.3	.3	.3	.4	.4
Sept.	.5	.5	.4	.4	.3	.3	.3	.3	.3	.3	.3	.3
Oct.	0.7	0.7	0.6	0.6	0.5	0.4	0.4	0.3	0.3	0.3	0.3	0.2
Nov.	0.9	0.9	0.8	0.7	.7	.6	.5	.5	.4	.3	.3	.3
Dec.	1.0	1.0	1.0	0.9	0.8	0.8	0.7	0.6	0.6	0.5	0.4	0.3

Figure 25-4. *Excerpt from Polaris Tables*, Nautical Almanac. (*Note: These values vary year to year, so this excerpt can only be used for the example problems in this book.*)

Body POLARIS	POLARIS
GMT	01-12-09 Dec16
IC	+.5
D (Ht 44')	-6.4'
Sum	-5.9'
hs	31° 33.4'
ha	31° 27.5'
Alt Corr	-1.6'
Add'l Corr	
Ho	31° 25.9'
GHA γ (h)	99° 15.8'
Incre. (m/s)	3° 02.7'
Total GHA γ	102° 18.5'
±360°	
DRλ (+E,-W)	67° 37.2'W
LHA γ	34° 41.3'

Figure 25-5A. *Polaris sight form, LHAγ computed.*

proper GHA increment; in this case a GHAγ of 102° 18.5' results. Since the longitude is west and is less than the GHAγ in this instance, the exact LHAγ is found by subtracting the DR longitude from the GHAγ. The form now appears as in Figure 25-5A.

Next, the Polaris tables shown in Figure 25-4 are entered to find the three parts of the Polaris correction. The entering argument for the tables is the LHA of Aries, 34° 41.3'. The appropriate column is therefore the third one, headed 30°–39°. To find the a_0 part, the upper third of the table containing corrections for each integral degree of LHA from 30° to 40° is used, with interpolation if necessary to arrive at the a_0 correction corresponding to LHAγ 34° 41.3'. In this case, the tabulated a_0 values for both LHAγ 34° and 35° are the same, 0° 06.8', so no interpolation is needed. After recording this 0° 06.8' a_0 correction on the form, we proceed to the middle third of the table, in the same LHAγ column, to obtain the a_1 part. The entering argument in the left margin is the tabulated latitude closest to the DR latitude, or 30° in this case. The corresponding a_1 value is 0.6', which is recorded onto the form. The a_2 part is obtained from the lower third of the table, staying in the same column, opposite the month of observation, December in this case; it is 0.9'. As the final step in finding the total correction to be applied to Ho, a_0, a_1, and a_2 are added algebraically to form the sum of +8.3', which is then added algebraically to the required –60' constant, to yield a total correction in this example of –53.3'. Applying this correction to Ho yields the desired latitude line, 30° 34.2'N. The sight form as it appears at this point is pictured in Figure 25-5B.

A plot of the resulting latitude line by Polaris is shown in Figure 25-5C on page 451.

Body POLARIS	POLARIS
GMT	01-12-09 Dec16
IC	+.5
D (Ht 44')	-6.4'
Sum	-5.9'
hs	31° 33.4'
ha	31° 27.5'
Alt Corr	-1.6'
Add'l Corr	
Ho	31° 25.9'
GHA γ (h)	99° 15.8'
Incre. (m/s)	3° 02.7'
Total GHA γ	102° 18.5'
±360°	
DRλ (+E,-W)	67° 37.2'W
LHA γ	34° 41.3'
a0	+6.8'
a1	+.6'
a2	+.9'
Sum	+8.3'
1° 00' (-)	-60.0'
Total Corr	-51.7'
Ho	31° 25.9'
LAT	30° 34.2'

Figure 25-5B. *Completed determination of latitude by Polaris.*

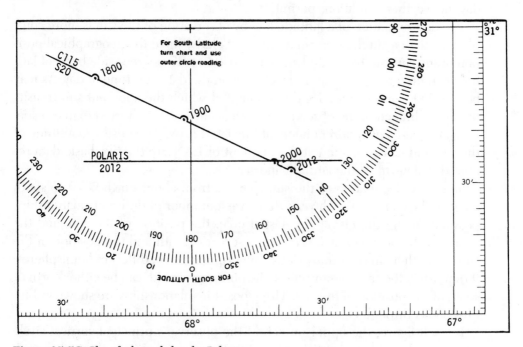

Figure 25-5C. *Plot of a latitude line by Polaris.*

After the latitude line determined by an observation of Polaris has been plotted, it may be combined with other LOPs to form a celestial fix or running fix by methods discussed in chapter 23. If no other LOP is available, a perpendicular may be dropped to the latitude line from the DR position at the time of the Polaris observation (or in other words, the DR longitude may be crossed with the Polaris latitude) to obtain an estimated position of high reliability.

Because the star Polaris is below the observer's celestial horizon when south of the terrestrial equator, it follows that the determination of latitude by Polaris is not possible when the observer is in south latitudes. When Polaris is not available for observation, either during daylight hours or because of the position of the observer, there is an excellent alternative body available for determination of a latitude line—the sun.

OBTAINING A LATITUDE LINE BY THE SUN

Because the sun always completes upper transit above the observer's celestial horizon in the midlatitudes of the world, observation of the sun at this meridian transit is a very convenient method of determining a latitude line. The sun latitude line thus obtained is considered to be one of the most accurate LOPs available in the course of a typical day's work in navigation at sea, except possibly when the sun's altitude is extremely high. In practice, most navigators observe the sun at meridian transit as a matter of routine at sea every day the weather conditions permit.

The determination of a latitude line by observation of the sun is more complicated than a similar determination by Polaris, since the geographical position of the sun can be located anywhere within a 47-degree-wide band of latitude centered on the equator, depending on the date and time of day. As may be recalled from chapter 18, the moment at which the apparent sun transits the upper branch of the observer's meridian each day is referred to in celestial navigation as *local apparent noon,* abbreviated *LAN.* The observed altitude of the sun and its declination at the moment of LAN are the two basic data required to determine the latitude line at LAN.

Since the declination of the sun changes from about north 23½° to south 23½° in the course of each year, there are a number of different relationships possible among the elevated celestial pole, the position of the sun, and the zenith of the observer at LAN. The observer's zenith and the position of the sun can both lie in the same celestial hemisphere, or in different hemispheres; if they are in the same hemisphere, the observer's zenith can be either north or south of the position of the sun. Three possible relationships are shown in Figures 25-6A through C.

In each figure, the upper branch of the observer's principal vertical circle (the vertical circle passing through the north and south points of the observer's

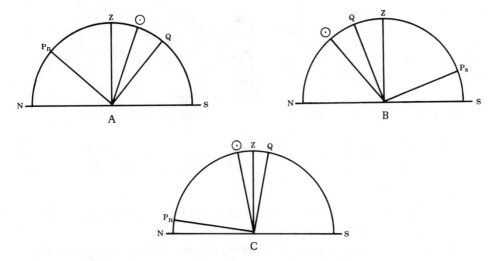

Figure 25-6. *(A) Latitude N, Dec. N; (B) Latitude S, Dec. N; (C) Latitude N, Dec. greater than Lat.*

celestial horizon, coincident with the celestial meridian) is pictured at the moment of LAN. The horizontal line represents the observer's celestial horizon, with the northernmost point to the left. The observer's zenith (directly over the ship's DR position) is represented by a vertical line labeled Z. The sun's position is indicated by a line labeled with the sun symbol ☉, the elevated pole by a line labeled P_n or P_s, and the celestial equator by a line labeled Q. The vertex from which all lines originate represents the center of the earth.

In the figures, the latitude of the observer is identical to the angle formed at the earth's center between the celestial equator Q and the observer's zenith Z. The angle formed between the sun's position and the observer's zenith may be recognized as the 90° – Ho side of the celestial triangle; for purposes of the latitude determination at LAN, this angle is referred to by its alternate name, the *zenith distance*. It is abbreviated by a lower case *z*, and labeled with a suffix to indicate the direction of the observer's zenith from the body at the moment of meridian transit. The suffix N is used to denote north, and the suffix S, south.

Consider the diagram in Figure 25-6A. If the zenith distance z(N) is indicated, it would be apparent that it is equal to 90° – Ho. Furthermore, if the declination of the sun is indicated, it would be seen that the latitude of the observer in this case is equal to the sum of the zenith distance z plus the declination of the sun. These angles are all identified in Figure 25-7A. For clarity, the relationships in Figure 25-7A are reproduced in Figure 25-7B as they would appear on a Mercator chart projection.

If these same angles were indicated on Figures 25-6B and C and on other diagrams representing other possible relationships among the sun, the observer's zenith, and the celestial equator at LAN, it would be seen that the latitude of the observer at LAN is always yielded by the sum or difference of the

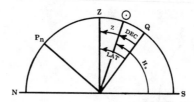

Figure 25-7A. *Relationship of observer's latitude to zenith distance and declination.*

zenith distance z and the declination of the sun. In general, the latitude of the observer can be obtained by an observation of the sun at LAN by applying the following two rules:

If the zenith distance z and declination are of the *same* name, *add* them for the latitude of the observer.

If the zenith distance z and declination are of *contrary* name, *subtract* the smaller from the larger.

In the latter case, the latitude will have the same name as the remainder.

As an example, if the declination of the sun in Figures 25-6A and 25-7A were N20° and the observed altitude Ho at LAN were 70°, the zenith distance z

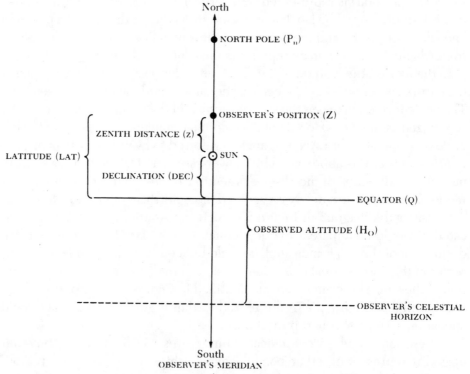

Figure 25-7B. *Relationship of observer's latitude to zenith distance and declination on a Mercator projection.*

would be 90° − 70° or 20°N, and the observer's latitude would be 20°N + N20°, or 40°N. In Figure 25-6B, if the declination of the sun were N20° and the Ho were 50°, the zenith distance would be 90° − 50°, or 40°S, since the observer's zenith is located south of the sun. Hence, the latitude would be 40°S − N20°, or 20°S. In Figure 25-6C, if the declination of the sun were N20° and the observed altitude Ho were 80°, the latitude of the observer would be 10°N.

Determination of the Time of LAN

Because the declination of the sun at the moment of meridian transit and the zenith distance are the two quantities combined to yield the latitude of the observer, it follows that both the declination and observed altitude of the sun must be determined as precisely as possible for an accurate latitude line. As an aid in obtaining the correct value for the declination of the sun and also to determine when to plan to make a LAN observation, the navigator will normally precompute the zone time at which meridian transit of the sun should occur, based on either the current position, if the ship is not under way with way on, or on the projected DR track, if the ship is moving. In other words, the navigator will usually *estimate the zone time of local apparent noon* for the position at which he or she expects to be located at meridian transit of the sun.

The first step in the estimation of the time of LAN at the ship's stationary or anticipated DR position is to obtain the local mean time of meridian passage of the apparent sun by referring to the lower right-hand side of the daily pages of the *Nautical Almanac*. Although the times of meridian passage listed are computed for the Greenwich meridian (and hence are Greenwich Mean Times), they can also be used with insignificant error as the local mean times of meridian passage on that date for that year at every other meridian, because of the nearly constant rate of motion of the sun about the earth over the period of a day. The data for 15, 16, and 17 December of one particular year are reproduced in Figure 25-8.

After plotting a DR position on a suitably prepared plotting sheet for the local mean time of meridian passage, the navigator must then convert the LMT to the zone time of the phenomenon at the plotted DR position. To perform

Day	SUN Eqn. of Time 00ʰ	12ʰ	Mer. Pass.
	ᵐ ˢ	ᵐ ˢ	ʰ ᵐ
15	05 12	04 58	11 55
16	04 43	04 29	11 56
17	04 14	04 00	11 56

Figure 25-8. *Times of meridian passage of the sun,* Nautical Almanac, *15, 16, 17 December. Caution: These data are correct only for the year of publication.*

this conversion, the navigator determines the arc difference between the DR longitude and the central meridian of the time zone by which the ship's clocks are set. Usually but not always the ship will be within this zone. Next, this arc difference is converted to a time difference using the tables "Conversion of Arc to Time" in the almanac. Since the sun will cross the observer's meridian before the central meridian of the time zone when the ship is located east of the central meridian, the time difference is *subtracted* if the initial DR position is to the *east* of the time zone central meridian. Conversely, if the initial DR position is to the *west* of the central meridian of the time zone being used, the time difference is *added*, since the sun will arrive over the observer's meridian after it has crossed the time zone central meridian. The resulting zone time of meridian passage is referred to as the *first estimate* of the zone time of LAN.

If the ship is stationary, this first estimate of the zone time of LAN should be the time at which the sun will transit the navigator's meridian. If the ship is moving, a second estimate of the time of LAN must be made, to account for the motion of the ship during the time required for the sun to travel between the time zone central meridian and the meridian passing through the first estimate DR position.

To obtain the second and final estimate of the time of LAN, the process described in the preceding paragraph is repeated, using a DR position plotted for the first estimate zone time as a starting point. The longitude difference for this position is found and converted to a time difference, and this time difference is again applied to the tabulated LMT of meridian passage to obtain the second and final estimate.

The following is a summary of the steps necessary to determine the zone time of LAN:

1. Obtain the LMT of meridian passage from the *Nautical Almanac*.
2. Plot a DR position for this time.
3. Determine the difference in longitude between this position and the central meridian of the time zone being used; convert this arc difference to a time difference.
4. Apply this time difference to the LMT of meridian passage, adding if west of the time zone central meridian, subtracting if east. The resulting time is the first estimate of the zone time of LAN.
5. If the ship is stationary, this first estimate should be the actual zone time of LAN.
6. If the ship is moving, plot another DR position for the first estimate zone time of LAN.
7. Compute and apply the time difference for this position to the tabulated LMT of meridian passage. The resulting time is the second and final estimate of the zone time of LAN.

456

In practice, the navigator will usually begin observing the sextant altitude of the sun about five minutes prior to the second estimated zone time of LAN, to account for any possible errors in the estimating procedure. The altitude recorded is normally the highest altitude reached by the sun as it transits the local meridian. It should be repeated here that if the navigator's vessel is proceeding on a heading having a large north-south velocity component, the motion of the vessel toward or away from the GP of the transiting sun will cause its altitude to appear to be changing at LAN. If the velocity component is in a direction toward the sun, the vessel is approaching the subsolar point, thereby causing the apparent altitude to seem to be increasing at meridian transit. Conversely, if the velocity component is away from the subsolar point, the sun will appear to be sinking at transit. Thus, if the greatest altitude is taken as the observed altitude at LAN under these circumstances, an error is introduced. When such a large north-south velocity component is present, the navigator should either record the altitude observed at the precomputed time of LAN, or record the altitude observed when the true azimuth of the sun is due north or south.

After the actual meridian transit of the sun has been observed and the sextant altitude and exact zone time of LAN have been recorded, the declination of the sun for this time is determined from either the *Nautical Almanac* or *Air Almanac*, the LAN diagram is drawn, and the observer's latitude is determined. As was the case with the observation of Polaris, no sight reduction tables are required to obtain a latitude line by observation of the sun. Nevertheless, a standard sight form for estimating the time of LAN and the reduction of the LAN sun line can be of considerable assistance to the navigator. The following example demonstrates the use of such a form.

Determining the Latitude at LAN: An Example

As an example of determining a latitude line by observation of the sun at local apparent noon, suppose that at 1100 ZT on 16 December a ship was located at a DR position L 35° 45.0′N, λ 54° 45.0′W, steaming on course 080°T at a speed of 20 knots. The navigator desires to obtain a second estimate of the zone time of LAN.

Working through the procedure set forth in the preceding section, the LMT of meridian passage is first obtained from the *Nautical Almanac* excerpt shown in Figure 25-8; for 16 December, it is 1156. A DR position is then plotted for this time as shown in Figure 25-10 (p. 459), and the DR longitude of this position is recorded on the LAN sight form. The difference between the DR longitude, 54° 22.0′W, and the central meridian of the time zone, 60°W, is 5° 38.0′:

$$60° - 54° 22.0′ = 5° 38.0′$$

DRλ	54° 22.0' W
STD Meridian	60° W
dλ (arc)	5° 38.0'
dλ (time)	23'
LMT Mer Pass	1156
ZT LAN (1st Est)	1133

Figure 25-9A. *Sun LAN form, first estimate completed.*

DRλ	54° 22.0' W
STD Meridian	60° W
dλ (arc)	5° 38.0'
dλ (time)	23'
LMT Mer Pass	1156
ZT LAN (1st Est)	1133
Rev DRλ	54° 31.8' W
STD Meridian	60° W
dλ (arc)	5° 28.2'
dλ (time)	22'
LMT Mer Pass	1156
ZT LAN (2nd Est)	1134
LAT BY LAN	
ZT LAN (OBS)	11-35-05 Dec16
ZD	+4
GMT	15-35-05 Dec16
Tab Dec	S 23° 19.2'
d#/d Corr	+.1'/+.1'
True Dec	S 23° 19.3'
IC	-.5'
D (Ht 44')	-6.4'
Sum	-6.9'
hs	30° 52.1'
ha	30° 45.2'
Alt Corr	+14.6'
Ho	30° 59.8'
89° 60'	89° 60.0'
Ho (-)	-30° 59.8'
Z Dist	59° 00.2' N
True Dec	S 23° 19.3'
LAT	35° 40.9' N

Figure 25-9B. *Completed sun LAN form.*

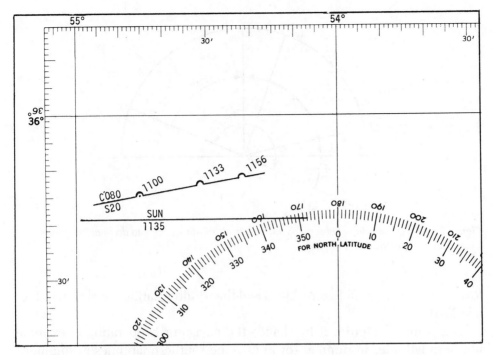

Figure 25-10. *Plot of a sun LAN line.*

Conversion of this arc to time by the tables in the *Nautical Almanac* yields 22 minutes 32 seconds, which may be rounded to the nearest whole minute, 23 minutes. Since the 1156 DR position is to the east of the 60° central meridian, the first estimate of the zone time of LAN is 1156 − 23, or ZT 1133. At this point, the partially completed sight form appears as in Figure 25-9A.

To obtain the second estimate, a DR position for 1133 is plotted, and the DR longitude 54° 31.8′W is picked off and subtracted from 60° to form the second estimate longitude difference, 5° 28.2′. Converting this arc to time and again subtracting from the LMT of meridian passage yields the second and final estimate of the zone time of LAN, 1134.

To complete the example, suppose that the actual zone time of the observation recorded when the sun appeared at its greatest altitude was 11-35-05 and the sextant altitude hs of the lower limb was 30° 52.1′. The sextant IC was −0.5′, and height of eye is 44′. The true declination for the zone time of observation is found from the *Nautical Almanac* to be S23° 19.3′. After applying the IC, dip, and altitude corrections to hs, an Ho of 30° 59.8′ results. Subtracting this Ho from 90° yields the zenith distance 59° 00.2′N, labeled "north" because the DR position is north of the GP of the sun. Inasmuch as the zenith distance and declination are of contrary name, the true declination is subtracted from the larger zenith distance to yield the latitude 35° 40.9′N. The completed sight

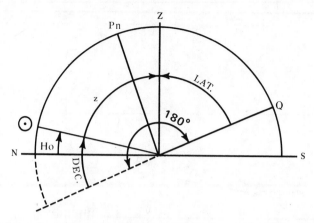

Figure 25-11. *Relationship of observer's latitude, zenith distance (z), and declination of the sun at lower transit in northern latitudes.*

form now appears as in Figure 25-9B and the resulting latitude is shown in Figure 25-10.

As a point of interest, if by chance the navigator is operating in extreme northern latitudes in summer (or in extreme southern latitudes in summer), the sun may be observable across the pole at low altitudes at *lower transit,* that is, local midnight. In such regions of constant daylight, the equation of time may be used to determine the local mean time of lower transit, from which the local zone time of the transit can be derived by using the same procedure as that for finding the zone time of LAN. The lowest observed altitude is used to determine zenith distance, which must then be added to the declination, and the sum subtracted from 180° to obtain the latitude, as indicated in Figure 25-11.

Just as was the case with the Polaris line discussed earlier, the sun latitude line may be combined with other simultaneous or nonsimultaneous LOPs to form a fix or running fix, or the line can be crossed with the longitude of the DR position corresponding to the time of the observation to produce an estimated position. In practice, it is standard procedure whenever the weather permits to obtain a running fix at the time of LAN by advancing a morning sun line, and to obtain a second afternoon running fix by advancing the LAN line. These running fixes, together with any electronic fixes, the morning and evening celestial fixes obtained at twilight, and the determination of compass error as set forth in the next chapter, constitute a basic celestial "day's work in navigation" for the navigator at sea. This subject will be discussed in greater detail in chapter 32.

SUMMARY

This chapter has presented the basic theory and procedures for determining a special type of celestial LOP at sea known as the latitude line, by observing either the star Polaris or the sun. Observation of any celestial body at meridian passage can produce a highly accurate latitude LOP, but Polaris and the sun are particularly well suited for this purpose. Polaris is perpetually near meridian transit, while the sun is the only body always observable at upper transit above the celestial horizon by day in midlatitudes. Once the latitude line has been derived, it may be combined with other LOPs to form a celestial fix or running fix of high accuracy, or it may be used in conjunction with the DR plot to obtain an estimated position of unusual reliability.

DETERMINING THE GYRO
ERROR AT SEA

26

Among the duties of the navigator prescribed in *OPNAVINST 3120.32* and specified by Navy fleet commanders' instructions is the requirement that the error of the ship's gyrocompass and magnetic compass be determined and recorded daily. The procedures for determining the gyro and compass error in piloting waters are set forth in chapter 9. In this chapter, the method of obtaining a gyro error at sea by observation of a celestial body will be discussed. In practice, once the error of the ship's gyrocompass system has been established, the ship's magnetic compasses are compared with the gyro to obtain the magnetic compass errors.

Essentially, gyro error at sea is determined by first observing the azimuth of a selected celestial body by means of the gyro repeater fitted with an azimuth circle, and then comparing this observed azimuth to a true azimuth computed by means of a special case solution of the celestial triangle. Although any celestial body could be used for this purpose, in practice most navigators prefer one of only two bodies—the star Polaris or the sun. As was the case with the latitude line discussed in the last chapter, Polaris is a favorite body for compass error determinations in north latitudes because of its location near the north celestial pole, while the sun is usually the only body available for observation for this purpose during daylight hours.

OBSERVING THE AZIMUTH OF POLARIS OR THE SUN

To observe the azimuth of Polaris, the sun, or any other celestial body, an instrument introduced in chapter 7—the *azimuth circle*—is used. The inexperienced navigator should review the description of the design and function of this device before proceeding further.

When the azimuth of the star Polaris is to be recorded, the procedure used is somewhat similar to that for the observation of the bearing of a terrestrial object with the azimuth circle. The star is first aligned in the black-coated reflector incorporated in the far sight vane, then the gyro azimuth is read from the portion of the compass card directly beneath the sight vane. Care should be taken to ensure that the azimuth circle is level at the moment the sight is recorded, to obtain an accurate reading.

To observe the azimuth of the sun, the mirror and prism assemblies at right angles to the sight vanes on the instrument are used, as explained in chapter 7. A narrow reflected beam of sunlight is cast across the proper portion of the compass card, and the bearing illuminated on the card by the beam is the gyro azimuth of the sun recorded. An alternative method of recording the azimuth of the sun, if its face is partially obscured by clouds, is to orient the azimuth circle in such a way that the image of the sun appears centered in the far sight vane reflector. The gyro bearing read beneath the vane then represents the sun's azimuth. As before, the azimuth circle should be level at the time of the observation.

Because of the construction of the azimuth circle, azimuths of celestial bodies are most accurately read when the body under observation is less than 20° in altitude; the most accurate observation possible occurs when the body is located on the celestial horizon of the observer. Hence, when the sun is to be used to determine gyro error, the optimum time of observation is near the time of sunrise and sunset. Polaris, because of its location near the north celestial pole, is always at about the same position both in altitude and azimuth above the observer's horizon in north latitudes. Obtaining a meaningful gyro error by Polaris in latitudes above 60° north is very difficult because of its high altitude in these higher latitudes.

DETERMINING THE GYRO ERROR BY POLARIS

The true azimuth of Polaris is tabulated in the *Nautical Almanac* "Polaris Tables" for latitudes between 0° and 65° north, in a horizontal section located beneath the a_0, a_1, and a_2 correction sections discussed in the last chapter. An extract from this Polaris azimuth section for LHAΥ between 0° and 119° appears in Figure 26-1.

Entering arguments for this azimuth section are the same LHAΥ used as a basis for entering the top three sections of the table, and the computed latitude by Polaris determined earlier. Mental interpolation is used where necessary to obtain the true azimuth of Polaris precise to the nearest tenth of a degree.

After the true azimuth of Polaris has been extracted from the Polaris azimuth table, it is then compared with the observed azimuth. The difference between the observed and true azimuth is the gyro error. The error is usually rounded to the nearest half degree, in keeping with the precision of the gyro repeater, and it is labeled either E (east) or W (west), depending on whether the true azimuth is greater than or less than the observed azimuth. The last few

POLARIS (POLE STAR) TABLES, 19__

FOR DETERMINING LATITUDE FROM SEXTANT ALTITUDE AND FOR AZIMUTH

L.H.A. ARIES	0°– 9°	10°– 19°	20°– 29°	30°– 39°	40°– 49°	50°– 59°	60°– 69°	70°– 79°	80°– 89°	90°– 99°	100°– 109°	110°– 119°
Lat.						AZIMUTH						
°	°	°	°	°	°	°	°	°	°	°	°	°
0	0·4	0·2	0·1	359·9	359·8	359·6	359·5	359·4	359·3	359·2	359·2	359·1
20	0·4	0·3	0·1	359·9	359·8	359·6	359·5	359·4	359·2	359·2	359·1	359·1
40	0·5	0·3	0·1	359·9	359·7	359·5	359·4	359·2	359·1	359·0	358·9	358·9
50	0·6	0·4	0·1	359·9	359·7	359·4	359·2	359·0	358·9	358·8	358·7	358·7
55	0·7	0·4	0·2	359·9	359·6	359·4	359·1	358·9	358·8	358·6	358·5	358·5
60	0·8	0·5	0·2	359·9	359·6	359·3	359·0	358·8	358·6	358·4	358·3	358·3
65	0·9	0·6	0·2	359·8	359·5	359·1	358·8	358·5	358·3	358·1	358·0	357·9

Latitude = Apparent altitude (corrected for refraction) − 1° + a_0 + c_1 + a_2

The table is entered with L.H.A. Aries to determine the column to be used; each column refers to a range of 10°. a_0 is taken, with mental interpolation, from the upper table with the units of L.H.A. Aries in degrees as argument; a_1, a_2 are taken, without interpolation, from the second and third tables with arguments latitude and month respectively. a_0, a_1, a_2 are always positive. The final table gives the azimuth of *Polaris*.

Figure 26-1. *Excerpt from Polaris Azimuth Tables,* Nautical Almanac.

spaces of the Polaris sight reduction form introduced in the last chapter are intended for the computation of the gyro error by Polaris.

As an example of the determination of a gyro error by Polaris, suppose that at the moment of the observation of Polaris (see Figure 25-5B), the gyro bearing of the star was observed and recorded on the sight form as 000.5° per gyrocompass (pgc). To find the true azimuth of Polaris, the LHA♈ 30°–39° column of the Polaris Tables is followed down into the bottom "Azimuth" section, shown in Figure 26-1. Since the tabulated azimuth corresponding with all latitudes except 65°N is 359.9°T, the exact azimuth corresponding to the computed altitude 30° 34.2′N is 359.9°T. Comparison of this value with 000.5° pgc results in a gyro error of 0.6°W, since the true azimuth is less than the observed azimuth. In practice, this value would be rounded off to 0.5°W. The completed gyro error determination appears on the sight form in Figure 26-2.

As was mentioned earlier, after the gyro error has been found the magnetic compass can be compared with the gyrocompass to ascertain the magnetic compass error.

GYRO ERROR BY POLARIS	
Zn POLARIS	359.9° T
Gyro Brg	000.5° pgc
Gyro Error	.6° W

Figure 26-2. *Completed gyro error computation, Polaris sight form.*

DETERMINING THE GYRO ERROR BY THE SUN

The conditions under which the star Polaris must be observed are not always conducive to accuracy in the determination of the gyro error. Polaris is a second-magnitude star, and is therefore not easily observed with the azimuth circle during twilight. Furthermore, dusk or darkness makes it difficult to level the azimuth circle, and the altitude of the star is often in excess of 20°, which is considered the maximum altitude for an accurate azimuth measurement. Hence, the determination of the gyro error by observation of the sun is generally preferred by most navigators. The two methods by which this may be accomplished—the sun amplitude sight and the sun azimuth sight—will be discussed below.

The Sun Amplitude Sight

As was stated earlier in this chapter, observation of the sun for the purpose of obtaining a gyro error is best performed when it is centered on the celestial horizon of the observer, either in the act of rising or setting. An observation of any celestial body at this position is termed an *amplitude sight.* There are two main advantages in observing the sun at this position. Not only is the bearing of the sun easy to observe at this time using an azimuth circle, but its true azimuth can be found without having to use a sight reduction table, by the use of either an *amplitude table* or the amplitude angle formula described below. Because of the combined effects of refraction and dip, in practice the assumption is usually made that the center of the sun is on the celestial horizon when its lower limb is about two-thirds of a solar diameter above the visible sea horizon. Although in some cases this assumption leads to a small error in the observed azimuth, any such error is insignificantly small in low and midlatitudes. In high latitudes, the sun must be observed when its disk is centered on the visible horizon to ensure accuracy in the observed azimuth. In this case, a correction to the amplitude angle to compensate for the dip of the visible horizon is obtained from a separate table.

The sun, like all celestial bodies, appears to rise in the east and set in the west, as a result of the effects of the earth's rotation. When the declination of the sun is zero at the time of the vernal or autumnal equinoxes, it seems to an observer anywhere on earth that the sun is located at the easternmost point of the celestial horizon at sunrise, bearing 090°T. This effect is illustrated in Figure 26-3.

In this figure, AP_1, AP_2, and AP_3 are assumed positions lying along a meridian experiencing sunrise at the time of equinox. Because of the great distance of the earth from the sun, it would seem to an observer at each of the three positions that the sun were bearing 090°T at the moment at which this body ap-

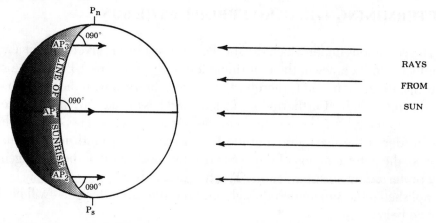

Figure 26-3. *Sun at 0° declination bears 090°T at sunrise everywhere on earth.*

peared above the celestial horizon. By the same reasoning, it could be shown that at sunset the true bearing of the sun as it crossed the celestial horizon to the west would appear to be 270°T from all three positions.

If the sun were not at equinox, but rather had a declination of, say, 20° north, it should be obvious that the sun would no longer appear to rise bearing 090°T and set bearing 270°T. It would in fact subtend an angle between its apparent position and the easternmost or westernmost point of the observer's celestial horizon called the *amplitude angle*. This amplitude angle may be defined as the horizontal angular distance measured in a northerly or southerly direction from the prime vertical—that vertical circle passing through the east and west points of the observer's celestial horizon—to the apparent position of the body on the celestial horizon. The angle is given the prefix E (east) if the body is rising, or W (west) if it is setting, and the suffix N (north) if it has northerly declination, or S (south) if it has southerly declination. If the observer were located on the equator, the amplitude angle would be identical in size to the declination; thus, if the sun had 20° north declination at sunrise, the amplitude angle at the equator would be E20°N. As the distance of the observer from the equator increases, however, the size of the amplitude angle also increases, reaching a maximum of 90° when the sum of the degrees of declination and the observer's latitude equal 90. In the case of the sun having declination N20° at sunrise, the maximum value E90°N occurs at latitude 70° north and south. Figures 26-4A and 26-4B should help to visualize this.

Figure 26-4A depicts the earth as it might be viewed from a position in space directly opposite the great circle line along which sunrise is being observed from three different positions on earth, AP_1, AP_2, and AP_3; Figure 26-4B represents the plane of the celestial horizon for each of these positions at sunrise. At AP_1, located on the equator, the sun appears to form an amplitude angle equal

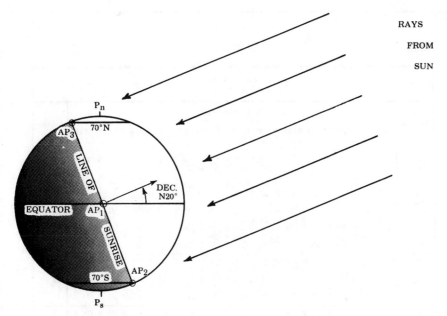

Figure 26-4A. *Representation of sunrise great circle, with sun at declination N20°.*

in size to its declination, E20°N, at sunrise, which occurs here at about 0600 ZT. At AP₃ located at 70°N, however, sunrise occurs much earlier, at about 0000 ZT. Because the sun is visible across the north pole at that time, its amplitude angle is E90°N, equivalent to a true bearing of due north. At AP₂, sunrise does not occur until about 1200 ZT, when the sun again forms an amplitude angle of E90°N, or 000° true azimuth. Consequently, all latitudes north of 70°N on this date would experience 24 hours of sunlight, while all positions south of 70°S would experience no daylight whatsoever. The subject of sunrise and sunset will be discussed in greater detail in the next chapter.

From the preceding discussion, it follows that the size of the amplitude angle depends on the declination of the sun and the latitude of the observer. If

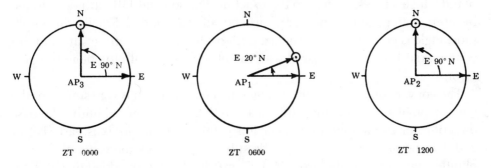

Figure 26-4B. *Plane of celestial horizon from three positions at sunrise.*

Figure 26-5. *Excerpt from Amplitude Table 22,* Bowditch.

TABLE 27
Amplitudes

Latitude	18°0	18°5	19°0	19°5	20°0	20°5	21°0	21°5	22°0	22°5	23°0	23°5	24°0	Latitude
0	18.0	18.5	19.0	19.5	20.0	20.5	21.0	21.5	22.0	22.5	23.0	23.5	24.0	0
10	18.3	18.8	19.3	19.8	20.3	20.8	21.3	21.8	22.4	22.9	23.4	23.9	24.4	10
15	18.7	19.2	19.7	20.2	20.7	21.3	21.8	22.3	22.8	23.3	23.9	24.4	24.9	15
20	19.2	19.7	20.3	20.8	21.3	21.9	22.4	23.0	23.5	24.0	24.6	25.1	25.6	20
25	19.9	20.5	21.1	21.6	22.2	22.7	23.3	23.9	24.4	25.0	25.5	26.1	26.7	25
30	20.9	21.5	22.1	22.7	23.3	23.9	24.4	25.0	25.6	26.2	26.8	27.4	28.0	30
32	21.4	22.0	22.6	23.2	23.8	24.4	25.0	25.6	26.2	26.8	27.4	28.0	28.7	32
34	21.9	22.5	23.1	23.7	24.4	25.0	25.6	26.2	26.9	27.5	28.1	28.7	29.4	34
36	22.5	23.1	23.7	24.4	25.0	25.7	26.3	26.9	27.6	28.2	28.9	29.5	30.2	36
38	23.1	23.7	24.4	25.1	25.7	26.4	27.1	27.7	28.4	29.1	29.7	30.4	31.1	38
40	23.8	24.5	25.2	25.8	26.5	27.2	27.9	28.6	29.3	30.0	30.7	31.4	32.1	40
41	24.2	24.9	25.6	26.3	26.9	27.6	28.3	29.1	29.8	30.5	31.2	31.9	32.6	41
42	24.6	25.3	26.0	26.7	27.4	28.1	28.8	29.5	30.3	31.0	31.7	32.5	33.2	42
43	25.0	25.7	26.4	27.2	27.9	28.6	29.3	30.1	30.8	31.6	32.3	33.0	33.8	43
44	25.4	26.2	26.9	27.6	28.4	29.1	29.9	30.6	31.4	32.1	32.9	33.7	34.4	44
45	25.9	26.7	27.4	28.2	28.9	29.7	30.5	31.2	32.0	32.8	33.5	34.3	35.1	45
46	26.4	27.2	27.9	28.7	29.5	30.3	31.1	31.8	32.6	33.4	34.2	35.0	35.8	46
47	26.9	27.7	28.5	29.3	30.1	30.9	31.7	32.5	33.3	34.1	35.0	35.8	36.6	47
48	27.5	28.3	29.1	29.9	30.7		32.4	33.2	34.0	34.9	35.7	36.6	37.4	48

the sun is observed when its center is located on the celestial horizon, Table 22 in the *American Practical Navigator (Bowditch)* may be used to obtain the value of the amplitude angle. An excerpt from this table for declinations between 18° and 24° appears in Figure 26-5; the complete table is reproduced in appendix F of this book.

The table is entered using as arguments the declination of the sun at the time of the amplitude sight, as determined from an almanac, and the DR latitude of the observer for the time of the observation. The magnitude of the amplitude angle corresponding to the exact declination and DR latitude is found by interpolation, and the proper prefix and suffix are added. Finally, the amplitude angle is converted to a true azimuth of the rising or setting sun, and this computed azimuth is compared with the observed azimuth to obtain the gyro error.

The conversion of an amplitude angle to a true azimuth is very similar to the conversion of a prefixed and suffixed azimuth angle to true azimuth. The base direction for the amplitude conversion is 090°T if the prefix is east (E), or 270°T if the prefix is west (W). If the suffix is north (N), the value of the amplitude angle is subtracted from 090°T for rising bodies, or added to 270°T for setting bodies. Conversely, if the suffix is south (S), the value of the amplitude angle is added to 090°T or subtracted from 270°T.

As an example of the use of the amplitude sight of the sun to obtain a gyro error, suppose that at 0600 ZT on 16 December the gyro bearing of the rising sun observed when the lower limb was two-thirds of a diameter above the visible horizon was 116° pgc. The 0600 ZT DR position was L 22° 30.0′N, λ 58° 30.0′W.

The first step in obtaining the gyro error is to obtain the declination of the sun for the time of the observation. The DR longitude places the ship in the +4 time zone, so the GMT of the observation is 1000 16 December. The declination of the sun for this time from the *Nautical Almanac* (see Figure 20-1B) is S23° 18.6′. The amplitude angle can now be determined by interpolation in Table 22 of *Bowditch*, using this declination and the DR latitude as entering arguments. To simplify the computations, both arguments can be expressed to the nearest tenth of a degree with no loss in the accuracy of the result.

For purposes of the interpolation, the four values of the amplitude angle for the tabulated arguments bracketing the exact calculated arguments are extracted from the excerpt shown in Figure 26-5 and written as shown below:

		Declination		
		23.0°	S23.3°	23.5°
	20°	24.6		25.1
Latitude	22.5°			
	25°	25.5		26.1

The interpolation is conducted in two steps, once for latitude, and again for declination; the order of the steps is immaterial. Thus, interpolating first for the exact latitude, we have:

		Declination		
		23.0°	S23.3°	23.5°
	20°	24.6		25.1
Latitude	22.5°	*25.1*		*25.6*
	25°	25.5		26.1

Next, for declination:

		Declination		
		23.0°	S23.3°	23.5°
Latitude	22.5°	25.1	*25.4*	25.6

Since it is sunrise, the prefix is E, and because the declination is south, the suffix is S. Thus, the resultant amplitude angle is E25.4°S. Converting this amplitude angle to a true azimuth yields a value of 90° + 25.4° or 115.4°T. Finally, this true azimuth is compared to the observed azimuth 116.0° pgc to obtain the resultant gyro error, 0.6°W. In practice, the error is usually rounded to the nearest half degree or 0.5°W in this case.

TABLE 28

Correction of Amplitude as Observed on the Visible Horizon

| Latitude | Declination | | | | | | | | | | | | | Latitude |
	0°	2°	4°	6°	8°	10°	12°	14°	16°	18°	20°	22°	24°	
0	0.0	0.0	0.0	0.0	0.0	0.0	0.0	0.0	0.0	0.0	0.0	0.0	0.0	0
10	0.1	0.1	0.1	0.1	0.1	0.1	0.1	0.1	0.1	0.1	0.1	0.1	0.1	10
15	0.2	0.2	0.2	0.2	0.2	0.2	0.2	0.2	0.2	0.2	0.2	0.2	0.2	15
20	0.3	0.3	0.3	0.3	0.3	0.3	0.3	0.3	0.3	0.3	0.3	0.3	0.3	20
25	0.3	0.3	0.3	0.3	0.3	0.4	0.3	0.3	0.3	0.3	0.3	0.3	0.3	25
30	0.4	0.4	0.4	0.4	0.5	0.4	0.4	0.4	0.4	0.4	0.4	0.5	0.5	30
32	0.4	0.4	0.4	0.4	0.5	0.4	0.4	0.4	0.4	0.4	0.5	0.5	0.5	32
34	0.5	0.5	0.5	0.5	0.5	0.5	0.5	0.5	0.5	0.5	0.5	0.5	0.5	34
36	0.5	0.5	0.5	0.5	0.5	0.5	0.5	0.5	0.6	0.5	0.6	0.6	0.6	36
38	0.6	0.6	0.6	0.6	0.6	0.6	0.6	0.6	0.6	0.6	0.6	0.6	0.6	38
40	0.6	0.6	0.6	0.6	0.6	0.6	0.6	0.6	0.6	0.6	0.7	0.7	0.7	40
42	0.6	0.6	0.6	0.6	0.7	0.7	0.7	0.7	0.7	0.7	0.7	0.7	0.7	42
44	0.7	0.7	0.7	0.6	0.6	0.7	0.7	0.7	0.8	0.8	0.8	0.8	0.9	44
46	0.7	0.7	0.7	0.7	0.7	0.8	0.8	0.8	0.8	0.8	0.8	0.9	0.9	46
48	0.8	0.8	0.8	0.8	0.8	0.8	0.8	0.8	0.9	0.9	1.0	1.0	1.0	48
50	0.8	0.8	0.8	0.8	0.9	0.9	0.9	0.9	0.9	1.0	1.0	1.1	1.0	50
51	0.8	0.8	0.8	0.8	0.9	0.9	0.9	0.9	0.9	1.0	1.1	1.1	1.1	51
52	0.9	0.9	0.9	0.9	0.9	0.9	1.0	1.0	1.0	1.1	1.1	1.1	1.3	52
53	0.9	0.9	0.9	0.9	0.9	0.9	1.0	1.0	1.0	1.1	1.2	1.2	1.3	53
54	1.0	1.0	1.0	1.0	1.0	1.0	1.1	1.1	1.1	1.2	1.2	1.3	1.3	54
55	1.0	1.0	1.0	1.0	1.1	1.1	1.0	1.2	1.2	1.2	1.3	1.3	1.4	55
56	1.0	1.0	1.0	1.0	1.1	1.1	1.2	1.2	1.2	1.3	1.3	1.4	1.5	56
57		1.1	1.1	1.1	1.1		1.2	1.2	1.3					57

Figure 26-6. *Correction of Amplitude Table 23,* Bowditch.

Had the amplitude observation of the sun been made when its center was located on the visible horizon, a small correction to the resulting observed azimuth would have been necessary. This correction is tabulated in Table 23 of *Bowditch*, reproduced in part in Figure 26-6. The correction extracted for sun observations is always applied in the direction away from the elevated pole, as per the instructions at the bottom of the table.

As an alternative to the use of the amplitude table, if an electronic calculator with trigonometric functions is available, the magnitude of the amplitude angle A can be easily calculated from the formula

$$\sin A = \frac{\sin \text{Dec}}{\cos \text{Lat}}$$

where Dec is the declination of the sun to the nearest tenth of a degree at the time of the observation, and Lat is the DR latitude. For the example given in this section,

$$\sin A = \frac{\sin 23.3°}{\cos 22.5°} \; ; \qquad A = 25.35° \cong 25.4°$$

470
PART II. CELESTIAL NAVIGATION

The amplitude angle found by this method can be slightly more accurate than that found by the tabular interpolation method, and using the formula is certainly faster. Once calculated, the amplitude angle is then converted to true azimuth as before.

The Sun Azimuth Sight

At times between sunrise and sunset, the gyro error may be obtained by computing the true azimuth of the sun by means of any one of a number of special tables designed for azimuth computations, or by use of the *Sight Reduction Tables for Marine Navigation, No. 229*. In practice, *Tables No. 229* are generally preferred for obtaining the true azimuth, inasmuch as these tables are widely available, and they are useful for reducing a standard sight as well. It would also be possible to obtain a true azimuth accurate to the nearest whole degree using volumes 2 and 3 of *Tables No. 249*, but this precision is insufficient to yield a gyro error accurate to the nearest half degree. *Tables No. 249*, therefore, is not suitable for the gyro error determination.

To use *Tables No. 229* for computation of the exact azimuth of the sun, it is necessary to interpolate in the tables for the exact values of DR latitude, LHA, and declination of the sun for the time of the azimuth observations. The DR latitude is, of course, obtained from the DR plot, while the LHA and declination of the sun are derived from either the *Air Almanac* or, preferably, the *Nautical Almanac*. A sight form designed for use in performing the required triple interpolation in *Tables No. 229* is helpful; a form prepared for use at the U.S. Naval Academy appears in Figure 26-7.

If preferred, a strip-type form for computing exact azimuth may also be used; such a form is included in appendix D in the back of this book. The data entered and the computations made are the same.

Entering arguments for the tables for the exact azimuth determination are the whole degrees of DR latitude, LHA, and declination lower than the exact calculated values. The tabulated azimuth angle Z corresponding with these integral arguments is first extracted and entered on the form as "Tab. Z." Next, the amount of change in the value of the tabulated azimuth angle, or "Z difference," is found between this tabulated azimuth angle and the value corresponding to the next higher integral degree of each of the three entering arguments. The Z difference for 1-degree increase of latitude, for example, is found by comparing the tabulated azimuth with the value contained in the next adjacent latitude column to the right, while keeping the LHA and declination the same. Likewise, the other two Z differences are found by successive comparisons of the tabulated azimuth angle for the next higher degrees of LHA and declination. All three Z differences are recorded on the exact azimuth form.

Using the exact minutes of each of the three entering arguments, each Z difference is then interpolated for the exact value of its corresponding argument;

EXACT AZIMUTH USING TABLES 229

Body	_____		EXACT		Z DIFF.	CORR.
			Deg : Min		(+ or –)	(+ or –)
DR L	_____	LAT	:			
DR λ	_____	LHA	:			
Date (L)	_____	DEC	:			
ZT	_____			Total (±)		
ZD (+ or –)	_____			Tab Z	_____	
GMT	_____			Exact Z	_____	
Date (G)	_____			Exact Zn	_____	
Tab GHA	_____			Gyro/Compass Brg	_____	
Inc'mt	_____			Gyro/Compass Error	_____	
GHA	_____					

DR λ _____

LHA _____

        d(+/–)

Tab Dec _____

d corr _____

Dec _____

NORTH LAT

LHA greater than 180° . . . . Zn = Z
LHA less than 180° . . . . Zn = 360° – Z

SOUTH LAT

LHA greater than 180° . . . . Zn = 180° – Z
LHA less than 180° . . . . Zn = 180° + Z

Figure 26-7. USNA form Exact Azimuth using Tables No. 229.

the result of each such interpolation is the correction necessary to the tabulated azimuth angle for the minutes of the argument. All three corrections are added algebraically to form the total correction to the tabulated azimuth angle. This total correction is then applied to obtain the exact azimuth angle corresponding to the exact values of the three entering arguments. As the final step, this exact azimuth angle is converted to a true azimuth, using the conversion formula printed on the azimuth sight form. Comparison of the true azimuth thus computed with the observed azimuth of the sun yields the gyro error.

The above procedure for determining the gyro error by observation of the azimuth of the sun may be summarized by the following steps:

1. Obtain and record the DR latitude, exact LHA, and declination of the sun for the time of the azimuth observation.
2. Using the integral degrees of these three quantities as entering arguments in *Tables No. 229*, extract the corresponding tabulated azimuth angle Z.
3. Obtain and record the three Z differences between this tabulated Z and the tabulated values for the next higher degree of each of the entering arguments.

Body	SUN
DR L	30-15.0 N
DR λ (+E-W)	64-33.0 W
Date (L)	16 DEC
ZT	11-12-06
ZD (+ or -)	+4
GMT	15-12-06
Date (G)	16 DEC
Tab GHA	46-06.1
Inc'mt	3-01.5
GHA	49-07.6
DR λ	64-33.0 W
LHA	344-34.6

	EXACT Deg	EXACT Min	Z DIFF. (+ or -)	CORR. (+ or -)
LAT	30	15.0 N		
LHA	344	34.6		
DEC	S 23	19.2		

Total (±) ____

Tab Z _____

Exact Z _____

Exact Zn _____

Gyro/Compass Brg _____

Gyro/Compass Error _____

	d(+/-)
Tab Dec	S 23-19.2	+.1
d corr	0
Dec	S 23-19.2

NORTH LAT

LHA greater than 180° .... Zn = Z
LHA less than 180° .. .. Zn = 360° - Z

SOUTH LAT

LHA greater than 180° .... Zn = 180° - Z
LHA less than 180° .... Zn = 180° + Z

Figure 26-8A. Exact azimuth form prepared for use with Tables No. 229.

4. Interpolate each Z difference for the correction corresponding to the exact minutes of its entering argument.
5. Add all three corrections algebraically to form the total correction to tabulated Z.
6. Apply the correction to obtain the exact azimuth angle.
7. Convert the exact azimuth angle to a true azimuth.
8. Compare the true azimuth with the observed azimuth to obtain the gyro error.

As an example of the determination of a gyro error by the sun azimuth sight, suppose that at 11-12-06 on 16 December the azimuth of the sun was observed bearing 161° pgc. The ship's DR position for this time was L 30° 15.0'N, λ 64° 33.0'W.

Working through the stepwise procedure enumerated above, the *Nautical Almanac* is first consulted to obtain the LHA and declination of the sun for the GMT of the observation, 15-12-06 on 16 December. After the resulting LHA and declination of the sun have been entered on the form along with the DR latitude, it should appear as in Figure 26-8A, ready for use with *Tables No. 229.*

EXACT AZIMUTH USING TABLES 229

Body	SUN		EXACT		Z DIFF.	CORR.
			Deg	Min	(+ or –)	(+ or –)
DR L	30-15.0 N	LAT	30	15.0 N	+.2	+.1
DR λ(+E-W)	64-33.0 W	LHA	344	34.6	+1.1	+.6
Date (L)	16 DEC	DEC	S 23	19.2	+.3	+.1
ZT	11-12-06			Total (±)		+.8
ZD (+ or –)	+4			Tab Z		162.0
GMT	15-12-06			Exact Z		162.8
Date (G)	16 DEC			Exact Zn		162.8° T
Tab GHA	46-06.1			Gyro/Compass Brg		161° pgc
Inc'mt	3-01.5			Gyro/Compass Error		1.8° E
GHA	49-07.6					
DR λ	64-33.0 W		**NORTH LAT**			
LHA	344-34.6					

d(+/–)

Tab Dec	S 23-19.2	+.1
d corr	0	
Dec	S 23-19.2	

NORTH LAT

LHA greater than 180° . . . . Zn = Z
LHA less than 180° . . . . Zn = 360° – Z

SOUTH LAT

LHA greater than 180° . . . . Zn = 180° – Z
LHA less than 180° . . . . Zn = 180° +Z

Figure 26-8B. *Completed exact azimuth form.*

Note that the GHA in this case was less than the westerly DR longitude, necessitating the addition of 360° to the GHA to produce a positive LHA.

The tabulated azimuth angle Z from *Tables No. 229* corresponding with the integral arguments latitude 30°N, LHA 344°, and declination S23° is 162.0. The Z difference between this value and the tabulated value for the same LHA and declination but a latitude of 31°N is +0.2. The Z difference for 30°N latitude, S23° declination, but LHA 345° is +1.1, and for 30°N latitude, LHA 344°, and declination S24°, the Z difference is +0.3. Interpolation of the +0.2 latitude Z difference for 15.0′ of latitude yields a latitude Z correction of +0.1. In similar fashion, interpolation for the remaining LHA and declination correction results in respective values of +0.6 and +0.1. Adding the three corrections yields a total Z correction of +0.8. Applying this +0.8 to the tabulated azimuth figure forms the exact azimuth angle 162.8°, which is converted to a true azimuth 162.8°T. As the final step, comparison with the observed azimuth, 161° pgc, yields a gyro error of 1.8°E, which for practical purposes is rounded off to 2.0°E. The completed azimuth form appears in Figure 26-8B.

Before leaving the subject of the azimuth sight, it should be stressed that the true azimuth of *any* celestial body can be found using *Tables No. 229*, for comparison with the observed azimuth of the body to obtain a compass error. The

PART II. CELESTIAL NAVIGATION

sun, however, is the body most commonly used for this purpose, since observation of its azimuth is very convenient in the latitude regions in which most ships operate.

SUMMARY

This chapter has examined the procedures most commonly used to determine the gyro error of a ship operating at sea beyond piloting waters. In lower-northern latitudes, observation of the azimuth of Polaris at twilight is a very convenient method of obtaining the gyro error, while an amplitude or azimuth observation of the sun yields excellent results by day in either northern or southern latitudes. Computation of the true azimuth of Polaris requires only the Polaris tables in the *Nautical Almanac.* The amplitude and thence the true azimuth of the sun at sunrise or sunset can be quickly obtained using either Tables 27 and 28 in the *American Practical Navigator (Bowditch)* or a trigonometric calculation, and the true azimuth of the sun at other times may be derived using the *Sight Reduction Tables for Marine Navigation, No. 229.*

27

TWILIGHT AND OTHER RISING AND SETTING PHENOMENA

There are several phenomena associated with the rising and setting of the sun and other celestial bodies of significance and consequent interest to the navigator. The most important of these are twilight, sunrise and sunset, and moonrise and moonset. *Twilight* is the period before sunrise when darkness is giving way to daylight, and after sunset, when the opposite progression takes place. Morning twilight ends at *sunrise,* defined as the first appearance of the sun's upper limb above the visible horizon, and evening twilight begins at *sunset,* or the disappearance of the sun's upper limb below the horizon. *Moonrise* and *moonset* are defined similarly to sunrise and sunset, by the contact of the upper limb of the moon with the visible horizon.

Twilight is of special interest to the navigator, as this is the only time when the visible horizon is still light enough to be clearly defined, while the navigational stars and planets are bright enough to be observed with a marine sextant. Sunrise and sunset are only slightly less important. Not only do the times of these respective events signify the ending and beginning of twilight, but as was discussed in the last chapter, they also indicate the approximate time at which a sun amplitude sight may be observed. Moonrise and moonset do not have much navigational significance, but these phenomena can be of interest to the Navy navigator for tactical planning applications.

Because of the regular rate of increase of the Greenwich hour angle (GHA) of the sun and the very small rate of change in its declination with time, phenomena related to the rising and setting of the sun are tabulated in the *Nautical Almanac* for the central day of each 3-day period covered. Mean times of sunrise and sunset and the darker limits of civil and nautical twilight are given as they occur along the Greenwich meridian at various latitudes between 72°N and 60°S. On the other hand, because the motion of the moon is very irregular both in GHA and declination, tabulations of the mean times of moonrise and moonset along the Greenwich meridian between latitudes 72°N and 60°S are required for each day. Because of the regular rate of increase of the GHA of the sun (approximately 15° per hour) the tabulated mean times of sun-associated phenomena at Greenwich (i.e., the GMTs) can be used as the local mean times (LMTs) of the phenomena at all other meridians. But because of the nonuniform motion of the moon, the tabulated GMTs of moonrise and moonset are valid without correction only along the Greenwich meridian.

Because of the inclination of the earth's axis, at certain times of the year the sun is either continually above or continually below the celestial horizon at the higher tabulated latitudes in the *Nautical Almanac*. This results in either constant daylight, twilight, or night being experienced at these latitudes throughout the 3-day period. The symbol //// is shown in place of a time to indicate twilight throughout the night, the symbol ⬜ indicates constant daylight, and the symbol ⬛ indicates constant night. If the moon is continually above or below the celestial horizon at a tabulated latitude, this fact is also denoted by the symbols ⬜ and ⬛, respectively. The phase and age of the moon are given at the bottom of each right-hand daily page, along with its time of meridian passage at Greenwich. A set of tabulated data for twilight, sunrise and sunset, and moonrise and moonset from the *Nautical Almanac* for 15, 16, and 17 December of a typical year appears in Figure 27-1.

In the *Air Almanac*, tables for predicting the darker limits of civil twilight and sunrise and sunset are located in the back of each volume, while tables for moonrise and moonset similar to those in the *Nautical Almanac* appear in the daily pages. The *Nautical Almanac* tabulations are generally preferred by most surface navigators, however, and their use will be demonstrated in this chapter.

PREDICTING THE DARKER LIMIT TIME OF TWILIGHT

As mentioned above, twilight is the period of incomplete darkness occurring just before sunrise or just after sunset. There are two kinds of twilight of concern in celestial navigation, differentiated by the position of the sun below the horizon at the darker limit:

Civil twilight is the period extending from sunrise or sunset at the lighter limit to the time at which the center of the sun is 6° below the celestial horizon at the darker limit.

Nautical twilight is the period extending from sunrise or sunset at the lighter limit to the time at which the center of the sun is 12° below the celestial horizon at the darker limit.

Civil twilight is characterized by a fairly sharp horizon and a light sky wherein only the brighter first- and second-magnitude stars and navigational planets are visible. During the darker stages of nautical twilight, the horizon may become very vague and the sky quite dark, with the dimmer second- and third-magnitude stars and planets becoming visible.

Lat.	Twilight Naut.	Twilight Civil	Sun- rise	Moonrise 15	Moonrise 16	Moonrise 17	Moonrise 18
°	h m	h m	h m	h m	h m	h m	h m
N 72	08 20	10 48	■	▭	▭	18 22	20 42
N 70	08 01	09 48	■	▭	16 01	18 55	20 56
68	07 45	09 14	■	▭	17 08	19 18	21 08
66	07 32	08 49	10 28	15 14	17 44	19 36	21 17
64	07 21	08 29	09 47	16 14	18 09	19 51	21 25
62	07 12	08 15	09 19	16 47	18 29	20 03	21 31
60	07 03	08 00	08 58	17 11	18 45	20 13	21 37
N 58	06 56	07 48	08 40	17 31	18 58	20 22	21 42
56	06 49	07 38	08 26	17 47	19 10	20 30	21 47
54	06 43	07 29	08 13	18 01	19 20	20 36	21 51
52	06 38	07 21	08 02	18 13	19 29	20 43	21 54
50	06 32	07 14	07 52	18 23	19 37	20 48	21 58
45	06 21	06 58	07 32	18 45	19 54	21 00	22 05
N 40	06 10	06 44	07 15	19 03	20 07	21 10	22 10
35	06 01	06 33	07 01	19 18	20 19	21 18	22 16
30	05 52	06 22	06 49	19 31	20 29	21 26	22 20
20	05 36	06 04	06 28	19 52	20 47	21 38	22 28
N 10	05 20	05 46	06 09	20 11	21 02	21 49	22 34
0	05 03	05 29	05 52	20 29	21 16	22 00	22 41
S 10	04 44	05 11	05 34	20 46	21 30	22 10	22 47
20	04 22	04 51	05 15	21 05	21 45	22 21	22 53
30	03 53	04 26	04 53	21 26	22 02	22 33	23 01
35	03 34	04 11	04 41	21 39	22 12	22 40	23 05
40	03 11	03 53	04 26	21 53	22 23	22 48	23 10
45	02 41	03 30	04 08	22 10	22 36	22 58	23 16
S 50	01 56	03 01	03 45	22 31	22 52	23 09	23 22
52	01 28	02 46	03 34	22 41	23 00	23 14	23 26
54	00 44	02 28	03 22	22 52	23 08	23 20	23 29
56	////	02 06	03 08	23 04	23 17	23 26	23 33
58	////	01 37	02 52	23 19	23 28	23 33	23 37
S 60	////	00 49	02 31	23 36	23 40	23 41	23 41

Lat.	Sun- set	Twilight Civil	Twilight Naut.	Moonset 15	Moonset 16	Moonset 17	Moonset 18
°	h m	h m	h m	h m	h m	h m	h m
N 72	■	13 02	15 30	▭	▭	13 37	12 45
N 70	■	14 02	15 49	▭	14 25	13 03	12 28
68	■	14 36	16 05	▭	13 16	12 38	12 15
66	13 22	15 01	16 18	13 29	12 40	12 19	12 04
64	14 04	15 21	16 29	12 29	12 13	12 03	11 55
62	14 31	15 37	16 39	11 55	11 53	11 50	11 47
60	14 53	15 50	16 47	11 30	11 36	11 39	11 40
N 58	15 10	16 02	16 54	11 10	11 22	11 29	11 34
56	15 25	16 12	17 01	10 53	11 10	11 20	11 28
54	15 38	16 21	17 07	10 39	10 59	11 13	11 23
52	15 49	16 29	17 13	10 27	10 49	11 06	11 19
50	15 58	16 37	17 18	10 16	10 41	11 00	11 15
45	16 19	16 53	17 30	09 53	10 23	10 46	11 06
N 40	16 36	17 06	17 41	09 34	10 08	10 35	10 58
35	16 50	17 18	17 50	09 19	09 55	10 25	10 52
30	17 02	17 29	17 59	09 05	09 44	10 17	10 46
20	17 23	17 48	18 15	08 42	09 25	10 02	10 36
N 10	17 42	18 05	18 31	08 22	09 08	09 49	10 28
0	17 59	18 22	18 48	08 03	08 52	09 37	10 20
S 10	18 17	18 40	19 07	07 44	08 36	09 25	10 11
20	18 36	19 00	19 29	07 23	08 19	09 12	10 02
30	18 58	19 25	19 58	06 59	07 59	08 57	09 52
35	19 11	19 40	20 17	06 45	07 47	08 48	09 46
40	19 26	19 58	20 40	06 29	07 34	08 38	09 39
45	19 44	20 21	21 10	06 09	07 18	08 26	09 31
S 50	20 06	20 50	21 56	05 45	06 58	08 11	09 22
52	20 17	21 06	22 23	05 33	06 49	08 04	09 17
54	20 29	21 23	23 08	05 19	06 39	07 57	09 12
56	20 43	21 45	////	05 03	06 26	07 48	09 07
58	21 00	22 15	////	04 44	06 13	07 39	09 01
S 60	21 20	23 04	////	04 21	05 56	07 27	08 54

Day	SUN Eqn. of Time 00ʰ	SUN Eqn. of Time 12ʰ	SUN Mer. Pass.	MOON Mer. Pass. Upper	MOON Mer. Pass. Lower	MOON Age	MOON Phase
	m s	m s	h m	h m	h m	d	
15	05 12	04 58	11 55	01 50	14 16	17	●
16	04 43	04 29	11 56	02 40	15 04	18	
17	04 14	04 00	11 56	03 27	15 49	19	

Figure 27-1. *Sample rising and setting phenomena predictions,* Nautical Almanac, *15, 16, 17 December.*

The prediction of the zone time of the beginning of morning civil or nautical twilight and the ending of evening civil or nautical twilight for either a stationary or moving ship is somewhat similar to the estimation of the zone time of LAN discussed in chapter 25. Because the LMT of the darker limits of twilight varies with the latitude of the observer, it is necessary to adjust the tabulated mean time of twilight phenomena for the difference in latitude between the latitude tabulated in the *Nautical Almanac* and the latitude of the observer. The adjusted mean time of the twilight phenomenon of interest is then converted to local zone time by applying the arc-time difference in longitude between the observer's longitude and the standard meridian of the local time zone, as was done in the LAN problem.

In practice, the navigator is usually most concerned with the prediction of the zone time of the darker limit of civil twilight, inasmuch as this time occurs about midway in the period suitable for celestial observations. If the ship is stationary in a position such as an anchorage, the time of the twilight phenomena can be determined with great accuracy using the *Nautical Almanac.* If the ship is moving, the time can be predicted within an accuracy of one or two minutes based on the DR plot.

Regardless of whether the ship is stationary or moving, the procedure for estimating the zone time of a darker limit of civil or nautical twilight is basically the same. As the first step, the local mean time of the event of interest at the nearest tabulated latitude in the daily pages of the almanac smaller than the DR latitude (for a moving ship) or the ship's position (for a stationary ship) is extracted. Next, a correction to this time for the difference between the ship's latitude and the lower tabulated latitude is obtained from Table I of "Tables for Interpolating Sunrise, Moonrise, Etc." in the back of the *Nautical Almanac.* This table is reproduced in Figure 27-2.

The entering arguments for Table I are the difference in latitude between the two tabulated values bracketing the ship's latitude (called the *tabular interval* in Table I), the closest tabulated difference to the actual difference between the ship's latitude and the lower of the two tabulated latitudes, and the closest tabulated time difference between the mean times of the event of interest at the upper and lower tabulated latitudes. The sign of the resulting LMT correction is determined by inspection of the daily predictions.

Application of this latitude LMT correction to the listed mean time of the phenomenon at the smaller tabulated latitude yields the LMT of the phenomenon at all three meridians of interest: Greenwich, the observer's meridian, and the standard meridian of the local time zone.

Next, this initial estimate of the LMT of the phenomenon is converted to a zone time by application of an arc-time correction corresponding to the difference between the observer's longitude and the standard meridian of the local time zone, just as was done in estimating the time of LAN. The resulting zone

TABLES FOR INTERPOLATING SUNRISE, MOONRISE, ETC.
TABLE I—FOR LATITUDE

Tabular Interval			Difference between the times for consecutive latitudes															
10°	5°	2°	5m	10m	15m	20m	25m	30m	35m	40m	45m	50m	55m	60m	1h 05m	1h 10m	1h 15m	1h 20m
° ′	° ′	° ′	m	m	m	m	m	m	m	m	m	m	m	m	h m	h m	h m	h m
0 30	0 15	0 06	0	0	1	1	1	1	1	2	2	2	2	2	0 02	0 02	0 02	0 02
1 00	0 30	0 12	0	1	1	2	2	3	3	3	4	4	4	5	05	05	05	05
1 30	0 45	0 18	1	1	2	3	3	4	4	5	5	6	7	7	07	07	07	07
2 00	1 00	0 24	1	2	3	4	5	5	6	7	7	8	9	10	10	10	10	10
2 30	1 15	0 30	1	2	4	5	6	7	8	9	9	10	11	12	12	13	13	13
3 00	1 30	0 36	1	3	4	6	7	8	9	10	11	12	13	14	0 15	0 15	0 16	0 16
3 30	1 45	0 42	2	3	5	7	8	10	11	12	13	14	16	17	18	18	19	19
4 00	2 00	0 48	2	4	6	8	9	11	13	14	15	16	18	19	20	21	22	22
4 30	2 15	0 54	2	4	7	9	11	13	15	16	18	19	21	22	23	24	25	26
5 00	2 30	1 00	2	5	7	10	12	14	16	18	20	22	23	25	26	27	28	29
5 30	2 45	1 06	3	5	8	11	13	16	18	20	22	24	26	28	0 29	0 30	0 31	0 32
6 00	3 00	1 12	3	6	9	12	14	17	20	22	24	26	29	31	32	33	34	36
6 30	3 15	1 18	3	6	10	13	16	19	22	24	26	29	31	34	36	37	38	40
7 00	3 30	1 24	3	7	10	14	17	20	23	26	29	31	34	37	39	41	42	44
7 30	3 45	1 30	4	7	11	15	18	22	25	28	31	34	37	40	43	44	46	48
8 00	4 00	1 36	4	8	12	16	20	23	27	30	34	37	41	44	0 47	0 48	0 51	0 53
8 30	4 15	1 42	4	8	13	17	21	25	29	33	36	40	44	48	0 51	0 53	0 56	0 58
9 00	4 30	1 48	4	9	13	18	22	27	31	35	39	43	47	52	0 55	0 58	1 01	1 04
9 30	4 45	1 54	5	9	14	19	24	28	33	38	42	47	51	56	1 00	1 04	1 08	1 12
10 00	5 00	2 00	5	10	15	20	25	30	35	40	45	50	55	60	1 05	1 10	1 15	1 20

Table I is for interpolating the L.M.T. of sunrise, twilight, moonrise, etc., for latitude. It is to be entered, in the appropriate column on the left, with the difference between true latitude and the nearest tabular latitude which is *less* than the true latitude; and with the argument at the top which is the nearest value of the difference between the times for the tabular latitude and the next higher one; the correction so obtained is applied to the time for the tabular latitude; the sign of the correction can be seen by inspection. It is to be noted that the interpolation is not linear, so that when using this table it is essential to take out the tabular phenomenon for the latitude *less* than the true latitude.

Figure 27-2. *Table I, sunrise, moonrise interpolation tables,* Nautical Almanac.

time is the time of the darker limit of twilight if the observer's ship is stationary; if it is moving, this time becomes the "first estimate" of the zone time of the event.

In the case of a moving ship, a DR position for the first estimate time is plotted. A second and final estimate of the zone time of the twilight phenomenon is then made by repeating the procedure described above, using the first estimate DR position as the observer's position.

Inasmuch as the procedure for predicting the zone time of the darker limit of twilight is virtually identical to the procedure used for predicting the time of sunrise or sunset, an example of which will be given in the next section of this chapter, no example of a twilight solution will be given herein.

PREDICTING THE TIME OF SUNRISE AND SUNSET

As was mentioned at the beginning of this chapter, the prediction of the zone time of sunrise and sunset is of importance not only because these phenomena mark the end of morning and the beginning of evening twilight, but also because one of the most accurate types of azimuth sights possible—the amplitude sight—can be made near these times.

The prediction of sunrise and sunset is accomplished in virtually the same manner as the prediction of the time of a darker limit of twilight. As is the case with predictions of the latter type, the determination of times of sunrise and sunset can be made either for a stationary position or a moving ship. As an aid to the inexperienced navigator, the procedure for predicting the zone time of a darker limit of twilight or the zone time of sunrise or sunset using the *Nautical Almanac* is summarized as follows:

1. Obtain the local mean time of the phenomenon of interest at Greenwich on the day in question for the tabulated latitude smaller than the DR latitude (moving ship) or the ship's position (stationary ship). (Remember that the local mean time of the phenomenon at Greenwich can be used with insignificant error as the local mean time at all other meridians.)
2. In the case of a moving ship, plot and record a DR position for this time.
3. Using this position, determine a first estimate of zone time of the phenomenon as follows:
   A. Find the latitude LMT correction from Table I corresponding to the difference between the ship's latitude and the lower tabulated latitude.
   B. Apply this correction to the tabulated mean time of the phenomenon.
   C. Convert the resulting adjusted LMT of the phenomenon to zone time by applying the arc-time difference between the ship's longitude and the standard meridian of the time zone by which the ship's clocks are set. In so doing, add the difference if west of the standard meridian, subtract if east.

For a stationary ship, this first estimate is the predicted zone time of the phenomenon. In the case of a moving ship, the following additional steps to find a second estimated time are required:

4. Plot a DR position for the zone time of the first estimate.
5. Using this position, determine the second estimate of zone time of the phenomenon:
   A. Find a second latitude correction from Table I.
   B. Apply this correction to the tabulated time of the phenomenon at the lower tabulated latitude.
   C. Convert the resulting second estimate of LMT to zone time by determining and applying a second longitude arc-time difference for the first estimate DR longitude.

If the first estimate of the zone time of the phenomenon for a moving ship places the ship above the next higher latitude for which tabulations are printed in the *Nautical Almanac,* care must be taken when finding the second estimate to use the listed time corresponding to this next higher latitude as the base time to which the second estimate Table I and arc-time corrections are applied. Conversely, if the first estimate places the ship below the latitude used as an entering argument for the initial estimate of the time of the phenomenon, the second estimate Table I and longitude corrections should be applied to the tabulated mean time corresponding to the next lower tabulated latitude.

Although the computation of time of twilight or sunrise and sunset can be done without one by following the steps given above, a form such as the one prepared by the Navy for the purpose helps to keep the data organized; its use will be demonstrated below.

An Example Sunrise Computation

As an example of the application of this procedure to find the predicted zone time of sunrise for a moving ship, suppose that at 0600 ZT on 16 December a ship is located at a DR position L 31° 42.0′N, λ 65° 26.3′W, steaming on course 140°T at 20 knots. The navigator desires to predict the time of sunrise, assuming the ship remains on the same course and speed.

Using the procedure outlined above, the first step is to extract the tabulated LMT of sunrise at the approximate DR position of the ship. Since the DR plot

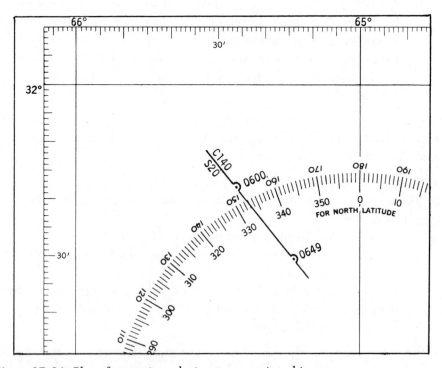

Figure 27–3A. *Plot of a sunrise solution on a moving ship.*

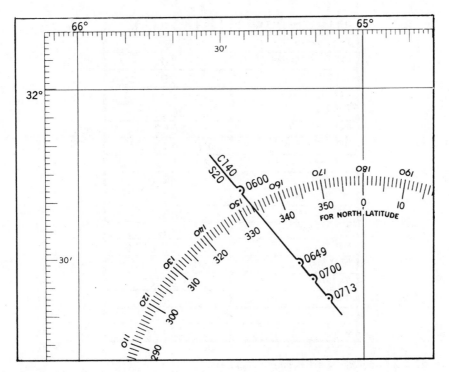

Figure 27-3B. *Completed plot of a sunrise solution on a moving ship.*

indicates that the ship is between 30° and 35°N, the mean time tabulated in the *Nautical Almanac* for 16 December for latitude 30°N is extracted; it is 0649. A DR position is plotted and recorded for this time at L 31° 29.3′N, λ 65° 14.0′W. At this point, the DR plot appears as shown in Figure 27-3A.

For this DR position, the tabular interval is 5°, the difference in latitude is 31° 29.3′ − 30° = 1° 29.3′, and the difference in time between the tabulations for 30° and 35°N is 12 minutes. Entering Table I with the closest tabulated arguments, a latitude LMT correction of +3 minutes is extracted. Applying this correction to the tabulated mean time for 30°N, 0649, yields an adjusted LMT of 0652.

Next, this adjusted LMT of sunrise is converted to zone time by application of the arc-time difference corresponding to the difference in longitude between the 0649 DR position and 60°W, the standard meridian of the time zone in this case. The arc difference is 5° 14.0′, so the corresponding time correction taken from the "Conversion of Arc to Time" tables in the almanac is 21 minutes; the sign of the correction is positive, since the ship is located to the west of the standard meridian. Application of this correction to the LMT adjusted for latitude, 0652, yields a first estimate of zone time of sunrise of 0713.

To find the second estimate of the zone time of sunrise, a DR position is first plotted for 0713; it is L 31° 23.0′N, λ 65° 07.8′W. Now the correction proce-

DR Lat (Tab LMT)	31° 29.3′ N
Tab Interval	5°
Lat Time Interval	10′
Lat Diff	1° 29.3′
Corr Table 1	+3′
Tab LMT	0649
Corr LMT	0652
DRλ (Tab LMT)	65° 14.0′ W
STD Mer	60° W
dλ (arc)	5° 14.0′
dλ (time)	+21
LMT	0652
ZT (1st est)	0713
Rev DR λ	65° 07.8′ W
STD Mer	60° W
dλ (arc)	5° 07.8′
dλ (time)	+21
LMT	0652
ZT (2nd est)	0713

Figure 27-3C. *Completed sunrise computation form.*

dure is repeated using this position as a basis. The LMT latitude correction from Table I is +3, resulting in a second estimate of LMT of sunrise identical to the first, 0652. The arc-time longitude correction for the 0713 DR position is again 21 minutes, so the second and final estimate of zone time of sunrise is 0713. The completed DR plot appears in Figure 27-3B, and the completed sunrise computation form appears in Figure 27-3C.

PREDICTING THE TIME OF MOONRISE AND MOONSET

Although the times of moonrise and moonset do not have much navigational significance, the navigator occasionally will have need to predict the times of these phenomena for other purposes, such as operational or tactical planning. Additionally, when the moon is at or near its full phase and its altitude is low, the horizon beneath it may be illuminated to a degree sufficient to permit sextant altitude observations of the moon itself or other celestial bodies in the same vicinity. Prediction of the times of moonrise and moonset may indicate when the possibility for such an auxiliary sight opportunity may exist during a given night.

Prediction of the time of moonrise or moonset is more complicated than a similar prediction for phenomena associated with the sun because the rate of

TABLE II—FOR LONGITUDE

Long. East or West	Difference between the times for given date and preceding date (for east longitude) or for given date and following date (for west longitude)																	
	10ᵐ	20ᵐ	30ᵐ	40ᵐ	50ᵐ	60ᵐ	1ʰ + 10ᵐ	1ʰ + 20ᵐ	1ʰ + 30ᵐ	1ʰ + 40ᵐ	1ʰ + 50ᵐ	1ʰ + 60ᵐ	2ʰ 10ᵐ	2ʰ 20ᵐ	2ʰ 30ᵐ	2ʰ 40ᵐ	2ʰ 50ᵐ	3ʰ 00ᵐ
°	m	m	m	m	m	m	m	m	m	m	m	m	h m	h m	h m	h m	h m	h m
0	0	0	0	0	0	0	0	0	0	0	0	0	0 00	0 00	0 00	0 00	0 00	0 00
10	0	1	1	1	1	2	2	2	2	3	3	3	04	04	04	04	05	05
20	1	1	2	2	3	3	4	4	5	6	6	7	07	08	08	09	09	10
30	1	2	2	3	4	5	6	7	7	8	9	10	11	12	12	13	14	15
40	1	2	3	4	6	7	8	9	10	11	12	13	14	16	17	18	19	20
50	1	3	4	6	7	8	10	11	12	14	15	17	0 18	0 19	0 21	0 22	0 24	0 25
60	2	3	5	7	8	10	12	13	15	17	18	20	22	23	25	27	28	30
70	2	4	6	8	10	12	14	16	17	19	21	23	25	27	29	31	33	35
80	2	4	7	9	11	13	16	18	20	22	24	27	29	31	33	36	38	40
90	2	5	7	10	12	15	17	20	22	25	27	30	32	35	37	40	42	45
100	3	6	8	11	14	17	19	22	25	28	31	33	0 36	0 39	0 42	0 44	0 47	0 50
110	3	6	9	12	15	18	21	24	27	31	34	37	40	43	46	49	0 52	0 55
120	3	7	10	13	17	20	23	27	30	33	37	40	43	47	50	53	0 57	1 00
130	4	7	11	14	18	22	25	29	32	36	40	43	47	51	54	0 58	1 01	1 05
140	4	8	12	16	19	23	27	31	35	39	43	47	51	54	0 58	1 02	1 06	1 10
150	4	8	13	17	21	25	29	33	38	42	46	50	0 54	0 58	1 03	1 07	1 11	1 15
160	4	9	13	18	22	27	31	36	40	44	49	53	0 58	1 02	1 07	1 11	1 16	1 20
170	5	9	14	19	24	28	33	38	42	47	52	57	1 01	1 06	1 11	1 16	1 20	1 25
180	5	10	15	20	25	30	35	40	45	50	55	60	1 05	1 10	1 15	1 20	1 25	1 30

Table II is for interpolating the L.M.T. of moonrise, moonset and the Moon's meridian passage for longitude. It is entered with longitude and with the difference between the times for the given date and for the preceding date (in east longitudes) or following date (in west longitudes). The correction is normally *added* for west longitudes and *subtracted* for east longitudes, but if, as occasionally happens, the times become earlier each day instead of later, the signs of the corrections must be reversed.

xxxii

Figure 27-4. *Table II, moonrise/moonset longitude interpolations,* Nautical Almanac.

change of GHA and to a lesser extent the declination of the moon are not constant. Whereas the GHA of the sun increases at the rate of approximately 15° per hour, the GHA of the moon increases at an irregular rate, usually faster but at times slower than that of the sun. Thus, in the *Nautical Almanac* daily pages, predictions of the mean times of moonrise and moonset along the Greenwich meridian must be given for each date covered by the 3-day period. For convenience, the data for the first day on the following page are also given (Figure 27-1). Because the rate of change of the GHA of the moon is so irregular, the tabulated mean times of its rising and setting at the Greenwich meridian cannot be considered the same as the LMTs of these phenomena at other meridians as in the case of the sun-related phenomena. Hence, after the tabulated GMT of moonrise or moonset has been adjusted for the exact latitude of the observer by use of Table I, shown in Figure 27-2, the resulting LMT at the Greenwich meridian must then be corrected for the longitude of the observer by use of a special longitude LMT correction table designed for use with the moon. This table, designated Table II, is located in the back of the *Nautical Almanac* beneath Table I, and is shown in Figure 27-4. Its use will be explained below. After the tabulated time of the lunar event of interest at Greenwich has been corrected for the exact latitude and longitude of the observer using Tables I and II, the resulting adjusted LMT must then be converted to zone time in the usual fashion.

As was the case with the other rising and setting phenomena discussed earlier, the time of moonrise or moonset may be predicted for either a stationary or moving ship. The first step in the prediction procedure is to choose the proper columns of tabulations in the *Nautical Almanac*; two adjacent days' columns must always be used because of the irregular motion of the moon. If the observer is located in *east* longitudes on the day for which the moonrise or moonset prediction is to be made, the columns containing tabulations for the day in question and for the *preceding* day are used. Conversely, if the observer is in *west* longitudes, the columns containing tabulations for the day in question and for the *following* day are used.

For a stationary ship, after the proper data columns have been selected, the difference in mean time of the phenomenon resulting from the difference between the observer's latitude and the closest tabulated latitude smaller than the observer's latitude is first taken into account by the use of Table I of Figure 27-2. Corrections are found from Table I both for the mean time at the tabulated latitude on the day in question and on the preceding (λE) or following (λW) day. After these corrections have been applied to both days' times, the correction to LMT for the observer's longitude is then found from Table II in Figure 27-4. Entering arguments for this table are the observer's longitude and the difference between the two LMTs of the phenomenon corrected for latitude as described above. Interpolation to the nearest minute may be required in some cases in which the time difference is large. The correction obtained from Table II is normally added for west longitudes and subtracted for east longitudes, but if the tabulated times on the successive days of interest become earlier instead of later, the sign of the correction must be reversed.

After the LMT correction for longitude from Table II has been obtained, it is applied to the LMT adjusted for latitude on the day in question to obtain the LMT of the phenomenon at the stationary position. This LMT must then be converted to zone time by application of the arc-time difference between the observer's meridian and the standard meridian of the observer's time zone.

In the case of a moving ship, the tabulated time of moonrise or moonset at the tabulated latitude smaller than the approximate DR latitude of the ship on the day in question is used as a starting point. The exact DR position corresponding to this LMT is plotted and used as the basis for finding the two latitude corrections as explained before. The longitude correction from Table II is then found and applied to the DR position time corrected for latitude to form a "first estimate" of the LMT of moonrise or moonset. This first estimate in

LMT is then converted to zone time, using as a basis for this conversion the DR position plotted for the tabulated mean time of the phenomenon on the given day. A second estimate is then found by repeating the interpolation procedure, using as a basis the first estimate zone time DR position. This second estimate of zone time of the phenomenon is then considered the approximate time of the event at the projected position of the moving ship.

The procedure for predicting the time of moonrise or moonset may be summarized as follows:

1. Obtain the GMT of the phenomenon of interest on the day in question for the tabulated latitude smaller than the DR latitude (moving ship) or the ship's position (stationary ship).
2. In the case of a moving ship, plot a DR position for this time, using the GMT as local mean time.
3. Using this position, determine a first estimate of zone time of moonrise or moonset as follows:
   A. Find a latitude correction to LMT from Table I for both the day in question and the preceding day, if in east longitude, and the following day, if in west longitude.
   B. Apply these corrections to the tabulated times corresponding to the nearest tabulated latitudes smaller than the ship's latitude.
   C. Find the difference between these two adjusted LMTs.
   D. Find the longitude correction to LMT from Table II using as entering arguments this time difference and the ship's longitude.
   E. Apply the longitude correction to the LMT adjusted for latitude on the day in question to find the LMT of moonrise or moonset at the ship's stationary or DR position.
   F. Convert this LMT to zone time, using the arc-time difference between the initial DR longitude or the ship's stationary position and the standard meridian of the observer's time zone.

The result at this point is the first estimate of the zone time of the moon phenomenon. For a stationary ship, this first estimate is the predicted time of moonrise or moonset. For a moving ship, the following additional steps are required to determine a second and final estimate:

4. Plot a DR position for the zone time of the first estimate.
5. Using this position, repeat all parts of step 3 above. The result is the predicted zone time of moonrise or moonset for the moving ship.

A moonrise/moonset computation form similar to the one used by the Navy and illustrated in the following section is extremely helpful during such computations.

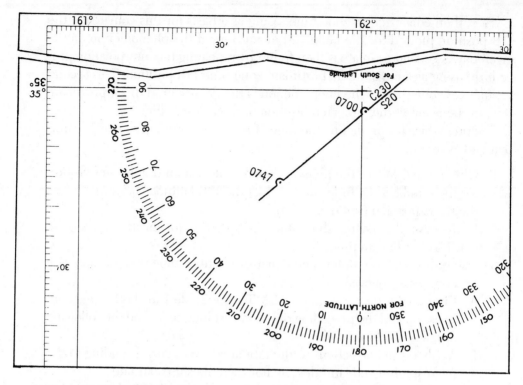

Figure 27-5A. *Initial plot of a moonset solution on a moving ship.*

An Example Moonset Computation

As an example of the prediction of the time of moonset for a moving ship, suppose that in the morning of 16 December a ship's 0700 DR position was L 35° 03.5'S, λ 162° 01.0'E, steering on course 230°T at a speed of 20 knots. The navigator desires to predict the time of moonset if the ship remains on the same course and speed.

Following through the procedure set forth above, the first step is to determine the LMT of moonset on 16 December at the approximate DR position. Since the time of moonset tabulated for latitude 35° south (the lower tabulated latitude) on 16 December in the *Nautical Almanac* is 0747, a DR position is plotted for this time for use in obtaining the first estimate of the zone time of moonset; the plot is shown in Figure 27-5A.

The coordinates of the 0747 DR position are L 35° 15.8'S, λ 161° 43.0'E. Since the ship is in east longitude, the columns of the almanac containing moonset time for 15 and 16 December will be used to obtain latitude and longitude LMT corrections. The times of moonset at 35°S on 15 and 16 December are recorded (see Figure 27-1) as 0645 and 0747, respectively. The latitude difference between the 0747 DR latitude and 35°S is 15.8', the tabular interval

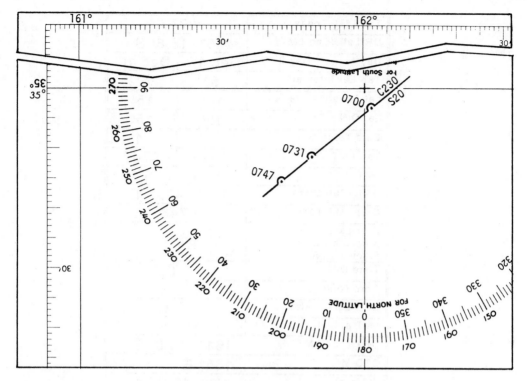

Figure 27-5B. *Completed plot of a moonset solution on a moving ship.*

is 5°, and the differences between the listed times at 35°S and 40°S on 15 and 16 December are –16 minutes and –13 minutes, respectively. The resulting latitude LMT corrections from Table I of Figure 27-2 are each –1 minute. Applying these corrections to the tabulated times for 15 and 16 December yields values of LMT corrected for latitude of 0644 and 0746, respectively.

Next, Table II is entered for the longitude LMT correction, using as arguments the difference between the two partially corrected LMTs, 0746 – 0644, or 62 minutes, and the 0747 DR longitude. After mental interpolation in the table, a correction of –28 minutes results. Application of this correction to the adjusted LMT 0746 results in a first estimate of the LMT of moonrise of 0746 – 28, or 0718. To convert this LMT to zone time, an arc-time difference is found for the difference in longitude between the 0747 DR position and the time zone standard meridian, 165°E in this case. The arc difference is 3° 17.0′, and the corresponding time correction is 13 minutes; it is added, since the ship is west of the time zone standard meridian. Thus, the first estimate of the zone time of moonset is 0731.

To find the second estimate, a DR position is plotted for 0731, L 35° 12.0′S, λ 161° 48.0′E, and the procedure is repeated. After applying the latitude LMT corrections from Table I to the tabulated times for 35°S for 15 and 16 Decem-

Tab LMT Today	0747	
DR Lat (Tab LMT)	35° 15.8' S	
Day 1	Day 2	Dec 15	Dec 16
Tab Lat Interval	5°	5°
Lat Time Interval	15'	15'
Lat Diff	15'	15'
Corr Table 1	-1'	-1'
Tab LMT	0645	0747
LMT	0644	0746
DRλ (Tab LMT)	161° 43.0' E	
Corr LMT Today	0746	
Corr LMT preceding(E) following (W)	0644	
Time Diff	62'	
Corr Table 2	-28	
Corr LMT Today	0746	
LMT actual	0718	
DRλ (Tab LMT)	161° 43.0' E	
STD Mer	165° E	
dλ (arc)	3° 17.0'	
dλ (time)	+13'	
LMT actual	0718	
ZT (1st est)	0731	
Rev DRλ	161° 48.0' E	
STD Mer	165° E	
dλ (arc)	3° 12.0'	
dλ (time)	+13'	
LMT actual	0718	
ZT (2nd est)	0731	

Figure 27-5C. *Completed moonset computation form.*

ber, partially corrected LMTs of 0644 and 0746 again result. The longitude LMT correction is again −28 minutes, resulting in an LMT corrected for latitude and longitude of 0718. To correct this LMT to zone time, the arc-time correction for the difference between the 0731 DR longitude and 165°E is found; it is again +13 minutes. Hence, the second and final estimate of the zone time of moonset in the morning of 16 December for this ship is 0718 + 13 or ZT 0731. The completed plot is shown in Figure 27-5B, and the completed computation form for this example is given in Figure 27-5C.

SUMMARY

This chapter concludes part 2 of *Marine Navigation* dealing with the practice of celestial navigation by discussing the procedures used to predict the times of certain rising and setting phenomena. The determination of the times of the darker limits of morning and evening celestial twilight are of greatest concern to the navigator, followed by the prediction of the times of sunrise and sunset and, finally, moonrise and moonset. The procedures for prediction of the times of sun-related rising and setting phenomena are relatively uncomplicated, while the procedures for estimating the times of the phenomena of least significance in celestial navigation—moonrise and moonset—are fairly involved because of the irregular rate of change of the GHA and declination of the moon.

PART 3.
ELECTRONIC NAVIGATION

INTRODUCTION TO
ELECTRONIC NAVIGATION

28

The previous chapters of this book have each been concerned with some aspect of the practice of celestial navigation at sea. In the days of sail, the principles of celestial navigation just described were sufficient by themselves to safely direct the voyages of sailing vessels. Because of the relatively slow speed of even the fastest sailing ship, unfavorable weather conditions over part of a transit would not unduly hazard the vessel if a good dead reckoning plot were maintained. Even if several days elapsed without a good celestial fix being obtained, the ship could not stray an excessive distance from her planned track because of her slow speed. A divergence of a few tens of miles between the actual track and the DR track in times of overcast or stormy weather on the open sea was hardly significant for a sailing voyage of several thousand miles.

With the advent of the airplane and the deep-draft steamship, however, and more recently the missile-carrying submarine, the capability of a regular determination of position regardless of the environmental conditions prevailing over the open sea became more and more critical. Not only are the effects of any divergence from the planned track amplified by the increased speed of airborne travel, but the deep draft and huge displacement of modern merchant and combatant ships make close adherence to the preplanned track a necessity for both economy of operations and safe navigation. The precision of today's guided weapons also demands precision and accuracy in the navigation of the platforms from which they are launched, as well as in midcourse guidance, if their full capabilities are to be realized.

The first practical use of electronic navigation occurred in the early years of this century, when primitive shore-based radio direction-finding equipment was used to establish a bearing to a ship, and later to an airplane, transmitting a continuous radio signal. Once determined, the bearing was then transmitted to the ship or plane via radiotelegraph. At some locations, several of these shore stations were linked by telephone lines, enabling a rough determination of position to be made by crossed bearings, which position could then be transmitted to the ship or plane. This technique is still in use today in more refined form in many coastal areas of the world.

In the 1930s, improved radio direction-finding techniques and equipment led to the establishment of navigational aids for both contemporary air and surface navigation called *radiobeacons*. These aids consisted essentially of small radio transmitters located along coastal areas; they were designed to provide radio bearings that could be used in lieu of visual bearings at sea in foul weather. Position-finding by these beacons became known as *radionavigation*. Continued refinements in the state of the art of electronics technology and a better understanding of electromagnetic wave propagation led to the subsequent development of radar, sonar, and longer-range radio navigation systems, with the result that modern electronic navigational aids bear little resemblance to the early devices employed in radio navigation. The term *radionavigation* is still applied to any electronic navigation system employing radio waves as the means of position-finding; the term *electronic navigation* has come into wide use to describe not only radionavigation systems, but also all other systems dependent upon an electronic device for the establishment of a position.

Modern electronic navigation systems can be classified both by their range and their scope. The range classifications most generally used are short range, medium or intermediate range, and long range, with the exact limits between each being rather indefinite. The scope of a system is used here to mean either a self-contained or externally supported system, and either an active (i.e., transmitting) or passive (i.e., not transmitting) mode of operation.

The short-range systems in most common use at the present time are radiobeacons, radar, and Decca (considered to be both a short- and a medium-range system). Medium-range systems include Decca and certain types of extended-range radar. The only remaining long-range system in full operation is Loran-C. All of these systems depend on active radio-frequency (RF) transmissions, and with the exception of radar, all are externally supported with respect to a ship operating at sea. In addition to these systems, there is a category of systems that do not conveniently fit into any range classification. These are the so-called *advanced navigation systems,* of which three are of prime importance: the shipboard inertial navigation system, the Doppler sonar system, and the satellite-based Navstar Global Positioning System. The first of these is the only self-contained, passive electronic navigation system currently available.

Before specifically discussing any of these systems, it is necessary to review the basic theory of the electromagnetic wave upon which most of them depend for their operation. This chapter reviews the fundamental characteristics and behavior of the electromagnetic wave, and discusses the way in which these waves are employed in radio navigation. The radio navigation systems listed above are each the subject of a separate section of chapter 29, and chapter 30 deals with the three advanced navigation systems mentioned. Radar is discussed in chapter 10 in part 1.

THE ELECTROMAGNETIC WAVE

For navigational purposes it is not necessary to be more than generally aware of the methods by which an electromagnetic wave is propagated, but it is very important to understand what forces affect such a wave once it has left the generating antenna(s). Very briefly, an electromagnetic wave is produced by a rapidly expanding and collapsing magnetic field, which is in turn produced by alternately energizing and deenergizing an electronic circuit especially designed for the generation of such waves. In the science of electronics, such a generating circuit is referred to as an *oscillator.* An amplifier of some type is generally used to boost the power of the oscillator output, and an antenna is used to form the outgoing wave.

An electromagnetic wave, because of the methods by which it is propagated, always resembles a sine wave in appearance; it can be characterized by its wavelength, frequency, and amplitude, each of which is illustrated in Figure 28-1. In the figure, one complete electromagnetic wave or cycle is shown, and the terms used to describe it are defined as follows:

A *cycle* is one complete sequence of values of the strength of the wave as it passes through a point in space.

The *wavelength,* abbreviated in electronics by the symbol λ, is the length of a cycle expressed in distance units, usually either meters or centimeters.

The *amplitude* is the wave strength at any particular point along the wave.

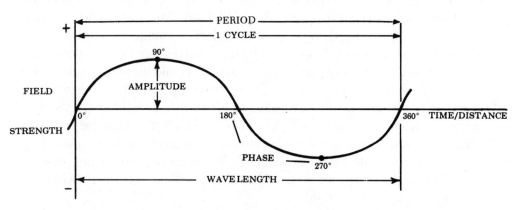

Figure 28-1. *Electromagnetic wave terminology.*

The *frequency,* abbreviated as *f,* is the number of cycles repeated during 1 second of time. If the time frame shown in Figure 28-1 were 1 second long, for example, it could be said that the frequency of the wave depicted is 1 cycle per second.

The *period* (τ) is the time required to complete one cycle of the wave; in the example above, the period would be 1 second. Period and frequency are related by the formula

$$\tau = \frac{1}{f}$$

In the vacuum of space, an electromagnetic wave is theorized to travel at a velocity approaching the speed of light, or 300,000,000 meters per second. Frequency and wavelength are related by the formula

$$\lambda = \frac{300,000,000}{f}$$

where λ is the wavelength in meters and *f* the frequency in cycles per second. Hence, every specific electromagnetic frequency is radiated at a specific wavelength.

In recent years, the term *Hertz,* abbreviated *Hz,* has come to be used in place of *cycles per second,* in honor of the German pioneer in electromagnetic radiation, Heinrich Hertz. One Hertz is defined as 1 cycle per second. Frequency is expressed in terms of numbers of thousands (kilo), millions (mega), or billions (giga) of Hertz. For example, 10,000 cycles per second is expressed as 10 kiloHertz, usually written 10 kHz, or 2.5 million cycles per second as 2.5 megaHertz, written as 2.5 MHz.

One additional term associated with the electromagnetic wave has great import in electronic navigation. The *phase* of a wave is the amount by which a cycle has progressed from a specified origin. For most purposes, it is stated in degrees, with a complete cycle being 360° in length. Two waves can be partially compared by measuring their phase difference. For example, two waves having crests one-quarter of a cycle apart can be described as being 90° "out of phase."

The behavior of an electromagnetic wave is dependent upon its frequency and corresponding wavelength. For descriptive purposes, electromagnetic frequencies can be arranged in ascending order to form a "frequency spectrum" diagram, shown in Figure 28-2. As can be seen, electromagnetic waves are classified as *audible waves** at the lower end of the spectrum, *radio waves* from about 5 kHz to 3×10^5 MHz, and *visible light* and various other types of *rays* at the upper end of the spectrum. In electronic navigation, frequencies within the radio wave spectrum are used.

---

* That is, audible *frequency* waves; to be heard, such electromagnetic waves must be transformed into sound waves by a receiver/speaker combination.

For ease of reference, the radio wave spectrum is further broken down into eight so-called *bands* of frequencies, as indicated in Figure 28-2. These eight bands are described as follows:

The *Very Low Frequency (VLF)* band includes all radio frequencies less than 30 kHz. The primary navigational use of this band is for the Omega system.

The *Low Frequency (LF)* band extends from 30 to 300 kHz. The navigational systems Decca and Loran-C, and most radiobeacons, use this band.

The *Medium Frequency (MF)* band extends from 300 kHz to 3 MHz, and is not used much for navigation at present.

The *High Frequency (HF)* band includes frequencies from 3 to 30 MHz, and is chiefly used for long-range communication.

The *Very High Frequency (VHF)* band extends from 30 to 300 MHz and is used for short- to medium-range communication.

The *Ultra High Frequency (UHF)* band includes frequencies between 300 and 3,000 MHz. It is used for satellite navigation and radar.

The *Super High Frequency (SHF)* band from 3,000 to 30,000 MHz and the *Extremely High Frequency (EHF)* band from 30,000 to 300,000 MHz are used almost exclusively for precise line-of-sight radar.

For purposes of classification and identification, especially in the military services, radio frequencies in the bands above 1,000 MHz are further broken down by letter designators, as follows:

L-band	1,000–2,000 MHz
S-band	2,000–4,000 MHz
C-band	4,000–8,000 MHz
X-band	8–12.5 GHz
K-band	12.5–40 GHz

Because severe absorption of radar waves occurs near the resonant frequency of water vapor (22.2 GHz), the K-band is further subdivided into the lower K-band (12.5–18 GHz) and upper K-band (26.5–40 GHz). Most marine navigation radars operate in the C- and X-bands, and many weapons fire control radars operate in the K-band range.

A series of electromagnetic waves transmitted at constant frequency and amplitude is called a *continuous wave,* abbreviated CW. This wave cannot be heard except at the very lowest frequencies, where it may produce a high-pitched hum in a receiver. Because an unmodified continuous wave cannot convey much information, in electronic navigation the wave is often modified or *modulated* in some way. When this is done, the basic continuous wave is referred to as a *carrier wave.*

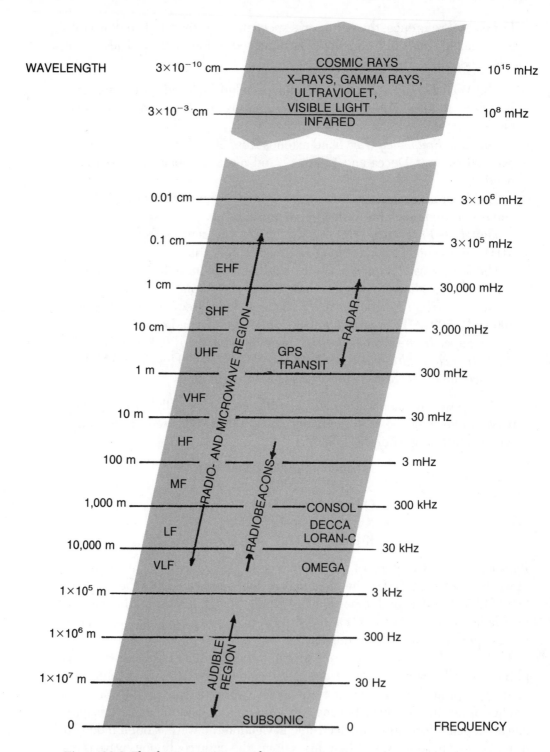

Figure 28-2. The electromagnetic wave frequency spectrum.

In practice there are three methods by which a carrier wave may be modulated to convey information; these are amplitude, frequency, and pulse modulation. In *amplitude modulation,* abbreviated AM, the amplitude of the carrier wave is modified in accordance with the amplitude of a modulating wave, usually but not always an audible frequency. The top drawing in Figure 28-3 illustrates this type of modulation. In the receiver, the signal is demodulated by removing the modulating wave, which in the case of a voice radio is then amplified and related to the listener by means of a speaker. This type of modulation is widely used in the commercial radio broadcast band.

In *frequency modulation,* abbreviated FM, the frequency of the carrier wave instead of the amplitude is altered in accordance with the frequency of the modulating wave, as shown in Figure 28-3 (center). This type of modulation is used for FM commercial radio broadcasts and the sound portion of television broadcasts.

Pulse modulation is different from either amplitude or frequency modulation in that there is usually no impressed modulating wave employed. In this form of modification, the continuous wave is actually broken up into very short bursts or "pulses," separated by relatively long periods of silence during which no wave is transmitted. As stated in chapter 10, this is the transmission used in most types of marine navigational and surface search radar; it is also used in some common long-range radio navigation aids, most notable of which is Loran. Figure 28-3 (bottom) depicts a pulse-modulated wave.

As was mentioned in the first part of this section, in electronics the device that generates an electromagnetic wave is termed an oscillator. In the case of radio waves, the output of the oscillator is boosted in power by an amplifier, then modulated as described above by a modulator unit. In voice radio, this modulator incorporates a microphone that converts an audible wave into a modulating electronic wave. The modulated radio wave is then passed through a second amplifier and finally transmitted into space by means of the antenna. These components, in addition to a power supply and a device to control the frequency of the wave originated by the oscillator, are collectively referred to as a *transmitter.* A block diagram of an AM radio transmitter appears in Figure 28-4.

A radio *receiver* is a device that is designed to convert a radio wave into a form suitable to convey information. It should be able to select carrier waves of a desired frequency, demodulate the wave, amplify it as necessary, and present the information conveyed in a usable form. The output of a receiver may be presented audibly by earphones or a loud speaker, or visually on a dial or cathode ray tube (CRT) display. A receiver can be thought of as incorporating three components—an antenna to convert an incoming radio wave into an electric current, a demodulator to separate a modulating wave from the carrier wave, and an output display device to present the output from the demodulator in a usable form.

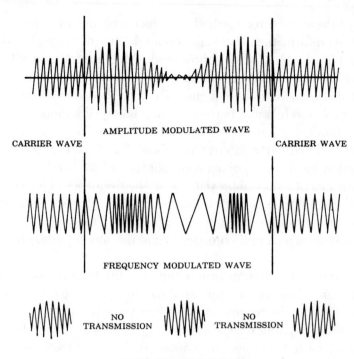

Figure 28-3. *Amplitude, frequency, and pulse modulation of a carrier wave.*

Radio receivers differ mainly in the six characteristics listed below:

Type: The type of signal that they will receive (i.e., AM, FM, or pulse)
Frequency range: The range of frequencies to which they can be tuned
Selectivity: The ability to separate waves of a desired frequency from others of nearly the same frequency
Sensitivity: The ability to detect and extract information from weak signals against a background of random radio-frequency noise

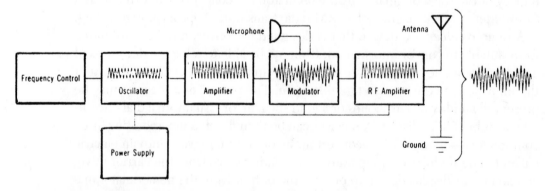

Figure 28-4. *Block diagram of components of an AM radio transmitter.*

Stability: The ability to resist drift from a frequency to which the receiver is tuned

Fidelity: The accuracy with which the characteristics of the original modulating wave are reproduced

THE BEHAVIOR OF RADIO WAVES IN THE EARTH'S ATMOSPHERE

When a radio wave is generated by a transmitter located within the earth's atmosphere, the wave travels outward in all directions much like a light wave from an unshielded light source. Because the radio and light wave are both forms of an electromagnetic wave, differing only in frequency and wavelength, their behavior is very similar in several respects.

Radio waves, like light waves, may be reflected from the surface of a material they strike. In both cases, the quality of the reflection depends on the irregularity of the surface as compared to the wavelength of the impinging wave and the density of the material. Thus, a sea of 10-foot waves would form a good reflecting surface for a radio wave hundreds of meters in length, but a very poor reflecting surface for a wave of a few centimeters. Whereas most of the energy within a wave striking a highly reflective surface is retained in the reflected wave, most of the energy striking a nonreflective surface is absorbed by the material in the form of heat.

Electrons within the molecules composing the earth's atmosphere and its crust are excited by the energy of a passing electromagnetic wave; as these electrons collide with each other and with other adjacent molecules, the energy is converted to heat and lost to the original wave. This effect, called *absorption,* is inversely proportional to the frequency of the wave; at lower radio frequencies the effect is maximized, due to the fact that at these frequencies the time available for each part of the wave to affect the electrons is relatively long. Because of the lower frequencies and longer wavelengths of a radio wave as compared to a light wave, the absorption effect is much more pronounced for the radio wave. The lower the frequency of the radio wave, the more energy it must have to proceed through the atmosphere to a given distance from the transmitter.

When an electromagnetic wave encounters an obstruction of opaque material (i.e., impervious to radio waves), the area behind the obstruction is shadowed, since the waves that would otherwise reach the area are blocked. This blockage is not complete, however, because the portion of the wave to either side of the obstruction begets a secondary series of waves that travel into the shadow zone. This effect, called *diffraction,* is illustrated in Figure 28-5. In the case of a light wave, it accounts for a shadow behind an obstruction being less than absolutely black, and in the case of a radio wave, for the reception of a weak radio signal within such a shadow zone.

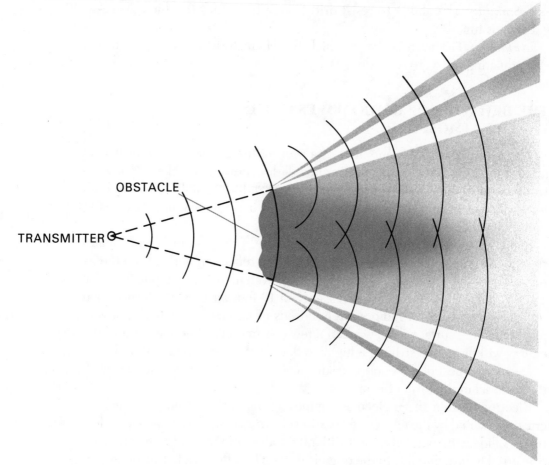

Figure 28-5. *Diffraction effect of an electromagnetic wave.*

If two or more electromagnetic waves arrive simultaneously at the same point in space, *interference* may result. The amount of this interference depends on the phase and frequency relationship of the waves involved. If two waves of the same frequency but 180° out of phase arrive simultaneously, each wave cancels the other, resulting in a *null* at that point.

Under certain conditions, a portion of the electromagnetic energy in a radio wave may be reflected back toward the earth's surface from the *ionosphere,* a layer of charged particles about 90 to 400 kilometers high; such a reflected wave is called a *sky wave.* The portion of the earth's upper atmosphere in daylight is subjected to continual bombardment by ultraviolet rays of the sun. These high-energy light waves cause electrons in the gas molecules of the upper atmosphere to become excited and to free themselves from their molecules, forming ionized layers. These layers reach their maximum intensity when the sun is highest.

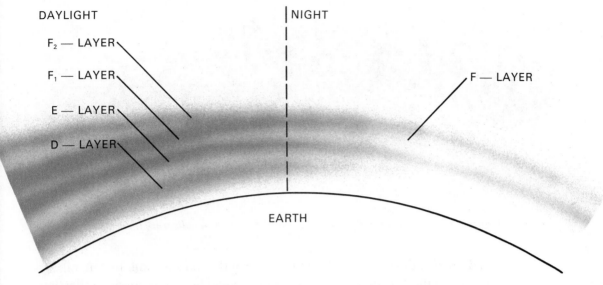

Figure 28-6. *The four layers of the ionosphere of importance in radio wave propagation.*

There are four such *ionospheric layers* of importance in the study of radio wave propagation, pictured in Figure 28-6; they are designated the D-layer, E-layer, and F_1- and F_2-layers, each of which is described below:

The *D-layer* is the ionized layer closest to the earth's surface, about 60 to 90 kilometers high. It is considerably less dense than the other layers, and appears to be formed only during the daylight hours.

The *E-layer* is located about 110 kilometers above the earth's surface. Its density is greatest in the area immediately beneath the sun, and it persists throughout the night with decreased intensity.

The *F_1-layer* occurs only in daylight regions of the upper atmosphere, usually between 175 and 200 kilometers above the earth's surface.

The *F_2-layer* is found at altitudes between 250 to 400 kilometers. Its strength is greatest by day, but due to the very low density of the atmosphere at this height, the freed electrons in this layer persist several hours after sunset. There is a tendency for the F_1- and F_2-layers to merge thereafter to form the so-called *F-layer*, which is ordinarily the only layer of importance to radio wave propagation after dark.

All layers of the ionosphere are somewhat variable, with the main patterns seeming to be diurnal, seasonal, and by sunspot cycle. The layers may either be conducive to the sky wave transmission of a radio wave to a desired area of reception, or they may hinder or even entirely prevent such transmission, depending on the frequency of the wave, its angle of incidence, and the height

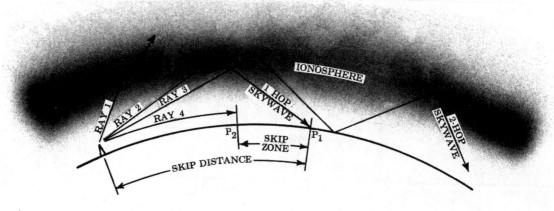

Figure 28-7. A sky wave and ground wave pattern for a given radio wave transmission.

and density of the various layers at the time of the transmission. In general, frequencies in the MF and HF bands are most suitable for ionospheric reflections during both day and night, with frequencies in the upper LF and lower VHF bands producing usable sky waves only at night. Frequencies outside of these limits either will not produce any sky waves or the sky waves they do form are so weak as to be unusable. In times of abnormal sunspot activity, the normal ionospheric reflection patterns may be disrupted to such an extent that no sky waves of any frequency are produced.

Because of the higher resistance of the earth's crust as compared to the atmosphere, the lower portion of a radio wave radiated parallel to the earth's surface is slowed somewhat, causing the wave to bend toward earth, just as a light wave from a star is bent upon entering the earth's atmosphere. A wave of this type that tends to follow the earth's curvature is termed a *ground wave*. Again, the amount of curvature resulting from this cause is inversely proportional to the frequency of the wave; the lower the frequency, the more it will tend to bend in conformity to the earth's shape. The ultimate range of such a ground wave is dependent upon the absorption effect.

Combining the effects of sky wave and ground wave transmission for a given radio wave of suitable frequency, a radiation pattern of the type depicted in Figure 28-7 may result. In this figure, ray 1 of the transmitted radio wave is radiated at an angle too great to permit its reflection by the ionosphere as a sky wave. Hence, it penetrates the ionosphere and escapes into space. Ray 2 impinges upon the ionosphere at the steepest angle permitting a reflected sky wave under these conditions. The resultant sky wave, called a "one-hop" wave, reaches the earth's surface at point P_1; no sky waves could be received within the distance of P_1 from the transmitter, since the angle of incidence of ray 2 on the ionosphere is the maximum possible angle that will produce a sky wave. Ray 3 strikes the ionosphere at a lesser angle, is reflected in turn by the earth's

506

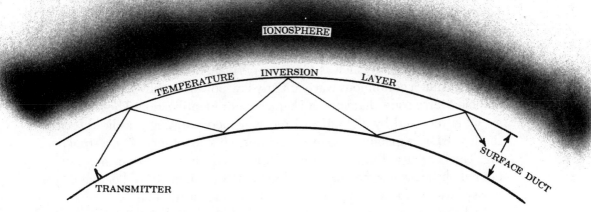

Figure 28-8. *Formation of a surface duct.*

surface, and is once again reflected by the ionosphere, such a doubly reflected wave is termed a "two-hop" sky wave. Ray 4 is a ground wave. It can penetrate the atmosphere only as far as point P_2 before absorption weakens the wave to such an extent that it is not receivable beyond this point. Hence, this particular radio wave cannot be received between points P_1 and P_2; such a zone between the maximum range of the ground wave and the minimum range of a sky wave is termed a *skip zone.* The term *skip distance* is sometimes applied to the minimum range at which a sky wave may be received.

Occasionally, under relatively rare atmospheric conditions, a phenomenon known as *surface ducting* may occur, extending the range of a ground wave well beyond its normal limits of reception. A surface duct is formed between the earth's surface and the bottom of a layer of air within which an extreme temperature inversion exists, shown in Figure 28-8. Because the width of the duct must be large as compared to the wavelength of the radio wave to be effective, surface ducting is usually associated with the higher radio and radar frequencies. The phenomenon is most common in tropical latitudes, especially in the Pacific regions, where a surface duct once formed may persist for several hours or even days.

SHORT- AND MEDIUM-RANGE NAVIGATION SYSTEMS

Most short- to medium-range radio navigation systems are designed to provide either a bearing to a shore-based transmitter site, as in the case of radiobeacons, or a range and bearing from the transmitter to a natural or manufactured navigation aid, in the case of radar. Decca, which may be considered both a short- and a medium-range system, employs a type of hyperbolic lattice pattern to determine position; principles of its operation are discussed in the following section of this chapter. Since radar as a navigation system is discussed in some detail in chapter 10, it will not be dealt with here.

Essentially, a radiobeacon is a single transmitter, frequently combined with visual navigational aids such as a buoy or permanent light structure. Radiobeacons transmit a continuous wave at fairly low power, usually modulated by an audible Morse code character or characters for identification. The transmitted signal is received by a shipboard receiver incorporating a radio direction-finding (RDF) capability, and used very much like an observed terrestrial bearing. The bearings of signals received simultaneously from two or more different radiobeacons may be crossed to form a fix, or a single RDF bearing may be crossed with any other type of LOP to obtain a fix or running fix.

Radiobeacons and their use will be discussed in further detail in the first section of the following chapter.

LONG-RANGE HYPERBOLIC NAVIGATION SYSTEMS

Most contemporary earth-based long-range electronic navigation systems are called *hyperbolic systems* because the line of position that they yield is a segment of a hyperbola rather than a radial line, as in the case of a terrestrial or radiobeacon LOP. Suppose that two transmitting stations designated M (*master*) and S (*secondary*) located some distance apart both transmit a short radio wave pulse simultaneously at fixed time intervals. If the speed of propagation of the ground waves were assumed constant, a series of equally spaced concentric circles could be drawn surrounding each of the stations to represent the distance a single pulse would travel during a fixed time interval; the spacing between each circle is constant, since the pulse would travel a uniform distance during each time interval. Figure 28-9 illustrates a series of such circles.

A line drawn connecting the two stations M and S is called the *base line*. If a perpendicular line AB were drawn bisecting this base line, simultaneously transmitted pulses from both stations would arrive along this bisector at the same time. Thus, the bisector would represent a locus of all points at which the time difference in receipt of the master and secondary station signals was zero.

Now, suppose that we wished to locate a second locus of points at which the secondary signal arrived one time unit ahead of the master pulse. Returning to the base line, if we moved one-half the spacing of a circle from the bisector toward the secondary station, we would also be moving one-half of the time interval away from the master. Hence, at this point on the base line, the secondary station pulse would be received one time unit before the master station pulse. The locus of all other points at which an identical time difference could be measured would form the hyperbola labeled CD in the figure.

Figure 28-9. *Construction of a hyperbolic navigation pattern.*

In similar fashion, other hyperbolas could be constructed, each of which would represent the locus of all points at which a specific time difference between the times of arrival of the master and secondary pulses would be measured. A family of such hyperbolas appears in Figure 28-9.

As can be seen by an inspection of the figure, the divergence between adjacent hyperbolas representing successive time differences increases as the distance of the receiver from the base line increases, and the curvature of the hyperbolas approaches a straight line. In fact, at extreme ranges, the branches of the hyperbolas become asymptotic to straight lines converging at the midpoint of the base line. Thus, any hyperbolic system is inherently more accurate on or near the base line, with the precision of any LOP obtained decreasing with increased distance from the base line. Near the base line, any change in the position of the receiver would correspond to a relatively large change in the time difference of receipt of the master and secondary pulses, but at longer ranges the change in time difference because of a similar change in position is relatively small.

The time difference measured anywhere on either base-line extension is a constant equal to the time required for a pulse to traverse the length of the base line. Moreover, the time difference is nearly constant in the vicinity of the extensions, with a large change in position in a direction perpendicular to the extensions being necessary to produce a measurable change. Hence, the hyperbolic system subtended by a single pair of stations is unusable in the parabolic areas centered on the base-line extensions.

Up to this point in the discussion of the hyperbolic system it has been assumed that two pulses were transmitted simultaneously by both the master and secondary stations. In actuality, in such a pulsed hyperbolic system the pulses are not transmitted simultaneously; rather, the secondary station transmits its pulse at a fixed interval called a *coding delay* after the arrival of the master pulse at the secondary station. This is done so that the master pulse will always precede the secondary pulse, no matter where the receiver is located in the hyperbolic grid pattern. An additional variable delay can also be included for security purposes during wartime. In such a system the time difference measured along the perpendicular bisector of the base line, line AB in Figure 28-9, is no longer zero but rather the interval corresponding to the sum of the time delay plus the transmission time between the master and secondary stations. This type of system is employed in Loran-C.

Until the late 1960s in hyperbolic navigation systems, the actual time of arrival of the master pulse was used to key the transmission of the secondary station pulse. Such secondary stations were therefore called *slave stations*. Since then, however, the long-range hyperbolic systems have employed atomic time standards to regulate both the master and secondary station transmissions. This was done in order to increase system accuracy by eliminating random timing errors arising from variable atmospheric propagation conditions along the base lines connecting the master and its associated slave stations. In such a system, the secondary stations are always referred to as such, rather than the old slave station designations, since their transmissions are no longer directly dependent on the arrival of the master pulse.

As an alternative to the use of a short pulse, the circular transmission patterns shown in Figure 28-9 can be achieved by use of continuous waves, with the leading edges of each cycle representing a given time and corresponding distance interval. The corresponding hyperbolas then represent loci of constant phase differences called *isophase lines*, with zero phase differences occurring half a wavelength apart along the base line. In this type of system the spaces between the hyperbolas are referred to as *lanes*; the position of the receiver within a lane is determined by measuring the *phase difference* between the master and secondary signals. A disadvantage of this type of system is that it is not possible to distinguish one lane from another by use of the phase comparison alone. Hence, the lanes must either be counted as they are traversed from some fixed position, or the lane must be identified by other means. The Decca system is the only large-scale radionavigation system still in full operation that uses such continuous waves. The Omega system, currently being phased out of operation, uses the same principle to establish worldwide coverage with only eight operational stations.

Regardless of the type of transmission, in hyperbolic navigation systems the user's distance from the stations can seldom exceed about six times the length of the base line for good accuracy in the line of position. In systems in which the base line is very short compared to the distances over which the system is to be radiated, the pattern is referred to as a *collapsed* or *degenerated hyperbolic type,* and only the straight asymptotic part of the hyperbola is used. In general, the shorter the base line, the more the character of the LOP resembles a bearing. There are no collapsed hyperbolic systems currently in operation.

In electronic navigation, as in visual piloting and celestial navigation, a fix is obtained by the intersection of two or more simultaneous or near simultaneous LOPs. In the case of the hyperbolic system, at least two pairs of stations must be available to the user in order to obtain a fix. In cases in which only one hyperbolic LOP is available, it may be crossed with a simultaneous LOP obtained by any other means to determine a fix. Such single hyperbolic LOPs, when available, are often used to check the accuracy of a celestial fix, especially when only two or three bodies are observed.

ACCURACY OF AN ELECTRONIC FIX

In all types of electronic navigation systems, the determination of position is influenced by various types of random effects that can cause errors in the indicated position. These position errors can be caused by such things as atmospheric disturbances in the transmission path, errors in the transmitter or receiver clocks, inaccuracy in the electronic circuitry, and gyro error, among others. As a result, a series of positions determined at a given time or location will usually result in a cluster of points around or near the true position. It is hoped that most of these will be fairly close, but some may be farther away. In electronic navigation, then, it is frequently desirable to use some standard measures by which the accuracy and precision of a fix produced by a given system can be expressed.

There are two such measures in common use to describe the accuracy and precision of electronic navigation systems. One technique is called *circular error probable* (CEP). To determine the CEP, a circle is drawn, centered on the true position, whose circumference encompasses 50 percent of all the indicated positions. The CEP value is then the radius of this circle (see Figure 28-10A). Using this measure, it can be said that there is a 50 percent probability that any new position determined will fall within the CEP distance from the true position. Circular error probable is often used by the military services, not only as a measure of navigational accuracy, but also to express the accuracy of weapons systems like bombs or guns relative to a desired target.

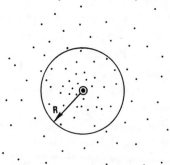

Figure 28-10A. *The CEP radius defines a circle that encompasses 50 percent of all indicated positions.*

A second measure of accuracy and precision in common use especially in scientific applications is the *root mean square* (rms). In words, this is the square root of the sum of the squares of all the errors (E), divided by the number of measurements (N). Mathematically, it can be expressed by the formula

$$\text{rms} = \sqrt{\sum_{n=1}^{N} (E_n)^2/N}$$

It is analogous to the *standard deviation* in statistics. A circle centered on the true position with radius equal to the rms value would be expected to contain 68 percent of all the indicated positions; such a circle of radius equal to 2 × rms should contain 95 to 98 percent of all the indicated positions (see Figure 28-10B). The rms error measure is preferred by scientists because it is more sensitive to the presence of even a small number of large errors, while the CEP is not. Thus, the rms error is a more conservative measure of accuracy and precision than is CEP. Roughly, the relationship between the two is

$$2 \text{ rms} = 2.5 \times \text{CEP}$$

In electronic navigation systems three *types* of accuracy are important; they are defined as follows:

1. *Predictable* (sometimes called *absolute*) *accuracy*: the accuracy of an indicated position with respect to its true geographic coordinates.
2. *Repeatable accuracy*: the accuracy with which the user can return to a position whose coordinates have been determined at a previous time with the same navigation system.

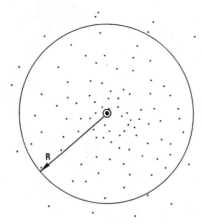

Figure 28-10B. *The 2 rms radius defines a circle that encompasses 95 to 98 percent of all indicated positions.*

3. *Relative accuracy*: the accuracy with which a user can measure position relative to that of another user of the same system at the same time.

The importance of predictable accuracy in navigation is obvious. Since the position of one ship or plane relative to another is also often a prime concern in military maneuvers, the repeatable and relative accuracies are significant as well. To a yachtsman or fisherman who may have marked a favorite fishing spot

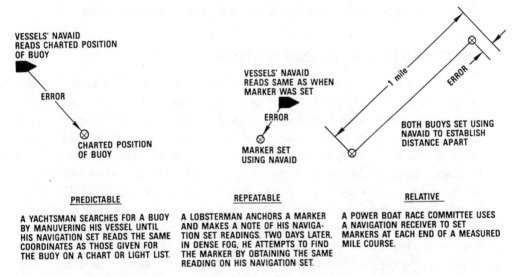

Figure 28-11. *Three examples of the distinctions among predictable, repeatable, and relative errors. (Courtesy of Magnavox Corporation)*

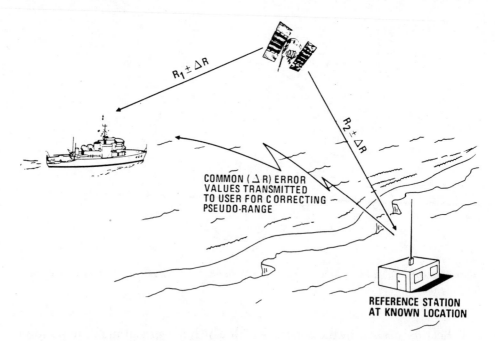

Figure 28-12. *An example of the differential technique for increasing electronic fix accuracy. Here, a common error (ΔR) in the ranges to a GPS satellite from a reference station and a nearby ship have been computed by the reference station, for separate transmission to the ship. (Courtesy of Magnavox Corporation)*

by the coordinates given by his or her radionavigation set, the repeatable accuracy may be even more important than the predictable accuracy. Figure 28-11 gives a few more examples of the distinctions between these terms.

DIFFERENTIAL RADIONAVIGATION

In certain applications requiring an exceptionally high degree of accuracy relative to the true geographic position, a technique called *differential radionavigation* can be employed. In differential navigation a reference receiver is placed at a known location, usually on shore, whose true position has been accurately verified, usually by means of land surveying techniques. The position as determined by the radionavigation system of interest is then continually compared with this known position. Any resulting correction or difference (symbolized by the Greek letter Δ) that must be applied to the data in order to obtain the true position is then transmitted to designated mobile receivers using that navigation system aboard ships, planes, or ground vehicles operating nearby, in the region within which the correction is the same or nearly so—often to a radius of several hundred kilometers or more from the reference location (see Figure 28-12). Positional data determined and thus adjusted at the mobile receivers can yield highly precise fixes accurate to 1–2 meters rms or less.

The differential technique has been successfully applied to all the hyperbolic and satellite-based radionavigation systems discussed in the following two chapters, to yield highly accurate positional information for many applications, including piloting in restricted waters, airport approach control, and geodetic surveying and oil exploration. It has come into increasing use in recent years as the cost of the more capable electronic equipment required to use it has decreased.

SUMMARY

This chapter has described the characteristics and behavior of the electromagnetic wave upon which the radionavigation systems to be discussed in the following chapter are based. In general, the behavior of any particular radiated RF signal is dependent on its power, frequency, and corresponding wavelength. The type of modulation used to produce meaningful positioning information to the user depends on the purpose for which the system is intended. Radar and Loran-C use pulsed transmissions, while most radiobeacons and Decca employ continuous waves. Radionavigation systems can provide bearings to the transmitter site, as in the case of radiobeacons, or hyperbolic lattice patterns, as in the case of Loran-C and Decca; radar provides both a range and a bearing from the transmitter to a desired natural or artificial navigation aid. Each of the former systems will be discussed in the following chapter, while radar is the subject of chapter 10. A certain amount of random error is inherent in all electronic navigation systems, necessitating the use of measures of probable error and precision, the most common of which are circular error probable (CEP) and root mean square (rms). Differential radionavigation involving the use of a receiver at a known location ashore is a technique coming into increased use to eliminate the effects of random error when high precision and accuracy are required.

RADIONAVIGATION SYSTEMS
29

The preceding chapter reviews the principles of behavior of the electromagnetic wave in the earth's atmosphere, and discusses the general characteristics of the most widely used electronic radionavigation systems. This chapter specifically describes radiobeacons, Loran-C, and Decca in order to acquaint the inexperienced navigator with the practical methods of position-finding through the use of each of these systems. However, in electronic navigation, just as in celestial navigation, there is no substitute for practical experience. Working with receivers under actual conditions is the only way to attain proficiency in the use of any electronic navigation system.

RADIOBEACONS

As previously mentioned, radiobeacons were first employed as coastal navigational aids along the shores of the United States in the 1930s. They consist of single transmitters, often co-located with visual navigational aids. Radiobeacon signals are transmitted at rather low power on frequencies in the LF and MF bands between about 250 to 400 kHz. Most signals are limited in range to less than 200 miles, with the majority not receivable beyond about 20 miles. The signals are continuous waves modulated by Morse code characters for identification. Often radiobeacons located in a given area are grouped on a common frequency, and each transmitter transmits only during a segment of a time-sharing plan. Radio bearings to the site of the transmitter are determined by the use of a radio receiver equipped with a radio direction-finding (RDF) capability. There are a number of moderately priced manually operated RDF receivers and several more expensive fully automatic models available commercially for private users. Many Navy navigators use the services of the ECM (electronic countermeasures) personnel in CIC during piloting to obtain radiobeacon bearings.

All RDF receivers, whether manual or automatic, operate on the principle that when a coil is rotated in the field of radiation of a transmitter, the received signal strength is greatest when the coil face is oriented 90° away from the direction of the incoming signal, and weakest, or *nulled,* when the coil is normal to the signal, facing either directly toward or directly away from it. Thus, particularly on older models of receivers, it is very easy for the inexperienced operator to read an RDF bearing as 180° away from the true bearing of the transmitter. Most newer models, however, have features that tend to reduce the possibility of such error. Other potential sources of error include improper calibration of the receiver, the effect of coastal refraction if the path of the signal passes over land or along the shoreline between the beacon and the receiving vessel, and varying amounts of ionospheric interference at night. As a general rule, RDF bearings are normally considered accurate only to within ±2° for distances under 150 miles to the transmitter in favorable conditions, and ±5–10° when conditions are unfavorable.

Descriptions of the characteristic signals and normal ranges of all permanent radiobeacons in U.S. and foreign waters are contained in the Coast Guard *Light List* and the DMAHTC *List of Lights* volumes (see Figure 29-1A). Amplifying information on how to use radio bearings for navigation is contained in *Publication No. 117, Radio Navigational Aids.*

Since radio waves transmitted by radiobeacons travel outward along great circles, radio bearings to the beacons should not be directly read from a receiver and plotted on a Mercator chart. All RDF bearings should first be converted to rhumb lines before they can be plotted, using a conversion table included in *Publication 117.* A portion of this table appears in Figure 29-1B.

Entering arguments for the table are the difference in longitude to the nearest half degree between the positions of the radiobeacon and the receiver, and the midlatitude to the nearest whole degree between the two positions. The sign of the correction is determined by the rules included at the bottom of the table.

Corrected radiobeacon bearings are plotted on a Mercator chart using a reciprocal drawn from the radiobeacon to the approximate position of the ship. Additional lines representing the probable limits of error at the time the bearings were obtained can also be drawn, as illustrated in Figure 29-1C. The intersection of two or more such bearing envelopes indicates the probable ship's position as well as the area within which the ship might lie given the probable bearing errors, as shown in the figure.

Because of the relative inaccuracy of RDF bearing LOPs as compared to LOPs obtained visually or by the use of radar, radiobeacons are not widely used by U.S. Navy navigators. Many navigators of small boats and merchant ships not equipped with effective navigational radars, however, often make extensive use of these aids whenever they are engaged in coastal navigation and piloting.

The U.S. Coast Guard is currently in the first stages of a project that will eventually result in the broadcasting of differential GPS signals from radiobea-

(1) No.	(2) Name	(3) Position	(4) Characteristic	(5) Range	(6) Sequence	(7) Frequency	(8) Remarks
			CANADA AND UNITED STATES -- GREAT LAKES				
		N/W					
	Lake Superior						
1070	La Pointe, Wi.	46°43'44" 90°47'05"	V (•••—).	20	288 A2A,		Seasonal. Carrier signal.
1080	Superior Entry South Breakwater Light Station, Wi.	46°42'37" 92°00'22"	SN (••• —•).	10	316 A2A.		Seasonal. Carrier signal.
			UNITED STATES - ATLANTIC AND GULF COASTS				
1110	Martinicus Rock Light Station, Me.	43°47'01" 68°51'19"	MR (—— •—•).	20	314 A2A.		
1120	The Cuckolds Light Station, Me.	43°46'46" 69°39'02"	CU (—•—• ••—).	10	320 A2A.		
1130	Halfway Rock Light Station, Me.	43°39'21" 70°02'15"	HR (•••• •—•).	10	291 A2A.		
1140	Portland Lighted Horn Buoy "P", Me.	43°31'37" 70°05'31"	PH (•——• ••••).	30	301 A2A.		
1150	Portsmouth Harbor New Castle Light, N.H.	43°04'15" 70°42'36"	NCE (—• —•—• •).	10	322 A2A.		
1160	Eastern Point Light Station, Ma.	42°34'50" 70°39'54"	EP (• •——•).	10	325 A2A.		
1164.5	Great Inagua.	20°57'37" 73°40'27"	ZIN (—••• •• —•).	50	376 A2A.		AERO.
1170	Eastern Point Light Station Calibration Radiobeacon, Ma.	42°34'48" 70°39'54"	EP (• •——•). period 60ˢ tr(2) 10ˢ (—) 20ˢ repeats(1) 30ˢ	5	298, 480 A2A.		Transmits upon request through Coast Guard Radio Marshfield (NMF) or via Gloucester Coast Guard Station on 2182 kHz.
1180	Lyndy, Ma.	42°27'00" 70°58'00"	LO (•—•• ———).	100	382 A2A.		AERO.
1190	Boston Lighted Horn Buoy "B", Ma.	42°22'42" 70°47'00"	BH (—••• ••••).	30	304 A2A.		
1200	Scituate Harbor, Ma.	42°11'54" 70°43'12"	SH (••• ••••). period 60ˢ tr 50ˢ si 10ˢ	10	295 A2A.		
1210	Highland Light Station, Ma.	42°02'24" 70°03'40"	HI (•••• ••). period 360ˢ tr 50ˢ (—) 10ˢ si 300ˢ	100	I	286 A2A.	
1212	Nantucket Shoals Lighted Horn Buoy "N", Ma.	40°30'00" 69°26'00"	NS (—• •••). period 360ˢ tr 50ˢ (—) 10ˢ si 300ˢ	50	II	286 A2A.	
1214	Montauk Point Light Station, N.Y.	41°04'02" 71°51'46"	MP (—— •——•). period 360ˢ tr 50ˢ (—) 10ˢ si 300ˢ	125	III	286 A2A.	

Figure 29-1A. *Sample page from the radiobeacon section of the DMAHTC List of Lights.*

cons located throughout the coastal and inland waterways of the United States, including Puerto Rico, Alaska, and Hawaii. As will be elaborated upon in chapter 30, these signals will enable properly equipped users to navigate with 10-meter accuracy in all U.S. harbors and approach areas.

RADIO NAVIGATIONAL AIDS

200F.   Radio Bearing Conversion Table

Correction to be applied to radio bearing to convert to Mercator bearing

Difference of longitude

Mid. lat.	0.5°	1°	1.5°	2°	2.5°	3°	3.5°	4°	4.5°	5°	5.5°	6°	6.5°	7°	7.5°	Mid. lat.
4	...	...	...	...	0.1	0.1	0.1	0.1	0.2	0.2	0.2	0.2	0.2	0.2	0.3	4
5	...	0.1	0.1	.1	.1	.1	.2	.2	.2	.2	.2	.3	.3	.3	.3	5
6	...	.1	.1	.1	.1	.2	.2	.2	.2	.3	.3	.3	.3	.4	.4	6
7	...	.1	.1	.1	.2	.2	.2	.3	.3	.3	.3	.4	.4	.4	.5	7
8	...	.1	.1	.1	.2	.2	.2	.3	.3	.4	.4	.4	.5	.5	.5	8
9	...	.1	.1	.1	.2	.2	.2	.3	.3	.3			.5	.5	.6	.6	9
31	.1	.2	.4	.5			1.0	1.2	1.3	1.4	1.6		1.8	1.9		31
32	.1	.3	.4	.5	.7		.9	1.1	1.2	1.3	1.4	1.6	1.7	1.8	2.0	32
33	.1	.3	.4	.6	.7	.8	1.0	1.1	1.2	1.4	1.5	1.6	1.8	1.9	2.1	33
34	.1	.3	.4	.6	.7	.8	1.0	1.1	1.2	1.4	1.5	1.7	1.8	2.0	2.1	34
35	.1	.3	.4	.6	.7	.9	1.0	1.2	1.3	1.4	1.6	1.7	1.9	2.0	2.2	35
36	.1	.3	.4	.6	.7	.9	1.0	1.2	1.3	1.5	1.6	1.8	1.9	2.1	2.2	36
37	.2	.3	.4	.6	.8	.9	1.1	1.2	1.4	1.5	1.6	1.8	2.0	2.1	2.2	37
38	.2	.3	.5	.6	.8	.9	1.1	1.2	1.4	1.5	1.7	1.8	2.0	2.2	2.3	38
39	.2	.3	.5	.6	.8	1.0	1.1	1.2	1.4	1.6	1.7	1.9	2.1	2.2	2.4	39
40	.2	.3	.5	.6	.8	1.0	1.1	1.3	1.4	1.6	1.8	1.9	2.1	2.2	2.4	40
41	.2	.3	.5	.6	.8	1.0	1.2	1.3	1.5	1.6	1.8	2.0	2.1	2.3	2.5	41
42	.2	.3	.5	.7	.8	1.0	1.2	1.3	1.5	1.7	1.8	2.0	2.2	2.3	2.5	42
43	.2	.3	.5	.7	.8	1.0	1.2	1.4	1.5	1.7	1.9	2.1	2.2	2.4	2.6	43
44	.2	.4	.5	.7	.9	1.1	1.2	1.4	1.6	1.7	1.9	2.1	2.2	2.4	2.6	44
45	.2	.4	.5	.7	.9	1.1	1.2	1.4	1.6	1.8	2.0	2.1	2.3	2.5	2.6	45
46	.2	.4	.5	.7	.9	1.1	1.3	1.4	1.6	1.8	2.0	2.2	2.3	2.5	2.7	46
47	.2	.4	.6	.7	.9	1.1	1.3	1.5	1.7	1.8	2.0	2.2	2.4	2.6	2.8	47
48	.2	.4	.6	.8	.9	1.1	1.3	1.5	1.7	1.8	2.1	2.2	2.4	2.6	2.8	48
49	.2	.4	.6	.8	1.0	1.1	1.3	1.5	1.7	1.9	2.1	2.3	2.5	2.6	2.8	49
50	.2	.4	.6	.8	1.0	1.1	1.3	1.5	1.7	1.9	2.1	2.3	2.5	2.7	2.9	50
51	.2	.4	.6	.8	1.0	1.2	1.4	1.6	1.8	2.0	2.1	2.3	2.5	2.7	2.9	51
52	.2	.4	.6	.8	1.0	1.2	1.4	1.6	1.8	2.0	2.2	2.4	2.6	2.8	3.0	52
53	.2	.4	.6	.8	1.0	1.2	1.4	1.6	1.8	2.0	2.2	2.4	2.6	2.8	3.0	53
54	.2	.4	.6	.8	1.0	1.2	1.4	1.6	1.8	2.0	2.2	2.4	2.6	2.8	3.0	54
55	.2	.4	.6	.8	1.0	1.2	1.4	1.6	1.8	2.1	2.2	2.4	2.7	2.9	3.1	55
56	.2	.4	.6	.8	1.0	1.2	1.4	1.7	1.9	2.1	2.3	2.5	2.7	2.9	3.1	56
57	.2	.4	.6	.8	1.1	1.2	1.5	1.7	1.9	2.1	2.3	2.5	2.7	2.9	3.2	57
58	.2	.4	.6	.8	1.1	1.3	1.5	1.7	1.9	2.1	2.3	2.6	2.8	3.0	3.2	58
59	.2	.4	.6	.8	1.1	1.3	1.5	1.7	1.9	2.2	2.4	2.6	2.8	3.0	3.2	59
60	.2	.4	.6	.9	1.1	1.3	1.5	1.7	2.0	2.2	2.4	2.6	2.8	3.0	3.2	60

Receiver (latitude)	Transmitter (direction from receiver)	Correction Sign	Receiver (latitude)	Transmitter (direction from receiver)	Correction Sign
North	Eastward	+	South	Eastward	−
North	Westward	−	South	Westward	+

2–5

Figure 29-1B. *Portion of the Radio Bearing Conversion Table,* Publication 117.

LORAN-C

The Loran (for *Long-Range Navigation*) system was originally developed by the United States in 1940. It was one of the first attempts to implement a long-range hyperbolic navigation system that could provide all-weather fixing information for both ships and aircraft at sea. World War II and the Korean War spurred the establishment of the Loran system. As originally conceived, the system used pairs of master and slave stations transmitting sequential pulsed radio waves in the upper MF band, with frequencies between 1,850 and 1,950 kHz, to establish a hyperbolic lattice pattern in the area of coverage based on time differences of receipt of the master and the slave station pulse. As described in the previous chapter, in such a system each hyperbolic LOP in the

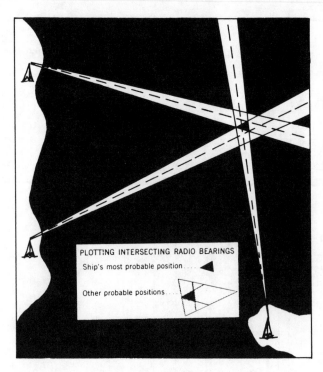

Figure 29-1C. A plot of a ship's position as determined by three radiobeacon bearings.

pattern represents a locus of points at which a unique time difference can be recorded between receipt of the master station pulse and the pulse of its associated slave. The original system, which came to be called *Loran-A*, featured ground-wave coverage out to between 450 and 800 miles from the base line by day, and sky-wave ranges to 1,400 miles by night. Between about 1955 and 1970, it was deployed throughout the coastal regions of the northern Atlantic and Pacific Oceans, where it became probably the most widely used electronic navigation aid in the world up to that time.

In the early 1950s, the need for a longer-range electronic navigation system that would extend coverage farther out into the midocean regions began to be felt. Initial attempts to extend the range of Loran-A by simply lowering its carrier frequencies were abandoned because of difficulties encountered in station synchronization and in distinguishing between ground and sky waves as a result of the shorter reflection time of the sky wave at lower frequencies. To overcome these difficulties, a system employing synchronized pulses for both time-difference and phase-comparison measurements was developed. This system became known as Loran-C. The first stations in the system became operational in 1957. In contrast to Loran-A, the characteristics of the Loran-C signal enabled a master station to operate with more than one slave station, usually a group of three arranged in a rough geographic triangle with the master in the center. Each group of master and associated slave stations was called a *chain*.

The Loran-C system has been continually improved and expanded since its inception, to the point where today its ground-wave coverage extends over most of the continental and coastal regions of the United States, as well as much of the northern Atlantic and Pacific Oceans and the Mediterranean, Red Sea, and Persian Gulf, with sky waves receivable over most of the northern hemisphere with the exception of the Indian Ocean and northwest Pacific. Loran-A was thus rendered redundant with the result that it was phased out of existence during the 1980s. The U.S. Coast Guard assumed managerial control of all U.S.-operated Loran-C stations in 1978.

Both Loran-A and Loran-C originally operated under a concept wherein the receipt of the master pulse at the associated slave stations would trigger the sequential transmission of the "slave" pulse. In the late sixties and early seventies, however, atomic time standards were introduced to regulate all transmissions within each chain more precisely, and the slave stations in Loran-C were consequently redesignated *secondary* stations. All stations in the system transmit a signal on a common carrier frequency in the mid-LF band of 100 kHz, with a band width extending 10 kHz to either side. The ground-wave range of Loran-C is about 1,200 miles; one-hop sky waves have a range of about 2,300 miles, and two-hop signals have been received as far as 4,000 miles from the transmitting stations. One-hop sky waves are produced both by day and by night, while two-hop sky waves are formed only at night. The increased range of the Loran-C ground wave, coupled with the use of atomic time standards instead of the receipt of the master pulse to regulate the transmission of the secondary station pulses within each chain, has permitted base-line distances to be extended to between 1,000 and 1,500 miles for present-day Loran-C chains. The absolute accuracy of the system varies from about ±200 yards rms near the base line to ±500 yards rms near the extreme range of the system, but the repeatable accuracy is much better. Differential techniques as explained in the last chapter have been used with great success to reduce the system error in situations requiring greater precision. The low frequencies and high power (over 1,500 kw in some cases) at which Loran-C signals are transmitted allow ground waves to penetrate the surface layers of seawater, enabling them to be received by submerged submarines.

Unfortunately, the increased advantages of Loran-C as compared to the old Loran-A are achieved at the cost of increased complexity of both the system itself and the receiver required to use it. This complexity is necessary for two reasons. First, at the frequency at which the Loran-C signal is transmitted, a great deal of power is required to propagate the wave for the long range for which the system is designed. Second, at this frequency sky waves are propagated that arrive a very short time after the arrival of the ground wave at any particular location.

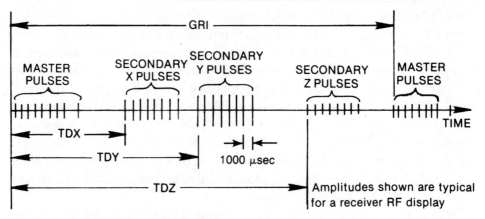

Figure 29-2. *Loran-C pulse sequence, for a four-station chain.*

To lessen the large power requirements, in Loran-C a so-called *multipulsed* signal is used. Each pulse transmitted by a master station actually consists of nine separate comparatively weak pulses; the first eight are separated by an interval of 1,000 μs, and the ninth by 2,000 μs. Each secondary station transmits eight such pulses at 1,000 μs intervals, with the extra pulse in the master signal being used for visual master station identification and trouble blink. Within the Loran-C receiver, each of the basic eight weak pulses is electronically superimposed or "integrated" to form "strong" master and secondary pulses of approximately 320 μs duration, which can then be electronically compared to obtain the time-difference measurement.

To eliminate sky-wave interference, the integrated master and secondary pulses of the ground wave are compared at a sampling point 30 μs from their leading edges. The comparison is thus made before any reflected sky waves of the pulse can reach the receiver. Another pulse feature designed to eliminate sky-wave interference is called *phase coding,* wherein the phase of the carrier wave (i.e., the 100 kHz signal) is changed systematically from pulse to pulse, so that any sky waves that reach the receiver will be out of phase with ground waves, and thus be rejected.

The multipulsed signals of each master station and its associated secondary stations are transmitted in a predetermined sequence, as shown in Figure 29-2. The master station signal is allowed to arrive at each of its secondary stations before the secondary station signals are commenced, and a coding delay is included between the secondary transmissions to ensure that all transmissions will always be received in the same sequence throughout the area covered by the chain.

In 1992 the Coast Guard transferred to foreign ownership or closed down all U.S.-operated Loran-C stations located on foreign soil. As a result, several new foreign Loran-C chains have been established in areas of the world previously covered by U.S. chains, bringing the current total to twenty fully operational chains.

The countries of Norway, Denmark, Germany, Ireland, the Netherlands, and France have established a common Loran-C system known as the Northwest European Loran-C System (NELS). The developing system will eventually be comprised of nine stations forming four chains. Two of these at Edje, Denmark, and Bo, Norway, are scheduled to become operational soon, with others at Lessay, France, and Sylt, Germany, to follow later.

The countries of Japan, the People's Republic of China, the Republic of Korea, and the Russian Federation have established a Loran-C network known as the Far East Radionavigation Service (FERNS). Japan took over operation of the former U.S. Coast Guard stations in its territory; they are currently operated by the Japanese Maritime Safety Agency. In 1996, five chains (Korea, North China Sea, East China Sea, South China Sea, and Russia) became operational.

Each of the more than twenty Loran-C chains presently in operation uses a unique pulse repetition rate (PRR)/pulse repetition interval (PRI) combination, with the PRRs currently between 10 to 20 μs, and the PRIs between 40,000 to 100,000 μs. The specific PRI used in a given chain is referred to as its *group repetition interval (GRI)*, often called a *rate*. The first four digits of the GRI are used as the chain designator; the secondary station is identified by suffixing its letter (W, X, Y, or Z) to the GRI. For example, the code 9960-X designates the master–secondary-X station pair of the Northeast U.S. chain, which uses a basic PRR of 10 pulses per second, transmitted at a specific PRI of 99,600 μs. An observed time difference in μs, corresponding to a hyperbolic LOP, is added as a suffix to the basic code, as for example 9960-X-11300.

LORAN-C Receivers

In the 1960s and early 1970s, a typical Loran-C receiver was a cubic meter of large, complex, and bulky electronics with readouts consisting of difficult-to-operate oscilloscope-type CRT displays requiring manual signal matching to obtain sequential time-difference (TD) readings for each Loran-C master-secondary station pair within a given area of coverage. Two or more TDs would then have to be plotted manually to obtain a fix, either on a Loran-C chart or, in some cases, by the use of a set of Loran-C tables. Consequently, mainly only relatively large naval warships, submarines, and merchant ships carried Loran-C receivers, and often these were little used because of the fairly extensive operator training required.

The advent of space-age technology in the 1970s, however, had a tremendous impact on the Loran-C system, and the receivers designed to use it. In recent years, solid-state technology has resulted in a proliferation of relatively inexpensive ($200 to $300 for some models), compact, and virtually fully automatic receivers for both marine and aircraft applications. Most receivers can

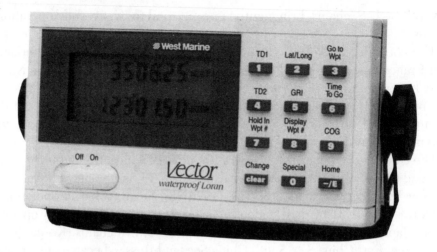

Figure 29-3. *A modern Loran-C receiver.*

now provide direct lat/long digital readouts, precise to a tenth of a minute of arc, and many provide as standard auxiliary features multiple waypoint (intermediate destination) selection, course and speed over the ground, and course and distance to next waypoint, all available at the touch of a button. Once turned on and initialized, these receivers will automatically select the best chain to use based on signal strength in the area, and the best combination of master-secondary pulses to use for a fix based on optimum fix geometry.

Several models of receivers are now approved for use on Navy ships, submarines, and aircraft. All are fully automatic, easily operated with minimal operator training, and highly reliable.

Plotting the Loran-C Fix

After the time-difference readings for two or more Loran-C station pairs receivable at a particular location have been obtained and recorded, it remains only to plot the corresponding portions of the time-difference hyperbolas on a chart to obtain the Loran-C fix. Of course, if the user's receiver has a direct lat/long output feature, only the fix position need be plotted. Even in these cases, however, it is prudent to check the actual time-difference readouts against the Loran-C chart occasionally, to be certain that the position being read out is consistent with the charted time-difference hyperbolas in the area in which the position plots. If by chance only one Loran-C line is obtainable in the area in which the vessel is operating, it may be combined with LOPs obtained by other means to form a fix or running fix, in the usual fashion.

The Defense Mapping Agency and the National Ocean Survey both issue series of Loran-C charts upon which the hyperbolic patterns covering the area represented on the chart are overprinted. There is a separate set of color-

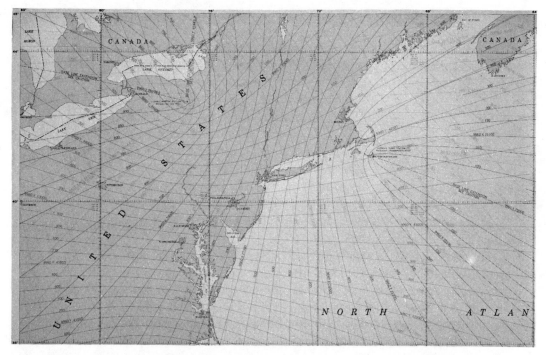

Figure 29-4A. *Portion of a Loran-C chart of an area covered by the Northeast U.S. chain (GRI 9960).*

coded time-difference hyperbolas for each master-secondary station pair receivable within that area. The hyperbolas are printed at intervals of 25, 50, 100, or 200 μs, depending on the scale of the chart and the degree of convergence of the hyperbolic patterns depicted. A portion of a typical Loran-C chart appears in Figure 29-4A.

It is fairly simple to plot a Loran-C LOP on a chart overprinted with Loran hyperbolas. The navigator merely selects the two time-difference hyperbolas printed on the chart near this DR position that bracket the observed reading for the secondary station in question, and plots a short segment of the hyperbola corresponding to this reading in the appropriate position between them. The segment is a straight line plotted in rough conformance with the trend of the hyperbolas to either side.

To assist in the correct positioning of the segment between the two printed hyperbolas, an aid called a *linear interpolator* is printed in the margin of every Loran chart. An example appears in Figure 29-4B.

The interpolator is used in conjunction with a pair of dividers. First, the dividers are spread to the interval appearing between the two printed hyperbolas between which the observed segment is to be drawn in the vicinity of the DR position. Without changing the spread thus established, the dividers are then moved to the linear interpolator, where they are held at a right angle to

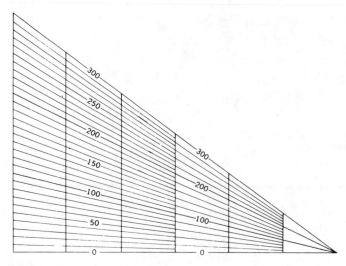

Figure 29-4B. *A Loran-C linear interpolator.*

the bottom base line with one point resting thereon. The dividers are then shifted to the right or left until the top point rests on the top curve of the interpolator. The portion of the interpolator between the divider points now represents the charted interval. The top point of the dividers is then compressed downward across the interpolator scale until the point rests on the scale marking corresponding to the exact value of the observed reading. The dividers are then shifted back to the site at which the segment is to be drawn on the chart, with care being taken not to disturb the spread determined by the interpolator, and a point is located at the spread distance from the lower-valued overprinted hyperbola. The segment is then drawn through this point, and labeled with the rate above the line and the time of observation to the nearest minute below the line, as shown in Figure 29-4C.

Two or more such Loran-C LOPs obtained from different master-secondary station pairs within a half hour can be combined to form a Loran-C fix. If the readings on which the LOPs are based were made within a reasonably short time span, such as two or three minutes, the intersection of the two plotted LOPs can be considered the Loran-C fix position. If more than a few minutes has elapsed between the readings, however, the earlier LOP(s) should be advanced to the time of the later LOP in the same manner as is done in plotting a running fix. A single Loran-C LOP cannot usually be used to obtain a running fix with a later Loran-C LOP from the same secondary station in surface navigation, since the angles between successive hyperbolas are fairly small in most areas of coverage. It can, however, be crossed with an LOP from another source obtained either simultaneously or within three hours of the Loran-C LOP to determine a fix or running fix, respectively.

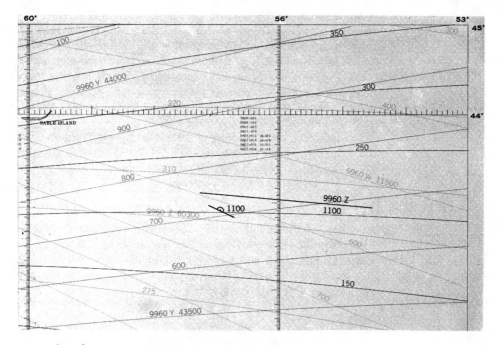

Figure 29-4C. Loran-C LOP plotted for a reading of 9960-Z-60306.0.

If sky waves are used to obtain a Loran-C time difference, the reading obtained from the receiver must be corrected to an equivalent ground-wave reading by the application of a sky-wave correction. On Loran-C charts, sky-wave corrections for day and night for each rate appearing on the chart are printed at various intersections of the latitude-longitude lines over the areas where they are applicable. The correction to be applied is determined by double interpolation based on the ship's DR position.

The Future of Loran-C

For many years, Loran-C was primarily used by mariners operating in the coastal waters around the United States and in the Great Lakes. In the 1980s, however, the reliability and consistency of the system, coupled with the growing availability of newer, more compact, affordable, and highly automated solid-state receivers, made it increasingly attractive not only to the marine community, but also to the commercial and civil aviation communities as an excellent means of airborne position-fixing in all areas of coverage over the continental United States. In fact, it is estimated that there are currently many times more aviation users of Loran-C than marine users.

At one time it was thought that when the satellite-based GPS system became fully operational in the mid-1990s Loran-C would be shut down, but this has not happened. Instead, the system has in fact been expanded in recent years, as described above. Given the proliferation of users in both the marine and aviation communities, it is probable that Loran-C will continue to operate at least until the year 2015 and perhaps even beyond.

DECCA

The Decca system, more properly called the Decca Navigator System, is unique in that it was owned and operated for many years, until the late 1980s, by a private enterprise, the Racal-Decca Navigator Company Limited, based in London, England. The system was originally conceived in 1937 by an American engineer, W. J. O'Brien, and developed by the British Admiralty Signals Establishment. Its first practical use was in guiding the leading minesweepers and landing craft in the Allied invasion of Normandy in 1944. The Decca Navigator Company, formed in 1945, further refined the system and established the first commercial Decca chain in southeast England the following year. The present coverage of the system is depicted in Figure 29-5.

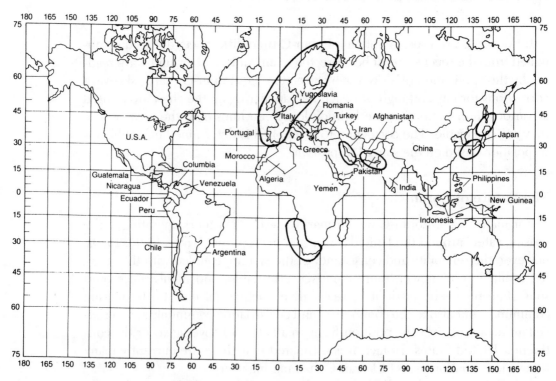

Figure 29-5. *Areas of the world covered by Decca. Currently there are thirty chains still in operation.*

Decca is similar to Loran-C in that each chain uses a master station in combination with up to three slave stations, but the systems differ in that Decca employs unmodulated continuous waves (CW) rather than the pulsed waves of Loran. The characteristic hyperbolic grid pattern is formed by phase comparisons of the transmitted master and slave signals.

All stations in the Decca system transmit on frequencies in the LF band between 70 and 130 kHz. The nominal maximum range of Decca is considered to be 240 miles from the master station both by day and by night, with sky-wave interference rendering the system unusable beyond this limit; Decca can thus be described as both a short- and a medium-range system. While the range of the system is somewhat limited, this disadvantage is compensated for by the extreme accuracy and relative simplicity of obtaining Decca LOPs within the area of coverage. The average maximum rms LOP error within the area covered by a Decca chain is given in the following table.

Nautical miles from master station	rms error in yards	
	Day	Night
100	30	100
150	60	350
200	100	700
240	150	1,200

The average Decca chain is composed of a master and three slave stations arranged in a so-called *star* pattern, with an angle of about 120° between each master-slave base line. The signals transmitted by each of the stations in a given chain are all harmonics of a single fundamental frequency (f), and chains are differentiated by assigning each a separate fundamental in the range from 14.00 to 14.33 kHz; harmonics are used to simplify the phase comparison process by which Decca LOPs are obtained. For purposes of identification and reference, each of the slave stations is designated by a color—purple, red, or green. All master stations transmit on a frequency of $6f$, all purple stations on a frequency of $5f$, all reds on $8f$, and all greens on $9f$.

The Decca receiver actually consists of four separate receivers, each of which can be set to receive one of the four signals transmitted by a given chain by simply selecting the correct fundamental frequency for that chain. Within the receiver, the signals for each master-slave pair are then electronically multiplied up to a single comparison frequency. The $6f$ master frequency is multiplied by 4 and the red $8f$ frequency by 3 to obtain a comparison frequency for the master–red-slave pair of $24f$; at the same time, in another part of the receiver the $6f$ master is multiplied by 3 and the green $9f$ slave by 2 to produce a comparison frequency for the master–green-slave pair of $18f$. In similar fashion the master–purple-slave comparison frequency $30f$ is formed. Phase com-

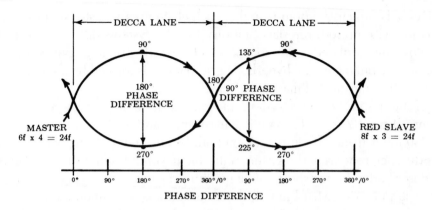

Figure 29-6A. *Phase comparison of Decca master/red-slave signals along the base line.*

parisons for each pair of stations within the selected chain are then made at the three comparison frequencies to yield three Decca LOPs.

Since all signals in each Decca chain are phase-locked along the base line between, say, a master and red-slave station, the comparisons in Figure 29-6A might be made at the comparison frequency $24f$. In the figure, the distance between each 0° phase-difference measurement, which occurs every half wavelength, is referred to as a *Decca lane*. If the fundamental f were 14.00 kHz for this example, the wavelength would be 880 meters, and the Decca lane width measured along the base line would be 440 meters.

In the left-hand lane, a receiver is located such that a phase difference of 180° is measured between the master and red-slave comparison frequencies. Thus, the receiver must be located at a point 0.50 of the distance in the lane away from the master station. In the right-hand lane, a phase comparison of 90° is obtained, so the receiver is located 0.25 of the distance between the lanes, again measured from the master toward the slave station.

If only a phase comparison were available between two such continuous waves, it would be impossible to determine in which of the two lanes shown in Figure 29-6A the receiver was located without some additional information. In the Decca system, lane identification is accomplished by the transmission of a *lane identification signal* by each master and slave station within a given chain once each 20 seconds for a duration of 0.6 second. The normal signals transmitted by the chain are interrupted during these times. Briefly, during the lane identification signal period all four harmonics are radiated simultaneously, and within the receiver the harmonics are summed to derive a pulse train having the fundamental frequency f for that chain. Using this frequency, a half-wavelength *zone* is thus created encompassing 18 green lanes, 24 red lanes, and 30 purple lanes. Phase comparisons are performed on this identification signal for each of the station pairs, which in effect indicates in which lane

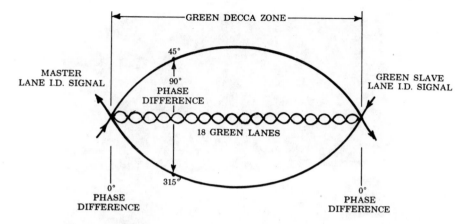

Figure 29-6B. *Phase comparison of Decca master and green-slave-lane identification signals.*

within each of the colored zones the receiver is located. This operation is illustrated in Figure 29-6B for a master–green-slave station pair. In this figure, phase comparison of the master and green-slave-lane identification signals indicates that the receiver is located in the fifth green lane contained by the zone, measured from the master toward the slave station.

In addition to the lane identification signal described above, each station periodically transmits a *zone identification signal* during each 20-second cycle on a frequency of $8.2f$. Called the orange frequency, it is combined within suitably equipped receivers with the $8f$ frequency to form a coarse hyperbolic pattern in which each 360° difference cycle embraces five zones. The receiver performs a phase difference measurement using this derived zone identification pattern, and indicates by a two-character code in which of the five zones the receiver is located. The orange frequency is also used for system monitoring and control functions within each chain.

For identification purposes, each Decca zone is assigned a letter from *A* through *J* either clockwise or counterclockwise from the base-line extension, depending on the slave station color; the lettering is repeated every tenth zone. Each lane within a zone is identified by numbers; the numbers 0–23 are used for red lanes, 30–47 for the green lanes, and 50–79 for purple lanes.

Obtaining and Plotting the Decca Fix

In the Decca receiver, the zone letter, lane number, and fractional position within the lane to the nearest 0.01 are continuously indicated for each master-slave station pair on dials called *decometers*. The face of each dial indicates the fractional position within the lane, while the zone letter and lane number appear in a window just above the dial. The newest model of shipboard Decca receiver, designated the Mark 21, is pictured in Figure 29-7. It incorporates

Figure 29-7. *The Mark 21 Decca receiver. (Courtesy Racal-Decca Navigator Co., Ltd.)*

three decometers, one for each color-coded pair in the chain, plus an LED readout that displays the lane identification signal for each of the chain pairs three times each minute.

Once the receiver has been set up for a particular Decca chain, the decometers provide continuous readings of the zone, lane, and fractional lane position for each of the station pairs until the vessel passes out of range of that chain. Once obtained, the position information is used to plot segments of hyperbolic LOPs on a suitable chart overprinted with color-coded Decca hyperbolic patterns in the same fashion as is done in the case of Loran. An example Decca fix appears in Figure 29-8.

A variety of receivers is available for use with the Decca system, each designed to fill the needs of a particular application. There are also several automatic graphic display devices available, designed to plot a continuous track over a chartlet, using information automatically supplied by the Decca receiver. This system is especially valuable in aircraft, in which it constitutes a type of automatic flight log.

Because of the limited areas of the world covered by Decca (see Figure 29-5), U.S. Navy ships are normally outfitted with Decca receivers and charts only when extended operations in an area covered by Decca are anticipated. Heavy use is routinely made of the system by merchant and fishing vessels and aircraft

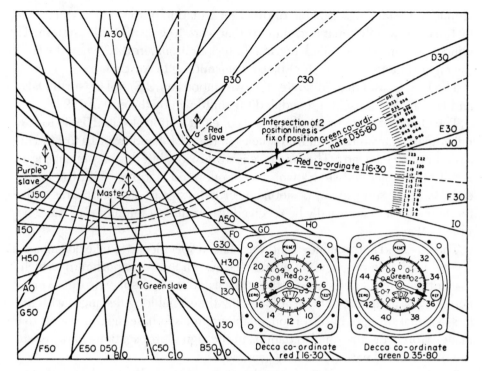

Figure 29-8. *A plot of a Decca fix. (Courtesy Racal-Decca Navigator Co., Ltd.)*

operating in the North Sea and the English Channel, and in the Sea of Japan. Although the operation of Decca is now subsidized by the British government, local usage in the United Kingdom is so heavy that it is currently anticipated that the system will continue in use at least through the year 2000, notwithstanding the availability of GPS.

OMEGA

One of the major disadvantages of all of the hyperbolic navigation systems developed and deployed in the years between 1940 and 1970 was the relatively small areas of the world that they cover. Moreover, the precision of the longest-range of these systems, Loran-C, falls off rapidly beyond about 1,200 miles, although one-hop sky-wave signals are often obtainable out to about 2,300 miles. To overcome these disadvantages, in 1947 the U.S. Navy began a research and development program with the goal of developing an electronic navigation system that could achieve worldwide coverage through the use of frequencies in the LF or VLF bands. Gradually over the next twenty years, the various technical problems associated with such transmissions were solved, and in the late 1960s, the last major technical obstacle was overcome with the development of atomic time standards capable of precise regulation of the signals transmitted by the eight widely separated stations that were to constitute the system. The Omega system, as it is called, became fully operational in 1982 when the last of the eight stations in the system began transmitting in Australia.

The U.S. Coast Guard operates the two stations on U.S. soil in North Dakota and Hawaii, but these are scheduled for shutdown by the end of 1997.

The Omega system presently consists of eight stations located 5,000 to 6,000 miles apart, transmitting on frequencies in the VLF band from 10 to 14 kHz at a power of 10 kilowatts. The coverage is worldwide, but this may change as the result of the planned shutdown of U.S.-operated stations. With the current eight-station system, at any position on earth the signals of at least three and often four stations can usually be received, to produce a minimum of three possible LOPs.

The 10–14 kHz frequency band was chosen specifically to take advantage of several favorable propagation characteristics, the most important of which is that at these frequencies, the earth's surface and the ionosphere form a very efficient wave guide capable of propagating the signal long distances with little loss of signal strength or distortion, enabling precise phase measurements to be made. Moreover, any small variations in the signal propagation characteristics can be predicted with reasonable accuracy.

The Omega system is similar to the Decca system discussed previously, in that for most applications the same principle of phase comparison of two CW transmissions is used to obtain an LOP, but it is different in that any two Omega stations receivable in a given geographic area may be considered as a pair for the phase comparison. It is also possible with specially designed equipment to obtain ranges to two or more Omega stations, and then determine an Omega position by plotting the intersection of the resulting range LOPs. In practice, the phase comparison method is used for the most part.

The basic frequency at which all eight stations transmit is 10.2 kHz, with atomic frequency standards being used to ensure that the transmissions of all eight stations are kept exactly in phase. Lines of zero phase differences (*isophase lines*) form a hyperbolic pattern between each pair of stations. For comparison purposes, a typical hyperbolic pattern formed by a station pair having a baseline length of 1,000 miles, such as Loran-C, is contrasted to a similar pattern formed by an Omega station pair with a base line of 6,000 miles in Figure 29-9.

The small divergence of successive Omega hyperbolas, coupled with the accuracy with which propagation corrections may be predicted, normally results in Omega fixes being accurate to within ±1 mile rms by day and ±2 miles rms by night anywhere in the world, except in areas experiencing unusual propagation effects.

Two 10.2 kHz Omega continuous waves transmitted exactly in phase but traveling in opposite directions produce a series of Omega lanes; within each,

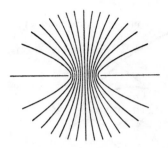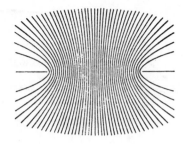

Figure 29-9. Hyperbolic patterns for 1,000-mile (left) and 6,000-mile (right) base lines.

a phase-difference measurement would progress from 0° to 360° as the receiver moved across the lane. Two such Omega lanes, produced by one complete cycle of two Omega signals, are shown in Figure 29-10A. Position of a receiver within an Omega lane is expressed in terms of so-called *centicycle* (cec) or *centilane* (cel) units, each defined as 0.01 of the width of an Omega lane. A reading of 50 centicycles, for example, would indicate a position at which a phase-difference reading of 180° was obtained.

To alleviate the problem of lane identification, three other signals in addition to the basic 10.2 kHz Omega signal are transmitted by each station on a time-sharing or "multiplexed" basis. These frequencies are 11.05, 11.33, and 13.6 kHz. Within an appropriately designed Omega receiver, these three signals can be electronically combined with the basic 11.33 and 10.2 kHz signals to form so-called *difference frequencies* of 0.283, 1.133, and 3.4 kHz, respectively, which are $\frac{1}{36}$, $\frac{1}{9}$, and $\frac{1}{3}$ the frequency of the 10.2 kHz signal. These three difference frequencies can then be compared to establish three broader or "coarse" lane widths of 288, 72, and 24 miles, as measured along the base line. Each 3.4 kHz broad lane contains three 10.2 kHz lanes, each 1.133 kHz coarse lane contains three 3.4 broad lanes, and each 0.283 kHz (283 Hz) lane four 1.133 coarse lanes. Hence, users need only be able to establish position within

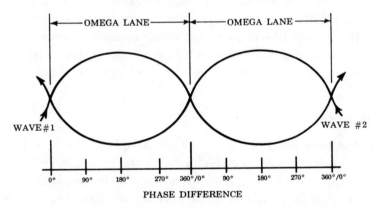

Figure 29-10A. Two Omega lanes formed by phase comparison of two 10.2 kHz signals.

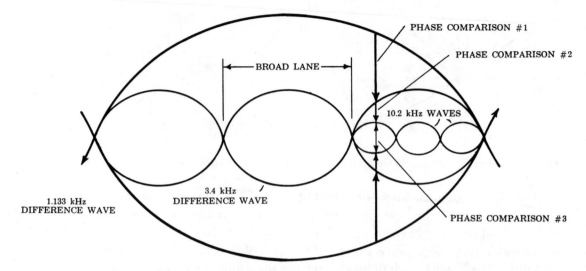

Figure 29-10B. *Three successive Omega phase comparisons for lane resolution on surface ships.*

±144 miles (half the width of a 283 Hz coarse lane) to determine in which 10.2 kHz lane they are located, through four successive phase comparisons as indicated in Figure 29-10B.

If all four Omega frequencies were transmitted simultaneously in the same order at each of the eight Omega stations, it would be difficult if not impossible to differentiate the various station signals. Hence, the stations in the Omega system transmit in a 10-second format called a *commutation pattern*, shown in Figure 29-11.

Each station in the Omega system is designated by a letter from A through H; the current locations of the stations are indicated. Each station transmits its frequencies as shown; note that the transmission timing pattern is different for each station. These time patterns and the location and duration of the signal within the repeating 10-second commutation pattern serve to identify the individual stations. Each transmission period is interrupted by a 0.2-second pause, to provide the desired 10-second period for the complete commutation cycle.

In addition to the four navigation signals, each station also transmits a timing signal on a frequency unique to it four times each cycle, as indicated in Figure 29-11.

Omega Receivers

Omega receivers, like Loran-C receivers, have been the beneficiaries of a great deal of improved solid-state technology in recent years. Fully automated sets that provide direct lat/long readouts are now widely available at moderate cost (around $500) for both marine and aviation use. However, the system has not enjoyed much popularity in either of these communities, primarily because

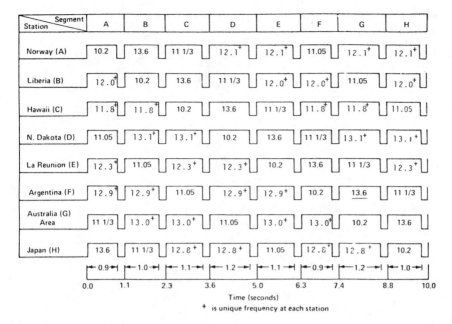

Figure 29-11. *Omega station commutation pattern.*

its precision has not proven sufficient for use except as a backup in areas requiring fairly precise position-keeping, such as coastal waters or near airports. The primary users, therefore, have been vessels and larger aircraft making ocean transits in areas not covered by Loran-C.

Plotting the Omega Fix

Inasmuch as differing ionospheric reflection and atmospheric wave propagation conditions affect the Omega signals, it is always necessary to adjust each basic phase difference reading obtained from an Omega pair by a *propagation correction*. The value of this correction varies with the location of the Omega receiver within the hyperbolic pattern of the station pair being used, with the season of the year, and with the time of day. Newer receivers apply these corrections automatically, but for older less capable receivers they must be computed and applied manually. The corrections are predicted by means of a computer model that is updated periodically by empirical data gathered in the course of system usage. The DMAHTC publishes as a set the *Omega Propagation Correction Tables* for the 10.2, 11.33, and 13.6 kHz Omega frequencies transmitted by each of the eight Omega stations under *Publication Number 224*.

Charts for Omega navigation similar to those designed for use with Loran are issued by the DMAHTC. They depict families of hyperbolas for 28 different Omega pairs for the basic 10.2 kHz Omega frequency. Most Omega charts are printed only with every third 10.2 kHz isophase hyperbola, due to the chart

537

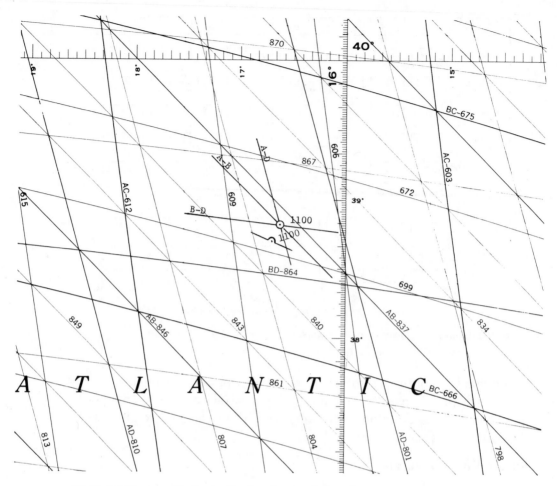

Figure 29-12. *A sample Omega fix plotted on an Omega chart.*

scale used. Only the hyperbolas for station pairs providing the best intercept angles within the geographic area covered by the chart are overprinted, with a separate color for each pair. The plot of a particular centicycle reading within the printed pattern is accomplished using a linear interpolator in the same manner as is done on Loran and Decca charts.

A typical Omega fix is shown plotted in Figure 29-12. Note especially the labeling of the three LOPs and the fix position.

Future of Omega

Omega was scheduled for phaseout in the United States in 1997, but it is probable that sufficient interest in the system exists in the worldwide radio-navigation community to keep the system in at least partial operation for the next several years.

SUMMARY

When the Omega navigation system is compared with the other radionavigation systems discussed in this chapter, the fact that worldwide coverage is achieved with an accuracy of ±1 to 2 miles rms with only eight stations makes it unique. For air and coastal marine navigation, however, the greater precision afforded by the other systems close to their base lines has led to their more widespread use. Because they are range-sensitive, it is not meaningful to attempt to compare the various systems by system accuracy and precision alone, inasmuch as the rms accuracy of all systems with the exception of Omega varies from a few hundred yards to several miles within the area of system coverage. The table below, however, may be useful in highlighting the various characteristics of the systems described herein.

System	Frequency Band kHz	Principle of Operation	Maximum Range Day/Night (Mi)
Radiobeacon	30–3,000	CW transmission & RDF	20–200
Loran-C	90–110	Pulse time difference and phase comparison	1,200/2,300
Decca	70–130	CW phase comparison	240
Omega	10–14	CW phase comparison	Worldwide

ADVANCED NAVIGATION SYSTEMS

30

The continued rapid pace of advances in electronics technology since the 1960s has had a significant impact in the routine practice of navigation, just as it has in most other areas of modern life. The last chapter described several systems wholly dependent upon modern electronics technology for their effectiveness. In addition to these systems, there have appeared in recent years several more esoteric navigational systems. These systems have come to be called *advanced navigation systems*, because of their degree of sophistication and precision in comparison with the other electronic systems discussed thus far.

There are currently three advanced navigation systems of paramount importance in operation as of this writing. They are the Ship's Inertial Navigation System, the doppler sonar system, and the Navstar Global Positioning System. This chapter will give a brief description of each; further information may be obtained from appropriate technical manuals.

THE SHIP'S INERTIAL NAVIGATION SYSTEM

All of the electronic navigation systems discussed previously, as well as the satellite-based and sonar-based advanced systems described later in this chapter, are dependent upon an electromagnetic or electromechanical (sonar) wave or series of waves transmitted either by the ship herself or by an external transmitter. In this section, a system unique in that it is independent of any such externally or internally transmitted signal will be discussed—the Ship's Inertial Navigation System (SINS).

The SINS system, like several of the other conventional and advanced navigation systems now operational, was originally developed in the late 1950s and early 1960s to meet the precision position-finding needs of the ballistic missile submarine. Since its first deployment on board the submarine USS *George Washington* in 1960, however, the SINS system has been continually refined, upgraded, and reduced in size, so that today its use has been extended to submarines, aircraft carriers, Aegis cruisers, many destroyers, and helicopter/dock landing ships.

Inertial navigation is defined as the process of directing the movements of a vessel based on sensed accelerations in known spatial directions by means of instruments that mechanize the Newtonian laws of motion, integrating such accelerations to determine velocity and position. The basic instruments used in all inertial navigation systems are gyros, accelerometers, and the electronic computer.

Gyros were introduced in chapter 9 in connection with the shipboard gyro-compass. Essentially, a classical gyro consists of a rapidly spinning mass that in accordance with Newton's laws maintains the orientation of its spin axis with respect to a universal reference system unless disturbed by some outside force, such as gravity or friction (see Figure 9-6, page 160). An *accelerometer* is a device designed to measure acceleration (A) along a given axis by measuring the force (F) exerted along this axis upon a given mass (M), using Newton's Second Law of Motion, $F = MA$.

Basically, in the SINS system a platform upon which two accelerometers are mounted is stabilized by a system of three gyros in such a way that it is constantly coincident with a plane tangential to the earth's surface. The two accelerometers are continually oriented in a north-south and east-west direction. Hence, they are sensitive only to horizontal north-south and east-west accelerations. Integration of these accelerations with respect to time yields corresponding north-south and east-west velocity components, which when vectorially summed produce true ship's velocity. From this velocity, the ship's position expressed in terms of rectangular coordinates is continually converted to the spherical coordinates latitude and longitude by the computer and read out on a suitable output device.

The accuracy of the SINS system is dependent in large measure on the precision and reliability of its instruments and supporting equipment. Some of the more significant sources of potential error in the system include the following:

- Errors caused by the daily rotational motion of the earth
- Friction in the gyro support systems
- Misalignment of the stabilized platform, causing vertical components of the earth's gravitational field to be falsely interpreted as horizontal components
- Imperfections in construction of the gyros and accelerometers

Because the combined error caused by these and other sources cannot be completely eliminated, all SINS systems experience some degree of cumulative error that increases with the passage of time. Thus, the position furnished by the system must be periodically compared to positions established by external means, and the system must be occasionally updated and reset. The current system installed in U.S. ballistic missile and attack submarines, however, requires far less frequent updates than did the initial SINS system of the early Polaris submarines. This is important since position determination by external means often necessitates cruising near the surface, where vulnerability to detection is greatly increased. When a SINS position update is needed, Loran-C, Omega, and the satellite-based GPS system described in this chapter can be used for this purpose, as can positions determined by celestial or bathymetric navigation.

Two of the more interesting advances that have occurred during the continued efforts to refine the SINS system in recent years have been the development of the *electrostatic gyro* (ESG) and the *laser gyro.* In the electrostatic gyro, the rotor consists of a 1-centimeter diameter solid beryllium sphere spinning at 216,000 RPM in a near-perfect vacuum. The rotor is supported solely by an electrostatic field, which holds the sphere suspended a few thousandths of an inch from the internal surface of the evacuated case containing it. The ESG is thus freed from the classical gyro bearing friction, as well as many of the associated random torques that a mechanical support can introduce. Hence, it represents the best approximation to a theoretically perfect gyroscope yet devised. In current submarine SINS systems, called ESGN systems, two independent ESG-based systems monitor each other, thus producing positions of such reliability that external fixes for resets are required only once every 30 days. In actuality they are seldom required to go this long between resets because of input from the GPS satellite system whenever the submarine is close enough to the surface to receive it.

The laser gyro has been incorporated in many of the newer SINS systems developed in recent years, especially those designed for aircraft. In reality, this device is not a gyro in the traditional sense, since there is no spinning central mass, but rather a closed geometric laser path (usually triangular) centered on an expected spin axis. Identically phased laser beams are continuously generated, which travel in opposite directions around the closed path. Any rotation about the spin axis causes an apparent phase difference in the two beams at a measurement point at one vertex of the triangle, since the path of the laser beam traveling in the direction of the rotation is effectively lengthened, while the path of the beam traveling in the opposite direction is shortened (see Figure 30-1). The amount of the phase difference thus measured is directly proportional to the speed of rotation. Because it does not depend for its operation on a spinning mass, laser gyro SINS systems can be even more precise than the ESG-based SINS systems.

The position produced by modern SINS systems should normally be much more accurate than a DR position derived by conventional means, but the navigator must still keep in mind that the SINS position is not a fix. The navigator, therefore, should refrain from placing complete confidence in any SINS position.

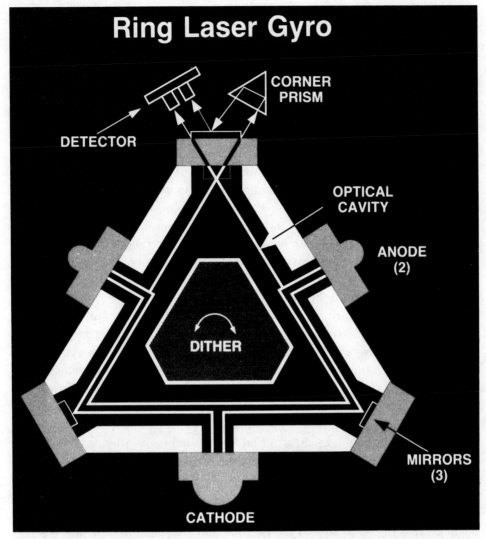

Ring Laser Gyro

CORNER PRISM

DETECTOR

OPTICAL CAVITY

ANODE (2)

DITHER

MIRRORS (3)

CATHODE

Figure 30-1. *A schematic of a ring laser gyro. When the gyro is rotated or dithered around an axis perpendicular to the plane of the two laser beams traveling in opposite directions, the resulting different path lengths cause frequency shifts that are proportional to the angular rotational rate.*

THE SHIPBOARD DOPPLER SONAR SYSTEM

The *doppler sonar system* is a relatively recent development, that can determine water depth and vessel speed over the ground with a high degree of precision and accuracy. The system is based on the doppler shift of a continuously transmitted sonar pattern.

A complication inherent in any doppler navigation system is that the propagation characteristics of a sonar signal tend to vary with the temperature, density, and salinity of the water into which the signal is transmitted. Inasmuch as the system depends on small frequency shifts of returning sonar echoes caused by the vessel's horizontal motion, for good accuracy in deeper water a method must be incorporated into any such system to compensate for random frequency shifts caused by variable water conditions. Furthermore, since only the horizontal component of a vessel's velocity is of interest, any random vertical errors that might be introduced as a result of pitching or rolling must also be eliminated.

There are both military and commercial versions of the doppler sonar system for large ships. In its simplest configuration, the system consists of a fixed hull-mounted transducer that transmits pulsed sonar signals in two beams, one oriented forward and the other aft, at about a 30° angle from the vertical. This dual-beam pattern is called the Janus configuration, after the two-faced Roman god of war who could look both forward into the future and back at the past at the same time. The system provides a continuous readout of ship's fore-and-aft speed precise to 0.1 knot; water depth to the nearest meter, foot, or fathom; and nautical miles run to the nearest hundredth.

The dual-beam pattern permits the forward and aft doppler shifts to be continuously compared in order to eliminate any vertical motion error, as well as error caused by irregular bottom features, thus allowing a very precise determination of horizontal velocity and depth to be made. In more advanced configurations, the doppler sonar system employs a dual axis four-beam array, with the beams transmitted fore and aft and port and starboard 90° apart as shown in Figure 30-2. For many military applications, the transducer is mounted on a gyro-stabilized horizontal platform that keeps the beams oriented in the four cardinal directions—north, south, east, and west—to allow true ship's course and speed over the ground to be computed, for input into other shipboard navigational systems and dead reckoning plotters. For commercial applications in highly stable deep-draft ships, such as large tankers and RoRo ships, the four-beam array is rigidly affixed to the bottom, and stabilization is achieved internally by electronic means. The precise determination of athwartship's speed provided by the doppler system is of great importance to these huge ships, as this speed component is critical when maneuvering in restricted waters and while docking. The safe docking speed for a vessel exceeding 100,000 deadweight tons is about 0.2 feet per second; the berthing facility might collapse if contact were made with a speed greater than about 1 foot per second.

Most doppler sonar systems use a bottom-bounce mode in water depths up to about 300 meters (1,000 feet), and a volume reverberation mode wherein the signal is reflected from the water mass itself in depths exceeding this limit. The maximum error of current systems has been demonstrated to be not more than 0.17 percent of the cumulative distance run since the last reset—about 0.85 mile after a run of 500 miles. At present most doppler sonar systems are

544

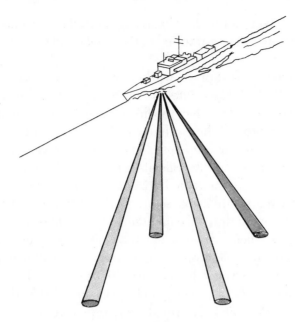

Figure 30-2. A dual-axis four-beam Janus sonar array of a doppler sonar system.

installed in commercial ships such as large tankers and freighters. They are be-ing used to great advantage both to maintain highly accurate DR plots in mid-ocean regions, especially in those areas not having reliable Omega or Loran-C coverage, and during docking evolutions and piloting in restricted coastal wa-ters and channels. System velocity data are used as input via digital and analog interfaces to a variety of other systems, including fin stabilizers, integrated steering systems, and satellite navigation systems.

THE NAVSTAR GLOBAL POSITIONING SYSTEM

In the late 1950s, with the advent of the Polaris SSBN submarine, the need for a highly reliable and precise position-fixing system that would be available worldwide regardless of weather conditions or proximity of shore transmitting stations became apparent to the U.S. Navy. At that time, some efforts to extend Loran-C coverage were meeting with limited success, but Omega was still a concept in a few scientists' minds, and midocean position-fixing depended largely on celestial navigation. With the launching of the first artificial satellite *Sputnik I* in 1957, however, and the rapid advances in space-age electronics and missile technology that quickly followed, it was soon realized that the means to establish the desired comprehensive navigation system were at hand.

A satellite navigation system based on the doppler shift of signals transmit-ted by orbiting satellites was first proposed by a team of scientists of the Ap-plied Physics Laboratory of Johns Hopkins University in late 1957, as a result of their studies of the behavior of the signals transmitted by *Sputnik.* In the following year, a Navy contract was awarded to the Applied Physics Laboratory to develop such a system.

This led to the development of the world's first large-scale satellite navigation system, called the U.S. Navy Satellite System (NAVSAT), which became operational in 1964 with the launch of the first of an eventual eight operational Transit satellites. They orbited the earth in roughly equidistant circular orbits 600 miles high. The eight-satellite constellation could provide periodic two-dimensional fixes worldwide at intervals of from 35 to 95 minutes, as one by one the satellites passed over the radio horizons of users on the earth below.

For the next thirty years until its phaseout in late 1996, the NAVSAT system (called the Transit system in the civil community) proved to be one of the most cost-effective and beneficial government-sponsored programs of modern times, providing highly reliable positioning information to all manner of both military and civil users . However, by the early 1970s, the need for a precision satellite-based navigation system that would be available worldwide at all times had become acute within all U.S. armed services. Moreover, a continuous three-dimensional position-finding capability (i.e., latitude, longitude, and altitude) was stipulated as a prime system objective, in contrast to the periodic two-dimensional capability of the NAVSAT system. Such a system would be usable not only by military ships, submarines, aircraft, and ground vehicles, but would also be of great benefit to the civil sector in a wide variety of applications, from precise topographic mapping to aircraft and ship collision-avoidance systems. In April of 1973, the Department of Defense formally initiated the development program for this second generation navigation satellite system, called the Navstar Global Positioning System, or GPS. Later that same year, the GPS Joint Program Office was established and staffed by military and civilian representatives from all four U.S. armed services, the Coast Guard, the Defense Mapping Agency, and NATO.

Several possible satellite configurations were considered during the developmental phases in the early 1980s. The arrangement finally chosen was four satellites in each of six orbits. Each satellite orbit is circular, about 10,900 nautical miles high (20,200 kilometers), inclined at angles of 55 degrees with respect to the earth's axis. Each satellite has a 12-hour period. Three of the satellites are spares, maintained in orbit for quick replacement of any operational satellite that may go "down" unexpectedly. The goal of the system configuration is to provide for at least a 0.95 probability of a minimum of four satellites being above the 5° minimum altitude required for good reception at any time at any position on earth. However, there are times in the midlatitudes of both hemispheres when the fourth satellite is so low above the user's horizon that a temporary degradation of three-dimensional accuracy occurs. But these times are infrequent and are a maximum of one-half hour in duration.

546

Figure 30-3. *A block-2 GPS satellite being assembled at the Navstar assembly line at Seal Beach, California.*

The GPS control segment consists of five *monitor stations* located at Colorado Springs, Colorado; Hawaii and Kwajalein in the Pacific; Ascension Island in the Atlantic; and Diego Garcia in the Indian Ocean. The last three of these can upload data to the GPS satellites. The *master control station* (MCS) for the system is located at Falcon Air Force Base in Colorado. The monitor stations passively track all satellites as they come over the radio horizon, accumulating ranging data that are then passed to the MCS. There, this information is processed to determine satellite orbits and to update each satellite's navigation message. The updated messages are then transmitted to the satellites via the uplinks at Kwajalein, Ascension Island, and Diego Garcia.

As was mentioned in chapter 4 in conjunction with the nautical chart, the geodetic reference plane used for the GPS system is WGS-84. Users should be careful to ascertain that this reference plane is also used on any charts on which they might plot GPS fixes, since a significant variation in the actual versus the plotted position might otherwise result.

The GPS system was declared fully operational in mid-1995. Updates on satellite status and other information of interest to system users is published in the *Notice to Mariners,* and status reports are also available via the NAVINFONET.

GPS System Operation

Position determination using the GPS system is based on the ability of the user's receiver to accurately determine the distance to the GPS satellites above the user's horizon at the time of the fix. If accurate distances or *ranges* to two

2-RANGE FIX

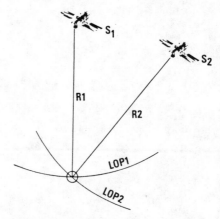

Figure 30-4A. *If ranges R1 and R2 to two satellites could be measured directly, they would produce two LOPs sufficient to determine a point fix on the earth's surface. (Courtesy of Magnavox Corporation)*

such satellites can be found, a fix can then be determined on the earth's surface, as shown in Figure 30-4A. In order to do this, the receiver needs to know the exact time at which the signal was broadcast by the satellite and the exact time that it was received. Then, if the propagation speed down through the atmosphere were known, the resulting range could be easily calculated.

Unfortunately, in practice the problem is more complicated, because only atomic time standards can provide the required accuracy to determine the ranges this way, and they would make the receivers prohibitively expensive. Most GPS receivers employ instead crystal oscillator clocks, which introduce a common time bias error into the signal time-of-arrival measurements. These, plus any propagation speed errors, result in a small range error, which is common to all the GPS ranges measured at a particular time. So, the measured ranges thus determined are called *pseudoranges*. Initially, three pseudoranges to three GPS satellites will not converge at a single point. Within the receiver the computer section adjusts the ranges in equal increments until the resulting LOPs do converge at a single point, in effect solving three simultaneous equations (one for each pseudorange) for three unknowns—the clock error, latitude, and longitude (see Figure 30-4B). A fourth satellite is required if altitude is to be determined.

GPS Satellite Signals

Each GPS satellite, like the earlier NAVSAT Transit satellites, broadcasts simultaneously on two frequencies, in order that high-precision receivers can directly determine and eliminate the effects of ionospheric and atmospheric

3-PSEUDORANGE FIX

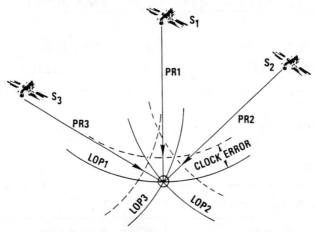

Figure 30-4B. *If pseudoranges (PRs) to three satellites are available, they can be adjusted in equal amounts until a fix can be produced from three converging LOPs. (Courtesy of Magnavox Corporation)*

refraction on the signal, thus allowing for a more accurate computation of the propagation speed. However, the Navstar frequencies are higher, at 1,575.42 MHz and 1,227.6 MHz, and they are broadcast continually, in contrast to a 2-minute format used by NAVSAT. Because they are in the L-band subdivision of the UHF frequency range, they are designated L1 and L2.

Both these signals are modulated by 30-second navigation messages transmitted at 50 bits per second (bps). The first 18 seconds of each 30-second frame are ephemeris data for that particular satellite, which precisely define the position of the satellite as a function of time. The remaining 12 seconds are almanac data, which define the orbits and operational status of all satellites in the system. The GPS receivers store and use the ephemeris data to determine the pseudorange, and the almanac data to help receivers determine the four best satellites to use for positional data at any given time.

The L1 and L2 satellite navigational signals are also modulated by two additional binary sequences, one called the *C/A code*, for acquisition and coarse navigation, and the other, the *P-code*, for precision ranging after acquisition and synchronization have been achieved by the receiver (see Figure 30-5). The L1 signal is modulated both by the C/A and P-code, and the L2 only by the P-code. Positional accuracies of better than 50 meters rms are possible using the C/A code alone, and 5–10 meters rms are possible using both the C/A and P-code. However, for security reasons it has been deemed undesirable to provide full system capability indiscriminately to all users. Therefore, the P-code is not routinely made available, but is encrypted such that it is decipherable

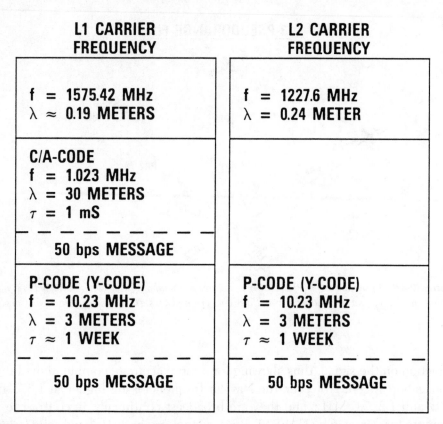

L1 CARRIER FREQUENCY	L2 CARRIER FREQUENCY
f = 1575.42 MHz λ ≈ 0.19 METERS	f = 1227.6 MHz λ = 0.24 METER
C/A-CODE f = 1.023 MHz λ = 30 METERS τ = 1 mS ─ ─ ─ ─ ─ ─ ─ ─ ─ ─ 50 bps MESSAGE	
P-CODE (Y-CODE) f = 10.23 MHz λ = 3 METERS τ ≈ 1 WEEK ─ ─ ─ ─ ─ ─ ─ ─ ─ ─ 50 bps MESSAGE	P-CODE (Y-CODE) f = 10.23 MHz λ = 3 METERS τ ≈ 1 WEEK ─ ─ ─ ─ ─ ─ ─ ─ ─ ─ 50 bps MESSAGE

Figure 30-5. *The characteristics of the GPS satellite navigation signals. (Courtesy of Magnavox Corporation)*

only by U.S. and NATO military users equipped with special receivers; the encrypted P-code is designated the Y-code. Moreover, C/A-code accuracy can also be degraded by a second cryptographic technique called selective availability (SA) encryption, wherein random errors would be inserted into the ephemeris data, thereby decreasing system accuracy to as low as 2,500 meters rms or less for receivers using the C/A code alone. Selective availability requires a presidential order to invoke, and would only be done in wartime to deny an enemy the tactical advantage of accurate GPS positioning.

Thus, civil and foreign users have available a so-called *Standard Positioning Service* (SPS) with a stated horizontal positioning accuracy of not less than 100 meters 95 percent of the time, and 150 meters vertically, while U.S. and NATO military users have available *Precise Positioning Service* (PPS) accurate to at least 16 meters horizontally 95 percent of the time, and 25 meters vertically.

Differential GPS

As a means of enhancing the SPS accuracy of the GPS system in selective areas, the *differential technique* described in chapter 28 can be applied to the

GPS signals (see Figure 28-12). In differential GPS, the reference station receiver measures the pseudoranges from at least five, and often more, Navstar satellites. Since the reference receiver is at a known surveyed location, it can calculate precisely what the pseudoranges should be at any time. The differences between these calculated values and those obtained from the GPS transmissions are then transmitted as corrections to remote GPS receivers in the surrounding area, to produce more accurate fixes than could otherwise be obtained.

Tests have indicated that accuracies approaching 2–5 meters rms are achievable using differential GPS on moving vessels, and under 1 meter rms in stationary situations. As previously mentioned in chapter 28, the Coast Guard is transmitting differential GPS data from various radiobeacons located in U.S. coastal and inland waters and the coastal waters of Puerto Rico, Hawaii, and Alaska, so that commercial ships and pleasure craft equipped with differential-capable GPS receivers can use the GPS system to navigate in confined harbors and approach channels, where the required positional accuracy is under 10 meters rms. Differential techniques can also be applied for airport approach control and surveying applications. These techniques could be defeated in wartime by changing the C/A and Y-codes and/or the L1 and L2 carrier frequencies randomly with time.

GPS Receivers

There are currently three basic types of GPS receivers being designed and built for the various user communities. These are called *slow-sequencing, fast-sequencing,* and *continuous-tracking* receivers. The least complicated and lowest-cost receiver for most applications is the slow-sequencing type, wherein only one measurement channel is used to dwell sequentially on the L1 C/A code from each satellite in use for periods of about 1.2 seconds, with occasional interrupts to collect ephemeris and almanac data. Because the pseudoranges thus determined are not computed and processed simultaneously but rather over an approximate 5-second interval, some error is introduced, but this is negligible compared to the 50-meter rms error intentionally inserted into the encrypted satellite signals. Most of the lower-cost receivers designed for low-dynamic environments (e.g., ground troops and civil marine applications) are presently of this type.

Fast-sequencing receivers have two channels—one for making continuous pseudorange measurements, and the other for collection of the ephemeris and almanac data. The pseudorange measurements in this type of receiver are also sequential, but are done at a much faster rate than the slow-sequencing-type receiver, making the tracking appear to be continuous. This type of receiver is used in medium-dynamic applications such as ground vehicles and marine applications requiring higher precision.

FIGURE OF MERIT	ESTIMATED POSITION ERROR	
	METERS	FEET
1	Less than 25	Less than 82
2	Less than 50	Less than 164
3	Less than 75	Less than 246
4	Less than 100	Less than 328
5	Less than 200	Less than 656
6	Less than 500	Less than 1640
7	Less than 1000	Less than 3280
8	Less than 1000	Less than 16400
9	Unknown	Unknown

Figure 30-6. *Correlation of figure of merit to estimated position error.*

Continuous-tracking receivers employ multiple channels (preferably at least five) to track, compute, and process the pseudoranges to the various satellites being utilized simultaneously, thus obtaining the highest possible degree of accuracy. Four channels are continuously locked onto the satellite signals, making simultaneous pseudorange and doppler measurements, while the fifth channel gathers ephemeris and almanac data from all visible satellites not otherwise being tracked. Because this type of receiver can provide highly precise, continuous, and virtually instantaneous positioning data, it is the type carried by highly dynamic platforms such as high-performance aircraft and most naval ships and submarines, where high precision is required for weapons guidance purposes.

Receivers of all three types are compact, highly automated, and easy to use. Handheld slow-sequencing receivers are about the size of a large transistor radio, while aircraft and shipboard types are not much larger. Current prices for civil models range from under $300 for low-dynamic sets to $1,000 or more for the highly precise high-dynamic sets capable of being used with any of several optional geodetic reference planes besides the standard WGS-84. Many will accept differential GPS inputs, and some, like the AN/WRN-6 receiver used aboard most naval ships, have a display called the *figure of merit,* an accuracy rating scale ranging from 1 to 9, with 1 being most accurate, and 9 least. The figure of merit is calculated by the receiver based primarily on possible error caused by the geometry between the receiver and the satellites in view plus any known timing errors (called geometric dilution of position, or GDOP). Other possible sources of error could include ionospheric and atmospheric delays, mathematical and/or timing errors within the receiver, and satellite signals taking a circuitous path to the receiver (known as multipath error). Typically, a figure of merit of greater than 4 (indicating an estimated position error of less than 100 meters—see Figure 30-6) would render the indicated fix position unacceptable for use by Navy ships, and would probably indicate erroneous signal decryption.

GPS receiver prices have gone down dramatically in recent years as solid-state technology has continued to advance and production costs have decreased. As a point of interest, several automobile manufacturers have begun to offer GPS sets with digital map displays of position as standard equipment on their luxury models, and optional equipment on others. Undoubtedly the popularity of the GPS system for everyday use in an ever-expanding variety of air, land, and marine applications will continue to increase greatly over the next several years (see Figure 30-7).

THE RUSSIAN GLONASS SYSTEM

Because of the many advantages offered by the continuous availability of satellite-based positioning data, the United States is not the only nation currently involved in satellite radionavigation. The Russian Federation is presently in the final stages of deployment of a system very similar to GPS called *Glonass.* Presently there are some 24 operational Glonass satellites plus a number of spares in roughly circular 10,300 nautical mile orbits in three orbital planes 120° apart. The spacing of the satellites is such that a minimum of 5 satellites are in view at all times to users worldwide.

Although not much has been officially released about Glonass by Russian authorities to date, technical data gathered by civil scientists in this country and Europe indicate that there are many similarities between it and the GPS system. Like GPS, the Glonass satellites transmit on two L-band frequencies, L1 presently at $1,602$ MHz $+ 0.5625\, n$, where n equals the number of the satellite, and L2 around $2,150$ MHz; they are modulated with a 30-second data message, transmitted in two codes very similar to the GPS P- and C/A codes. Accuracy of the Glonass system has been reported to be better than standard GPS with selective availability (SA) off, that is, under 10 meters rms. One point of difference between the two systems is that GPS uses WGS-84 as its reference datum for satellite positions and fixes, whereas Glonass uses a different reference called PZ-90. The difference between terrestrial coordinates expressed in the two frames is estimated to be about 20 meters, which must be compensated for when comparing positional data obtained by the two systems.

Among the goals for the Glonass system that have been announced by the Russians are an improvement of satellite life from 3 years in the case of the current prototypes now in orbit to 5 years, improved stability of on-board clocks, a reduction in the user range error to under 5 meters rms, and the addition of a provision to transmit the offset in the time scales of the GPS and Glonass systems.

One of the major problems concerning Glonass has been a lack of availability of Glonass receivers, which are not as yet manufactured outside of Russia. They are currently almost impossible to obtain in the West, though this may

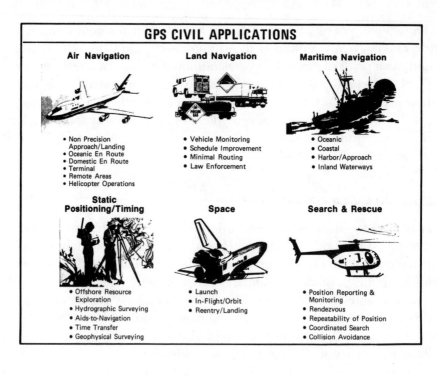

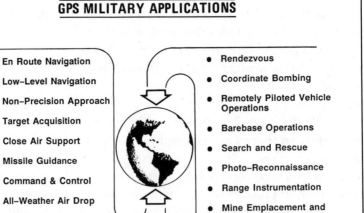

Figure 30-7. *GPS military and civil applications.*

change in the next few years. If a receiver were developed that allowed simultaneous reception and analysis of the positioning information from both systems, it is thought that positional accuracy comparable to that obtained by the use of differential GPS could well result.

SUMMARY

This chapter has introduced three advanced electronic navigation systems currently of major importance in the field of marine navigation. Two of these, the ship's inertial navigation system (SINS) and the doppler sonar system, have been fully operational for some time, and are now in their second and third generations of equipment. The Navstar Global Positioning System (GPS) achieved fully operational status in mid-1995. All of these use state-of-the-art electronics technology, and have proven to be highly effective for a variety of military and civil applications.

The Russian Glonass satellite navigation system, similar to GPS in operation, may also be used more for air and marine navigation in the coming years if suitable receivers can be developed and marketed for it.

The GPS and Glonass systems have rendered other medium- to long-range terrestrial and satellite-based radionavigation systems obsolete for most applications, with the result that many if not all of these will probably be shut down during the next decade.

BATHYMETRIC NAVIGATION
31

Bathymetric navigation can be defined as the art of establishing a geographic position on the open sea by use of geological features of the ocean floor. These features are located by means of an instrument called the *echo sounder*; in practice, this device is more commonly known as the *fathometer*, the name applied to an early, widely adopted model of this device by its manufacturer, the Raytheon Company.

The inexperienced navigator may question the need for bathymetric navigation in the present day when advanced techniques of celestial and electronic navigation are available. It must be remembered, however, that the professional navigator will use all means available to establish the ship's position and safely direct her movements. Even in the best circumstances, bathymetric position-fixing can provide valuable backup information. In the worst circumstances, the method may be the only means of determining the ship's position. Among the advantages of bathymetric navigation are that it uses fixed features of the ocean bottom, it is impervious to electronic jamming or weather, its application is nearly worldwide, and it is easily learned and applied.

This chapter briefly describes the nature of the ocean bottom features and the charts used to portray them, and examines the characteristics of the echo sounder and its output information. Finally, the techniques of bathymetric navigation presently in common use on many U.S. Navy surface ships are discussed, along with their advantages and limitations.

GEOLOGICAL FEATURES OF THE OCEAN BOTTOM

In order to make use of information received from the echo sounder, the navigator must first be aware of the basic characteristics of the geological features present on the ocean floor. In general, ocean bottom features are equivalent to topographical features seen on land, with the difference that undersea geology is usually much more subdued and gentle than land geology, owing to the more subtle erosion forces present within the oceans. Some predominant features of the ocean floor are pictured in Figure 31-1 and are described as follows:

An *escarpment* is a long, steep face of rock, similar to a cliff on land.

A *seamount* is an elevation rising steeply from its surroundings to a height of 500 fathoms (1,000 meters) or more; it is generally of relatively small horizontal extent.

A *guyot* is a flat-topped seamount, rather similar to the mesas found in the southwestern United States. Guyots are especially plentiful in the Pacific Ocean region.

Submarine canyons are similar to their dry-land counterparts and are found off most continental slopes.

A *trench* is a relatively narrow canyon, distinctive because of its great depth.

The ocean bottom areas contiguous to the continents are generally devoid of any of the distinguishing features described above, but they are nonetheless useful for navigation by means of their *depth contours,* which are lines on a chart representing points of equal depth with respect to the surface datum. The zone between the emergence of a continent from the sea and the deep-sea bottom is called the *continental margin.* Within this margin three different subdivisions can generally be identified:

The *continental shelf* is the zone immediately adjacent to a continent or island, extending from the low-water shoreline and sloping gently to an

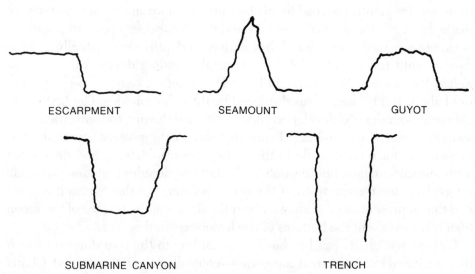

ESCARPMENT   SEAMOUNT   GUYOT

SUBMARINE CANYON   TRENCH

Figure 31-1. *Features of the ocean bottom.*

area of steeper slope. Its depth ranges from 10 or 20 fathoms down to 300 fathoms, and it may extend seaward from beyond the shoreline for a widely varying distance.

The *continental slope* is the area extending from the edge of the continental shelf into greater depth.

The *continental rise* is a gentle slope with a generally smooth surface, rising from the deep-sea floor to the foot of the continental slope.

Beyond the continental margin, the deep ocean floor generally consists of expansive plains and gently rolling hills, occasionally interrupted by distinguishing features, such as those listed above. Hence, as will be explained in more detail later, when such features can be identified by means of the echo sounder, they provide excellent positional information to the navigator, especially when used in combination with depth contours of the surrounding region.

THE BATHYMETRIC CHART

Bottom soundings and associated bottom sketches and charts have long been regularly used by the mariner. The use of the hand lead line in the Nile River well before the birth of Christ has been recorded. The first formal hydrographic office was established in Spain in 1508, and although intended primarily as a regulatory enterprise, a department was tasked with the charting of routes to and from the New World. Unfortunately, much of the information gathered by this and similar agencies of other contemporary governments was considered a secret to be closely guarded, inasmuch as a safe and rapid sea route was frequently the road to great wealth. As more and more voyages were made, however, the depths in and around the major ocean ports of the world became fairly well documented, but midocean depths were virtually unmeasurable until the invention of a mechanical sounding device known as the *Kelvin-White sounding machine* in 1878. This sounding system was in wide use until about 1930, when it was displaced by the echo sounder. The first workable echo sounder was developed about 1920, and the first transatlantic line of soundings was made in 1922. From that day to the present, continual improvements have been made in the techniques of obtaining and presenting sounding information, but it is only in the last few decades that any systematic efforts have been made to chart the ocean bottom. For this reason it is often said that at present more is known about the details of the surface of the moon than is known about the bottoms of the lesser-explored oceans.

Bathymetric charts can be classified according to the soundings on which they are based into two broad categories—controlled and uncontrolled. Charts in the former category are produced as the result of a systematic survey of selected areas, while charts in the latter category are based on random data collected from ships that have traversed an area while transiting from one port to another. Today, controlled bathymetric charts exist for nearly all undersea areas within the midlatitude continental margins, and for deep ocean areas beneath the more heavily traveled ocean routes between continents. Many infrequently traveled ocean areas, however, are covered only by uncontrolled charts based on random soundings made many years ago.

The DMAHTC issues a series of bathymetric charts referred to as bottom contour charts. Bottom contours also appear on most five-digit coastal charts issued both by DMAHTC and the National Ocean Survey. Depth contours shown on charts produced by both agencies are based on an assumed sound propagation velocity of 4,800 feet per second, identical to the standard velocity to which most echo sounders are calibrated. Thus, no correction is necessary to echo sounder depths, other than the addition of the ship's keel depth in cases in which the installed equipment does not incorporate an automatic draft adjustment feature.

The contour interval of a bathymetric chart depends on the accuracy of the sounding data available and the chart scale. In general, controlled charts are produced on a scale between 1:10,000 to 1:100,000, and have contour intervals from 5 to 50 fathoms, depending on the scale. Uncontrolled charts ranging in scale from 1:1,000,000 to 1:1,500,000 are contoured at 100-fathom intervals. The importance of the proper contour interval cannot be overemphasized. The left-hand portion of Figure 31-2A depicts three 100-fathom contours, while the right-hand portion depicts 10-fathom contours in the area of the inset. If the scale of the chart were fairly large, the 100-fathom interval would be impractical for bathymetric navigation, while the 10-fathom interval provides a much more complete picture of the bottom for navigational purposes.

It should be apparent that if a very small contour interval were chosen for a small-scale chart, a morass of unintelligible contour lines would result. On the other hand, if too large an interval were selected, the chart would not be detailed enough to permit its use in bathymetric navigation.

In general, in bathymetric navigation as in piloting, the largest scale chart available depicting a given bottom area or geological feature of interest should be used. In the left-hand side of Figure 31-2B, a portion of an ocean area containing a seamount is contoured at a 250-fathom interval, such as might be found on a three-digit chart of the area; in the right-hand side, the same feature is depicted using 50-fathom intervals. It should be apparent that the closer interval spacing provides a much more accurate picture of the shape and size of the seamount. In fact, it is doubtful that the feature could be recognized as suitable for position-fixing from an inspection of the left-hand representation, while it is readily identifiable in the right-hand drawing.

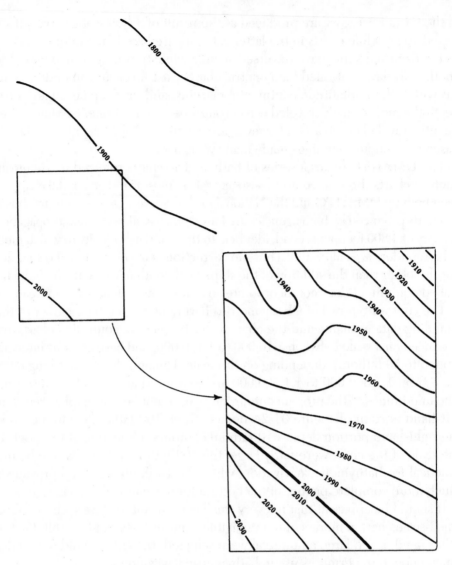

Figure 31-2A. *Contrast between 100- and 10-fathom contour intervals.*

CHARACTERISTICS OF THE ECHO SOUNDER

The echo sounder or fathometer consists of two basic components—the transducer and the recorder. The *transducer* used on most ships is located at or near the keel, and contains both a *projector* for transmission of a sound signal into the water and a *hydrophone* for reception of the returning echoes. The *recorder* is located where required, usually in the chart room. An echo sounder widely used in the U.S. Navy is the AN/UQN-4, pictured in Figure 7-4 on page 117.

The AN/UQN-4 transmits on a frequency of 12 kHz. Both the pulse duration and the pulse repetition rate are variable, and several different combina-

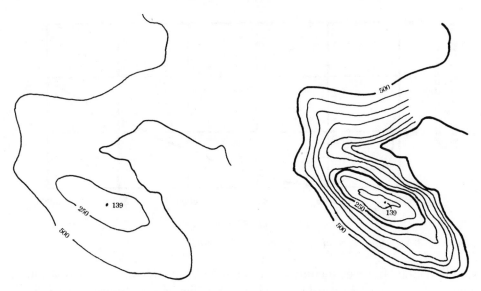

Figure 31-2B. Seamount depicted using 250- and 50-fathom contour intervals.

tions can be selected depending upon the depth of water in which the ship is operating; a digital readout indicates the depth in either feet or fathoms. The UQN-4 is also equipped with a continuous-feed graphic recorder, which produces a graphic trace of the depths encountered, and two additional very useful features: a *draft adjustment,* which automatically adds the ship's draft to the transducer depth, as well as a so-called *lost tracking indicator,* which is illuminated whenever a depth varying more than 200 feet from a preset depth is recorded. The AN/UQN-4 subtends a cone 30 degrees in width, in contrast to the 60-degree cone of most other fathometers.

The signal transmitted by an echo sounder is electromechanical in nature. The transmitter, which is usually physically located in the recorder, emits a pulsed CW electromagnetic signal that is translated into a sound pulse by the transducer. The sound pulse train is radiated into the water in the shape of a cone, the dimensions of which are governed by the frequency of the pulsed wave and diameter of the transducer; as mentioned above, the normal cone width is 60 degrees. When a pulse strikes any surface or boundary layer of a region within which the sound propagation characteristics are different from those of the water into which the pulse was transmitted, an echo is returned. The strength of the echo depends on the quality of the reflective surface the pulse strikes. The returning echo is converted into an electromagnetic signal by the transducer, and in the recorder the depth is obtained by taking into consideration the elapsed time between the transmission of the pulse and the return of its echo and the transmission speed of sound in seawater. For purposes of the depth computation, most Navy echo sounders are calibrated to use a

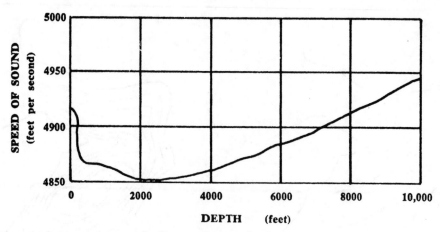

Figure 31-3. *Variation in sound velocity with ocean depth.*

standard 4,800 feet per second as the sound velocity. Because the actual sound velocity is both variable and somewhat faster than this standard, varying with salinity, temperature, and pressure as shown in Figure 31-3, some difference almost always exists between the actual depth and the indicated depth. The actual depth is always greater than the indicated depth, thus providing a small variable safety factor.

There are other sources of error in the echo sounder depth as well. Since the transducer pulses propagate outward in the shape of a cone, the first reflective surface that the cone encounters will produce a reflection interpreted by the recorder as the depth immediately beneath the ship. In actuality, however, the surface may lie off to one side, as shown in Figure 31-4. Such an echo returned from an object at the edge of the sound cone is termed a *side echo.*

A third possible source of error is the rolling and to a lesser extent the pitching motion of a ship while under way. The position of the transducer on most ships is fixed, causing the transmission cone to be canted at an angle to the vertical as the ship proceeds. This effect is a factor that must be taken into account regardless of the cone width, as shown in Figure 31-5.

INTERPRETATION OF THE ECHO SOUNDER RECORDING

When an echo sounder is to be used for bathymetric navigation, the continuous graphic trace is used to display and record the depth information. Most types of trace paper are marked with several scales, corresponding to the various scales on the recorder that the operator can select. As the paper feeds through the recorder at a constant rate, an electrostatic pen moves rapidly over the surface of the paper at a right angle to its direction of movement, shading the paper at locations corresponding to the depth measurement. Thus, a depth profile is plotted versus time.

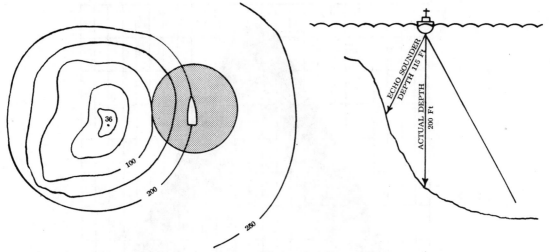

Figure 31-4. *Echo sounder depth error resulting from a side echo.*

Because of the conic sound propagation pattern of the echo sounder, a bottom trace of an object protruding above an otherwise level ocean floor appears hyperbolic in shape, as, for example, the seamount trace pictured in Figure 31-6. The shape of the hyperbola is a function of the beam width of the transducer, the depth of the feature, and the speed of the ship.

If the ocean bottom is fairly irregular, a pattern of several hyperbolas superimposed on one another as in Figure 31-7 may result. The multiple hyperbolas are due in large measure to the side echoes returned by the edges of the propagation cone; the effect is even more pronounced if the ship is rolling from side to side. When interpreting such a trace, the navigator must bear in mind that the minimum depths recorded to the top of each hyperbolic trace are not necessarily those directly beneath the ship. Hence, the trace cannot be regarded as a profile of the bottom along the route of the ship, but rather it is a repre-

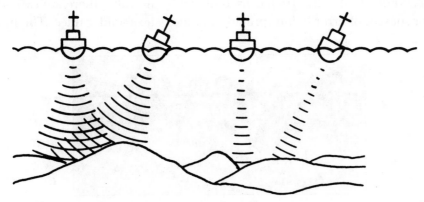

Figure 31-5. *Echo sounder depth errors resulting from ship motion.*

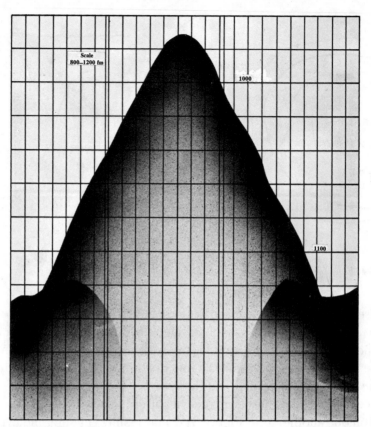

Figure 31-6. *Echo sounder trace of a seamount.*

sentation of the average depths over the area swept by the sound propagation cone as the ship moves along her path.

As will be seen in the following section, the disadvantages of the conic propagation effect described above are balanced by certain other beneficial effects.

In addition to the distortion caused by the beam width, there are two other major sources of error in interpretation of the echo sounder trace. The first is

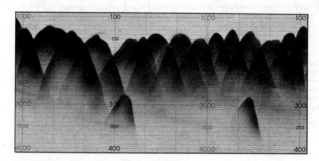

Figure 31-7. *Echo sounder trace of an irregular bottom.*

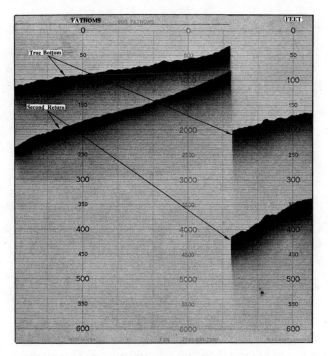

Figure 31-8A. *Echo sounder trace on two scales, showing multiple bottom returns.*

the *multiple bottom return,* an example of which is pictured in Figure 31-8A. The effect is most common in relatively shallow water with a highly reflective bottom such as sand or gravel, and is caused by reverberation of the sound pulses between the bottom and the water surface. A similar phenomenon sometimes occurs if the bottom is covered to a depth of several fathoms by a material of poor reflective qualities, such as soft mud. One return is formed by the weak reflection from the surface of the mud, while a second return results from the actual bottom reflection.

A second possible source of error in echo sounder trace interpretation is the *deep scattering layer,* a suspension of biological matter such as plankton in a layer between the surface and the bottom. The effect of this layer is illustrated on the trace pictured in Figure 31-8B.

When a layer of this type is present, it usually rises toward the surface at night and sinks somewhat by day. Often the effect is very persistent in a given area, and may lead to numerous reports of shallow water at locations where the actual depth is very great. A feature called the American Scout Seamount with depths of 30 to 90 fathoms was reported east of Newfoundland so often between the years 1948 to 1964 that it was printed on several contemporary charts of the area. Controlled surveys conducted in the 1960s, however, found no depths in the area less than 2,350 fathoms. The fictitious seamount reports are believed to have been caused solely by the deep scattering layer.

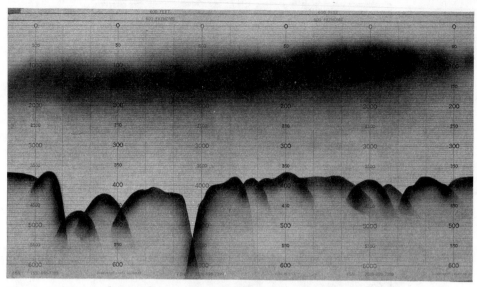

Figure 31-8B. *An echo sounder trace showing the deep scattering layer.*

TECHNIQUES OF BATHYMETRIC POSITION-FINDING

There are several techniques available for determining the ship's position by bathymetric methods. The particular technique employed depends largely on the general geological pattern of the bottom, the type of echo sounder installed, and the skill of the navigator in interpreting the recorder output.

If the ship passes directly over an easily recognizable bottom feature such as a seamount or submarine canyon, the ship's position can be determined with a high degree of accuracy, particularly if the surrounding ocean floor is fairly uniform. Any of the features depicted in Figure 31-1 are readily identifiable on the fathometer trace, even with the distortion effects due to the propagation cone present. Seamounts are particularly well suited for position-finding, as many have distinguishable depths, sizes, and shapes. When the ship passes over such an unmistakable feature, the navigator has only to note the time of passage and commence a new DR plot from that point on the chart.

If a uniquely identifiable feature either does not exist in the area over which the ship is operating, or if the ship does not pass directly over such a feature, a more complicated technique is then necessary to determine the ship's position. There are three such procedures used for the most part—the "line-of-soundings" technique, "contour advancement," and the "side-echo" technique.

The *line-of-soundings technique* has been widely used by both commercial and Navy navigators for many years. This procedure depends to a great extent on the accuracy of the depth contours depicted on the chart in use, and is essentially independent of the chart scale. In this technique, an overlay of thin

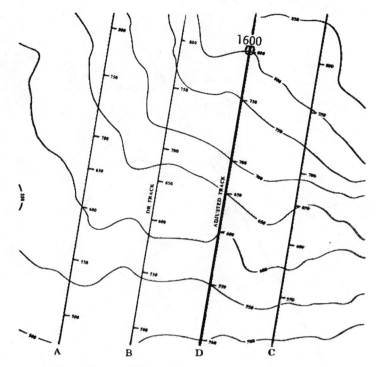

Figure 31-9. *Line-of-soundings technique.*

transparent paper or plastic acetate is laid down over the bathymetric chart, and the DR plot is constructed on the overlay in the usual manner. As the ship proceeds, the times at which she passes over the depths corresponding to the charted contour lines are noted, and DR positions for these times are plotted on the DR course line; each is labeled with the depth recorded at the time, as shown in Figure 31-9. The length of time during which these depth recordings are made and plotted depends on the scale of the chart and contour interval.

After a number of depth recordings have been plotted, the overlay is detached and shifted about in the vicinity of the original DR track until a match between the plotted depths and the underlying charted depth contour lines is made. This matching process is represented on Figure 31-9 by lines A, B, and C; line D is the final position of the template. After the match has been made, the overlay is secured in position, and the last depth mark on the overlay DR plot is taken as either a fix or an estimated position, depending upon the degree of confidence the navigator places in the position thus determined.

It should be apparent that this method is not suitable if the ship is proceeding in a direction roughly parallel to the depth contours, or in areas where adjacent contours are parallel or nearly so, as for example where the bottom is flat, or rising or falling at a constant slope.

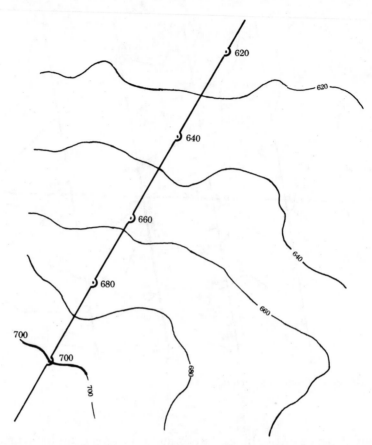

Figure 31-10A. *DR plot on bathymetric chart, contour advancement technique.*

The *contour advancement technique* also involves the use of an overlay, and is closely related to the line-of-soundings method just discussed. First, a DR plot is laid down on the bathymetric chart and a DR position corresponding to an initial recorded depth is placed on the course line directly over the corresponding depth contour; this initial contour is referred to as the reference contour. Next, DR positions for times at which the ship passes depths corresponding to succeeding depth contours are plotted relative to the initial DR position, as shown in Figure 31-10A. In this figure, the 700-fathom curve is the reference contour.

After a suitable number of DR positions have been so plotted—usually four or five are sufficient—an overlay is placed over the DR plot and the reference contour is traced thereon, similar to the heavy line in Figure 31-10A.

Next, the overlay is detached and moved along the DR course line until the reference contour appears directly over the next DR position. In the case of Figure 31-10A, the overlay is advanced until the traced 700-fathom curve on the overlay crosses over the 680-fathom DR position. The contour corre-

PART III. ELECTRONIC NAVIGATION

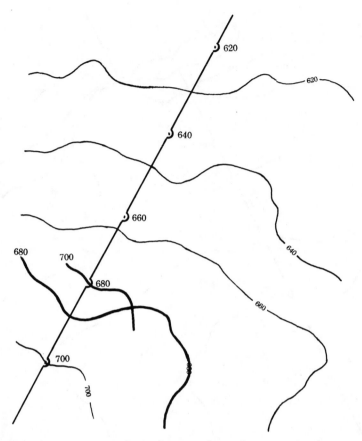

Figure 31-10B. *Reference contour advanced to time of second contour recording.*

sponding to this DR position is now traced onto the overlay, as shown in Figure 31-10B.

At this point there are now two contours drawn on the overlay, one atop the other. Now the overlay is again moved in the direction of the DR course line until the reference contour lies over the next DR position, and the third contour is traced. After the reference contour has been so aligned with every DR position plotted and the corresponding depth contours traced, the overlay is fixed in position. The point on the chart over which all traced contours seem to intersect is the ship's position corresponding to the time of the last DR position plotted, as shown in Figure 31-10C. To complete the plot, the position is noted, the overlay is removed, and a new DR plot can be commenced at that position.

Again, this technique does not lend itself to use in areas where the direction of travel is parallel to the depth contours or where the contours are parallel to one another and spaced equal distances apart on the chart.

The two techniques just discussed are very useful in situations in which the

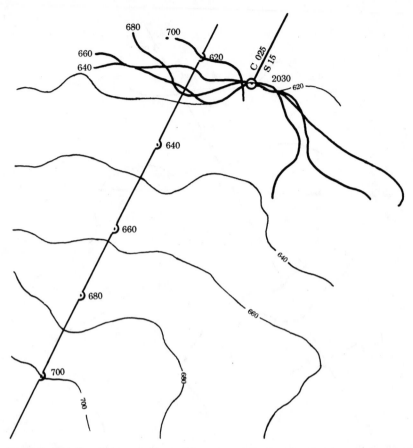

Figure 31-10C. *Completed contour advancement overlay.*

bottom is fairly uneven, the depth contours are closely spaced on the chart, and the direction of travel is approximately at a right angle to the contour lines. If an isolated seamount or guyot is to be used on an otherwise level ocean floor, however, the *side-echo technique* is used if the ship's path does not take her directly over the feature.

If a seamount is to be used to determine the ship's position, a DR track is first laid down from a position within 20 or 30 miles of the feature directly across its top. As the ship approaches, continual depth readings are recorded. If the minimum depth agrees with the charted minimum depth over the top of the feature, it can be assumed that the ship has passed directly over it; a fix symbol is plotted, and the DR plot is continued from this point. If the minimum depth is more than the charted depth, however, it can be assumed that the ship passed near but not directly over the top of the seamount. In this case, the side-echo technique is applied as described below.

First, a construction line is drawn perpendicular to the DR course line through a DR position corresponding to the time of the minimum depth reading

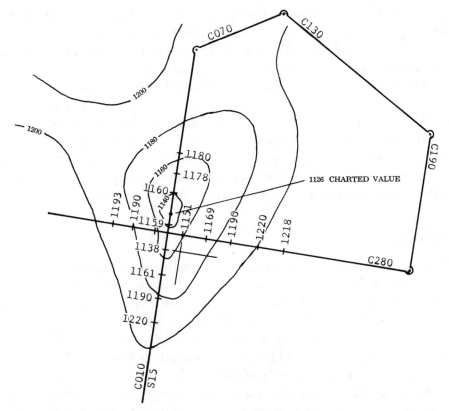

Figure 31-11. Side-echo technique.

obtained. After a downward trend in the depths being recorded is observed, the ship's course is changed in a wide 270° turn to either side, such that the projected ship's track will cross the original DR course line at a right angle near the minimum depth reading. Echo-sounder readings are again carefully recorded, and a DR position for the time at which a minimum depth is again recorded is plotted. A perpendicular construction line is then drawn through this point, so that the plot now resembles that depicted in Figure 31-11.

The point at which the two construction lines intersect locates the seamount top relative to the DR plot. An estimated position can then be plotted by shifting the DR position used for the second construction line through the distance separating the intersection of the two construction lines and the charted position of the top of the seamount. The ship's course is then adjusted as desired, based on the estimated position.

Although the use of the side-echo technique is disadvantageous because of the course change required, it has the compensating benefit that use of the technique takes full advantage of the conic propagation zone of the echo sounder. If the cone were fairly narrow like a radar beam, the ship would have to pass directly over a raised feature to be able to find and use it for navigational purposes. The expanding cone allows features to be readily located, and the side-echo technique makes possible the use of the feature for position-fixing without the necessity of passing directly over its top.

New techniques of position-finding by means of undersea topographical features are now being developed that incorporate the electronic computer to resolve difficulties arising in situations more ambiguous than the rather straightforward examples given in this section. Computers will provide the ability to perform statistical analysis upon a large amount of bathymetric data that the navigator now has the ability to gather, but cannot at the present time conveniently analyze for positional information over complicated bottom topography.

SUMMARY

This chapter has discussed techniques of bathymetric navigation used on surface ships of the U.S. Navy. Although the techniques described here are not widely used when celestial observations and electronic position-finding are available, they nonetheless represent another potential source of positional information available to the navigator of any type of vessel. The techniques of bathymetric navigation should not be overlooked, as positional information supplied from this source can be a valuable backup to information obtained by other means at sea. Bathymetric information is one of the few kinds of navigational data obtainable by the navigator in any weather, independent of outside facilities.

In this final chapter of part 3, the routine of a typical U.S. Navy navigator's day at sea will be briefly outlined, with special emphasis on the interplay of celestial and electronic means of position-finding. While nonmilitary readers may find that some of the auxiliary duties and activities described here are not applicable to their past experience and probable future circumstances at sea, nevertheless most of the actual work of navigating will be common to almost all vessels operating on the high seas. Many practicing navigators of commercial and private vessels, in fact, have often been heard to remark that the only real difference between Navy and civilian or Merchant Marine navigators at sea is that the former usually have many more assistants. Be this as it may, there is undeniably a great deal of similarity in the routines of all professional navigators of oceangoing vessels.

The transit from Norfolk, Virginia, to Naples, Italy, preplanned as an example in chapter 15, will form the framework of the discussion in this chapter. Before examining the events of a typical day en route on this voyage, it is first necessary to outline the normal routine of a Navy navigator's day's work at sea.

THE ROUTINE OF A DAY'S WORK

For purposes of this discussion, the typical day's work in navigation while on an extended voyage can be considered to commence with the preparation of the rough draft of the Captain's Night Orders, which were briefly described in chapter 15. In actual practice, the navigator usually begins this task after the evening celestial fix has been plotted and the ship's DR track for the night has been laid down from it.

The following is a fairly comprehensive listing of the tasks generally accomplished by the navigator of a U.S. Navy ship, with the help of the ship's staff of quartermasters, in the course of a typical day's work in navigation at sea:

1. Prepare the rough draft of the Captain's Night Orders.
2. Compute the commencement of morning star-time (civil twilight), the time of sunrise, and compile a list of the most favorable stars and planets suitable for observation.
3. If no unusual events are anticipated during the night, turn in until morning star-time.
4. Arise prior to morning star-time, and make the sextant ready for the day's observations.
5. Observe, plot, and record the morning celestial fix.
6. Observe sunrise, obtain and record a sun amplitude observation, and compute and record the resulting gyro error.
7. Construct the day's DR track, incorporating any corrections necessary to regain the preplanned intended track based on the morning fix.
8. Prepare the 0800 ship's position report slip (see chapter 15) based on the morning celestial fix.
9. Wind the ship's chronometers and determine and record chronometer error by radio time signals.
10. While the sun is at low altitude, observe its azimuth to determine the gyro error if necessary.
11. Obtain a midmorning sun observation for a morning sun line.
12. Compute the time of LAN.
13. Observe the sun at LAN and plot a running fix by advancing the midmorning sun line.
14. Prepare the noon position report slip.
15. Obtain a midafternoon sun observation for an afternoon sun line and an azimuth for determination of the gyro error if necessary.
16. Compute the time of sunset and the ending of evening star-time (ending of civil twilight) and compile a list of the most favorable stars and planets for observation.
17. Observe, plot, and record the evening celestial fix.
18. Prepare the 2000 position report slip based on the evening celestial fix.
19. Lay down the ship's DR track for the night from the evening fix.
20. Prepare the rough draft of the Captain's Night Orders.

Additional responsibilities that are interspersed throughout the routine tasks listed above are as follows:

1. Obtain, plot, and record additional celestial LOPs and fixes of opportunity, such as additional sun lines and daytime observations of the moon and Venus when available.
2. Obtain, plot, and record electronic fixes as available.
3. Obtain, plot, and record any bathymetric fixes of opportunity.
4. Perform routine administrative duties.

As mentioned above, the navigator does not personally have to accomplish each of the tasks listed, as the ship's quartermasters are skilled enough to perform many if not all of them. Exactly which tasks will be carried out and by whom is in the last analysis the prerogative of the individual navigator, based on experience and the ability of subordinates.

THE CAPTAIN'S NIGHT ORDERS

The Captain's Night Orders have already been introduced in "Voyage Planning," chapter 15. As was explained at that time, the night orders are written primarily for the benefit of the night OODs, to ensure that they will be informed as to the important details of an operational and navigational nature expected to occur during their watches. Although there is no established format for the night orders, most include the following information:

1. The authority under which the ship is operating
2. The status of the ship, including:
   a. Material condition (X-ray, Yoke, Zebra)
   b. Status of the engineering plant
   c. Condition of readiness set
   d. Major equipment casualties
   e. Times of anticipated changes to ship's status
3. Formation status and command structure, including:
   a. Range and bearing to formation guide
   b. OTC, SOPA, and screen commander
   c. Vessels joining or departing formation during the night
   d. Base course and speed
   e. Anticipated changes to formation or command structure
4. Navigational information, including:
   a. Times of planned base course and speed changes
   b. Times and approximate bearings and ranges at which aids to navigation should be observed either visually or on radar
   c. Frequency at which fixes are to be obtained and plotted
5. Pertinent tactical information
6. Times at which the navigator and commanding officer are to be called the following morning

The navigator's staff can normally fill in the more routine items included as part of the night orders, but the navigator usually adds the pertinent navigational and operational data. After the rough draft of the night orders has been completed, they are submitted to the commanding officer for annotation and final approval. On most ships, after the night orders have been signed by the captain, the OOD is authorized to make the listed course and speed changes at

CAPTAINS NIGHT ORDERS

SHIP USS EXAMPLE	TIME ZONE +1 N	DATE 24 - 25 JUNE

ENROUTE FROM   NORFOLK , VA   TO   NAPLES , ITALY

OPERATING WITH N/A	AREA NORTH ATLANTIC

OTC C.O. THIS SHIP	FLAG SHIP N/A

STANDARD TACTICAL DATA

FORMATION   N/A

BASE COURSE 102 °T 118 °PSTGC	SPEED 16 KTS 118 RPM

FORMATION AXIS N/A °T	GUIDE N/A	BEARING N/A °T	DISTANCE N/A YARDS

SCREEN DATA

TYPE SCREEN	SCREEN	CIRCLE	NO	SHIPS UNASSIGNED

OWN SHIP DATA

ENGINES ON THE LINE 2	GENERATORS ON THE LINE 2	PLANT 2+4	SHIP DARKENED YES ( ) NO (X)

EQUIPMENT CASUALTIES	ETR		ETR
SPN 40 AIR - SEARCH RADAR	UKN		

WEATHER DATA

SUNRISE 250447	SUNSET 241920	MOONRISE 242334	MOONSET 251115

NAVIGATION AND WEATHER REMARKS

1. SHIP IS 3 HRS. 17 MINUTES AHEAD OF PIM AT 242000N JUN.

2. NO UNUSUAL WEATHER CONDITIONS EXPECTED 24-25 JUN.

3. CALL NAVIGATOR FOR MORNING STARS AT 0400.

NIGHT INTENTIONS
1. CHANGE COURSE TO 109°T AT PT "G" AT MIDNIGHT (250000N).

1. CARRY OUT STANDING NIGHT ORDERS. CHECK   2. ETERNAL VIGILANCE IS THE PRICE
   THEM OVER TO REFRESH YOUR MEMORY.   OF SAFETY
              3. CALL ME WHEN IN DOUBT AND IN ANY EVENT AT ___0630___

SIGNATURE (COMMANDING OFFICER) *T. F. Epley*

SIGNATURE (NAVIGATOR) *R R Hobb* SIGNATURE (EXECUTIVE OFFICER) *D. C. Gaberti*

WATCH	OOD	JOOD
20-24	*P.S*	*C.T.*
00-04		
04-08		

WATCH REMARKS

Figure 32-1. Captain's Night Orders, Norfolk-Naples transit, 24–25 June.

the times indicated, using the night orders as the authority to do so. The OOD will then notify the CO of the action completed.

Most well-run commercial ships and private vessels will use some written equivalent of the Captain's Night Orders, so that the night watch officers will have a clear idea of any action intended to be accomplished during their watches. On sailing craft, the ship's log will often be used as the medium for recording and relaying this information.

A typical example of the Captain's Night Orders that might have been written for the night of 24–25 June, while en route on the transit preplanned in chapter 15, appears in Figure 32-1.

LAYING DOWN THE SHIP'S DR TRACK

After the navigator or his staff has obtained and plotted a celestial or electronic fix, a DR track should be laid down from this fix to take the ship back to the preplanned track, if in fact the fix shows the ship to be displaced from it. As was mentioned in the voyage planning chapter, if the ship is en route on an extended transit, the navigator will normally desire to maintain a position about three to four hours ahead of the position of intended movement (PIM) along the track until the ship is near the end of the voyage. The slack time thus built up allows for delays en route resulting from such causes as maneuvering exercises, engineering drills, equipment casualties, and rough weather.

THE INTERPLAY OF CELESTIAL, ELECTRONIC, AND BATHYMETRIC FIXING INFORMATION

The modern-day navigator has available a wide variety of sources from which to obtain position-fixing information at sea. As recently as the 1950s, the navigator had to rely almost exclusively on celestial observations to direct the movements of a vessel on the open sea. With the advent of worldwide navigation systems such as Omega and Navstar GPS, however, celestial navigation has for many naval and merchant vessels become for the first time in history a backup technique rather than the primary method of position-finding at sea. Even under these circumstances, however, the truly professional navigator should continue to observe morning and evening stars and the sun at local apparent noon, if for no other reason than to remain proficient in celestial navigation in the event of malfunction or destruction in wartime of the electronic aids. Furthermore, observation of a celestial body either as an amplitude or azimuth sight remains the primary method of determining gyrocompass error at sea, an essential daily task of the navigator. The prudent navigator should always take full advantage of all means at his or her disposal to determine and verify a vessel's position at sea.

AN EXAMPLE OF A DAY'S WORK IN NAVIGATION AT SEA

As an example of a day's work in navigation at sea, let us return to the mock voyage discussed in chapter 15. The voyage planning process was illustrated by means of an example based on an independent transit from Norfolk, Virginia, to Naples, Italy. It was determined that the ship should depart Norfolk at 1000R 17 June and arrive at Naples at 0800A 30 June. The preplanned track is shown plotted in Figure 15-5 on page 265.

For purposes of this discussion, it will be assumed that all applicable bathymetric, Omega plotting sheets, and large-scale charts are on board, and that the ship is equipped with an echo sounder and an Omega receiver. Furthermore, it will be assumed that all Omega stations receivable in the North Atlantic area are operational on the 10.2 kHz transmission frequency. As several different examples of the plotting of positions determined by celestial, electronic, and bathymetric methods have already been presented in earlier chapters, detailed solutions of the various fixes obtained in the course of this example will not be presented. Rather, the LOPs and fix positions are plotted as they might appear on the chart at the end of the day's work. For purposes of clarity in this example, all fixes are labeled to show the source from which they were derived, and Omega LOPs are labeled with the corrected readings received. In practice, celestial and bathymetric fixes would be labeled only with the time in the usual manner, and Omega LOPs would be labeled only with the station pair letters.

Let us begin the example of the day's work at midnight (0000N) 25 June. From the Captain's Night Orders shown in Figure 32-1, the ship is steaming approximately 3 hours ahead of PIM, to allow for any unexpected delays that might occur during the remainder of the transit.

The navigator has been using as the ship's primary working charts the bathymetric chart series covering the route of transit across the North Atlantic. In addition to these, the 7600-series Omega charts of the area have been used to plot Omega fixes. Once plotted, the Omega fix positions have been transferred to the bathymetric charts. The navigator has also been recording the ship's progress every four hours on the applicable intermediate-scale, three-digit general-purpose charts of the North Atlantic, for use by bridge and CIC watchstanders. A portion of Chart No. 126 indicating the ship's position at 250000N June appears in Figure 32-2. Note that at this time the ship has just arrived at Point G after completing the F-G segment of her track. She has turned to course 109°T at a speed of 16 knots, the track and SOA of the G-H segment, as per the instructions in the night orders.

The previous evening the navigator computed the time of the commencement of civil twilight to be 0413, and sunrise to be 0447 (unless otherwise specified, all times given henceforth will be time zone N times). Consequently, the

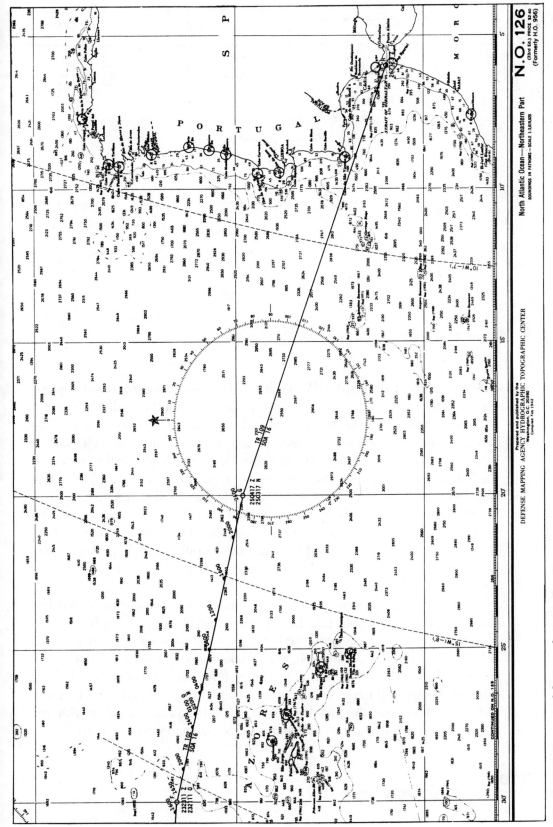

Figure 32-2. *Segment of ship's track, Points F to H.*

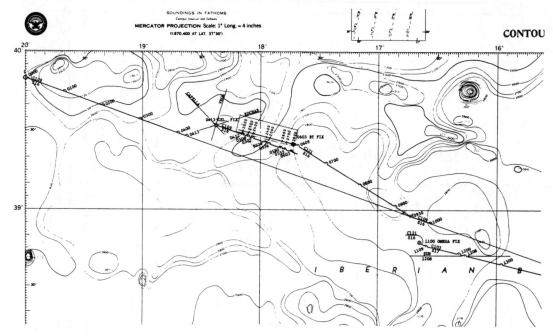

Figure 32-3. *Ship's navigation plot, 0000–1200N 25 June.*

navigator instructed the QMOW to call at 0400 so the navigator could observe the morning stars, which were selected earlier by use of the Rude Starfinder.

After observation of the morning stars, three of those listed in volume 1, *Tables No. 249*, were plotted and used to determine the morning celestial fix, recorded at 0413. This fix is shown in Figure 32-3, which depicts the portion of the bathymetric chart used for the navigation plot from 0000 to 1200N. The ship's position is about 13 miles northeast of the 0413 DR position.

At sunrise, the azimuth of the sun was observed on the gyrocompass repeater and compared with the computed amplitude to determine the gyro error. The gyro azimuth was 059° pgc, and the amplitude angle perhaps computed using Table 28 of *Bowditch* was E31.1°N or 058.9°T. Thus, the gyro error was considered negligible.

Shortly after plotting the 0413 celestial fix, the navigator noticed that the ship's course 109°T would take her over a series of charted depth contours ideal for obtaining a bathymetric (BT) fix. Therefore, the navigator decided to delay recommending a course change to take the ship back to the intended track until after the BT fix was obtained. A DR position was plotted for the times at which the ship crossed over each charted contour, and the BT fix was plotted at 0603 using the line-of-soundings overlay technique. Shortly thereafter at 0609 the navigator recommended a course change to 121°T to take the

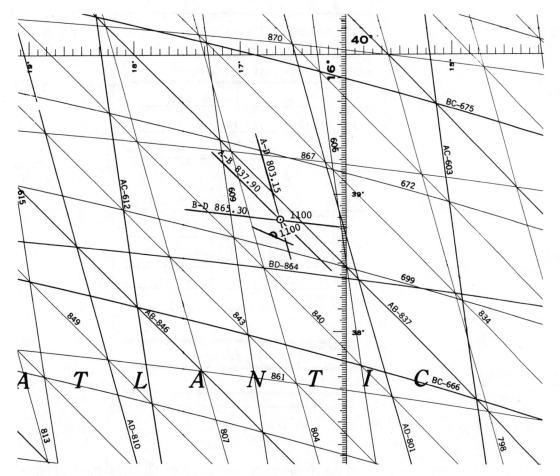

Figure 32-4. *The 1100 Omega fix plotted on Omega chart No. 7624.*

ship back onto track by 0930. The 0800 position report slip was then made up based on the DR plot extended from the 0603 BT fix.

At 1000 the ship went to general quarters and commenced a series of maneuvering and engineering casualty drills. During this time, the navigator shifted the DR plot to a small-area plotting sheet, and obtained several fixes using the Omega system. At 1100 the ship secured from general quarters and resumed course 121°T at a speed of 16 knots. At 1109 the navigator recommended a course change to 103°T based on an 1100 Omega fix and a speed increase to 17 knots to make up some of the distance lost during the general quarters drills. The 1100 Omega fix is illustrated as it was plotted on the Omega chart No. 7624 in Figure 32-4.

Between 1100 and 1200, the ship's radio communication facility patched a radio time signal into the chartroom for a chronometer time check, and the re-

TO:

COMMANDING OFFICER, USS **EXAMPLE**

AT (Time of day)	DATE
1200 N	25 JUN

LATITUDE	LONGITUDE	DETERMINED AT
38° 43.0' N	16° 19.2' W	1100

BY (Indicate by check in box,

☐ CELESTIAL   ☒ D. R.   ☒ OMEGA ~~LORAN~~   ☐ RADAR   ☐ VISUAL

SET	DRIFT	DISTANCE MADE GOOD SINCE (time) (mi.)
070°	1.5 kn	0800 N - 46.8

DISTANCE TO	MILES	ETA
PT H	524	262148Z

TRUE HDG.	ERROR	VARIATION
103°	FOR: GYRO 0° AFT GYRO 0°	12° W

MAGNETIC COMPASS HEADING (Check one)

☐ STD   ☒ STEER-ING   ☐ REMOTE IND   ☐ OTHER   117°

DEVIATION	1104 TABLE DEVIATION	DG: (Indicate by check in box)
2° W	2° W	☐ ON ☒ OFF

REMARKS

1 HR. 35 MIN AHEAD OF PIM

RESPECTFULLY SUBMITTED

R R Hobbs

CC:

Figure 32-5. *The 1200 position report, 25 June.*

sults were entered into the chronometer record book. The 1200 position report shown in Figure 32-5 was made up based on the 1100 Omega fix and projected 1200 DR position. At 1208 the sun was observed at LAN; the resulting sun line passed through the 1208 DR position.

At 1500 an afternoon sun line was observed, and the 1208 LAN line was advanced to form the 1500 running fix. This is shown in Figure 32-6, which depicts the navigation plot from 1200 to 2400N. An Omega fix obtained at 1530 was coincident with the 1530 DR position; it is shown plotted on the Omega chart in Figure 32-7 on page 584.

At 1600 the navigator noted that the intended track would again take the ship through an area of bottom contours suitable for a BT fix. From 1649 to 1903' (19-03-30), DR positions were plotted for the times at which the contoured depths were indicated on the echo sounder trace, and the 1903' BT fix was plotted using the contour advancement overlay technique. While these depth marks were being taken, the navigator computed the times of sunset and ending of civil twilight as 1922 and 1951 respectively, and made up a list of selected stars for the evening celestial sight using the Rude Starfinder. Since the

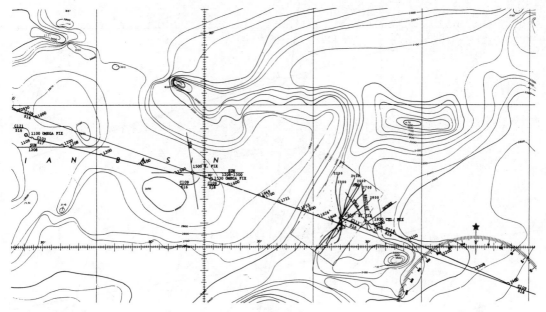

Figure 32-6. *Ship's navigation plot, 1200–2400N 25 June.*

1903′ BT fix showed the ship's position to be north of the intended track, the navigator recommended a course adjustment to 113°T.

Just before sunset a second sun amplitude observation was made to check the gyro error once again. The evening stars were then observed and the complete solutions of three of them were worked using the *Nautical Almanac* and *Tables No. 249*, volume 1. The resulting 1950 celestial fix was plotted and used as the basis for the 2000 position report. Because this position also placed the ship north of the intended track, the navigator decided to remain on course 113°T at 16 knots until midnight.

As a final action in the day's work prior to retiring for the night, the navigator made up the rough draft of the Captain's Night Orders for the night of 25–26 June. The planned course change at midnight was noted, as well as the fact that the ship would be approaching the coast of Spain and the Straits of Gibraltar the following day.

NOTES ON THE DAY'S WORK

In the example of a day's work in navigation just given, only the more noteworthy events were described in the interest of brevity. A great many other routine items accomplished daily by the navigator and the ship's staff of quartermasters were omitted, among which were checking the ship's clocks, preparing the sextant for use, and obtaining and reporting weather observa-

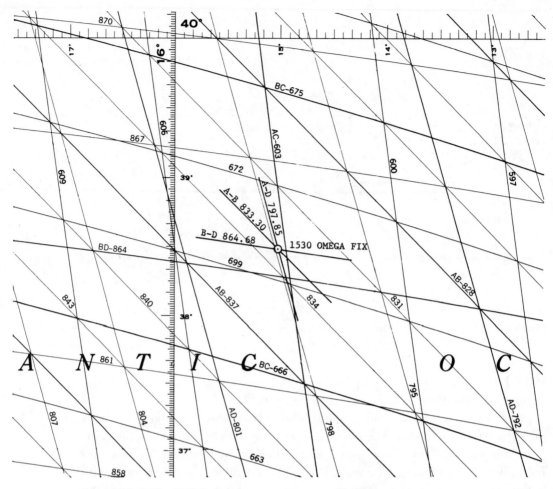

Figure 32-7. *The 1530 Omega fix.*

tions. Although only two Omega fixes were plotted, in practice such a widely available and accurate means of position-fixing would be used much more often. Other electronic aids to navigation also available in the area were disregarded, the most notable of these being GPS. The example given, however, should provide the inexperienced navigator with a good feel for the pace of the navigator's activity at sea.

SUMMARY

This chapter concludes parts 2 and 3 on celestial and electronic navigation with a brief outline of a typical day's work in navigation at sea. While the primary means of navigating at sea has undergone a historic shift from celestial to electronic and satellite navigation during the last decade, celestial observations are still the principal method of determining gyrocompass error at sea, and celestial navigation, with the possible exception of the ship's inertial navigation system, remains the only reliable method of determining the ship's position at sea using information independent of externally transmitted electromagnetic or electromechanical radiations. Celestial navigation remains the sole reliable method of navigation at sea that does not require an electrical power supply—a point of prime importance to all sailboat navigators, most of whose craft can barely generate and store enough power to keep their running lights shining for the duration of a voyage at sea.

If they have mastered the material presented in *Marine Navigation*, students of navigation can feel confident that they are familiar with the basic knowledge needed to safely direct the movements of a surface vessel anywhere in the world. However, in navigation, be it visual piloting, celestial, or electronic, there is no substitute for experience in developing the expertise and fine judgment required and expected of the professional navigator.

ABBREVIATIONS AND SYMBOLS COMMONLY USED IN PILOTING

A

ATP	Allied Tactical Publication	kn	knot (of speed)
Bn	beacon	L	latitude
C	course	λ	longitude
CIC	combat information center	LMT	local mean time
CO	commanding officer	LOGREQ	logistics requirement message
COG	course over the ground		
D	drift	LOP	line of position
DB	danger bearing	LPO	leading petty officer
DG	degaussing	M	magnetic
DMA	Defense Mapping Agency	m	meter
DMAHTC	Defense Mapping Agency Hydrographic/ Topographic Center	MHW	mean high water
		Mi	mile
		MLW	mean low water
DNC	digital nautical chart	N	north
DR	dead reckoning	NIMA	National Imagery and Mapping Agency
E	east		
ec.	eclipse (of a light)	NOS	National Ocean Survey
ECM	electronic countermeasures	OOD	officer of the deck
ENC	electronic navigation chart	OTSR	optimum track ship routing
EP	estimated position	P_n	north pole
ETA	estimated time of arrival	P_s	south pole
ETD	estimated time of departure	pgc	per gyro compass
fm	fathom	PIM	position of intended movement
ft	foot		
GMT	Greenwich Mean Time	PMP	parallel motion protractor
GPS	Global Positioning System	PPI	plan position indicator
IALA	International Ass'n of Lighthouse Authorities	psc	per standard compass
		pstgc	per steering compass
JA	captain's battle sound-powered telephone circuit	QMOW	quartermaster of the watch
		R	relative
JW	navigation sound-powered telephone circuit	RF	radio frequency
		RPM	revolution per minute

S	speed; south; set	yd	yard (unit of measure)
SI	International System of Units	ZD	zone description
		ZT	zone time
SOA	speed of advance		
SOG	speed over ground	**Miscellaneous Symbols**	
SOP	standard operating procedure	°	degrees
		′	minutes of arc
SOPA	senior officer present afloat	″	seconds of arc
T	true	□	estimated position
TR	track	⊙	fix, running fix
W	west	⌒	DR position
XO	executive officer		

ABBREVIATIONS AND SYMBOLS USED IN CELESTIAL AND ELECTRONIC NAVIGATION

A	amplitude; away (altitude intercept)	d corr^n	correction for change in declination
a	altitude intercept	dec	declination
a_0, a_1, a_2	Polaris sight corrections	DG	degaussing
ADR	average daily rate (of chronometer)	DMA	Defense Mapping Agency
		DMAHTC	Defense Mapping Agency Hydrographic/
aL	assumed latitude		Topographic Center
AM	amplitude modulation; ante meridian (before noon)	DNC	digital nautical chart
AP	assumed position	DR	dead reckoning
aλ	assumed longitude	DS corr^n	double-second difference correction
BT	bathymetric		
C	Centigrade (Celsius); chronometer time; compass; course	DST	daylight savings time
		DUT1	UTC/UT1 ($\cong$GMT) difference correction
CE	chronometer error; compass error (magnetic)	dλ	longitude difference
		E	east
cec	centicycle	EHF	extremely high frequency
cel	centilane	ENC	electronic navigation chart
CEP	circular error probable	EP	estimated position
CIC	Combat Information Center	ETA	estimated time of arrival
		ETD	estimated time of departure
cm	centimeter		
CO	commanding officer	F	Fahrenheit; fast frequency
COG	course over the ground		
corr^n	correction	f	multiplication factor (*Air Almanac*)
CPA	closest point of approach		
cps	cycles per second	FM	frequency modulation
CRT	cathode ray tube	fm	fathom
CW	continuous wave	ft	foot
D	deviation; drift (of current)	G	Greenwich; Greenwich meridian (upper branch)
d	altitude difference		

g	Greenwich meridian (lower branch)	OTC	Officer in Tactical Command
GE	gyro error	P	pole; parallax
GHA	Greenwich hour angle	pgc	per gyrocompass
GMT	Greenwich Mean Time	P in A	parallax in altitude
GP	geographic position	PM	post meridian (after noon); pulse modulation
GPS	Global Positioning System		
GRI	group repetition interval	P_n	north pole; north celestial pole
h	hour		
HA	hour angle	PRI	pulse repetition interval
ha	apparent altitude	PRR	pulse repetition rate
Hc	computed altitude	P_s	south pole; south celestial pole
HF	high frequency		
Ho	observed altitude	psc	per standard compass
HP	horizontal parallax	p stg c	per steering compass
hs	sextant altitude	Pub	publication
ht	height	Q	celestial equator
Hz	Hertz	QMOW	Quartermaster of the Watch
IC	index correction	R	relative
in	inch	RA	right ascension
kHz	kilohertz	RDF	radio direction finding
kn	knot	RF	radio frequency
L	latitude	R Fix	running fix
λ	longitude; wave length	rms	root mean square error
LAN	local apparent noon	RPM	revolution per minute
LAPS	large area plotting sheet	S	slow; south; speed
LF	low frequency	s	second
LL	lower limb	SAPS	small area plotting sheet
LMT	local mean time	SD	semidiameter
LOP	line of position	SHA	sidereal hour angle
M	magnetic; observer's meridian (upper branch); nautical mile	SHF	super high frequency
		SINS	ships inertial navigation system
m	observer's meridian (lower branch); meters; minutes	SOA	speed of advance
		SOG	speed over ground
MF	medium frequency	SOPA	senior officer present afloat
MHz	megahertz	T	toward (altitude intercept); true
μs	microsecond		
N	north	t	meridian angle
NIMA	National Imagery and Mapping Agency	TB	temperature-barometric correction
NOS	National Ocean Survey	Tg	ground wave
NWP	Naval Warfare Publication	TR	track
OOD	Officer of the Deck	Ts	sky wave

UHF	ultra high frequency	**Miscellaneous Symbols**	
UL	upper limb	Δ	delta (unit change)
UTC	Coordinated Universal Time	°	degrees
		′	minutes of arc
V	variation	″	seconds of arc
v corr^n	correction for irregular orbital motion	■	remains below horizon
		☐	remains above horizon
VHF	very high frequency	////	twilight lasts all night
VLF	very low frequency	♀	Venus
W	watch time; west	☉	sun
WE	watch error	☽	moon
yd	yard	☆	star
Z	azimuth angle	♈	First Point of Aries
z	zenith distance	⊡	estimated position
ZD	zone description	☉	fix, running fix
Zn	true azimuth	⌓	DR position
ZT	zone time		

OUTLINE OF A TYPICAL NAVIGATION BRIEF

C

<u>NAVIGATION BRIEF</u>

REF: COMNAVSURFLANT 3530.4

I. <u>ARRIVAL/DEPARTURE</u>: GOOD MORNING/AFTERNOON/EVENING CAPTAIN, XO, GENTLEMEN. TODAY'S NAVIGATION BRIEF IS FOR OUR _____. WE EXPECT TO SET NAVIGATION DETAIL AT _____, SEA AND ANCHOR DETAIL AT _____, AND GET UNDERWAY/MOOR FROM/AT PIER _____, AT TIME _____, THE UNIFORM WILL BE _____.

 <u>WATCHBILL</u>: THE WATCHBILL IS AS FOLLOWS . . . (FROM SWO)

 <u>TIDES/CURRENTS</u>: THE TIDES AT _____ WILL BE _____ FT ABOVE MEAN LOW WATER, GOING TOWARDS _____ AT _____. CURRENTS WILL BE _____ KTS EBB/FLOOD, GOING TOWARDS _____ KTS AT _____.

 <u>OPERATIONAL REQUIREMENTS</u>: OPERATIONAL REQUIREMENTS ARE AS FOLLOWS: _____
_____
_____.

 <u>SPEED RESTRICTIONS</u>: THE FOLLOWING SPEED RESTRICTIONS APPLY TO OUR TRACK: MAX SPEED IS _____ KTS. WE WILL SLOW IN THE AREA(S) OF:_____
_____.

II. <u>WEATHER</u>: THE WEATHER WILL BE AS FOLLOWS: _____
_____
_____.

III.   <u>TIDES/CURRENTS</u>: SEE ATTACHED GRAPHS:

IV. <u>ASTRONOMICAL DATA:</u> SUNRISE WILL BE AT _____. SUNSET WILL BE AT _____. MOONRISE IS AT _____, AND MOONSET IS AT _____.

V. <u>CHARTS:</u> ALL CHARTS HAVE BEEN CORRECTED UP TO _____. ALL CHARTS AND THEIR TRACKS FOR THEIR USE IN CIC AND ON THE BRIDGE HAVE BEEN COMPARED. THEY ARE COMPLETE AND ACCURATE. THE BUOYAGE SYSTEM IN EFFECT IS IALA BRAVO/IALA ALPHA. WE WILL TAKE GREEN BUOYS TO _____ AND RED BUOYS TO _____.

VI. <u>TRACK:</u> THE FOLLOWING SHALL BE COVERED BY THE CONNING OFFICER:

       A. COURSES . . .
       B. TURN/DANGER BEARINGS AND RANGES . . .
       C. SHOAL WATER . . .
       D. DEPTH OF WATER FOR CHANNEL, TURNING BASIN . . .
       E. VISUAL AND RADAR POINTS . . .
       F. VESSEL TRAFFIC SEPARATION SCHEME . . .
       G. LINE OF DEMARCATION . . .
       H. DEGAUSSING AREA . . .
       I. ANTICIPATED TRAFFIC . . .
       J. PIER HEADING . . .
       K. DESCRIPTION OF ANCHORAGE OR MOORING . . .
       L. TYPE OF BOTTOM . . .
       M. HEAD/DROP BEARING . . .
       N. AMOUNT OF ANCHOR CHAIN REQUIRED . . .

VII. <u>GROUND TACKLE:</u> THE FOLLOWING SHALL BE COVERED BY THE FIRST LIEUTENANT:

       A. READY ANCHOR PORT/STBD
       B. SCOPE OF CHAIN _____ SHOTS
       C. STATUS OF ANCHOR WINDLASS/WINCHES SAT/UNSAT
       D. SPECIAL MOORING BUOY PROCEDURES _____
       E. MOORING PLAN _____
       _____.

       F. LET GO OR WALK OUT TO _____ SCOPE.

VIII. <u>SIGNIFICANT TRAFFIC:</u> THE FOLLOWING WILL BE COVERED BY THE OPERATIONS OFFICER:

       A. ENTERING/DEPARTING MOVEMENTS _____
       _____.

       B. HARBOR SPECIAL EVENTS _____
       _____.

       C. MEDIA COVERAGE _____
       _____.

IX. <u>TUGS AND PILOTS:</u> WE WILL/WILL NOT BE TAKING ON A PILOT IN THE VICINITY OF _____. COMMUNICATIONS WITH THE PILOT

WILL BE MADE _____, VIA B/B CHANNEL(S) _____. WE EXPECT TO HAVE _____ TUGS TO ASSIST US AND MEET US IN THE VICINITY OF _____.

X. <u>STATUS OF NAVIGATIONAL EQUIPMENT:</u> COMPASS/REPEATER ERROR IS _____. WRN-6 IS _____, WITH A FIGURE OF MERIT OF _____. THE _____ RADARS ARE UP AND ARE OPERATIONAL. WRN-5 IS _____.

XI. <u>ENGINEERING PLANT STATUS:</u> THE FOLLOWING SHALL BE BRIEFED BY CHIEF ENGINEER:

    A. THE FOLLOWING EQUIPMENT IS OOC/OUT OF SERVICE:

       _____

       _____

       _____

       _____

       _____.

    B. DEGAUSSING EQUIPMENT IS _____.

XII. <u>SPECIAL CONSIDERATIONS:</u> THE FOLLOWING WILL BE BRIEFED BY THE NAVIGATOR IF APPLICABLE:

    A. HONORS
    B. FLAG OFFICER MOVEMENTS
    C. VISITORS
    D. HELO OPS
    E. BOATS IN THE WATER
    F. HARBOR EXERCISES
    G. ACCOMMODATION LADDERS
    H. DEBRIEF SCHEDULE

XIII. <u>EMERGENCIES:</u> THE FOLLOWING EMERGENCY CASUALTIES SHALL BE DISCUSSED BY THE OOD:

    _____

    _____

    _____

    _____

    _____

_____   _____

      SUBMITTED BY   APPROVED BY

SIGHT REDUCTION - TABLES 229 AND NAUTICAL ALMANAC

Body			
GMT			
IC			
D (Ht)			
Sum			
hs			
ha			
Alt Corr			
Add'l Corr Moon HP/Corr			
Ho			
GHA (h)			
Incre. (m/s)			
v/v Corr SHA			
Total GHA			
±360°			
aλ (+E,-W)			
LHA			
Tab Dec			
d#/d Corr			
True Dec			
a LAT Same Contrary			
Dec Inc/d			
Tens/DSD			
Units/DSD Corr			
Total Corr			
Hc (Tab)			
Hc (Comp)			
Ho			
a			
Z			
Zn			

Figure D-1. *Sight reduction by* Nautical Almanac *and* Tables 229.

SIGHT REDUCTION - TABLES 249 VOL 1 AND NAUTICAL ALMANAC

Body			
GMT			
IC			
D (Ht)			
Sum			
hs			
ha			
Alt Corr			
Add'l Corr			
Ho			
GHA γ (h)			
Incre. (m/s)			
Total GHA γ			
±360°			
aλ (+E,-W)			
LHA γ			
a LAT			
Hc			
Ho			
a			
Zn			
P and N Corr			

Figure D-2A. *Sight reduction by* Nautical Almanac *and* Tables 249, *vol. 1.*

SIGHT REDUCTION - TABLES 249 VOL 1 AND AIR ALMANAC

Body			
GMT			
IC			
D (Ht)			
R			
Total Corr			
hs			
Ho			
GHA γ (h, 10m)			
incre. (m/s)			
Total GHA γ			
±360°			
aλ (+E,-W)			
LHA γ			
a LAT			
Hc			
Ho			
a			
Zn			
P and N Corr			

Figure D-2B. *Sight reduction by* Air Almanac *and* Tables 249, *vol. 1.*

SIGHT REDUCTION - TABLES 249 VOLS 2 AND 3 AND AIR ALMANAC

Body			
GMT			
IC			
D (Ht)			
R			
SD			
P. in A. (Moon)			
Total Corr			
hs			
Ho			
GHA (h, 10m)			
Incre. (m/s)			
SHA (star)			
Total GHA			
±360°			
aλ (+E,-W)			
LHA			
Tab Dec			
a LAT Same Contrary			
Dec Inc/d			
Dec Corr			
Hc (Tab)			
Hc (Comp)			
Ho			
a			
Z			
Zn			

Figure D-3. *Sight reduction by* Air Almanac *and* Tables 249, *vols. 2 and 3.*

SIGHT REDUCTION - ELECTRONIC CALCULATOR

Body	
IC	
D (Ht)	
Sum	
hs	
ha	
Alt Corr	
Add'l Corr	
Moon HP Corr	
Total Corr to hs	
Ho (°+m/60)	
DR Lat (°+m/60)	
DRλ (°+m/60)	
GMT (h + m/60 + s/3600) /Date	
Tab GHA (h) = °+m/60 /v* (v/60)	
GHA$_{fract}$ = GMT x 15.0 (sun) 15.04 (γ) (14.32 + v*) (moon) (15.0 + v*) (planet)	
GHA (Tab GHA + GHA$_{fract}$)	
SHA (°+m/60)	
Total GHA (GHA + SHA)	
±360	
DRλ (+E,-W)	
LHA	
Tab Dec (°+m/60)	
d# (d/60)/d Corr (GMT$_{fract}$ x d#)	
True Dec (Tab Dec + d Corr)	
θ_1 (LHA) θ_2 (True Dec) θ_3 (DR Lat) sin (θ_n) A C E cos (θ_n) B D F	A____ C____ E____ B____ D____ F____
Hc (sin^{-1} {(C x E) + (D x B x F)})	
Ho	
a	
Z (tan^{-1}[A/{(B x E) - (F x C/D)}|)	
Zn LHA < 180, Z(+), Zn = Z + 180 LHA < 180, Z(-), Zn = 360 + (-Z) LHA > 180, Z(+), Zn = Z LHA > 180, Z(-), Zn = 180 + (-Z)	

Figure D-4. *Sight reduction by electronic calculator.*

LATITUDE BY LAN

DRλ			
STD Meridian			
dλ (arc)			
dλ (time)			
LMT Mer Pass			
ZT LAN (1st Est)			
Rev DRλ			
STD Meridian			
dλ (arc)			
dλ (time)			
LMT Mer Pass			
ZT LAN (2nd Est)			
LAT BY LAN			
ZT LAN (OBS)			
ZD			
GMT			
Tab Dec			
d#/d Corr			
True Dec			
IC			
D (Ht)			
Sum			
hs			
ha			
Alt Corr			
Ho			
89° 60′			
Ho (-)			
Z Dist			
True Dec			
LAT			

Figure D-5. Latitude by LAN.

LATITUDE/GYRO ERROR BY POLARIS

Body POLARIS			
GMT			
IC			
D (Ht)			
Sum			
hs			
ha			
Alt Corr			
Add'l Corr			
Ho			
GHA γ (h)			
Incre. (m/s)			
Total GHA γ			
±360°			
DRλ (+E,-W)			
LHA γ			
a0			
a1			
a2			
Sum			
1° 00′ (-)			
Total Corr			
Ho			
LAT			
GYRO ERROR BY POLARIS			
Zn POLARIS			
Gyro Brg			
Gyro Error			

Figure D-6. *Latitude/gyro error by Polaris.*

EXACT AZIMUTH BY TABLES 229

Body _____

DR L _____

DR λ _____

Date (L) _____

ZT _____

ZD (+ or -) _____

GMT _____

Date (G) _____

Tab GHA _____

Inc'mt _____

GHA _____

DR λ _____

LHA _____ d(+/-)

Tab Dec _____

d corr _____

Dec _____

	EXACT Deg : Min	Z DIFF. (+ or -)	CORR. (+ or -)
LAT			
LHA			
DEC			
	Total (±)		

Tab Z _____

Exact Z _____

Exact Zn _____

Gyro/Compass Brg _____

Gyro/Compass Error _____

NORTH LAT

LHA greater than 180°..... Zn = Z.

LHA less than 180°........ Zn = 360°-Z

SOUTH LAT

LHA greater than 180°..... Zn = 180°-Z

LHA less than 180°........ Zn = 180°+Z

Figure D-7A. *Exact azimuth by* Tables 229, *USNA form.*

EXACT AZIMUTH BY TABLES 229

Body			
GMT			
GHA (h)			
Incre. (m/s)			
Total GHA			
±360°			
DRλ (+E,-W)			
LHA			
Tab Dec			
d#/d Corr			
True Dec			
DR LAT Same Contrary			
Tab Z			
Dec Inc/Z Diff			
Dec Corr			
Lat Inc/Z Diff			
Lat Corr			
LHA Inc/Z Diff			
LHA Corr			
Dec Corr			
Lat Corr			
Total Corr			
Tab Z			
Exact Z			
Exact Zn			
Gyro Brg			
Gyro Error			

Figure D-7B. *Exact azimuth by* Tables 229, *Navy form.*

TIME OF SUNRISE/SUNSET/TWILIGHT

DR Lat (Tab LMT)			
Tab Interval			
Lat Time Interval			
Lat Diff			
Corr Table 1			
Tab LMT			
Corr LMT			
DRλ (Tab LMT)			
STD Mer			
dλ (arc)			
dλ (time)			
LMT			
ZT (1st est)			
Rev DR			
STD Mer			
dλ (arc)			
dλ (time)			
LMT			
ZT (2nd est)			

Figure D-8. *Time of sunrise/sunset/twilight.*

TIME OF MOONRISE/MOONSET

Tab LMT Today					
DR Lat (Tab LMT)					
Day 1	Day 2					
Tab Lat Interval					
Lat Time Interval					
Lat Diff					
Corr Table 1					
Tab LMT					
LMT					
DRλ (Tab LMT)					
Corr LMT Today					
Corr LMT preceding(E) following (W)					
Time Diff					
Corr Table 2					
Corr LMT Today					
STD Mer					
dλ (arc)					
dλ (time)					
LMT actual					
ZT (1st est)					
Rev DRλ					
STD Mer					
dλ (arc)					
dλ (time)					
LMT actual					
ZT (2nd est)					

Figure D-9. *Time of moonrise/moonset.*

GEOGRAPHIC RANGE TABLE 13, *THE AMERICAN PRACTICAL NAVIGATOR (BOWDITCH)*

E

TABLE 13
Geographic Range

Feet	Meters	7 (2)	10 (3)	13 (4)	16 (5)	20 (6)	23 (7)	26 (8)	30 (9)	33 (10)	36 (11)	Meters	Feet
		Miles	Miles	Miles	Miles	Miles	Miles	Miles	Miles	Miles	Miles		
0	0	3.1	3.7	4.2	4.7	5.2	5.6	6.0	6.4	6.7	7.0	0	0
3	1	5.1	5.7	6.2	6.7	7.3	7.6	8.0	8.4	8.7	9.0	1	3
7	2	6.2	6.8	7.3	7.8	8.3	8.7	9.1	9.5	9.8	10.1	2	7
10	3	6.8	7.4	7.9	8.4	8.9	9.3	9.7	10.1	10.4	10.7	3	10
13	4	7.3	7.9	8.4	8.9	9.5	9.8	10.2	10.6	10.9	11.2	4	13
16	5	7.8	8.4	8.9	9.4	9.9	10.3	10.6	11.1	11.4	11.7	5	16
20	6	8.3	8.9	9.5	9.9	10.5	10.8	11.2	11.6	12.0	12.3	6	20
23	7	8.7	9.3	9.8	10.3	10.8	11.2	11.6	12.0	12.3	12.6	7	23
26	8	9.1	9.7	10.2	10.6	11.2	11.6	11.9	12.4	12.7	13.0	8	26
30	9	9.5	10.1	10.6	11.1	11.6	12.0	12.4	12.8	13.1	13.4	9	30
33	10	9.8	10.4	10.9	11.4	12.0	12.3	12.7	13.1	13.4	13.7	10	33
36	11	10.1	10.7	11.2	11.7	12.3	12.6	13.0	13.4	13.7	14.0	11	36
39	12	10.4	11.0	11.5	12.0	12.5	12.9	13.3	13.7	14.0	14.3	12	39
43	13	10.8	11.4	11.9	12.4	12.9	13.3	13.6	14.1	14.4	14.7	13	43
46	14	11.0	11.6	12.2	12.6	13.2	13.5	13.9	14.3	14.7	15.0	14	46
49	15	11.3	11.9	12.4	12.9	13.4	13.8	14.2	14.6	14.9	15.2	15	49
52	16	11.5	12.1	12.7	13.1	13.7	14.0	14.4	14.8	15.2	15.5	16	52
56	17	11.9	12.5	13.0	13.4	14.0	14.4	14.7	15.2	15.5	15.8	17	56
59	18	12.1	12.7	13.2	13.7	14.2	14.6	15.0	15.4	15.7	16.0	18	59
62	19	12.3	12.9	13.4	13.9	14.4	14.8	15.2	15.6	15.9	16.2	19	62
66	20	12.6	13.2	13.7	14.2	14.7	15.1	15.5	15.9	16.2	16.5	20	66
72	22	13.0	13.6	14.1	14.6	15.2	15.5	15.9	16.3	16.6	16.9	22	72
79	24	13.5	14.1	14.6	15.1	15.6	16.0	16.4	16.8	17.1	17.4	24	79
85	26	13.9	14.5	15.0	15.5	16.0	16.4	16.8	17.2	17.5	17.8	26	85
92	28	14.3	14.9	15.4	15.9	16.5	16.8	17.2	17.6	17.9	18.2	28	92
98	30	14.7	15.3	15.8	16.3	16.8	17.2	17.5	18.0	18.3	18.6	30	98
115	35	15.6	16.2	16.8	17.2	17.8	18.2	18.5	19.0	19.3	19.6	35	115
131	40	16.5	17.1	17.6	18.1	18.6	19.0	19.4	19.8	20.1	20.4	40	131
148	45	17.3	17.9	18.5	18.9	19.5	19.8	20.2	20.6	21.0	21.3	45	148
164	50	18.1	18.7	19.2	19.7	20.2	20.6	20.9	21.4	21.7	22.0	50	164
180	55	18.8	19.4	19.9	20.4	20.9	21.3	21.7	22.1	22.4	22.7	55	180
197	60	19.5	20.1	20.6	21.1	21.7	22.0	22.4	22.8	23.1	23.4	60	197
213	65	20.2	20.8	21.3	21.8	22.3	22.7	23.0	23.5	23.8	24.1	65	213
230	70	20.8	21.4	22.0	22.4	23.0	23.4	23.7	24.2	24.5	24.8	70	230
246	75	21.4	22.1	22.6	23.0	23.6	24.0	24.3	24.8	25.1	25.4	75	246
262	80	22.0	22.6	23.2	23.6	24.2	24.5	24.9	25.3	25.7	26.0	80	262
279	85	22.6	23.2	23.8	24.2	24.8	25.2	25.5	26.0	26.3	26.6	85	279
295	90	23.2	23.8	24.3	24.8	25.3	25.7	26.1	26.5	26.8	27.1	90	295
312	95	23.8	24.4	24.9	25.3	25.9	26.3	26.6	27.1	27.4	27.7	95	312
328	100	24.3	24.9	25.4	25.9	26.4	26.8	27.2	27.6	27.9	28.2	100	328
361	110	25.3	25.9	26.4	26.9	27.5	27.8	28.2	28.6	29.0	29.3	110	361
394	120	26.3	26.9	27.4	27.9	28.5	28.8	29.2	29.6	29.9	30.2	120	394
427	130	27.3	27.9	28.4	28.9	29.4	29.8	30.1	30.6	30.9	31.2	130	427
459	140	28.2	28.8	29.3	29.7	30.3	30.7	31.0	31.5	31.8	32.1	140	459
492	150	29.0	29.7	30.2	30.6	31.2	31.6	31.9	32.4	32.7	33.0	150	492
525	160	29.9	30.5	31.0	31.5	32.0	32.4	32.8	33.2	33.5	33.8	160	525
558	170	30.7	31.3	31.9	32.3	32.9	33.2	33.6	34.0	34.4	34.7	170	558
591	180	31.5	32.1	32.7	33.1	33.7	34.1	34.4	34.9	35.2	35.5	180	591
623	190	32.3	32.9	33.4	33.9	34.4	34.8	35.2	35.6	35.9	36.2	190	623
656	200	33.1	33.7	34.2	34.6	35.2	35.6	35.9	36.4	36.7	37.0	200	656
722	220	34.5	35.1	35.7	36.1	36.7	37.0	37.4	37.8	38.2	38.5	220	722
787	240	35.9	36.5	37.0	37.5	38.1	38.4	38.8	39.2	39.5	39.8	240	787
853	260	37.3	37.9	38.4	38.9	39.4	39.8	40.1	40.6	40.9	41.2	260	853
919	280	38.6	39.2	39.7	40.1	40.7	41.1	41.4	41.9	42.2	42.5	280	919
984	300	39.8	40.4	40.9	41.4	41.9	42.3	42.7	43.1	43.4	43.7	300	984

Object height (Feet/Meters) at left and right. Height of eye of observer in feet (top number) and meters (bottom number) across the top.

TABLE 13

Geographic Range

Object height (Feet)	(Meters)	39 / 12	43 / 13	46 / 14	49 / 15	52 / 16	56 / 17	59 / 18	62 / 19	66 / 20	69 / 21	Object height (Meters)	(Feet)
		Miles	Miles	Miles	Miles	Miles	Miles	Miles	Miles	Miles	Miles		
0	0	7.3	7.7	7.9	8.2	8.4	8.8	9.0	9.2	9.5	9.7	0	0
3	1	9.3	9.7	10.0	10.2	10.5	10.8	11.0	11.2	11.5	11.7	1	3
7	2	10.4	10.8	11.0	11.3	11.5	11.9	12.1	12.3	12.6	12.8	2	7
10	3	11.0	11.4	11.6	11.9	12.1	12.5	12.7	12.9	13.2	13.4	3	10
13	4	11.5	11.9	12.2	12.4	12.7	13.0	13.2	13.4	13.7	13.9	4	13
16	5	12.0	12.4	12.6	12.9	13.1	13.4	13.7	13.9	14.2	14.4	5	16
20	6	12.5	12.9	13.2	13.4	13.7	14.0	14.2	14.4	14.7	15.0	6	20
23	7	12.9	13.3	13.5	13.8	14.0	14.4	14.6	14.8	15.1	15.3	7	23
26	8	13.3	13.6	13.9	14.2	14.4	14.7	15.0	15.2	15.5	15.7	8	26
30	9	13.7	14.1	14.3	14.6	14.8	15.2	15.4	15.6	15.9	16.1	9	30
33	10	14.0	14.4	14.7	14.9	15.2	15.5	15.7	15.9	16.2	16.4	10	33
36	11	14.3	14.7	15.0	15.2	15.5	15.8	16.0	16.2	16.5	16.7	11	36
39	12	14.6	15.0	15.2	15.5	15.7	16.1	16.3	16.5	16.8	17.0	12	39
43	13	15.0	15.3	15.6	15.9	16.1	16.4	16.7	16.9	17.2	17.4	13	43
46	14	15.2	15.6	15.9	16.1	16.4	16.7	16.9	17.1	17.4	17.7	14	46
49	15	15.5	15.9	16.1	16.4	16.6	16.9	17.2	17.4	17.7	17.9	15	49
52	16	15.7	16.1	16.4	16.6	16.9	17.2	17.4	17.6	17.9	18.2	16	52
56	17	16.1	16.4	16.7	16.9	17.2	17.5	17.7	18.0	18.3	18.5	17	56
59	18	16.3	16.7	16.9	17.2	17.4	17.7	18.0	18.2	18.5	18.7	18	59
62	19	16.5	16.9	17.1	17.4	17.6	18.0	18.2	18.4	18.7	18.9	19	62
66	20	16.8	17.2	17.4	17.7	17.9	18.3	18.5	18.7	19.0	19.2	20	66
72	22	17.2	17.6	17.9	18.1	18.4	18.7	18.9	19.1	19.4	19.6	22	72
79	24	17.7	18.1	18.3	18.6	18.8	19.2	19.4	19.6	19.9	20.1	24	79
85	26	18.1	18.5	18.7	19.0	19.2	19.5	19.8	20.0	20.3	20.5	26	85
92	28	18.5	18.9	19.2	19.4	19.7	20.0	20.2	20.4	20.7	20.9	28	92
98	30	18.9	19.3	19.5	19.8	20.0	20.3	20.6	20.8	21.1	21.3	30	98
115	35	19.9	20.2	20.5	20.7	21.0	21.3	21.5	21.8	22.1	22.3	35	115
131	40	20.7	21.1	21.3	21.6	21.8	22.1	22.4	22.6	22.9	23.1	40	131
148	45	21.5	21.9	22.2	22.4	22.7	23.0	23.2	23.4	23.7	24.0	45	148
164	50	22.3	22.7	22.9	23.2	23.4	23.7	24.0	24.2	24.5	24.7	50	164
180	55	23.0	23.4	23.6	23.9	24.1	24.5	24.7	24.9	25.2	25.4	55	180
197	60	23.7	24.1	24.4	24.6	24.9	25.2	25.4	25.6	25.9	26.1	60	197
213	65	24.4	24.7	25.0	25.3	25.5	25.8	26.1	26.3	26.6	26.8	65	213
230	70	25.1	25.4	25.7	25.9	26.2	26.5	26.7	27.0	27.2	27.5	70	230
246	75	25.7	26.0	26.3	26.5	26.8	27.1	27.3	27.6	27.9	28.1	75	246
262	80	26.2	26.6	26.9	27.1	27.4	27.7	27.9	28.2	28.4	28.7	80	262
279	85	26.8	27.2	27.5	27.7	28.0	28.3	28.5	28.8	29.0	29.3	85	279
295	90	27.4	27.8	28.0	28.3	28.5	28.9	29.1	29.3	29.6	29.8	90	295
312	95	28.0	28.3	28.6	28.9	29.1	29.4	29.7	29.9	30.2	30.4	95	312
328	100	28.5	28.9	29.1	29.4	29.6	29.9	30.2	30.4	30.7	30.9	100	328
361	110	29.5	29.9	30.2	30.4	30.7	31.0	31.2	31.4	31.7	31.9	110	361
394	120	30.5	30.9	31.2	31.4	31.7	32.0	32.2	32.4	32.7	32.9	120	394
427	130	31.5	31.8	32.1	32.4	32.6	32.9	33.2	33.4	33.7	33.9	130	427
459	140	32.4	32.7	33.0	33.3	33.5	33.8	34.1	34.3	34.6	34.8	140	459
492	150	33.3	33.6	33.9	34.1	34.4	34.7	34.9	35.2	35.5	35.7	150	492
525	160	34.1	34.5	34.7	35.0	35.2	35.6	35.8	36.0	36.3	36.5	160	525
558	170	34.9	35.3	35.6	35.8	36.1	36.4	36.6	36.9	37.1	37.4	170	558
591	180	35.7	36.1	36.4	36.6	36.9	37.2	37.4	37.7	37.9	38.2	180	591
623	190	36.5	36.9	37.1	37.4	37.6	38.0	38.2	38.4	38.7	38.9	190	623
656	200	37.3	37.6	37.9	38.2	38.4	38.7	39.0	39.2	39.5	39.7	200	656
722	220	38.7	39.1	39.4	39.6	39.9	40.2	40.4	40.7	40.9	41.2	220	722
787	240	40.1	40.5	40.8	41.0	41.3	41.6	41.8	42.0	42.3	42.5	240	787
853	260	41.5	41.8	42.1	42.4	42.6	42.9	43.2	43.4	43.7	43.9	260	853
919	280	42.8	43.1	43.4	43.7	43.9	44.2	44.5	44.7	45.0	45.2	270	919
984	300	44.0	44.4	44.6	44.9	45.1	45.5	45.7	45.9	46.2	46.4	300	984

Object height (Feet, Meters) · Height of eye of observer in feet and meters · Object height (Meters, Feet)

TABLE 13
Geographic Range

Object Height Feet	Meters	72 / 22 Miles	75 / 23 Miles	79 / 24 Miles	82 / 25 Miles	85 / 26 Miles	89 / 27 Miles	92 / 28 Miles	95 / 29 Miles	98 / 30 Miles	115 / 35 Miles	Object Height Meters	Feet
0	0	9.9	10.2	10.4	10.6	10.8	11.0	11.2	11.4	11.6	12.5	0	0
3	1	12.0	12.2	12.4	12.6	12.8	13.1	13.2	13.4	13.6	14.6	1	3
7	2	13.0	13.3	13.5	13.7	13.9	14.1	14.3	14.5	14.7	15.6	2	7
10	3	13.6	13.9	14.1	14.3	14.5	14.7	14.9	15.1	15.3	16.2	3	10
13	4	14.1	14.4	14.6	14.8	15.0	15.3	15.4	15.6	15.8	16.8	4	13
16	5	14.6	14.9	15.1	15.3	15.5	15.7	15.9	16.1	16.3	17.2	5	16
20	6	15.2	15.4	15.6	15.8	16.0	16.3	16.5	16.6	16.8	17.8	6	20
23	7	15.5	15.8	16.0	16.2	16.4	16.6	16.8	17.0	17.2	18.2	7	23
26	8	15.9	16.2	16.4	16.6	16.8	17.0	17.2	17.4	17.5	18.5	8	26
30	9	16.3	16.6	16.8	17.0	17.2	17.4	17.6	17.8	18.0	19.0	9	30
33	10	16.6	16.9	17.1	17.3	17.5	17.8	17.9	18.1	18.3	19.3	10	33
36	11	16.9	17.2	17.4	17.6	17.8	18.1	18.2	18.4	18.6	19.6	11	36
39	12	17.2	17.5	17.7	17.9	18.1	18.3	18.5	18.7	18.9	19.9	12	39
43	13	17.6	17.9	18.1	18.3	18.5	18.7	18.9	19.1	19.3	20.2	13	43
46	14	17.9	18.1	18.3	18.5	18.7	19.0	19.2	19.3	19.5	20.5	14	46
49	15	18.1	18.4	18.6	18.8	19.0	19.2	19.4	19.6	19.8	20.7	15	49
52	16	18.4	18.6	18.8	19.0	19.2	19.5	19.7	19.8	20.0	21.0	16	52
56	17	18.7	19.0	19.2	19.4	19.5	19.8	20.0	20.2	20.3	21.3	17	56
59	18	18.9	19.2	19.4	19.6	19.8	20.0	20.2	20.4	20.6	21.5	18	59
62	19	19.1	19.4	19.6	19.8	20.0	20.3	20.4	20.6	20.8	21.8	19	62
66	20	19.4	19.7	19.9	20.1	20.3	20.5	20.7	20.9	21.1	22.1	20	66
72	22	19.9	20.1	20.3	20.5	20.7	21.0	21.2	21.3	21.5	22.5	22	72
79	24	20.3	20.6	20.8	21.0	21.2	21.4	21.6	21.8	22.0	22.9	24	79
85	26	20.7	21.0	21.2	21.4	21.6	21.8	22.0	22.2	22.4	23.3	26	85
92	28	21.2	21.4	21.6	21.8	22.0	22.3	22.4	22.6	22.8	23.8	28	92
98	30	21.5	21.8	22.0	22.2	22.4	22.6	22.8	23.0	23.2	24.1	30	98
115	35	22.5	22.7	22.9	23.1	23.3	23.6	23.8	24.0	24.1	25.1	35	115
131	40	23.3	23.6	23.8	24.0	24.2	24.4	24.6	24.8	25.0	25.9	40	131
148	45	24.2	24.4	24.6	24.8	25.0	25.3	25.5	25.6	25.8	26.8	45	148
164	50	24.9	25.2	25.4	25.6	25.8	26.0	26.2	26.4	26.6	27.5	50	164
180	55	25.6	25.9	26.1	26.3	26.5	26.7	26.9	27.1	27.3	28.2	55	180
197	60	26.3	26.6	26.8	27.0	27.2	27.5	27.6	27.8	28.0	29.0	60	197
213	65	27.0	27.3	27.5	27.7	27.9	28.1	28.3	28.5	28.7	29.6	65	213
230	70	27.7	27.9	28.1	28.3	28.5	28.8	29.0	29.1	29.3	30.3	70	230
246	75	28.3	28.6	28.7	28.9	29.1	29.4	29.6	29.8	29.9	30.9	75	246
262	80	28.9	29.1	29.3	29.5	29.7	30.0	30.1	30.3	30.5	31.5	80	262
279	85	29.5	29.7	29.9	30.1	30.3	30.6	30.8	30.9	31.1	32.1	85	279
295	90	30.0	30.3	30.5	30.7	30.9	31.1	31.3	31.5	31.7	32.6	90	295
312	95	30.6	30.9	31.1	31.3	31.4	31.7	31.9	32.1	32.2	33.2	95	312
328	100	31.1	31.4	31.6	31.8	32.0	32.2	32.4	32.6	32.8	33.7	100	328
361	110	32.2	32.4	32.6	32.8	33.0	33.3	33.5	33.6	33.8	34.8	110	361
394	120	33.2	33.4	33.6	33.8	34.0	34.4	34.4	34.6	34.8	35.8	120	394
427	130	34.1	34.4	34.6	34.8	35.0	35.2	35.4	35.6	35.8	36.7	130	427
459	140	35.0	35.3	35.5	35.7	35.9	36.1	36.3	36.5	36.6	37.6	140	459
492	150	35.9	36.2	36.4	36.5	36.7	37.0	37.2	37.4	37.5	38.5	150	492
525	160	36.7	37.0	37.2	37.4	37.6	37.8	38.0	38.2	38.4	39.4	160	525
558	170	37.6	37.8	38.0	38.2	38.4	38.7	38.9	39.0	39.2	40.2	170	558
591	180	38.4	38.6	38.8	39.0	39.2	39.5	39.7	39.8	40.0	41.0	180	591
623	190	39.1	39.4	39.6	39.8	40.0	40.2	40.4	40.6	40.8	41.8	190	623
656	200	39.9	40.2	40.4	40.6	40.8	41.0	41.2	41.4	41.5	42.5	200	656
722	220	41.4	41.6	41.8	42.0	42.2	42.5	42.7	42.8	43.0	44.0	220	722
787	240	42.8	43.0	43.2	43.4	43.6	43.9	44.0	44.2	44.4	45.4	240	787
853	260	44.1	44.4	44.6	44.8	45.0	45.2	45.4	45.6	45.8	46.7	260	853
919	280	45.4	45.7	45.9	46.1	46.3	46.5	46.7	46.9	47.1	48.0	280	919
984	300	46.6	46.9	47.1	47.3	47.5	47.7	47.9	48.1	48.3	49.2	300	984

TABLE 22
Amplitudes

Latitude	0°0	0°5	1°0	1°5	2°0	2°5	3°0	3°5	4°0	4°5	5°0	5°5	6°0	Latitude
0	0.0	0.5	1.0	1.5	2.0	2.5	3.0	3.5	4.0	4.5	5.0	5.5	6.0	0
10	0.0	0.5	1.0	1.5	2.0	2.5	3.0	3.6	4.1	4.6	5.1	5.6	6.1	10
15	0.0	0.5	1.0	1.6	2.1	2.6	3.1	3.6	4.1	4.7	5.2	5.7	6.2	15
20	0.0	0.5	1.1	1.6	2.1	2.7	3.2	3.7	4.3	4.8	5.3	5.9	6.4	20
25	0.0	0.6	1.1	1.7	2.2	2.8	3.3	3.9	4.4	5.0	5.5	6.1	6.6	25
30	0.0	0.6	1.2	1.7	2.3	2.9	3.5	4.0	4.6	5.2	5.8	6.4	6.9	30
32	0.0	0.6	1.2	1.8	2.4	2.9	3.5	4.1	4.7	5.3	5.9	6.5	7.1	32
34	0.0	0.6	1.2	1.8	2.4	3.0	3.6	4.2	4.8	5.4	6.0	6.6	7.2	34
36	0.0	0.6	1.2	1.9	2.5	3.1	3.7	4.3	4.9	5.6	6.2	6.8	7.4	36
38	0.0	0.6	1.3	1.9	2.5	3.2	3.8	4.4	5.1	5.7	6.4	7.0	7.6	38
40	0.0	0.7	1.3	2.0	2.6	3.3	3.9	4.6	5.2	5.9	6.5	7.2	7.8	40
42	0.0	0.7	1.3	2.0	2.7	3.4	4.0	4.7	5.4	6.1	6.7	7.4	8.1	42
44	0.0	0.7	1.4	2.1	2.8	3.5	4.2	4.9	5.6	6.3	7.0	7.7	8.4	44
46	0.0	0.7	1.4	2.2	2.9	3.6	4.3	5.0	5.8	6.5	7.2	7.9	8.7	46
48	0.0	0.7	1.5	2.2	3.0	3.7	4.5	5.2	6.0	6.7	7.5	8.2	9.0	48
50	0.0	0.8	1.6	2.3	3.1	3.9	4.7	5.4	6.2	7.0	7.8	8.6	9.4	50
51	0.0	0.8	1.6	2.4	3.2	4.0	4.8	5.6	6.4	7.2	8.0	8.8	9.6	51
52	0.0	0.8	1.6	2.4	3.2	4.1	4.9	5.7	6.5	7.3	8.1	9.0	9.8	52
53	0.0	0.8	1.7	2.5	3.3	4.2	5.0	5.8	6.7	7.5	8.3	9.2	10.0	53
54	0.0	0.9	1.7	2.6	3.4	4.3	5.1	6.0	6.8	7.7	8.5	9.4	10.2	54
55	0.0	0.9	1.7	2.6	3.5	4.4	5.2	6.1	7.0	7.9	8.7	9.6	10.5	55
56	0.0	0.9	1.8	2.7	3.6	4.5	5.4	6.3	7.2	8.1	9.0	9.9	10.8	56
57	0.0	0.9	1.8	2.8	3.7	4.6	5.5	6.4	7.4	8.3	9.2	10.1	11.1	57
58	0.0	0.9	1.9	2.8	3.8	4.7	5.7	6.6	7.6	8.5	9.5	10.4	11.4	58
59	0.0	1.0	1.9	2.9	3.9	4.9	5.8	6.8	7.8	8.8	9.7	10.7	11.7	59
60	0.0	1.0	2.0	3.0	4.0	5.0	6.0	7.0	8.0	9.0	10.0	11.1	12.1	60
61	0.0	1.0	2.1	3.1	4.1	5.2	6.2	7.2	8.3	9.3	10.3	11.4	12.5	61
62	0.0	1.1	2.1	3.2	4.3	5.3	6.4	7.5	8.5	9.6	10.7	11.8	12.9	62
63	0.0	1.1	2.2	3.3	4.4	5.5	6.6	7.7	8.8	10.0	11.1	12.2	13.3	63
64	0.0	1.1	2.3	3.4	4.6	5.7	6.9	8.0	9.2	10.3	11.5	12.6	13.8	64
65.0	0.0	1.2	2.4	3.6	4.7	5.9	7.1	8.3	9.5	10.7	11.9	13.1	14.3	65.0
65.5	0.0	1.2	2.4	3.6	4.8	6.0	7.3	8.5	9.7	10.9	12.1	13.4	14.6	65.5
66.0	0.0	1.2	2.5	3.7	4.9	6.2	7.4	8.6	9.9	11.1	12.4	13.6	14.9	66.0
66.5	0.0	1.3	2.5	3.8	5.0	6.3	7.5	8.8	10.1	11.3	12.6	13.9	15.2	66.5
67.0	0.0	1.3	2.6	3.8	5.1	6.4	7.7	9.0	10.3	11.6	12.9	14.2	15.5	67.0
67.5	0.0	1.3	2.6	3.9	5.2	6.5	7.9	9.2	10.5	11.8	13.2	14.5	15.9	67.5
68.0	0.0	1.3	2.7	4.0	5.3	6.7	8.0	9.4	10.7	12.1	13.5	14.8	16.2	68.0
68.5	0.0	1.4	2.7	4.1	5.5	6.8	8.2	9.6	11.0	12.4	13.8	15.2	16.6	68.5
69.0	0.0	1.4	2.8	4.2	5.6	7.0	8.4	9.8	11.2	12.6	14.1	15.5	17.0	69.0
69.5	0.0	1.4	2.9	4.3	5.7	7.2	8.6	10.0	11.5	12.9	14.4	15.9	17.4	69.5
70.0	0.0	1.5	2.9	4.4	5.9	7.3	8.8	10.3	11.8	13.3	14.8	16.3	17.8	70.0
70.5	0.0	1.5	3.0	4.5	6.0	7.5	9.0	10.5	12.1	13.6	15.1	16.7	18.2	70.5
71.0	0.0	1.5	3.1	4.6	6.2	7.7	9.3	10.8	12.4	13.9	15.5	17.1	18.7	71.0
71.5	0.0	1.6	3.2	4.7	6.3	7.9	9.5	11.1	12.7	14.3	15.9	17.6	19.2	71.5
72.0	0.0	1.6	3.2	4.9	6.5	8.1	9.8	11.4	13.0	14.7	16.4	18.1	19.8	72.0
72.5	0.0	1.7	3.3	5.0	6.7	8.3	10.0	11.7	13.4	15.1	16.8	18.6	20.3	72.5
73.0	0.0	1.7	3.4	5.1	6.9	8.6	10.3	12.1	13.8	15.6	17.3	19.1	20.9	73.0
73.5	0.0	1.8	3.5	5.3	7.1	8.8	10.6	12.4	14.2	16.0	17.9	19.7	21.6	73.5
74.0	0.0	1.8	3.6	5.4	7.3	9.1	10.9	12.8	14.7	16.5	18.4	20.3	22.3	74.0
74.5	0.0	1.9	3.7	5.6	7.5	9.4	11.3	13.2	15.1	17.1	19.0	21.0	23.0	74.5
75.0	0.0	1.9	3.9	5.8	7.7	9.7	11.7	13.6	15.6	17.6	19.7	21.7	23.8	75.0
75.5	0.0	2.0	4.0	6.0	8.0	10.0	12.1	14.1	16.2	18.3	20.4	22.5	24.7	75.5
76.0	0.0	2.0	4.1	6.2	8.3	10.4	12.5	14.6	16.8	18.9	21.1	23.3	25.6	76.0
76.5	0.0	2.1	4.3	6.4	8.6	10.8	13.0	15.2	17.4	19.6	21.9	24.2	26.6	76.5
77.0	0.0	2.2	4.4	6.7	8.9	11.2	13.5	15.7	18.1	20.4	22.8	25.2	27.7	77.0

TABLE 22
Amplitudes

Latitude	Declination													Latitude
	6°0	6°5	7°0	7°5	8°0	8°5	9°0	9°5	10°0	10°5	11°0	11°5	12°0	
°	°	°	°	°	°	°	°	°	°	°	°	°	°	°
0	6.0	6.5	7.0	7.5	8.0	8.5	9.0	9.5	10.0	10.5	11.0	1..5	12.0	0
10	6.1	6.6	7.1	7.6	8.1	8.6	9.1	9.6	10.2	10.7	11.2	11.7	12.2	10
15	6.2	6.7	7.2	7.8	8.3	8.8	9.3	9.8	10.4	10.9	11.4	11.9	12.4	15
20	6.4	6.9	7.5	8.0	8.5	9.0	9.6	10.1	10.6	11.2	11.7	12.2	12.8	20
25	6.6	7.2	7.7	8.3	8.8	9.4	9.9	10.5	11.0	11.6	12.2	12.7	13.3	25
30	6.9	7.5	8.1	8.7	9.2	9.8	10.4	11.0	11.6	12.1	12.7	13.3	13.9	30
32	7.1	7.7	8.3	8.9	9.4	10.0	10.6	11.2	11.8	12.4	13.0	13.6	14.2	32
34	7.2	7.8	8.5	9.1	9.7	10.3	10.9	11.5	12.1	12.7	13.3	13.9	14.5	34
36	7.4	8.0	8.7	9.3	9.9	10.5	11.1	11.8	12.4	13.0	13.6	14.3	14.9	36
38	7.6	8.3	8.9	9.5	10.2	10.8	11.5	12.1	12.7	13.4	14.0	14.7	15.3	38
40	7.8	8.5	9.2	9.8	10.5	11.1	11.8	12.4	13.1	13.8	14.4	15.1	15.7	40
42	8.1	8.8	9.4	10.1	10.8	11.5	12.1	12.8	13.5	14.2	14.9	15.6	16.2	42
44	8.4	9.1	9.8	10.5	11.2	11.9	12.6	13.3	14.0	14.7	15.4	16.1	16.8	44
46	8.7	9.4	10.1	10.8	11.6	12.3	13.0	13.7	14.5	15.2	15.9	16.7	17.4	46
48	9.0	9.7	10.5	11.2	12.0	12.8	13.5	14.3	15.0	15.8	16.6	17.3	18.1	48
50	9.4	10.1	10.9	11.7	12.5	13.3	14.1	14.9	15.7	16.5	17.3	18.1	18.9	50
51	9.6	10.4	11.2	12.0	12.8	13.6	14.4	15.2	16.0	16.8	17.7	18.5	19.3	51
52	9.8	10.6	11.4	12.2	13.1	13.9	14.7	15.6	16.4	17.2	18.1	18.9	19.7	52
53	10.0	10.8	11.7	12.5	13.4	14.2	15.1	15.9	16.8	17.6	18.5	19.3	20.2	53
54	10.2	11.1	12.0	12.8	13.7	14.6	15.4	16.3	17.2	18.1	18.9	19.8	20.7	54
55	10.5	11.4	12.3	13.2	14.0	14.9	15.8	16.7	17.6	18.5	19.4	20.3	21.3	55
56	10.8	11.7	12.6	13.5	14.4	15.3	16.2	17.2	18.1	19.0	20.0	20.9	21.8	56
57	11.1	12.0	12.9	13.9	14.8	15.7	16.7	17.6	18.6	19.6	20.5	21.5	22.4	57
58	11.4	12.3	13.3	14.3	15.2	16.2	17.2	18.1	19.1	20.1	21.1	22.1	23.1	58
59	11.7	12.7	13.7	14.7	15.7	16.7	17.7	18.7	19.7	20.7	21.7	22.8	23.8	59
60	12.1	13.1	14.1	15.1	16.2	17.2	18.2	19.3	20.3	21.4	22.4	23.5	24.6	60
61	12.5	13.5	14.6	15.6	16.7	17.8	18.8	19.9	21.0	22.1	23.2	24.3	25.4	61
62	12.9	14.0	15.0	16.1	17.2	18.4	19.5	20.6	21.7	22.8	24.0	25.1	26.3	62
63	13.3	14.4	15.6	16.7	17.9	19.0	20.2	21.3	22.5	23.7	24.9	26.0	27.3	63
64	13.8	15.0	16.2	17.3	18.5	19.7	20.9	22.1	23.3	24.6	25.8	27.1	28.3	64
65.0	14.3	15.5	16.8	18.0	19.2	20.5	21.7	23.0	24.3	25.5	26.8	28.1	29.5	65.0
65.5	14.6	15.8	17.1	18.3	19.6	20.9	22.2	23.5	24.8	26.1	27.4	28.7	30.1	65.5
66.0	14.9	16.2	17.4	18.7	20.0	21.3	22.6	23.9	25.3	26.6	28.0	29.4	30.7	66.0
66.5	15.2	16.5	17.8	19.1	20.4	21.8	23.1	24.5	25.8	27.2	28.6	30.0	31.4	66.5
67.0	15.5	16.8	18.2	19.5	20.9	22.2	23.6	25.0	26.4	27.8	29.2	30.7	32.1	67.0
67.5	15.9	17.2	18.6	19.9	21.3	22.7	24.1	25.5	27.0	28.4	29.9	31.4	32.9	67.5
68.0	16.2	17.6	19.0	20.4	21.8	23.2	24.7	26.1	27.6	29.1	30.6	32.2	33.7	68.0
68.5	16.6	18.0	19.4	20.9	22.3	23.8	25.3	26.8	28.3	29.8	31.4	33.0	34.6	68.5
69.0	17.0	18.4	19.9	21.4	22.9	24.4	25.9	27.4	29.0	30.6	32.2	33.8	35.5	69.0
69.5	17.4	18.9	20.4	21.9	23.4	25.0	26.5	28.1	29.7	31.4	33.0	34.7	36.4	69.5
70.0	17.8	19.3	20.9	22.4	24.0	25.6	27.2	28.9	30.5	32.2	33.9	35.7	37.4	70.0
70.5	18.2	19.8	21.4	23.0	24.6	26.3	27.9	29.6	31.3	33.1	34.9	36.7	38.5	70.5
71.0	18.7	20.3	22.0	23.6	25.3	27.0	28.7	30.5	32.2	34.0	35.9	37.8	39.7	71.0
71.5	19.2	20.9	22.6	24.3	26.0	27.8	29.5	31.3	33.2	35.1	37.0	38.9	40.9	71.5
72.0	19.8	21.5	23.2	25.0	26.8	28.6	30.4	32.3	34.2	36.1	38.1	40.2	42.3	72.0
72.5	20.3	22.1	23.9	25.7	27.6	29.4	31.3	33.3	35.3	37.3	39.4	41.5	43.7	72.5
73.0	20.9	22.8	24.6	26.5	28.4	30.4	32.3	34.4	36.4	38.6	40.7	43.0	45.3	73.0
73.5	21.6	23.5	25.4	27.4	29.3	31.4	33.4	35.5	37.7	39.9	42.2	44.6	47.1	73.5
74.0	22.3	24.2	26.2	28.3	30.3	32.4	34.6	36.8	39.0	41.4	43.8	46.3	49.0	74.0
74.5	23.0	25.1	27.1	29.3	31.4	33.6	35.8	38.1	40.5	43.0	45.6	48.2	51.1	74.5
75.0	23.8	25.9	28.1	30.3	32.5	34.8	37.2	39.6	42.1	44.8	47.5	50.4	53.4	75.0
75.5	24.7	26.9	29.1	31.4	33.8	36.2	38.7	41.2	43.9	46.7	49.6	52.8	56.1	75.5
76.0	25.6	27.9	30.2	32.7	35.1	37.7	40.3	43.0	45.9	48.9	52.1	55.5	59.3	76.0
76.5	26.6	29.0	31.5	34.0	36.6	39.3	42.1	45.0	48.1	51.3	54.8	58.7	63.0	76.5
77.0	27.7	30.2	32.8	35.5	38.2	41.1	44.1	47.2	50.5	54.1	58.0	62.4	67.6	77.0

TABLE 22
Amplitudes

Latitude	12°0	12°5	13°0	13°5	14°0	14°5	15°0	15°5	16°0	16°5	17°0	17°5	18°0	Latitude
0	12.0	12.5	13.0	13.5	14.0	14.5	15.0	15.5	16.0	16.5	17.0	17.5	18.0	0
10	12.2	12.7	13.2	13.7	14.2	14.7	15.2	15.7	16.3	16.8	17.3	17.8	18.3	10
15	12.4	12.9	13.5	14.0	14.5	15.0	15.5	16.1	16.6	17.1	17.6	18.1	18.7	15
20	12.8	13.3	13.9	14.4	14.9	15.5	16.0	16.5	17.1	17.6	18.1	18.7	19.2	20
25	13.3	13.8	14.4	14.9	15.5	16.0	16.6	17.1	17.7	18.3	18.8	19.4	19.9	25
30	13.9	14.5	15.1	15.6	16.2	16.8	17.4	18.0	18.6	19.1	19.7	20.3	20.9	30
32	14.2	14.8	15.4	16.0	16.6	17.2	17.8	18.4	19.0	19.6	20.2	20.8	21.4	32
34	14.5	15.1	15.7	16.4	17.0	17.6	18.2	18.8	19.4	20.0	20.7	21.3	21.9	34
36	14.9	15.5	16.1	16.8	17.4	18.0	18.7	19.3	19.9	20.6	21.2	21.8	22.5	36
38	15.3	15.9	16.6	17.2	17.9	18.5	19.2	19.8	20.5	21.1	21.8	22.4	23.1	38
40	15.7	16.4	17.1	17.7	18.4	19.1	19.7	20.4	21.1	21.8	22.4	23.1	23.8	40
41	16.0	16.7	17.3	18.0	18.7	19.4	20.1	20.8	21.4	22.1	22.8	23.5	24.2	41
42	16.2	16.9	17.6	18.3	19.0	19.7	20.4	21.1	21.8	22.5	23.2	23.9	24.6	42
43	16.5	17.2	17.9	18.6	19.3	20.0	20.7	21.4	22.1	22.9	23.6	24.3	25.0	43
44	16.8	17.5	18.2	18.9	19.7	20.4	21.1	21.8	22.5	23.3	24.0	24.7	25.4	44
45	17.1	17.8	18.5	19.3	20.0	20.7	21.5	22.2	22.9	23.7	24.4	25.2	25.9	45
46	17.4	18.2	18.9	19.6	20.4	21.1	21.9	22.6	23.4	24.1	24.9	25.7	26.4	46
47	17.7	18.5	19.3	20.0	20.8	21.5	22.3	23.1	23.8	24.6	25.4	26.2	26.9	47
48	18.1	18.9	19.6	20.4	21.2	22.0	22.8	23.5	24.3	25.1	25.9	26.7	27.5	48
49	18.5	19.3	20.1	20.8	21.6	22.4	23.2	24.0	24.8	25.7	26.5	27.3	28.1	49
50	18.9	19.7	20.5	21.3	22.1	22.9	23.7	24.6	25.4	26.2	27.1	27.9	28.7	50
51	19.3	20.1	20.9	21.8	22.6	23.4	24.3	25.1	26.0	26.8	27.7	28.5	29.4	51
52	19.7	20.6	21.4	22.3	23.1	24.0	24.9	25.7	26.6	27.5	28.3	29.2	30.1	52
53	20.2	21.1	21.9	22.8	23.7	24.6	25.5	26.4	27.3	28.2	29.1	30.0	30.9	53
54	20.7	21.6	22.5	23.4	24.3	25.2	26.1	27.0	28.0	28.9	29.8	30.8	31.7	54
55	21.3	22.2	23.1	24.0	24.9	25.9	26.8	27.8	28.7	29.7	30.6	31.6	32.6	55
56	21.8	22.8	23.7	24.7	25.6	26.6	27.6	28.5	29.5	30.5	31.5	32.5	33.5	56
57	22.4	23.4	24.4	25.4	26.4	27.4	28.4	29.4	30.4	31.4	32.5	33.5	34.6	57
58	23.1	24.1	25.1	26.1	27.2	28.2	29.2	30.3	31.3	32.4	33.5	34.6	35.7	58
59	23.8	24.8	25.9	27.0	28.0	29.1	30.2	31.3	32.4	33.5	34.6	35.7	36.9	59
60	24.6	25.7	26.7	27.8	28.9	30.1	31.2	32.3	33.5	34.6	35.8	37.0	38.2	60
61	25.4	26.5	27.6	28.8	29.9	31.1	32.3	33.5	34.6	35.9	37.1	38.3	39.6	61
62	26.3	27.5	28.6	29.8	31.0	32.2	33.5	34.7	36.0	37.2	38.5	39.8	41.2	62
63	27.3	28.5	29.7	30.9	32.2	33.5	34.8	36.1	37.4	38.7	40.1	41.5	42.9	63
64	28.3	29.6	30.9	32.2	33.5	34.8	36.2	37.6	39.0	40.4	41.8	43.3	44.8	64
65.0	29.5	30.8	32.2	33.5	34.9	36.3	37.8	39.2	40.7	42.2	43.8	45.4	47.0	65.0
65.5	30.1	31.5	32.9	34.3	35.7	37.1	38.6	40.1	41.7	43.2	44.8	46.5	48.2	65.5
66.0	30.7	32.1	33.6	35.0	36.5	38.0	39.5	41.1	42.7	44.3	46.0	47.7	49.4	66.0
66.5	31.4	32.9	34.3	35.8	37.3	38.9	40.5	42.1	43.7	45.4	47.2	48.9	50.8	66.5
67.0	32.1	33.6	35.1	36.7	38.3	39.9	41.5	43.2	44.9	46.6	48.4	50.3	52.3	67.0
67.5	32.9	34.4	36.0	37.6	39.2	40.9	42.6	44.3	46.1	47.9	49.8	51.8	53.9	67.5
68.0	33.7	35.3	36.9	38.6	40.2	41.9	43.7	45.5	47.4	49.3	51.3	53.4	55.6	68.0
68.5	34.6	36.2	37.9	39.6	41.3	43.1	44.9	46.8	48.8	50.8	52.9	55.1	57.5	68.5
69.0	35.5	37.2	38.9	40.6	42.5	44.3	46.2	48.2	50.3	52.4	54.7	57.0	59.6	69.0
69.5	36.4	38.2	40.0	41.8	43.7	45.6	47.7	49.7	51.9	54.2	56.6	59.2	61.9	69.5
70.0	37.4	39.3	41.1	43.0	45.0	47.1	49.2	51.4	53.7	56.1	58.7	61.5	64.6	70.0
70.5	38.5	40.4	42.4	44.4	46.4	48.6	50.8	53.2	55.7	58.3	61.1	64.3	67.8	70.5
71.0	39.7	41.7	43.7	45.8	48.0	50.3	52.7	55.2	57.8	60.7	63.9	67.5	71.7	71.0
71.5	40.9	43.0	45.1	47.4	49.7	52.1	54.7	57.4	60.3	63.5	67.1	71.4	76.9	71.5
72.0	42.3	44.5	46.7	49.1	51.5	54.1	56.9	59.9	63.1	66.8	71.1	76.7	90.0	72.0
72.5	43.7	46.0	48.4	50.9	53.6	56.4	59.4	62.7	66.4	70.8	76.5	90.0		72.5
73.0	45.3	47.8	50.3	53.0	55.8	58.9	62.3	66.1	70.5	76.3	90.0			73.0
73.5	47.1	49.6	52.4	55.3	58.4	61.8	65.7	70.2	76.0	90.0				73.5
74.0	49.0	51.7	54.7	57.9	61.4	65.3	69.9	75.8	90.0					74.0
74.5	51.1	54.1	57.3	60.9	64.9	69.5	75.6	90.0						74.5

TABLE 22
Amplitudes

Latitude	Declination													Latitude
	18°0	18°5	19°0	19°5	20°0	20°5	21°0	21°5	22°0	22°5	23°0	23°5	24°0	
0	18.0	18.5	19.0	19.5	20.0	20.5	21.0	21.5	22.0	22.5	23.0	23.5	24.0	0
10	18.3	18.8	19.3	19.8	20.3	20.8	21.3	21.8	22.4	22.9	23.4	23.9	24.4	10
15	18.7	19.2	19.7	20.2	20.7	21.3	21.8	22.3	22.8	23.3	23.9	24.4	24.9	15
20	19.2	19.7	20.3	20.8	21.3	21.9	22.4	23.0	23.5	24.0	24.6	25.1	25.6	20
25	19.9	20.5	21.1	21.6	22.2	22.7	23.3	23.9	24.4	25.0	25.5	26.1	26.7	25
30	20.9	21.5	22.1	22.7	23.3	23.9	24.4	25.0	25.6	26.2	26.8	27.4	28.0	30
32	21.4	22.0	22.6	23.2	23.8	24.4	25.0	25.6	26.2	26.8	27.4	28.0	28.7	32
34	21.9	22.5	23.1	23.7	24.4	25.0	25.6	26.2	26.9	27.5	28.1	28.7	29.4	34
36	22.5	23.1	23.7	24.4	25.0	25.7	26.3	26.9	27.6	28.2	28.9	29.5	30.2	36
38	23.1	23.7	24.4	25.1	25.7	26.4	27.1	27.7	28.4	29.1	29.7	30.4	31.1	38
40	23.8	24.5	25.2	25.8	26.5	27.2	27.9	28.6	29.3	30.0	30.7	31.4	32.1	40
41	24.2	24.9	25.6	26.3	26.9	27.6	28.3	29.1	29.8	30.5	31.2	31.9	32.6	41
42	24.6	25.3	26.0	26.7	27.4	28.1	28.8	29.5	30.3	31.0	31.7	32.5	33.2	42
43	25.0	25.7	26.4	27.2	27.9	28.6	29.3	30.1	30.8	31.6	32.3	33.0	33.8	43
44	25.4	26.2	26.9	27.6	28.4	29.1	29.9	30.6	31.4	32.1	32.9	33.7	34.4	44
45	25.9	26.7	27.4	28.2	28.9	29.7	30.5	31.2	32.0	32.8	33.5	34.3	35.1	45
46	26.4	27.2	27.9	28.7	29.5	30.3	31.1	31.8	32.6	33.4	34.2	35.0	35.8	46
47	26.9	27.7	28.5	29.3	30.1	30.9	31.7	32.5	33.3	34.1	35.0	35.8	36.6	47
48	27.5	28.3	29.1	29.9	30.7	31.6	32.4	33.2	34.0	34.9	35.7	36.6	37.4	48
49	28.1	28.9	29.8	30.6	31.4	32.3	33.1	34.0	34.8	35.7	36.6	37.4	38.3	49
50	28.7	29.6	30.4	31.3	32.1	33.0	33.9	34.8	35.6	36.5	37.4	38.3	39.3	50
51	29.4	30.3	31.2	32.0	32.9	33.8	34.7	35.6	36.5	37.5	38.4	39.3	40.3	51
52	30.1	31.0	31.9	32.8	33.7	34.7	35.6	36.5	37.5	38.4	39.4	40.4	41.3	52
53	30.9	31.8	32.8	33.7	34.6	35.6	36.5	37.5	38.5	39.5	40.5	41.5	42.5	53
54	31.7	32.7	33.6	34.6	35.6	36.6	37.6	38.6	39.6	40.6	41.7	42.7	43.8	54
55	32.6	33.6	34.6	35.6	36.6	37.6	38.7	39.7	40.8	41.9	42.9	44.0	45.2	55
56	33.5	34.6	35.6	36.7	37.7	38.8	39.9	41.0	42.1	43.2	44.3	45.5	46.7	56
57	34.6	35.6	36.7	37.8	38.9	40.0	41.1	42.3	43.5	44.6	45.8	47.1	48.3	57
58	35.7	36.8	37.9	39.1	40.2	41.4	42.6	43.8	45.0	46.2	47.5	48.8	50.1	58
59	36.9	38.0	39.2	40.4	41.6	42.8	44.1	45.4	46.7	48.0	49.3	50.7	52.2	59
60.0	38.2	39.4	40.6	41.9	43.2	44.5	45.8	47.1	48.5	49.9	51.4	52.9	54.4	60.0
60.5	38.9	40.1	41.4	42.7	44.0	45.3	46.7	48.1	49.5	51.0	52.5	54.1	55.7	60.5
61.0	39.6	40.9	42.2	43.5	44.9	46.3	47.7	49.1	50.6	52.1	53.7	55.3	57.0	61.0
61.5	40.4	41.7	43.0	44.4	45.8	47.2	48.7	50.2	51.7	53.3	55.0	56.7	58.5	61.5
62.0	41.2	42.5	43.9	45.3	46.8	48.2	49.8	51.3	52.9	54.6	56.3	58.1	60.0	62.0
62.5	42.0	43.4	44.8	46.3	47.8	49.3	50.9	52.5	54.2	56.0	57.8	59.7	61.7	62.5
63.0	42.9	44.3	45.8	47.3	48.9	50.5	52.1	53.8	55.6	57.5	59.4	61.4	63.6	63.0
63.5	43.8	45.3	46.9	48.4	50.0	51.7	53.4	55.2	57.1	59.1	61.1	63.4	65.7	63.5
64.0	44.8	46.4	48.0	49.6	51.3	53.0	54.8	56.7	58.7	60.8	63.0	65.3	68.1	64.0
64.5	45.9	47.5	49.1	50.8	52.6	54.4	56.3	58.4	60.5	62.7	65.2	67.9	70.9	64.5
65.0	47.0	48.7	50.4	52.2	54.0	56.0	58.0	60.1	62.4	64.9	67.6	70.7	74.2	65.0
65.5	48.2	49.9	51.7	53.6	55.6	57.6	59.8	62.1	64.6	67.3	70.4	74.1	78.8	65.5
66.0	49.4	51.3	53.2	55.2	57.2	59.4	61.8	64.3	67.1	70.0	73.9	78.6	90.0	66.0
66.5	50.8	52.7	54.7	56.8	59.1	61.4	64.0	66.8	70.0	73.7	78.5	90.0		66.5
67.0	52.3	54.3	56.4	58.7	61.1	63.7	66.5	69.7	73.5	78.4	90.0			67.0
67.5	53.9	56.0	58.3	60.7	63.3	66.2	69.5	73.3	78.2	90.0				67.5
68.0	55.6	57.9	60.4	63.0	65.9	69.2	73.1	78.1	90.0					68.0
68.5	57.5	60.0	62.7	65.6	68.9	72.9	77.9	90.0						68.5
69.0	59.6	62.3	65.3	68.7	72.6	77.7	90.0							69.0
69.5	61.9	65.0	68.4	72.4	77.6	90.0								69.5
70.0	64.6	68.1	72.2	77.4	90.0									70.0
70.5	67.8	71.9	77.2	90.0										70.5
71.0	71.7	77.1	90.0											71.0
71.5	76.9	90.0												71.5
72.0	90.0													72.0

Chart No.1

United States of America

G

Nautical Chart Symbols Abbreviations and Terms

Ninth Edition
JANUARY 1990

Prepared Jointly by

DEPARTMENT OF COMMERCE
National Oceanic and Atmospheric Administration
National Ocean Service

DEPARTMENT OF DEFENSE
Defense Mapping Agency
Hydrographic/Topographic Center

Published at Washington, D.C.
DEPARTMENT OF COMMERCE
National Oceanic and Atmospheric Administration
National Ocean Service

SELECTED CONTENTS

INTRODUCTION

General Remarks—The ninth edition of Chart No. 1, Nautical Chart Symbols and Abbreviations is presented in a completely new format. A major change is the incorporation of the symbols contained in the International Hydrographic Organization (IHO) Chart 1 (INT1). The various sections comprising the Table of Contents follow the sequence presented in INT1; therefore the numbering system in this publication follows the standard format approved and adopted by the IHO.

Where appropriate, each page lists separately the current preferred U.S. symbols shown on charts of the National Ocean Service (NOS) and Defense Mapping Agency Hydrographic/Topographic Center (DMAHTC). Also shown in separate columns are the IHO symbols and symbols used on foreign charts reproduced by DMAHTC.

New with this edition is a schematic layout of a typical page showing what kind of information each column presents. In addition a typical layout of a U.S. chart is shown (Section A); a new page outlining tidal levels and other charted tidal data has also been added (Section H).

For more information on the use of the chart, the practice of navigation, chart sounding datum, and visual and audible aids to navigation, the user should refer to DMAHTC Pub. No. 9, American Practical Navigator (Bowditch), Volumes I and II.

Tide and current data can be found in NOS publications Tide Tables and Tidal Current Tables. Detailed information on lights, buoys, and beacons is available in the Coast Guard Light List and DMAHTC List of Lights. In addition, color plates of the U.S. Aids to Navigation System and the Uniform State Waterway Marking System are contained in the Coast Guard Light Lists.

Other important information that cannot be shown conveniently on the nautical chart can be found in the U.S. Coast Pilots and DMAHTC Sailing Directions.

Metric Charts and Feet/Fathom Charts—In January, 1972 the United States began producing a limited number of nautical charts in meters. Since then, some charts have been issued with soundings and contours in meters; however, for some time to come there will still be many charts on issue depicting sounding units in feet or fathoms. Modified reproductions of foreign charts are being produced retaining the sounding unit value of the country of origin. The sounding unit is stated in bold type outside the border of every chart and in the chart title.

Soundings—The sounding datum reference is stated in the chart title. In all cases the unit of depth used is shown in the chart title and in the border of the chart in bold type.

Drying Heights—On rocks and banks that cover and uncover, the elevations shown are above the sounding datum, as stated in the chart title.

Shoreline—Shoreline shown on charts represents the line of contact between the land and a selected water elevation. In areas affected by tidal fluctuation, this line of contact is usually the mean high-water line. In confined coastal waters of diminished tidal influence, a mean water level line may be used. The shoreline of interior waters (rivers, lakes) is usually a line representing a specified elevation above a selected datum. Shoreline is symbolized by a heavy line (Section C1).

Apparent Shoreline is used on charts to show the outer edge of marine vegetation where that limit would reasonably appear as the shoreline to the mariner or where it prevents the shoreline from being clearly defined. Apparent shoreline is symbolized by a light line (Section C32).

Landmarks—A conspicuous feature on a building may be shown by a landmark symbol with a descriptive label (Section E10 and E22). Prominent buildings that are of assistance to the mariner may be shown by actual shape as viewed from above (Section D5, D6, and D34). Legends associated with landmarks, when shown in capital letters, indicate conspicuous; the landmark may also be labeled "CONSPI" or "CONSPICUOUS."

Buoys—The buoyage systems used by other countries often vary from that used by the United States. U.S. Charts show the colors, lights and other characteristics in use for the area of the individual chart. In the U.S. system, on entering a channel from seaward, buoys on the starboard side are red with even numbers, on the port side, black or green with odd numbers. Lights on buoys on starboard side of the channel are red or white, on the port side, white or green. Mid-channel buoys have red and white or black and white vertical stripes and may be passed on either side. Junction or obstruction buoys have red and green or red and black horizontal bands, the top band color indicating the preferred side of passage. This system does not apply to foreign waters.

Light Visibility (Range)—(Other than on the Great Lakes and adjacent waterways.) A light's visibility (range) is given in nautical miles. Where the visibility (range) is shown as x/x M for a two (2) color light, the first number indicates the visibility (range) of the first color, while the second number indicates the visibility (range) of the second color. For example, Fl W G 12/8M indicates the visibility (range) of the white light to be 12 nautical miles and the green light to be 8 nautical miles. Where a light has three (3) colors, only the longest and shortest visibilities (ranges) may be given, in which case the middle visibility (range) is represented by a hyphen. For example, Fl W R G 12–8M indicates the visibility (range) of the white light to be 12 nautical miles, the green light to be 8 miles, and the red light to be between 12 and 8 nautical miles.

IALA Buoyage System—The International Association of Lighthouse Authorities (IALA) Maritime Buoyage System (combined Cardinal-Lateral System) is being implemented by nearly every maritime buoyage jurisdiction worldwide as either REGION A buoyage (red to port) or REGION B buoyage (red to starboard). The terms "REGION A" and "REGION B" will be used to determine which type of buoyage is in effect or undergoing conversion in a particular area. The major difference in the two buoyage regions will be in the lateral marks. In REGION A they will be red to port; in REGION B they will be red to starboard. Shapes of lateral marks will be the same in both REGIONS, can to port; cone (nun) to starboard. Cardinal and other marks will continue to follow current guidelines and may be found in both REGIONS. A modified lateral mark, indicating the preferred channel where a channel divides, will be introduced for use in both REGIONS. Section Q and the color plates at the back of this publication illustrate the IALA buoyage system for both REGIONS A and B.

Aids to Navigation Positioning—The aids to navigation depicted on charts comprise a system consisting of fixed and floating aids with varying degrees of reliability. Therefore, prudent mariners will not rely solely on any single aid to navigation, particularly a floating aid.

The buoy symbol is used to indicate the approximate position of the buoy body and the sinker which secures the buoy to the seabed. The approximate position is used because of practical limitations in positioning and maintaining buoys and their sinkers in precise geographical locations. These limitations include, but are not limited to, inherent imprecisions in position fixing methods, prevailing atmospheric and sea conditions, the slope of and the material making up the seabed, the fact that buoys are moored to sinkers by varying lengths of chain, and the fact that buoy body and/or sinker positions are not under continuous surveillance but are normally checked only during periodic maintenance visits which often occur more than a year apart. The position of the buoy body can be expected to shift inside and outside the charting symbol due to the forces of nature. The mariner is also cautioned that buoys are liable to be carried away, shifted, capsized, sunk, etc. Lighted buoys may be extinguished or sound signals may not function as the result of ice, running ice, other natural causes, collisions, or other accidents.

For the foregoing reasons a prudent mariner must not rely completely upon the position or operation of floating aids to navigation, but will also utilize bearings from fixed objects and aids to navigation on shore. Further, a vessel attempting to pass close aboard always risks collision with a yawing buoy or with the obstruction the buoy marks.

Colors—Colors are optional for characterizing various features and areas on the charts. For instance the land tint in this publication is gold as used on charts of the NOS; however, charts of the DMAHTC show land and tint as gray.

Heights—Heights of lights, landmarks, structures, etc. are referred to the shoreline plane of reference. Heights of small islets or offshore rocks, which due to space limitations must be placed in the water area, are bracketed. The unit of height is shown in the chart title.

Conversion Scales-Depth conversion scales are provided on all charts to enable the user to work in meters, fathoms, or feet.

Traffic Separation Schemes—Traffic separation schemes show established routes to increase safety of navigation, particularly in areas of high density shipping. These schemes were established by the International Maritime Organization (IMO) and are described in the IMO publication "Ships Routing".

Traffic separation schemes are generally shown on nautical charts at scales of 1:600,000 and larger. When possible, traffic separation schemes are plotted to scale and shown as depicted in Section M.

Correction Dates—The date of each edition is shown below the lower left border of the chart. This includes the date of the latest Notice to Mariners applied to the chart.

U.S. Coast Pilots, Sailing Directions, Light Lists, Lists of Lights—These related publications furnish information required by the navigator that cannot be shown conveniently on the nautical charts.

U.S. Nautical Chart Catalogs and Indexes—These list nautical charts, auxiliary maps, and related publications and include general information relative to the use and ordering of charts.

Corrections and Comments—Notices of Corrections for this publication will appear in the weekly Notice to Mariners. Users should refer corrections, additions, and comments for improving this product to DIRECTOR, DEFENSE MAPPING AGENCY/HYDROGRAPHIC TOPOGRAPHIC CENTER, ATTN: PR, WASHINGTON, D.C. 20315-0030.

C *Natural Features*

			Coastline	Supplementary national symbols: a — e	

Foreshore → I, J

1		Coastline, surveyed		
2		Coastline, unsurveyed		
3	high low	Steep coast, Steep coast with rock cliffs, Cliffs		
4		Coastal hillocks, elevation not determined		
5		Flat coast		
6		Sandy shore		
7		Stony shore, Shingly shore		Stones
8		Sandhills, Dunes		Dunes

C Natural Features

Relief

Plane of Reference for Heights → H

10		Contour lines with spot height	
11		Spot heights	
12		Approximate contour lines with approximate height	
13		Form lines with spot height	
14		Approximate height of top of trees (above height datum)	

Water Features, Lava

20		River, Stream	
21		Intermittent river	
22		Rapids, Waterfalls	
23		Lakes	

C Natural Features

24		Salt pans		
25		Glacier		
26		Lava flow		

Vegetation			Supplementary national symbols: 0	
30	Wooded	Wood, in general		Wooded
31		Prominent trees (in groups or isolated)		
31.1		Deciduous tree		
31.2		Evergreen (except conifer)		
31.3		Conifer		
31.4		Palm		
31.5		Nipa palm		
31.6		Casuarina		
31.7		Filao		
31.8		Eucalytus		

C *Natural Features*

32	Mangrove		Mangrove	
33	Marsh (used in small areas)		Marsh	Marsh
	Swamp		Swamp	

	Supplementary National Symbols			
a	Uncovers		Chart sounding datum line	
b			Approximate sounding datum line	
c	Mud		Foreshore; Strand (in general) Stones; Shingle; Gravel; Mud ; Sand	
d	Breakers Breakers (if extensive)		Breakers along a shore	

C *Natural Features*

e		Rubble	
f	610 606	Hachures	
g		Shading	
h		Lagoon	
i	Wooded	Deciduous woodland	
j	Wooded	Coniferous woodland	
k		Tree plantation	
l	Cultivated	Cultivated fields	
m	Grass	Grass fields	
n	Rice	Paddy (rice) fields	
o	Bushes	Bushes	

E Landmarks

Plane of reference for Heights → H   Lighthouses → P   Beacons → Q

General

1	⊙TANK ○ Tk ⊕ ⊘	Examples of landmarks	◆Building ⊙ Hotel 🏛	
2	⊙ CAPITOL DOME ⊙ WORLD TRADE CENTER	Examples of conspicuous landmarks	◆ BUILDING ⊙ HOTEL 🏛 WATER TOWER	
3.1		Pictorial symbols (in true position)		
3.2		Sketches, Views (out of position)		
4		Height of top of a structure above plane of reference for heights	🏛 (30)	
5	(30)	Height of structure above ground level	🏛 (30)	

Landmarks

10.1	✠ Ch	Church	✠ ⊞ ｜ Ch.	⚓ ▪
10.2		Church tower	✠ Tr. ⊞ Tr.	
10.3	⊙SPIRE ○ Spire	Church spire	✠ Sp. ⊞ Sp.	● ♦ ⚲
10.4	⊙CUPOLA ○Cup	Church cupola	✠ Cup. ⊞ Cup.	♀
11	✚ Ch	Chapel		⚲
12	⚲	Cross, Calvary		+ ⊥
13	⊠	Temple	⊠	⊕
14	⊠	Pagoda	⊠	
15	⊠	Shinto shrine, Josshouse	⊠	卍

16	⋈		Buddhist temple	⋈　　卍	
17	⚲ ⚲ ⌇		Mosque, Minaret	⚲	⌇ ⚲
18	⊡ ⚲		Marabout	⊙ Marabout	
19	⌐ Cem ¬	†⸸†† / Cem	Cemetery (for all religious denominations)	L L L L / L L L L / L L L L	⌐ ᴜ ᴜ ᴜ ¬
20	⊙TOWER ○ Tr		Tower	⌷	Tr	
21	STANDPIPE ⊙ ○S'pipe		Water tower, Water tank on a tower Standpipe	⌷		
22	⊙CHIMNEY ○ Chy		Chimney	⌸	◀ Chy	⌷
23	⊙FLARE ○ Flare		Flare stack (on land)	⌻		
24	⊙MONUMENT ○ Mon		Monument	⌾	Mon	⌿ ⊡
25.1	⊙WINDMILL ○ Windmill	⊙WINDMILL ⊗	Windmill	⤬		⚹ ⤴
25.2			Windmill (wingless)	⤬ Ru		
26	⊙ WINDMOTOR ○ Windmotor		Windmotor	⚵		⊗ ⤬
27	⊙F S ⊙F P ○ FS ○ F P		Flagstaff, Flagpole	⚑	FS	
28	⊙ R MAST ⊙ MAST ○ R Mast ○ TV Mast		Radio mast, Television mast	⟨ɐ⟩		
29	⊙ R TR ⊙ TV TR ○ R Tr ○ TV Tr		Radio tower, Television tower	⟨ɐ⟩		
30.1			Radar mast:	⊙ Radar Mast	

30.2			Radar tower	⊙ Radar Tr.	
30.3			Radar scanner	⊙ Radar Sc.	
30.4	⊙DOME (RADAR) °Dome (Radar)		Radar dome	⊙ Radome	
31	⊙ANT (RADAR) ° Ant (Radar)		Dish aerial		
32	⊙TANK ⊘ ○ Tk ⊕	⊙TANK ○ Tk	Tanks	• ⊕	T a n k s
33	⊙ SILO ○ Silo		Silo , Elevator	○ Silo ⊙ Silo	
34.1			Fortified structure (on large-scale charts)		
34.2	▚ Cas		Castle, Fort, Blockhouse (on smaller-scale charts)	▚	✛
34.3	ᴗ̄		Battery, Small fort (on smaller-scale charts)	✚	
35.1			Quarry (on large-scale charts)		
35.2	⚒		Quarry (on smaller-scale charts)	⚒	Ψ
36	⚒		Mine	⚒	

Supplementary National Symbols					
a		⚲	Moslem Shrine		
b		⟙	Tomb		

c	⌣	Watermill		☼
d	▨ ▮ ▷ Facty	Factory		
e	○ Well	Well		
f	⚑ Sch	School		
g	⚑ HS	High school		
h	⚑ Univ	University		
i	⊙ GAB ○ Gab	Gable		
j	⛺	Camping site		
k	Tel Tel Off	Telegraph Telegraph office		
l	Magz	Magazine		
m	Govt Ho	Government house		
n	Inst	Institute		
o	Ct Ho	Courthouse		
p	Pav	Pavilion		
q	T	Telephone		
r	Ltd	Limited		
s	Apt	Apartment		
t	Cap	Capitol		
u	Co	Company		
v	Corp	Corporation		

H Tides, Currents

		Terms Relating to Tidal Levels	Supplementary national symbols: a−k
1		Chart Datum, Datum for sounding reduction	CD
2		Lowest Astronomical Tide	LAT
3		Highest Astronomical Tide	HAT
4	MLW	Mean Low Water	MLW
5	MHW	Mean High Water	MHW
6	MSL	Mean Sea Level	MSL
7		Land survey datum	
8	MLWS	Mean Low Water Springs	MLWS
9	MHWS	Mean High Water Springs	MHWS
10	MLWN	Mean Low Water Neaps	MLWN
11	MHWN	Mean High Water Neaps	MHWN
12	MLLW	Mean Lower Low Water	MLLW
13	MHHW	Mean Higher High Water	MHHW
14		Mean Higher Low Water	MHLW
15		Mean Lower High Water	MLHW
16	Sp	Spring tide	
17	Np	Neap tide	

H *Tides, Currents*

Tidal Levels and charted Data   Tide gauge → T

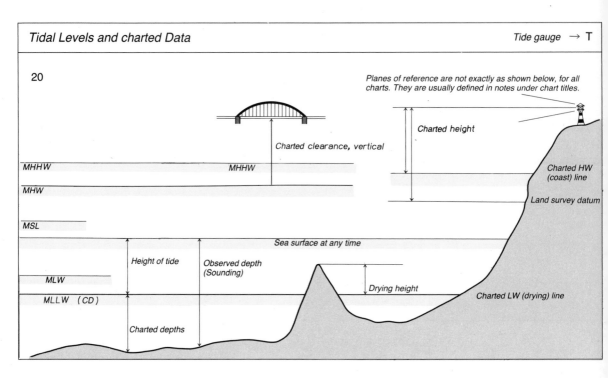

20

Planes of reference are not exactly as shown below, for all charts. They are usually defined in notes under chart titles.

Charted height

Charted clearance, vertical

MHHW   MHHW   *Charted HW (coast) line*

MHW   *Land survey datum*

MSL

Sea surface at any time

Height of tide   *Observed depth (Sounding)*

MLW   *Drying height*

MLLW (CD)   *Charted LW (drying) line*

Charted depths

Tide Tables

Tidal Levels referred to Datum of Soundings

30	Place	Lat N	Long E	Heights in meters above datum			
				MHWS	MHWN	MLWN	MLWS
				MHHW	MLHW	MHLW	MLLW

Tabular statement of semi-diurnal or diurnal tides

Note:
The order of the columns of levels will be the same as that used in national tables of tidal predictions.

---

31 *Tidal stream table*

Tidal streams referred to ..

Hours	◇ Geographical Position	Directions of streams (degrees)	Rates at spring tides (knots)	Rates at neap tides (knots)
Before High Water { 6 5 4 3 2 1				
High Water				
After High Water { 1 2 3 4 5 6				

H *Tides, Currents*

Tidal Streams and Currents

Supplementary national symbols: m — t

Breakers → K   Tide Gauge → T

40	→ 2 kn →	Flood stream (current) with rate	2,5 kn →
41	· 2kn →	Ebb stream (current) with rate	2,5 kn →
42		Current in restricted waters	»»»» →
43		Ocean current with rates and seasons	2,5−4,5 kn Jan−Mar (see Note)
44	Tide rips ⌃⌃⌃ ～～ ⌃⌃ Symbol used only in small areas	Overfalls, tide rips, races	～～ ～～ ～～ ～～ ～～
45	⌒⌒⌒ Eddies ⌒⌒⌒ Symbol used only in small areas	Eddies	⌒ ⌒ ⌒ ⌒ ⌒
46	Ⓐ Ⓑ	Position of tabulated tidal data with designation	Ⓐ

Supplementary National Symbols

a	HW	High water	
b	HHW	Higher high water	
c	LW	Low water	
d	LWD	Low-water datum	
e	LLW	Lower low water	
f	MTL	Mean tide level	
g	ISLW	Indian spring low water	
h	HWF&C	High-water full and change (vulgar establishment of the port)	
i	LWF&C	Low-water full and change	

j	CRD	Columbia River Datum	
k	GCLWD	Gulf Coast Low Water Datum	
l	Str	Stream	
m	2kn	Current, general, with rate	
n	vel	Velocity; Rate	
o	kn	Knots	
p	ht	Height	
q	fl	Flood	
r		New moon	
s		Full moon	
t		Current diagram	

General

1	ED	Existence doubtful	ED
2	SD	Sounding doubtful	SD
3.1	Rep	Reported, but not surveyed	Rep
3.2	⋮3⋮ Rep (1983)	Reported with year of report, but not surveyed	Rep(1973)
4	⋮3⋮ Rep	Reported but not confirmed sounding or danger.	⋮7⋮ ⋮4$_6$⋮

Soundings

Plane of Reference for Depths → H   Plane of Reference for Heights → H

10	19 8$_2$ 7$\frac{3}{4}$ 8$_2$ 19	Sounding in true position (Upright soundings are used on English unit charts and Sloping soundings are used on Metric charts).	12 9$_7$
11	(23) ∙⟋1036	Sounding out of position	+ (12) 3375
12	(5)	Least depth in narrow channel	(9$_7$)
13	$\overline{65}$	No bottom found at depth shown	$\overline{200}$
14	8$_2$ 19 8$_2$ 19	Soundings which are unreliable or taken from a smaller-scale source (Sloping soundings are used on English unit charts and upright soundings are used on Metric charts).	12 9$_7$
15	6	Drying heights above chart datum	1$_8$ 0 3

I Depths

Depths in Fairways and Areas			*Supplementary national symbols:* **a, b**

Plane of Reference for Depths → H

20		Limit of dredged area	----------	
21		Dredged channel or area with depth of dredging in meters	7,2 m.	7,2 meters
22	24 FEET OCT 1983 / 30 FEET APR 1984	Dredged channel or area with depth and year of the latest control survey	Dredged to 7,2m(1978)	7,2 m (1978)
23		Dredged channel or area with maintained depth	Maintained depth 7,2m	7,2 m
24		Depth at chart datum, to which an area has been swept by wire drag. The latest date of sweeping may be shown in parentheses	10₈ 10₂ 9₆ (1980) 11 9₈	
25		Unsurveyed or inadequately surveyed area; area with inadequate depth information	Inadequately Surveyed / Inadequately Surveyed	

Depths

Depth Contours

	Feet	Fm/Meters					
	0	0				Low water line	
	6	1					
	12	2					
	18	3					
	24	4					
	30	5					
	36	6					
	60	10					
	120	20				One or two lighter	
	180	30				blue tints may be	
	240	40				used instead of the	
	300	50				'ribbons' of tint at	
	600	100				10 or 20 m	
	1,200	200					
	1,800	300					
	2,400	400					
	3,000	500					
	6,000	1,000					

30

(Depth contour tint scale: 0, 1, 2, 3, 4, 5, 6, 10, 20, 30, 40, 50, 100, 200, 300, 400, 500, 1000)

(Contour values at right: 0, 2, 3, 5, 8, 10, 15, 20, 25, 30, 40, 50, 75, 100, 200, 300, 400, 500, 600, 700, 800, 900, 1000, 2000, 3000, 4000, 5000, 6000, 7000, 8000, 9000, 10000)

31	Approximate depth contour Continuous lines, with values	— 5 — (blue or black) —100—	Approximate depth contours	— — 20 — — — — — 50 — — —

Note: The extent of the blue tint varies with the scale and purpose of the chart, or its sources. On some charts, contours and figures are printed in blue.

Supplementary National Symbols

a	6	Swept channel	
b	89 17 119 15	Swept area, not adequately sounded (shown by purple or green tint)	
c	2 ft 6 5	Stream	

J Nature of the Seabed

Types of Seabed

Rocks → K

Supplementary national abbreviations: a−ag

1	S	Sand	S
2	M	Mud	M
3	Cy; Cl	Clay	Cy
4	Si	Silt	Si
5	St	Stones	St
6	G	Gravel	G
7	P	Pebbles	P
8	Cb	Cobbles	Cb
9	Rk; rky	Rock; Rocky	R
10	Co	Coral and Coralline algae	Co
11	Sh	Shells	Sh
12	S/M	Two layers, eg. Sand over mud	S/M
13.1	Wd	Weed (including Kelp)	Wd
13.2	Kelp	Kelp, Seaweed	
14	Sandwaves	Mobile bottom (sand waves)	
15	Spring	Freshwater springs in seabed	

Types of Seabed, Intertidal Areas

20.1	Gravel	Area with stones, gravel or shingle	G St
20.2		Small area with stones or gravel	
21	Rock	Rocky area, which covers and uncovers	
22	Coral	Coral reef, which covers and uncovers	

J *Nature of the Seabed*

Qualifying Terms

30	f; fne	fine		f
31	m	medium	only used in relation to sand	m
32	c; crs	coarse		c
33	bk; brk	broken		bk
34	sy; stk	sticky		sy
35	so; sft	soft		so
36	stf	stiff		sf
37	Vol	volcanic		v
38	Ca	calcareous		ca
39	h; hrd	hard		h

Supplementary National Abbreviations

a	Grd	Ground	
b	Oz	Ooze	
c	Ml	Marl	
d	Sn	Shingle	
e	Blds	Boulders	
f	Ck	Chalk	
g	Qz	Quartz	
h	Sch	Schist	
i	Co Hd	Coral head	
j	Mds	Madrepores	
k	Vol Ash	Volcanic ash	
l	La	Lava	
m	Pm	Pumice	
n	T	Tufa	
o	Sc	Scoriae	
p	Cn	Cinders	
q	Mn	Manganese	
r	Oys	Oysters	
s	Ms	Mussels	
t	Spg	Sponge	
u	K	Kelp	
v	Grs	Grass	
w	Stg	Sea-tangle	
x	Spi	Spicules	
y	Fr	Foraminifera	
z	Gl	Globigerina	
aa	Di	Diatoms	

J   *Nature of the Seabed*

ab	Rd	Radiolaria	
ac	Pt	Pteropods	
ad	Po	Polyzoa	
ae	Cir	Cirripedia	
af	Fu	Fucus	
ag	Ma	Mattes	
ah	sml	Small	
ai	lrg	Large	
aj	rt	Rotten	
ak	str	Streaky	
al	spk	Speckled	
am	gty	Gritty	
an	dec	Decayed	
ao	fly	Flinty	
ap	glac	Glacial	
aq	ten	Tenacious	
ar	wh	White	
as	bl; bk	Black	
at	vi	Violet	
au	bu	Blue	
av	gn	Green	
aw	yl	Yellow	
ax	or	Orange	
ay	rd	Red	
az	br	Brown	
ba	ch	Chocolate	
bb	gy	Gray	
bc	lt	Light	
bd	dk	Dark	
be	vard	Varied	
bf	unev	Uneven	

K Rocks, Wrecks, Obstructions

General

No.	Symbol	Description	
1		Danger line, in general	
2	21, Rk 21, Obstr 3₂ 3₂ ⁰5	Swept by wire drag or diver	

Rocks

Plane of Reference for Heights → H   Plane of Reference for Depths → H

No.	Symbol		Description		
10	²⁵ ⁰(21)		Rock (islet) which does not cover, height above height datum	(4,1) (3,1) (1,7)	▲ (4 m)
11	*(2) ⁰(2) 4	*Uncov 2 ft ⁰Uncov 2 ft *(2) ⁰(2)	Rock which covers and uncovers, height above chart datum	(3,7) (2,7) *(1,6) *(1,6)	
12	* ⊛	* ⊛ (0) ⁰(0)	Rock awash at the level of chart datum	* ⊛ ⊛	⊛
13	+ ⊛		Dangerous underwater rock of uncertain depth	+ ⊛ ⊛	
14			Dangerous underwater rock of known depth	+(4₈) +(12₁)	
14.1		12 Rk	in the corresponding depth area		
14.2		5 Rk	outside the corresponding depth area	⊛ (4₈) ⊛ (12₁)	3₂

15	+ 35 Rk	35 Rk	Non-dangerous rock, depth known	21 R	35R. 35R +(35)
16	+Co 3₁+ Reef Line		Coral reef which covers	Co Co	
17	Breakers	Br	Breakers	5₈ 19 Br 18	

Wrecks					
Plane of Reference for Depths → H					
20	Hk	Hk	Wreck, hull always dry, on large-scale charts	Wk	
21		Hk	Wreck, covers and uncovers, on large-scale charts	Wk	Wk Wk Wk Wk
22			Submerged wreck, depth known, on large-scale charts	5½ Wk	9 Wk
23		Hk	Submerged wreck, depth unknown, on large-scale charts	Wk	Wk Wk
24	PA		Wreck showing any portion of hull or superstructure at level of chart datum		Wk Wk Wk
25	Masts	Mast (10 ft) Funnel	Wreck showing mast or masts above chart datum only	Mast	
26	5½ Wk		Wreck, least depth known by sounding only	4₆ Wk 25 Wk	(9)
27	2₁ Wk 5 Wk	2₁ Wk 5½ 5 Wk	Wreck, least depth known, swept by wire drag or diver	4₆ Wk 25 Wk	+++ 2₁ Rk
28			Dangerous wreck, depth unknown		
29	+++		Sunken wreck, not dangerous to surface navigation	+++	
30	8 Wk		Wreck, least depth unknown, but considered to have a safe clearance to the depth shown	25 Wk	5

K *Rocks, Wrecks, Obstructions*

31	(Foul) (Wks) #	(Foul) (Wks) (Wreckage)	Remains of a wreck or other foul area, non-dangerous to navigation but to be avoided by vessels anchoring, trawling etc.	#	Foul	# ⁄ ᶠᴮ

Obstructions

Plane of Reference for Depths → H   Kelp, Sea-Weed → J

40	⊙ Obstn	◯ Obstn	Obstruction, depth unknown	◯ Obstn Obstn	#
41	⑤ Obstn	⑤ Obstn	Obstruction, least depth known	4₆ Obstn 16₈ Obstn	
42	(21) Obstn (5) Obstn	(21) Obstn (5) Obstn	Obstruction, least depth known, swept by wire drag or diver	4₆ Obstn 16₈ Obstn	
43.1	Subm piles Stakes, Perches	∘∘ Subm piles ∘-------∘ Subm piling	Stumps of posts or piles, wholly submerged	◯ Obstn ⊤ ⊤ ⊤	⌅ Ⓣ Subm piles
43.2	∘∘ Snags	∘∘ Stumps	Submerged pile, stake, snag, well or stump (with exact position)	⌅	⌅ ⌅ ⓣ
44.1	⊔⊔⊔⊔⊔⊔⊔⊔⊔⊔ Fsh stks		Fishing stakes	⊔⊔⊔⊔⊔ ⊔⊔⊔⊔⊔	
44.2	⬚━		Fish trap, fish weirs, tunny nets	⬚	
45	━ ━ ━		Fish trap area, tunny nets area	⌐ Fish traps ⌐ ⌐ Tunny nets ⌐	
46.1	⊙ Obstruction (fish haven) ▭ (actual shape)	⊙ Obstruction (fish haven) ▭	Fish haven (artificial fishing reef)	⊂⊃ ⬭	
46.2	▭ } Obstn ▭ Fish haven (Auth min 42 ft)		Fish haven with minimum depth	⊂⊃ (2₄) ⬭ 2₄	
47	◠ Oys ◡		Shellfish cultivation (stakes visible)	⌐ Shellfish ⌐	

K Rocks, Wrecks, Obstructions

Supplementary National Symbols				
a	∗ ⊛		Rock awash (height unknown)		
b	ⓢ Rk 5 Rks		Shoal sounding on isolated rock or rocks		9R 2r 2P ⊕(8)
c			Sunken wreck covered 20 to 30 meters		⊹⊹⊹
d	⟨⟩Sub vol		Submarine volcano		
e	⟨⟩Discol water		Discolored water		
f	2₁ Rk 3₂ 3₂		Sunken danger with depth cleared (swept) by wire drag		Obst.n 2₁ 5
g	Reef		Reef of unknown extent		
h	● ⊛Co	Coral Co Co ⊛Co	Coral reef, detached (uncovers at sounding datum)		
i	⊡ Subm Crib		Submerged Crib		□
j	□ Crib (above water)		Crib (above water)		
k			Submerged Crib with depth		□('2')
l			Crib with drying height		□ (2)

Dredged and Swept Areas → I	Submarine Cables, Submarine Pipelines → L	Tracks, Routes → M

General

1.1		Maritime limit in general usually implying: Permanent obstructions	
1.2		Maritime limit in general usually implying: No permanent obstructions	
2.1	Restricted Area	Limit of restricted area	
2.2	PROHIBITED AREA / PROHIB AREA	(Screen optional) / Limit of prohibited area (no unauthorized entry)	Entry Prohibited

Anchorages, Anchorage Areas

10		Anchorage (large vessels)		
		Anchorage (small vessels)		
11.1	(14)	Anchor berths	A N 53 (14)	(6) No1
11.2	3	Anchor berths, swinging circle may be shown	(A) (N 53) (14)	
12.1		Anchorage area in general		
12.2		Numbered anchorage area	No1	
12.3		Named anchorage area	Neufeld	
12.4		Deep Water Anchorage area, Anchorage area for Deep Draft Vessels	DW	
12.5		Tanker anchorage area	Tanker	

12.6			Anchorage for periods up to 24 hours	24 h ⚓	
12.7	*Explosives Anchorage*		Explosives anchorage area		
12.8	*QUAR ANCH* / *QUARANTINE ANCHORAGE*		Quarantine anchorage area		
12.9			Reserved anchorage	*Reserved* ⚓ (see Caution)	*Anch Reserved*

Note: Anchors as part of the limit symbol are not shown for small areas. Other types of anchorage area may be shown.

13			Sea-plane landing area		⚓ Υ ⚓
14			Anchorage for sea-planes	⚓	

Restricted Areas

20	*ANCH PROHIB* / *Anch Prohibited*	*ANCH PROHIB* / *ANCH PROHIBITED*	Anchoring prohibited		
21	*Fish Prohibited*	*FISH PROHIB*	Fishing prohibited		
22	— — — —		Limit of nature reserve: Nature reserve, Bird sanctuary, Game preserve, Seal sanctuary		
23.1	*Explosives Dumping Ground*	*Explosives Dumping Ground*	Explosives dumping ground	*Explosives Dumping Ground*	
23.2	*Explosives Dumping Ground (Discontd)*	*Explosives Dumping Ground (Discont)*	Explosives dumping ground (disused) Foul (explosives)	*Explosives Dumping Ground (disused)*	

24	r ‒ ‒ ‒ ‒ ‒ ┐ | Dump Site | |	⊢ T T T T T T ⊣ ⊢ Dump Site ⊢	Dumping ground for chemical waste	⊢ T T T T T T T T T T T T ⊢ Dumping Ground for ⊢ Chemical waste ⊢	
25	r ‒ ‒ ‒ ‒ ‒ ┐ | Degaussing Range |	⊢ T T T T T T ⊢ Degaussing Range ⊢	Degaussing range	⊢ T T⌇⌇⌇⌇⌇⌇⌇T T T T ⊢ ⊢ Degaussing range	
26			Historic wreck and restricted area	Historic Wk	

Military Practice Areas					
30			Firing danger area		
31	T T T T T T T T T T T T PROHIBITED AREA		Military area, entry prohibited	⊢ T T T T T T T T T T ⊢ ⊢ Prohibited Area ⊢	
32			Mine-laying practice area	r ‒ ‒ Ω ‒ ‒ Ω ‒ ‒ ‒ | Ωᵨ	
33			Submarine transit lane and exercise area	‒ ‒ ‒ ‒ ‒ ‒ ‒ ‒ ‒ ‒ ‒ ‒ ‒ ‒ ‒ ‒ ‒ ‒ ‒ ‒	
34			Mine field	r T T T T T T T T T T ┐ ⊢ ⊢ Minefield ⊢ (see Caution) ⊢	r ‒ ‒ ‒ ┐ | ⌐ ‒ ‒

International Boundaries and National Limits				Supplementary national symbols: a, b	
40	+ + + + + + + +	——— – – ———	International boundary on land	FEDERAL REPUBLIC OF GERMANY + + + + + + + + + DENMARK	
41	+ + + + + + + +	——— – – ———	International maritime boundary	FEDERAL REPUBLIC OF GERMANY + — + — + — + — + DENMARK	
42	——————————		Straight territorial sea baseline	⟋_⟍○⟋‾⟍	
43	——————————		Seaward limit of territorial sea	——— + + ———	
44	——————————		Seaward limit of contiguous zone	——— + ———	

45			Limits of fishery zones		
46			Limit of continental shelf		
47			Limit of Exclusive Economic Zone	EEZ	
48			Customs limit		
49			Harbour limit	Harbour limit	

Various Limits

Supplementary national symbols: c— j

60.1		Limit of fast ice, Ice front		
60.2		Limit of sea ice (pack ice)-seasonal		
61	Log boom	Log pond	Log Pond	
62.1	Spoil Area	Spoil ground	Spoil Ground	
62.2	Spoil Area Discontinued	Spoil ground (disused)	Spoil Ground (disused)	
63		Dredging area	Dredging Area	
64		Cargo transhipment area	Cargo Transhipment Area	
65		Incineration area	Incineration Area	

Supplementary National Symbols

a		COLREGS demarcation line	
b		Limit of fishing areas (fish trap areas)	
c	Dumping Ground	Dumping ground	

d	Disposal Area 92 depths from survey of JUNE 1972 85	Disposal area (Dump Site)	
e	-------------------	Limit of airport	
f	—— · —— 	Reservation line (Options)	
g	Dump Site	Dump site	

Light Structures, Major Floating Lights

1		Major light, minor light light, lighthouse	☆ ★ Lt Lt Ho	✦ ● ✪ ·
2	■ PLATFORM (lighted)	Lighted offshore platform	⊡	
3	○ Marker (lighted)	Lighted beacon tower	BY ☆ Bn Tr	
4		Lighted beacon	R BRB ☆ Bn	
5	○ Art Articulated light (floating light)	Articulated light Buoyant beacon, resilient beacon	R ☆ Bn	
6		Light vessel; Lightship Normally manned light-vessel	LtV	
7	FLOAT FLOAT	Unmanned light-vessel; light float		FLOAT
8		LANBY, superbuoy as navigational aid		

P Lights

Light Characters					

Light Characters on Light Buoys → Q

	Abbreviation		Class of light	Illustration　　　Period shown ⊢——⊣	
	International	National			
10.1	**F**	F	*Fixed*		
	Occulting (total duration of light longer than total duration of darkness)				
10.2	**Oc**	Oc; Occ	*Single-occulting*		Oc; Occ
	Oc(2) Example	Oc (2); Gp Occ	*Group-occulting*		Oc (2); Gp Occ
	Oc(2+3) Example	Oc(2+3)	*Composite group-occulting*		Oc(2+3)
10.3	*Isophase (duration of light and darkness equal)*				
	Iso	Iso; E Int	*Isophase*		Iso; E Int
	Flashing (total duration of light shorter than total duration of darkness)				
10.4	**Fl**	Fl	*Single-flashing*		Fl
	Fl(3) Example	Fl (2); Gp Fl	*Group-flashing*		Fl (2); Gp Fl
	Fl(2+1) Example	Fl (2+I)	*Composite group-flashing*		Fl (2+I)
10.5	**LFl**	L Fl	*Long-flashing* *(flash 2 s or longer)*		L Fl
	Quick (repetition rate of 50 to 79 – usually either 50 or 60 – flashes per minute)				
10.6	**Q**	Q; Qk Fl	*Continuous quick*		Q; Qk Fl
	Q(3) Example	Q(3)	*Group quick*		Q(3)
	IQ	IQ; Int Qk Fl; I Qk Fl	*Interrupted quick*		IQ; Int Qk Fl; I Qk Fl

	Abbreviation		Class of light	Illustration	Period shown ⊢————┤	
	International	National				
	Very quick (repetition rate of 80 to 159 – usually either 100 or 120 – flashes per min)					
10.7	**VQ**	VQ; V Qk Fl	Continuous very quick			VQ; V Qk Fl
	VQ(3) Example	VQ (3)	Group very quick			
	IVQ	IVQ	Interrupted very quick			
10.8	**Ultra quick** (repetition rate of 160 or more – usually 240 to 300 – flashes per min)					
	UQ	UQ	Continuous ultra quick			
	IUQ	IUQ	Interrupted ultra quick			
10.9	**Mo (A)** Example	Mo (A)	Morse Code			
10.10	**FFl**	F Fl	Fixed and flashing			F Fl
10.11	**Al.WR**	Al; Alt	Alternating			Al; Alt

Colors of Lights

	International	National	Class of light		Colors of lights shown on standard charts
11.1	**W**	W	White (only on sector- and alternating lights)		
11.2	**R**	R	Red		
11.3	**G**	G	Green		on multicolored charts
11.4	**Bu**	Bu; Bl	Blue		
11.5	**Vi**	Vi	Violet		
11.6	**Y**	Y	Yellow		on multicolored charts at sector lights
11.7	**Y** **Or**	Y Or	Orange		
11.8	**Y** **Am**	Y Am	Amber		

P Lights

Period			
12	90s	Period in seconds	90s

Elevation			
Plane of Reference for Heights → H			Tidal Levels → H
13	12m 36ft	Elevation of light given in meters or feet	12m

Range			
Note: Charted ranges are nominal ranges given in Nautical miles			
14	15M	15M	Light with single range	15M
	10M	10M	Light with two different ranges Note: only lesser of two ranges is charted	15/10M
	7M	7M	Light with three or more ranges Note: only least of three ranges is charted	15-7M

Disposition			
15	Hor	horizontally disposed	(hor)
	Vert	vertically disposed	(vert)

Example of a full Light Description

16		Name Fl (3) WRG 15s 21ft 11M		Name ☆ Fl(3)WRG.15s 21m 15-11M
	Fl(3)	Class of light: group flashing repeating a group of three flashes	Fl(3)	Class of light: group flashing repeating a group of three flashes
	WRG	Colors: white, red, green, exhibiting the different colors in defined sectors	WRG	Colors: white, red, green, exhibiting the different colors in defined sectors
	15s	Period: the time taken to exhibit one full sequence of 3 flashes and eclipses: 15 seconds	15s	Period: the time taken to exhibit one full sequence of 3 flashes and eclipses: 15 seconds
	21ft	Elevation of focal plane above datum: 21 feet	21m	Elevation of focal plane above datum: 21 meters
	11M	Nominal range	15-11M	Nominal range: white 15 M, green 11 M, red between 15 and 11 M

Lights Marking Fairways

Leading Lights and Lights in Line

20.1	Lts in line 270°	Leading lights with leading line (firm line is fairway) and arcs of visibility Bearing given in degrees and tenths of a degree	Name Oc.3s 8m12M Name Oc.6s 24m15M Oc.3s Oc.3s 225.3°
20.2		Leading lights ‡: any two objects in line Bearing given in degrees and minutes	Oc.4s12M Oc.R 4s10M Oc.R & Oc ‡ 269°18'
20.3		Leading lights on small-scale charts	Ldg Oc.R & F.R
21		Lights in line, marking the sides of a channel	Fl.G Fl.G 2 Fl.R 270° 270°
22		Rear or upper light	Rear Lt or Upper Lt
23		Front or lower light	Front Lt or Lower Lt

Direction Lights

30.1	RED GREEN	Direction light with narrow sector and course to be followed, flanked by darkness or unintensified light	Fl(2)5s10m11M Dir 269°
30.2		Direction light with course to be followed, uncharted sector is flanked by darkness or unintensified light	Oc.12s 6M Dir 299° Dir 255,5° Fl(2)5s11M
30.3		Direction light with narrow fairway sector flanked by sectors of different character	F.G Al.WG Oc.W.4s Al.WR Dir WRG 15-5M F.R
31		Moiré effect light (day and night) Arrows show when course alteration needed	Dir 295°

Note: Quoted bearings are always from seaward.

P Lights

Sector Lights			
40		Sector light on standard charts	Fl.WRG.4s 21m 18-12M
41.1		Sector lights on standard charts, the white sector limits marking the sides of the fairway	Oc.WRG. 10-6M
41.2		Sector lights on multicoloured charts, the white sector limits marking the sides of the fairway	Oc.WRG. 10-6M
42		Main light visible all-round with red subsidiary light seen over danger	Fl(3)10s 62m 25M F.R.55m 12M
43		All-round light with obscured sector	Fl.5s 41m 30M
44		Light with arc of visibility deliberately restricted	Iso.WRG
45		Light with faint sector	Q.14m 5M
46		Light with intensified sector	Oc.R.8s 7M / Oc.R.8s

P Lights

Lights with limited Times of Exhibition

50	Occas		Lights exhibited only when specially needed (tor fishing vessels, ferries) and some private lights	☆ F.R. (occas)	
51			Daytime light (charted only where the character shown by day differs from that shown at night)	Fl.10s40m27M ☆ (F.37m11M Day)	
52			Fog light (exhibited only in fog, or character changes in fog)	Name ☆ Q.WRG.5m10-3M Fl.5s (in fog)	
53			Unwatched (unmanned) light with no standby or emergency arrangements	☆ Fl.5s(U)	
54		Temp	Temporary light	(temp)	
55		Exting	Extinguished light	(exting)	

Special Lights

Flare Stack (at Sea) → L   Flare Stack (on Land) → E   Signal Stations → T

60	•AERO		Aero light	✧ Aero Al.Fl.WG.7,5s11M	★ AERO
61.1			Air obstruction light of high intensity	✧ Aero F.R.313m11M RADIO MAST (353)	
61.2			Air obstruction lights of low intensity	(89) ⌁ (R Lts)	
62		Fog Det.Lt	Fog detector light	Fog Det Lt	
63			Floodlight, floodlighting of a structure	◁▽▷	(Illuminated)
64			Strip light		
65	• Priv		Private light other than one exhibited occasionally	★ F.R.(priv)	★ ● Priv maintd

Supplementary National Symbols

a			Riprap surrounding light		
b	S-L Fl		Short-Long Flashing		S-L Fl
c			Group-Short Flashing		
d	F Gp Fl		Fixed and Group Flashing		F Gp Fl

Buoys and Beacons

IALA Maritime Buoyage System, which includes Beacons → Q 130		
1	Position of buoy	—○—

Colors of Buoys and Beacon Topmarks

Abbreviations for Colors → P		
2	Green and black	G B G G G
3	Single colors other than green and black	R R Y Y R
4	Multiple colors in horizontal bands, the color sequence is from top to bottom	BY GRG BRB
5	Multiple colors in vertical or diagonal stripes, the darker color is given first	RW RW RW
6	Retroreflecting material	

Note: Retroreflecting material may be fitted to some unlit marks. Charts do not usually show it. Under IALA Recommendations, black bands will appear blue under a spotlight.

Lighted Marks

Marks with Fog Signals → R		
7	Lighted marks on standard charts	Fl.G G Fl.R R
8	Lighted marks on multicolored charts	Fl.R R Iso RW Fl.G G

Topmarks and Radar Reflectors

For Application of Topmarks within the IALA-System → Q 130	Topmarks on Special Purpose Buoys and Beacons → Q		
9	IALA System buoy topmarks (beacon topmarks shown upright)	
10	Beacon with topmark, color, radar reflector and designation	No 2 Name R
11	Buoy with topmark, color, radar reflector and designation	No 3 G

Note: Radar reflectors on floating marks are usually not charted.

Q Buoys, Beacons

	Buoys	Features Common to Buoys and Beacons → Q 1–11		
	Shapes of Buoys			
20	Ɋ N ⌂	Conical buoy, nun buoy	⌂	
21	Ɋ C ⊿	Can or cylindrical buoy	⊿	
22	Ɋ SP ⌂	Spherical buoy	⌂	
23	Ɋ P ⊿	Pillar buoy	⊿	
24	Ɋ S /	Spar buoy, spindle buoy	/	
25	Ɋ ⌂	Barrel buoy	⌂	
26	⊥	Super buoy	⊥	
	Light Floats			
30	✳	Light float as part of IALA System	Fl.G.3s No 3 Name	✳
31		Light float (unmanned light-vessel) not part of IALA System	Fl.10s12m 26 M	
	Mooring Buoys			
	Oil or Gas Installation Buoy → L		Small Craft Mooring → U	
40	▬	Mooring buoys	⌂ ⌂ ⌂ ⌂	
41	▬	Lighted mooring buoy (Example)	Fl.Y.2,5s No 1	
42		Trot, mooring buoys with ground tackle and berth numbers	(1) (2)	
43	See Supplementary national symbols S, t	Mooring buoy with telegraphic or telephonic communication	⌂〜〜〜〜〜	
44		Numerous moorings (example)	Small Craft Moorings	

Q Buoys, Beacons

Special Purpose Buoys		
Note: Shapes of buoys are variable. Lateral or Cardinal buoys may be used in some situations.		
50		Firing danger area (Danger Zone) buoy	⚲ DZ
51		Target	⚲ Target
52		Marker Ship	⚲ Marker Ship
53		Barge	⚲ Barge
54		Degaussing Range buoy	⚲
55	⚲ Tel	Cable buoy	⚲
56	⚲	Spoil ground buoy	⚲
57	⚲	Buoy marking outfall	⚲
58	⏚ ODAS	ODAS-buoy (Ocean-Data-Acquisition System). Data-Collecting buoy of superbuoy size	⏚ ODAS
	⚲W or ⚲ ⚲ ⚲W or ⚲ ⚲	Special-purpose buoys	
59		Wave recorder, current meter	⚲
60	⚲ AERO	Seaplane anchorage buoy	
61		Buoy marking traffic separation scheme	
62		Buoy marking recreation zone	⚲
Seasonal Buoys		
70	⚲ Priv (maintained by private interests, use with caution)	Buoy privately maintained (example)	⚲ Priv
71		Temporary buoy (example)	⚲ (Apr–Oct)

Q Buoys, Beacons

Beacons

Lighted Beacons → P   *Features Common to Beacons and Buoys* → Q 1–11

80	□Bn	⊥ Bn	Beacon in general, characteristics unknown or chart scale too small to show	⊥ ⊙ **Bn**	
81	□ RW ▲ ▪		Beacon with color, no distinctive topmark	⊥ BW	
82			Beacons with colors and topmarks (examples)	R BY BRB	
83			Beacon on submerged rock (topmark as appropriate)	BRB	BRB

Minor impermanent Marks usually in drying Areas (Lateral Mark of Minor Channel)

Minor Pile → F

				PORT HAND	STARBOARD HAND
90	∘ Pole • Pole		Stake, pole	⊥	
91	∘Stake •Stake		Perch, stake	Ψ	↑
92			Withy	Ψ̵	⅄̵

Minor Marks, usually on Land

Landmarks → E

100	⊙CAIRN ℓcairn △ ⊕	⊙CAIRN ℓcairn △ ⊕	Cairn	⊕
101			Colored or white mark	□ **Mk**

Beacon Towers

110	□RW	♤ ✦	Beacon towers without and with topmarks and colors (examples)	R G R G BY BRB
111			Lattice beacon	♯

Q *Buoys, Beacons*

Special Purpose Beacons			
Leaaing Lines, Clearing Lines → M			
Note: Topmarks and colors shown where scale permits.			
120		Leading beacons	
121		Beacons marking a clearing line	
122	COURSE 053° 00' TRUE MARKERS MARKERS	Beacons marking measured distance with quoted bearings	Measured Distance 1852 m 090°–270°
123		Cable landing beacon (example)	
124		Refuge beacon	Ref. Ref.
125		Firing danger area.beacons	
126		Notice board	

Q *Buoys, Beacons*

---

130 *IALA Maritime Buoyage System*

IALA   International Association of Lighthouse Authorities

Where in force, the IALA System applies to all fixed and floating marks except lighthouses, sector lights, leading lights and leading marks, light-vessels and lanbys. The standard buoy shapes are cylindrical (can) ⌷ . conical △ , spherical ⌂ , pillar ⌁ , and spar ⌶ , but variations may occur, for example: light-floats ⇱ . In the illustrations below, only the standard buoy shapes are used. In the case of fixed beacons (lit or unlit) only the shape of the topmark is of navigational significance.

130.1 *Lateral marks* are generally for well-defined channels. There are two international Buoyage Regions – A and B – where Lateral marks differ.

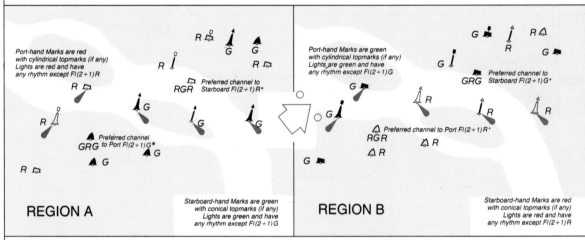

Port-hand Marks are red with cylindrical topmarks (if any) Lights are red and have any rhythm except Fl(2+1)R

Preferred channel to Starboard Fl(2+1)R*

Preferred channel to Port Fl(2+1)G*

Port-hand Marks are green with cylindrical topmarks (if any) Lights are green and have any rhythm except Fl(2+1)G

Preferred channel to Starboard Fl(2+1)G*

Preferred channel to Port Fl(2+1)R*

REGION A

Starboard-hand Marks are green with conical topmarks (if any) Lights are green and have any rhythm except Fl(2+1)G

REGION B

Starboard-hand Marks are red with conical topmarks (if any) Lights are red and have any rhythm except Fl(2+1)R

A preferred channel buoy may also be a pillar or a spar. All preferred channel marks have horizontal bands of color.
Where for exceptional reasons an Authority considers that a green color for buoys is not satisfactory, black may be used.

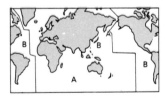

IALA Buoyage Regions A and B

130.2 *Direction of Buoyage*
The direction of buoyage is that taken when approaching a harbor from seaward or along coasts, the direction determined by buoyage authorities, normally clockwise around land masses.

    Symbol showing direction of buoyage where not obvious.

    Symbol showing direction of buoyage on multicolored charts.

Q *Buoys, Beacons*

In the illustrations below all marks are the same in Regions A and B.

130.3 *Cardinal Marks* indicating navigable water to the named side of the marks.

UNLIT MARKS

Topmark: 2 black cones

N

NW — NE

North Mark
Black above yellow

BY — BY

West Mark
YBY — YBY
Point of interest

W ——— E

Yellow with black band

East Mark
BYB — BYB
Black with yellow band

South Mark
YB — YB
Yellow above black

SW — SE

S

LIGHTED MARKS

White Light

Time (seconds)
0   5   10   15

North Mark	N BY	VQ or Q
East Mark	E BYB	VQ(3)5s or Q(3)10s
South Mark	S YB	VQ(6)+LFl 10s or Q(6)+LFl 15s
West Mark	W YBY	VQ(9)10s or Q(9)15s

The same abbreviations are used for lights on spar buoys and beacons. The periods 5s, 10s and 15s, may not always be charted.

130.4 *Isolated Danger Marks* stationed over dangers with navigable water around them.

Body: black with red horizontal band(s)
Topmark: 2 black spheres

BRB — BRB   BRB — BRB   Fl(2)   white light

130.5 *Safe Water Marks* such as mid-channel and landfall marks.

Body: red and white vertical stripes
Topmark (if any): red sphere

RW — RW — RW   RW — RW — RW   Iso Oc or L Fl 10s Mo (A)   Mo (A)   white light

---

130.6 *Special Marks* not primarily to assist navigation but to indicate special features.

Body (shape optional): yellow ‡
Topmark (if any): yellow ×

Y — Y — Y — (Y)   Y — Y — Y — (Y) Fl.Y etc.   yellow light

‡ In special cases yellow can be in conjunction with another color.

BEACONS with IALA System topmarks are charted by upright symbols, eg. BYB R bRB R (minor beacon) or, on smaller-scale charts: Bn R Bn G

Beacon towers are charted: R G BRB BYB etc. (occasionally lighted)

RADAR REFLECTORS on buoys and beacons are not generally charted.

COLOR ABBREVIATIONS under symbols, especially those of spar buoys, may be omitted, or may be at variance with symbols shown above.

LIGHT FLOATS: The IALA System is not usually applied to large light floats (replacing manned lightships) but may be applied to smaller light floats.

Q *Buoys, Beacons*

	Supplementary National Symbols			
a	⚲BELL ⚲BELL		Bell buoy	
b	⚲GONG ⚲GONG		Gong buoy	
c	⚲WHIS ⚲WHIS		Whistle buoy	
d	⚲RW ⚲BW		Fairway buoy (RWVS; BWVS)	
e	⚲RW ⚲BW		Midchannel buoy (RWVS; BWVS)	
f	⚲R "2"		Starboard-hand buoy (entering from seaward – US waters)	
g	⚲"1" ⚲"1"		Port-hand buoy (entering from seaward – US waters)	
h	⚲RB ⚲BR ⚲RG ⚲GR ⚲G		Bifurcation, Junction, Isolated danger, Wreck and Obstruction buoys	
i		⚲	Warping buoy	
j		⚲Y	Quarantine buoy	
k		⚲Explos Anch	Explosive anchorage buoy	
l		⚲Deviation	Compass adjustment buoy	
m		⚲BW	Fish trap (area) buoy (BWHB)	
n		⚲W	Anchorage buoy (marks limits)	
o	⚲ Chec		Checkered	
p	⚲ Diag		Diagonal bands	
q	B		Black	
r	▲R Bn △RG Bn		Triangular beacon	
	■B Bn		Black beacon	
	▪G Bn □GR Bn □W Bn		Square and other shaped beacons	
	□Bn		Color unknown	
s		⬤Tel ⬤Tel	Mooring buoy with telegraphic communications	
t		⬤T ⬤T	Mooring buoy with telephonic communications	
u	⚲		Lighted beacon	•Bn ⚲

AERO, Aero	*Aero light*	P 60
AERO RBn	*Aeronautical radiobeacon*	S 16
Aero RC	*Aeronautical radiobeacon*	S 16
Al	*Alternating*	P 10.11
ALP	*Articulated Loading Platform*	L 12
Alt	*Alternating*	P 10.11
Am	*Amber*	P 11.8
anc	*Ancient*	O 84
ANCH, Anch	*Anchorage*	N 20, O 21
approx	*Approximate*	O 90
Apprs	*Approaches*	O 22
B	*Bay, bayou*	O 4
Bdy Mon	*Boundary monument*	B 24
bk	*Broken*	J 33
Bkw	*Breakwater*	F 4.1
Bl	*Blue*	P 11.4
BM	*Bench mark*	B 23
Bn	*Beacon*	O 4
Bn Tr	*Beacon tower*	O 3
Br	*Breakers*	K 17
brg	*Bearing*	B 62
brk	*Broken*	J 33
Bu	*Blue*	P 11.4
c	*Course*	J 32
C	*Can, cylindrical*	Q 21
C	*Cove*	O 9
CALM	*Centenary Anchor Leg Mooring*	L 16
Cas	*Castle*	E 34.2
Cb	*Cobbles*	J 8
cbl	*Cable*	B 46
cd	*Candela*	B 54
CD	*Chart datum*	H 1
Cem	*Cemetery*	E 19
CG	*Coast Guard station*	T 10
Chan	*Channel*	O 14
Ch.	*Church*	E 10.1
Chy	*Chimney*	E 22
Cl	*Clay*	J 3
CL	*Clearance*	D 20, D 21
cm	*Centimeter(s)*	B 43
Co	*Coral*	J 10
Co rf	*Coral reef*	O 26
Cr	*Creek*	O 7
crs	*Course*	J 32

Cup, Cup.	*Cupola*	E 10.4
Cus Ho	*Customs house*	F 61
Cy	*Clay*	J 3
D	*Destroyed*	O 94
Destr	*Destroyed*	O 94
dev	*Deviation*	B 67
DIA, Dia	*Diaphone*	R 11
Dir	*Direction*	P 30, P 31
dist	*Distant*	O 85
dm	*Decimeter(s)*	B 42
Dn.	*Dolphin*	F 20
Dol.	*Dolphin*	F 20
DW	*Deep Water route*	M 27.1, N 12.4
DZ	*Danger Zone*	Q 50
E	*East, eastern*	B 10
ED	*Existence doubtful*	I 1
EEZ	*Exclusive Economic Zone*	N 47
E Int	*Equal interval, isophase*	P 10.3
Entr	*Entrance*	O 16
Est	*Estuary*	O 17
exper	*Experimental*	O 93
Explos	*Explosive*	R 10
Exting, exting	*Extinguished*	P 55
f	*Fine*	J 30
F	*Fixed*	P 10.1
Fd	*Fjord*	O 5
F Fl	*Fixed and flashing*	P 10.10
FISH	*Fishing*	N 21
Fl	*Flashing*	P 10.4
Fla	*Flare stack*	L 11
fm	*Fathom*	B 48
fms	*Fathoms*	B 48
fne	*Fine*	J 30
Fog Det Lt	*Fog detector light*	P 62
Fog Sig	*Fog signal*	R 1
FP	*Flagpole*	E 27
FS, FS.	*Flagstaff*	E 27
ft	*Foot, feet*	B 47
G	*Gravel*	J 6
G	*Green*	P 11.3
G	*Gulf*	O 3

Gp Fl	Group flashing	P 10.4
Gp Occ	Group occulting	P 10.2
h	Hard	J 39
h	Hour	B 49
H	Pilot transferred by helicopter	T 1.4
HAT	Highest astronomical tide	H 3
Hbr Mr	Harbormaster	F 60
Historic Wk	Historic wreck	N 26
Hk	Hulk	F 34
Hor	Horizontally disposed	P 15
Hor Cl	Horizontal clearance	D 21
Hosp	Hospital	F 62.2
hr	Hour	B 49
hrd	Hard	J 39
IALA	International Association of Lighthouse Authorities	Q 130
In	Inlet	O 10
Intens	Intensified	P 45
Int Qk Fl	Interrupted quick flashing	P 10.6
IQ	Interrupted quick flashing	P 10.6
I Qk Fl	Interrupted quick flashing	P 10.6
Iso	Isophase	P 10.3
IUQ	Interrupted ultra quick	P 10.8
km	Kilometer(s)	B 40
kn	Knot(s)	B 52
L	Loch, lough, lake	O 6
Lag	Lagoon	O 8
LANBY	Large Automatic Navigational Buoy	P 8
Lat, lat	Latitude	B 1
LASH	Lighter aboard ship	G 184
LAT	Lowest astronomical tide	H 2
Ldg	Landing	F 17
Ldg	Leading	P 21
Le	Ledge	O 28
L Fl	Long flashing	P 10.5
Lndg	Landing	F 17
LNG	Liquified natural gas	G 185
Long, long	Longitude	B 2
LOP	Line of position	S 21, S 31, S 41
LPG	Liquified petroleum gas	G 186
LSS	Life saving station	T 12
Lt	Light	P 1

Lt Ho	Lighthouse	P 1
Lt V	Light vessel	O 6
m	Meter(s)	B 41
m	Minute(s) of time	B 50
m	Medium (in relation to sand)	J 31
M	Mud, muddy	J 2
M	Nautical mile(s)	B 45
mag	Magnetic	B 61
MHHW	Mean higher high water	H 13
MHLW	Mean higher low water	H 14
MHW	Mean high water	H 5
MHWN	Mean high water neaps	H 11
MHWS	Mean high water springs	H 9
Mi	Nautical mile(s)	B 45
min	Minute of time	B 50
Mk	Mark	Q 101
MLHW	Mean lower high water	H 15
MLLW	Mean lower low water	H 12
MLW	Mean low water	H 4
MLWN	Mean low water neaps	H 10
MLWS	Mean low water springs	H 8
mm	Millimeter(s)	B 44
Mo	Morse	P 10.9
MON, Mon, Mon.	Monument	B 24, E 24
MSL	Mean sea level	H 6
Mt	Mountain	O 32
Mth	Mouth	O 19
N	North, northern	B 9
N	Nun	Q 20
NE	Northeast	B 13
NM	Nautical mile(s)	B 45
N Mi	Nautical mile(s)	B 45
No	Number	N 12.2
Np	Neap tide	H 17
NW	Northwest	B 15
NWS SIG STA	Weather signal station	T 29
Obsc	Obscured	P 43
Obscd	Obscured	P 43
Obs spot	Observation spot	B 21
Obstn	Obstruction	K 40, K 41, K 42
Obstr	Obstruction	K 41
Oc	Occulting	P 10.2
Occ	Occulting	P 10.2

V Index of Abbreviations

Occas	Occasional	P 50
ODAS	Ocean Data Acquisition System	Q 58
Or	Orange	P 11.7
P	Pebbles	J 7
P	Pillar	Q 23
PA	Position approximate	B 7
Pass	Passage, pass	O 13
PD	Position doubtful	B 8
PTL STA	Pilot station	T 3
Pk	Peak	O 35
Post Off	Post office	F 63
Priv, priv	Private	P 65, Q 70
Prod. well	Production well	L 20
PROHIB	Prohibited	N 2.2, N 20, N 21
Pyl	Pylon	D 26
Q	Quick	P 10.6
Qk Fl	Quick flashing	P 10.6
R	Coast radio station providing QTG services	S 15
R	Red	P 11.2
R	Rocky	J 9
Ra	Radar reference line	M 32
Ra (conspic)	Radar conspicuous object	S 5
Ra Antenna	Dish aerial	E 31
Racon	Radar transponder beacon	S 3
Radar Sc.	Radar scanner	E 30.3
Radar Tr.	Radar tower	E 30.2
Radome, Ra Dome	Radar dome	E 30.4
Ra Ref	Radar reflector	S 4
RBn	Circular radiobeacon	S 10
RC	Circular radiobeacon	S 10
Rd	Roads, roadstead	O 22
RD	Directional radiobeacon	S 11
RDF	Radio direction finding station	S 14
Ref.	Refuge	Q 124
Rep	Reported	I 3
Rf	Reef	O 26
RG	Radio direction finding station	S 14
Rk	Rocky	J 9
Rky	Rocky	J 9
R Mast	Radio mast	E 28
Ra Ro	Roll on Roll off	F 50
R Sta	Coast radio station providing QTG services	S 15

R Tower	Radio tower	E 29
Ru	Ruins	D 8, F 33.1
RW	Rotating radiobeacon	S 12
S	Sand	J 1
S	South, southern	B 11
S	Spar, spindle	Q 24
s	Second of time	B 51
SALM	Single Anchor Leg Mooring	L 12
SBM	Single Buoy Mooring	L 16
Sc	Scanner	E 30.3
Sd	Sound	O 12
SD	Sounding doubtful	I 2
SE	Southeast	B 14
sec	Second of time	B 51
sf	Stiff	J 36
sft	Soft	J 35
Sh	Shells	J 12
Shl	Shoal	O 25
Si	Silt	J 4
so	Soft	J 35
Sp	Spring tide	H 16
SP	Spherical	Q 22
Sp.	Spire	E 10.3
Spipe	Standpipe	E 21
SPM	Single point mooring	L 12
SS	Signal station	T 20
st	Stones	J 5
stf	Stiff	J 36
stk	Sticky	J 34
Str	Strait	O 11
Subm	Submerged	O 93
Subm piles	Submerged piles	K 43.1
Subm ruins	Submerged ruins	F 33.2
sy	Sticky	J 34
SW	Southwest	B 16
T	True	B 63
t	Metric ton(s)	B 53
Tel	Telephone, telegraph	D 27
Temp, temp	Temporary	P 54
Tk	Tank	E 32
Tr, Tr., TR	Tower	E 10.2, E 20
TT	Tree tops	C 14
TV Mast	Television mast	E 28
TV Tower	Television tower	E 29

Uncov	*Uncovers*	K 11
UQ	*Ultra quick*	P 10.8
v	*Volcanic*	J 37
var	*Variation*	B 60
Vert	*Vertically disposed*	P 15
Vert Cl	*Vertical clearance*	D 20
Vi	*Violet*	P 11.5
Vil	*Village*	D 4
VLCC	*Very large crude carrier*	G 187
vol	*Volcanic*	J 37
VQ	*Very quick*	P 10.7
V Qk Fl	*Very quick flash*	P 10.7
W	*West, western*	B 12
W	*White*	P 11.1
Wd	*Weed*	J 13.1
WGS	*World Geodetic System*	S 50
Whf	*Wharf*	F 13
WHIS, Whis	*Whistle*	R 15
Wk	*Wreck*	K 20–23, K 26–27, K 30
Y	*Yellow*	P 11.6

V Index of Abbreviations

Supplementary National Abbreviations:

Apt	Apartment	Es		Grd	Ground	Ja
				Grs	Grass	Jv
				gty	Gritty	Jam
B	Black	Qq		GUN	Fog gun	Rd
bk	Black	Jas		gy	Gray	Jbb
bl	Black	Jas				
Blds	Boulders	Je				
br	Brown	Jaz		HECP	Harbor entrance control point	Tb
bu	Blue	Jau		HHW	Higher high water	Hb
				HS	High school	Eg
				ht	Height	Hp
Cap	Capitol	Et		HW	High water	Hq
ch	Chocolate	Jba		HWF & C	High water full and change	Hh
Chec	Checkered	Qo		Hz	Hertz	Bg
Ck	Chalk	Jf				
Cn	Cinders	Jp				
Co	Company	Eu		in	Inch	Bc
Co Hd	Coral head	Ji		ins	Inches	Bc
COLREGS	Collision regulations	Na		Inst	Institute	En
Corp	Corporation	Ev		ISLW	Indian springs low water	Hg
cps	Cycles per second	Bj				
CRD	Columbia River Datum	Hj				
c/s	Cycles per second	Bj		K	Kelp	Ju
Ct Ho	Court house	Eo		kc	Kilocycle	Bk
				kHz	Kilohertz	Bh
				kn	Knot(s)	Ho
dec	Decayed	Jan				
deg	Degree(s)	Bn				
Di	Diatoms	Jaa		La	Lava	Jl
Diag	Diagonal bands	Qp		LLW	Lower low water	He
Discol water	Discolored water	Ke		LOOK TR	Lookout tower	Tf
dk	Dark	Jbd		lrg	Large	Jai
				lt	Light	Jbc
				Ltd	Limited	Er
Explos Anch	Explosives anchorage	Qk		LW	Low water	Hc
				LWD	Low water datum	Hd
				LWF & C	Low water full and change	Hi
Facty	Factory	Ed				
F Gp Fl	Fixed and group flashing	Pd				
fl	Flood	Hq		m²	Square meter(s)	Ba
fly	Flinty	Jao		m³	Cubic meter(s)	Bb
Fr	Foraminifera	Jy		Ma	Mattes	Jag
Fu	Fucus	Jaf		Magz	Magazine	El
				Mc	Megacycle(s)	Bl
				Mds	Madrepores	Jj
GAB, Gab	Gable	Ei		MHz	Megahertz	Bi
GCLWD	Gulf Coast Low Water Datum	Hk		Ml	Marl	Jc
Gl	Globigerina	Jz		Mn	Manganese	Jq
glac	Glacial	Jap		Mo	Morse code	Rf
gn	Green	Jav		Ms	Mussels	Js
Govt Ho	Government house	Em		MTL	Mean Tide Level	Hf

Index of Abbreviations

Supplementary National Abbreviations:

NAUTO	Nautophone	Rc
or	Orange	Jax
Oys	Oysters	Jr
Oz	Ooze	Jb
Pav	Pavilion	Ep
Pm	Pumice	Jm
Po	Polyzoa	Jad
Pt	Pteropods	Jac
Quar	Quarantine	Fd
Qz	Quartz	Jg
Rd	Radiolaria	Jab
rd	Red	Jay
rt	Rotten	Jaj
Ry	Railway, railroad	Db
Sc	Scoriae	Jo
Sch	Schist	Jh
Sch	School	Ef
Sem	Semaphore	Tg
Sh	Shingle	Jd
S–LFl	Short–long flashing	Pb
sml	Small	Jah
Spg	Sponge	Jt
Spi	Spicules	Jx
spk	Speckled	Jal
Stg	Seatangle	Jw
St M	Statute mile(s)	Be
St Mi	Statute mile(s)	Be
Str	Stream	Hl
str	Streaky	Jak
SUB–BELL	Submarine fog bell	Ra
Subm crib	Submerged crib	Ki
SUB–OSC	Submarine oscillator	Rb.
Sub vol	Submarine volcano	Kd
T	Telephone	Eq,Qt
T	Short ton(s)	Bm
T	Tufa	Jn
Tel	Telegraph	Qs
Tel off	Telegraph office	Ek
ten	Tenacious	Jaq

unev	Uneven	Jbf
Univ	University	Eh
us	Microsecond(s)	Bf
usec	Microsecond(s)	Bf
vard	Varied	Jbe
vel	Velocity	Hn
vi	Violet	Jat
Vol Ash	Volcanic ash	Jk
wh	White	Jar
WHIS	Whistle	Qc
yd	Yard	Bd
yds	Yards	Bd
yl	Yellow	Jaw

IALA MARITIME BUOYAGE SYSTEM
LATERAL MARKS REGION A

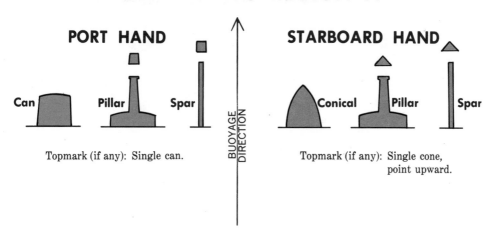

PORT HAND

Can   Pillar   Spar

Topmark (if any): Single can.

BUOYAGE DIRECTION

STARBOARD HAND

Conical   Pillar   Spar

Topmark (if any): Single cone,
point upward.

Lights, when fitted, may have any phase
characteristic other than that used
for preferred channels.

Examples

Quick Flashing
Flashing
Long Flashing
Group Flashing

PREFERRED CHANNEL
TO STARBOARD

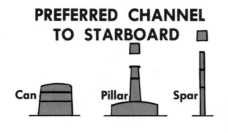

Can   Pillar   Spar

Topmark (if any): Single can.

BUOYAGE DIRECTION

PREFERRED CHANNEL
TO PORT

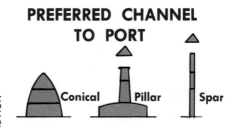

Conical   Pillar   Spar

Topmark (if any): Single cone,
point upward.

Lights, when fitted, are composite
group flashing Fl (2 + 1).

IALA MARITIME BUOYAGE SYSTEM
LATERAL MARKS REGION B

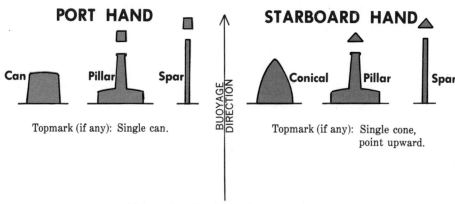

PORT HAND

Can · Pillar · Spar

Topmark (if any): Single can.

STARBOARD HAND

Conical · Pillar · Spar

Topmark (if any): Single cone, point upward.

BUOYAGE DIRECTION

Lights, when fitted, may have any phase
characteristic other than that used
for preferred channels.

Examples
Quick Flashing
Flashing
Long Flashing
Group Flashing

PREFERRED CHANNEL TO STARBOARD

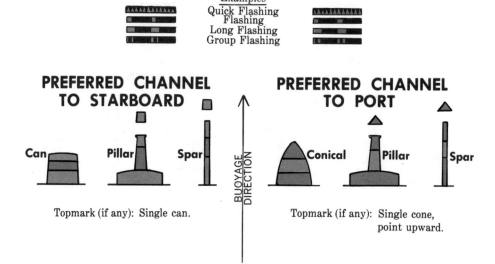

Can · Pillar · Spar

Topmark (if any): Single can.

PREFERRED CHANNEL TO PORT

Conical · Pillar · Spar

Topmark (if any): Single cone, point upward.

BUOYAGE DIRECTION

Lights, when fitted, are composite
group flashing Fl (2+1).

IALA MARITIME BUOYAGE SYSTEM
CARDINAL MARKS REGIONS A AND B

Topmarks are always fitted (when practicable).
Buoy shapes are pillar or spar.

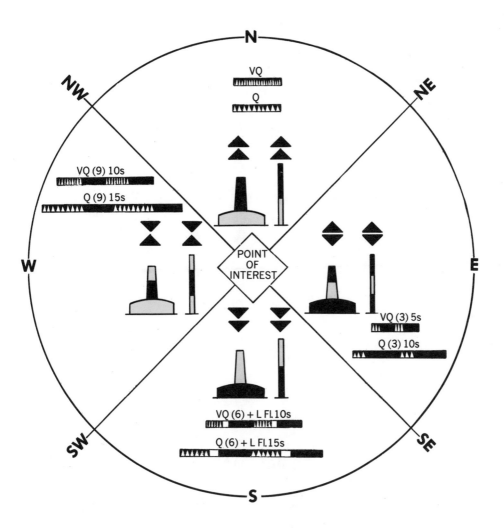

Lights, when fitted, are **white** , Very Quick Flashing
or Quick Flashing; a South mark also has a
Long Flash immediately following the quick flashes.

IALA MARITIME BUOYAGE SYSTEM
REGIONS A AND B

ISOLATED DANGER MARKS

Topmarks are
always fitted
(when practicable).

Shape: Optional, but not
conflicting with lateral
marks; pillar or spar
preferred.

Light, when fitted, is
white
Group Flashing (2)

Fl (2)

SAFE WATER MARKS

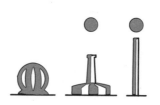

Topmark (if any):
Single sphere.

Shape: Spherical
or
pillar or spar.

Light, when fitted,
is **white**
Isophase or Occulting,
or one Long Flash
every 10 seconds or
Morse "A".

Iso

Occ

L Fl. 10s

Morse "A"

SPECIAL MARKS

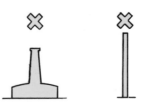

Topmark (if any):
Single X shape.

Shape: Optional, but not
conflicting with
navigational marks.

Light (when fitted) is
yellow and may have
any phase characteristic
not used for white lights.

Examples

Fl Y

Fl(4) Y

CONTENTS

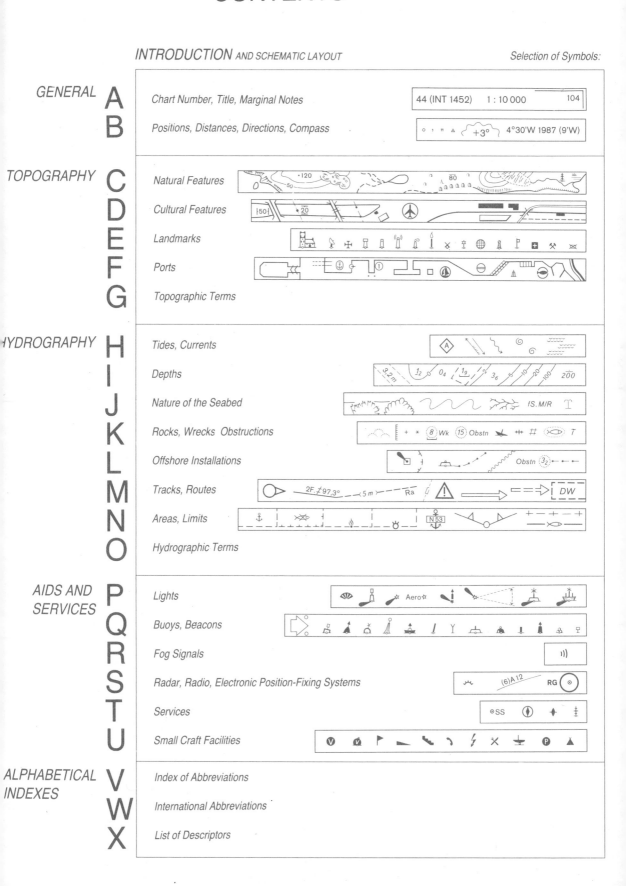

INTRODUCTION AND SCHEMATIC LAYOUT   Selection of Symbols:

GENERAL   A B

Chart Number, Title, Marginal Notes   44 (INT 1452)   1 : 10 000   104

Positions, Distances, Directions, Compass   +3°   4°30'W 1987 (9'W)

TOPOGRAPHY   C D E F G

Natural Features

Cultural Features

Landmarks

Ports

Topographic Terms

HYDROGRAPHY   H I J K L M N O

Tides, Currents

Depths

Nature of the Seabed

Rocks, Wrecks Obstructions

Offshore Installations

Tracks, Routes

Areas, Limits

Hydrographic Terms

AIDS AND
SERVICES   P Q R S T U

Lights

Buoys, Beacons

Fog Signals

Radar, Radio, Electronic Position-Fixing Systems

Services

Small Craft Facilities

ALPHABETICAL
INDEXES   V W X

Index of Abbreviations

International Abbreviations

List of Descriptors

INDEX

sextant altitude corrections. *See* sextant, marine

sextant, marine, 4–5, 329–62
  altitude corrections for, 340–62
  care of, 340
  characteristics of, 329–32
  early, 4–5
  index error of, 334–36
  optical principle of, 332
  techniques for use, 336–40
Ship's Deck Log, 9, 17, 19
ship's handling characteristics, 234–37
Ship's Position Log, 270
sight forms, blank, 597–605
Sight Reduction Tables No. 229, 374, 377–86
Sight Reduction Tables No. 249, 398–408
  use of as starfinder, 408
six-minute rule, 18, 123
six rules of DR, 134–35
slide lines, 241
solar eclipse, 282–83
solar system, 280–82
speed, measurement of, 114–17
  by use of electronic nav systems, 116
  by use of shaft RPM, 116–17
speed over the ground (SOG), def., 114
stadimeter, 113–14
  Brandon type, 113
  Fisk type, 113
Summary of Corrections, 52
sunrise, -set, predicting times of, 480–84
swing circle, 244, 245, 249–51
swinging ship, 155

telescopic alidade, 112
terrestrial coordinate system, 283–84
  equator, def., 283
  latitude, def., 283–84
  longitude, def., 284
  meridian, def., 283
    prime meridian, 283
three-minute rule, 18, 123
Tidal Current Tables, 74, 203–217
  use of, 208–217
  See also current; tide
tide, 177–98
  bridge problem, 195–97

causes of, 177–79
constructing graph of, 191–92
def., 177
prediction of, 183–95
reference planes, 180–83
shoal problem, 197
types of, 179–80
Tide Tables, 72–73, 183–97
  use of, 188–97
  See also tide
time, 252–58, 310–28
  bases of, 310–14
    apparent solar, 311
    atomic, 314
    mean solar, 311–12
    sidereal, 313–14
  conversions, 255–58
  diagrams, 318–21
  equation of, 312–13
  format of written, 348
  Greenwich Mean (GMT), 254–55
  and longitude, 314–18
  zone (ZT), 253–55
time diagrams, 318–21
timepieces, 126–28
  chronometer, 126–27
timing celestial observations, 321–25
track, ship's, 143–47, 262–69, 577
  labeling of, 143
  plotting of, 263–69
transfer, 235–38
  use of during piloting, 237–38
true azimuth, def., 298
turn bearings, 238–41
twilight, predicting times of, 476–80

variation, 150–53
vectorizing, electronic charts, 48
visibility computations, 88–98
  plotting, 95–98
  procedures for, 91–95
  range terms, 89–90
voyage planning, 252, 259–70

weather reports, 9
WGS-84, 33–35, 48, 547, 553
World Port Index, Pub. No. 150, 65